Communications in Computer and Information Science 1966

Rationale

The CCIS series is devoted to the publication of proceedings of computer science conferences. Its aim is to efficiently disseminate original research results in informatics in printed and electronic form. While the focus is on publication of peer-reviewed full papers presenting mature work, inclusion of reviewed short papers reporting on work in progress is welcome, too. Besides globally relevant meetings with internationally representative program committees guaranteeing a strict peer-reviewing and paper selection process, conferences run by societies or of high regional or national relevance are also considered for publication.

Topics

The topical scope of CCIS spans the entire spectrum of informatics ranging from foundational topics in the theory of computing to information and communications science and technology and a broad variety of interdisciplinary application fields.

Information for Volume Editors and Authors

Publication in CCIS is free of charge. No royalties are paid, however, we offer registered conference participants temporary free access to the online version of the conference proceedings on SpringerLink (http://link.springer.com) by means of an http referrer from the conference website and/or a number of complimentary printed copies, as specified in the official acceptance email of the event.

CCIS proceedings can be published in time for distribution at conferences or as post-proceedings, and delivered in the form of printed books and/or electronically as USBs and/or e-content licenses for accessing proceedings at SpringerLink. Furthermore, CCIS proceedings are included in the CCIS electronic book series hosted in the SpringerLink digital library at http://link.springer.com/bookseries/7899. Conferences publishing in CCIS are allowed to use Online Conference Service (OCS) for managing the whole proceedings lifecycle (from submission and reviewing to preparing for publication) free of charge.

Publication process

The language of publication is exclusively English. Authors publishing in CCIS have to sign the Springer CCIS copyright transfer form, however, they are free to use their material published in CCIS for substantially changed, more elaborate subsequent publications elsewhere. For the preparation of the camera-ready papers/files, authors have to strictly adhere to the Springer CCIS Authors' Instructions and are strongly encouraged to use the CCIS LaTeX style files or templates.

Abstracting/Indexing

CCIS is abstracted/indexed in DBLP, Google Scholar, EI-Compendex, Mathematical Reviews, SCImago, Scopus. CCIS volumes are also submitted for the inclusion in ISI Proceedings.

How to start

To start the evaluation of your proposal for inclusion in the CCIS series, please send an e-mail to ccis@springer.com.

Biao Luo · Long Cheng · Zheng-Guang Wu ·
Hongyi Li · Chaojie Li
Editors

Neural Information Processing

30th International Conference, ICONIP 2023
Changsha, China, November 20–23, 2023
Proceedings, Part XII

Editors
Biao Luo
School of Automation
Central South University
Changsha, China

Long Cheng
Institute of Automation
Chinese Academy of Sciences
Beijing, China

Zheng-Guang Wu
Institute of Cyber-Systems and Control
Zhejiang University
Hangzhou, China

Hongyi Li
School of Automation
Guangdong University of Technology
Guangzhou, China

Chaojie Li
School of Electrical Engineering
and Telecommunications
UNSW Sydney
Sydney, NSW, Australia

ISSN 1865-0929 ISSN 1865-0937 (electronic)
Communications in Computer and Information Science
ISBN 978-981-99-8147-2 ISBN 978-981-99-8148-9 (eBook)
https://doi.org/10.1007/978-981-99-8148-9

This Springer imprint is published by the registered company Springer Nature Singapore Pte Ltd.
The registered company address is: 152 Beach Road, #21-01/04 Gateway East, Singapore 189721, Singapore

Preface

Welcome to the 30th International Conference on Neural Information Processing (ICONIP2023) of the Asia-Pacific Neural Network Society (APNNS), held in Changsha, China, November 20–23, 2023.

The mission of the Asia-Pacific Neural Network Society is to promote active interactions among researchers, scientists, and industry professionals who are working in neural networks and related fields in the Asia-Pacific region. APNNS has Governing Board Members from 13 countries/regions – Australia, China, Hong Kong, India, Japan, Malaysia, New Zealand, Singapore, South Korea, Qatar, Taiwan, Thailand, and Turkey. The society's flagship annual conference is the International Conference of Neural Information Processing (ICONIP). The ICONIP conference aims to provide a leading international forum for researchers, scientists, and industry professionals who are working in neuroscience, neural networks, deep learning, and related fields to share their new ideas, progress, and achievements.

ICONIP2023 received 1274 papers, of which 394 papers were accepted for publication in Communications in Computer and Information Science (CCIS), representing an acceptance rate of 30.93% and reflecting the increasingly high quality of research in neural networks and related areas. The conference focused on four main areas, i.e., "Theory and Algorithms", "Cognitive Neurosciences", "Human-Centered Computing", and "Applications". All the submissions were rigorously reviewed by the conference Program Committee (PC), comprising 258 PC members, and they ensured that every paper had at least two high-quality single-blind reviews. In fact, 5270 reviews were provided by 2145 reviewers. On average, each paper received 4.14 reviews.

We would like to take this opportunity to thank all the authors for submitting their papers to our conference, and our great appreciation goes to the Program Committee members and the reviewers who devoted their time and effort to our rigorous peer-review process; their insightful reviews and timely feedback ensured the high quality of the papers accepted for publication. We hope you enjoyed the research program at the conference.

October 2023

Biao Luo
Long Cheng
Zheng-Guang Wu
Hongyi Li
Chaojie Li

Organization

Honorary Chair

Weihua Gui	Central South University, China

Advisory Chairs

Jonathan Chan	King Mongkut's University of Technology Thonburi, Thailand
Zeng-Guang Hou	Chinese Academy of Sciences, China
Nikola Kasabov	Auckland University of Technology, New Zealand
Derong Liu	Southern University of Science and Technology, China
Seiichi Ozawa	Kobe University, Japan
Kevin Wong	Murdoch University, Australia

General Chairs

Tingwen Huang	Texas A&M University at Qatar, Qatar
Chunhua Yang	Central South University, China

Program Chairs

Biao Luo	Central South University, China
Long Cheng	Chinese Academy of Sciences, China
Zheng-Guang Wu	Zhejiang University, China
Hongyi Li	Guangdong University of Technology, China
Chaojie Li	University of New South Wales, Australia

Technical Chairs

Xing He	Southwest University, China
Keke Huang	Central South University, China
Huaqing Li	Southwest University, China
Qi Zhou	Guangdong University of Technology, China

Local Arrangement Chairs

Wenfeng Hu	Central South University, China
Bei Sun	Central South University, China

Finance Chairs

Fanbiao Li	Central South University, China
Hayaru Shouno	University of Electro-Communications, Japan
Xiaojun Zhou	Central South University, China

Special Session Chairs

Hongjing Liang	University of Electronic Science and Technology, China
Paul S. Pang	Federation University, Australia
Qiankun Song	Chongqing Jiaotong University, China
Lin Xiao	Hunan Normal University, China

Tutorial Chairs

Min Liu	Hunan University, China
M. Tanveer	Indian Institute of Technology Indore, India
Guanghui Wen	Southeast University, China

Publicity Chairs

Sabri Arik	Istanbul University-Cerrahpaşa, Turkey
Sung-Bae Cho	Yonsei University, South Korea
Maryam Doborjeh	Auckland University of Technology, New Zealand
El-Sayed M. El-Alfy	King Fahd University of Petroleum and Minerals, Saudi Arabia
Ashish Ghosh	Indian Statistical Institute, India
Chuandong Li	Southwest University, China
Weng Kin Lai	Tunku Abdul Rahman University of Management & Technology, Malaysia
Chu Kiong Loo	University of Malaya, Malaysia
Qinmin Yang	Zhejiang University, China
Zhigang Zeng	Huazhong University of Science and Technology, China

Publication Chairs

Zhiwen Chen	Central South University, China
Andrew Chi-Sing Leung	City University of Hong Kong, China
Xin Wang	Southwest University, China
Xiaofeng Yuan	Central South University, China

Secretaries

Yun Feng	Hunan University, China
Bingchuan Wang	Central South University, China

Webmasters

Tianmeng Hu	Central South University, China
Xianzhe Liu	Xiangtan University, China

Program Committee

Rohit Agarwal	UiT The Arctic University of Norway, Norway
Hasin Ahmed	Gauhati University, India
Harith Al-Sahaf	Victoria University of Wellington, New Zealand
Brad Alexander	University of Adelaide, Australia
Mashaan Alshammari	Independent Researcher, Saudi Arabia
Sabri Arik	Istanbul University, Turkey
Ravneet Singh Arora	Block Inc., USA
Zeyar Aung	Khalifa University of Science and Technology, UAE
Monowar Bhuyan	Umeå University, Sweden
Jingguo Bi	Beijing University of Posts and Telecommunications, China
Xu Bin	Northwestern Polytechnical University, China
Marcin Blachnik	Silesian University of Technology, Poland
Paul Black	Federation University, Australia
Anoop C. S.	Govt. Engineering College, India
Ning Cai	Beijing University of Posts and Telecommunications, China
Siripinyo Chantamunee	Walailak University, Thailand
Hangjun Che	City University of Hong Kong, China

Wei-Wei Che	Qingdao University, China
Huabin Chen	Nanchang University, China
Jinpeng Chen	Beijing University of Posts & Telecommunications, China
Ke-Jia Chen	Nanjing University of Posts and Telecommunications, China
Lv Chen	Shandong Normal University, China
Qiuyuan Chen	Tencent Technology, China
Wei-Neng Chen	South China University of Technology, China
Yufei Chen	Tongji University, China
Long Cheng	Institute of Automation, China
Yongli Cheng	Fuzhou University, China
Sung-Bae Cho	Yonsei University, South Korea
Ruikai Cui	Australian National University, Australia
Jianhua Dai	Hunan Normal University, China
Tao Dai	Tsinghua University, China
Yuxin Ding	Harbin Institute of Technology, China
Bo Dong	Xi'an Jiaotong University, China
Shanling Dong	Zhejiang University, China
Sidong Feng	Monash University, Australia
Yuming Feng	Chongqing Three Gorges University, China
Yun Feng	Hunan University, China
Junjie Fu	Southeast University, China
Yanggeng Fu	Fuzhou University, China
Ninnart Fuengfusin	Kyushu Institute of Technology, Japan
Thippa Reddy Gadekallu	VIT University, India
Ruobin Gao	Nanyang Technological University, Singapore
Tom Gedeon	Curtin University, Australia
Kam Meng Goh	Tunku Abdul Rahman University of Management and Technology, Malaysia
Zbigniew Gomolka	University of Rzeszow, Poland
Shengrong Gong	Changshu Institute of Technology, China
Xiaodong Gu	Fudan University, China
Zhihao Gu	Shanghai Jiao Tong University, China
Changlu Guo	Budapest University of Technology and Economics, Hungary
Weixin Han	Northwestern Polytechnical University, China
Xing He	Southwest University, China
Akira Hirose	University of Tokyo, Japan
Yin Hongwei	Huzhou Normal University, China
Md Zakir Hossain	Curtin University, Australia
Zengguang Hou	Chinese Academy of Sciences, China

Lu Hu	Jiangsu University, China
Zeke Zexi Hu	University of Sydney, Australia
He Huang	Soochow University, China
Junjian Huang	Chongqing University of Education, China
Kaizhu Huang	Duke Kunshan University, China
David Iclanzan	Sapientia University, Romania
Radu Tudor Ionescu	University of Bucharest, Romania
Asim Iqbal	Cornell University, USA
Syed Islam	Edith Cowan University, Australia
Kazunori Iwata	Hiroshima City University, Japan
Junkai Ji	Shenzhen University, China
Yi Ji	Soochow University, China
Canghong Jin	Zhejiang University, China
Xiaoyang Kang	Fudan University, China
Mutsumi Kimura	Ryukoku University, Japan
Masahiro Kohjima	NTT, Japan
Damian Kordos	Rzeszow University of Technology, Poland
Marek Kraft	Poznań University of Technology, Poland
Lov Kumar	NIT Kurukshetra, India
Weng Kin Lai	Tunku Abdul Rahman University of Management & Technology, Malaysia
Xinyi Le	Shanghai Jiao Tong University, China
Bin Li	University of Science and Technology of China, China
Hongfei Li	Xinjiang University, China
Houcheng Li	Chinese Academy of Sciences, China
Huaqing Li	Southwest University, China
Jianfeng Li	Southwest University, China
Jun Li	Nanjing Normal University, China
Kan Li	Beijing Institute of Technology, China
Peifeng Li	Soochow University, China
Wenye Li	Chinese University of Hong Kong, China
Xiangyu Li	Beijing Jiaotong University, China
Yantao Li	Chongqing University, China
Yaoman Li	Chinese University of Hong Kong, China
Yinlin Li	Chinese Academy of Sciences, China
Yuan Li	Academy of Military Science, China
Yun Li	Nanjing University of Posts and Telecommunications, China
Zhidong Li	University of Technology Sydney, Australia
Zhixin Li	Guangxi Normal University, China
Zhongyi Li	Beihang University, China

Ziqiang Li	University of Tokyo, Japan
Xianghong Lin	Northwest Normal University, China
Yang Lin	University of Sydney, Australia
Huawen Liu	Zhejiang Normal University, China
Jian-Wei Liu	China University of Petroleum, China
Jun Liu	Chengdu University of Information Technology, China
Junxiu Liu	Guangxi Normal University, China
Tommy Liu	Australian National University, Australia
Wen Liu	Chinese University of Hong Kong, China
Yan Liu	Taikang Insurance Group, China
Yang Liu	Guangdong University of Technology, China
Yaozhong Liu	Australian National University, Australia
Yong Liu	Heilongjiang University, China
Yubao Liu	Sun Yat-sen University, China
Yunlong Liu	Xiamen University, China
Zhe Liu	Jiangsu University, China
Zhen Liu	Chinese Academy of Sciences, China
Zhi-Yong Liu	Chinese Academy of Sciences, China
Ma Lizhuang	Shanghai Jiao Tong University, China
Chu-Kiong Loo	University of Malaya, Malaysia
Vasco Lopes	Universidade da Beira Interior, Portugal
Hongtao Lu	Shanghai Jiao Tong University, China
Wenpeng Lu	Qilu University of Technology, China
Biao Luo	Central South University, China
Ye Luo	Tongji University, China
Jiancheng Lv	Sichuan University, China
Yuezu Lv	Beijing Institute of Technology, China
Huifang Ma	Northwest Normal University, China
Jinwen Ma	Peking University, China
Jyoti Maggu	Thapar Institute of Engineering and Technology Patiala, India
Adnan Mahmood	Macquarie University, Australia
Mufti Mahmud	University of Padova, Italy
Krishanu Maity	Indian Institute of Technology Patna, India
Srimanta Mandal	DA-IICT, India
Wang Manning	Fudan University, China
Piotr Milczarski	Lodz University of Technology, Poland
Malek Mouhoub	University of Regina, Canada
Nankun Mu	Chongqing University, China
Wenlong Ni	Jiangxi Normal University, China
Anupiya Nugaliyadde	Murdoch University, Australia

Xinyue Xu	Hong Kong University of Science and Technology, China
Yong Xu	Beijing Institute of Technology, China
Ngo Xuan Bach	Posts and Telecommunications Institute of Technology, Vietnam
Hao Xue	University of New South Wales, Australia
Yang Xujun	Chongqing Jiaotong University, China
Haitian Yang	Chinese Academy of Sciences, China
Jie Yang	Shanghai Jiao Tong University, China
Minghao Yang	Chinese Academy of Sciences, China
Peipei Yang	Chinese Academy of Science, China
Zhiyuan Yang	City University of Hong Kong, China
Wangshu Yao	Soochow University, China
Ming Yin	Guangdong University of Technology, China
Qiang Yu	Tianjin University, China
Wenxin Yu	Southwest University of Science and Technology, China
Yun-Hao Yuan	Yangzhou University, China
Xiaodong Yue	Shanghai University, China
Paweł Zawistowski	Warsaw University of Technology, Poland
Hui Zeng	Southwest University of Science and Technology, China
Wang Zengyunwang	Hunan First Normal University, China
Daren Zha	Institute of Information Engineering, China
Zhi-Hui Zhan	South China University of Technology, China
Baojie Zhang	Chongqing Three Gorges University, China
Canlong Zhang	Guangxi Normal University, China
Guixuan Zhang	Chinese Academy of Science, China
Jianming Zhang	Changsha University of Science and Technology, China
Li Zhang	Soochow University, China
Wei Zhang	Southwest University, China
Wenbing Zhang	Yangzhou University, China
Xiang Zhang	National University of Defense Technology, China
Xiaofang Zhang	Soochow University, China
Xiaowang Zhang	Tianjin University, China
Xinglong Zhang	National University of Defense Technology, China
Dongdong Zhao	Wuhan University of Technology, China
Xiang Zhao	National University of Defense Technology, China
Xu Zhao	Shanghai Jiao Tong University, China

Contents – Part XII

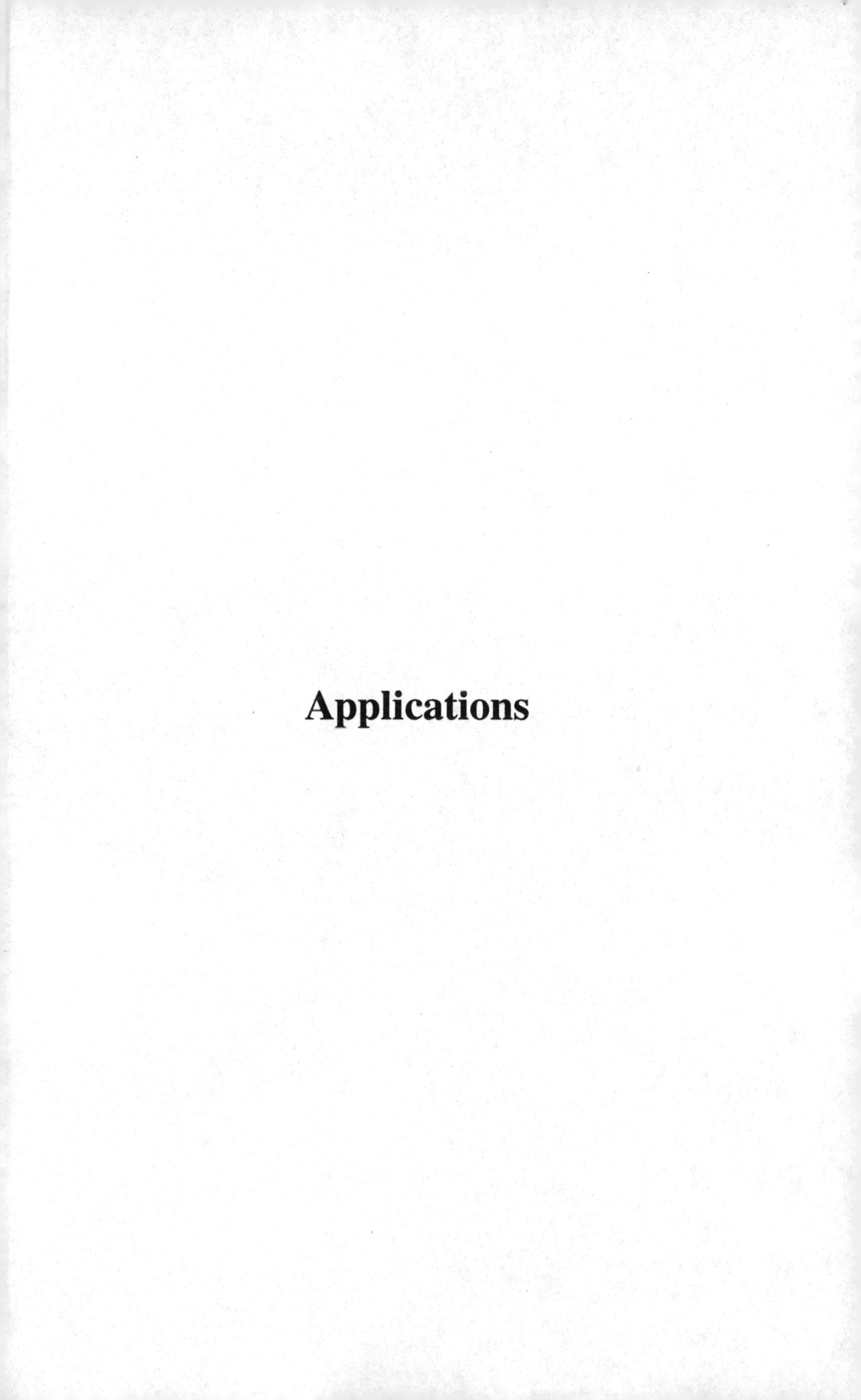

Applications

PBTR: Pre-training and Bidirectional Semantic Enhanced Trajectory Recovery

Qiming Zhang, Tianxi Liao(✉), Tongyu Zhu, Leilei Sun, and Weifeng Lv

State Key Laboratory of Software Development Environment, Beihang University, Beijing 100191, China
{qimingzhang,tx_liao,zhutongyu,leileisun,lwf}@buaa.edu.cn

Abstract. The advancement of position acquisition technology has enabled the study based on vehicle trajectories. However, limitations in equipment and environmental factors often result in missing track records, significantly impacting the trajectory data quality. It is a fundamental task to restore the missing vehicle tracks within the traffic network structure. Existing research has attempted to address this issue through the construction of neural network models. However, these methods neglect the significance of the bidirectional information of the trajectory and the embedded representation of the trajectory unit. In view of the above problems, we propose a Seq2Seq-based trajectory recovery model that effectively utilizes bidirectional information and generates embedded representations of trajectory units to enhance trajectory recovery performance, which is a **P**re-Training and **B**idirectional Semantic enhanced **T**rajectory **R**ecovery model, namely **PBTR**. Specifically, the road network's representations extracting time factors are captured by a pre-training technique and a bidirectional semantics encoder is employed to enhance the expressiveness of the model followed by an attentive recurrent network to reconstruct the trajectory. The efficacy of our model is demonstrated through its superior performance on two real-world datasets.

Keywords: Trajectory recovery · Route Representation Learning · Sequence to sequence model · Bidirectional information

1 Introduction

With the widespread use of GPS devices, a large amount of vehicle trajectory data has been collected. Many studies have shown that complete vehicle trajectories can provide high-quality input for various downstream tasks, such as travel time estimation [5,20], route optimization [21] and location prediction [18]. However, in real life, due to limitations in equipment and signal transmission, there are often missing points in trajectory data, which increases the uncertainty of trajectories. Therefore, recovering the missing information in the trajectory is very important.

B. Luo et al. (Eds.): ICONIP 2023, CCIS 1966, pp. 3–16, 2024.
https://doi.org/10.1007/978-981-99-8148-9_1

To tackle this problem, many deep learning-based models [4,14,19] have been proposed. For example, DHTR [17] extends the classic sequence-to-sequence model by implementing a subsequence-to-sequence recovery model, which predicts the coarse-grained grid of high-sampling-rate points by integrating a sequence-to-sequence model with a calibration component of the Kalman Filter. MTrajRec [12] devises a multi-task sequence-to-sequence learning architecture to predict road segment and moving ratio simultaneously, which utilizes the segment ID sequences and GPS coordinate's grid sequences to learn the semantic meanings and spatial context features.

However, these methods often focus on the good designs of modules to capture complex sequential or data correlations, and rarely take the complete representations of road networks into consideration, which employ simplistic encoding techniques such as one-hot encoding to embed trajectory location points and one-way encoders to encode trajectory sequences, ignoring the bidirectional information inherent in the trajectory sequences. As the discrete IDs of segments in the trajectories are first need to be transformed into constructive vectors with a full mapping from the physical world, trajectory recovery tasks bear resemblance to sentence completion tasks in the field of Natural Language Processing (NLP). Extensive research [3,6] has demonstrated that effective word embedding vectors can significantly enhance model performance, and pre-training alone improves over techniques devised for a specific task. Consequently, a pre-training task based on the historical trajectories is designed to learn the representations of road networks for the downstream task. Though the trajectory traveler goes through is directional, the effect of contextual segments is mutual. It is natural to capture the underlying inter-dependencies of contextual segments within the trajectory by implementing a bidirectional encoder-decoder for the specific task.

Based on the above discussion, this paper proposes a novel model that addresses the aforementioned issues, which is called **P**re-Training and **B**idirectional Semantic enhanced **T**rajectory **R**ecovery model (PBTR). Subsequently, The model first pre-trains the input historical trajectories to obtain high-quality embedded representations. Then, a Seq2Seq-based bidirectional encoder is employed to extract the spatiotemporal semantics of the trajectory sequence. Finally, an attentive recurrent decoder is employed to decode the hidden states of the encoder, which obtains the reconstructed trajectory sequence.

In summary, the main contributions of the proposed method can be summarized as follows:

- We propose a novel Seq2Seq-based model that can effectively leverage the semantic information of trajectories to accomplish the task of trajectory restoration with pre-trained representations of road networks and enhanced bidirectional semantic information of contextual road segments.
- Based on the above model, We design an Encoder-Decoder module, consisting of a pre-training and bidirectional road representation encoder and an attentive recurrent decoder, which effectively extracts the underlying dependency of trajectories in the road network to further improve the performance.

- We conduct substantial experiments using two real-world taxi trajectory datasets. Extensive results have shown the superiority of the proposed PBTR in both effectiveness and efficiency.

2 Related Works

2.1 Trajectory Recovery

With the emergence of a large amount of trajectory data and the rise of deep learning, many studies have adopted deep learning models for trajectory restoration. Some studies have conducted trajectory restoration by mining the periodic patterns of trajectories. Feng et al. [4] proposed a historical attention model to capture multiple periodic patterns of trajectories. Xia et al. [19] further developed multiple attention mechanisms to fully explore the movement patterns of trajectories. Sun et al. [14] introduced graph neural networks to capture the shifting periodicity of human mobility. However, these studies divide time into fixed time slices and cannot be directly applied to trajectory restoration in free space. Wang et al. [17] introduced Kalman filtering and subseq2seq module to restore trajectories in free space. Ren et al. [12] designed a multitasking learning model based on Seq2Seq, which maps trajectory points in free space onto road segments and simultaneously completes trajectory restoration and map matching in an end-to-end manner.

2.2 Sequence-to-Sequence Models

The sequence-to-sequence model (Seq2Seq) was originally used for machine translation problems [1]. Due to the similarity between text sequences and trajectory sequences, Seq2Seq and its variants are often used for trajectory generation [9], trajectory similarity learning [7], and other trajectory sequence tasks. Park et al. [11] proposed a Seq2Seq model based on LSTM encoder-decoder for real-time generation of surrounding vehicle trajectory sequences. Liu et al. [9] detected anonymous trajectories through the Seq2Seq model. However, to our best knowledge, this work is the first attempt to employ a bidirectional decoder in the Seq2Seq model to recover trajectories in road networks.

2.3 Pre-training Embedding

The importance of pre-training in natural language processing has been proved [10]. Due to the similarity between the trajectory sequence and the text sequence, some studies use the pre-training method in natural language processing to generate the vector representation of the trajectory. Feng et al. introduced Skip Gram [10] to model location points. However, the time information in trajectory points is equally important, and some studies have attempted to introduce time information for pre-training. Lin et al. [8] proposed a pre-training method based on the BERT model, which learns the embedding representation of position points and time, and evaluates the quality of the embedding vector in the user's next position prediction task. Zhao et al. [22] proposed the temporal POI embedding model to capture temporal characteristics on different days. Wan et al. [16] proposed a Time-Aware Location Embedding method with a tree structure to incorporate the temporal information. Shimizu et al. [13] designed a novel method that combines visited time and stayed duration to improve the quality of generated place embeddings.

3 Preliminaries

In this section, we first introduce the notations throughout the manuscript, and then formally define our task.

Definition 1. GPS Trajectory: A GPS Trajectory $\widetilde{\tau}$ can be defined as a sequence of GPS positions with timestamps, i.e., $\widetilde{\tau} = (p_1, p_2, ..., p_n)$, where $p_i = (lat, lng, t), \forall i, 1 \leq i \leq n$.

Definition 2. Map-Matched Trajectory: A Map-matched Trajectory τ can be defined as a sequence of road segments with timestamps, i.e., $\tau = (a_1, a_2, ..., a_m)$, where $a_j = (\hat{q}_j, t_j), \forall j, 1 \leq j \leq m$.

Definition 3. Road Network: A road network is a directed graph $\mathcal{G} = (\mathbb{V}, \mathbb{E})$, where $\mathbb{V} = (v_1, v_2, ..., v_k)$ is a set of nodes representing intersections of road segments, and $\mathbb{E} = (e_1, e_2, ..., e_l)$ refers a set of edges representing the road segments which connect nodes v in $\mathbb{V}$. It contains two properties: 1) start and end nodes, indicating the start and end GPS position of a road segment; 2) length, which refers to the distance of a road segment in meters.

Definition 4. Keeping Ratio: A keeping ratio ϵ is the retention rate of the map-matched trajectory τ. For example, a keeping ratio of 10% represents randomly removing 90% of the road sections from the τ and retaining the remaining 10%.

Problem Definition: Given an $\epsilon-$keeping ratio trajectory τ_ϵ that misses some road segments and $\mathcal{G} = (\mathbb{V}, \mathbb{E})$ that represents the road network. We aim to recover the map-matched trajectory $\tau = (a_1, a_2, ..., a_m)$. The road segment in τ will be projected onto the road network.

4 Methodology

In this section, we introduce the proposed Pre-training and Bidirectional Semantic enhanced Trajectory Recovery Model, named PBTR. Figure 1 illustrates the architecture of our PBTR model. It mainly consists of two parts: 1) Pre-Training and Bidirectional Road Representation Encoder which divides into a pre-training embedding module and a bidirectional encoder; 2) Attentive Recurrent Decoder.

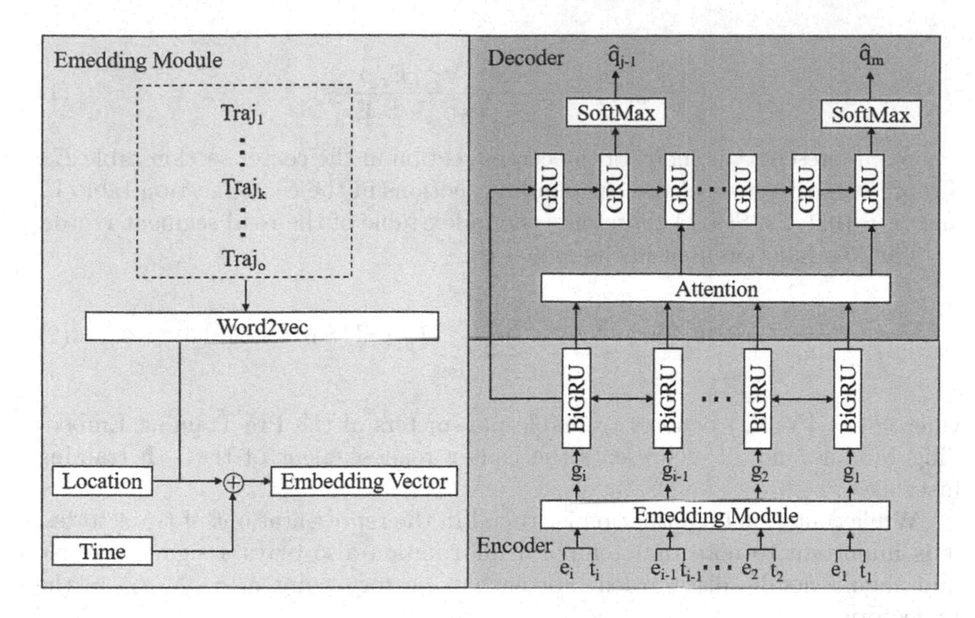

Fig. 1. Structure of PBTR.

4.1 Pre-training and Bidirectional Road Representation Encoder

The key to trajectory recovery is first to learn the fine-grained representation vectors of segments within the road networks, which helps learn the inter-dependencies underlying the segments and further improves the performance of recovery. An intuitive solution is to design an Encoder-Decoder architecture, which takes the historical trajectories as input and outputs the recovered trajectories. However, the objective of the task is to minimize the gap between target sequences and recovered sequences, which deviates from the original purpose of considering the co-occurrence between segments within the road networks. Meanwhile, prior work [6] has demonstrated a tendency that pre-training alone could improve techniques devised for a specific task. A pre-training-based sequential encoder is designed here.

Pre-training Embedding Module. Let $\mathcal{R}$ denotes the set of road segments in trajectories. To obtain a denser representation of road segment vector $\mathbf{E_r} \in \mathbb{R}^{|\mathcal{R}|\times d}$, we designed an embedding module that adopts the skip-gram idea in Word2vec [10], treating road segments as words and trajectories as sentences. The input of the embedding module is the set of map-matched trajectories Γ, and the output is the representation of road segment vector $\mathbf{E_r}$. Firstly, the input trajectory is processed into training samples according to the window size u. Then, we slide the center segment of the window to predict the surrounding segments, in order to train the surrounding segment vector table $\mathbf{V}$ and the center word vector table $\mathbf{E_r}$:

$$P(r_o|r_c) = \frac{\exp\langle \mathbf{V}_o^\top, \mathbf{E}_{\mathbf{r_c}} \rangle}{\sum_{i \in \mathcal{S}} \exp\langle \mathbf{V}_i^\top, \mathbf{E}_{\mathbf{r_c}} \rangle}, \tag{1}$$

where c represents the index of the center section in the center section table E_r, o represents the index of the surrounding sections in the center section table V, and $S = \{0, 1, ..., |S| - 1\}$ represents the index table of the road segment vector.

The loss function used in the model is:.

$$\mathcal{L}_1(\theta_1) = -\sum_{t=1}^{|\mathcal{D}_b|} \sum_{-u \le j \le u, j \ne 0} P_{\theta_1}(r^{(t+j)}|r^{(t)}), \tag{2}$$

where $\theta_1 = \{\mathbf{V}, \mathbf{E_r}\}$ denotes trainable parameters of the Pre-Training Embedding Module and $r^{(t)}$ represents the center road segment of the t-th training instance.

While contextual information is crucial in the representations of trajectories, it is important to note that temporal information also plays a significant role and should not be disregarded. For each trajectory point $a_i = (r_i, t_i)$ in the trajectory:

$$e_{t_i}(2j) = \sin(t_i/10000^{2j/d}), \tag{3}$$

$$e_{t_i}(2j+1) = \cos(t_i/10000^{2j/d}), \tag{4}$$

$$g_i = \mathbf{E_{r_i}} + e_{t_i} \tag{5}$$

where j denotes the j-th dimension, $\mathbf{E_{r_i}}$ represents the embedded representation corresponding to road segment r_i.

Algorithm 1: Training Algorithm for Embedding Module

Input: Set of trajectories : $\Gamma = \{\tau_1, \tau_2, ..., \tau_{n_t}\}$;
Window size: u

Output: Embedding of road segments : $\mathbf{E_r} \in \mathbb{R}^{|\mathcal{R}| \times d}$

1 //construct training instances: $\mathcal{D} \longleftarrow \emptyset$

2 **for** $i = 1$ *to* n_t **do**

3   **for** $j = 1$ *to* n_r-u **do**

4   Put a training instance $(r_j, r_{j+1}, ..., r_{j+u-1})$ from τ_i into $\mathcal{D}$

5 //Train the model: initialize the embedding of road segments $\mathbf{E_r}$
for $k \in \{1, 2, ..., EPOCH\}$ **do**

6   Select one batch $\mathcal{D}_b$ from $\mathcal{D}$;
    Update $\mathbf{E_r}$ by minimizing the objective $\mathcal{L}_1(\theta_1)$ with $\mathcal{D}_b$

7 return $\mathbf{E_r}$

The training process of the embedding module is shown in Algorithm 1.

Bidirectional Encoder. Encoder learns bidirectional spatiotemporal information from embedded ϵ-keeping ratio trajectory $\tau_\epsilon = (g_1, g_2, ..., g_{m_\epsilon})$ and generates context vector $H = (h_1, h_2, ..., h_n)$. Because of the similarity between trajectory restoration and machine translation, we use the classical Seq2Seq model [15] to restore the trajectory. Firstly, the encoder part is introduced. Different from the prediction task, the bidirectional Semantic information of the trajectory is used to encode in the trajectory restoration task to help the model better understand the trajectory pattern. Therefore, the encoder uses bidirectional Gate Recurrent Unit (GRU) [2] to encode the input trajectory, and the specific formula is as follows:

$$z_i{}^{(f)} = \sigma(\mathbf{W_z}{}^{(f)} \cdot [h_{i-1}{}^{(f)}, g_i] + \mathbf{b_z}{}^{(f)}), \tag{6}$$

$$r_i{}^{(f)} = \sigma(\mathbf{W_r}{}^{(f)} \cdot [h_{i-1}{}^{(f)}, g_i] + \mathbf{b_r}{}^{(f)}), \tag{7}$$

$$\overline{h}_{i-1}{}^{(f)} = \tanh(\mathbf{W_s}{}^{(f)} \cdot [r_i{}^{(f)} * g_{i-1}, g_i] + \mathbf{b_s}{}^{(f)}), \tag{8}$$

$$h_i{}^{(f)} = (1 - \mathbf{z_i}{}^{(f)}) * h_{i-1}{}^{(f)} + \mathbf{z_i}{}^{(f)} * \overline{h}_{i-1}{}^{(f)} \tag{9}$$

where $\mathbf{W_x}{}^{(f)}$ represents the weights for the forward respective gate(x) neurons and $\mathbf{b_x}{}^{(f)}$ is the bias for the respective gate(x).

$$z_i{}^{(b)} = \sigma(\mathbf{W_z}{}^{(b)} \cdot [h_{i+1}{}^{(b)}, g_i] + \mathbf{b_z}{}^{(b)}), \tag{10}$$

$$r_i{}^{(b)} = \sigma(\mathbf{W_r}{}^{(b)} \cdot [h_{i+1}{}^{(b)}, g_i] + \mathbf{b_r}{}^{(b)}), \tag{11}$$

$$\overline{h}_{i+1}{}^{(b)} = \tanh(\mathbf{W_s}{}^{(b)} \cdot [r_i{}^{(b)} * g_{i+1}, g_i] + \mathbf{b_s}{}^{(b)}), \tag{12}$$

$$h_i{}^{(b)} = (1 - \mathbf{z_i}{}^{(b)}) * h_{i+1}{}^{(b)} + \mathbf{z_i}{}^{(b)} * \overline{h}_{i+1}{}^{(b)} \tag{13}$$

where $\mathbf{W_x}{}^{(b)}$ represents the weights for the reverse respective gate(x) neurons and $\mathbf{b_x}{}^{(b)}$ is the bias for the respective gate(x). To simplify, the encoder derives the hidden state h_i as:

$$h_i{}^{(f)} = GRU^{(f)}(h_{i-1}, g_i), \tag{14}$$

$$h_i{}^{(b)} = GRU^{(b)}(h_{i+1}, g_i), \tag{15}$$

$$h_i = Concat\{h_i{}^{(f)}, h_i{}^{(b)}\} \tag{16}$$

where the last state h_i will be considered as the context vector as well as the initial hidden state for the decoder.

4.2 Attentive Recurrent Decoder

After obtaining the fixed-dimensional representation of the input historical trajectory sequence given by the last hidden state of the Encoder module above, the conditional probability of the output recovered sequence is needed to be computed. The Decoder takes the output H of the Encoder module above as input and outputs the recovered map-matched trajectory τ. The decoder uses GRU for decoding while introducing an attention mechanism to help the model understand the spatiotemporal dependence of trajectories. The hidden state s_j is determined by the following formula:

$$s_j = GRU^{(f)}(s_{j-1}, Concat\{Emb(\hat{q}_{j-1}), m_j\}) \tag{17}$$

where the context vector m_j is computed by a weighted sum of all the output vectors h from the encoder and m_j is formulated as:

$$m_j = \sum_{k=1}^{n} \alpha_{j,k} h_j, \tag{18}$$

$$\alpha_{j,k} = \frac{\exp(f_{j,k})}{\sum_{k'}^{n} \exp(f_{j,k'})}, \tag{19}$$

$$f_{j,k} = l^{\top} \cdot \tanh(\mathbf{W_h} s_j + \mathbf{W_s} h_k) \tag{20}$$

where l, $\mathbf{W_h}$ and $\mathbf{W_s}$ are the learnable parameters, and s_j denotes the current mobility status from the decoder. For the road segment recovery, we adopt the cross entropy as the loss function:

$$\mathcal{L}_2(\theta_2) = -\sum_{n \in \mathcal{N}} \sum_{t \in \mathcal{T}^{\mathcal{P}}} \langle y^{n,t}, \hat{q}^{n,t} \rangle + \lambda |\theta_2|^2 \tag{21}$$

where $\theta_2 = \{\mathbf{W_x}^{(f)}, \mathbf{W_x}^{(b)}, \mathbf{b_x}^{(f)}, \mathbf{b_x}^{(b)}, \mathbf{W_h}, \mathbf{W_s}\}$, $\langle , \rangle$is the inner product, $y^{n,t}$ is the one-hot representation of the segment in n-th trajectory's t-th time step, $\mathcal{T}^{\mathcal{P}}$ denotes the missing time steps and λ is a parameter to control the power of regularization.

4.3 Model Training

To fully train the model and improve its performance, we perform a two-stage training process. Algorithm 2 illustrates the training process of the PBTR model. During the training process, we apply the gradient descent approach to update parameters $\theta = \{\theta_1, \theta_2\}$, with learning rate lr and a pre-defined $Epoch_{max}$.

Algorithm 2: Two-stage training process of PBTR

Input: Trajectories $\mathcal{T}$, ϵ-keeping ratio Trajectories $\mathcal{T}_\epsilon$, initialized parameters $\theta = \{\theta_1, \theta_2\}$, learning rate lr, max iteration $Epoch_{max}$

Output: A well trained PBTR with parameters θ

1 Training θ_1 using Algorithm 1;
2 **for** $i \in \{1, 2, ..., Epoch_{max}\}$ **do**
3   Calculate gradient $\nabla L_2(\theta_2)$using Eq. 21;
    Update $\theta_2 \leftarrow \theta_2 - \nabla L_2(\theta_2)$

5 Experiments

In this section, we first set up the experiments, and then present the performance comparison and result analysis.

5.1 Experimental Settings

Data Description. To measure the performance of our proposed model, we use two real-world trajectory datasets collected from Guangxi, and Porto respectively. The trajectory data from Guangxi is sampled from gantries on the highway in Guangxi Province, while the dataset from Porto is a public trajectory dataset, and was originally released for a taxi trajectory prediction competition on Kaggle. The Guangxi dataset contains 133,274 drivers and 5.42 million road segment records over a period of 1 month, in September 2021. The Porto dataset contains 225,472 drivers and 6.76 million GPS records during one year. In the two datasets, we split the dataset into training set, validation set, and test set with a splitting ratio of 7: 2: 1 and conduct experiments with keeping ratio = 10%, 20%, 30%.

Baseline Algorithms. We consider using the following successful methods for comparison:

- **Top:** It is a simple counting-based method. The most popular road segments in the training set are used as recovery for all trajectories.
- **Seq2Seq** [15]**:** It is a deep learning model that converts one sequence into another under the encoder-decoder structure.
- **DeepMove** [4]**:** It is a multi-modal embedding recurrent neural network that can capture the complicated sequential transitions by jointly embedding the multiple factors that govern human mobility.
- **AttnMove** [19]**:** This is the latest trajectory recovery method which leverages various attention mechanisms to model the regularity and periodical patterns of the user's mobility.
- **MTrajRec** [12]**:** It implements a multi-task sequence-to-sequence learning architecture to predict road segment and moving ratio simultaneously.

Evaluation Metrics. We use *Accuracy*, *Recall*, and *Precision* to evaluate the performance of trajectory recovery by comparing the recovered road segments ζ_R to the ground truth ζ_G:

- **Accuracy** is the proportion of correctly recovered road segments to the total road segment.
- **Recall** is denoted as $recall = \frac{|\zeta_R \cap \zeta_G|}{|\zeta_G|}$.
- **Precision** is defined as $precision = \frac{|\zeta_R \cap \zeta_G|}{|\zeta_R|}$

Table 1. Experimental results on two datasets.

Metric	Datasets	Porto			Guangxi		
	Keeping Ratio	30%	20%	10%	30%	20%	10%
Accuracy	Top	0.4997	0.4719	0.4534	0.6122	0.5974	0.5827
	Seq2Seq	0.5465	0.5289	0.5256	0.6748	0.6462	0.6275
	DeepMove	0.5225	0.5066	0.4891	0.7078	0.6723	0.6667
	AttnMove	0.5454	0.5423	0.5296	0.7221	0.6861	0.6778
	MTrajRec	0.6335	0.6271	0.5835	0.7314	0.6749	0.6573
	PBTR	**0.6951**	**0.6364**	**0.6089**	**0.7750**	**0.7575**	**0.7188**
Recall	Top	0.5802	0.5550	0.5378	0.6586	0.6519	0.6348
	Seq2Seq	0.6237	0.6170	0.6122	0.7195	0.6939	0.6742
	DeepMove	0.6024	0.5998	0.5951	0.7476	0.7179	0.7082
	AttnMove	0.6194	0.6169	0.6097	0.7542	0.7231	0.7151
	MTrajRec	0.6785	0.6665	0.6443	0.7611	0.7122	0.6950
	PBTR	**0.7186**	**0.6741**	**0.6585**	**0.7993**	**0.7851**	**0.7518**
Precision	Top	0.6549	0.6204	0.5958	0.7808	0.7532	0.7406
	Seq2Seq	0.7090	0.7063	0.7050	0.8153	0.7934	0.7737
	DeepMove	0.6948	0.6850	0.6737	0.8416	0.8106	0.8038
	AttnMove	0.7114	0.7156	0.7068	0.8500	0.8263	0.8194
	MTrajRec	0.7711	0.7668	0.7366	0.8530	0.8066	0.7887
	PBTR	**0.8138**	**0.7785**	**0.7553**	**0.8798**	**0.8674**	**0.8358**

5.2 Result and Analysis

Overall Performance. We compare our PBTR with the baseline models in terms of *Accuracy*, *Recall*, and *Precision*. The performance of different approaches with different keeping ratios for trajectory recovery is presented in Table 1. We have the following observations:

- Statistical models perform worse than deep learning models. This is because statistical models only restore trajectories based on historical probabilities and cannot capture complex spatiotemporal relationships, while recurrent neural networks in deep learning models are better at capturing the spatiotemporal dependencies of sequences.
- DeepMove and Attnmove both use the historical trajectories of specific users to enhance the performance of the model. But as shown in the table, the

indicators of these two models on both datasets are not as good as PBTR. This is because learning from the historical trajectories of specific users does not improve the generalization of the model, while our pre-training embedding module learns a universal road segment embedding representation from all trajectories, resulting in better performance.
- PBTR outperforms all the baselines for all the metrics. Specifically, the *Accuracy* of PBTR outperforms the best baseline by 6.2% when the keeping ratio is 30%. *Recall* and *Precision* are improved by 4.0% and 4.3% respectively. This proves the effectiveness of our PBTR in trajectory recovery.

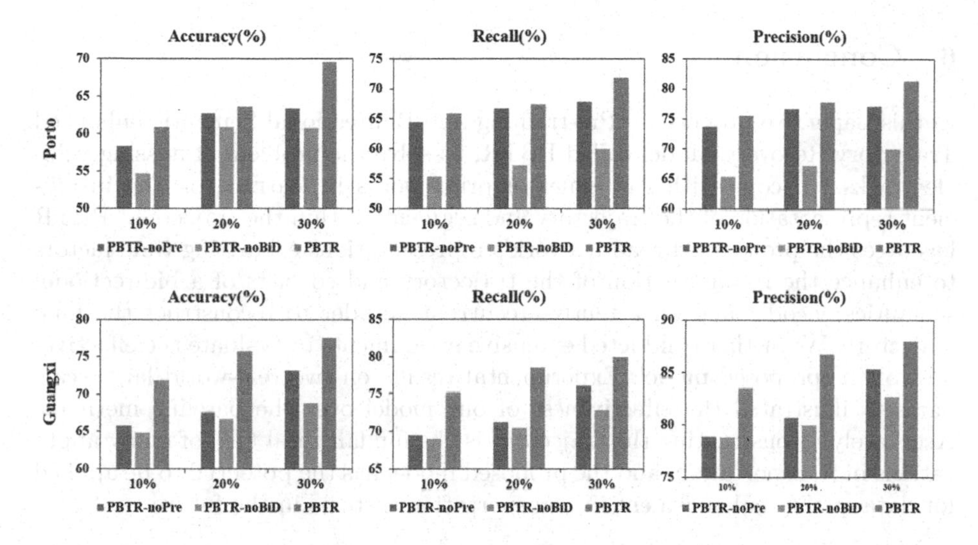

Fig. 2. Ablation study results.

5.3 Ablation Study

We have introduced a pre-training module and bidirectional encoding module to improve the performance of our PBTR. To explore the role of these modules, we conducted experiments on two datasets among the following variants of PBTR:

- **PBTR-noPre:** We remove the embedding module from PBTR and use one-hot encoding to represent road segments.
- **PBTR-noBiD:** We remove the bidirectional encoder from PBTR and use GRU to encode trajectories.
- **PBTR:** The model uses both embedding module and bidirectional encoder.

Importance of the Embedding Module: As shown in Fig. 2, both PBTR-noPre and PBTR get better performance as the keeping ratio increases. But when the pre-training module is removed, the performance declines significantly.

Especially, it causes *Accuracy* to shrink by 6.2% and *Recall* to shrink by 4.0% with a 30% keeping ratio on the Porto dataset. This is because the pre-training embedding module introduces contextual information into the embedded representation of road segments, which improves the training efficiency of the model.

Importance of the Bidirectional Encoder: As illustrated in Fig. 2, the results of PBTR-noBiD fall obviously compared with PBTR. Particularly, *Accuracy* decreases by 8.7% and *Precision* decreases by 11.3% after removing the bidirectional encoding module at the keeping ratio of 30% on the Porto dataset. One intuitive reason is that the bidirectional encoder utilizes the bidirectional semantics of trajectories, providing more information to the decoder.

6 Conclusion

In this paper, we proposed a Pre-training and Bidirectional Semantic enhanced Trajectory Recovery Model called PBTR, to solve the problem of missing vehicle tracks. To cope with the issues of prior works in learning on the insufficient representation of the trajectory and segments within the trajectory, PBTR leverages the pre-trained road network's representations extracting time factors to enhance the reconstruction of the trajectory and consists of a bidirectional semantics encoder and an attentive recurrent decoder to reconstruct the final trajectory. We further conducted extensive experiments to evaluate the effectiveness of the proposed model. Experimental results on two real-world large-scale datasets illustrated the effectiveness of our model over the baseline methods. Accurately reconstructing the trajectory is the fundamental task of many applications in transportation, and the proposed model has the potential to be applied for data quality enhancement in transportation systems in the future.

Acknowledgement. This work was supported by the National Natural Science Foundation of China (No. 62272023) and the Fundamental Research Funds for the Central Universities (No. YWF-23-L-1203).

References

1. Bahdanau, D., Cho, K., Bengio, Y.: Neural machine translation by jointly learning to align and translate. arXiv preprint arXiv:1409.0473 (2014)
2. Cho, K., et al.: Learning phrase representations using RNN encoder-decoder XGBoostfor statistical machine translation. arXiv preprint arXiv:1406.1078 (2014)
3. Erhan, D., Courville, A., Bengio, Y., Vincent, P.: Why does unsupervised pre-training help deep learning? In: Proceedings of the Thirteenth International Conference on Artificial Intelligence and Statistics, pp. 201–208. JMLR Workshop and Conference Proceedings (2010)
4. Feng, J., et al.: DeepMove: predicting human mobility with attentional recurrent networks. In: Proceedings of the 2018 World Wide Web Conference, pp. 1459–1468 (2018)

5. Han, L., Du, B., Lin, J., Sun, L., Li, X., Peng, Y.: Multi-semantic path representation learning for travel time estimation. IEEE Trans. Intell. Transp. Syst. **23**(8), 13108–13117 (2021)
6. Hendrycks, D., Lee, K., Mazeika, M.: Using pre-training can improve model robustness and uncertainty. In: Chaudhuri, K., Salakhutdinov, R. (eds.) Proceedings of the 36th International Conference on Machine Learning. Proceedings of Machine Learning Research, vol. 97, pp. 2712–2721. PMLR (2019)
7. Li, X., Zhao, K., Cong, G., Jensen, C.S., Wei, W.: Deep representation learning for trajectory similarity computation. In: 2018 IEEE 34th International Conference on Data Engineering (ICDE), pp. 617–628. IEEE (2018)
8. Lin, Y., Wan, H., Guo, S., Lin, Y.: Pre-training context and time aware location embeddings from spatial-temporal trajectories for user next location prediction. In: Proceedings of the AAAI Conference on Artificial Intelligence, vol. 35, pp. 4241–4248 (2021)
9. Liu, Y., Zhao, K., Cong, G., Bao, Z.: Online anomalous trajectory detection with deep generative sequence modeling. In: 2020 IEEE 36th International Conference on Data Engineering (ICDE), pp. 949–960. IEEE (2020)
10. Mikolov, T., Chen, K., Corrado, G., Dean, J.: Efficient estimation of word representations in vector space. arXiv preprint arXiv:1301.3781 (2013)
11. Park, S.H., Kim, B., Kang, C.M., Chung, C.C., Choi, J.W.: Sequence-to-sequence prediction of vehicle trajectory via LSTM encoder-decoder architecture. In: 2018 IEEE Intelligent Vehicles Symposium (IV), pp. 1672–1678. IEEE (2018)
12. Ren, H., Ruan, S., Li, Y., Bao, J., Meng, C., Li, R., Zheng, Y.: MTrajRec: map-constrained trajectory recovery via seq2seq multi-task learning. In: Proceedings of the 27th ACM SIGKDD Conference on Knowledge Discovery & Data Mining, pp. 1410–1419 (2021)
13. Shimizu, T., Yabe, T., Tsubouchi, K.: Learning fine grained place embeddings with spatial hierarchy from human mobility trajectories. arXiv preprint arXiv:2002.02058 (2020)
14. Sun, H., Yang, C., Deng, L., Zhou, F., Huang, F., Zheng, K.: PeriodicMove: shift-aware human mobility recovery with graph neural network. In: Proceedings of the 30th ACM International Conference on Information & Knowledge Management, pp. 1734–1743 (2021)
15. Sutskever, I., Vinyals, O., Le, Q.V.: Sequence to sequence learning with neural networks. In: Advances in Neural Information Processing Systems, vol. 27 (2014)
16. Wan, H., Li, F., Guo, S., Cao, Z., Lin, Y.: Learning time-aware distributed representations of locations from spatio-temporal trajectories. In: Li, G., Yang, J., Gama, J., Natwichai, J., Tong, Y. (eds.) DASFAA 2019. LNCS, vol. 11448, pp. 268–272. Springer, Cham (2019). https://doi.org/10.1007/978-3-030-18590-9_26
17. Wang, J., Wu, N., Lu, X., Zhao, W.X., Feng, K.: Deep trajectory recovery with fine-grained calibration using Kalman filter. IEEE Trans. Knowl. Data Eng. **33**(3), 921–934 (2019)
18. Wu, R., Luo, G., Shao, J., Tian, L., Peng, C.: Location prediction on trajectory data: a review. Big Data Min. Anal. **1**(2), 108–127 (2018)
19. Xia, T., et al.: AttnMove: history enhanced trajectory recovery via attentional network. In: Proceedings of the AAAI Conference on Artificial Intelligence, vol. 35, pp. 4494–4502 (2021)
20. Xu, Y., Sun, L., Du, B., Han, L.: Spatial semantic learning for travel time estimation. In: Memmi, G., Yang, B., Kong, L., Zhang, T., Qiu, M. (eds.) KSEM 2022, Part III. LNCS, vol. 13370, pp. 15–26. Springer, Cham (2022). https://doi.org/10.1007/978-3-031-10989-8_2

21. Yuan, J., Zheng, Y., Xie, X., Sun, G.: T-drive: enhancing driving directions with taxi drivers' intelligence. IEEE Trans. Knowl. Data Eng. **25**(1), 220–232 (2011)
22. Zhao, S., Zhao, T., King, I., Lyu, M.R.: Geo-teaser: geo-temporal sequential embedding rank for point-of-interest recommendation. In: Proceedings of the 26th International Conference on World Wide Web Companion, pp. 153–162 (2017)

Event-Aware Document-Level Event Extraction via Multi-granularity Event Encoder

Zetai Jiang[1,2], Sanchuan Tian[1,2], and Fang Kong[1,2](✉)

[1] Laboratory for Natural Language Processing, Soochow University, Suzhou, China
{20215227035,20195227074}@stu.suda.edu.cn, kongfang@suda.edu.cn
[2] School of Computer Science and Technology, Soochow University, Suzhou, China

Abstract. Event extraction (EE) is a crucial task in natural language processing that entails identifying and extracting events from unstructured text. However, the prior research has largely concentrated on sentence-level event extraction (SEE), while disregarding the increasing requirements for document-level event extraction (DEE) in real-world scenarios. The latter presents two significant challenges, namely the arguments scattering problem and the multi-event problem, which are more frequently observed in documents. In this paper, we propose an event-aware document-level event extraction framework, which can accurately detect event locations throughout the entire document without triggers and encode information at three different granularities (i.e., event-level, document-level, and sentence-level) via a multi-granularity event encoder. The resulting event-related holistic representation is then utilized for subsequent event record generation, thereby improving the accuracy of argument classification. Our proposed models effectiveness is demonstrated through experimental results obtained from a large Chinese financial dataset.

Keywords: Event extraction · Document-level · Multi-event · Event-aware

1 Introduction

Event extraction (EE) is a natural language processing (NLP) task that involves identifying and extracting structured information about events from unstructured text data. In recent years, event extraction has garnered considerable attention in the NLP community, owing to its potential to support downstream tasks like knowledge graph construction, dialogue systems, question answering, etc. Additionally, the task has demonstrated broad usefulness across diverse domains, such as finance, law, medicine, etc. The rise of widespread digitization across various domains has led to an increasing need for document-level event

Z. Jiang and S. Tian—Equal contribution.

B. Luo et al. (Eds.): ICONIP 2023, CCIS 1966, pp. 17–29, 2024.
https://doi.org/10.1007/978-981-99-8148-9_2

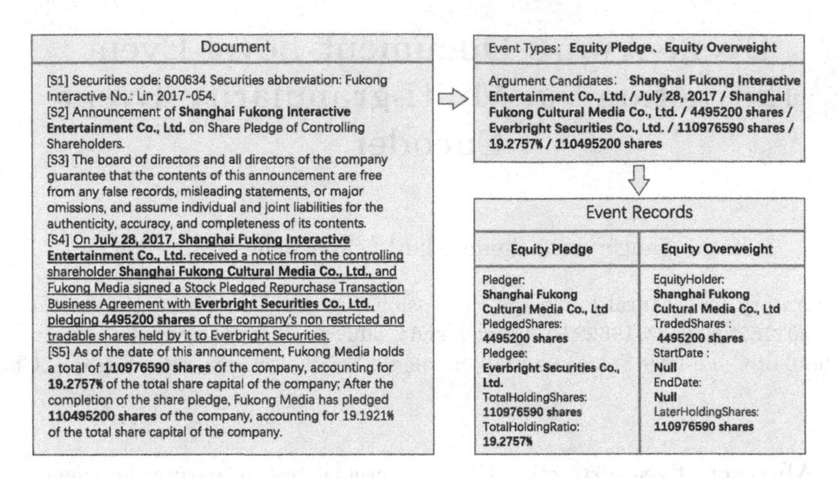

Fig. 1. An example of document-level event extraction, contains multiple event records, and the arguments of these events are scattered in different sentences of the document.

extraction (DEE). However, existing researches [3,4,6,8,9,12] have primarily focused on sentence-level event extraction (SEE), which may not be suitable for DEE due to the aggravation of the arguments-scattering problem and the multi-event problem in longer documents.

The provided illustration in Fig. 1 exemplifies two critical challenges related to document-level event extraction: the multi-event problem and the arguments-scattering problem. Regarding the former challenge, the presented example comprises two event records, namely `Equity Pledge` and `Equity Overweight`, which lack clear boundaries in the text. Therefore, a capable model must identify these events comprehensively throughout the entire document. Concurrently, as the number of events increases, the number of arguments to extract also increases linearly. Figure 1 further demonstrates that these arguments are dispersed across the sentences in the document. Thus, it is crucial to identify all arguments accurately and assign them to their respective roles.

To address the aforementioned challenges in DEE tasks, Yang et al. [14] proposed the DCFEE model, which utilizes sequence labeling to identify arguments and completes events based on the central sentence of the triggers. However, the simplistic approach used for central sentence detection and argument completion may not fully capture the complex relationship between events and their corresponding arguments. Zheng et al. [15] made an innovative contribution to the DEE task by considering it as a direct event table filling task without triggers, decomposing it into two parts: argument candidate extraction and entity-based directed acyclic graph (EDAG) generation. Nonetheless, due to the absence of trigger information, the model must rely on the entire document to obtain complete event records, which can be excessively long and contain a considerable amount of redundant and irrelevant information.

As shown in Fig. 1, the document is composed of five sentences, with S1 and S3 being unhelpful for the event extraction task and possibly hindering its performance. In contrast, S4 contains two event mentions simultaneously. These sentences not only contain event mentions but also a substantial amount of relevant arguments, we refer to such sentences as event-aware sentences, which we can utilize to effectively capture relationships between and within multiple events, thereby overcoming the challenges associated with multi-event and arguments-scattering. By integrating event-aware sentences, we can obtain more comprehensive information that not only facilitates the accurate determination of event numbers but also enables the capture of dispersed arguments based on specific event mentions rather than the entire document. Thus, our primary objective is to identify event-aware sentences in lengthy documents and encode it comprehensively as a cue for downstream event record generation tasks.

Motivated by this observation, this paper proposes a new framework called Event-aware Document-level Event Extraction (EADEE). EADEE introduces a novel multi-granularity event encoder that can detect events at both document-level and sentence-level and encode information at different granularities. Additionally, to explicitly incorporate the structural and semantic information of event ontology, which is often neglected in previous research, we incorporate the event-level representation as a new granularity into the encoder. The resulting comprehensive representations are leveraged for downstream event record generation tasks. By utilizing these event-aware multi-granularity representations, the decoder can better capture the connections between different events and arguments, thus improving the accuracy of argument classification. EADEE effectively identifies event-aware sentences and integrates representations from three different granularities, namely event-level, document-level, and sentence-level, thereby enhancing its ability to address the challenges posed by multi-event and arguments-scattering.

This paper makes the following contributions:

- We propose EADEE, a novel document-level event extraction framework that can effectively detect event mentions in trigger-free scenarios and utilize them as clues, instead of the entire document, to complete the event extraction task.
- We design a multi-granularity encoder that can detect events in both documents and sentences simultaneously and integrate multi-granularity information for the event record decoder.
- We conduct extensive experiments on a widely used document-level dataset, and the experimental results demonstrate the effectiveness of EADEE compared to other competitive baselines.

2 Related Work

Event extraction has been extensively studied due to its numerous applications. For instance, Chen et al. [3] proposed DMCNN, an event extraction model of convolutional neural network by Dynamic Pooling. Liu et al. [6] introduced syntactic quick arcs to jointly extract events. Recently, Ma et al. [7] attempted to use prompt tuning to make better use of pre-trained language models (PLMs) for argument extraction. However, these methods often focus on the sentence-level, which may not be sufficient for the increasing amount of document-level data in real-world scenarios.

Document-level event extraction has gained increasing attention in recent years. Yang and Mitchell [13] introduced contextual information by modeling event structures, while Yang et al. [14] proposed a DEE framework based on key-event sentences and a corresponding argument filling strategy to complete event extraction from the document. Zheng et al. [15] treated DEE as a direct event table filling task without triggers and generated event records by constructing an entity-based directed acyclic graph. However, these approaches often treat the entire document as a single unit, which can result in redundant information and lack of identification of semantically richer sentences containing event mentions. Moreover, they often neglect the structural and semantic information provided by event ontology, which has been proven to be effective in assisting trigger and argument identification.

3 Methodology

3.1 Task Definition

In order to present our approach, it is necessary to first clarify the formalization of the DEE task. This task involves processing an input document $\mathcal{D} = \{S_1, S_2, ..., S_{N_s}\}$ with N_s sentences, as well as a predefined set of event types $\mathcal{T}$ and argument roles $\mathcal{R}$. Our objective is to extract all possible event records $\mathcal{E} = \{e_1, e_k, ..., e_k\}$, which consists of k records. Each event record e_i contains a corresponding event type $t_i \subset \mathcal{T}$ and a set of roles $(r_i^1, r_i^2, ..., r_i^n) \in \mathcal{R}$ that must be extracted based on t_i. We need to specify the event type t_i of each event record e_i and fill in the roles with the arguments $(a_i^1, a_i^2, ..., a_i^n)$, which are derived from the candidate arguments $\mathcal{A}$ in the document.

Our proposed model integrates various levels of information granularity to facilitate event record generation. Figure 2 showcases the architecture of our model, which comprises three primary components: (1) candidate argument extractor, (2) multi-granularity event encoder, and (3) event record decoder. We will provide a detailed explanation of each component in the following sections.

3.2 Candidate Argument Extractor

Given a document $\mathcal{D} = \{S_1, S_2, ..., S_{N_s}\}$ consisting of N_s sentences, we first capture the semantic representation of each sentence by encoding it separately using a sentence-level Transformer [11] encoder. The Transformer architecture has been shown to effectively capture contextual representations in various NLP tasks, making it a suitable choice for our model. Denoting the Transformer encoder as *Sent-Encoder*$(\cdot)$, we obtain the contextual representation $\mathbf{C}_i = \{\mathbf{c}_i^1, \mathbf{c}_i^2, ..., \ \mathbf{c}_i^l\} \in \mathbb{R}^{l \times d_h}$ of sentence $S_i = \{w_i^1, w_i^2, ..., w_i^l\}$ using the following equation:

$$\{\mathbf{c}_i^1, \mathbf{c}_i^2, ..., \mathbf{c}_i^l\} = Sent\text{-}Encoder(\{w_i^1, w_i^2, ..., w_i^l\}) \quad (1)$$

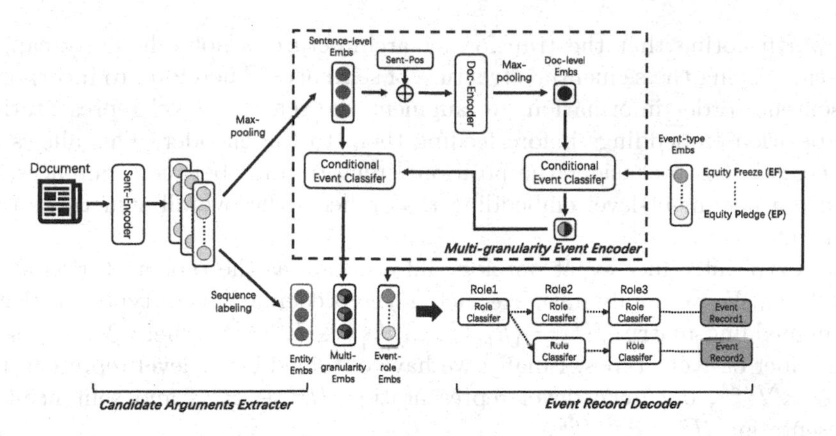

Fig. 2. The overall architecture EADEE. Firstly, each sentence in the document is encoded separately using a sentence-level encoder. Subsequently, these encoded representations are passed through a multi-granularity event encoder, which captures the semantic information at multiple levels of granularity to obtain a more comprehensive representation. After combining the encoded candidate arguments, multi-granularity event representation, and argument ontology information, the event record decoder generates the final event record in role order.

Where l denotes sentence length, and d_h denotes hidden layer dimension.

Next, we perform a sequence labeling task on each sentence using conditional random fields (CRF) to identify entity mentions. We consider all N_a entities in the document as candidate arguments $\mathcal{A} = \{a_1, a_2, ..., a_{N_a}\}$, and for any argument a_i with its span covering the j-th to k-th words in sentence S_i, we obtain its representation $h_i^a \in \mathbb{R}^{d_h}$ by performing max-pooling over $[\mathbf{c}_i^j, ..\mathbf{c}_i^k]$. This yields the representation of all candidate arguments $H^{\mathcal{A}} = \{h_1^a, h_2^a, ..., h_{N_a}^a\} \in \mathbb{R}^{N_a \times d_h}$. Similarly, we can obtain the sentence-level representation $H^S = \{h_1^s, h_2^s, ..., h_{N_s}^s\} \in \mathbb{R}^{N_s \times d_h}$ by performing max-pooling on the contextual representation $\mathbf{C}_i$ of each sentence S_i.

3.3 Multi-granularity Event Encoder

Multi-granularity Representations. Through the candidate argument extractor, we obtain sentence-level representations H^S. However, the scope of these representations is limited to the local sentence. To achieve a document-level representation, we leverage the Transformer architecture to construct a document-level encoder, known as *Doc-Encoder*$(\cdot)$. we first input the sentence-level representations into the encoder, after which we apply a max-pooling operation to combine the most salient features from the sentence-level embeddings into a single document-level embedding $H^D \in \mathbb{R}^{d_h}$. This approach allows us to distill the essential information from the sentence-level representations and produce a comprehensive representation of the whole document.

$$H^D = max\text{-}pooling(Doc\text{-}Encoder(H^S)) \tag{2}$$

It is worth noting that the transformer architecture is not inherently capable of distinguishing the sequential structure of sentences. Therefore, to incorporate the sentence order information, we augment the sentence-level representations with position embeddings before feeding them to the encoder. This allows the *Doc-Encoder*$(\cdot)$ to consider the positional relationships between sentences and produce a document-level embedding that reflects the overall structure of the document.

We extra introduce event ontology information as the representation at the event-level. We randomly initialize each of the predefined event types to a learnable embedding matrix $H^T = \{h_1^t, h_2^t, ..., h_{N_t}^t\} \in \mathbb{R}^{N_t \times d_h}$, where N_t represents the number of event types. Finally, we have obtained event-level representation $H^T \in \mathbb{R}^{N_t \times d_h}$, document-level representation $H^D \in \mathbb{R}^{d_h}$, and sentence-level representation $H^S \in \mathbb{R}^{N_s \times d_h}$.

Conditional Event Classifier. To synthesize representations from different granularities, it is necessary to detect and encode every possible event at multiple levels. To this end, we introduce a conditional event classifier (CEC) that can incorporate conditional dependencies into textual representations and classify them to achieve event detection at both the document-level and sentence-level.

The conditional event classifier integrate representations of different granularities through conditional layer regularization (CLN) [2,9,10], which is derived from traditional layer normalization [1] but can dynamically generates α and β based on previous conditional information. Given a contextual representation h and a conditional embedding c, we can combine their information into a single embedding $\overline{h}$ via CLN. The formulation for this process is as follows:

$$\begin{aligned} \overline{h} = CLN(h, c) = \alpha_c \odot (\frac{h-\mu}{\sigma}) + \beta_c \\ \alpha_c = W_\alpha c + b_\alpha, \beta_c = W_\beta c + b_\beta \end{aligned} \quad (3)$$

where $\mu \in \mathbb{R}$ and $\sigma \in \mathbb{R}$ are the mean and standard deviation of the elements of h, and $\alpha_c \in \mathbb{R}^{d_h}$ and $\beta_c \in \mathbb{R}^{d_h}$ are the gain and bias generated according to the conditions c.

To improve classification accuracy, the conditional event classifier further enhances the representation using self-attention mechanism. This mechanism calculates the dot product to obtain a similarity weight distribution of values, which helps to improve the quality of the representation:

$$S = Self\text{-}Attention\left(\overline{H}\right) \quad (4)$$

where $\overline{H}$ is the representation matrix constructed from $\overline{h}$.

The resulting representation is then fed into a classifier to complete the event detection task. It is important to note that we concatenate the resulting representation $s \subset S$ and conditional embedding c to capture the similarity information from different dimensions more comprehensively. The details are as follows:

$$p = Classifier(s, c) = Softmax(W[s; c; |s - c|; s \odot c]) \quad (5)$$

where $W \in \mathbb{R}^{2\times 4d_h}$ is learnable parameter, $[\cdot;\cdot]$ denotes concatenation, $|\cdot|$ denotes an absolute value operator, $\odot$ denotes the element-wise production.

Equations (3), (4), and (5) form a complete CEC module, we cascade the utilization of conditional event classifier, we first incorporates event ontology information $h_i^t \in \mathbb{R}^{d_h}$, which represents the event $t_i \subset \mathcal{T}$, into the document-level representation $H^D \in \mathbb{R}^{d_h}$ as event-level conditions, as detailed below:

$$S^D, p^D = CEC(H^D, h_i^t) \tag{6}$$

where $CEC(\cdot)$ stands for conditional event classifier, S^D denotes the representation that fuses the event ontology and the document information, and p^D is the document event detection. Note that since the document-level representation H^D has only one dimension, we use a multi-layer perceptron (MLP) instead of the self-attention mechanism during encoding.

If the event t_i exists, we incorporate the newly obtained representation $S^D \in \mathbb{R}^{d_h}$ into the sentence-level representation $H^S \in \mathbb{R}^{N_s\times d_h}$ as the document-level condition:

$$S^S, p^S = CEC(H^S, S^D) \tag{7}$$

We end up with a representation S^S that combines information at three different granularities and a prediction p^S for each sentence. We filter event-aware sentences based on p^S and obtain the multi-granularity representation $H^M \in \mathbb{R}^{N_e\times d_h}$ from $S^S \in \mathbb{R}^{N_s\times d_h}$, which is finally used for event record generation.

Event-Aware Sentence Detection Strategy. In contrast to traditional event extraction methods, this paper addresses the challenge of accurately detecting event-aware sentences in a document without explicit trigger. We propose a novel approach inspired by DCFEE [14], which leverages the distribution of arguments to identify key-event sentences. However, unlike DCFEE, which uses a heuristic strategy to pad the missing arguments from the vicinity of the key-event sentence, we assume that the sentence with the most arguments is the event-aware sentence. By doing so, we can obtain pseudo-sentence-level labels of these sentences without requiring additional manual annotation.

Furthermore, we argue that the CEC module utilized for document-level event detection and encoding can be efficiently transferred to the sentence-level. Therefore, we only need to train a singular CEC module that can proficiently handle tasks at multiple levels in the multi-granularity event encoder. Finally, by training on both ground-truth document-level labels and pseudo-sentence-level labels, we obtain a solitary CEC module with enhanced robustness, capable of effectively performing event detection and encoding at both document and sentence levels.

3.4 Event Record Decoder

To handle the variable number of event records, we adopt the same method as previous work [15] by constructing a directed acyclic graph for each event type. This results in a tree structure with leaf nodes corresponding to the records for each event type. However, in contrast to previous work, which used every sentence in the document to generate the final event record, we employ a more effective approach. Specifically, we replace the sentence representations with the event-aware multi-granularity representation H^M, obtained from our multi-granularity event encoder. This approach eliminates redundant or irrelevant information from the text, leading to more accurate event record generation.

The event record generation task can be viewed as a sequence of event role classification tasks for candidate arguments $H^{\mathcal{A}}$. To perform each event role classification, we begin by concatenating the comprehensive representation H^M with the candidate argument representation $H^{\mathcal{A}}$, which are then preliminarily encoded by the record encoder. This results in a candidate argument representation $\tilde{H}^{\mathcal{A}}$ that incorporates the event information, thus facilitating more effective event perception:

$$[\tilde{H}^M, \tilde{H}^{\mathcal{A}}] = \textit{Rec-Encoder}([H^M, H^{\mathcal{A}}]) \tag{8}$$

where *Rec-Encoder*$(\cdot)$ stands for record encoder, again using transformer architecture.

To enhance the discrimination of event roles, we map them to a randomly initialized matrix $H^R = \{h_1^r, h_2^r, ..., h_{N_r}^r\} \in \mathbb{R}^{N_r \times d_h}$ based on the event ontology information. This mapping is specific to each event type. To classify a given event role, we feed the role embedding $h_i^r \subset H^R$ as a condition into the conditional event classifier together with the candidate argument representation $\tilde{H}^{\mathcal{A}}$ as follows:

$$S^{\mathcal{A}}, p^{\mathcal{A}} = CEC(h_i^r, \tilde{H}^{\mathcal{A}}) \tag{9}$$

Using the conditional event classifier, we obtain a probability distribution $p^{\mathcal{A}}$ over the possible roles for a given candidate argument. We can assign the argument to its corresponding event role according to the probability. We repeat this process for each event role in order, until all roles have been assigned, resulting in complete records of the event.

3.5 Model Training

During the training phase, our model's overall loss function $\mathcal{L}$ is divided into four components: $\mathcal{L}_{ae}$, $\mathcal{L}_{ded}$, $\mathcal{L}_{sed}$, and $\mathcal{L}_{rg}$. These correspond to the argument extraction, document-level event detection, sentence-level event detection, and event record generation tasks described above, respectively.

$$\mathcal{L} = \lambda_1 \mathcal{L}_{ae} + \lambda_2 \mathcal{L}_{ded} + \lambda_3 \mathcal{L}_{sed} + \lambda_4 \mathcal{L}_{rg} \tag{10}$$

where $\mathcal{L}_{ae}$, $\mathcal{L}_{ded}$, $\mathcal{L}_{sed}$, and $\mathcal{L}_{rg}$ are all cross-entropy loss function, λ_1, λ_2, λ_3, and λ_4 are hyper-parameters.

4 Experiments

4.1 Experimental Setup

Dataset and Metrics. To assess the performance of our model, we utilized the ChFinAnn dataset, proposed by Zheng et al. [15], as a benchmark. This large-scale dataset comprises 32,040 Chinese financial documents and encompasses five prevalent event types in the financial domain, including Equity Freeze (EF), Equity Repurchase (ER), Equity Underweight (EU), Equity Overweight (EO), and Equity Pledge (EP). We partitioned the dataset into training, validation, and testing sets according to an 8:1:1 ratio.

To ensure impartial evaluations, we followed the evaluation metrics prescribed by Zheng et al. [15]. Specifically, we selected the most similar predicted event record when the predicted event type was correct for each document. We then presented the comparison results between the predicted event record and the ground truth in the form of Precision (P), Recall (R), and F1 score (F1).

Implementation Details. To prepare the input for our model, we limited the maximum number of sentences to 64 with a maximum sentence length of 128 characters. We utilized the basic transformer in each encoding stage, which consists of four layers with 768 hidden units each. During training, we employed the Adam optimizer [5] with a learning rate of $1e-4$. We selected the best performance on the validation set within 100 epochs to determine the testing set performance. For our loss function, we set $\lambda_1 = 0.05$, $\lambda_2 = 0.95$, $\lambda_3 = 0.01$, and $\lambda_4 = 0.95$. We also trained all event ontology embeddings from scratch.

4.2 Baselines

We compared our model with several baseline models, including **DCFEE** [14], which utilizes a sentence-level extraction model to extract arguments and fill event records with arguments surrounding key-event sentences. This model has two variants, **DCFEE-O** and **DCFEE-M**; **DCFEE-O** generates one record at a time, while **DCFEE-M** generates multiple possible argument combinations based on the closest distance to the key sentence. **DOC2EDAG** [15], a novel end-to-end model that considers DEE as a direct event table filling task without triggers. This model generates an entity-based directed acyclic graph to efficiently accomplish table filling tasks. Additionally, we compared our model with **GreedyDec**, a simplified variant of **DOC2EDAG** that fills only one event table entry in a greedy manner.

Table 1. Overall performance of all methods on ChFinAnn dataset, where P, R, and F1 denote precision, recall, and F1 score, respectively.

	EF			ER			EU			EO			EP		
	P.	R.	F1.	P.	R.	F1.	P.	R.	F1.	P.	R.	F1.	P.	R.	F1.
DCFEE-O	66.0	41.6	51.1	84.5	81.8	83.1	62.7	35.4	45.3	51.4	42.6	46.6	64.3	63.6	63.9
DCFEE-M	51.8	40.7	45.6	83.7	78.0	80.8	49.5	39.9	44.2	42.5	47.5	44.9	59.8	66.4	62.9
GreedyDec	**78.7**	46.1	58.1	86.0	76.3	80.9	67.2	40.4	50.5	69.7	41.4	51.9	**81.8**	40.3	54.0
Doc2EDAG	72.9	63.4	**67.8**	91.4	84.2	87.6	72.6	59.9	65.6	78.1	65.6	71.3	79.7	70.0	74.5
EADEE	66.5	**67.4**	67.0	**91.4**	**85.1**	**88.1**	**74.3**	**63.7**	**68.6**	**79.5**	**66.2**	**72.2**	77.5	**72.6**	**75.0**

Table 2. F1 scores for all event types on single-event (S.) and multi-event (M.) sets.

	EF		ER		EU		EO		EP	
	S.	M.	S.	M.	S.	M.	S.	M.	S.	M.
DCFEE-O	56.0	46.5	86.7	54.1	48.5	41.2	47.7	45.2	68.4	61.1
DCFEE-M	48.4	43.1	83.8	53.4	48.1	39.6	47.1	42.0	67.0	60.6
GreedyDec	75.7	39.6	83.9	49.1	61.3	34.1	65.8	30.6	78.4	36.4
Doc2EDAG	**76.4**	59.2	89.8	68.8	71.0	58.4	75.7	65.3	**84.3**	68.6
EADEE	71.7	**64.4**	**90.0**	**71.8**	**74.4**	**61.2**	**76.4**	**66.9**	82.8	**69.1**

4.3 Experiment Results

Main Results. Table 1 demonstrates that our proposed EADEE model has significantly improved upon other baselines across five distinct event types. In comparison to the previous state-of-the-art model, DOC2EDAG, our model has achieved higher F1 scores for all event types except for EF, with increases of 0.5, 3.0, 0.9, and 0.5 for ER, EU, EO, and EP, respectively. We attribute this success to our use of a multi-granularity event encoder, which can incorporate information from multiple perspectives and provide a more comprehensive representation of event-aware sentences for downstream event record generation.

Our EADEE model shows a marked improvement over DCFEE for all five event types, including both DCFEE-O and DCFEE-M. Our work introduces the concept of event-aware sentences, building on the notion of key-event sentences proposed in DCFEE. We determine the position of event-aware sentences through argument distribution and consider this position as a pseudo-labels for training purposes. Our experimental results show that our approach to using these pseudo-labels during training can integrate event-related information more smoothly and effectively than directly using the argument around the key-event sentence to fill the event record. Additionally, we have addressed the issue of insufficient information from pseudo-labels by transferring the document-level CEC module for use in the sentence-level task.

Table 3. F1 scores of ablation experiments on EADEE variants for each event type and the averaged (Avg.).

Model	EF	ER	EU	EO	EP	Avg.
EADEE	67.0	88.1	68.6	72.2	74.2	74.0
-MgEnc	−3.3	−3.2	−1.2	−2.7	−1.0	−2.3
-EaSent	−0.9	−1.8	−0.6	−1.0	−1.4	−1.1

Single-Event vs. Multi-event. To assess the performance of our model in more complex multi-event scenarios, we divided the test set into two parts: a single-event set containing only one event record per document and a multi-event set containing multiple records per document. As detailed in Table 2, our model achieved superior results in the multi-event set compared to other methods and performed competitively in the single-event set. We attribute this outcome to our event-aware sentence filtering mechanism. Unlike other baselines, our approach enables us to not only detect events in the document but also filter each event-aware sentence without triggers that are strongly related to the event record to be generated. Furthermore, our use of event-aware representations, in contrast to DOC2EDAG's employment of the whole document representation, provides more accurate and effective information for the event record decoder.

4.4 Ablation Studies

Ablation studies were conducted on EADEE to assess the efficacy of each component in our approach, namely: 1) *-MgEnc*, which replaces the multi-granularity encoding with the representation obtained directly from Eq. (1), and 2) *-EaSent*, which replaces the representation of event-aware sentences with the representation of the entire document. The results presented in Table 3 indicate that: 1) Both multi-granularity encoding and detection of event-aware statements improve the model's performance in complex document-level event extraction tasks, with an average improvement of 2.3 and 1.1 in F1 scores, respectively. 2) The fusion of multi-granularity information is more beneficial for the model's performance than detecting event-aware sentence representations.

5 Conclusion

In this paper, we propose a novel event-aware document-level event extraction framework via multi-granularity event encoder, named EADEE, that utilizes a multi-granularity event encoder to address the challenges of argument scattering and multi-event problems in DEE. Through the multi-granularity event encoder, EADEE can comprehensively integrate the representations of different granularities, and effectively identify the event-aware sentences in the document without triggers. By using these event-aware holistic representations, the accuracy of event record generation can be effectively improved. We conducted extensive experiments to demonstrate the effectiveness of our proposed model. In the future, we plan to investigate how to incorporate more fine-grained token-level information into our model to further improve its performance.

Acknowledgements. This work was supported by Projects 62276178 under the National Natural Science Foundation of China, the National Key RD Program of China under Grant No. 2020AAA0108600 and the Priority Academic Program Development of Jiangsu Higher Education Institutions.

References

1. Ba, J.L., Kiros, J.R., Hinton, G.E.: Layer normalization (2016)
2. Yu, B., et al.: Semi-open information extraction. In: Proceedings of the Web Conference, pp. 1661–1672 (2021)
3. Chen, Y., Xu, L., Liu, K., Zeng, D., Zhao, J.: Event extraction via dynamic multi-pooling convolutional neural networks. In: Proceedings of the Annual Meeting of the Association for Computational Linguistics, pp. 167–176 (2015)
4. Feng, Y., Li, C., Ng, V.: Legal judgment prediction via event extraction with constraints. In: Proceedings of the Annual Meeting of the Association for Computational Linguistics, pp. 648–664 (2022)
5. Kingma, D.P., Ba, J.: Adam: a method for stochastic optimization. In: Bengio, Y., LeCun, Y. (eds.) 3rd International Conference on Learning Representations, ICLR 2015, San Diego, CA, USA, 7–9 May 2015. Conference Track Proceedings (2015)
6. Liu, X., Luo, Z., Huang, H.: Jointly multiple events extraction via attention-based graph information aggregation. In: Proceedings of the Conference on Empirical Methods in Natural Language Processing (EMNLP), pp. 1247–1256 (2018)
7. Ma, Y., et al.: Prompt for extraction? PAIE: prompting argument interaction for event argument extraction. In: Proceedings of the 60th Annual Meeting of the Association for Computational Linguistics (Volume 1: Long Papers), pp. 6759–6774 (2022)
8. Nguyen, T.H., Cho, K., Grishman, R.: Joint event extraction via recurrent neural networks. In: Proceedings of the Conference of the North American Chapter of the Association for Computational Linguistics: Human Language Technologies, pp. 300–309 (2016)
9. Sheng, J., et al.: CasEE: a joint learning framework with cascade decoding for overlapping event extraction. In: Proceedings of the Annual Meeting of the Association for Computational Linguistics, pp. 164–174 (2021)
10. Su, J.: Conditional text generation based on conditional layer normalization (2019)
11. Vaswani, A., et al.: Attention is all you need. In: Advances in Neural Information Processing Systems, pp. 5998–6008 (2017)
12. Wang, S., Yu, M., Chang, S., Sun, L., Huang, L.: Query and extract: refining event extraction as type-oriented binary decoding. In: Findings of the Association for Computational Linguistics: ACL, pp. 169–182 (2022)
13. Yang, B., Mitchell, T.M.: Joint extraction of events and entities within a document context. In: Proceedings of the 2016 Conference of the North American Chapter of the Association for Computational Linguistics: Human Language Technologies, pp. 289–299. Association for Computational Linguistics (2016)

14. Yang, H., Chen, Y., Liu, K., Xiao, Y., Zhao, J.: DCFEE: a document-level Chinese financial event extraction system based on automatically labeled training data. In: Proceedings of ACL 2018, System Demonstrations, pp. 50–55. Association for Computational Linguistics, Melbourne (2018)
15. Zheng, S., Cao, W., Xu, W., Bian, J.: Doc2EDAG: an end-to-end document-level framework for Chinese financial event extraction. In: Proceedings of the 2019 Conference on Empirical Methods in Natural Language Processing and the 9th International Joint Conference on Natural Language Processing (EMNLP-IJCNLP), pp. 337–346 (2019)

Curve Enhancement: A No-Reference Method for Low-Light Image Enhancement

Zhixiang Zhang[1,2] and Shan Jiang[1,2(✉)]

[1] Key Laboratory of Ethnic Language Intelligent Analysis and Security Governance of MOE, Minzu University of China, Beijing, China
{liyoh,jshan}@muc.edu.cn

[2] School of Information Engineering, Minzu University of China, Beijing 100081, China

Abstract. In this paper, we introduce an end-to-end method for enhancing low-light images without relying on paired datasets. Our solution is reference-free and unsupervised, addressing the lack of real-world low-light paired datasets effectively. Specifically, we design a Brightness Boost Curve (BB-Curve) that enhances the brightness of image pixels in a finely mapped form. Additionally, we propose a lightweight deep neural network that can estimate the curve parameters and evaluate the quality of the enhanced images using a series of no-reference loss functions. We validate our method through experiments conducted on several datasets and provide both subjective and quantitative evaluations to demonstrate its significant brightness enhancement capabilities, free from smearing and artifacts. Notably, our approach displays a strong ability to generalize while retaining details that are crucial for image interpretation. With the reduced network structure and simple curve mapping, our model yields superior training speed and the best prediction performance among comparative methods.

Keywords: Low-light image enhancement · Deep learning · No-reference · Curve mapping

1 Introduction

Low-light images are a frequently encountered form of degraded images, which can be attributed to unfavorable shooting conditions, bad shooting equipment, inappropriate camera parameter settings, etc. The primary goal of this task is to increase the brightness and sharpness of images captured in low-light conditions while concurrently preserving their color and contrast. This process enhances the image's visual appeal, while simultaneously reducing distraction and highlighting useful features.

In recent years, remarkable achievements have been made in low-light image enhancement utilizing deep learning techniques. Among them, LLNet [1] pioneered the deep learning-based solution. Although Retinex models [2] were further developed by Yang *et al.* [3] until recently, these models typically disregard

B. Luo et al. (Eds.): ICONIP 2023, CCIS 1966, pp. 30–41, 2024.
https://doi.org/10.1007/978-981-99-8148-9_3

the potential impact of noise. The current low-light image enhancement methods, mainly utilize supervised learning methods. However, datasets containing corresponding well-lit natural lighting images for low-light images do not exist, and the current paired datasets are achieved by synthesizing or adjusting camera exposure, ISO, and other parameters. EnGAN [4] is the first to use Generative Adversarial Network (GAN) for unsupervised learning without the need for paired training data. Zero-DCE [5] proposed a novel method to transform the image enhancement problem into a specific curve estimation problem, which does not require paired or unpaired images for training. Kar *et al.* [6] combined Zero-shot learning with Koschmieder's Model [7] to improve both low light image enhancement and underwater image recovery. Moreover, some scholars use reinforcement learning (RL) [8] methods to address supervised learning limitations. RL, however, demands intricate reward mechanisms.

In this paper, we design a no-reference/unsupervised low-light image enhancement method that can get rid of the dependence on paired datasets. Specifically, the contributions of this paper can be summarized as follows:

- We designed a brightness boost curve (BB-Curve) that can efficiently enhance the brightness of image pixels in a refined mapping form.
- Additionally, we designed a lightweight deep neural network that achieved excellent training speed and predictive performance.
- Furthermore, we have redesigned a series of no-reference loss functions to replace the labels in supervised learning methods.

2 Methodology

2.1 Brightness Boost Curve

We normalize the pixel values of the input image to [0, 1] and then consider designing a curve that can map lower pixel values to higher values, subject to the following conditions: (1) The curve should be differentiable within the closed interval [0, 1] to prevent extreme variations in enhancement effects, especially in intervals near non-differentiable points. (2) The curve should be strictly monotonic within [0, 1]. This prevents inconsistent enhancement effects of similar pixels and maintains the contrast between adjacent pixels. (3) The curve should be iterable over itself. This achieves multiple enhancements.

Propose BB-Curve. The following curve was meticulously designed to meet the three aforementioned requirements perfectly, which can be defined by

$$f(x;\alpha) = \frac{x}{x + \alpha\,(1 - x)} \tag{1}$$

where x represents the original pixel value, $f(x;\alpha)$ represents the enhanced pixel value, and α is a trainable parameter. As shown in Fig. 1(a), this curve has the ability to enhance and darken pixels simultaneously. However, the enhancement and suppression intervals are not symmetrical. In order to maintain consistency, Eq. (1) is adopted as the enhancement curve, and the inverse function f^{-1} is applied as the suppression curve, which can be defined as

$$f^{-1}(x;\alpha) = \frac{\alpha x}{1 + (\alpha - 1)\, x} \tag{2}$$

To prevent overexposure and brightness imbalance, we set an upper limit $\beta = 0.4$, whereas α is constrained to the interval $[\beta, 1]$.

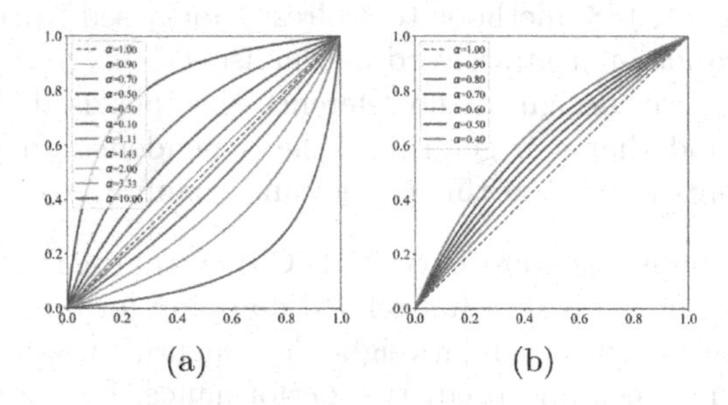

Fig. 1. (a) The graph of the curve $f(x;\alpha)$ with different α. (b) The graph of the curve $f(x;\alpha)$ after introducing the homogenization factor $\mathcal{H}$. The abscissa of all subfigures represents the input pixel values, and the ordinate represents the output pixel values.

Introduce a Homogenization Factor. In Fig. 1(a), it can be observed that the change in curve amplitude is not uniform when α changes by the same unit. This non-uniformity is not conducive to the learning of deep neural networks. So we propose the homogeneity factor $\mathcal{H}$, which can be represented as

$$\mathcal{H}(\alpha;\beta,\xi) = \xi(\alpha - \beta)(\alpha - 1) + 1 \tag{3}$$

where ξ represents the homogeneity coefficient, adjusted according to β. In this paper, ξ is set as 1.04. The purpose of $\mathcal{H}$ is to act as a factor for modifying α so that α falls onto a more uniform position on the curve. Specifically, α can be reassigned to $\alpha \cdot \mathcal{H}$. Figure 1(b) demonstrates the effectiveness of the homogenization factor.

Extended BB-Curve. Since our application scenario involves RGB images, the above formulas, need to be expanded to cover the entire image. In other words, we can rewrite Eq. (1) and Eq. (2) as

$$\begin{aligned} BBC\left((i,j);\mathcal{A}\right) &= \frac{I\left(i,j\right)}{I\left(i,j\right) + \mathcal{A}\left(i,j\right)\left(1 - I\left(i,j\right)\right)}, \\ BBC^{-1}\left((i,j);\mathcal{A}\right) &= \frac{\mathcal{A}\left(i,j\right) I\left(i,j\right)}{1 + \left(\mathcal{A}\left(i,j\right) - 1\right) I\left(i,j\right)} \end{aligned} \tag{4}$$

For a given input image I, $BBC(\cdot)$ corresponds to its enhanced version, whereby (i, j) determines the coordinate of the pixel and is applied to RGB channels separately. $\mathcal{A}$ is the curve parameter map that is consistent with the size of the input image, and $\mathcal{A}(i, j)$ stands for the parameter α of the particular curve corresponding to the pixel, i.e., enhancing pixels in this manner is carried out channel by channel and pixel by pixel.

Iterable BB-Curve. Too large curvature tends to produce greater brightness deviation when dealing with complex lighting scenes. Hence, we defined β to limit the maximum curvature of the curve. This limitation, though, also restricts the curve's adjustment ability. To address this contradiction, we modified the BB-Curve into an iterative form, which means that Eq. (4) can be reformulated as

$$\begin{aligned} BBC_n\left((i,j);\mathcal{A}\right) &= \frac{BBC_{n-1}\left(i,j\right)}{BBC_{n-1}\left(i,j\right)+\mathcal{A}_n\left(i,j\right)\left(1-BBC_{n-1}\left(i,j\right)\right)}, \\ BBC_n^{-1}\left((i,j);\mathcal{A}\right) &= \frac{\mathcal{A}_n\left(i,j\right)BBC_{n-1}^{-1}\left(i,j\right)}{1+\left(\mathcal{A}_n\left(i,j\right)-1\right)BBC_{n-1}^{-1}\left(i,j\right)} \end{aligned} \tag{5}$$

where the variable n denotes the number of iterations. In this paper, we set n to 8, taking into account algorithmic efficiency and performance. In particular, when n equals 1, Eq. (5) degenerates to Eq. (4). Up to this point, we can conclusively consider Eq. (5) to be the final version of the BB-Curve.

2.2 RV-Net

In this study, a lightweight convolutional neural network, known as RV-Net for its Reduced structure and V-shaped, was developed (Fig. 2(a)). RV-Net inputs RGB images and outputs $3 \times n$ curve parameter estimation maps. RV-Net has 9 layers of full convolution with symmetric skip connections. Each layer applies 28 3×3 convolution kernels with a stride of 1. ReLu is the activation function used in the first 8 layers, while the last layer uses Hardtanh to converge the output range to $\mathcal{A} \in [-1, 1]$. In fact, this is not a suitable range for the BB-Curve, thus the range mapping function is needed

$$RM(\mathcal{A}) = \frac{d-c}{b-a}(\mathcal{A}-a)+c \tag{6}$$

Equation (6) enables a remapping of the numerical values within $[a, b]$ to $[c, d]$. To match Eq. (5), set $a = 0, b = 1, c = \beta = 0.4, d = 1$. Specifically, for the curve parameter map $\mathcal{A}$ produced from RV-Net, different $\mathcal{A}(i, j)$ are remapped differently and then homogenized and fed into BB-Curve or its inverse function. Further details are available in Fig. 2(b). Overall, the RV-Net architecture is designed as a lightweight neural network with only 83,660 trainable parameters and 5.47G FLOPs[1] intended for processing a $3 \times 256 \times 256$ input image.

[1] FLOPs: Floating Point Operations, refers to the count of floating-point arithmetic operations (additions, multiplications, divisions, etc.) that a model or computation involves.

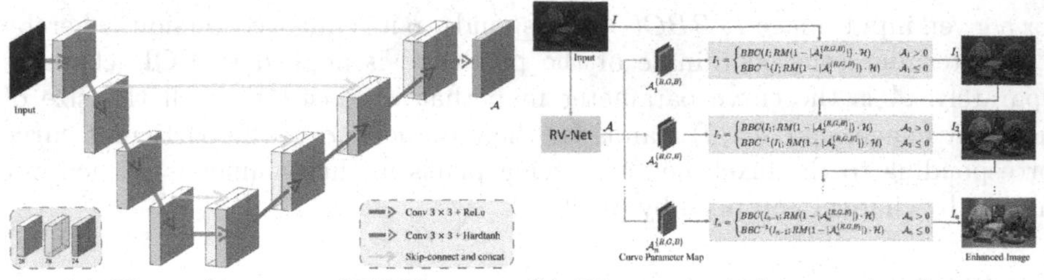

(a) The architecture of RV-Net. (b) The general framework of our approach.

Fig. 2. Details of the neural network and the framework of the proposed method. The RV-Net is designed to estimate the curve parameter map $\mathcal{A}$, which is then remapped and homogenized for use in iteratively enhancing the input image using Brightness Boost Curve (BB-Curve).

2.3 No-Reference Loss Functions

Our approach does not require paired images for reference in evaluating enhanced images. Several no-reference loss functions are used to assess the quality of enhanced images and to guide RV-Net in learning the correspondence between low-light images and curve parameter map $\mathcal{A}$. Inspired by the method of no-reference low-light image enhancement proposed by Guo *et al.* [5] and Kar *et al.* [6], the following four loss functions are designed and applied to RV-Net in this paper.

Neighborhood Consistency Loss (L_{nbh}). L_{nbh} maintains contrast by preventing the values of a pixel and its adjacent pixels from changing too much. L_{nbh} can be defined as follows:

$$L_{nbh} = \frac{1}{K}\sum_{i=1}^{K}\sum_{j\in\Omega(i)} |\mu(Y_i - Y_j)^2 - \mu(I_i - I_j)^2| \tag{7}$$

Let i represent a local region, where it is set to 4×4. K is the number of i, and $\Omega(i)$ is the 4-neighborhood centered on i. $\mu(Y)$ and $\mu(I)$ denote the mean value of the local region in the enhanced image and the input image, respectively.

Exposure Level Loss (L_{expo}). L_{expo} is introduced to regulate areas with under- or over-exposure by measuring the square of the distance between the average intensity of a local region and a preset well-exposed intensity E. L_{expo} can pull up the local average value of the image to a better value. According to an existing study [9], the exposure level of an image depends on the luminance space, and here E is empirically set to 0.64. L_{expo} can be calculated as

$$L_{expo} = \frac{1}{M}\sum_{k=1}^{M} |\mu(Y_k) - E|^2 \tag{8}$$

where k is empirically set to 16×16 non-overlapping regions, M denotes its number, and $\mu(Y)$ represents the average brightness of the locally non-overlapping regions in the enhanced image.

Gray World Assumption Loss (L_{gw}). The mean value of each channel of the RGB image tends to be gray based on the gray world assumption [10] of natural image statistics. L_{gw} is very useful for maintaining the original color of the enhanced image, and it is calculated as

$$L_{gw} = \frac{1}{P} \sum_{(c_1,c_2)\in\Theta} |\mu(Y^{c_1}) - \mu(Y^{c_2})| \tag{9}$$

where $\Theta = \{(R,G), (R,B), (G,B)\}$ is the color channel pair, P is its number, i.e., 3, and $\mu(Y^c)$ denotes the mean value of channel c of the enhanced image.

Parametric Map Variation Loss. Through our experiments, we found that the difference in enhancement magnitude between neighboring pixels should not be too large, i.e., neighboring pixels need to have similar curve parameters to maintain the original difference, otherwise, severe color artifacts may occur. Based on the literature [11], we set $L_{\mathcal{A}v}$ to act on the curve parameter map $\mathcal{A}$ as a way to control its variation. $L_{\mathcal{A}v}$ can be defined as

$$L_{\mathcal{A}v} = \frac{1}{N} \sum_{n=1}^{N} \sum_{c\in\theta} (|\nabla_h \mathcal{A}_n^c|^2 + |\nabla_v \mathcal{A}_n^c|^2) \tag{10}$$

where $\theta = \{R, G, B\}$ denotes the main color channel, N denotes the number of iterations of BB-Curve, ∇_h and ∇_v denote the gradients in the horizontal and vertical directions, respectively.

Total Loss. Finally, the total loss is a weighted sum of the above losses, namely

$$L = W_{nba}L_{nbh} + W_{expo}L_{expo} + W_{gw}L_{gw} + W_{\mathcal{A}v}L_{\mathcal{A}v} \tag{11}$$

where $W_{nba}, W_{expo}, W_{gw}, W_{av}$ are the corresponding weights, which are set to 3, 8, 5, and 220, respectively in this paper.

3 Experiment

In this section, we first present the details of the parameter settings and experimental platform, followed by the datasets and comparison methods used in this paper, and finally show the experimental results.

3.1 Implementation Details

Our method is implemented using the PyTorch deep learning framework. The training is performed with a batch size of 16, and the model weights are initialized using a normal distribution with a mean of 0 and a standard deviation of 0.02. Additionally, a default constant is used for bias. The optimization process employs the ADAM [12] optimizer, and the learning rate is set to 0.0001. Our primary hardware configuration includes an AMD EPYC 7302P CPU and a Titan Xp 12 GB graphics card. Based on this platform and parameter settings, our method only requires training for less than 5 epochs to yield good results on the data settings in Sect. 3.2, with each epoch only taking approximately 56 s to train.

3.2 Datasets and Comparison Methods

Training Datasets. To fully exploit the bi-directional adjustment capability of BB-Curve, this paper's training dataset must include both low-light and high-exposure images. The SICE dataset [13] fits the needs of this method perfectly. 589 scenes were captured in this dataset, each of which contains multiple exposure images of the same scene from underexposed to overexposed, and we remove the labels from them to keep only the degraded images that are improperly exposed, for a total of 4800 images, and we resize the resolution to 256×256 to reduce the performance overhead and expedite the training process.

Comparison Methods. Since our method belongs to unsupervised learning based on deep learning, this paper also selects the state-of-the-art (SOTA) unsupervised learning method based on deep learning as a comparative experiment. There are SSIE [14] representing self-supervised learning, RUAS [15] using the Zero-shot method, and representative unsupervised/no-reference learning methods for EnGAN [4] and Zero-DCE [5]. All comparison methods use official codes, default models and default settings.

Test Datasets. We selected multiple datasets of low-light images, comprising a total of 697 images in total, to verify the effectiveness of the proposed method in this paper. There are: 1) the LOL dataset [16]; 2) the DICM dataset [17], of which the first 41 were dark; 3) the LIME dataset [18]; 4) 105 night_real_clean images from the D2N dataset [19]; 5) the VV dataset[2]; and finally, 6) the MEF [20] dataset.

3.3 Experiment Results

Benchmark Metrics. Image quality assessment is widely used in the field of digital image processing. We select full-reference (FR) metrics including PSNR,

[2] https://sites.google.com/site/vonikakis/datasets.

SSIM [21], LOE [22], and NRMSE. The no-reference (NR) metrics we have selected to assess image quality include NIQMC [23] and CEIQ [24], PIQE [25], NIQE [26], BRISQUE [27], and ENIQA [28].

We evaluated the 10 aforementioned metrics on three datasets: LOL, DICM, and VV. However, the FR metrics were exclusively used on the LOL dataset due to the lack of paired reference images in the remaining datasets. The results of the objective evaluation metrics are displayed in Table 1. Our method earned the most first and second places, signifying that the multidimensional combined quality of our approach outperforms all comparison methods on the tested dataset.

Table 1. Comparison of objective evaluation metrics. Red is the best and blue is the second best.

Dataset		LOL (500 samples)			
Metrics		NIQMC/CEIQ↑	PIQE/NIQE/BRISQUE/ENIQA↓	PSNR/SSIM↑	LOE/NRMSE↓
Zoro-shot	RUAS	3.43/2.48	21.26/7.83/28.34/0.16	10.71/0.44	165.38/0.63
Self-supervised	SSIE	4.62/3.12	23.46/3.88/21.22/0.10	16.80/0.72	256.92/0.36
Unsupervised	Zero-DCE	4.44/2.97	44.49/8.85/33.2/0.10	16.39/0.62	289.05/0.34
	EnGAN	4.54/3.13	32.13/4.97/24.41/0.12	16.25/0.68	371.06/0.35
	Ours	4.48/3.02	42.02/8.64/31.96/0.04	16.82/0.65	218.05/0.33

Dataset		DICM (41 samples)		VV (24 samples)	
Metrics		NIQMC/CEIQ↑	PIQE/NIQE/BRISQUE/ENIQA↓	NIQMC/CEIQ↑	PIQE/NIQE/BRISQUE/ENIQA↓
Zoro-shot	RUAS	4.69/2.81	31.62/4.45/33.13/0.09	4.37/2.86	30.3/3.51/38,92/0.15
Self-supervised	SSIE	5.16/3.10	48.95/3.61/29.67/0.28	5.20/3.23	47.53/3.04/31.03/0.28
Unsupervised	Zero-DCE	5.22/3.22	36.18/3.09/28.06/0.12	5.31/3.36	25.84/2.91/32.59/0.19
	EnGAN	5.25/3.26	38.97/3.19/36.26/0.22	5.44/3.45	52.05/3.85/44.74/0.33
	Ours	5.24/3.19	35.36/2.99/29.35/0.11	5.31/3.36	25.65/2.75/32.63/0.19

Subjective Evaluation. Figure 3 illustrates a visual comparison between our method and other SOTA approaches on some images from our test datasets. It can be seen that our method undergoes significant improvements in brightness as well as preserves image details satisfactorily. From observations on Fig. 3(b), we can find that RUAS produces a weak enhancement of the extremely dark areas and overexposure of the brightest local areas of the input image on many samples. Figure 4(a) reveals two major flaws of RUAS at the same time. SSIE shows poor generalization ability, for example, the background color of the clock image in the top of Fig. 3(c) shows a significant deviation. SSIE even performs poorly in subjective visual effects on some of the LOL datasets(training dataset), as shown in Fig. 4(b). As can be seen in Fig. 3(d), Zero-DCE is not very different from ours and our method is slightly brighter, but our method nearly surpasses Zero-DCE across the board in terms of the objective evaluation metrics demonstrated in Table 1. EnGAN is balanced in the samples, but it also has the problem that the details are not obvious enough, for example, the details of the man's face at the bottom of Fig. 3(e) are not as clear as that of ours.

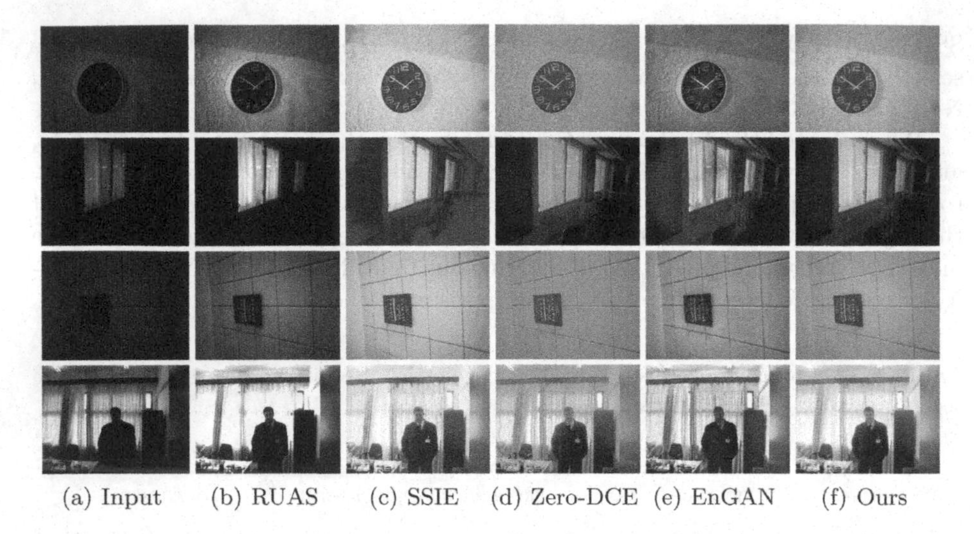

(a) Input (b) RUAS (c) SSIE (d) Zero-DCE (e) EnGAN (f) Ours

Fig. 3. Subjective visual comparison on the test datasets among SOTA low-light image enhancement approaches. RUAS has difficulty controlling the overall luminance balance of the image; SSIE is mediocre and the colors are prone to bias; and EnGAN tends to favor more intense colors, with some details under-represented.

3.4 Performance Comparison

Based on the hardware platforms in Sect. 3.1, we counted the average frame generation rate of each method on all test datasets, and all methods were run

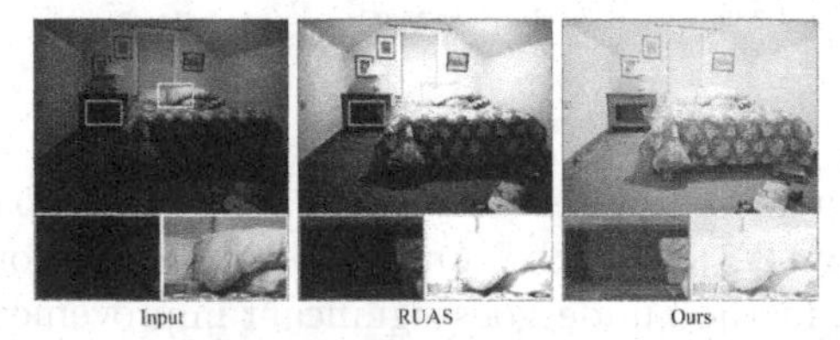

(a) Comparison of visual quality details of RUAS and our method on D2N dataset.

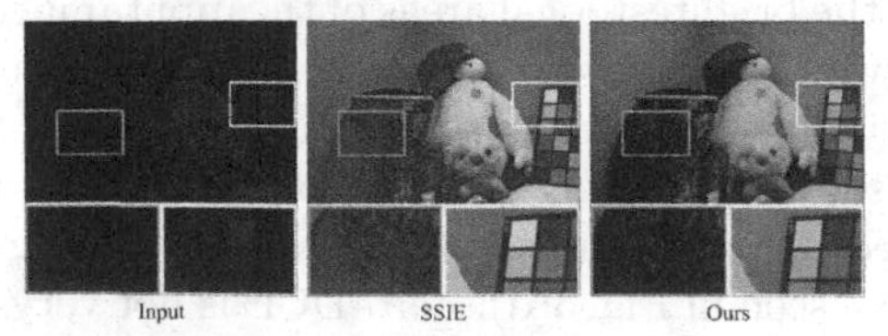

(b) Comparison of visual quality details of SSIE and our method on LOL dataset.

Fig. 4. Visual quality comparison. (a) RUAS is prone to underexposure and overexposure. (b) SSIE is prone to smearing and insufficient picture sharpness.

on GPU. For the deep learning framework, Pytorch was used for all except SSIE, which used Tensorflow. Figure 5 visualizes the prediction performance of our method and the comparison methods. Our paper achieved the best result by utilizing a reduced neural network and an efficient boost curve, which outperformed the worst-performing EnGAN by 3.5 times, and Zero-DCE delivers commendable performance with its also lightweight network architecture.

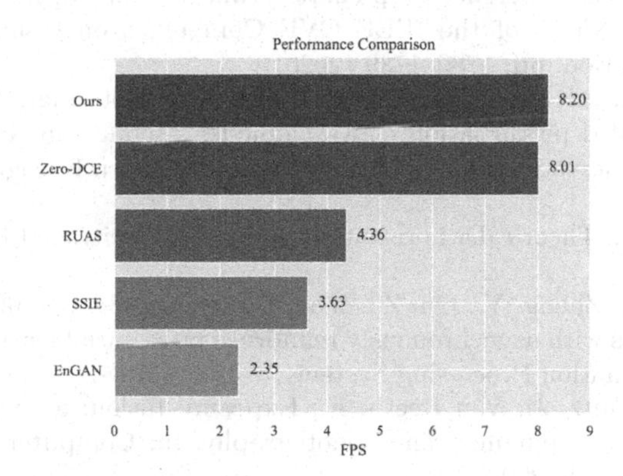

Fig. 5. Frame generation rate of each method on all test data.

4 Conclusion

In this paper, we propose a no-reference low-light image enhancement method to address the challenge of lacking paired datasets. We enhance the image brightness using BB-Curve in a rapid and efficient manner. We apply deep learning methods to evaluate curve parameters and redesign a series of no-reference loss functions to assess the quality of the enhanced image. Hence, our approach can also be seen as a hybrid of traditional and deep learning method. The experiments demonstrate that the comprehensive quality of our approach outperforms the existing unsupervised low-light image enhancement methods, and also obtains the best prediction performance. Follow-up works consider post-processing of the enhanced images to seek less noise and better contrast.

Acknowledgements. This research was supported by the National Natural Science Foundation of China (52105167) and the Fundamental Research Funds for the Central Universities (2022QNPY80).

References

1. Lore, K.G., Akintayo, A., Sarkar, S.: LLNet: a deep autoencoder approach to natural low-light image enhancement. Pattern Recogn. **61**, 650–662 (2017)

2. Rahman, Z.U., Jobson, D.J., Woodell, G.A.: Retinex processing for automatic image enhancement. J. Electron. Imaging **13**(1), 100–110 (2004)
3. Yang, W., Wang, W., Huang, H., Wang, S., Liu, J.: Sparse gradient regularized deep retinex network for robust low-light image enhancement. IEEE Trans. Image Process. **30**, 2072–2086 (2021)
4. Jiang, Y., et al.: EnlightenGAN: deep light enhancement without paired supervision. IEEE Trans. Image Process. **30**, 2340–2349 (2021)
5. Guo, C., et al.: Zero-reference deep curve estimation for low-light image enhancement. In: Proceedings of the IEEE/CVF Conference on Computer Vision and Pattern Recognition, pp. 1780–1789 (2020)
6. Kar, A., Dhara, S.K., Sen, D., Biswas, P.K.: Zero-shot single image restoration through controlled perturbation of Koschmieder's model. In: Proceedings of the IEEE/CVF Conference on Computer Vision and Pattern Recognition, pp. 16205–16215 (2021)
7. Koschmieder, H.: Theorie der horizontalen sichtweite. Beitrage Phys. Freien Atmos. 33–53 (1924)
8. Yu, R., Liu, W., Zhang, Y., Qu, Z., Zhao, D., Zhang, B.: DeepExposure: learning to expose photos with asynchronously reinforced adversarial learning. In: Advances in Neural Information Processing Systems, vol. 31 (2018)
9. Mertens, T., Kautz, J., Van Reeth, F.: Exposure fusion: a simple and practical alternative to high dynamic range photography. In: Computer Graphics Forum, vol. 28, pp. 161–171 (2009)
10. Buchsbaum, G.: A spatial processor model for object colour perception. J. Franklin Inst. **310**, 1–26 (2002)
11. Rodríguez, P.: Total variation regularization algorithms for images corrupted with different noise models: a review. J. Electr. Comput. Eng. **2013**, 1–18 (2013)
12. Kingma, D., Ba, J.: Adam: a method for stochastic optimization. In: International Conference on Learning Representations (2014)
13. Cai, J., Gu, S., Zhang, L.: Learning a deep single image contrast enhancer from multi-exposure images. IEEE Trans. Image Process. **27**(4), 2049–2062 (2018)
14. Zhang, Y., Di, X., Zhang, B., Wang, C.: Self-supervised image enhancement network: training with low light images only. arXiv preprint arXiv:2002.11300 (2020)
15. Liu, R., Ma, L., Zhang, J., Fan, X., Luo, Z.: Retinex-inspired unrolling with cooperative prior architecture search for low-light image enhancement. In: Proceedings of the IEEE/CVF Conference on Computer Vision and Pattern Recognition, pp. 10561–10570 (2021)
16. Wei, C., Wang, W., Yang, W., Liu, J.: Deep retinex decomposition for low-light enhancement (2018)
17. Lee, C., Lee, C., Kim, C.S.: Contrast enhancement based on layered difference representation. In: 2012 19th IEEE International Conference on Image Processing, pp. 965–968 (2012)
18. Guo, X., Li, Y., Ling, H.: LIME: low-light image enhancement via illumination map estimation. IEEE Trans. Image Process. **26**(2), 982–993 (2016)
19. Punnappurath, A., Abuolaim, A., Abdelhamed, A., Levinshtein, A., Brown, M.S.: Day-to-night image synthesis for training nighttime neural ISPS. In: Proceedings of the IEEE/CVF Conference on Computer Vision and Pattern Recognition (2022)
20. Ma, K., Zeng, K., Wang, Z.: Perceptual quality assessment for multi-exposure image fusion. IEEE Trans. Image Process. **24**(11), 3345–3356 (2015)
21. Wang, Z., Bovik, A., Sheikh, H., Simoncelli, E.: Image quality assessment: from error visibility to structural similarity. IEEE Trans. Image Process. **13**(4), 600–612 (2004)

22. Wang, S., Zheng, J., Hu, H.M., Li, B.: Naturalness preserved enhancement algorithm for non-uniform illumination images. IEEE Trans. Image Process. **22**(9), 3538–3548 (2013)
23. Gu, K., Lin, W., Zhai, G., Yang, X., Zhang, W., Chen, C.W.: No-reference quality metric of contrast-distorted images based on information maximization. IEEE Trans. Cybern. **47**(12), 4559–4565 (2017)
24. Yan, J., Li, J., Fu, X.: No-reference quality assessment of contrast-distorted images using contrast enhancement. arXiv preprint arXiv:1904.08879 (2019)
25. Venkatanath, N., Praneeth, D., Bh, M.C., Channappayya, S.S., Medasani, S.S.: Blind image quality evaluation using perception based features. In: 2015 Twenty First National Conference on Communications, pp. 1–6 (2015)
26. Mittal, A., Soundararajan, R., Bovik, A.C.: Making a "completely blind" image quality analyzer. IEEE Signal Process. Lett. **20**(3), 209–212 (2013)
27. Mittal, A., Moorthy, A.K., Bovik, A.C.: No-reference image quality assessment in the spatial domain. IEEE Trans. Image Process. **21**(12), 4695–4708 (2012)
28. Chen, X., Zhang, Q., Lin, M., Yang, G., He, C.: No-reference color image quality assessment: from entropy to perceptual quality. EURASIP J. Image Video Process. **2019**(1), 1–14 (2019)

A Deep Joint Model of Multi-scale Intent-Slots Interaction with Second-Order Gate for SLU

Qingpeng Wen[1], Bi Zeng[1], Pengfei Wei[1(✉)], and Huiting Hu[2]

[1] School of Computer Science and Technology, Guangdong University of Technology, Guangzhou 510006, China
wpf@gdut.edu.cn
[2] School of Information Science and Technology, Zhongkai University of Agriculture and Engineering, Guangzhou 510006, China

Abstract. Slot filling and intent detection are crucial tasks of Spoken Language Understanding (SLU). However, most existing joint models establish shallow connections between intent and slot by sharing parameters, which cannot fully utilize their rich interaction information. Meanwhile, the character and word fusion methods used in the Chinese SLU simply combines the initial information without appropriate guidance, making it easy to introduce a large amount of noisy information. In this paper, we propose a deep joint model of Multi-Scale intent-slots Interaction with Second-Order Gate for Chinese SLU **(MSIM-SOG)**. The model consists of two main modules: (1) the Multi-Scale intent-slots Interaction Module (MSIM), which enables cyclic updating the multi-scale information to achieve deep bi-directional interaction of intent and slots; (2) the Second-Order Gate Module (SOG), which controls the propagation of valuable information through the gate with second-order weights, reduces the noise information of fusion, accelerates model convergence, and alleviates model overfitting. Experiments on two public datasets demonstrate that our model outperforms the baseline and achieves state-of-the-art performance compared to previous models.

Keywords: Intent Detection · Slot Filling · Multi-Scale intent-slots Interaction Module (MSIM) · Second-Order Gate (SOG)

1 Introduction

For task-oriented dialogue systems, Spoken Language Understanding (SLU) is a critical component [17], it includes two subtasks Intent Detection (ID) and Slot Filling (SF) [4]. SF is a sequence labeling task to obtain the slot information of the utterance; ID is a classification task to identify the intent of the utterance. An example of a simple Chinese SLU is shown in Fig. 1.

B. Luo et al. (Eds.): ICONIP 2023, CCIS 1966, pp. 42–54, 2024.
https://doi.org/10.1007/978-981-99-8148-9_4

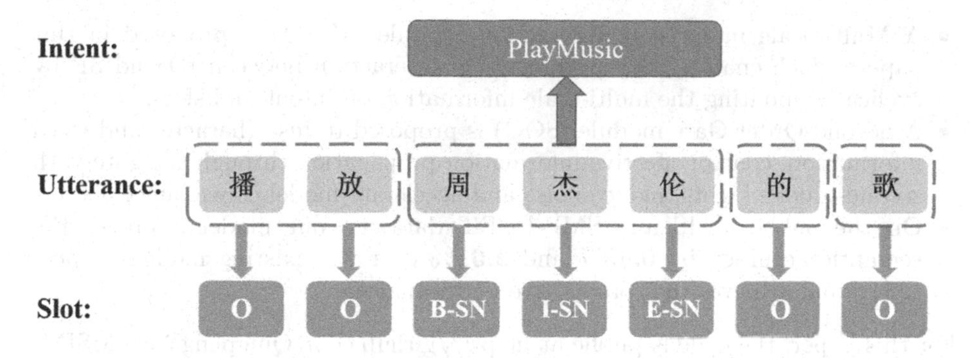

Fig. 1. An example of Chinese SLU, where `B-SN` denotes `B-singer_name`, `I-SN` denotes `I-singer_name` and `E-SN` denotes `E-singer_name`, the blue dashed box denotes word segmentation and the yellow box denotes character segmentation.

The main challenge in English SLU research is correlating ID and SF effectively. In response, Xu et al. [18] proposed a variable-length attention encoder-decoder model in 2020, in which SF is guided by intent information and achieves intent-enhanced association, but it lacks a bi-directional correlation between ID and SF. Recent research [11,14,16] has demonstrated that ID and SF tasks can mutually reinforce each other. Accordingly, Li et al. [7] proposed a bi-directional correlation BiLSTM-CRF model in 2022, updating ID and SF in both directions, but the deep interaction remains unestablished.

Compared to English SLU, Chinese SLU also faces challenges in segmenting Chinese utterances and effectively integrating character information. Unlike English, Chinese lacks natural word separators, rendering character segmentation techniques unreliable. As shown in Fig. 1, the segmentation of '周杰伦(Jay Chou)' into '周(week)-杰(Jay)-伦(Aron)' by characters incorrectly predicts '周(week)' as 'Datetime_date'. However, we expect the model to correctly segment it into '周杰伦(Jay Chou)' and predict it as the slot label 'singer_name' by using a suitable Chinese Word Segmentation (CWS) system. To address this, Teng et al. [15] improved the CWS system using Multi-level Word Adapter to fuse character and word information, but it lacks bi-directional interaction between ID and SF and introduces noise and overfitting problems in the fusion mechanism. This paper proposes a deep joint model of Multi-Scale intent-slots Interaction with Second-Order Gate (MSIM-SOG) to better fuse character and word information and establish a deep bi-directional interaction between two tasks. Experimental results on two publicly datasets called CAIS and SMP-ECDT show that our model outperforms all other models and achieves SOTA performance.

To summarize, the following are the contributions of this paper:

- In this paper, we propose a deep joint model of Multi-Scale intent-slots Interaction with Second-Order Gate for Chinese SLU (MSIM-SOG), which optimizes the performance of Chinese SLU tasks and improves current joint model.

- A Multi-Scale intent-slots Interaction Module (MSIM) is proposed in this paper, which enables deep bi-directional interaction between ID and SF by cyclically updating the multi-scale information on intent and slots.
- A Second-Order Gate module (SOG) is proposed to fuse character and word information, control effective information propagation through the gate with second-order weights, reduce noise and accelerate model convergence.
- On the public CAIS and SMP-ECDT datasets, our model improves the semantic accuracy by **0.49%** and **2.61%** over the existing models respectively, and achieves the competitive performance.

For this paper, the code is public at https://github.com/QingpengWen/MSIM-SOG.

2 Related Work

English SLU Task: The Spoken Language Understanding (SLU) task consists of two main tasks: Intent Detection (ID) and Slot Filling (SF). Early research in ID often utilized common classification methods like SVM [2] and RNN [6]. While SF extracts semantic information through sequence labeling such as CRF [19] and LSTM [20]. However, these approaches commonly cause error propagation as they lack the interaction of ID and SF. Ma et al. [10] proposed a two-stage selective fusion framework that explored intent-enhanced models. However, it simply guided slot by intent, and the bi-directional relationship was still not established. Sun et al. [13] designed a bi-directional interaction model based on a gate mechanism to achieve bi-directional association between ID and SF.

Chinese SLU Task: Although these approaches have made great progress in English SLU, there are still some challenges in Chinese SLU include dealing with ambiguous words, the effective fusion of word and character information, and lacks of deep bi-directional interaction models for ID and SF. As a result, existing English SLU models cannot be directly applied to the Chinese SLU task. To address this, Zhu et al. [21] proposed a two-stage Graph Attention Interaction Refine framework to mitigate ambiguity in Chinese SLU, but it may incorrectly identify slot boundaries due to the absence of CWS system. Teng et al. [15] proposed a Multi-level Word Adapter to fuse character and word information, but it only used the intent guidance slot, while the fusion mechanism they used introduces noisy information and risks losing critical information.

3 Approach

In this section, we will introduce the MSIM-SOG model proposed in this paper in detail. The general model framework is illustrated in Fig. 2.

3.1 Char-Word Channel Layer

Based on MLWA [15], we construct a character-level and word-level channel layer (Char-Word Channel Layer), which obtains the complete character sequence information and utterance representation information for SF and ID tasks. The Char-Word Channel Layer consists of the Self-Attentive Encoder module, the LSTM module, and the MLP Attention module. Among them, the Self-Attentive Encoder module extracts the character and word encoding representation. Then the LSTM module is utilized to extract the contextual and sequence information for the SF task, while the MLP Attention module extracts the complete representation of the utterance for the ID task.

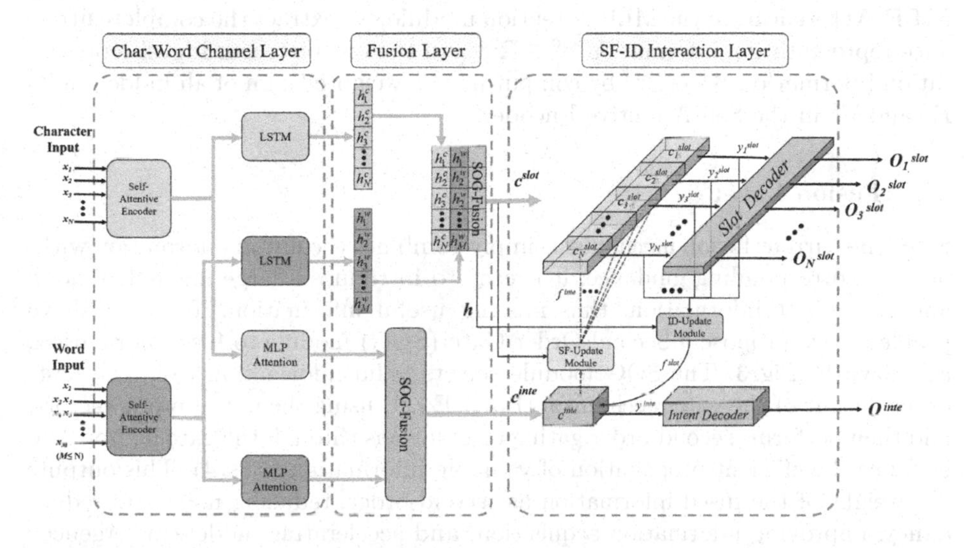

Fig. 2. The MSIM-SOG model proposed in this paper. The model includes our Char-Word Channel Layer, Fusion Layer and SF-ID interaction layer. The internal structure of the SOG Fusion module is shown in Fig. 3.

Self-attentive Encoder: The Self-Attentive Encoder mainly consists of an Embedding encoder, a Self-Attention encoder, and a BiLSTM encoder [3]. For a given Chinese utterance $c = \{c_1, c_2, ..., c_N\}$ containing N characters. The Embedding encoder converts it into the character vector $\mathbf{E}^c_{emb} \in \mathbb{R}^{N\times d} = \{\mathbf{e}^{c,e}_1, \mathbf{e}^{c,e}_2, \cdots, \mathbf{e}^{c,e}_N\}$. The BiLSTM encoder loops the input utterance forward and backward to obtain context-aware sequence feature information $H^c \in \mathbb{R}^{N\times d} = \{h^c_1, h^c_2, ..., h^c_N\}$, where $h^c_j \in \mathbb{R}^d = BiLSTM(e^{c,e}_j, h^c_{j-1}, h^c_{j+1})$ and the Self-Attention encoder captures the contextual information of each character in a valid sequence as $A^x \in \mathbb{R}^{N\times d} = softmax(\frac{Q\cdot K^T}{\sqrt{d^k}}) \cdot V$, where Q, K and V are matrices acquired by the application of different linear projections to the input vectors and d^k denotes the vector dimension. Subsequently, we concatenate these outputs to obtain the final character-level encoding representation as $\mathbf{E}^c \in \mathbb{R}^{N\times 2d} = \{\mathbf{e}^c_1, \mathbf{e}^c_2, \cdots, \mathbf{e}^c_N\}$.

For word-level encoding, we adopt the CWS system to capture the word segmentation sequence $w = \{w_1, w_2, ..., w_M\}$ (M $\leq$ N) by segmenting the utterance. And the final word-level encoding is denoted as $\mathbf{E}^w \in \mathbb{R}^{M\times 2d} = \{\mathbf{e}_1^w, \mathbf{e}_2^w, \cdots, \mathbf{e}_M^w\}$.

LSTM: In the LSTM module, we extract the contextual information of the character-level encoding $\mathbf{E}^c$ and capture the character sequence information to obtain the hidden state output $\mathbf{H}^c \in \mathbb{R}^{N\times 2d} = \{\mathbf{h}_1^c, \mathbf{h}_2^c, \mathbf{h}_3^c, \cdots, \mathbf{h}_N^c\}$, and use it for SF task, where $h_j^c = LSTM(e_j^c, h_{j-1}^c)$.

Equally, by extracting the word-level encoding information $\mathbf{E}^w$, we obtain the output of the hidden state is $\mathbf{H}^w \in \mathbb{R}^{M\times 2d} = \{\mathbf{h}_1^w, \mathbf{h}_2^w, \cdots, \mathbf{h}_M^w\}$.

MLP Attention: In the MLP Attention module, we extract the complete utterance representation information $\mathbf{S}^c \in \mathbb{R}^{2d}$ and the complete word-level representation information $\mathbf{S}^w \in \mathbb{R}^{2d}$ by computing the weighted sum of all hidden units $\mathbf{E}^c$ and $\mathbf{E}^c$ in the Self-Attentive Encoder.

3.2 Fusion Layer

Since the current fusion mechanism simply combines the initial information without the corresponding guidance, it is easy to introduce a large amount of noise and redundant information, thus missing useful information. To solve above problems, we propose a Second-Order Gate (SOG) module to fuse information, as shown in Fig. 3. The SOG module selects valid information from the first-order output of the gate mechanism (Eq. 2–Eq. 3) using the initial input vectors, and then performs second-order gating calculations through the gate neuron λ to enhance the efficient propagation of valuable information (Eqs. 4). This outputs the weight of the fused information as second-order, reducing noise and redundancy, improving information acquisition, and accelerating model convergence.

Given the input vectors $\mathbf{x} \in \mathbb{R}^d$ and $\mathbf{y} \in \mathbb{R}^d$ and the output $\mathbf{h} \in \mathbb{R}^d$, then the SOG fusion is calculated as follows:

$$\lambda = f\left[W_x \cdot \tanh(\mathbf{x}) + W_y \cdot \tanh(\mathbf{y})\right] \tag{1}$$

$$\mathbf{h}_x = \lambda \cdot tanh(\mathbf{x}) + \mathbf{x} \tag{2}$$

$$\mathbf{h}_y = (1 - \lambda) \cdot tanh(\mathbf{y}) + \mathbf{y} \tag{3}$$

$$\mathbf{h} = SOG(\mathbf{x}, \mathbf{y}) = \lambda \cdot \mathbf{h}_x + (1 - \lambda) \cdot \mathbf{h}_y \tag{4}$$

where $\mathbf{W}_x$ and $\mathbf{W}_y$ are trainable parameters, $f(\cdot)$ denotes the activation function and λ is the gate neuron that controls the Fusion information weighting.

Subsequently, we use the fusion output $\mathbf{c}^{slot} \in \mathbb{R}^{N\times 2d} = \left\{\mathbf{c}_1^{slot}, \mathbf{c}_2^{slot}, .., \mathbf{c}_N^{slot}\right\}$ of the hidden information $\mathbf{H}^c$ and $\mathbf{H}^w$ as the input of the SF task, apply the output $\mathbf{h} \in \mathbb{R}^{N\times 2d} = \{\mathbf{h}_1, \mathbf{h}_2, ..., \mathbf{h}_N\}$ to update the intent information, and use the fusion output $\mathbf{c}^{inte} \in \mathbb{R}^{2d}$ of the representation information $\mathbf{S}^c$ and $\mathbf{S}^w$ as the input of the ID task. The calculation formula is as follows:

$$\mathbf{c}_j^{slot} = SOG(\mathbf{H}_j^c, \mathbf{H}_{f_{align}(j,\mathbf{w})}^w) \tag{5}$$

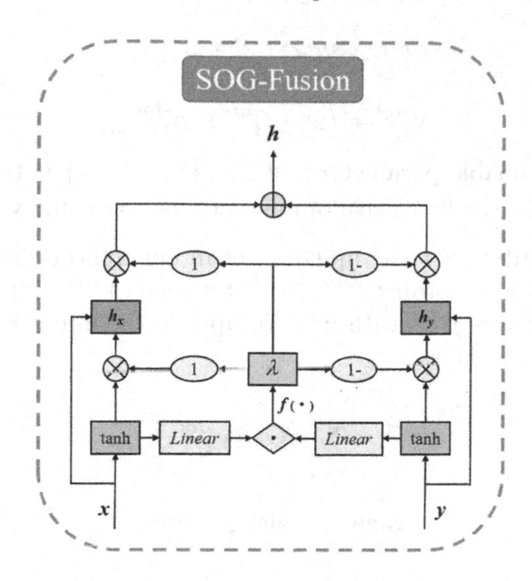

Fig. 3. SOG Fusion module, where $\mathbf{x}$ and $\mathbf{y}$ are fusion input vectors, $f(\cdot)$ denotes the activation function, λ is the gate neuron that controls the weight of the fused information, $\mathbf{h}_x$ and $\mathbf{h}_y$ are the selection information and $\mathbf{h}$ is the fusion output.

$$\mathbf{h}_j = SOG(\mathbf{H}^w_{f_{align}(j,\mathbf{w})}, \mathbf{H}^c_j) \tag{6}$$

$$\mathbf{c}^{intc} = SOG(\mathbf{S}^c, \mathbf{S}^w) \tag{7}$$

$$f_{align}(j, \mathbf{w}) = \begin{cases} 1 & j \leq len(\mathbf{w}_1) \\ \sum_{i=2}^{|\mathbf{w}|} i \cdot \mathbb{I}(\sum_{k=1}^{i-1} len(\mathbf{w}_k) < j \leq \sum_{k=1}^{i} len(\mathbf{w}_k)) & other \end{cases} \tag{8}$$

where $\mathbf{w}$ is the word sequence, $len(\cdot)$ counts the number of characters in a word, $\mathbb{I}(\cdot)$ is the indicator function. $j = \{1, 2, ..., N\}$ is each character's position index.

3.3 SF-ID Interaction Layer

To fully exploit the rich interaction information of intent and slots, we propose the Multi-Scale intent-slots Interaction Module (MSIM), which consists of SF-Update Module, ID-Update Module and Decoder Module. The MSIM module first uses the intent information to update the multi-scale information of slots obtained from the fusion layer and then uses them to guide the previous intent information. Finally, the deep bi-directional interaction between SF and ID is achieved through multiple interactions. A specific interaction is as follows.

SF-Update Module: For the updating of multi-scale slots information, we first obtains the update information $f^{inte} \in \mathbb{R}^{N\times 2d} = \{f_1^{inte}, f_2^{inte}, ..., f_N^{inte}\}$ by fusing $\mathbf{y}^{inte} \in \mathbb{R}^{2d}$ and $\mathbf{c}^{slot}$ using the SOG module, then the update information f^{inte} is calculated with $\mathbf{c}^{slot}$ to update the multi-scale slots information $\mathbf{y}^{slot} \in \mathbb{R}^{N\times 2d} = \{\mathbf{y}_1^{slot}, \mathbf{y}_2^{slot}, ..., \mathbf{y}_N^{slot}\}$, which is calculated as follows:

$$f_j^{inte} = SOG\left(\mathbf{c}_j^{slot}, \mathbf{y}^{inte}\right) \tag{9}$$

$$\mathbf{y}_j^{slot} = \left(w_j^I \cdot f_j^{inte}\right) \cdot \mathbf{c}_j^{slot} \tag{10}$$

where w_j^I is the trainable parameter and $j = \{1, 2, ..., N\}$ is the position index of each character. In the first cycle of interactions, we define $\mathbf{y}^{inte} = \mathbf{c}^{inte}$.

ID-Update Module: For the updating of intent information, similar to SF-Update module, we first obtains $f^{slot} \in \mathbb{R}^{2d}$ by fusing $\mathbf{y}^{slot}$ and $\mathbf{h}$ using the SOG module, then calculates f^{slot} with $\mathbf{c}^{inte}$ to update the intent information $\mathbf{y}^{inte}$. The calculation is as follows:

$$f^{slot} = \sum_{j=1}^{N} SOG(\mathbf{y}_j^{slot}, \mathbf{h}_j) \tag{11}$$

$$\mathbf{y}^{inte} = f^{slot} + \mathbf{c}^{inte} \tag{12}$$

Decoder Module: After the cyclic interaction, we decode the final information $\mathbf{y}^{slot}$ and $\mathbf{y}^{inte}$ by the Decoder module to obtain the final multi-scale slots output $\mathbf{O}^{slot} = \left\{\mathbf{O}_1^{slot}, \mathbf{O}_2^{slot}, ..., \mathbf{O}_N^{slot}\right\}$ and intent output $\mathbf{O}^{inte}$, which is calculated as follows.

$$P(\tilde{\mathbf{y}}^{slot} = j|\mathbf{c}^{slot}) = softmax[w^{S\text{-}O} \cdot (\mathbf{h}_N \oplus \mathbf{y}_j^{slot})] \tag{13}$$

$$P(\widetilde{\mathbf{y}}^{inte}|\mathbf{c}^{inte}) = softmax[w^{I\text{-}O} \cdot \mathbf{y}^{inte}] \tag{14}$$

$$\mathbf{O}_j^{slot} = argmax\ \left[P\left(\widetilde{\mathbf{y}}^{slot} = j\middle|\, \mathbf{c}^{slot}\right)\right] \tag{15}$$

$$\mathbf{O}^{inte} = argmax\ [P(\widetilde{\mathbf{y}}^{inte}|\mathbf{c}^{inte})] \tag{16}$$

where $w^{S\text{-}O}$ and $w^{I\text{-}O}$ are trainable parameters, $\oplus$ denotes concatenation operation, $j = \{1, 2, ..., N\}$ is the position index of each character.

3.4 Joint Loss Function

According to Goo et al. [1], a joint training scheme with NLLOSS is used for optimization in this paper, and the joint loss function is calculated as follows:

$$\mathcal{L} = -logP\left(\hat{\mathbf{y}}^{inte}\middle|\mathbf{c}^{inte}\right) - \sum_{i=1}^{N} logP(\hat{\mathbf{y}}_i^{slot}|\mathbf{c}_i^{slot}) \tag{17}$$

4 Experiments

4.1 Datasets and Evaluation Metrics

We conduct experiments on two public Chinese SLU datasets, CAIS [15] and SMP-ECDT [21], to evaluate the model validity. The CAIS dataset contains

7,995 training sets, 994 validation sets, and 1,024 test sets. The SMP-ECDT dataset contains 1,832 training sets, 352 validation sets, and 395 test sets.

In this paper, we use F1 score and accuracy to evaluate the accuracy of SF and ID, respectively. Moreover, we use sentence-level semantic accuracy to indicate that the output of this utterance is considered as a correct prediction when and only when the intent and all slots are perfectly matched.

4.2 Experimental Setting

In this paper, we set the dropout rate to 0.5, the initial learning rate is set to 0.001, the learning rate is adjusted dynamically by using the warmup strategy [8], and the Adam optimizer [5] is used to optimize the parameters of the model. For the CAIS dataset, we set the number_cycles to 3, while the SMP-ECDT dataset, we set it to 4. The model is trained on a Linux system using the PyTorch framework and Tesla A100, and multiple experiments are conducted with different random seeds to select the model parameter for evaluation on the test dataset that perform the best on the validation dataset.

4.3 Baseline Models

In this section, we select the following models for comparison, which are Slot-Gated Full Atten [1], a slot-oriented gating model to improve the semantic accuracy; CM-Net [9], a collaborative memory network to augment the local contextual representation; Stack-Propagation [12], a stack-propagation model to capture semantic knowledge of intent; MLWA [15], a multi-level word adapter that fuses word information with character information; GAIR [21], a two-stage Graph Attention Interaction Refine framework that leverages SF and ID information.

On the CAIS dataset, we uses the model performance from the paper GAIR [21]. On the SMP-ECDT dataset, we compare the published code of the model by running experiments separately, while the CM-Net [9] cannot be compared with this model due to the fact that the official codes are not provided.

4.4 Main Results

Table 1 presents the experimental results of the proposed MSIM-SOG model and the baseline models on the CAIS and SMP-ECDT datasets. From the analysis of the experimental results, we give the following experimental conclusions.

1. The MSIM-SOG model proposed in this paper outperforms the above model in all metrics and achieves state-of-the-art performance
2. Compared with the baseline model MLWA [15], our model achieves larger improvements. In detail, on the CAIS and SMP-ECDT datasets, our model improved Slot F1 Score by 1.66% and 2.95%, Intent Acc by 0.49% and 1.58%, and Semantic Acc by 1.18% and 3.78%, respectively. These results indicate that our model effectively fuses character and word information and enhances performance through a deep bi-directional interaction between ID and SF.

Table 1. The main results of the above models on the CAIS and SMP-ECDT datasets. The numbers with * indicate that the improvement of the model in this paper is statistically significant at all baselines, with $p < 0.05$.

Model	CAIS dataset			SMP-ECDT dataset		
	Slot F1 Score	Intent Acc	Semantic Acc	Slot F1 Score	Intent Acc	Semantic Acc
Slot-Gated Full Atten [1]	81.13	94.37	80.83	60.91	86.02	53.75
CM-Net [9]	86.16	94.56	–	–	–	–
Stack-Propagation [12]	87.64	94.37	84.68	71.32	91.06	63.75
MLWA [15]	88.61	95.16	86.17	75.76	94.65	68.58
GAIR [21]	88.92	95.45	86.86	77.68	95.45	69.75
MSIM-SOG	**90.27***	**95.65***	**87.35***	**78.71***	**96.23***	**72.36***

3. Compared with the current SOTA model GAIR [21], our model improved Slot F1 Score by 1.35% and 1.03%, Intent Acc by 0.20% and 0.78%, and Semantic Acc by 0.49% and 2.61% on the CAIS and SMP-ECDT datasets, respectively. These results show that our model, when utilizing a suitable CWS system and incorporating character information, outperforms the GAIR [21] model without CWS system.

The aforementioned outcomes demonstrate the advancement of the MSIM-SOG model proposed in this paper. We attribute these results to the following reasons: (1) The SOG module effectively fuses word and character information, enhancing model accuracy. (2) The deep interaction of ID and SF in MSIM improves performance by selecting effective multi-scale slots and intent information. (3) The use of a suitable CWS system and character information prevents incorrect slot identification and predictions.

4.5 Ablation Study

In this section, we conducted an ablation study to investigate the impact of the MSIM and SOG module on the performance enhancement of the MSIM-SOG model. We analyzed the effects by ablating four important modules and employing different approaches in the experiment (Table 2).

Table 2. Main results of ablation experiments on CAIS and SMP-ECDT datasets.

Model	CAIS dataset			SMP-ECDT dataset		
	Slot F1 Score	Intent Acc	Semantic Acc	Slot F1 Score	Intent Acc	Semantic Acc
MSIM w/o joint learning	87.61 (↓ 2.66)	94.56 (↓ 1.09)	85.27 (↓ 2.08)	76.83 (↓ 1.88)	95.02 (↓ 1.21)	70.75 (↓ 1.61)
MSIM w/o intent→ slot	88.75 (↓ 1.52)	95.35 (↓ 0.30)	86.61 (↓ 0.74)	77.69 (↓ 1.02)	95.67 (↓ 0.56)	71.79 (↓ 0.57)
MSIM w/o slot→ intent	89.69 (↓ 0.58)	95.15 (↓ 0.50)	86.75 (↓ 0.60)	78.35 (↓ 0.36)	95.47 (↓ 0.76)	71.68 (↓ 0.68)
MSIM-SOG w/o SOG	88.96 (↓ 1.31)	94.95 (↓ 0.70)	86.36 (↓ 0.99)	77.61 (↓ 1.10)	95.35 (↓ 0.88)	71.08 (↓ 1.28)
MSIM-SOG	**90.27**	**95.65**	**87.35**	**78.71**	**96.23**	**72.36**

Effect on MSIM: To demonstrate the advancement of the MSIM module, we first ablated the joint learning strategy, directly feeding intent and slot information from the fusion layer into the decoder. The experimental results indicated a significant drop in performance on both datasets compared to the original model, due to the lack of explicit interaction between intent and slot information. Subsequently, we conducted an ablation study on the unidirectional interaction of intent and slot, removing the SF-Update Module and ID-Update Module separately. The results indicated that the unidirectional interaction model had higher accuracy than the model without joint learning, but it performed significantly worse than the MSIM with deep bi-directional interaction. This confirms the mutual enhancement of multi-scale slots and intent information through deep interaction, aligning with previous studies.

Effect on SOG: To verify the advancement of the SOG module, we remove the SOG module and use MLWA [15] instead. The aforementioned experimental results demonstrate that the performance of SF and ID both decreased significantly. This indicates that the SOG module has a significant contribution in improving information acquisition, reducing the impact of model noise information and improving the learning ability of the model.

4.6 Convergence Analysis

To analyze the contribution of the SOG module in accelerating model convergence and reducing overfitting, we compared the semantic accuracy and loss curves of the model with and without the SOG module (replaced by MLWA [15]) after 300 epochs of training on the test set, as shown in Fig. 4a and Fig. 4b.

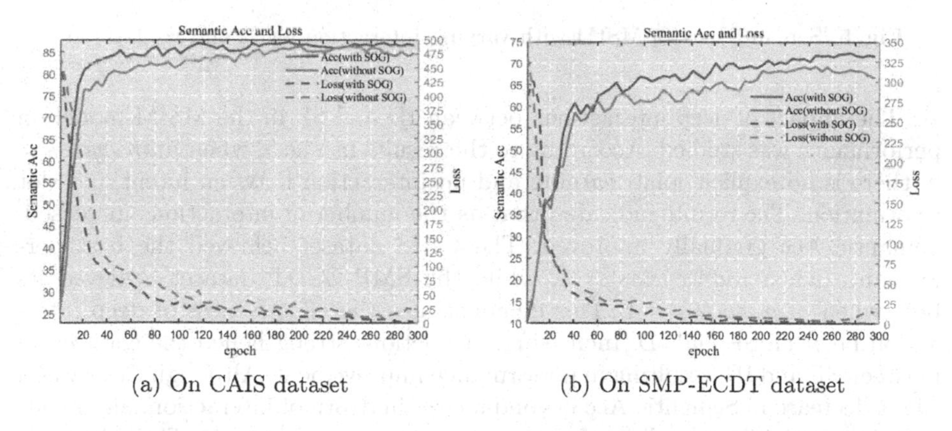

(a) On CAIS dataset (b) On SMP-ECDT dataset

Fig. 4. Semantic Acc and Loss Overall on CAIS and SMP-ECDT Dataset.

The results in Fig. 4a and Fig. 4b demonstrate that the model with the SOG module achieved convergence at around 117 and 160 epochs on the CAIS and SMP-ECDT datasets respectively, while the model without the SOG module reached convergence at 170 and 200 epochs. This indicates that the SOG module effectively accelerates model convergence and improves accuracy. On the loss curve, the model with the SOG module maintains relatively stable loss after 200 epochs of training, whereas the model without the SOG module shows an increase in loss after 270 epochs on the CAIS dataset and 280 epochs on the SMP-ECDT dataset, suggesting that the SOG module effectively alleviates model overfitting. These results highlight the effectiveness of the SOG module in accelerating convergence and reducing overfitting.

4.7 Effect of Interation

To assess the impact of deep interactions between ID and SF in the MSIM model, we evaluated its performance with different depths of interaction levels on the CAIS and SMP-ECDT datasets using Semantic Acc.

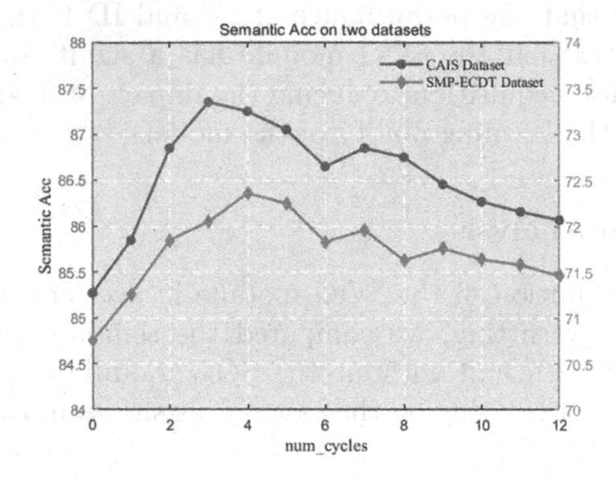

Fig. 5. Semantic Acc of MSIM with varying interaction levels on two datasets.

The impact of deep interactions between ID and SF in the MSIM model on performance was studied. According to the results in Fig. 5, when *num_cycles* = 0, there is no explicit joint learning and no interaction between intent and slot information. The results indicated that as the number of interactions increased, Semantic Acc gradually improved. The CAIS dataset achieved the best performance when *num_cycles* = 3, while the SMP-ECDT dataset achieved its best when *num_cycles* = 4. This demonstrates the effectiveness of deep interaction between SF and ID. Increasing interactions strengthened the connection between SF and ID, resulting in performance improvement. Although there was a slight decrease in Semantic Acc beyond a certain depth of interaction, all models with interactions outperformed the model without interactions. These findings emphasize the significance of deep interaction between ID and SF in enhancing model performance and validating the mutual reinforcement of SF and ID tasks.

5 Conclusion and Future Work

This paper introduces the MSIM-SOG model to address the challenges of fusing Chinese word and character information in the Chinese SLU domain while studying the deep interaction between ID and SF. The model consists of two modules:

MSIM enables deep bi-directional interaction between ID and SF by updating multi-scale slots and intent information cyclically. The SOG module enhances fusion by selecting the first-order gate output and performing second-order gating calculation. Experimental results on Chinese SLU datasets demonstrate significant performance improvement compared to existing models, achieving state-of-the-art results. Future work includes applying the MSIM-SOG model to multi-intent Chinese datasets to assess its generalization ability, as well as exploring the applicability of the SOG fusion mechanism in other NLP tasks such as sentiment analysis, recommendation systems, and semantic segmentation.

Acknowledgements. This work was supported in part by the National Science Foundation of China under Grant 62172111, in part by the Natural Science Foundation of Guangdong Province under Grant 2019A1515011056, in part by the Key technology project of Shunde District under Grant 2130218003002.

References

1. Goo, C.W., Gao, G., Hsu, Y.K., Huo, C.L., Chen, T.C.: Slot-gated modeling for joint slot filling and intent prediction. In: NAACL, pp. 753–757 (2018)
2. Haffner, P., Tür, G., Wright, J.H.: Optimizing SVMs for complex call classification. In: ICASSP (2003)
3. Hochreiter, S., Schmidhuber, J.: Long short-term memory. Neural Comput. **9**, 1735–1780 (1997)
4. Kim, S., D'Haro, L.F., Banchs, R.E., Williams, J.D., Henderson, M.: The fourth dialog state tracking challenge. In: Jokinen, K., Wilcock, G. (eds.) Dialogues with Social Robots. LNEE, vol. 999, pp. 435–449. Springer, Singapore (2017). https://doi.org/10.1007/978-981-10-2585-3_36
5. Kingma, D.P., Ba, J.: Adam: a method for stochastic optimization. CoRR, pp. 1–11 (2014)
6. Lai, S., Xu, L., Liu, K., Zhao, J.: Recurrent convolutional neural networks for text classification. In: AAAI 2015, pp. 2267–2273 (2015)
7. Li, C., Zhou, Y., Chao, G., Chu, D.: Understanding users' requirements precisely: a double bi-LSTM-CRF joint model for detecting user's intentions and slot tags. Neural Comput. Appl. **34**, 13639–13648 (2022)
8. Liu, L., et al.: On the variance of the adaptive learning rate and beyond. ArXiv, pp. 1–13 (2019)
9. Liu, Y., Meng, F., Zhang, J., Zhou, J.: CM-net: a novel collaborative memory network for spoken language understanding. In: EMNLP-ICJNLP, pp. 1051–1060
10. Ma, Z., Sun, B., Li, S.: A two-stage selective fusion framework for joint intent detection and slot filling. IEEE Trans. Neural Netw. Learn. 1–12 (2022)
11. Ni, P., Li, Y., Li, G., Chang, V.I.: Natural language understanding approaches based on joint task of intent detection and slot filling for IoT voice interaction. Neural Comput. Appl. 1–18 (2020)
12. Qin, L., Che, W., Li, Y., Wen, H., Liu, T.: A stack-propagation framework with token-level intent detection for spoken language understanding. In: EMNLP-IJCNLP, pp. 2078–2087 (2019)
13. Sun, C., Lv, L., Liu, T., Li, T.: A joint model based on interactive gate mechanism for spoken language understanding. Appl. Intell. **52**, 6057–6064 (2021)

14. Tang, H., Ji, D.H., Zhou, Q.: End-to-end masked graph-based CRF for joint slot filling and intent detection. Neurocomputing **413**, 348–359 (2020)
15. Teng, D., Qin, L., Che, W., Liu, T.: Injecting word information with multi-level word adapter for Chinese spoken language understanding. In: ICASSP, pp. 8188–8192 (2021)
16. Wei, P., Zeng, B., Liao, W.: Joint intent detection and slot filling with wheel-graph attention networks. J. Intell. Fuzzy Syst. **42**, 2409–2420 (2021)
17. Weld, H., Huang, X., Long, S., Poon, J., Han, S.C.: A survey of joint intent detection and slot filling models in natural language understanding. ACM Comput. Surv. **55**, 1–38 (2021)
18. Xu, C., Li, Q., Zhang, D., Cui, J.: A model with length-variable attention for spoken language understanding. Neurocomputing **379**, 197–202 (2020)
19. Xu, P., Sarikaya, R.: Convolutional neural network based triangular CRF for joint intent detection and slot filling. In: 2013 IEEE Workshop on Automatic Speech Recognition and Understanding, pp. 78–83 (2013)
20. Yao, K., Peng, B., Zhang, Y., Yu, D., Zweig, G.: Spoken language understanding using long short-term memory neural networks. In: IEEE-SLT, pp. 189–194 (2014)
21. Zhu, Z., Huang, P., Huang, H., Liu, S., Lao, L.: A graph attention interactive refine framework with contextual regularization for jointing intent detection and slot filling. In: ICASSP, pp. 7617–7621 (2022)

Instance-Aware and Semantic-Guided Prompt for Few-Shot Learning in Large Language Models

Jinta Weng[1,2], Donghao Li[1,2], Yifan Deng[1,2], Jie Zhang[1,2], Yue Hu[2(✉)], and Heyan Huang[3(✉)]

[1] School of Cyber Security, University of Chinese Academy of Sciences, Beijing, China
{wengjinta,lidonghao,dengyifan,zhangjie,huyue}@iie.ac.com
[2] Institute of Information Engineering, Chinese Academy of Sciences, Beijing, China
[3] School of Computer Science and Technology, Beijing Institute of Technology, Beijing, China
hhy63@bit.edu.cn

Abstract. The effectiveness of large language models (LLMs) and instruction learning has been demonstrated in different pre-trained language models (such as ChatGPT). However, current prompt learning methods usually use a unified template for the same tasks, and the template is difficult to capture significant information from different instances. To integrate the semantic attention dynamically on the instance level, We propose ISPrompt, an instance-semantic-aware prompt learning model. Specifically, the instance-driven prompt generated from the semantic dependency tree is introduced. Then, the proposed model would select a suitable semantic prompt from the prompt selection pool to motivate the prompt-based fine-tuning process. Our results show that the proposed model achieves state-of-the-art performance on few-shot learning tasks, which proves that ISPrompt integrating the instance semantics dynamically could assume as a better knowledge-mining tool for PLMs.

Keywords: Large Language Models · instruction learning · AIGC · prompt learning · deep learning

1 Introduction

ChatGPT [8] and GPT-4 [16] have made Large Language Models (LLMs) become the apple of the AI domain. With the exponential increase in model parameters and high-quality training corpus, the intelligence emergent phenomena is certificated by these neural networks containing trillions of parameters [27]. On the one hand, the large language model seems an intelligent generative robot integrating the nature knowledge and the human inference. On the other hand, it also brings up a novel knowledge-embedding way, allowing users to retrieve knowledge in natural language expression [2,6,11]. However, it is still a challenge to construct a

B. Luo et al. (Eds.): ICONIP 2023, CCIS 1966, pp. 55–67, 2024.
https://doi.org/10.1007/978-981-99-8148-9_5

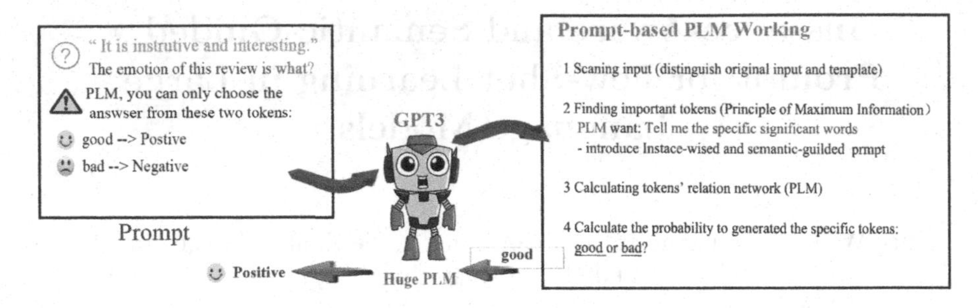

Fig. 1. The semantic viewpoint of PLM working process with prompt. The blue text shows the process for emotional analysis task, user first construct a prompt (left) with the question and specif label mapping tokens. Then, the prompt will insert into the PLM and motivate the prompt-based PLM working process to find the specif significant words (right) in the label mapping. (Color figure online)

clear prompt (e.g., question, instruction) to query the specific knowledge embedded in the LLMs. A minor change in the model size, training data, and prompt design, can make the LLMs generate an inconsistent answer [26]. In addition, limited model size and few-shot training samples are cover with numerous realistic tasks. Thus, the design of prompts has become a new research area, and different prompt paradigms like in-context learning [1], demonstration learning [21], and Chain of thought (COT) [28] have been developed.

In these paradigms, these queries are mainly constructed in hard prompt(textual prompt) or soft prompt format. The hard prompt is constructed by textual query and information about the current question [3]. By manually defining a template for the current input, the input would transit to a textual query with the "[Mask]" token. For example, in the emotion analysis task, the input X *'I think the AI is out of life'* with the template *'X. The emotion of the above sentence is [MASK].'* will generate a new sentence *'I think the AI is out of life. The emotion of the above sentence is [MASK]'*, and the [MASK] position is the generative goal of the LLMs.
Despite the above-mentioned textual construction [7,9], it still exists another effective soft prompt representation, which uses a continual meaningless vector to constitute a prompt embedding representation. For example, prompt-tuning, P-tuning [12,13], prefix-tuning [10], and perfect tuning [15]. To increase the diversity of prompts, different prompt strategies, like PET [19], KPT [5], UPT (Unfied prompt tuning) [24], and AutoPrompt [20], have also been introduced to enhance the multi-dimensional information of the prompt's design. However, there still exist some limitations though these methods. First, the effectiveness of textual prompt inevitably requires a lot of human effort and pre-experiments. When it comes to soft prompt, its efficacy is based on the large-scale training and suitable initialization [17]. Moreover, the soft prompt still lacks the nature and meaningful language expression [7]. Finally, both these prompt designs use the similar prompt for the same tasks [4], which could not realize the instance-aware

change in prompt structure. A natural question is: can we construct a more interpretable and applicable prompt from the viewpoints of semantic enhancement and instance-aware prompt? Therefore, the proposed paper introduces a novel prompt design – ISPrompt, **I**nstance-aware and **S**emantic-guided prompt for few-shot Learning in large language models. To demonstrate the ISPrompt, we first offer a hypothetical semantic model of Pre-trained language models (PLMs) in Fig. 1, which certify that designing more instructive prompts and semantic tokens are essential for utilizing the knowledge of enormous PLMs [17]. Therefore, we introduce the semantic-guided prompt generated by dependency-tree filters to facilitate PLMs to focus on meaningful tokens. The detailed contributions of the proposed work are:

- We propose an instance-aware prompt to reinforce the prompt's semantics and thus realize dynamic prompt augmentation and knowledge activation.
- We generate the semantic-based prompt dynamically from the prompt pool and select a suitable prompt to alleviate the effort of the manual prompts.
- Our proposed model ISPrompt shows state-of-the-art results and more reliability in different semantic understanding tasks and few-shot settings.

2 Problem Definition

Our main work is based on the prompt-based fine-tuning strategy with the few-shot setting as [3].

Table 1. Different combination of template and label mappings in SST-2 emotion classification task. The i-th word of label mapping represent i-th label.

Template	Mapping	Acc.
The emotion it is [*mask*]	terrible,great	92.1
	terrible,good	90.9
This emotion of this movie is [*mask*]	terrible,great	92.9
	great,terrible	86.7

Prompt-based fine-tuning uses an intuitive approach by defining a specific label mapping (also named as label verbalizer) and task-oriented template. In this method, a cloze question containing a masked token is first added to the original input x. Then the hidden representation of this "[MASK]" token is used to generate a token-level distribution on PLM's vocabulary. Consequently, only the distribution oflabel-mapping tokens is selected to realize the final prediction (Table 1). In the few-shot setting, the training corpus used for fine-tuning are construed by k samples per class. We use D for the training set, and $D^{(i)} = (x^i, y^i)$ is the $i-th$ training pair of the current dataset.

3 ISPrompt: Instance-Aware and Semantic-Based Prompt

The ISPrompt is based on reinforcement of the task's semantics since some task goals are more related to specific dependencies or word classes. For example, adjectives "beautiful" or adverbs "very" may contain more sentimental expression in the sentence "I very like the beautiful actress". Therefore, if we could add the adjective "beautiful" before the *mask* token, the prediction of the 'mask' token would prefer the mapping token of 'positive' emotion. Similarly, text entailment tasks are dependency-based, as they can be simplified to compare particular entities and relations in the context.

The semantic tokens mentioned above (e.g., adjective token, noun token, entity token, and relational token) always contain more instance-related information. Therefore, using these semantic tokens in the fine-tuning process to construct an instance-aware prompt is considerable. In detail, as Table 2 depicted, we first create two types of semantic filters on dependency-relation level and POS (part of speech) types. By limiting the available dependency types and POS tags, we could select some specific tokens to enhance the semantic of the constructed prompt.

However, using all POS and dependency relation types can lead to lengthy prompt and information interference, we only select a few filters for different tasks. For example, use words belonging to the "amod" dependency filters, or

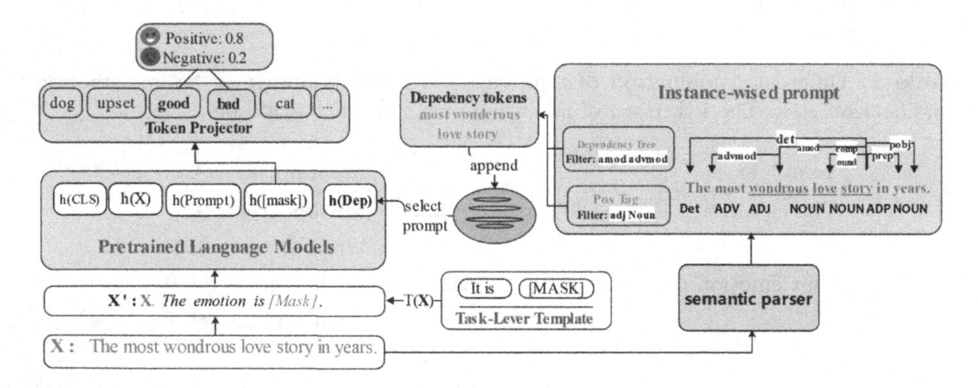

Fig. 2. The design of Dependency-based template. We design the template from the words containing the "amod" and "advmod" dependencies.

Table 2. Dep-filters. For more POS and dependency relations, please referring the Stanford dependency tree. For more about the dependency tree, please refer to Stanford NLP Tagger.

Filter Type	Filters Name
Word Class (POS)	NN | VBD | VBZ | VB
Dependency Tree	amod (Adj) | advmod (Adv) | ROOT | obj | nsubj

Table 3. Task-oriented dependencies filters. For more filter name, please referring stanford dependency tree.

Tasks	Dependency filters of Dep-promp
SST-2	amod (Adj) NN
SST-5	amod (Adj) | advmod (Adv) NN
TREC	amod | advmod | obj | NN | VBD | VBZ | VB
SNLI	amod (Adj) | advmod (Adv) NN
QNLI	amod (Adj) | advmod (Adv) NN

the "NN" POS filter to construct a prompt for the emotion classification task, for the reason that emotion classification is more related to the "amod" relation and noun-class word.

As Fig. 2 depicted, after defining POS filters and dependency tree filters, some specific tokens are selected to formulate a prompt and added to the original input. The ISPrompt template transition would be as follows:

$$T_{dep}(x^i) = x^i.\ \underline{[t^1_{dep}][t^2_{dep}][...][t^N_{dep}]}.[mask]. \tag{1}$$

The length of x^{dep} is up to the filter number. While using a POS filter, ISPrompt would add one token to the input and two for selecting a dependency tree filter [1]. Note that ISpromt is different and dynamically generated for each instance, thus realizing its instance-aware property.

3.1 Select Semantic Prompt

Different filter choices can generate various prompts. Thus, we introduce the prompt pool to store all augmented ISPrompts [25]. Note that the template transition $T(x)$ of the prompt largely determines the performance of the final prediction. Therefore, choosing a suitable template T is essential.

To find the right tokens that perform well in the current task, we offer manual and automatic selections, which allow users to select specific types of prompts or traverse the supporting prompt pool to select the best semantic tokens. In Table 3, we have listed some ideal prompt choices by Automatic Searching method for convenience [3], and the selection process is formulated as follows:

$$T_{select} = topk(\sum_{i}^{|D|} p\,(y|T_n(x_i), mask)) \tag{2}$$

Where |D| is the size of the task's training sample, n is the index of different templates, and y is the label token corresponding to the label of X.

[1] https://nlp.stanford.edu/software/tagger.shtml.

As the appropriate prompt is selected, a transition template of the current prompt is used to transform the original input x^i to prompting $\overline{x}^i$:

$$\overline{x}^i = T_{select}(x^i) = x^i.[t_1][t_2][t_{..}][t_N][MASK]. \tag{3}$$

where the *[mask]* token would later be used to generate the word distribution over PLM's vocabulary.

3.2 Model Formulation

Given these transition inputs, the target is predicting the predefined label-mapping tokens on *[mask]* position. Only the output of *[mask]* token $h_{[mask]}$ would be used to create a token-level distribution over PLM's vocabulary by introducing a learn-able linear projector W:

$$\delta = h_{[mask]} \otimes W_{proj} = PLM(\overline{x}^i_{mask}) \otimes W_{proj} \tag{4}$$

where the size of linear projector W is $R^{|h \times len(V)|}$, and h_{v_t} is the value of PLM token distribution in token v_t.

Subsequently, some tokens in PLM's vocabulary V are selected to represent each label and formed in a pre-defined label mapping. For example, in the emotion classification task, we could use the PLMs token "good" for the label positive and "bad" for the label negative.

$$F(y) : y_t \rightarrow v_t, y_t \in Y, v_t \in V \tag{5}$$

, where t is the label index of label set Y, and v represents the specif token of V

Note that the distributions of these represented tokens v_t would be retained and used to calculate the final prediction as the following equation:

$$\begin{aligned} p(y^i|\overline{x}^i) \Rightarrow p(F(y^i)|\delta) &= p(v^t|h_{[mask]} \otimes W_{proj}) \\ &= \ln \frac{\exp(h_{[mask]} \cdot w_t)}{\sum_{k=1}^{|Y|} \exp(h_{[mask]} \cdot w_k)} \end{aligned} \tag{6}$$

, where $|Y|$ is the label numbers, v_k is used to represent the token logit value of the k-th token in soft label mapping.

The target of task classification thus transits into the prediction of tokens in label mappings, and the final loss is formulated as:

$$L = -\frac{1}{N} \sum_{i} \sum_{t=1}^{|Y|} \log p(y^{it}|x^{it}) \tag{7}$$

, where i is the index of the training pair (x_i, y_i).

In this section, we propose an instance-aware prompt construction to reinforce the prompt's semantics and reduce the interference of invalid information. By integrating the semantic-aware prompts into the prompt-based fine-tuning process, it can realize efficient activation of PLMs knowledge by semantic tokens.

4 Experiments

We present the effectiveness of the proposed model ISPrompt in different semantic understanding tasks, for example, sentiment classification, question classification, and semantic entailment tasks. Note that We explore and analyze the generalization and transferability of the ISPrompt in the few-shot setting (Only few samples of each class are needed).

4.1 Datasets

We choose five different datasets from glue [23] and sentiment analysis benchmark. In sentiment classification, we choose two representative SST tasks, sst-2 and sst-5. The SST dataset aims to give a movie review to predict its emotion. The label of the sst-2 task is positive and negative tags, while in the sst-5 task are tags 'very positive', 'positive', 'neutral', 'negative', and 'very negative'. In the question classification dataset, the TREC-6 task is chosen to explore the task's understanding of QA [22]. TREC task aims to predict the six question types given an English question text. In the natural language inference task, we use SNLI and QNLI datasets to test the model's ability to judge whether the previous sentence could imply the semantics of the following sentence in the text [18].

4.2 Experiment Setting

We use RoBERTa-large [14] as our pre-training language model. Our experiments are developed in NVIDIA V100 32 GB (also could run in 1080ti with a small batch size). In the training process, we develop many experiments on different batch sizes *bs=4,8,16*, learning rate *lr=1e-5,2e-5,5e-5*. The reason that some study has shown the minor difference in the few-shot dataset could lead to differentiated results; we use five different sub-sets with the same sampling size. For example, the SST-2 emotion classification task with two classes needs to construct five different training sets with the size of 32 and a validation set with the same size of 32, while the testing size would use the original size of the SST-2 task without any other data setup. These sub-datasets are training and predicting singly. Their training process and prediction result would be recorded by different batch sizes and learning rates. All training pairs in the same task would use the same template transition in proposed prompting strategies to construct the real input. Finally, we choose the best result of each sub-dataset on different hyper-parameters and integrate the best result on the aspect of the sub-dataset. We chose the average accuracy and variance of different sub-datasets as the evaluation criteria because it reveals the overall performance and variation of the current task.

Table 4. The main result of the ISPrompt. The results of all experiments are evaluated by selecting the mean and variance of accuracy on five different segmented training datasets and the same testing dataset. Note that the results of existing baselines (⋆) using from the [3] to ensure fairness, and the results of prefix-tuning† and prompt-tuning† mode are based on the same experimental setting.

Baselines	TREC (acc)	SNLI (acc)	QNLI (acc)	SST-5 (acc)	SST-2 (acc)
Majority ⋆	18.8	33.8	49.5	50.9	23.1
prompt-based zero-shot learning⋆	32.0	49.5	50.8	35.0	83.6
GPT3-in-context-learning⋆	26.2(2.4)	47.1(0.6)	53.8(0.4)	30.6(0.9)	84.8
fine-tuning⋆	26.2(2.4)	48.4(4.8)	60.2(6.5)	43.9(2.0)	81.4(3.8)
Prefix-tuning†	36.0(1.1)	33.5(3.9)	54.5(2.2)	46.1(1.3)	88.1(2.3)
P-tuning†	40.2(1.3)	37.5(1.6)	57.6(3.9)	32.1(3.1)	90.1(1.2)
LMBFF⋆	84.8(5.1)	77.1(3.9)	63.7(4.2)	46.1(1.3)	92.1(1.1)
UPT	76.2	71.1	59.4	45.1	90.8
Jian et al., 2022	83.3 (1.5)	69.9 (2.4)	66.4 (3.5)	49.5 (1.1)	90.6 (0.1)
Our Methods					
ISPrompt	**87.2** (3.4)	**77.6 (1.7)**	**67.3 (3.3)**	48.1 **(1.0)**	91.8 **(0.7)**
w/o task template	**85.1(2.7)**	**76.1 (1.9)**	64.5 (3.1)	47.8 (0.9)	**92.0 (1.8)**

4.3 Baselines

We compare several baselines in related works and set our baselines with the Majority method (merely selecting the majority class as prediction), fine-tuning strategy [14], prompt-based zero-shot learning, GPT3-in-context-learning [1], UPT [29], and LMBFF model (The method details refer to Gao's work [3]).

As the main result described in Table 4, the proposed ISPrompt shows better performance on different types of tasks in the following aspects: (a) The proposed ISPrompt shows better overall effectiveness in few-shot setting, and integrate some semantic level information could assist downstream tasks; (b) Comparing the manual-designed and task-related prompt design (for example, LMBFF, and UPT methods), dynamically integrating instance-aware information for prompt-based fine-tuning is effective. (c) Compared to the soft prompt learning method (Prefix-tuning and P-tuning), ISPrompt's performance achieves higher performance and lower variance, and it also reveals that soft prompt is difficult to learn additional information with low resource data. (d) The effectiveness of SST-2 tasks shows that the proposed method appears insensitive to high-confidence tasks such as SST-2. However, it is still effective for binary entailment classification with low accuracy. We believe that there is no need to add a special augmented prompt for high-precision tasks or simple tasks, as longer prompts can bring a lot of distracting information. (e) We also conduct an ablation experiment in ISPrompt(-task) without any task pre-defined template and only use the instance-generative tokens, our results also achieve comparable result in most datasets.

However, using many ISPrompts choices it is difficult to surpass the results using a single prompt selection, as the longer prompt may introduce more semantic interference. Therefore, controlling the length of the prompt is a considered question in the ISPrompt fine-tuning process. We also consider that there is no need to add a special augmented prompt for high-precision tasks, for the reason that specific tokens may bring a lot of distracting information.

Overall, our results show that the proposed ISPrompt can get better improvement compared to other baselines and different semantic understanding tasks. Current prompt-based fine-tuning methods require a more semantic-driven template and dynamically utilize instance-aware information to motivate PLM.

Table 5. The comparison of different POS filter. The 'Orgin' mark the experiment is developed based on prompt-based fine-tuning without any change.

Task	POS Filter							
	Orgin	NN	WDT	NNP	WRB	WP	NNP	JJ
SST-2	89.3(1.3)	81.6 (6.4)	84.1 (4.7)	86.9 (2.6)	85.6 (4.7)	**88.1 (1.8)**	84.1 (4.7)	83.5 (5.6)
TREC	82.8(3.7)	83.2 (3.7)	84.4 (4.9)	**87.2 (3.4)**	84.0 (5.2)	84.0 (5.2)	**87.2 (3.4)**	79.4 (6.1)
QNLI	64.5(3.2)	66.5 (3.6)	67.7 (2.5)	68.0 (2.5)	68.0 (1.8)	**68.7 (2.1)**	68.0 (2.5)	68.0 (1.8)
SNLI	74.1(3.9)	74.4 (4.1)	77.4 (2.0)	**77.6(1.9)**	76.2(2.5)	75.7(2.1)	**77.6(1.9)**	74.9(2.4)
SST-5	43.1(1.7)	49.4 (0.8)	47.8 (3.1)	48.4 (1.6)	**50.4 (1.3)**	49.2 (1.3)	49.2 (2.7)	49.2 (2.7)

Table 6. The comparison of different Dependency Tree filters.

Task	Dependency Filter					
	Orgin	nsubj	amod	advmod	ROOT	obj
SST-2	89.3(1.3)	84.4(9.1)	82.1(10.5)	85.4(7.3)	86.6(6.6)	**87.9(4.3)**
TREC	82.8(3.7)	**85.5 (5.5)**	81.9 (5.1)	79.9 (7.1)	81.3 (7.5)	84.4 (6.6)
QNLI	64.5(3.2)	61.9 (3.9)	60.6 (2.6)	60.0 (3.6)	**64.1 (3.5)**	62.7 (3.3)
SNLI	74.1(3.9)	69.6 (6.3)	67.6 (4.7)	68.6 (6.5)	**70.8 (5.3)**	65.2 (8.4)
SST-5	43.1(1.7)	40.9 (5.0)	41.6 (2.4)	40.9 (2.7)	**42.4 (3.2)**	39.2 (6.1)

4.4 Select the Appropriate Dep-Filter for ISPrompt

Since ISPrompt is generated by tokens filtered by pre-defined filter names, such as specific word classes and dependency filters, we conducted several experiments on seven POS filters and five dependency filters to explore the influence of different filters. In POS Filter, we use the word class generated from the standford NLP POS Tagger and select 'NN', 'WDT', 'NNP', 'WRB', 'WP', 'NNP', and 'JJ' as POS filters. In the Dependency Filter, we choose dependency relationships 'nsubj', 'amod', 'advmod', 'ROOT', and 'obj'. For a more detailed description of these tags, see Sandford parse tagger (See Footnote 1).

As shown in Table 5 and Table 6, different types of filters are used to construct different augmented prompts. In Table 5, adding tokens with specific word classes to build an augmented prompt is effective. For example, the 'WP' filter (e.g., Wh-pronoun) is beneficial for SST-2 and QNLI tasks, while tokens with POS tag 'NNP' (e.g., noun) are effective in TREC and SNLI tasks. Also, tokens with the word class 'WRB' (Wh-adverb like 'how, where, and when') can enhance SST-5 performance. In dependency tree filters of Table 6, the 'nsubj' (nominal subject) relation is helpful for the TREC task, and the 'ROOT' (root Node of the dependency tree) relation is effective for QNLI, SNLI, and SST-5 tasks. Moreover, it appears that the tokens in our selecting dependencies are not sensitive to task SST-2, because SST-2 task is a high-accuracy task.

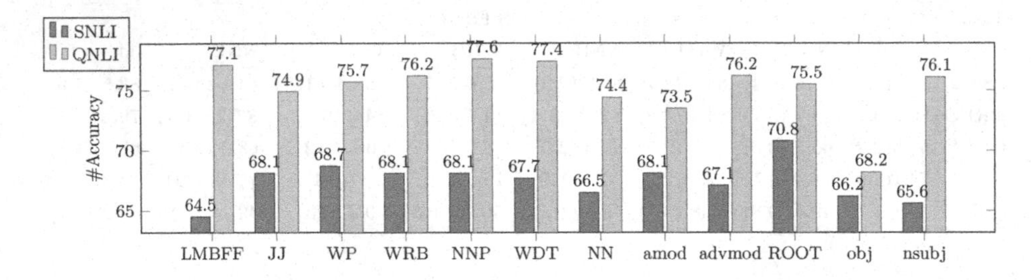

Fig. 3. The effectiveness of different using dependency filters in ISPrompt. The first column marked by **LMBFF** is the effectiveness of prompt-based fine-tuning baseline of [3].

4.5 Tasks Transferability

Using different dep filters may add new information to the original input. Therefore, we select text entailment tasks, QNLI, and SNLI tasks to explore the detailed effectiveness of dep-filter names. Both QNLI and SNLI are text-based tasks that depend on semantic understanding. Note that we still use the POS tags and dependencies mentioned above to evaluate the transferability of the same tasks.

As Fig. 3 depicted, the result shows that the POS-type filter with "NNP" and "'WDT' could achieve better results on QNLI and SNLI tasks compared to the LMBFF baseline. In detail, all the different types of proposed filters can achieve better results on text entailment tasks, which reveals that the fine-tuning process could learn more information from semantic tokens.

4.6 Different K-Shot Setting

Since the related research shows that the K-shot setting is the key point for a few-shot learning experiment, we also conduct a comparative experiment to explore the proposed ISPrompt on different K. In detail, we construct different datasets, which are K = 8,16,32,64,128, and 160 on TREC and SST-5 tasks. The result is

Table 7. The comparison of different K-shot settings. We use the average of accuracy and the gain of accuracy between different k-shot settings to select the best k-shot setting.

K-shot	Task (Filter Type)			
	TREC (Dependency Filter)	TREC (POS filter)	SST-5 (Dependency Filter)	SST-5 (POS filter)
8	74.2 (+0.0)	78. 3(+0.0)	46.7 (+0.0)	47.0 (+0.0)
16	**83.8 (+9.6)**	**85.8 (+7.5)**	50.3 (+3.6)	**50.2 (+3.2)**
32	91.8 (+8.0)	91.0 (+5.2)	48.5 (−1.8)	49.2 (−1.0)
64	95.0 (+4.8)	93.9 (+2.9)	**53.1 (+3.6)**	51.8 (+2.6)
128	96.4 (+1.4)	95.8 (+1.9)	54.3 (+3.2)	55.3 (+3.5)
160	96.3 (−0.1)	95.7 (−0.1)	54.3 (0.0)	54.5 (−1.2)

shown in Table 7, which reveals that K = 16 could gain more benefit than another K-shot setting. What is more, with the increase in instances, task accuracy is steadily increasing. In addition, without using all training data and limited K-show instances, ISPrompt could also achieve better comparative results.

5 Conclusion

Prompt tuning methods have shown their effectiveness to motivating the PLMs (like GPT and chatGPT) in different tasks. However, current prompt designs are primarily based on the task-level and manual-defined prompts [4], which can not dynamically change with different instances and reuse semantic information for few-shot learning tasks. To alleviate this, we propose an instance-aware and semantic-guided prompt to reinforce the prompt learning process and achieve efficient and dynamic prompt augmentation by the prompt selection pool. Our results show that the ISPrompt could achieve state-of-the-art effectiveness on different few-shot learning tasks, including emotion classification, question classification, and natural language inference tasks. It also reveals that ISPrompt could be assumed to be a better semantic augmentation tool for current prompt design and a better knowledge probe for prompt-based PLMs.

References

1. Brown, T., et al.: Language models are few-shot learners. In: Advances in Neural Information Processing Systems, vol. 33, pp. 1877–1901 (2020)
2. Devlin, J., Chang, M.W., Lee, K., Toutanova, K.: BERT: pre-training of deep bidirectional transformers for language understanding. arXiv e-prints arXiv:1810.04805 (2018)
3. Gao, T., Fisch, A., Chen, D.: Making pre-trained language models better few-shot learners. arXiv preprint arXiv:2012.15723 (2020)
4. Han, X., Zhao, W., Ding, N., Liu, Z., Sun, M.: PTR: prompt tuning with rules for text classification. arXiv preprint arXiv:2105.11259 (2021)
5. Hu, S., et al.: Knowledgeable prompt-tuning: Incorporating knowledge into prompt verbalizer for text classification. arXiv preprint arXiv:2108.02035 (2021)

6. Jiang, Z., Xu, F.F., Araki, J., Neubig, G.: How can we know what language models know? Trans. Assoc. Comput. Linguist. **8**, 423–438 (2020)
7. Kavumba, P., Takahashi, R., Oda, Y.: Are prompt-based models clueless? In: Proceedings of the 60th Annual Meeting of the Association for Computational Linguistics (vol. 1: Long Papers), pp. 2333–2352 (2022)
8. Kocoń, J., et al.: ChatGPT: jack of all trades, master of none. arXiv preprint arXiv:2302.10724 (2023)
9. Lester, B., Al-Rfou, R., Constant, N.: The power of scale for parameter-efficient prompt tuning. In: Proceedings of the 2021 Conference on Empirical Methods in Natural Language Processing, pp. 3045–3059 (2021)
10. Li, X.L., Liang, P.: Prefix-tuning: optimizing continuous prompts for generation. In: Proceedings of the 59th Annual Meeting of the Association for Computational Linguistics and the 11th International Joint Conference on Natural Language Processing (vol. 1: Long Papers), pp. 4582–4597. Association for Computational Linguistics (2021). https://doi.org/10.18653/v1/2021.acl-long.353. https://aclanthology.org/2021.acl-long.353
11. Lin, Y., Tan, Y.C., Frank, R.: Open sesame: getting inside BERT's linguistic knowledge (2019)
12. Liu, X., Ji, K., Fu, Y., Du, Z., Yang, Z., Tang, J.: P-Tuning v2: prompt tuning can be comparable to fine-tuning universally across scales and tasks (2021)
13. Liu, X., et al.: GPT understands, too. arXiv preprint arXiv:2103.10385 (2021)
14. Liu, Y., et al.: RoBERTa: a robustly optimized BERT pretraining approach. ArXiv abs/1907.11692 (2019)
15. Mahabadi, R.K., et al.: Perfect: prompt-free and efficient few-shot learning with language models. arXiv preprint arXiv:2204.01172 (2022)
16. OpenAI: Gpt-4 technical report (2023)
17. Qiu, X., Sun, T., Xu, Y., Shao, Y., Dai, N., Huang, X.: Pre-trained models for natural language processing: a survey. Sci. Chin. Technol. Sci. **63**(10), 1872–1897 (2020). https://doi.org/10.1007/s11431-020-1647-3
18. Rajpurkar, P., Zhang, J., Lopyrev, K., Liang, P.: Squad: 100,000+ questions for machine comprehension of text. In: Proceedings of the 2016 Conference on Empirical Methods in Natural Language Processing, pp. 2383–2392 (2016)
19. Schick, T., Schütze, H.: Exploiting cloze-questions for few-shot text classification and natural language inference. In: Proceedings of the 16th Conference of the European Chapter of the Association for Computational Linguistics: Main Volume, pp. 255–269. Association for Computational Linguistics, Online (2021). https://doi.org/10.18653/v1/2021.eacl-main.20. https://aclanthology.org/2021.eacl-main.20
20. Shin, T., Razeghi, Y., Logan IV, R.L., Wallace, E., Singh, S.: AutoPrompt: eliciting knowledge from language models with automatically generated prompts. arXiv preprint arXiv:2010.15980 (2020)
21. Sumers, T., Hawkins, R., Ho, M.K., Griffiths, T., Hadfield-Menell, D.: How to talk so AI will learn: instructions, descriptions, and autonomy. In: Advances in Neural Information Processing Systems, vol. 35, pp. 34762–34775 (2022)
22. Voorhees, E.M., Tice, D.M.: Building a question answering test collection. In: Proceedings of the 23rd Annual International ACM SIGIR Conference on Research and Development in Information Retrieval, pp. 200–207 (2000)
23. Wang, A., Singh, A., Michael, J., Hill, F., Levy, O., Bowman, S.: GLUE: a multi-task benchmark and analysis platform for natural language understanding. In: Proceedings of the 2018 EMNLP Workshop BlackboxNLP: Analyzing and Interpreting Neural Networks for NLP, pp. 353–355 (2018)

24. Wang, J., et al.: Towards unified prompt tuning for few-shot text classification. arXiv preprint arXiv:2205.05313 (2022)
25. Wang, Z., et al.: Learning to prompt for continual learning. In: Proceedings of the IEEE/CVF Conference on Computer Vision and Pattern Recognition, pp. 139–149 (2022)
26. Webson, A., Pavlick, E.: Do prompt-based models really understand the meaning of their prompts? arxiv abs/2109.01247 (2021)
27. Wei, J., et al.: Emergent abilities of large language models. arXiv preprint arXiv:2206.07682 (2022)
28. Wei, J., et al.: Chain of thought prompting elicits reasoning in large language models. arXiv preprint arXiv:2201.11903 (2022)
29. Zhong, W., et al.: Improving task generalization via unified schema prompt. arXiv preprint arXiv:2208.03229 (2022)

Graph Attention Network Knowledge Graph Completion Model Based on Relational Aggregation

Junkang Shi[1,2], Ming Li[3], and Jing Zhao[1,2](✉)

[1] Key Laboratory of Computing Power Network and Information Security, Ministry of Education, Shandong Computer Science Center (National Supercomputer Center in Jinan), Qilu University of Technology (Shandong Academy of Sciences), Jinan, China
zjstudent@126.com

[2] Shandong Provincial Key Laboratory of Computer Networks, Shandong Fundamental Research Center for Computer Science, Jinan, China

[3] School of Intelligence and Information Engineering, Shandong University of Traditional Chinese Medicine, Jinan, China
dianmail@sina.com

Abstract. Knowledge Graph Completion (KGC) refers to inferring missing links based on existing triples. Research has found that graph neural networks perform well in this task. By using the topology of the graph and the characteristics of the nodes to learn, the feature representation of the nodes can be updated, and more semantic information can be obtained from the surrounding entities and relationships. This paper aims to propose an end-to-end structured Graph Attention Network Enhanced Relationship Aggregation (GANERA) knowledge graph completion model. Firstly, entity aggregation is performed on the central entity by adding an entity attention mechanism. The addition of entity attention can distinguish the importance of different neighbor entities and screen more important entity embeddings. At the same time, the expression ability of the message function is enhanced through specific relational parameters, so that the model can extract richer relational information. Finally, the decoder chooses the convolutional network ConvR. We conduct experiments on standard datasets such as FB15k-237 and WN18RR, and the experimental results confirm the effectiveness of the model, while also achieving relative improvements in Hits@N and MRR values compared to other classical models.

Keywords: Knowledge graph completion · Relational Aggregation · Attention mechanism

1 Introduction

A knowledge graph is a structured semantic knowledge base that stores data in the form of triples (h, r, t), where h is the head entity, t is the tail entity,

B. Luo et al. (Eds.): ICONIP 2023, CCIS 1966, pp. 68–79, 2024.
https://doi.org/10.1007/978-981-99-8148-9_6

and r is the relationship between them [6]. Knowledge graph completion [15] has been successfully applied to many downstream tasks and applications such as intelligent question answering systems [3], recommendation systems [16] and visual relationship detection [18]. Figure 1 is a simple knowledge map related to movies. As we know, the amount of data that exists in reality is very large, so most knowledge graphs are incomplete, and the task of knowledge graph completion has attracted more and more attention.

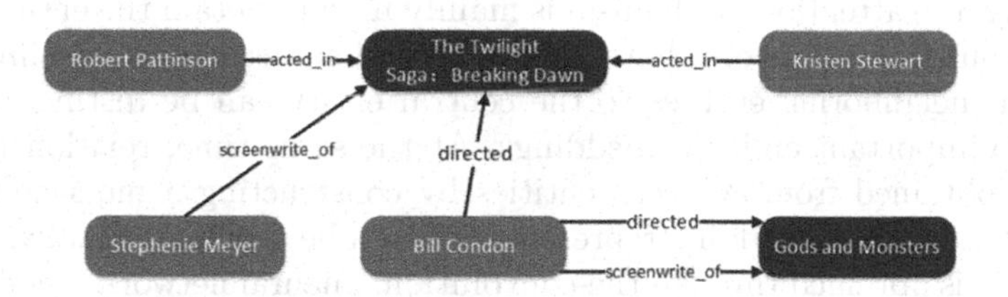

Fig. 1. A subgraph of the cast member knowledge graph. A subplot about the cast and crew of the movie Twilight.

Knowledge graph completion is to complete the missing links through the known entities and relationships existing in the knowledge graph. Its main task is to be able to predict missing entities (relationships) or confirm the relationship between any two entities and the correctness of existing relationships between entities based on existing entities and relationships in the knowledge graph, thereby achieving the integration, filtering and screening of existing knowledge, thereby improving the quality of entities (relationships) in the knowledge graph and solving many problems such as data missing in the knowledge graph. At present, the main research method is based on knowledge graph embedding methods, mainly embedding entities and relationships into continuous vector space for reasoning and prediction. The main research models include: translation model [5] and neural network model [8]. However, when dealing with triples, translation models treats triples independently, ignoring the hidden entity information and rich semantic relationships in the neighborhood around triples [1]. The model structure is relatively simple and has poor expressive power. The existing neural network model ignores the overall characteristics of the graph structure, does not make full use of the relationship between entities, and ignores the translation characteristics between entity relationships through too many convolution operations.

The emergence of graph neural networks [22] has realized the expansion of deep neural networks. We found that by using the characteristics of graph structure connectivity, it is possible to learn richer semantic information from neighboring entities and relationships. However, for heterogeneous graphs containing [19] different types of nodes and links, they have not been fully considered in existing graph neural network models. The amount of data that exists in reality is very large, and the attributes and associated relationships of entities are complex. Aggregating information from neighboring entities in a general way often causes message redundancy.

In order to effectively solve the problems existing in the existing model, we propose a graph attention network enhanced relationship aggregation knowledge graph completion model (GANERA) that uses an end-to-end structure. The graph neural network model with improved attention mechanism is used as an encoder to increase the interaction information between entities and relationships, and then the convolutional neural network model is used to predict scores. Specifically, the attention mechanism is mainly used to obtain the entity features of the surrounding neighbors. Through the attention mechanism, the importance of different neighboring entities to the central entity can be distinguished and filter more important entity embeddings. At the same time, relationship information is obtained from adjacent entities by constructing a message function, and then the entity embedding representation can be updated. Finally, the score of the triple is obtained through the convolutional neural network model ConvR. Our contributions are mainly as follows:

- In response to the problems existing in the current model, we propose the GANERA model, which is a graph attention-based method that mainly uses the attention mechanism to selectively integrate neighbor entity features and then updates the entity embedding representation through the message aggregation function.
- TBy introducing the attention mechanism, the ConvR model's ability to extract global features is enhanced, and the model's semantic understanding ability is increased.
- We verified the effect of the model on four different data sets. Experimental data shows that GANERA has good performance.

2 Related Work

2.1 Translation Model

The translation model mainly projects entities and relationships into a low-dimensional vector space, and is essentially a knowledge map completion through representation learning.

TransE is a classic knowledge graph translation model [2]. It maps entities and relationships to a low-dimensional vector space. The TransE model tries to make the head entity vector plus the relationship vector equal to the tail entity vector. This approach performs well when dealing with one-to-one relationships, but may encounter difficulties when dealing with one-to-many, many-to-one or many-to-many relationships.

2.2 Neural Network Model

Neural networks have powerful nonlinear fitting capabilities and can learn complex features and patterns from large amounts of data [21]. Convolutional neural network models mainly perform convolution operations on the embedded representations of entities and relationships to capture the interactions between entities and relationships [20]. Graph neural network models mainly learn node representations for graph data to capture the complex structure and semantics between entities and relationships.

The ConvR model is a multi-relation learning model based on adaptive convolution, which can effectively mine the interaction between entities and relationships, and improve the completeness of the knowledge graph [7]. The core idea of the ConvR model is to decompose and reconstruct the relationship representation into a set of filters, then perform convolution on the head entity representation, and finally compare it with the tail entity representation. Compared with the ConvE model, the ConvR model has stronger expressive power and higher parameter efficiency. However, the ConvR model only considers local interactions between entities and relationships, and does not fully consider global semantic information.

Graph Attention Network (GAT) can capture complex interactions between entities and relationships by performing a self-attention mechanism on the embedding representations of entities and relationships [14]. It can effectively capture important information in entity embeddings and relationship embeddings, thereby improving the performance of knowledge graph completion. However, because it needs to perform attention calculations on all neighbor nodes of each node, this can lead to memory and time consumption.

3 GANERA: Model Description

In this section, we will provide an in-depth introduction to the GANERA model. The overall structure of the model can be seen in Fig. 3, primarily consisting of two components: the encoder and the decoder. In the encoder part, we first use multi-head attention mechanism to fuse useful features of surrounding neighbor nodes, and then update the entity's embedding representation by aggregating relational features through constructing a relation-aware function. The input of the decoder is the updated entity embedding and vectorized relation embedding. To prevent feature loss caused by overfitting, we vectorize the original relationship as the input of the decoder. The dimension of the relationship embedding is the same as that of the output entity embedding.

We represent the knowledge graph as $\mathcal{G} = (\mathcal{V}, \mathcal{R}, \mathcal{E})$, where $\mathcal{G}$ represents the set of entities,V represents the set of entities, $\mathcal{R}$ represents the set of relations, and E represents the set of edges. Each triplet (s, r, o) denotes the presence of relation $r \in \mathcal{R}$ from entity s to entity o. Building upon previous research models, we permit information within directed knowledge graphs to traverse in three manners: original, inverse, and self-referential. As a result, the collection of edges and relations expands as follows:

$$\mathcal{E}' = \mathcal{E} \cup \{s, r, o^{-1} \mid (s, r, o) \in \{\mathcal{E}\}\} \cup \{s, \mathcal{E}, s \mid s \in \mathcal{E}\} \mathcal{R}' = \mathcal{R} \cup \mathcal{R}_{inv} \cup \{\mathcal{E}\} \quad (1)$$

Here, $\mathcal{R}_{inv} = \{r^{-1} | r \in \mathcal{R}\}$ and $\mathcal{E}$ represent the inverse relation and the self-referential relation respectively.

3.1 Entity Attention Fusion Layer

To acquire valuable information from the entity's neighborhood and alter the significance of nodes within each node's vicinity to prevent message redundancy, we initially employ a multi-head attention mechanism. This aids in adjusting the importance of nodes within each node's neighborhood while updating the embedded representation of entities effectively. In this layer, the input feature set of nodes is represented as $\mathbf{n} = \boldsymbol{n}1, \boldsymbol{n}2, ..., \boldsymbol{n}N$. A set of transformed node feature vectors, denoted as $\mathbf{n}' = \boldsymbol{n}1', \boldsymbol{n}2', ..., \boldsymbol{n}N'$, is generated, where $\boldsymbol{n}1$ and $\boldsymbol{n}1'$ represent the input and output embeddings of entity e_i respectively, and N is the number of entities (nodes). A single GAT layer can be described as:

$$e_{ij} = a(\mathbf{W}\boldsymbol{n}_i, \mathbf{W}\boldsymbol{n}_j) \tag{2}$$

where e_{ij} represents the attention weight of the edge (e_i, e_j) in $\mathcal{E}'$, $\mathbf{W}$ is a parameterized linear transformation matrix used to map input features into a high-dimensional output feature space, and $a()$ is the attention function we have chosen. The attention value of each edge is the importance of the feature of edge (e_i, e_j)'s of the source node e_i. Here, the softmax function is used to calculate the relative attention α_{ij} for all values in the neighborhood. The following equation shows the output of a layer (Fig. 2):

$$\boldsymbol{n}_i' = \sigma\left(\sum_{j \in \mathcal{N}_i} \alpha_{ij} \mathbf{W}\boldsymbol{n}_j\right) \tag{3}$$

We adopt multi-head attention to stabilize the learning process. The multi-head attention expression of concatenating K attention heads is shown below:

$$\boldsymbol{n}_i' = \overset{K}{\underset{k=1}{\Big\|}} \sigma\left(\sum_{j \in \mathcal{N}_i} \alpha_{ij}^k \mathbf{W}^k \boldsymbol{n}_j\right) \tag{4}$$

where $||$ signifies concatenation, σ represents any arbitrary non-linear function, α_{ij}^k denotes the normalized attention coefficient for the edge (e_i, e_j) obtained during the k-th iteration, and $\mathbf{W}^k$ stands for the linear transformation matrix associated with the k-th attention mechanism. In the final layer, the output embeddings are computed using an averaging operation instead of concatenation to achieve the multi-head attention effect. The formula is as follows:

$$\boldsymbol{n}_i' = \sigma\left(\frac{1}{K}\sum_{k=1}^{K}\sum_{j \in \mathcal{N}_i} \alpha_{ij}^k \mathbf{W}^k \boldsymbol{n}_j\right) \tag{5}$$

3.2 Relationship-Aware Function Layer

Subsequently, we utilize message functions to glean information from adjacent nodes and relationships. In the case where entity s is linked to entity o through

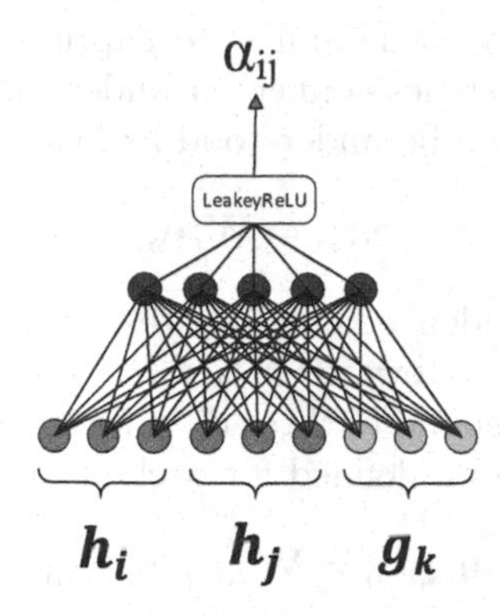

Fig. 2. Attention Mechanism

relationship r, their distributed representations are amalgamated as follows:

$$\mathbf{h}_{(s,r,o)} = \phi(\mathbf{e}_s, \mathbf{e}_r, \mathbf{e}_o) \tag{6}$$

Where $\mathbf{e}_s, \mathbf{e}_o \in \mathbb{R}^{d_0}$ stand for entity embeddings, while $\mathbf{e}_r \in \mathbb{R}^{d_0}$ denotes the relationship embedding. The function $\phi(\cdot)$ serves as a merging operator, integrating the relationship embedding into the entity embedding. Following this, weight coefficients $\boldsymbol{W}_g$ are learned, shared across the entire graph, for message generation:

$$\mathbf{m}_{(s,r,o)} = M(\mathbf{h}_{(s,r,o)}, \mathbf{w}_g) \tag{7}$$

Here, $\mathbf{m}_{(s,r,o)}$ symbolizes a message originating from entity s and directed towards entity o. The parameter $\mathbf{w}_g$ is a shared entity that transcends various positions and local structures, with the capacity to sift out prevalent features present across diverse topological arrangements. When an entity becomes connected through distinct relationships, it assumes diverse roles in relation to the central entity o.

$$\mathbf{h}^r_{(s,r,o)} = \phi_r(\mathbf{e}_s, \mathbf{e}_r, \mathbf{e}_o, \theta_r) \tag{8}$$

$$\mathbf{m}_{(s,r,o)} = M(\mathbf{h}^r_{(s,r,o)}, \boldsymbol{\theta}_g) \tag{9}$$

Where θr is acquired for the purpose of capturing features that are specific to each relationship. The embeddings of neighboring entities, along with the associated relationship information, are grouped together and subjected to various message functions. A straightforward method to derive θr involves training separate weight matrices for individual relationships. Considering the diverse range of encoders and decoders used in the experiment, there's a computational efficiency consideration that leads to the constraint of θ_r being a diagonal matrix.

$$\theta_r = \mathbf{W}_r = diag(\mathbf{w}_r) \tag{10}$$

Where $\mathbf{W}r \in \mathbb{R}^{d0 \times d_0}$ and $\mathbf{w}r \in \mathbb{R}^{d0 \times 1}$. The parameterization θ_r can take on more versatile forms. To put it simply, let's outline how $\mathbf{W}_r$ is applied within the present message function. The aim is to capture shared features linked to specific relationships. For entities connected under the same relationship r, the transformation procedure can be understood as follows:

$$\mathbf{e}_{(rs)} = \mathbf{W}_r \mathbf{e}_s \tag{11}$$

Here, $\mathbf{e}_{(rs)}$ denotes the hidden embedding resulting from the transformation associated with relationship r towards entity s'. Since the original, inverse, and self-loop relationships correspond to three distinct edge types with varying directions, distinct filters have been defined for each of these edge types.

$$\mathbf{m}_{(s,r,o)} = \mathbf{W}_{dir(r)} \mathbf{h}^r_{(s,r,o)} \tag{12}$$

here the relationship weight $\mathbf{W}_{dir(r)} \in \mathbb{R}^{d_1} \times \mathbb{R}^{d_0}$ is defined as follows:

$$\mathbf{W}_{dir(r)} = \begin{cases} \mathbf{W}_O & \text{if} r \in \mathcal{R} \\ \mathbf{W}_I & \text{if} r \in \mathcal{R}_{imv} \\ \mathbf{W}_S & \text{if} r \in \{\top\} \end{cases} \tag{13}$$

where O, I and S represent the original, reverse and self-loop directions. Therefore, the model can learn the entity and relationship features in the surrounding neighborhood in three directions. In order to obtain the new embedding of node o, The update function of the model can be expressed as:

$$\mathbf{e}'_o = f\left(\sum_{(s,r)\in\mathcal{N}(o)} a\mathbf{m}_{(s,r,o)}\right) \tag{14}$$

Here, $\mathcal{N}(o)$ refers to the collection of directly neighboring entities and the relationships connected to the central entity o, and $f()$ stands for a non-linear activation function. The sum of all messages is multiplied by a constant importance coefficient a, analogous to the application of $|\mathcal{N}(o)|$ in R-GCN. In addition, in order to prevent the loss of relationship features caused by overfitting, we transform the relationship embedding so that the relationship embedding has the same dimension as the entity embedding.

$$\mathbf{e}'_r = \mathbf{W}_{rel} \mathbf{e}_r \tag{15}$$

3.3 Decoder

In studying the ConvR model, we found that the ConvR model only considers local interactions between entities and relationships, and does not fully consider global semantic information. The updated entity features effectively enhance the model's semantic understanding ability. The ConvR model convolves the input entity embeddings with relation-specific filters that are adaptively constructed. For each filter $\mathbf{R}^{(\ell)}$, we generate a convolutional feature map

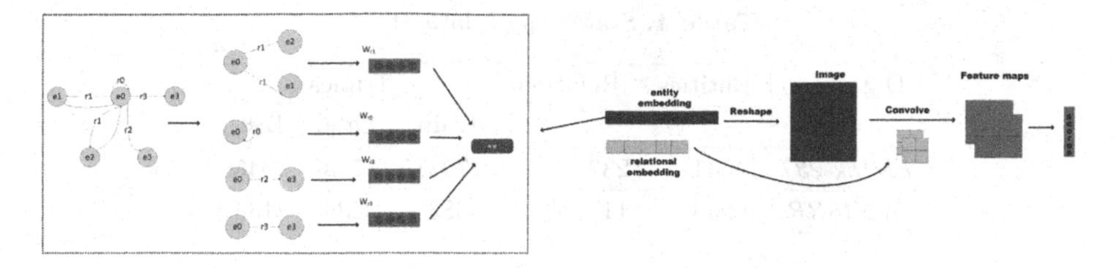

Fig. 3. The GANERA model is described as follows: GANERA is mainly composed of an encoder and a decoder. First, in order to integrate effective information from neighboring nodes, we update the entity embeddings through a multi-head attention mechanism. Then, through the message fusion function, we integrate the entity features under different relationship types. Finally, the decoder model scores the results.

$\mathbf{C}^{(\ell)} \in \mathbb{R}^{(d_e^h - h + 1) \times (d_e^w - w + 1)}$. Each element at position m and n within this map is computed using the following formula:

$$c_{m,n}^{(\ell)} = f\left(\sum_{i,j} s_{m+i-1,n+j-1} \times r_{i,j}^{(\ell)}\right) \tag{16}$$

where $f()$ is a non-linear function.In the experiment, the scoring function is as follows:

$$\psi(s, r, o) = f(\mathbf{W}\mathbf{c} + \mathbf{b})^{\top}\mathbf{o} \tag{17}$$

Here, $\mathbf{W} \in \mathbb{R}^{d_e \times c(d_e^h - h + 1)(d_e^w - w + 1)}$, and $\mathbf{b} \in \mathbb{R}^{d_e}$ represents the parameters of the fully connected layer.

4 Experiments

4.1 Benchmark Datasets

To comprehensively evaluate the proposed model, we used two standard benchmark datasets: WN18RR [4] and FB15k-237 [12]. Among them, WN18RR is a subset of WN18, which was created from the WN18 dataset extracted from a subset of WordNet. The WN18RR dataset removed the reversed relationships and retained the symmetric, asymmetric and composite relationships in the original dataset. FB15k-237 is a subset of the FB15k dataset inspired by observations of test leakage in FB15k. The FB15k-237 dataset retains mainly symmetric, asymmetric and composite relationships and also removes reversed relationships. The detailed information of these datasets is shown in Table 1.

4.2 Training Protocol

We use standard evaluation metrics MRR and Hits@N, where MRR is the mean reciprocal rank. The calculation method of MRR is for each test triple (h, r, t),

Table 1. Statistics of datasets

Dataset	Entities	Relations	Triples		
			Train	Valid	Test
FB15k-237	14541	237	272115	17535	20466
WN18RR	40943	11	86835	3034	3134

given h and r, score all possible tail entities t' and rank the triples according to the score, then calculate the reciprocal of the rank of the correct triple (h, r, t), and finally average the reciprocal ranks of all test triples. The larger the value of MRR, the better, indicating that the ranking of correct triples is closer to the front. Hits@N represents how many correct triples are in the top N items. The larger the value, the better the model is for knowledge graph completion. The initial learning rate of the model is set to 0.001 and we use Adam optimizer to optimize all parameters. The output embedding dimension of the graph attention network is set to 200, while the initial embedding dimension is set to 100 batch, the batch size for FB15k-237 is set to 1024 and the batch size for WN18RR is 256.

4.3 Baselines

To evaluate GANERA, we compared it with various baseline models, including translation models (TransE and TransR), tensor decomposition models (RESCAL and DistMult), and neural network models (ConvE, ConvR, and SACN). Among them, ConvE and ConvR are classic models based on convolutional neural networks, while SACN is a model based on graph convolutional neural networks with the same structure as the GANERA model.

4.4 Results and Analysis

Our test results are shown in Table 2. Table 2 reports the Hits@10, Hits@3, Hits@1 and MRR results of different baseline models and our model on standard datasets. The highest scores are shown in bold and the second highest scores are indicated by an underline.

The above table summarizes the comparison results of GANERA with existing knowledge graph completion models. The results clearly show that GANERA has achieved significant improvements in performance on all four datasets compared to other models. It has achieved better performance than traditional convolutional network models and translation models. The improvement in MRR values confirms our model's ability to accurately represent triple relationships. On the FB15k-237 dataset, all 4 evaluation indicators are better than other models. Compared to SACN with the same model architecture, GANERA's MRR increased by 3.9%, and its Hit@N values were also significantly higher than other models. Compared to the convolutional network model ConvE, GANERA's MRR increased by 15.2%, and its Hit@1 and Hit@3 values were also significantly higher than other models. On the WN18RR dataset, compared to the

Table 2. Results of the link prediction on WN18RR and FB15k-237 datasets.

	FB15k-237				WN18RR			
	Hits@1	Hits@3	Hits@10	MRR	Hits@1	Hits@3	Hits@10	MRR
TransE [2]	-	-	0.465	0.294	-	-	0.499	0.229
TransR [9]	–	–	0.472	0.304	–	–	0.503	0.235
ComplEX [13]	0.159	0.264	0.418	0.247	0.41	0.46	0.54	0.44
RESCAL [10]	–	–	0.413	0.334	–	–	0.483	0.452
DistMult [17]	0.156	0.274	0.428	0.231	0.393	0.442	0.498	0.433
ConvE [4]	0.235	0.353	0.499	0.315	0.44	0.468	0.532	0.439
ConvR [7]	0.264	0..384	0.513	0.353	**0.453**	0.483	0.534	0.476
SACN [11]	0.26	0.39	0.544	0.35	0.43	0.488	0.542	0.47
GANERA	**0.27**	**0.402**	**0.55**	**0.363**	0.412	**0.506**	**0.558**	**0.492**

tensor decomposition model DistMult, GANERA's MRR increased by 13.6%, and compared to SACN, GANERA's Hits@10 increased by 2.9%. GANERA outperformed other models in 3 out of 4 evaluation indicators on the WN18RR dataset. The experimental results on the other two datasets also show that our model is better than other baseline models. Overall, GANERA's overall performance is better than other models, confirming that after graph attention and message aggregation, the representation of entity embeddings is more accurate and can better achieve knowledge graph completion.

Table 3. Model ablation result on the validation sets

	Ablation	MRR	Hits@10
WN18RR	Full mode	**0.492**	**0.558**
	Without attention	0.479	0.503
	Without encoder	0.430	0.455
FB15K-237	Full mode	**0.363**	**0.55**
	Without attention	0.342	0.522
	Without encoder	0.324	0.512

4.5 Ablation Study

We conducted ablation experiments on the model, from the complete model to removing the attention part of the model and then to removing the encoder part, to see its impact on the final results. The ablation experiment results dataset is shown in Table 3. The data indicates that the complete model performs best on the FB15K-237 and WN18RR datasets. At the same time, by removing the attention mechanism, the results show that message fusion for relationships also improves the performance of the model. Overall, the ablation experiment proves the effectiveness of the model.

5 Conclusion and Future Work

We propose an end-to-end structured graph attention network enhanced relationship aggregation knowledge graph completion model, which adopts an end-to-end framework structure. Mainly using the structural characteristics of graph connectivity, selectively integrating useful information from neighboring entities through attention, increasing the feature representation of entities, and then performing feature fusion on entities under different relationships through specific message functions, integrating relationship features. In order to avoid the loss of features caused by overfitting, we use the original relationship embedding as the input for decoding. By introducing the attention mechanism, the ConvR model's ability to extract global features is enhanced, and the model's semantic understanding ability is increased.

In the future, we hope to incorporate the idea of neighbor selection and multi-relation knowledge graphs into our training framework, while considering the importance of neighbors when aggregating their vector representations and how to deal with the relationship embedding problem of multi-relation knowledge graphs. We also hope to expand our model to handle larger knowledge graphs

Acknowledgements. This work is supported in part by The Key R&D Program of Shandong Province (2021SFGC0101), The 20 Planned Projects in Jinan (202228120)

References

1. Arora, S.: A survey on graph neural networks for knowledge graph completion. arXiv preprint arXiv:2007.12374 (2020)
2. Bordes, A., Usunier, N., Garcia-Duran, A., Weston, J., Yakhnenko, O.: Translating embeddings for modeling multi-relational data. In: Advances in Neural Information Processing Systems, vol. 26 (2013)
3. Budiharto, W., Andreas, V., Gunawan, A.A.S.: Deep learning-based question answering system for intelligent humanoid robot. J. Big Data **7**, 1–10 (2020)
4. Dettmers, T., Minervini, P., Stenetorp, P., Riedel, S.: Convolutional 2D knowledge graph embeddings. In: Proceedings of the AAAI Conference on Artificial Intelligence, vol. 32 (2018)
5. Ebisu, T., Ichise, R.: Generalized translation-based embedding of knowledge graph. IEEE Trans. Knowl. Data Eng. **32**(5), 941–951 (2019)
6. Ji, S., Pan, S., Cambria, E., Marttinen, P., Philip, S.Y.: A survey on knowledge graphs: representation, acquisition, and applications. IEEE Trans. Neural Netw. Learn. Syst. **33**(2), 494–514 (2021)
7. Jiang, X., Wang, Q., Wang, B.: Adaptive convolution for multi-relational learning. In: Proceedings of the 2019 Conference of the North American Chapter of the Association for Computational Linguistics: Human Language Technologies, vol. 1 (Long and Short Papers), pp. 978–987 (2019)
8. Liang, S., Shao, J., Zhang, D., Zhang, J., Cui, B.: DRGI: deep relational graph infomax for knowledge graph completion. IEEE Trans. Knowl. Data Eng. (2021)
9. Lin, Y., Liu, Z., Sun, M., Liu, Y., Zhu, X.: Learning entity and relation embeddings for knowledge graph completion. In: Proceedings of the AAAI Conference on Artificial Intelligence, vol. 29 (2015)

10. Nickel, M., Tresp, V., Kriegel, H.P.: A three-way model for collective learning on multi-relational data. In: Icml, vol. 11, pp. 3104482–3104584 (2011)
11. Shang, C., Tang, Y., Huang, J., Bi, J., He, X., Zhou, B.: End-to-end structure-aware convolutional networks for knowledge base completion. In: Proceedings of the AAAI Conference on Artificial Intelligence, vol. 33, pp. 3060–3067 (2019)
12. Toutanova, K., Chen, D., Pantel, P., Poon, H., Choudhury, P., Gamon, M.: Representing text for joint embedding of text and knowledge bases. In: Proceedings of the 2015 Conference on Empirical Methods in Natural Language Processing, pp. 1499–1509 (2015)
13. Trouillon, T., Welbl, J., Riedel, S., Gaussier, É., Bouchard, G.: Complex embeddings for simple link prediction. In: International Conference on Machine Learning, pp. 2071–2080. PMLR (2016)
14. Veličković, P., Cucurull, G., Casanova, A., Romero, A., Lio, P., Bengio, Y.: Graph attention networks. arXiv preprint arXiv:1710.10903 (2017)
15. Wang, B., Shen, T., Long, G., Zhou, T., Wang, Y., Chang, Y.: Structure-augmented text representation learning for efficient knowledge graph completion. In: Proceedings of the Web Conference 2021, pp. 1737–1748 (2021)
16. Wu, S., Sun, F., Zhang, W., Xie, X., Cui, B.: Graph neural networks in recommender systems: a survey. ACM Comput. Surv. **55**(5), 1–37 (2022)
17. Yang, B., Yih, W.T., He, X., Gao, J., Deng, L.: Embedding entities and relations for learning and inference in knowledge bases. arXiv preprint arXiv:1412.6575 (2014)
18. Zhan, Y., Yu, J., Yu, T., Tao, D.: On exploring undetermined relationships for visual relationship detection. In: Proceedings of the IEEE/CVF Conference on Computer Vision and Pattern Recognition, pp. 5128–5137 (2019)
19. Zhang, C., Song, D., Huang, C., Swami, A., Chawla, N.V.: Heterogeneous graph neural network. In: Proceedings of the 25th ACM SIGKDD International Conference on Knowledge Discovery & Data Mining, pp. 793–803 (2019)
20. Zhang, J., Huang, J., Gao, J., Han, R., Zhou, C.: Knowledge graph embedding by logical-default attention graph convolution neural network for link prediction. Inf. Sci. **593**, 201–215 (2022)
21. Zheng, Y., Pan, S., Lee, V., Zheng, Y., Yu, P.S.: Rethinking and scaling up graph contrastive learning: an extremely efficient approach with group discrimination. In: Advances in Neural Information Processing Systems, vol. 35, pp. 10809–10820 (2022)
22. Zhou, J., et al.: Graph neural networks: a review of methods and applications. AI Open **1**, 57–81 (2020)

SODet: A LiDAR-Based Object Detector in Bird's-Eye View

Jin Pang and Yue Zhou(✉)

School of Electronic Information and Electrical Engineering,
Shanghai Jiao Tong University, Shanghai 200240, China
{sjtu.pj,zhouyue}@sjtu.edu.cn

Abstract. LiDAR-based object detection is of paramount significance in the realm of autonomous driving applications. Nevertheless, the detection of small objects from a bird's-eye view perspective remains challenging. To address this issue, the paper presents **SODet**, an efficient single-stage 3D object detector designed to enhance the perception of small objects like pedestrians and cyclists. SODet incorporates several key components and techniques. To capture broader context information and augment the capability of feature representation, the model constructs residual blocks comprising large-kernel depthwise convolutions and inverted bottleneck structures, forming the foundation of the CSP-based NeXtDark backbone network. Furthermore, the NeXtFPN feature extraction network is designed with the introduced SPPF module and the proposed special residual blocks, enabling the extraction and fusion of multi-scale information. Additionally, training strategies such as mosaic data augmentation and cosine annealing learning rate are employed to further improve small object detection accuracy. The effectiveness of SODet is demonstrated through experimental results on the KITTI dataset, showcasing a remarkable enhancement in detecting small objects from a bird's-eye perspective while maintaining a detection speed of 20.6 FPS.

Keywords: Object Detection · LiDAR Point Cloud · Bird's-Eye View · Autonomous Driving

1 Introduction

Within the domain of computer vision, the significance of 3D object detection has witnessed a steady rise in recent years, driven by the progressive advancements in areas like autonomous driving, robot navigation, and environmental perception. LiDAR, as an effective 3D sensing device, has found extensive applications in fields such as autonomous driving and robotics [1,29,30].

3D object detection based on LiDAR point clouds is a common method that extracts the 3D information and features of objects from processed LiDAR point cloud data, enabling object detection and recognition. Compared to image-based

B. Luo et al. (Eds.): ICONIP 2023, CCIS 1966, pp. 80–91, 2024.
https://doi.org/10.1007/978-981-99-8148-9_7

2D object detection, this approach can offer more precise and comprehensive spatial information, effectively mitigating challenges associated with occlusion and variations in lighting conditions [2].

However, 3D object detection based on LiDAR point clouds encounters a multitude of obstacles, including the sparsity, irregularity, and unordered characteristics of point cloud data [2,3]. Point clouds obtained from LiDAR sensors often exhibit sparsity, meaning there are gaps or missing points in the data, which can make it difficult to accurately detect and localize objects. Additionally, point clouds are typically irregular and unordered, lacking a predefined structure or grid-like arrangement. This irregularity poses challenges in feature extraction and modeling, as traditional 3D convolutional neural networks designed for regular grid data may not directly apply.

The bird's-eye view representation of LiDAR point clouds is a method that involves projecting three-dimensional point cloud data onto a two-dimensional plane from a top-down viewpoint. This technique finds frequent application in autonomous driving and robotics, specifically for object detection and recognition on the ground surface. Compared to other methods, the projection-based approach utilizing a bird's-eye view perspective offers significant advantages in terms of model design, deployment, and computational resources. Furthermore, the adoption of the bird's-eye view representation significantly alleviates the effects of line-of-sight occlusions and obstructions [6]. Simultaneously, it facilitates downstream multitask sharing within a unified bird's-eye view feature space. However, in the context of previous works [13,15,16,18–20], this method exhibits subpar performance in detecting small objects such as pedestrians and cyclists when compared to vehicles in the field of autonomous driving.

In response to the aforementioned challenges, the paper presents a high-efficiency single-stage 3D object detector named **SODet** (Small Object Detector) based on LiDAR point clouds, as shown in Fig. 1. The primary objective of this model is to enhance the detection precision of small objects, particularly pedestrians and cyclists, in the bird's-eye view perspective for autonomous driving applications, while maintaining rapid inference speed. The proposed SODet architecture comprises several key components. Firstly, the point cloud data is transformed into a bird's-eye view representation, enabling efficient processing and alleviating occlusion issues. Next, the bird's-eye view representation undergoes feature extraction and fusion stages, followed by classification and regression steps to perform object detection. To capture a wider range of contextual information and enhance feature representation capabilities, the paper constructs a residual block composed of large-kernel depthwise convolutions and inverted bottleneck structures. Building upon this, a backbone network called NeXtDark is designed, incorporating the Cross Stage Partial (CSP) structure. Furthermore, to effectively extract and integrate multi-scale information, a feature fusion network called NeXtFPN is designed based on the introduced Spatial Pyramid Pooling (SPPF) module and the proposed special residual block. Experimental results indicate that SODet achieves significant improvements in the detection accuracy in bird's-eye view, particularly for small objects, attaining a speed of 20.6 FPS on the KITTI dataset.

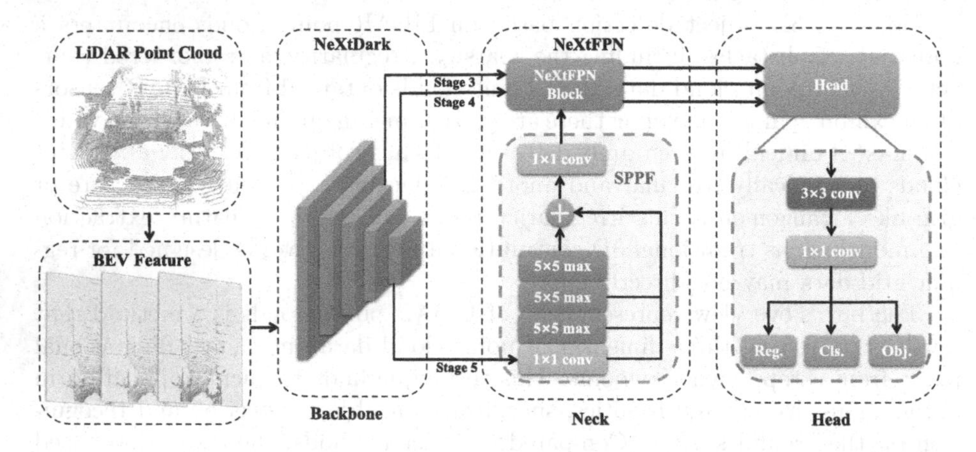

Fig. 1. Macro architecture of SODet. The point cloud data is encoded into a bird's-eye view representation, which is then fed into the network consisting of the backbone NeXtDark for feature extraction, the neck NeXtFPN for feature fusion, and the head for classification and regression.

In summary, this paper makes the following contributions:

- The introduction of SODet, an efficient 3D object detector operating on bird's-eye view LiDAR point cloud data, which achieves improved detection accuracy, particularly for small objects in autonomous driving scenarios.
- The development of NeXtDark, a feature extraction network based on special residual blocks incorporating large-kernel depthwise convolutions and inverted bottleneck structures.
- The proposal of NeXtFPN, a feature fusion network that effectively captures multi-scale information and enhances contextual understanding.
- The utilization of various training strategies to enhance the accuracy of small object detection, including data augmentation, learning rate optimization, and loss function.

2 Related Work

Thus far, a multitude of deep learning-based techniques have been extensively employed in the domain of 3D object detection using LiDAR point clouds. These methods have been specifically designed to address the distinctive attributes of point cloud data, incorporating tailored processing and optimization strategies.

Voxel-based methods in 3D object detection transform point cloud data into regular voxel grids and utilize 3D convolutional neural networks to process and analyze these voxels for object detection, such as VoxelNet [4]. However, the presence of 3D convolutions in this approach entails high computational complexity and time consumption. To address this issue, SECOND [5] employs sparse convolution operations to handle voxelized representations of LiDAR point cloud data, enhancing computational efficiency and accuracy. Sparse convolution solely considers the positions of point cloud data within voxels, significantly reducing computational burden and improving efficiency. However, it is not deployment-friendly, and compared to traditional convolutional operations, the implementation and tuning of sparse convolutions may be more complex, requiring advanced technical skills and computational resources. PointPillars [6] projects voxelized point cloud data onto a two-dimensional feature map, representing the distribution of points in the horizontal and vertical directions. Subsequently, 2D convolutional neural networks process the pillar features for object detection. Building upon the pillar-based point cloud representation, PillarNet [7] has further introduced a real-time and high-performance single-stage 3D object detector.

Point-based methods directly process LiDAR point cloud data without the need for additional data transformation or projection. PointNet [8] emerged as the pioneering end-to-end deep learning architecture specifically designed for direct manipulation of point cloud data. Through the utilization of shared-weight multilayer perceptrons, PointNet effectively captures unique features for individual points and subsequently combines them into a comprehensive global feature representation. Building upon PointNet, PointNet++ [9] extends its capabilities by addressing the inherent challenge of modeling local structures within point cloud data. Through a cascaded hierarchical aggregation scheme and multi-scale processing, PointNet++ effectively captures the hierarchical organization and local features of point cloud data. Nonetheless, point-based approaches encounter various challenges, including the sparse nature of point cloud, computational complexity, and sensitivity to occlusion.

Recently, there have been notable endeavors [10–12] that integrate the advantages of point-based methods, which enable fine-grained processing of point cloud data, with voxel-based methods, which facilitate holistic modeling of point cloud data.

Projection-based methods project 3D point cloud data onto a 2D plane for processing and analysis. Some approaches utilize a bird's-eye view perspective projection, such as [13–18]. Other methods leverage multi-view information to enhance the performance of 3D object detection. For example, MV3D [19] integrates features from the point cloud's top-view, front-view, and the RGB image, while AVOD [20] builds upon this by removing the front-view input. Multimodal approaches [21,22] aim to fuse information from LiDAR point clouds and camera images, thereby improving detection accuracy and robustness. While these approaches have demonstrated remarkable achievements, the calibration and alignment of diverse data types, as well as the computational complexity, pose substantial hurdles that still need to be overcome.

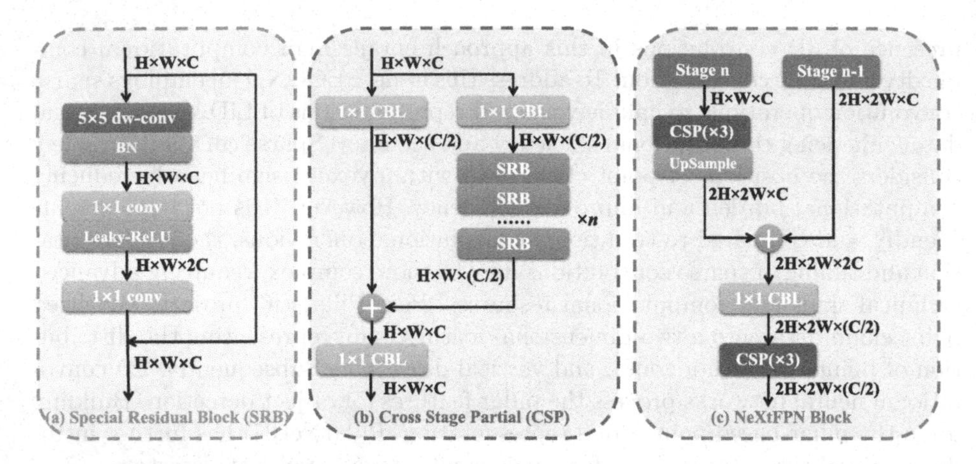

Fig. 2. (a) Special Residual Block (SRB) with a 5 × 5 depthwise convolution and inverted bottleneck structure, where BN stands for Batch Normalization. (b) Cross Stage Partial (CSP) structure with n SRBs, where CBL stands for Convolution+BN+Leaky-ReLU module. (c) NeXtFPN Block shows the fusion process of feature maps between Stage n and Stage $n-1$.

3 Methods

This section presents an efficient single-stage 3D object detector **SODet**, which focuses on enhancing the detection precision of small objects in bird's-eye view. As depicted in Fig. 1, the point cloud is first processed through bird's-eye view encoding (Sect. 3.1), and then fed into a network architecture (Sect. 3.2) consisting of the backbone NeXtDark for feature extraction, the neck NeXtFPN for feature fusion, and the head for classification and regression. By incorporating training strategies (Sect. 3.3), the model has achieved further improvements in the detection accuracy.

3.1 Point Cloud Encoding in Bird's-Eye View

Data Preprocessing. LiDAR is a type of sensor that utilizes laser beams to measure the distance to surrounding objects by measuring the time it takes for the laser beam to reflect back. This information is then used to generate point cloud data. In the KITTI dataset [23], point cloud data is stored in a floating-point binary file format. Each point in the point cloud is represented by four floating-point values, which correspond to the coordinates (x, y, z) of the point and an additional reflectance value (r). To enhance processing efficiency, the model selectively preserves the point cloud within the region of interest. This involves traversing each point $[x, y, z]^T$ and removing points that lie beyond the defined boundaries B, based on specified boundary conditions as illustrated in Eq. 1.

$$B = \left\{[x, y, z]^T \mid x \in [X_{min}, X_{max}], y \in [Y_{min}, Y_{max}], z \in [Z_{min}, Z_{max}]\right\} \quad (1)$$

Bird's-Eye View Representation. The advantages of the bird's-eye view lie in preserving the physical dimensions of objects and mitigating occlusion issues. Following the approach presented in [19], the cropped point cloud is projected onto a bird's-eye view representation and discretized into a two-dimensional grid. This grid is then encoded using three channels: height, intensity, and density. Specifically, assuming that the grid cell j contains n points $(p_0, p_1, p_2, ..., p_{n-1})$, where point i is represented as $p_i\left(x_i, y_i, z_i, r_i\right), i < n$, the height H_j, intensity I_j, and density D_j values of the grid cell j are calculated as shown in Eq. 2.

$$\begin{cases} H_j = max(z_0, z_1, z_2, ..., z_{n-1})/(Z_{max} - Z_{min}) \\ I_j = max(r_0, r_1, r_2, ..., r_{n-1}) \\ D_j = min(1.0, \ln(n+1)/\ln(64)) \end{cases} \tag{2}$$

3.2 Network Architecture

NeXtDark Backbone. The receptive field denotes the region within the input image that corresponds to an individual pixel in the feature map produced by a layer of a convolutional neural network. In the field of object detection, where small objects pose challenges in terms of limited information, increasing the visual receptive field can ensure the size of the feature map and enhance the network's ability to extract features. By capturing a broader context, the network can improve the accuracy of detecting small objects.

Motivated by recent advancements in large convolutional kernel research [24], a Special Residual Block (SRB) is constructed. This block comprises a combination of large-kernel depthwise convolutions and inverted bottleneck structures, as depicted in Fig. 2(a). Specifically, the proposed block is formed by a 5×5 depthwise convolution, followed by two 1×1 pointwise convolutions, and a residual connection. The introduction of large convolutional kernels expands the effective receptive field, while the depthwise convolution partially mitigates the computational cost associated with large kernels. To further enhance feature representation, the proposed block employs an inverted bottleneck structure, which was popularized by [25]. To avoid excessive parameterization, the hidden dimensions of the inverted bottleneck structure are set to twice the input dimensions, and two 1×1 pointwise convolutions are used to handle channel expansion and reduction separately.

Building upon the aforementioned special residual block, the backbone network NeXtDark is proposed in this paper. Following the principles of [26], NeXtDark employs a multi-level design with carefully designed feature map resolutions (1,3,9,9,3) at each downsampling stage. To address the parameter and computational efficiency of the network, NeXtDark introduces the Cross Stage Partial (CSP) structure [27]. As illustrated in Fig. 2(b), the CSP structure divides the input feature map into two branches using two 1×1 convolutions. The main branch extracts feature representations through n residual blocks, while the auxiliary branch, known as the transition branch, is concatenated with the main branch through cross-stage partial connections. These cross-stage partial connections facilitate feature propagation and utilization, improving the network's representation capacity and performance.

NeXtFPN Neck. In the neck network for feature fusion, the model introduces the spatial pyramid pooling module SPPF to fuse feature maps with different receptive fields, enriching the expressive power of the feature maps. The conventional SPP module utilizes a max pooling layer with pooling kernels of $k = 1 \times 1, 5 \times 5, 9 \times 9, 13 \times 13$ to perform multi-scale pooling operations on the input feature maps, capturing contextual information for objects of different scales, while the SPPF module replaces the single large-size pooling layer in the SPP module with multiple cascaded 5×5 maxpooling layers, as shown in Fig. 1. This modification retains the original functionality of the SPP module while further enhancing the operational speed of the model.

Based on the special residual block constructed in Sec. 3.2, this paper further designs the feature fusion network called NeXtFPN. Following the idea of [28], NeXtFPN aims to fuse semantic information at different scales by establishing a feature pyramid, improving the localization and recognition capabilities of the objects. As shown in Fig. 2(c), NeXtFPN block utilizes the CSP structure (consisting of three SRBs) and upsampling operation to propagate the semantic information from deep-level features (Stage n) to shallow-level features (Stage $n-1$), enhancing the expressive power of shallow-level features and thus improving the detection performance of small objects.

Detection Head. The detection head is responsible for converting the feature representations of the network into the detection results of the objects. As shown in Fig. 1, it takes the feature maps from the neck network and applies them to a 3×3 convolutional layer and a 1×1 convolutional layer to generate the regression predictions, classification predictions and object predictions. The 3×3 convolutional layer helps capture spatial relationships and local patterns within the feature maps, while the 1×1 convolutional layer is responsible for reducing the dimensionality of the feature maps and mapping them to the desired output space.

Specifically, the classification results have N channels, representing the predicted probabilities of N classes, ranging from 0 to 1. The object results output objectness scores to indicate the presence of an object, ranging from 0 to 1, where 0 represents the background and 1 represents the foreground. The regression results contain the positional information of the object bounding boxes, represented by the center coordinates (t_x, t_y), width t_w, height t_h, and angle θ of the boxes. Following the approach of [17], the orientation angle θ of each object is decomposed into regression parameters t_{im} and t_{re}, which are related by $\theta = arctan_2 (t_{im}, t_{re})$.

3.3 Training Strategy

Data Augmentation. Mosaic is a data augmentation technique that involves mixing four randomly selected training images by applying random cropping and scaling operations. This method significantly enriches the background of detected objects and helps enhances the detection precision of small objects particularly through the introduction of random scaling, which adds numerous small objects. Mosaic data augmentation has been broadly employed in the domain of RGB image detection, and this paper innovatively incorporates it into the training of bird's-eye view feature maps generated from LiDAR point clouds.

Learning Rate Optimization. Cosine annealing is a technique employed to dynamically adapt the learning rate during the training process of neural networks. In each training epoch, the learning rate is updated according to the Eq. 3, where η_t is the learning rate at the present training epoch, η_{max} is the maximum learning rate, η_{min} is the minimum learning rate, T_{cur} is the current training epoch, and T_{max} is the total number of training epochs. Using the cosine annealing method to decrease the learning rate is helpful in balancing the model's ability to converge quickly and fine-tune effectively during the training process.

$$\eta_t = \eta_{min} + \frac{1}{2}\left(\eta_{max} - \eta_{min}\right)\left(1 + \cos\left(\frac{T_{cur}}{T_{max}}\pi\right)\right) \tag{3}$$

Loss Function. The model's loss function L is obtained by taking the weighted total of the classification loss L_{cls}, regression loss L_{reg}, and object loss components L_{obj}. Specifically, L_{cls} is calculated for positive samples using the binary cross-entropy (BCE) loss, as shown in Eq. 4, where N represents the number of samples, $\hat{p}_i$ represents the predicted probability from the model (ranging from 0 to 1), and p_i denotes the true label. L_{obj} is obtained by summing the weighted BCE loss for both positive and negative samples. The introduction of weight factors balances the unequal quantity of positive and negative samples. The regression loss is computed for positive samples and defined as the sum of squared errors for the introduced multiple regression parameters. Specifically, the regression of the angle l_θ is calculated using Eq. 5, where S represents the discretized grid, B represents anchor boxes, t_{im} and t_{re} represent the ground truth labels, and $\hat{t}_{im}$ and $\hat{t}_{re}$ represent the predicted values.

$$L_{BCE} = -\frac{1}{N}\sum_{i=1}^{N}\left[p_i \log\left(\hat{p}_i\right) + \left(1 - p_i\right)\log\left(1 - \hat{p}_i\right)\right] \tag{4}$$

$$l_\theta = \sum_{i=0}^{S^2}\sum_{j=0}^{B}\left[\left(t_{im} - \hat{t}_{im}\right)^2 + \left(t_{re} - \hat{t}_{re}\right)^2 + \left(1 - \sqrt{\left(\hat{t}_{im}^2 + \hat{t}_{re}^2\right)}\right)^2\right] \tag{5}$$

4 Experiments

4.1 Implementation Details

The SODet is trained and evaluated on the challenging KITTI dataset [23], which includes object categories such as "car," "pedestrian," and "cyclist." The

Table 1. Performance comparison for birds-eye view detection. APs (in %) and frames per second (FPS) on KITTI dataset. E stands for easy, M for moderate, and H for hard.

Methods	Car			Cyclist			Pedestrian			Speed
	E	M	H	E	M	H	E	M	H	(FPS)
PIXOR [13]	81.70	77.05	72.95	–	–	–	–	–	–	**28.6**
SA-SSD [14]	95.03	91.03	85.96	–	–	–	–	–	–	25.0
AVOD [20]	86.80	85.44	77.73	63.66	47.74	46.55	42.51	35.24	33.97	12.5
BirdNet [15]	84.17	59.83	57.35	58.64	41.56	36.94	28.20	23.06	21.65	9.1
BirdNet+ [16]	84.80	63.33	61.23	72.45	52.15	46.57	45.53	38.28	35.37	10.0
C-YOLO [18]	74.23	66.07	65.70	36.12	30.16	26.01	22.00	20.88	20.81	15.6
SODet(ours)	**98.96**	**96.95**	**90.84**	**95.55**	**86.50**	**45.30**	**79.81**	**58.82**	**42.03**	20.6

detection range for point clouds is set to [−50 m, 50 m] along the X-axis, [−25 m, 25 m] along the Y-axis, and [−2.73 m, 1.27 m] along the Z-axis. The KITTI dataset provides annotations for 7,481 samples, which are split into a train-val set and a test set with a ratio of 6,000:1,481. The train-val set is further divided into a train subset and a validation subset with a ratio of 9:1. During the training phase, the SODet model utilizes the Adam optimizer with a maximum learning rate η_{max} of 10e−3 and a minimum learning rate η_{min} of 10e−7. The Leaky-ReLU activation function is employed to introduce non-linearity and improve the model's ability to capture complex patterns and features. During the evaluation phase, the detection results for each category are evaluated based on the KITTI evaluation criteria, considering "easy" (E), "moderate" (M), and "hard" (H) difficulty levels. The confidence threshold is set to 0.5, and the intersection over union (IoU) thresholds for "car," "pedestrian," and "cyclist" categories are set to 0.7, 0.5, and 0.5, respectively. In the post-processing stage, category-agnostic non-maximum suppression (NMS) is applied, with a NMS threshold of 0.3. The inference time is measured on an A100 GPU with a batch size of 1.

4.2 Results

Quantitative Evaluation. The quantitative evaluation results of SODet for bird's-eye view detection on the KITTI dataset are presented in Table 1, which is conducted based on the benchmark's utilization of bounding box overlap. When compared to previous BEV-based methods, the experimental results demonstrate that the proposed SODet significantly improves the detection accuracy in bird's-eye view. Specifically, it achieves a cyclist AP of 86.50% and a pedestrian AP of 58.82% on the moderate difficulty level with a detection speed of 20.6 FPS.

Ablation Experiments. Through ablation experiments as demonstrated in Table 2, the effects of different components and techniques in SODet on network performance were analyzed. Firstly, the effectiveness of the special residual

Table 2. Ablation experiments of SODet. APs (in %) on KITTI dataset. SRB stands for Special Residual Block, CSP represents Cross Stage Partial structure, and MDA refers to Mosaic Data Augmentation. E stands for easy, M for moderate, and H for hard.

SRB	CSP	SPPF	MDA	Car			Cyclist			Pedestrian		
				E	M	H	E	M	H	E	M	H
×	×	×	×	97.91	93.04	85.08	89.05	58.93	20.17	68.09	47.05	32.62
✓	×	×	×	98.21	94.72	88.26	90.74	65.36	34.83	75.76	54.52	37.71
✓	✓	×	×	98.39	95.25	87.56	90.79	72.13	35.43	74.79	47.51	34.09
✓	✓	✓	×	98.47	95.74	88.77	91.21	76.94	40.62	69.20	47.29	33.57
✓	✓	✓	✓	**98.96**	**96.95**	**90.84**	**95.55**	**86.50**	**45.30**	**79.81**	**58.82**	**42.03**

block (SRB) was evaluated. The utilization of large-kernel depthwise convolution allowed for an expanded receptive field, facilitating the capture of a broader context. Simultaneously, the specialized design of the inverted bottleneck structure mitigated the computational cost associated with the large-kernel convolution. The incorporation of special residual blocks enhanced the network's depth, resulting in improved feature representation and contextual understanding. Furthermore, the influence of the Cross Stage Partial (CSP) structure was assessed. The inter-stage connection enhanced feature propagation and utilization efficiency, thereby enhancing the network's representational capacity and performance. Through the introduction of the SPPF module, the model efficaciously integrated feature maps characterized by diverse receptive fields, thereby facilitating the comprehensive incorporation of contextual information across multiple scales and augmenting the representational capacity of the feature maps. Additionally, the impact of Mosaic Data Augmentation (MDA) on detection performance was investigated. This technique enriched the background of detected objects, thereby contributing to further enhancing the detection accuracy of small objects. The combined utilization of these key components and techniques enabled SODet to achieve superior performance.

5 Conclusion

The presented SODet in this study provides an efficient solution for addressing the challenges of detecting small objects in bird's-eye view LiDAR-based object detection. To capture a wider context and enhance the representation of features, the NeXtDark backbone network is constructed using residual blocks that consist of large-kernel depthwise convolutions and inverted bottlenecks, as well as the introduction of the CSP structure to facilitate information flow across stages. Additionally, the feature extraction network, NeXtFPN, is devised to extract and fuse multi-scale information by incorporating the introduced spatial pyramid pooling module SPPF and the specially designed residual blocks. During the training phase, mosaic data augmentation and cosine annealing learning rate are employed to further enhance the detection precision of small objects. The proposed approach, with its innovative network architecture design and effective training strategies, provides novel insights and methodologies for LiDAR-based object detection.

References

1. Qi, C., Liu, W., Wu, C., Su, H., Guibas, L. J.: Frustum PointNets for 3D object detection from RGB-D data. In: 2018 IEEE/CVF Conference on Computer Vision and Pattern Recognition, pp. 918–927 (2018)
2. Qi, C., Litany, O., He, K., Guibas, L. J.: Deep Hough voting for 3D object detection in point clouds. In: 2019 IEEE/CVF International Conference on Computer Vision (ICCV), pp. 9276–9285 (2019)
3. Chen, Y., Li, Y., Zhang, X., Sun, J., Jia, J.: Focal sparse convolutional networks for 3D object detection. In: 2022 IEEE/CVF Conference on Computer Vision and Pattern Recognition (CVPR), pp. 5418–5427 (2022)
4. Zhou, Y., Tuzel, O.: VoxelNet: end-to-end learning for point cloud based 3D object detection. In: 2018 IEEE/CVF Conference on Computer Vision and Pattern Recognition (CVPR), pp. 4490–4499 (2018)
5. Yan, Y., Mao, Y., Li, B.: SECOND: sparsely embedded convolutional detection. Sensors (Basel, Switzerland) **18**, 3337 (2018)
6. Lang, A. H., Vora, S., Caesar, H., Zhou, L., Yang, J., Beijbom, O.: PointPillars: fast encoders for object detection from point clouds. In: 2019 IEEE/CVF Conference on Computer Vision and Pattern Recognition (CVPR), pp. 12689–12697 (2019)
7. Shi, G.-H., Li, R., Ma, C.: PillarNet: real-time and high-performance pillar-based 3D object detection. In: Avidan, S., Brostow, G., Cissé, M., Farinella, G.M., Hassner, T. (eds.) ECCV 2022. LNCS, vol. 13670, pp. 35–52. Springer, Cham (2022). https://doi.org/10.1007/978-3-031-20080-9_3
8. Qi, C., Su, H., Mo, K., Guibas, L. J.: PointNet: deep learning on point sets for 3D classification and segmentation. In: 2017 IEEE Conference on Computer Vision and Pattern Recognition (CVPR), pp. 77–85 (2017)
9. Qi, C., Yi, L., Su, H., Guibas, L. J.: PointNet++: deep hierarchical feature learning on point sets in a metric space. In: NIPS (2017)
10. Shi, S., et al.: PV-RCNN: point-voxel feature set abstraction for 3D object detection. In: 2020 IEEE/CVF Conference on Computer Vision and Pattern Recognition (CVPR), pp. 10526–10535 (2020)
11. Shi, S., et al.: PV-RCNN++: point-voxel feature set abstraction with local vector representation for 3D object detection. Int. J. Comput. Vis. **131**, 531–551 (2021)
12. Noh, J., Lee, S., Ham, B.: HVPR: hybrid voxel-point representation for single-stage 3D object detection. In: 2021 IEEE/CVF Conference on Computer Vision and Pattern Recognition (CVPR), pp. 14600–14609 (2021)
13. Yang, B., Luo, W., Urtasun, R.: PIXOR: real-time 3D object detection from point clouds. In: 2018 IEEE/CVF Conference on Computer Vision and Pattern Recognition (CVPR), pp. 7652–7660 (2018)
14. He, C., Zeng, H., Huang, J., Hua, X., Zhang, L.: Structure aware single-stage 3D object detection from point cloud. In: 2020 IEEE/CVF Conference on Computer Vision and Pattern Recognition (CVPR), pp. 11870–11879 (2020)

15. Beltrán, J., Guindel, C., Moreno, F.M., Cruzado, D., Turrado García, F., de la Escalera, A.: BirdNet: a 3D object detection framework from LiDAR information. In: 2018 21st International Conference on Intelligent Transportation Systems (ITSC), pp. 3517–3523 (2018)
16. Barrera, A., Guindel, C., Beltrán, J., Abellán García, F.: BirdNet+: end-to-end 3D object detection in LiDAR bird's eye view. In: 2020 IEEE 23rd International Conference on Intelligent Transportation Systems (ITSC), pp. 1–6 (2020)
17. Simon, M., Milz, S., Amende, K., Groß, H.-M.: Complex-YOLO: an Euler-region-proposal for real-time 3D object detection on point clouds. In: ECCV Workshops (2018)
18. Simon, M., et al.: Complexer-YOLO: real-time 3D object detection and tracking on semantic point clouds. In: 2019 IEEE/CVF Conference on Computer Vision and Pattern Recognition Workshops (CVPRW), pp. 1190–1199 (2019)
19. Chen, X., Ma, H., Wan, J., Li, B., Xia, T.: Multi-view 3D object detection network for autonomous driving. In: 2017 IEEE Conference on Computer Vision and Pattern Recognition (CVPR), pp. 6526–6534 (2017)
20. Ku, J., Mozifian, M., Lee, J., Harakeh, A., Waslander, S.L.: Joint 3D proposal generation and object detection from view aggregation. In: 2018 IEEE/RSJ International Conference on Intelligent Robots and Systems (IROS), pp. 1–8 (2018)
21. Vora, S., Lang, A. H., Helou, B., Beijbom, O.: PointPainting: sequential fusion for 3d object detection. In: 2020 IEEE/CVF Conference on Computer Vision and Pattern Recognition (CVPR), pp. 4603–4611 (2020)
22. Li, Y., et al.:: DeepFusion: LiDAR-camera deep fusion for multi-modal 3D object detection. In: 2022 IEEE/CVF Conference on Computer Vision and Pattern Recognition (CVPR), pp. 17161–17170 (2022)
23. Geiger, A., Lenz, P., Stiller, C., Urtasun, R.: Vision meets robotics: the KITTI dataset. Int. J. Robot. Res. **32**(11), 1231–1237 (2013)
24. Liu, Z., Mao, H., Wu, C.-Y., Feichtenhofer, C., Darrell, T., Xie, S.: A ConvNet for the 2020s. In: 2022 IEEE/CVF Conference on Computer Vision and Pattern Recognition (CVPR), pp. 11966–11976 (2022)
25. Sandler, M., Howard, A.-G., Zhu, M., Zhmoginov, A., Chen, L.-C.: MobileNetV2: inverted residuals and linear bottlenecks. In: 2018 IEEE/CVF Conference on Computer Vision and Pattern Recognition (CVPR), pp. 4510–4520 (2018)
26. Redmon, J., Farhadi, A.: YOLOv3: an incremental improvement. ArXiv, abs/1804.02767 (2018)
27. Wang, C.-Y., Liao, H.-Y. M., Yeh, I.-H., Wu, Y.-H., Chen, P.-Y., Hsieh, J.-W.: CSPNet: a new backbone that can enhance learning capability of CNN. In: 2020 IEEE/CVF Conference on Computer Vision and Pattern Recognition Workshops (CVPRW), pp. 1571–1580 (2020)
28. Lin, T.-Y., Dollár, P., Girshick, R.B., He, K., Hariharan, B., Belongie, S. J.: Feature pyramid networks for object detection. In:2017 IEEE Conference on Computer Vision and Pattern Recognition (CVPR), pp. 936–944 (2017)
29. Wang, T., Zhu, X., Pang, J., Lin, D.: FCOS3D: fully convolutional one-stage monocular 3D object detection. In: 2021 IEEE/CVF International Conference on Computer Vision Workshops (ICCVW), pp. 913–922 (2021)
30. Reading, C., Harakeh, A., Chae, J., Waslander, S.L.: Categorical depth distribution network for monocular 3D object detection. In: 2021 IEEE/CVF Conference on Computer Vision and Pattern Recognition (CVPR), pp. 8551–8560 (2021)

Landmark-Assisted Facial Action Unit Detection with Optimal Attention and Contrastive Learning

Yi Yang[1], Qiaoping Hu[1], Hongtao Lu[1], Fei Jiang[2](✉), and Yaoyi Li[1]

[1] Department of Computer Science and Engineering, Shanghai Jiao Tong University, Shanghai 200240, China
{yiyang_cv,huqiaoping23,luht,dsamuel}@sjtu.edu.cn
[2] Shanghai Institute of AI Education, East China Normal University, Shanghai 200240, China
fjiang@mail.ecnu.edu.cn

Abstract. In this paper, we propose a weakly-supervised algorithm for facial action unit (AU) detection in the wild, which combines basic facial features and attention-based landmark features as well as contrastive learning to improve the performance of AU detection. Firstly, the backbone is a weakly-supervised algorithm since AU datasets in the wild are scarce and the utilization of other public datasets can capture robust basic facial features and landmark features. Secondly, we explore and select the optimal attention-based landmark encoder to capture facial landmark features that have been shown highly related to AUs. Then, we combine basic facial features and attention-based landmark features for AU detection. Finally, we propose a weighted multi-label contrastive loss function for the further improvement of AU detection. Extensive experiments on RFAU and BP4D demonstrate that our method outperforms or is comparable with state-of-the-art weakly-supervised and supervised AU detection methods.

Keywords: Facial Action Unit Detection · Facial Landmark · Attention · Contrastive Learning

1 Introduction

Action Units (AUs) [2] describe the movements of facial muscles. The goal of AU detection is to determine the occurrences of different AUs given a face image. AU detection in the wild is a difficult task because of the lack of AU datasets in the wild and the challenges including head poses, illumination, occlusion, etc. To detect AUs with limited datasets, several researchers have explored how to use the self-supervised or semi-supervised learning for AU detection. In TCAE

This paper is supported by NSFC (No. 62176155, 62007008, 62207014), Shanghai Municipal Science and Technology Major Project (2021SHZDZX0102). Hongtao Lu is also with MOE Key Lab of Artificial Intelligence, AI Institute, Shanghai Jiao Tong University.

B. Luo et al. (Eds.): ICONIP 2023, CCIS 1966, pp. 92–104, 2024.
https://doi.org/10.1007/978-981-99-8148-9_8

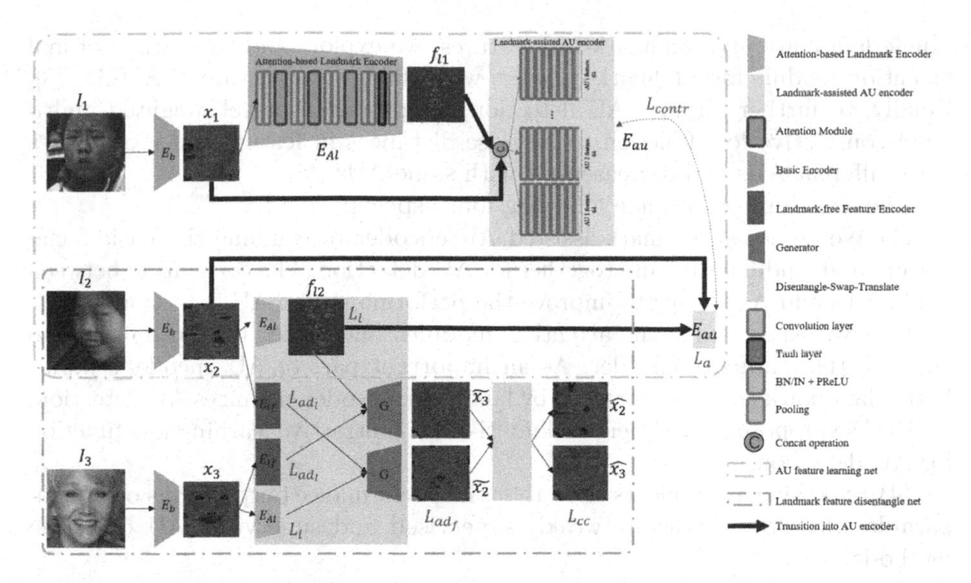

Fig. 1. The whole network contains an AU feature learning net and a landmark feature disentangling net. The image pair I_1 and I_2 in the AU feature learning net is used to extract AU features. The image pair I_2 and I_3 in the landmark feature disentangling net is used to disentangle landmark feature (useful for AU detection) and landmark-free feature. The AU feature learning net reuses the parameters of the landmark feature disentangling net to extract the basic facial and landmark features.

[15], facial changes are regarded as the combination of pose change and AU change. They proposed an autoencoder to disentangle AU features and head pose features. After training on the union of VoxCeleb1 [16] and VoxCeleb2 [1] datasets, they got two embeddings to encode the movements of AUs and head motions. The embedding of AUs was used for AU detection. The main limitation of TCAE is the assumption that AUs can be generated by moving pixels in faces. For example, TCAE failed on AU 9 because AU 9 is nose wrinkle and it cannot be generated by moving pixels in faces. Shao et al. [18] proposed a network named ADLD to disentangle the landmark-related feature and landmark-free feature. They combined the source landmark-related feature with the target landmark-free feature so that mapped a source image with AU label and a target image without AU label into a latent feature domain. ADLD was trained with BP4D [24] (or GFT [5]) as source domain dataset and EmotioNet [3] as target domain dataset to detect AUs on both BP4D (or GFT) and EmotioNet datasets. With the structure of ADLD, the target domain dataset can be replaced by any face datasets. However, ADLD directly uses the latent feature (which we call basic facial features in this paper) to extract AU features, without the fully use of landmark features. In this paper, to detect AUs in the wild, we propose a landmark-assisted AU detection algorithm with optimal attention and contrastive learning. Firstly, we combine basic features and landmark features for AU detection. Then, to extract better landmark features as well as disen-

tangle landmark and landmark-free features, we explore and select the optimal attention module for landmark encoder, with the basic structure of ADLD [18]. Finally, to further improve AU detection, we propose a novel weighted multi-label contrastive loss function to increase distances in feature space of images with different AUs and decrease that with same AUs.

The contributions include following four aspects.

(1) We design a landmark-assisted AU encoder to combine the basic facial feature and landmark feature together for AU detection. The correlation between landmarks and AUs helps to improve the performance for AU detection.

(2) We explore different attention modules and select the optima one to enhance the landmark encoder. As an important part of AU encoder's input, better landmark features extracted by landmark encoder enhances AU detection.

(3) We propose a novel weighted multi-label contrastive learning loss function for AU detection.

(4) Extensive experiments show that the performance outperforms or is comparable with state-of-the-art weakly-supervised and supervised AU detection methods.

2 Related Work

AU Detection with the Help of Landmarks. Facial landmarks are tightly related to AUs. For example, when AU 12 (Lip Corner Puller) occurs, landmarks in lip corners rise higher. Many previous works used landmarks to improve the performance for AU analysis. These works can be classified into two categories. The first category is directly utilizing facial landmarks as inputs for AU analysis. For example, Walecki et al. [22] used a set of facial landmarks that correspond to combinations of AU labels as input feature of the copula ordinal regression for jointly AU intensity estimation. The second category utilized landmarks as the localization of important regions on face to extract local features related to AUs. Li et al. [14] enhanced and cropped ROIs with roughly extracted facial landmark information for AU detection. Shao et al. [19] proposed a multi-task network to detection facial landmarks and AUs simultaneously, where predicted landmarks specify different ROIs for local AU feature learning. Motivated by the relationship between AUs and landmarks, we propose a landmark-assisted AU encoder to combine basic facial features and landmark features highly related to AU activations together for AU detection.

Attention Modules. Attention mechanism improves the representation of interests, the significance of which has been studied extensively in the literature [20]. The main idea of attention mechanism is focusing on important features and suppressing unnecessary ones. For example, Woo et al. [23] exploited two dimensional attention mechanism: channel and spatial attention. In this paper, we explore different forms of attention modules [4,9,23] and select the optima attention module to embed in landmark encoder, which focuses more on regions with activated AUs.

Contrastive Learning. Contrastive learning is essential in classification problems to make intra-class pairs to be closer to each other and inter-class pairs to be far apart from each other. Some existing works modified contrastive loss or triplet loss to solve facial expression analysis problems and achieved promising results. To train a compact embedding for facial expression similarity, Vemulapalli et al. [21] utilized a triplet loss function that encourages the distance between the two images that from the most similar pair to be smaller than that of these two images from the third image. For facial expression recognition, Hayale et al. [6] proposed a modified contrastive loss to train deep siamese neural networks, which encourages features of same facial expression images to be closer and that from different but close facial expressions (e.g., Disgust-Angry) to be further.

However, the above mentioned methods are designed for single-label multi-class problems, which cannot be directly used for the AU detection due to the multi-label features of AU data. The label of an image may contain several AUs. In this paper, we propose a novel contrastive loss for multi-label binary classification problem.

3 Proposed Framework

Overall Architecture Structure. The proposed framework includes a landmark feature disentangling net and an AU feature learning net, as shown in Fig. 1. The landmark feature disentangling net is used to learn the basic facial feature and the disentangled landmark feature with the landmark information supervision. The AU feature learning net reuses the parameters of the landmark feature disentangling net to extract the basic facial and landmark features. To enhance the features for AU detection, the basic facial and landmark features are concatenated to extract AU features with proposed landmark-assisted AU encoder. The final AU features are learned from the landmark-assisted AU encoder.

Concretely, for the AU feature learning net, first, I_1 and I_2 with AU labels from training dataset are extracted basic facial features (x_1 and x_2) and attention-based landmark features (f_{l1} and f_{l2}). Then, they are combined to extract AU features with landmark-assisted AU encoder E_{au}. Finally, it learns AU features with AU labels and the proposed contrastive loss. For the landmark feature disentangle net, we adopt an improved ADLD [18], where the landmark encoder is replaced by the proposed attention-based landmark encoder. Basic facial features of images I_2 and I_3 are entangled landmark features and landmark-free features, then we exchange and combine their features twice (see two Disentangle-Swap-Translate modules in the figure) to recover basic facial features $\widehat{x_2}$ and $\widehat{x_3}$. Landmark labels provide weak supervision information which helps to train the net.

Landmark-Assisted AU Encoder. We design this landmark-assisted AU encoder as shown in Fig. 2(a). The circle with a "C" represents a concat operation. We directly concat the basic facial feature (44 $\times$ 44, 64 channels) and the landmark feature (44 × 44, 49 channels since 49 landmarks are used in our experiments) as the input (44 × 44, 113 channels) for AU encoder. Considering we need every AU's feature for contrastive learning in Eq. 1, we capture each AU's feature with an independent module of a small convolution network. Each module extracts a 64-dimensional vector feature for the corresponding AU. These features are utilized in Eq. 1 for the computation of d_{nc}. Each module is followed by a small full connection layer to classify one AU. All the parameters are shown in Fig. 2(a).

Attention-Based Landmark Encoder. The serial channel and spatial attention CBAM module [23] is selected as our attention module. The landmark encoder is composed of several CNNs followed by attention modules as shown in Fig. 2(b). The structure and detailed parameters of the attention module are shown in Fig. 2(b) and 2(c). Explorations about different attention modules are shown in the section of Results Analysis.

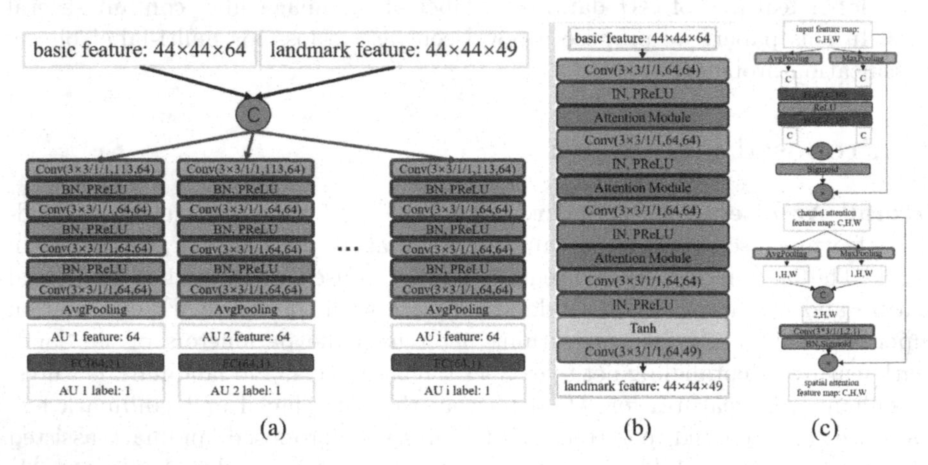

Fig. 2. (a) Landmark-assisted AU encoder. (b) Attention-based landmark encoder. (c) A serial channel and spatial attention module.

Weighted Multi-label Contrastive Learning. To apply contrastive learning into AU detection, we propose a novel weighted multi-label contrastive learning loss function:

$$
\begin{aligned}
L_{contr} = \frac{1}{NC} \sum_{n=1}^{N} \sum_{c=1}^{C} Y_{nc}^2 (\frac{1 - Y_{nc}}{2} d_{nc}^2 + \\
\frac{1 + Y_{nc}}{2} max\{margin - d_{nc}, 0\}^2) w_c,
\end{aligned}
\tag{1}
$$

where $Y_{nc} \in \{-1, 0, 1\}$, is the c-th AU contrastive label of the n-th image pair, N is the number of image pairs, C is the number of AU categories, and d_{nc} is the Euclidean distance of the image pair.

We set Y_{nc} to be -1 when the c-th AU both occurs in the image pair, i.e. both c-th AU labels in two images are 1. When $Y_{nc} = -1$, $L_{contr} = \frac{1}{NC} \sum \sum d_{nc}^2 w_c$.

The loss tries to reduce the distance of the AU features. We set Y_{nc} to be 1 when one image label is 1 and the other image label is 0. When $Y_{nc} = 1$, $L_{contr} = \frac{1}{NC}\sum\sum max\{margin - d_{nc}, 0\}^2 w_c$. The loss tries to increase the distance of features from different AU label images. There is a special states when Y_{nc} is set to be 0. When both c-th AU labels in two images are 0, the pair should be invalid and have no contribution to loss. Because the AU is not activated at this time. Just like background images (they have the same label of background) should not be selected to make closer in original contrastive learning, such distance of features without AU activation does not need to be reduced. For example, AU 4 is brow lowerer. For the face without AU 4, the brow may be upper or just natural. The upper brow and the natural brow do not need be the same or closer in feature space.

The weight w_c is used to balance different numbers of AU occurrence. We compute w_c as [19]:

$$w_c = \frac{C \cdot (1/f_c)}{\sum_{c=1}^{C} 1/f_c}, \tag{2}$$

where f_c is the frequency of the occurrence for c-th AU in the total training set. The lower the frequency of the AU occurrence, the bigger of the weight.

Model Training. To train the proposed model, seven loss functions are used, including the contrastive loss L_{contr}, AU detection loss L_a, landmark classification loss L_l, landmark adversarial loss L_{ad_l}, basic feature adversarial loss L_{ad_f}, self-reconstruction loss L_r, and cross-cycle consistency loss L_{cc}. Details about L_{contr} has been introduced in the last several paragraphs while other losses are directly follow the settings in [18]. We introduce them briefly in the following paragraph. More details can be found in [18].

The self-reconstruction loss L_r is used to encourage the generator G to generate similar basic facial features $G(f_{l2}, E_{lf}(x_2))$ and $G(E_{Al}(x_3), E_{lf}(x_3))$ to the original basic facial feature x_2 and x_3, where G is the generator with inputs of landmark features and landmark-free features. The cross-cycle consistency loss L_{cc} is used to encourage cross-cyclically reconstructed features $\widehat{x_2}$ and $\widehat{x_3}$ to be similar to basic facial features x_2 and x_3. Generated basic features $\tilde{x_2}$ and $\tilde{x_3}$ are encouraged to be indistinguishable from the rich feature x_2 and x_3. So two feature discriminators D_f^s and D_f^t are applied with a feature adversarial loss L_{ad_f} for x_2 (as real) and $\tilde{x_2}$ (as fake), x_3 (as real) and $\tilde{x_3}$ (as fake), respectively. The landmark-free feature encoder E_{lf} is encouraged to remove landmark information so a landmark adversarial loss L_{ad_l} after a landmark discriminator D_l is introduced as the adversary of E_{lf}. The facial landmark detection is regarded as a classification problem, where each output feature map of D_l can be regarded as a response map with a size of $44 \times 44 \times 1$ for each landmark. Each position in the response map is regarded as one of total 44^2 classes. The landmark classification loss L_l is used to utilize landmarks of images as the supervision information. It encourages the i-th response map of E_{Al} to have the highest response at the i-th landmark location while having near-zero responses at other locations.

4 Experiments

4.1 Datasets and Evaluation Metrics

In our experiments, RFAU [8] and BP4D [24] are both used as training datasets (providing I_1 and I_2 with their AU labels in Fig. 1) and evaluation datasets. EmotioNet [3] and AffWild2 [10–12] are only trained as auxiliary datasets, providing I_3 without AU labels in Fig. 1.

We evaluate our method with F1 score (F1-frame) for AU detection, as well as average results of all AUs. F1 score is the most popular metric for AU detection. It is computed as the harmonic mean of precision and recall. In following sections, F1-frame results are reported in percentages with "%" omitted.

4.2 Implementation Details

The total loss of the whole structure contains seven parts: the contrastive loss L_{contr}, landmark classification loss L_l, AU detection loss L_a, landmark adversarial loss L_{ad_l}, basic feature adversarial loss L_{ad_f}, self-reconstruction loss L_r, and cross-cycle consistency lossL_{cc}. The computation of L_{contr} has been introduced in the subsection of The Weighted Multi-label Contrastive Learning. Other losses are computed following the settings in ADLD [18]. The total loss is computed as:

$$L_{total} = \lambda_{contr} L_{contr} + \lambda_a L_a + \lambda_l L_l + \lambda_{ad_f} L_{ad_f} + \lambda_{ad_l} L_{ad_l} + \lambda_r L_r + \lambda_{cc} L_{cc}. \tag{3}$$

Table 1. F1 scores when training and testing on RFAU dataset with EmotioNet as auxiliary dataset. Bold ones indicate the best performance of supervised algorithms and weakly-supervised algorithms.

AU	Supervised					Weakly-Supervised			
	DRML [25]	ROINets [13]	AUNets [17]	CCT [7]	JAA [19]	ADLD [18]	**LA(ours)**	**ALA(ours)**	**ALAC(ours)**
1	28.3	**42.8**	33.6	33.5	38.8	32.6	**47.7**	44.8	46.8
2	25.2	**44.6**	34.7	30.8	31.6	22.8	30.6	**34.9**	31.9
4	34.2	43.9	**47.3**	42.5	47.0	33.2	47.3	46.4	**51.3**
6	59.7	**64.0**	48.2	60.0	58.7	53.8	55.5	**60.4**	55.6
10	32.9	**37.7**	21.0	36.0	27.6	26.8	24.8	**29.9**	29.8
12	81.7	84.3	83.4	85.3	**86.8**	73.2	**85.8**	81.5	84.0
14	31.1	**32.6**	23.7	29.7	26.3	3.9	24.0	24.0	**28.3**
17	28.9	**39.5**	32.8	36.1	33.4	30.3	40.5	36.1	**41.5**
23	25.5	**31.6**	21.7	25.1	18.4	25.0	26.8	**28.9**	28.0
24	22.7	17.6	8.1	**20.8**	13.8	4.5	20.8	**24.7**	20.5
27	87.3	91.4	87.8	91.8	**92.0**	75.2	82.2	83.2	**89.4**
28	57.6	**64.0**	58.3	61.5	54.3	56.1	47.5	**63.5**	59.2
Avg.	42.9	**49.5**	41.7	46.1	44.1	36.4	44.5	46.5	**47.2**

To evaluate how much the three proposed points enhance AU detection, we compare three designed models, LA, ALA, and ALAC. LA is the model with only the AU detector replaced by the proposed landmark-assisted AU detector comparing to ADLD [18]. ALA is the model with attention-based landmark encoder replacing original landmark encoder comparing to LA. ALAC is the model with contrastive learning comparing to ALA.

For the training of models, we set the batch size to be 16. Total epochs are 10 for LA and ALA. For ALAC, we train first 6 epochs on ALA then train 9 epochs on ALAC. We use the Adam solver ($\beta_1 = 0.95, \beta_2 = 0.999$) and an initial learning rate of 10^{-4} for E_b, E_{Al}, E_{au}. We use the Adam solver ($\beta_1 = 0.5, \beta_2 = 0.9$) and an initial learning rate of 5×10^{-5} for E_{fl}, G. All experiments are conducted with pytorch with GPU of Tesla P100.

5 Results Analysis

We compare our method with state-of-the-art supervised and weakly-supervised AU detection methods. Results on RFAU dataset and BP4D dataset are shown

Table 2. F1 scores when training and testing on BP4D dataset with EmotioNet as auxiliary dataset. Bold ones indicate the best performance of supervised algorithms and weakly-supervised algorithms.

AU	Supervised					Weakly-Supervised			
	DRML [25]	ROINets [13]	AUNets [17]	CCT [7]	JAA [19]	ADLD [18]	**LA (ours)**	**ALA (ours)**	**ALAC (ours)**
1	36.4	36.2	53.4	**60.0**	47.2	16.4	30.9	49.2	**50.8**
2	41.8	31.6	44.7	**60.7**	44.0	34.3	33.5	40.6	**44.5**
4	55.0	43.4	55.8	**59.6**	54.9	63.6	67.5	**67.6**	67.1
6	43.0	77.1	**79.2**	72.0	77.5	75.6	78.0	**80.5**	79.1
7	67.0	73.7	**78.2**	75.6	74.6	**78.5**	75.7	71.8	71.1
10	66.3	**85.0**	83.1	74.1	84.0	77.2	73.2	77.5	**79.5**
12	65.8	87.0	**88.4**	81.2	86.9	84.2	**87.9**	85.4	85.8
14	54.1	62.6	**66.6**	60.9	61.9	58.3	63.7	**71.1**	68.2
15	33.2	45.7	47.5	**56.6**	43.6	33.3	45.5	**54.5**	53.6
17	48.0	58.0	**62.0**	61.6	60.3	63.5	61.5	64.1	**65.8**
23	31.7	38.3	47.3	**63.1**	42.7	33.9	42.2	36.0	**42.4**
24	30.0	37.4	49.7	**73.7**	41.9	38.5	43.3	49.2	**50.1**
Avg.	48.3	56.4	63.0	**66.6**	60.0	54.8	58.6	62.3	**63.2**

Table 3. Comparison of different features for AU detection on BP4D with EmotioNet as auxiliary dataset. Let x, y correspondingly denote the input image and predicted results.

Model	AU							Test Models	Size (MB)
	1	2	4	6	12	17	Avg.		
bf [18]	50.5	35.7	61.8	74.1	75.2	69.0	61.0	$E_b, E_{au}^B : y = E_{au}^B(E_b(x))$	3.68
bf+e	47.8	31.0	64.0	75.4	82.3	65.3	61.0	$E_b, E_e^B, E_{au} : y = E_{au}(E_e^B(E_b(x)))$	4.35
lf	**54.6**	**42.3**	66.6	76.2	83.9	64.6	64.7	$E_b, E_l, E_{au}^L : y = E_{au}^L(E_l(E_b(x))$	4.15
lf+e	34.7	28.2	64.4	78.3	85.7	60.6	58.7	$E_b, E_l, E_e^L, E_{au} : y = E_{au}(E_e^L(E_l(E_b(x)))$	5.02
bf+lf	52.2	37.6	**66.7**	**79.7**	**86.6**	**70.2**	**65.5**	$E_b, E_l, E_{au} : y = E_{au}([E_b(x), E_l(E_b(x)])$	5.00

in Table 1 and Table 2. As shown in Table 1 and Table 2, comparing to ADLD, LA improves the average F1 score with 8.1 on RFAU and 3.8 on BP4D. This is because LA combines the landmark feature and the basic facial feature to detect AUs, which we call landmark-assisted AU detection, while ADLD only uses basic facial features for AU detection. ALA improves the average F1 score with 2 on RFAU and 3.7 on BP4D comparing to LA because of attention modules in landmark encoder. ALAC further improves the average F1 score with 0.7 on RFAU and 0.9 on BP4D comparing to ALA because of the contrastive learning. Besides, ALAC is even comparable with state-of-the-art supervised AU detection methods, where only ROINets outperforms ALAC on RFAU and only CCT outperforms ALAC on BP4D.

Different Features Comparison. To understand what feature is better for AU detection, we conduct an experiment on BP4D. Table 3 shows F1 score comparison for models using different features as the input for AU detection. Model "bf" means using only basic facial feature extracted by the basic feature encoder E_b for AU detection. The results of "bf" is directly copied from the paper [18]. To compare with "bf", we train other four models with the same 6 AUs on BP4D following the settings in ADLD [18]. For the model "lf", we use only the landmark feature extracted by the landmark encoder E_l for AU detection. For the model "bf+lf", we combine the basic facial feature and the landmark feature together for AU detection. F1 score results in Table 3 clearly show that combining the basic facial feature and the landmark feature is better than that of "bf" and "lf".

Table 4. F1 score comparison of different attention modules when training and testing on RFAU with EmotioNet as auxiliary dataset.

AU	1	2	4	6	10	12	14	17	23	24	27	28	Avg
LA	**47.7**	30.6	47.3	55.5	24.8	85.8	24.0	**40.5**	26.8	20.8	82.2	47.5	44.5
Single Self Attention	42.9	35.4	**54.1**	**64.5**	26.0	83.0	26.9	33.7	**32.6**	14.1	81.9	60.5	46.3
Multi Self Attention	39.1	**39.9**	53.4	56.4	25.8	82.7	18.6	31.9	25.8	17.0	81.4	**66.6**	44.9
Single CC Attention	40.9	37.8	41.2	56.8	28.8	76.6	**27.9**	33.8	27.1	24.5	83.7	61.2	45.0
Multi CC Attention	24.3	34.9	25.2	59.7	29.6	85.2	23.2	36.8	23.7	10.7	75.6	55.6	40.4
Single CBAM Attention	36.8	26.8	35.1	62.5	18.3	**86.4**	19.4	33.3	23.9	16.7	**87.3**	58.9	42.1
Multi CBAM Attention	44.8	34.8	46.4	60.4	**29.9**	81.5	24.0	36.1	28.9	**24.7**	83.2	63.5	**46.5**

Table 5. The testing F1 score results when training on BP4D and testing on RFAU with different auxiliary datasets.

Auxiliary Dataset	1	2	4	6	10	12	14	23	24	Avg
BP4D	13.2	5.2	17.7	22.7	7.3	33.6	14.8	**16.4**	3.9	15.0
AffWild2	12.8	**10.6**	**18.0**	30.4	**19.7**	40.0	17.3	14.9	14.3	19.8
EmotioNet	**14.9**	10.1	16.5	**37.0**	14.2	**50.5**	15.1	10.7	16.1	**20.6**
RFAU	12.8	10.5	16.9	31.1	15.1	48.4	**17.5**	16.0	**17.3**	**20.6**

Note that using both landmark features and basic facial features will increase the parameter number of AU encoder, since the input channel number increases from 64 (basic facial features, with E_{au}^B as AU encoder) or 49 (landmark features, with E_{au}^L as AU encoder) to 113 (64+49, with E_{au} as AU encoder). To explore whether the enhancement of performance comes from the increase of parameter number in AU encoder, we train two more models, "bf+e" and "lf+e". "e" represents an extra net, which contains a convolution layer of 1×1 kernel size, changing the channel from 64 (E_e^B) or 49 (E_e^L) to 113, and a Relu layer. The extra net changes the channel of basic feature or landmark feature and put them into AU encoder. The average F1 scores of "bf+e" and "bf" are very close. But the average F1 score of "lf+e" is obviously less than "lf". The reason may be the structures in landmark features are broken by the extra net. These results indicate the combining of basic facial features and landmark features but not the increasing number of parameters that improves the performance.

Different Attention Modules Comparison. To find the best attention module for AU detection based on LA model, we apply different attention modules in landmark encoders providing landmark features, then train models to find the best attention module. Table 4 shows F1 score comparison on RFAU dataset. "Single" means only applying an attention module after the third convolution layer in landmark encoder. "Multi" means embedding one attention module after each of first three convolution layers in landmark encoder. Self Attention represents the Dual Attention module [4], which includes a channel self attention module and a position self attention module connected in parallel. CC attention

Table 6. The testing F1 score results when both training and testing on RFAU dataset with different auxiliary datasets.

Auxiliary Dataset	1	2	4	6	10	12	14	17	23	24	27	28	Avg
BP4D	25.1	18.6	31.4	43.2	34.1	65.7	26.6	27.8	20.1	14.8	73.7	48.2	35.8
AffWild2	42.0	**35.3**	46.1	55.1	21.6	67.9	21.4	29.4	23.9	21.2	88.5	**66.8**	43.3
EmotioNet	**44.8**	34.8	46.4	**60.4**	29.9	81.5	24.0	**36.1**	28.9	**24.7**	83.2	63.5	46.5
RFAU	40.3	31.0	**50.9**	52.0	**34.6**	**84.6**	**30.1**	32.0	**32.5**	20.1	**90.6**	65.8	**47.0**

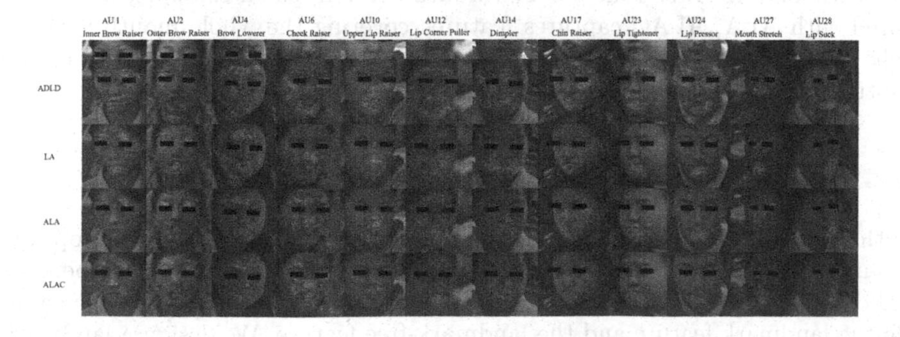

Fig. 3. Visualization of AU-specific saliency maps.

represents criss-cross Attention module [9], which claims to efficiently capture the contextual information. CBAM attention represents Convolutional Block Attention module [23], where we apply the serial connected channel and spatial attention module in [23] because it is reported the best connection way in different combinations of spatial and channel attention modules in [23]. Comparing these modules, we find the best way is to apply a serial channel and spatial attention [23] after each of first three layers in landmark encoder (Multi CBAM Attention), which we call attention-based landmark encoder. Besides, we can roughly find when applying attention modules in landmark encoder, self attention and criss-cross attention is better for single-layer use while CBAM attention is better for multi-layers use.

What's the Effect of the Auxiliary Dataset? We conduct two experiments to explore the effect of the auxiliary dataset. In the first experiment, we train ALA on BP4D with different auxiliary datasets including BP4D, AffWild2, EmotioNet, and RFAU, then test the models on RFAU dataset. The results are shown in Table 5. The cross-dataset testing result indicates that the network has captures as good features from BP4D and EmotioNet as that from BP4D and RFAU for the AU detection on RFAU. In the second experiment, we train ALA on RFAU with different auxiliary datasets including BP4D, AffWild2, EmotioNet, and RFAU, as shown in Table 6. When both the train dataset and auxiliary dataset are RFAU, the algorithm becomes a full-supervised algorithm and reaches the highest average F1 score of 47.0. Comparing Table 5 and Table 6, BP4D as the auxiliary dataset always reaches higher F1 score than others. This may indicate that using dataset in the wild as the auxiliary dataset is better for more robust feature extraction.

How does Contrastive Learning Help AU Detection? The last two columns in Table 1 and Table 2 can roughly demonstrate that contrastive learning can improve AU detection with a small degree. To further understand how does contrastive learning help AU detection and to analysis whether models can really capture features highly related to corresponding AUs, we compute saliency maps of those models on RFAU dataset. Figure 3 shows some examples. Generally, our proposed method (ALAC) shows more concentrating on regions related to corresponding AUs. Take AU 27 (Mouth Stretch) as an example, when compared with ALA, ALAC captures features covering the whole mouth with the help of contrastive learning, which indicates that the contrastive learning help capture features more related to AUs.

6 Conclusion and Future Work

In this paper, we present a landmark-assisted AU detection method with optimal attention and contrastive learning. This method is based on a weekly-supervised AU detection method with facial landmarks as supervision information to entangle the landmark feature and the landmark-free feature. We design a landmark-assisted AU encoder to use both basic facial features and landmark features to extract AU features. To extract better landmark features, we explore several different attention modules and finally select the best one (i.e., a serial channel and spatial attention module) to embed in landmark encoder. To further improve AU detection, we propose a weighted multi-label contrastive loss. Extensive experiments are conducted and show that our method outperforms or is comparable with the state-of-the-art weakly-supervised and supervised AU detection methods. Since our method sometimes may concentrate on regions not directly related to AUs, our future work will be constraining regions of feature extraction.

References

1. Chung, J.S., Nagrani, A., Zisserman, A.: VoxCeleb2: deep speaker recognition. Proc. Interspeech **2018**, 1086–1090 (2018)
2. Ekman, P., Friesen, W.V.: Facial action coding system: a technique for the measurement of facial movement. Consulting Psychologists Press (1978)
3. Fabian Benitez-Quiroz, C., Srinivasan, R., Martinez, A.M.: EmotioNet: an accurate, real-time algorithm for the automatic annotation of a million facial expressions in the wild. In: IEEE Conference on Computer Vision and Pattern Recognition (CVPR 2016), pp. 5562–5570 (2016)
4. Fu, J., et al.: Dual attention network for scene segmentation. In: IEEE International Conference on Computer Vision and Pattern Recognition (CVPR 2019), pp. 3146–3154 (2019)
5. Girard, J.M., Chu, W.S., Jeni, L.A., Cohn, J.F.: Sayette group formation task (GFT) spontaneous facial expression database. In: IEEE International Conference on Automatic Face and Gesture Recognition (FG 2017), pp. 581–588 (2017)
6. Hayale, W., Negi, P., Mahoor, M.: Facial expression recognition using deep Siamese neural networks with a supervised loss function. In: IEEE International Conference on Automatic Face and Gesture Recognition (FG 2019), pp. 1–7 (2019)
7. Hu, Q., Jiang, F., Mei, C., Shen, R.: CCT: a cross-concat and temporal neural network for multi-label action unit detection. In: IEEE International Conference on Multimedia and Expo (ICME 2018) (2018)
8. Hu, Q., et al.: RFAU: a database for facial action unit analysis in real classrooms. IEEE Trans. Affect. Comput. **13**, 1452–1465 (2020)
9. Huang, Z., et al.: CCNet: criss-cross attention for semantic segmentation. IEEE Trans. Pattern Anal. Mach. Intell. (2020). https://doi.org/10.1109/TPAMI.2020.3007032
10. Kollias, D., Zafeiriou, S.: Aff-wild2: extending the Aff-wild database for affect recognition. arXiv preprint arXiv:1811.07770 (2018)
11. Kollias, D., Zafeiriou, S.: A multi-task learning & generation framework: valence-arousal, action units & primary expressions. arXiv preprint arXiv:1811.07771 (2018)
12. Kollias, D., Zafeiriou, S.: Expression, affect, action unit recognition: Aff-wild2, multi-task learning and arcface. arXiv preprint arXiv:1910.04855 (2019)
13. Li, W., Abtahi, F., Zhu, Z.: Action unit detection with region adaptation, multi-labeling learning and optimal temporal fusing. In: IEEE International Conference on Computer Vision and Pattern Recognition (CVPR 2017) (2017)
14. Li, W., Abtahi, F., Zhu, Z., Yin, L.: EAC-Net: a region-based deep enhancing and cropping approach for facial action unit detection. In: IEEE International Conference on Automatic Face and Gesture Recognition (FG 2017), pp. 103–110 (2017)

15. Li, Y., Zeng, J., Shan, S., Chen, X.: Self-supervised representation learning from videos for facial action unit detection. In: IEEE International Conference on Computer Vision and Pattern Recognition (CVPR 2019) (2019)
16. Nagrani, A., Chung, J.S., Zisserman, A.: VoxCeleb: a large-scale speaker identification dataset. Proc. Interspeech **2017**, 2616–2620 (2017)
17. Romero, A., León, J., Arbeláez, P.: Multi-view dynamic facial action unit detection. Image Vis. Comput. **122**, 103723 (2018)
18. Shao, Z., Cai, J., Cham, T.J., Lu, X., Ma, L.: Unconstrained facial action unit detection via latent feature domain. IEEE Transa. Affect. Comput. **13**(2), 1111–1126 (2022)
19. Shao, Z., Liu, Z., Cai, J., Ma, L.: Deep adaptive attention for joint facial action unit detection and face alignment. In: European Conference on Computer Vision (ECCV 2018), pp. 725–740 (2018)
20. Vaswani, A., et al.: Attention is all you need. In: Advances in Neural Information Processing Systems (NIPS 2017), pp. 5998–6008 (2017)
21. Vemulapalli, R., Agarwala, A.: A compact embedding for facial expression similarity. In: IEEE International Conference on Computer Vision and Pattern Recognition (CVPR 2019), pp. 5676–5685 (2019)
22. Walecki, R., Rudovic, O., Pavlovic, V., Pantic, M.: Copula ordinal regression framework for joint estimation of facial action unit intensity. IEEE Trans. Affect. Comput. (TAC 2019) **10**(3), 297–312 (2019)
23. Woo, S., Park, J., Lee, J.Y., Kweon, I.S.: CBAM: convolutional block attention module. In: European Conference on Computer Vision (ECCV 2018) (2018)
24. Zhang, X., et al.: Bp4d-spontaneous: a high-resolution spontaneous 3D dynamic facial expression database. Image Vis. Comput. **32**(10), 692 – 706 (2014). http://www.sciencedirect.com/science/article/pii/S0262885614001012
25. Zhao, K., Chu, W.S., Zhang, H.: Deep region and multi-label learning for facial action unit detection. In: IEEE International Conference on Computer Vision and Pattern Recognition (CVPR 2016), pp. 3391–3399 (2016)

Multi-scale Local Region-Based Facial Action Unit Detection with Graph Convolutional Network

Yi Yang[1], Zhenchang Zhang[1], Hongtao Lu[1], and Fei Jiang[2](✉)

[1] Department of Computer Science and Engineering, Shanghai Jiao Tong University, Shanghai 200240, China
{yiyang_cv,zzc_salmon,luht}@sjtu.edu.cn
[2] Shanghai Institute of AI Education, East China Normal University, Shanghai 200240, China
fjiang@mail.ecnu.edu.cn

Abstract. Facial action unit (AU) detection is crucial for general facial expression analysis. Different AUs cause facial appearance changes over various regions at different scales, and may interact with each other. However, most existing methods fail to extract the multi-scale feature at local facial region, or consider the AU relationship in the classifiers. In this paper, we propose a novel multi-scale local region-based facial AU detection framework with Graph Convolutional Network (GCN). The proposed framework consists of two parts: multi-scale local region-based (MSLR) feature extraction and AU relationship modeling with GCN. Firstly, to extract the MSLR features, we build the improved AU centers for each AU, and then extract multi-scale feature around the centers with several predefined windows. Secondly, we employ the GCN framework to model the relationship between AUs. Specifically, we build the graph of AUs, then utilize two GCNs to update the MSLR feature and the AU classifiers respectively. Finally, the AU predicted probability is determined by both the multi-scale local feature and the relationship between AUs. Experimental results on two widely used AU detection datasets BP4D and DISFA show that the proposed algorithm outperforms the state-of-the-art methods.

Keywords: facial expression recognition · neural networks · multi-scale local feature extraction · graph convolutional network

1 Introduction

Facial expression is a natural and effective means for human interaction. Facial Action Coding System (FACS) developed by Ekman [5] describes the facial

---

This paper is supported by NSFC (No. 62176155, 62007008, 62207014), Shanghai Municipal Science and Technology Major Project (2021SHZDZX0102). Hongtao Lu is also with MOE Key Lab of Artificial Intelligence, AI Institute, Shanghai Jiao Tong University.

B. Luo et al. (Eds.): ICONIP 2023, CCIS 1966, pp. 105–116, 2024.
https://doi.org/10.1007/978-981-99-8148-9_9

behavior with a set of facial action units (AUs), each of which is anatomically related to a set of facial muscles. An automatic AU detection system has wide applications in diagnosing mental disorders [21], improving educational quality [19], and synthesizing human expression [4]. However, it is still a challenging task to recognize facial AUs from spontaneous facial displays, especially with large variations in facial appearance caused by head movements, occlusions, and illumination changes.

Since different AUs correspond to different muscular activation of the face, most existing methods either divided the image into equal parts [24], or extracted the same scale of feature near the facial landmarks [13]. However, different AUs cause facial appearance changes over various regions at different scales. A meticulous method is required to extract multi-scale AUs feature at local facial regions.

Another critical aspect of AU detection is the multi-label classification with the extracted feature. Each single local facial region defined in FACS can be ambiguous for AU detection because of the face variations in person-specific feature. Therefore, taking into account the relationship of various facial regions can provide more robustness than using a single local region. Some approaches tried to explicitly model the label dependencies by using Recurrent Neural Networks (RNNs) [18], while these methods predicted the labels in a sequential manner and required some pre-defined or learned orders. Recently, Graph Convolutional Networks (GCN) has shown great success in relation modeling [2], and provides alternative ways for the multi-label classification task.

In this paper, we propose a novel multi-scale local region-based AU detection framework with GCN, namely MSLR-GCN, to extract the multi-scale AU local features and implement multi-label classification with relationship modeling. This framework consists of two parts: multi-scale local region-based (MSLR) feature extraction and GCN-based AU relationship modeling. To extract the MSLR features, we first build the improved AU centers for each AU. The features are then extracted around the centers with several multi-scale windows on the last layer feature map. The window size is predefined to cover the whole AU, more details please see Sect. 3.2. With the extracted MSLR feature, we utilize the GCN framework to model the AU relationship. Specifically, we build the graph of AUs by the AU correlation matrix, and update the MSLR feature and the AU classifiers through GCN layers, more details please see Sect. 3.3. Finally, we get the AU predicted scores with the updated MSLR feature and the GCN-based classifiers. The main contributions of this paper are summarized as follows:

(i) We propose a novel end-to-end trainable framework for AU detection, which utilizes the multi-scale windows and the improved AU centers to extract multi-scale AUs local feature.
(ii) The framework employs GCN for multi-label classification, and the GCN-based AU classifiers can model the relationship between AUs to further improve the robustness of our AU detection algorithm.
(iii) Extensive experimental results have demonstrated the effectiveness of our proposed method, which outperforms the state-of-the-art methods on two widely used AU detection datasets BP4D and DISFA.

2 Related Work

We briefly review the related work of AU detection, especially for different feature extraction and the AUs relationship modeling algorithms. Moreover, the GCN used in our method is also introduced.

Feature Extraction for AU Detection. Most recently, Convolutional Neural Networks (CNNs) have been successfully applied to automatic facial action unit detection. Specifically, Zhao *et al.* [24] proposed a region layer to induce the CNN to adaptively focus on local facial regions; Li *et al.* [14] proposed a local feature learning method based on enhanced and cropped facial area; Shao *et al.* [20] apply several multi-scale CNN filters on the whole facial image, and compute an adaptive attention map for better local feature learning; Han *et al.* [6] proposed to learn the convolution filter size and capture the feature with different resolution. These feature extraction methods are related to our work and show good performance compared with traditional hand-crafted methods. However, they did not consider the significant similarity among human faces. Generally, they learn the position of the facial landmarks with deep neural network to obtain the local facial feature, or fine-tune the convolutional filters to obtain the multi-scale feature. Since the existed AU datasets (such as BP4D [22] and DISFA [17]) contain too many images of the same person, the complicated structure of neural network will make the model overfit to the training data.

AUs Multi-label Classification and GCNs. To model the relationship between AU local features, most existing methods just converted the multi-label problem into a set of binary classification problems, and only holistic feature representations were used. Graph Convolutional Network (GCN) was first introduced in [12] to perform semi-supervised classification. For the study of facial AU detection, Liu *et al.* [15] proposed the AU-GCN to update the AU local features through the GCN. The AU predicted probabilities are obtained from the AU features and the last fully-connected layer (regarded as the AU classifiers). Although the AU-GCN strengthened the relation between AU features, it did not consider the relationship between the classifiers. And the result enhancements between AUs with strong relationship are not consistent.

In this paper, we propose an end-to-end framework MSLR-GCN to implement AU detection. This framework takes the multi-scale local facial features into consideration, and learns the GCN-based classifiers to model the relationship between AUs.

3 Proposed Method

In this section, we first introduce the overall framework of the proposed algorithm. Then, we provide detailed descriptions of the two proposed strategies: multi-scale local region-based feature extraction and AU relationship modeling with GCN.

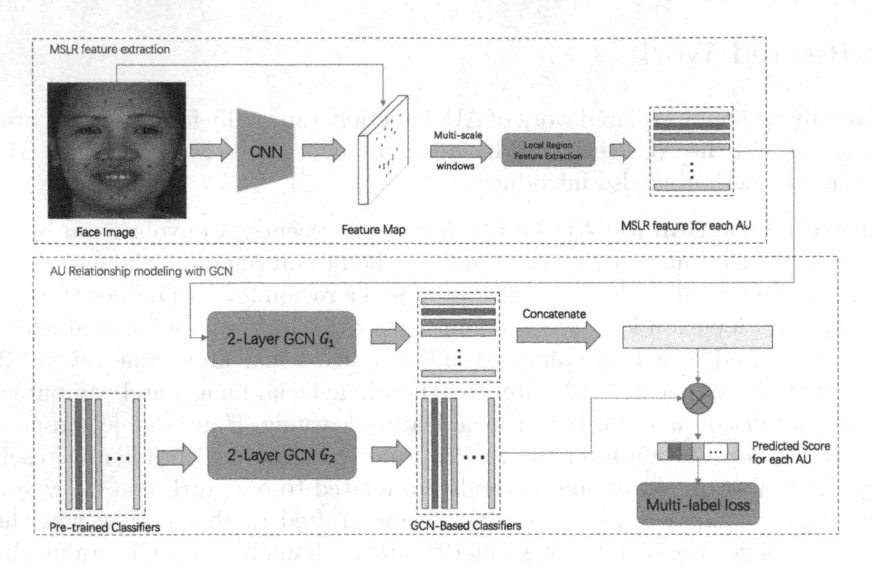

Fig. 1. Overview of the proposed MSLR-GCN framework for facial AU detection. MSLR-GCN consists of two parts: multi-scale local region-based (MSLR) feature extraction and AU relationship modeling with GCN.

3.1 Overview of MSLR-GCN

Feature extraction and AUs multi-label classification have been the two crucial aspects in facial AU detection. Most existing methods apply the same convolution operation to different facial region. However, different AUs cause facial appearance changes over various regions at different scales, therefore the multi-scale feature at local facial region should be considered. To overcome the above limitations, we propose the MSLR-GCN framework, as shown in Fig. 1.

Firstly, we choose ResNet-34 [9] as our backbone network to extract the image feature since we don't need too deep structure for this problem. The last layer of ResNet-34 contains 512 feature maps with 7×7 size. For each AU, we build the improved AU centers, and the MSLR feature is extracted around the AU centers with several multi-scale windows. Secondly, we update the MSLR feature as well as the AU classifiers by GCN, which can model the AUs relationship. Specifically, we build the graph of AUs using the AU correlation matrix, and the GCN is applied to update the graph node representation. The final AU predicted probability is obtained by the updated MSLR features and GCN-based classifiers.

Table 1. Positions of AU centers (the improved ones are in bold).

AU Index	AU Name	improved AU Centers
1	Inner Brow Raiser	1/2 scale above inner brow
2	Outer Brow Raiser	1/3 scale above outer brow
4	Brow Lowerer	**inner brow corner**
6	Cheek Raiser	1 scale below eye bottom
7	Lid Tightener	Eye center
10	Upper Lip Raiser	Upper lip center
12	Lip Corner Puller	**inner Lip corner**
14	Dimpler	**inner Lip corner**
15	Lip Corner Depressor	**inner Lip corner**
17	Chin Raiser	**midpoint of lip and chin**
23	Lip Tightener	**inner Lip center**
24	Lip Pressor	**inner Lip center**

3.2 MSLR Feature Extraction

Since different AUs correspond to different muscular activation of the face, a meticulous feature extraction method is required for effectively extracting multi-scale AUs feature at different local facial regions.

To get the AUs local feature, we first build the improved AU centers for each AU. According to [5], different AUs correspond to specific muscular activation of the face, therefore it is natural to define a set of activation centers for each AU. In this paper, we adopt the AU centers defined in [14], and make some modifications for our feature learning strategy. The detailed position of the improved AU centers is shown in Table 1.

The AU centers are previously calculated according to the positions of facial landmarks before the training/inference process, as detailed in [14]. Since many recent works can provide the facial landmarks positions in a fast and robust way, it's natural for us to calculate the improved AU center in advance, instead of obtaining it during training process [20], which will increase the model complexity as well as the training time.

Since different AUs cause facial appearance changes over various regions at different scales, the multi-scale region feature is extracted around the improved AU centers. Take the human face in Fig. 1 as example, we manually apply several multi-scale windows arround the improved AU centers, these windows are considered the region of interest (RoI) for AU detection. The sizes of RoIs are manually set to include the whole AU regions, and are mapped to the multi-scale of 2×2, 2×1.5, 1.5×2 in the last 7×7 feature map for ResNet-34. Note that the multi-scale windows do not need to be very accurately located, an approximated region covering the whole AU is enough to generate the MSLR feature.

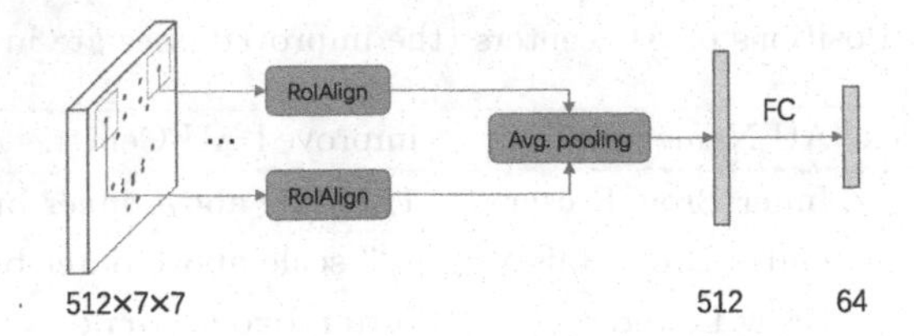

Fig. 2. Local region feature extraction for each AU(take AU2 for example here).

With various size of RoI features, the RoIAlign [8] is adopted to generate a fixed feature, which can help solve the misalignment problem between the RoI and the last feature map. Since the AU RoIs appear in pairs, we merge the fixed feature through an average pooling operation for each AU, as shown in Fig. 2. Finally, we obtain the MSLR features x as follows:

$$x_i = L_i(CNN(I; \theta_{cnn})) \in \mathbb{R}^d, \quad i = 0, 1, \cdots, n-1, \tag{1}$$

where θ_{cnn} denotes the backbone parameters, I is the original image, n denotes the number of AUs, L_i denotes the RoIAlign and average pooling operation for each AU, the dimension of each feature vector is $d = 64$.

3.3 AU Relationship Modeling with GCN

In this subsection, we introduce the AU relationship modeling strategy with GCN. In our model, the GCN is used for updating both the AU MSLR feature and the classifiers. The structure of the two GCNs are similar except for the input graph node embedding. Therefore, we will first introduce the framework of GCN, then show the difference of these input embeddings.

Essentially, the GCN works by updating the graph node representations based on the correlation matrix. In most applications, the correlation matrix is pre-defined. However, it is not provided in the AU multi-label classification. In this paper, we build the AU correlation matrix by statistics on the training set, which is exactly the frequency correlation matrix A that counting the co-occurrence of AU label pairs:

$$A_{i,j} = Q_{i,j}/N_i, \tag{2}$$

where $Q_{i,j}$ is the co-occurrence time for AU label i and j, N_i is the occurrence time for AU label i.

The visualization of the frequency correlation matrix is shown in Fig. 3, each element of which is the co-occurrence frequency of the AU labels. It can be seen from the figure that the frequency correlation matrix denotes the conditional probability between two AU labels, for instance, when AU6 appears, AU12 will also occur with a high probability. Through this AU correlation matrix, the directed graph is built over the AU labels.

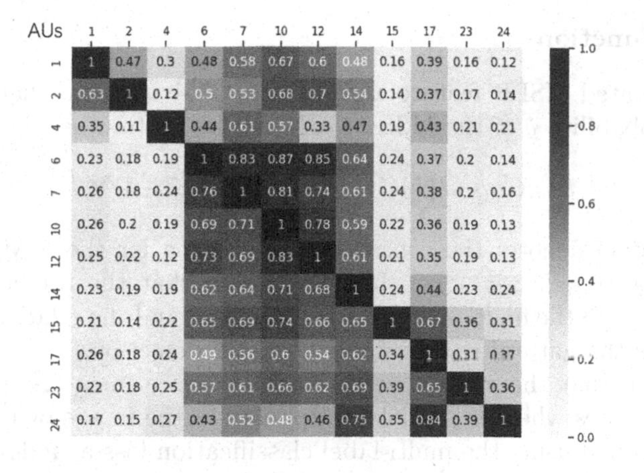

Fig. 3. Visualization of the frequency correlation matrix.

Based on the AU correlation matrix, the AU node representation will be updated by GCN layer. The update formula of the GCN layer is as below:

$$H^{(l+1)} = \sigma(D^{-\frac{1}{2}}(A+I)D^{-\frac{1}{2}}H^{(l)}W^{(l)}), \tag{3}$$

where $H^{(l)} \in \mathbb{R}^{n \times m}$ is the node representations in the l-th layer, m denotes the dimension of the node representations vector, $D^{-\frac{1}{2}}(A+I)D^{-\frac{1}{2}}$ is the normalized version of the AU correlation matrix, $D = \sum_j (A+I)_{ij}$ is the degree matrix, I is the identity matrix, σ denotes the LeakyReLU [16] activation function, and $W^{(l)} \in \mathbb{R}^{m \times m}$ is the trainable weight matrix for layer l. Note that the weight matrix is of square size, therefore the dimension of the input and output node representations will have same size.

The structures of the two GCNs are similar. In experiment, we empirically set 2 layers for each GCN. The dimensions of the node representation are equal in each GCN. For the GCN that updates the MSLR feature, the node vector dimension $m = d$ corresponds to the feature dimension; for the GCN that updates the AU classifiers, $m = d \times n$ corresponds to the concatenated feature size. The two GCNs are shown as below:

$$\hat{x} = G_1(x) \quad x, \hat{x} \in \mathbb{R}^{n \times d}, \tag{4}$$

$$\hat{C} = G_2(C) \quad C, \hat{C} \in \mathbb{R}^{n \times nd}, \tag{5}$$

where G_1 and G_2 are stacked by GCN layer, C is the pre-trained classifiers, $\hat{x}$ denotes the updated MSLR features, $\hat{C}$ denotes the updated AU classifiers. Note that G_1 is similar to the idea of [15] and models the AU feature relationship; G_2 is used for modeling the AU classifier relationship. Specifically, we first pre-train C by directly multiplying it with the concatenated feature to get the predicted score, then update C by G_2 to model the relationship of each single AU classifier. Since each single AU classifier is multiplied with the same concatenated feature, the classifiers with stronger relationship should have higher probability to generate similar predicted score.

3.4 Loss Function

With the updated MSLR feature $\hat{x}$ and the GCN-based classifiers $\hat{C}$, the final predicted probability $\hat{y}$ of the MSLR-GCN can be written as

$$\hat{y} = Concat(\hat{x})\hat{C}^T \quad \hat{C} \in \mathbb{R}^{n\times nd}, \hat{x} \in \mathbb{R}^{n\times d}, \tag{6}$$

where $Concat(*)$ denotes the concatenate operation for the n MSLR features, and $Concat(\hat{x}) \in \mathbb{R}^{nd}$, $\hat{y} \in \mathbb{R}^n$. It can be seen that the predicted output is influenced by both the multi-scale AUs local feature and the relationship between AUs, which is the natural way to implement AU detection.

We assume that the ground truth label of a face image is $y \in \mathbb{R}^n$, where $y_i \in \{1, 0\}$ denotes whether AU label i appears in the image or not. The whole network is trained using the multi-label classification loss as follows:

$$\mathcal{L} = \sum_{i=0}^{n-1} w_i[y_i log\hat{y}_i + (1 - y_i)log(1 - \hat{y}_i)], \tag{7}$$

where w_i is the balancing weight calculated in each batch using the Selective Learning strategy [7], which can help alleviate the problem of data imbalance.

4 Experiments

In this section, we demonstrate the effectiveness of the proposed MSLR-GCN through experiments on two widely used datasets: BP4D and DISFA. For fair comparisons, we run all our experiments in Ubuntu 18.04 System with eight NVIDIA 2080Ti GPUs.

4.1 Datasets and Implementation Details

Datasets. We evaluate our MSLR-GCN on two widely used public datasets: BP4D [22] and DISFA [17]. BP4D is a spontaneous facial expression database containing 328 videos for 41 participants (23 females and 18 males), each of which is involved in 8 tasks for emotional expression elicitation. There are over 140,000 frames with 12 AU labels of occurrence or absence. DISFA consists of 27 videos from 12 females and 15 males, each participant is asked to watch a 4-minute video to elicit facial AUs. About 130,000 frames are included in the dataset, each frame is annotated with AU intensities from 0 to 5. Following the settings in [24], 8 of the 12 AUs are used for evaluation in DISFA and the frames with intensities equal or greater than 2 are considered as positive, while others are treated as negative. We also adopt the subject-exclusive three-fold validation for the two datasets, with two folds for training and the remaining one for testing.

Table 2. Performance comparison with the state-of-the-art methods on the BP4D database in terms of F1 score.

Method	AU1	AU2	AU4	AU6	AU7	AU10	AU12	AU14	AU15	AU17	AU23	AU24	**Avg.**
JPML	32.6	25.6	37.4	42.3	50.5	72.2	74.1	65.7	38.1	40.0	30.4	42.3	45.9
DRML	36.4	[41.8]	43.0	55.0	67.0	66.3	65.8	54.1	33.2	48.0	31.7	30.0	48.3
ROI	36.2	31.6	43.4	77.1	73.7	**85.0**	87.0	62.6	45.7	58.0	38.3	37.4	56.4
EAC-Net	39.0	35.2	48.6	76.1	72.9	81.9	86.2	58.8	37.5	59.1	35.9	35.8	55.9
OFS-CNN	41.6	30.5	39.1	74.5	62.8	74.3	81.2	55.5	32.6	56.8	41.3	-	53.7
DSIN	[51.7]	40.4	56.0	76.1	73.5	79.9	85.4	62.7	37.3	62.9	38.8	41.6	58.9
LP-Net	43.4	38.0	54.2	77.1	76.7	83.8	87.2	63.3	45.3	60.5	**48.1**	**54.2**	61.0
AU-GCN	46.8	38.5	[60.1]	**80.1**	[79.5]	84.8	[88.0]	**67.3**	**52.0**	**63.2**	40.9	52.8	[62.8]
MSLR-GCN	**52.9**	**54.6**	**60.6**	[77.2]	**81.0**	**85.4**	**88.4**	[66.3]	[47.4]	[63.1]	[42.5]	[53.4]	**64.4**

Implementation Details. For each image, we detect the facial landmarks using dlib [10], so that the face alignment can be conducted to reduce the variations from scaling and in-plane rotation. The facial landmarks are also used to generate the improved AU centers, as detailed in Sect. 3.2. The images are cropped to 224×224 for the input of ResNet-34. Note that all the image cropping and the improved AU centers generation are implemented through CPU multi-threading, which can be done in around 5 h for all the raw images. Moreover, the backbone network ResNet-34 is pre-trained on VGG-face2 [1] to enhance the feature extraction ability.

For network optimization, we use Adam [11] as the optimizer. The initial learning rate is 0.0001, which decays by a factor of 10 for every 8 epochs and the network is trained for 20 epochs in total.

Evaluation Metrics. To evaluate the performance of all methods, we compute F1-frame score for 12 AUs in BP4D and 8 AUs in DISFA. The overall performance of the algorithm is described by the average F1 score (donated as **Avg.**).

4.2 Comparisons with the State-of-the-Art

In this subsection, we compare our MSLR-GCN with the state-of-the-art methods under the same subject-exclusive validation. The methods focused on representative feature extraction (JPML [23], DRML [24], ROI [13], EAC-Net [14], OFS-CNN [6]) and relationship modeling (DSIN [3], LP-Net [18], AU-GCN [15]) are compared. Table 2 and 3 report the F1 score of different methods on BP4D and DISFA databases. It can be seen from the tables that our MSLR-GCN outperforms all other compared methods by a large margin on average F1-score.

4.3 Ablation Study

We provide ablation study to investigate the effectiveness of each part in our MSLR-GCN. Table 4 shows the F1-scores on BP4D by individual ablation experiments. In the following paragraphs, we will detail the effectiveness of the MSLR feature extraction and the GCN-based classifiers.

Table 3. Performance comparison with the state-of-the-art methods on the DISFA database in terms of F1 score.

Method	AU1	AU2	AU4	AU6	AU9	AU12	AU25	AU26	**Avg.**
DRML	17.3	17.7	37.4	29.0	10.7	37.7	38.5	20.1	26.7
ROI	41.5	26.4	66.4	50.7	8.5	**89.3**	88.9	15.6	48.5
OFS-CNN	[43.7]	[40.0]	67.2	[59.0]	[49.7]	75.8	72.4	54.8	[57.8]
DSIN	42.4	39.0	[68.4]	28.6	46.8	70.8	90.4	42.2	53.6
LP-Net	29.9	24.7	**72.7**	46.8	49.6	72.9	**93.8**	**65.0**	56.9
AU-GCN	32.3	19.5	55.7	57.9	**61.4**	62.7	[90.9]	[60.0]	55.0
MSLR-GCN	**71.1**	**62.9**	45.1	**67.0**	31.8	[76.9]	83.3	40.8	**59.9**

Table 4. Ablation experiments on the BP4D database in terms of the average F1 score.

	Ablations									MSLR-GCN
+original AU centers?	–	✓	–	–	–	–	✓	–	–	–
+improved AU centers?	–	–	✓	✓	✓	–	–	✓	✓	✓
+multi-scale windows?	–	–	–	✓	✓	–	–	–	✓	✓
$+G_1$?	–	–	–	–	✓	–	–	–	–	✓
$+G_2$?	–	–	–	–	–	✓	✓	✓	✓	✓
Avg.	57.2	58.1	58.4	60.0	61.7	59.7	60.6	61.6	63.1	64.4

Effectiveness of MSLR. Since different AUs correspond to specific muscular activation of the face, we propose the MSLR feature extraction method to extract the multi-scale local feature in facial image. Specifically, the contribution of our MSLR is based on two aspects: the improved AU centers and the multi-scale windows. We first conduct the experiment of predicting AU probabilities by directly using ResNet-34 without any other tricks, which is the baseline of our experiment, and the average F1-score is 57.2. Then, we fuse the features around the original AU centers. This is exactly the idea of ROI [13] and the average F1-score is 58.1. While we replace the original AU centers by the improved ones, the average F1-score is enhanced to 60.0. In next experiment, we apply the multi-scale windows as detailed in Sect. 3.2, it can be seen that the average F1-score is further improved to 61.7.

Effectiveness of GCN. In our model, the GCN is used for updating the MSLR feature and the AU classifiers, corresponding to G_1 and G_2 in Sect. 3.3. Since G_1 is similar to the idea of AU-GCN [15], we will mainly discuss the effectiveness of G_2 in the following ablation study.

Based on the previous ablation study, we conduct the experiments of adding G_1 on the MSLR feature and adding G_2 on the pre-trained classifiers of each ablation model trained before. The AU classifiers are pre-trained from the ablation models, i.e. the baseline that directly uses RseNet-34. We update the pre-trained classifiers through a two-layer GCN G_2 to obtain the GCN-based classifiers. The AU correlation matrix that guides the information propagation is statistics from the training set, as shown in Fig. 3. It can be seen that the integrated method of $MSLR + G_1 + G_2$, i.e. MSLR-GCN, reaches the highest average F1-score of 63.1.

5 Conclusion

In this paper, we propose a novel end-to-end framework (MSLR-GCN) for facial action unit detection, which consists of two parts: MSLR feature extraction and GCN-based AU relationship modeling. The MSLR part utilizes the multi-scale windows and the improved AU centers to extract multi-scale AUs local feature. Then, we employ two GCNs to update the AU classifiers and the MSLR feature, which can model the relationship between AUs to further improve the robustness of our AU detection algorithm. Extensive experimental results show the effectiveness of our proposed method, which outperforms the state-of-the-art methods on two widely used AU detection datasets BP4D and DISFA. In our future work, we plan to build a dataset-independent AU detection model and applying it to detect facial expression in real scene applications.

References

1. Cao, Q., Shen, L., Xie, W., Parkhi, O.M., Zisserman, A.: Vggface2: a dataset for recognising faces across pose and age. In: 2018 13th IEEE International Conference on Automatic Face & Gesture Recognition (FG 2018), pp. 67–74. IEEE (2018)
2. Chen, Z.M., Wei, X.S., Wang, P., Guo, Y.: Multi-label image recognition with graph convolutional networks. In: Proceedings of the IEEE Conference on Computer Vision and Pattern Recognition, pp. 5177–5186 (2019)
3. Corneanu, C., Madadi, M., Escalera, S.: Deep structure inference network for facial action unit recognition. In: Proceedings of the European Conference on Computer Vision (ECCV), pp. 298–313 (2018)
4. Ding, H., Sricharan, K., Chellappa, R.: Exprgan: facial expression editing with controllable expression intensity. In: Thirty-Second AAAI Conference on Artificial Intelligence (2018)
5. Ekman, R.: What the Face Reveals: Basic and Applied Studies of Spontaneous Expression Using the Facial Action Coding System (FACS). Oxford University Press, Oxford (1997)
6. Han, S., et al.: Optimizing filter size in convolutional neural networks for facial action unit recognition. In: Proceedings of the IEEE Conference on Computer Vision and Pattern Recognition, pp. 5070–5078 (2018)
7. Hand, E.M., Castillo, C., Chellappa, R.: Doing the best we can with what we have: multi-label balancing with selective learning for attribute prediction. In: Thirty-Second AAAI Conference on Artificial Intelligence (2018)
8. He, K., Gkioxari, G., Dollár, P., Girshick, R.: Mask R-CNN. In: Proceedings of the IEEE International Conference on Computer Vision, pp. 2961–2969 (2017)
9. He, K., Zhang, X., Ren, S., Sun, J.: Deep residual learning for image recognition. In: Proceedings of the IEEE Conference on Computer Vision and Pattern Recognition, pp. 770–778 (2016)

10. King, D.E.: Dlib-ml: a machine learning toolkit. J. Mach. Learn. Res. **10**, 1755–1758 (2009)
11. Kingma, D.P., Ba, J.: Adam: a method for stochastic optimization. In: International Conference on Learning Representations (ICLR) (2015)
12. Kipf, T.N., Welling, M.: Semi-supervised classification with graph convolutional networks. In: International Conference on Learning Representations (ICLR) (2017)
13. Li, W., Abtahi, F., Zhu, Z.: Action unit detection with region adaptation, multi-labeling learning and optimal temporal fusing. In: Proceedings of the IEEE Conference on Computer Vision and Pattern Recognition, pp. 1841–1850 (2017)
14. Li, W., Abtahi, F., Zhu, Z., Yin, L.: EAC-Net: deep nets with enhancing and cropping for facial action unit detection. IEEE Trans. Pattern Anal. Mach. Intell. **40**(11), 2583–2596 (2018)
15. Liu, Z., Dong, J., Zhang, C., Wang, L., Dang, J.: Relation modeling with graph convolutional networks for facial action unit detection. In: Ro, Y.M., et al. (eds.) MMM 2020. LNCS, vol. 11962, pp. 489–501. Springer, Cham (2020). https://doi.org/10.1007/978-3-030-37734-2_40
16. Maas, A.L., Hannun, A.Y., Ng, A.Y.: Rectifier nonlinearities improve neural network acoustic models. In: Proceedings of ICML, vol. 30, p. 3 (2013)
17. Mavadati, S.M., Mahoor, M.H., Bartlett, K., Trinh, P., Cohn, J.F.: DISFA: a spontaneous facial action intensity database. IEEE Trans. Affect. Comput. **4**(2), 151–160 (2013)
18. Niu, X., Han, H., Yang, S., Huang, Y., Shan, S.: Local relationship learning with person-specific shape regularization for facial action unit detection. In: Proceedings of the IEEE Conference on Computer Vision and Pattern Recognition, pp. 11917–11926 (2019)
19. Otwell, K.: Facial expression recognition in educational learning systems, US Patent 10,319,249 (2019)
20. Shao, Z., Liu, Z., Cai, J., Ma, L.: Deep adaptive attention for joint facial action unit detection and face alignment. In: Proceedings of the European Conference on Computer Vision (ECCV), pp. 705–720 (2018)
21. Wearne, T., Osborne-Crowley, K., Rosenberg, H., Dethier, M., McDonald, S.: Emotion recognition depends on subjective emotional experience and not on facial expressivity: evidence from traumatic brain injury. Brain Inj. **33**(1), 12–22 (2019)
22. Zhang, X., et al.: BP4D-spontaneous: a high-resolution spontaneous 3D dynamic facial expression database. Image Vis. Comput. **32**(10), 692–706 (2014)
23. Zhao, K., Chu, W.S., De la Torre, F., Cohn, J.F., Zhang, H.: Joint patch and multi-label learning for facial action unit detection. In: Proceedings of the IEEE Conference on Computer Vision and Pattern Recognition, pp. 2207–2216 (2015)
24. Zhao, K., Chu, W.S., Zhang, H.: Deep region and multi-label learning for facial action unit detection. In: Proceedings of the IEEE Conference on Computer Vision and Pattern Recognition, pp. 3391–3399 (2016)

CRE: An Efficient Ciphertext Retrieval Scheme Based on Encoder

Kejun Zhang[1], Shaofei Xu[1(✉)], Pengcheng Li[2], Duo Zhang[3], Wenbin Wang[1(✉)], and Bing Zou[1]

[1] Beijing Electronic Science and Technology Institute, Beijing, China
{20222926,20212923}@mail.besti.edu.cn
[2] School of Computer Science and Technology, Xidian University, Xi'an, China
[3] School of Cyberspace Security, University of Science and Technology of China, Hefei, China

Abstract. Searchable Encryption is utilized to address the issue of searching for outsourced encrypted data on third-party untrusted cloud servers. Traditional approaches for ciphertext retrieval are limited to basic keyword-matching queries and fall short when it comes to handling complex semantic queries. Although several semantic retrieval schemes have been proposed in recent years, their performance is inadequate. This paper introduces a semantic retrieval scheme called CRE (Ciphertext Retrieval based on Encoder), which leverages the prompt-based RoBERTa pre-trained language model to generate precise embeddings for sentences in queries and documents. Moreover, to improve retrieval speed in the face of massive high-dimensional sentence embedding vectors, we introduce the HNSW algorithm. Through experimentation and theoretical analysis, this paper demonstrates that CRE outperforms $SSSW_2$ and $SSRB_2$ in terms of retrieval speed and accuracy.

Keywords: Ranked ciphertext retrieval · Semantic search · Pre-trained language model · Approximate nearest neighbor search

1 Introduction

The rapid advancement of cloud storage technology has led to a growing number of individual users choosing to store their data on cloud servers. To protect their private information from potential unauthorized access by the cloud server, users commonly encrypt their data before uploading. However, this practice introduces a new challenge: how can users efficiently and accurately retrieve the required documents from encrypted outsourced data? In response to this issue, the concept of Searchable Encryption has emerged.

In 2000, [25] introduced the concept of Searchable Encryption to address the need for retrieving encrypted data stored in the cloud. This concept involves the data owner uploading the encrypted documents along with the corresponding index to the cloud server. When a specific encrypted document needs to be retrieved, a query trapdoor is sent to the server. The server then matches the trapdoor with the

B. Luo et al. (Eds.): ICONIP 2023, CCIS 1966, pp. 117–130, 2024.
https://doi.org/10.1007/978-981-99-8148-9_10

encrypted index and returns the most relevant document. This pioneering idea has served as inspiration for numerous subsequent schemes [7,9,24], including multi-keyword retrieval schemes such as ranked multi-keyword search [5,6,22,26] and fuzzy multi-keyword search [1,12,28]. To meet the increasing demand for semantic search capabilities, [21] proposed the first semantic retrieval scheme based on Searchable Symmetric Encryption (SSE) for encrypted outsourced data. Since then, several semantic retrieval schemes under the encrypted environment have been proposed [8,10,13,29,30]. The rapid advancements in deep learning within the field of Natural Language Processing (NLP) have provided a novel approach for semantic search in the ciphertext. Consequently, several schemes leveraging deep learning technology for ciphertext retrieval have emerged in recent years [11,14,19]. However, these schemes still face challenges regarding retrieval accuracy and efficiency.

Drawing inspiration from the widespread utilization of Transformers in the field of Natural Language Processing, we propose a novel scheme CRE, which integrates the prompt-based sentence embedding method and self-attention mechanisms to convert sentences into embeddings that effectively capture the original semantic information. As a result, the efficiency and accuracy of ciphertext retrieval are significantly enhanced. Our main contributions are as follows:

(1) First, we utilize the abstractive text summarization model called BART to convert lengthy documents into concise summaries, thereby reducing the number of high-dimensional vectors. Unlike previous approaches that simply truncate excessively long documents, our scheme avoids the loss of semantic information.
(2) Second, we utilize RoBERTa pre-trained language model as the basis for our training. RoBERTa adopts dynamic masking instead of static masking in the MLM task, which enables RoBERTa to better capture the semantic information of words. Building upon RoBERTa, we utilize the prompt method to train our model. This approach enables us to convert sentences of documents and queries into 768-dimensional embeddings that effectively capture the original semantic information. In contrast, most previous approaches relied on keyword extraction methods to convey information in sentences, which posed challenges in accurately selecting the most representative keywords.
(3) Third, we utilize the HNSW algorithm, which is a hierarchical graph-based approximate nearest neighbor search algorithm. HNSW is specifically designed to handle the computation of distances among high-dimensional vectors efficiently. Unlike previous schemes that relied on constructing tree-based indexes for massive high-dimensional vectors, CRE leverages the advantages of HNSW to achieve faster and more accurate retrieval.

2 Related Work

2.1 Semantic Search over Encrypted Data

Traditional ciphertext retrieval schemes strictly rely on given keywords for retrieval. However, in many cases, users may not be able to determine specific keywords. Semantic search addresses this issue.

[13] argue that the syntactic relationships between keywords are crucial and propose the first scheme that considers the relationships between keywords. In paper [10], They construct a user interest model supported by the semantic ontology WordNet to accomplish personalized and accurate keyword searches. [29] constructed a Semantic Relationship Library for keyword sets, which is capable of identifying keywords that are semantically similar to the query keywords. Although the aforementioned schemes can achieve synonym recognition to some extent, they require the construction of significantly large semantic relationship libraries for synonyms. Some researchers have proposed the application of topic models such as LDA or BTM to ciphertext retrieval [8,30]. They match documents and query keywords by the same topic. However, this retrieval method fails to meet the fine-grained retrieval needs of users.

[2] employed the Vector Space Model (VSM) for retrieval and introduced the MRSE model, which vectorizes documents and queries. [15] demonstrated that deep learning models are unexplainable, making it impossible for adversaries to obtain private information even if they acquire the extracted feature vectors. Inspired by this, it is feasible to vectorize documents using deep learning techniques. [19] utilized the Word2vec model to transform keywords into embeddings, but the generated embeddings are fixed and unable to represent all senses of polysemous words. [11] employed pre-trained language model BERT to transform keywords into word embeddings. However, the original BERT model suffers from severe anisotropy issues when generating word embeddings. [14] utilized Sentence BERT to convert documents into vectors, but representing the entire document with a single vector may lead to the loss of semantic information.

2.2 BART (Bidirectional and Auto-Regressive Transformer)

[18] is primarily based on the Transformer architecture, comprising an encoder and a decoder. The encoder processes the corrupted text, capturing its underlying features and transmitting them to the decoder, then the decoder leverages these features to reconstruct the original text.

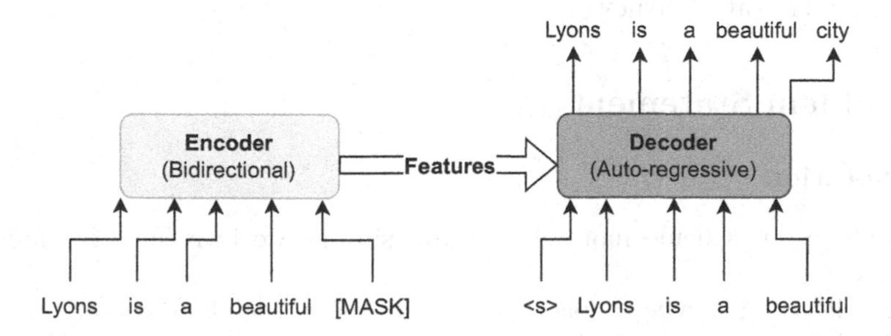

Fig. 1. Structure of the BART model, the encoder learns the features of damaged text and sends it to the decoder, then the decoder reconstructs the original text.

Figure 1 shows the architecture of the BART model. BART is trained by minimizing the reconstruction loss, which is the cross-entropy loss between the original text and generated text by the decoder [23]. To generate corrupted text, BART utilizes diverse noise-inducing methods that disrupt the original text and produce corrupted versions. These methods include token masking, token deletion, token padding, sentence shuffling, and document rotation.

2.3 Prompt-Based Bidirectional Encoder Representation from Transformer

The [CLS] token in the output of the BERT model [17] is typically used to represent the sentence embedding. However, due to the learned anisotropic token embedding space and static token embedding biases, this approach may inadequately capture the semantic information conveyed by sentences.

[16] leveraged the prompt method based on BERT to generate more accurate sentence embeddings. By reformulating the task of obtaining sentence embeddings as a masked language task, they demonstrated that using [MASK] tokens for sentence embeddings can effectively mitigate the biases caused by anisotropy and static token embedding biases. Jiang et al. used the prompt method to construct the template sentence and performed supervised and unsupervised training using the fine-tuned and unfine-tuned BERT models respectively. Their experiments showed that this method significantly enhanced the ability of sentence embedding to accurately capture original semantic information.

2.4 Hierarchical Navigable Small World Graphs

Hierarchical Navigable Small World [20] is a hierarchical graph-based approximate nearest neighbor search algorithm designed to facilitate rapid retrieval of vectors that closely resemble a given query vector in high-dimensional space. The algorithm leverages a hierarchical structure, where vector nodes are organized into multiple levels representing small-world networks. This hierarchical arrangement enables efficient localization of the most similar vector nodes, thereby enhancing retrieval efficiency.

3 Problem Statement

3.1 Notation

To introduce our scheme more clearly and simply, we introduce the following notations:

***DO*:** the data owner who uploads documents set to the cloud server.

***DU*:** the data user who gets the required encrypted documents from the server.

$\boldsymbol{D = \{D_1, D_2, ..., D_n\}}$: the set of plaintext document to be uploaded by *DO*.

$\boldsymbol{C = \{C_1, C_2, ..., C_n\}}$: the encrypted document set formed after the set D is encrypted.

$\boldsymbol{E = \{E_1, E_2, ..., E_n\}}$: the set of sentence embedding vectors for D, where

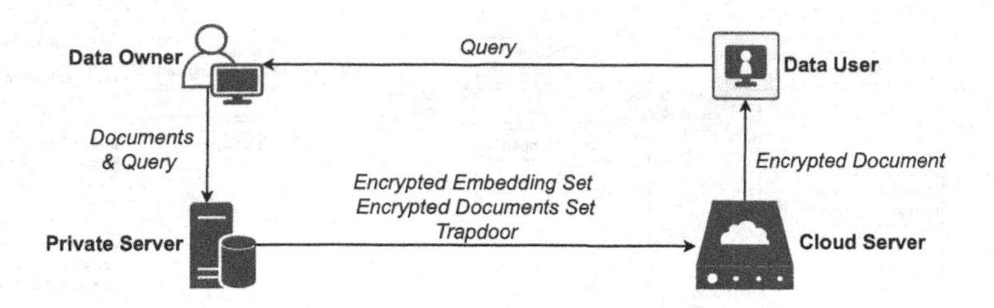

Fig. 2. CRE System Model

$E_i = \{s_1, s_2, ..., s_n\}$ represents the embedding set of all sentences in D_i.
$\widetilde{\boldsymbol{E}}$: the encrypted set E.
$\boldsymbol{Q}$: the query from DU.
$\boldsymbol{E_Q}$: the embedding vector of Q.
$\boldsymbol{T}$: the trapdoor formed by E_Q after encryption.

3.2 System Model

Our system model consists of four entities: the cloud server, the private server, the DO, and the DU. As shown in Fig. 2, DO first sends D to the private server, which transforms it into E. Then D and E are encrypted and uploaded to the cloud server. When DU wants to retrieve D_i from the cloud server, it needs to send Q to DO, DO verifies Q and forwards it to the private server, the private server converts Q into T and sends it to the cloud server. After receiving T, the cloud server uses the $HNSW$ algorithm to perform matching operations between T and $\widetilde{E}$. Finally, the cloud server returns C_i that meets $DU's$ requirements, then DU decrypts C_i locally to obtain D_i.

3.3 Threat Model

In our scheme, we assume that the cloud server is "honest-but-curious" [4]. Based on the known information of semi-honest public cloud servers, we investigate two models [14] with different threat levels, as follows:

Known Ciphertext Model: the cloud server only knows the encrypted document set C, the encrypted embedding set E, and the security query trapdoor T.

Known Background Model: The cloud server not only knows all the information in the first model but also can perform statistical analysis on the data. For instance, leveraging the query history, the cloud server can infer the relationship between trapdoors and their corresponding documents, thereby acquiring additional information.

3.4 Design Goal

According to the system model and threat model, our scheme should not only improve the accuracy and speed of document retrieval but also ensure the privacy and security of outsourced data. Our main design goals are as follows:

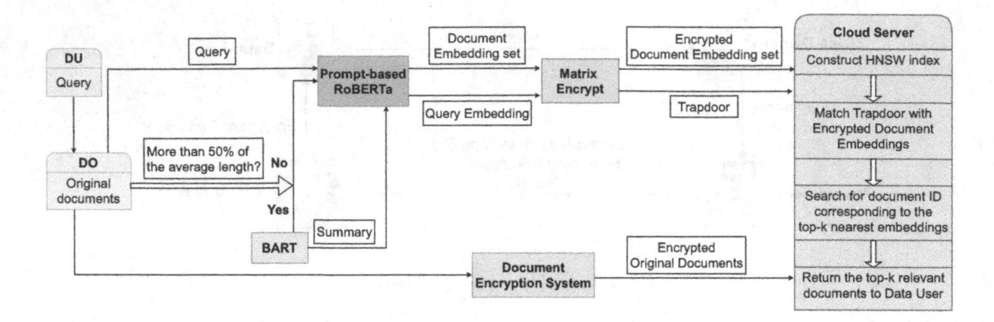

Fig. 3. CRE detailed structure. Initially, *DO* determines if the document length exceeds 50% of the average length. If it does, BART extracts the summary of the document. Subsequently, the document set is transmitted to a server equipped with the prompt-based RoBERTa model by *DO*. The server converts the document set into an embedding set, encrypts it using a matrix key, and uploads it to the cloud. Concurrently, *DO* encrypts the document set and uploads it to the cloud. When a search operation is required, the query from *DU* is transformed into an embedding and encrypted into a trapdoor. The cloud employs the HNSW algorithm to match the encrypted documents with the trapdoor, ultimately returning the top-k relevant results to *DU*.

Accuracy and Efficiency: The utilization of the prompt-based RoBERTa model enables the generation of sentence embeddings that accurately capture the semantic information, thus significantly enhancing the precision of retrieval results. Additionally, the incorporation of the HNSW algorithm in the retrieval process effectively improves the retrieval speed.

Security and Privacy: We encrypt the document embedding set with an invertible matrix, then use cosine similarity to calculate the distance between encrypted embeddings and query trapdoor. In addition, *DO* will add a random number every time it generates a new trapdoor. This ensures that even if two queries are entirely consistent, their respective trapdoors differ, improving the security of the access mode.

4 Ciphertext Retrieval Based on Encoder

In this section, we introduce our scheme against the background of the threat model presented in the previous section. The detailed architecture of the CRE scheme is shown in Fig. 3. Our scheme is divided into an embedding generation module and a retrieval module, which has a good performance in terms of retrieval efficiency and security. The main steps of our scheme will be introduced in detail below.

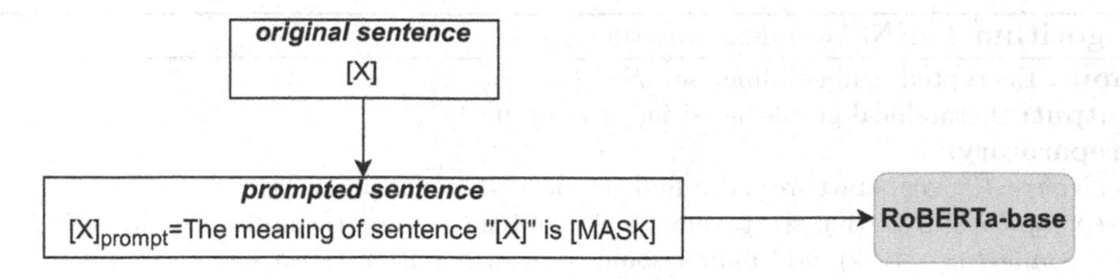

Fig. 4. Prompt-based RoBERTa model, input original sentence $[X]$, map it to $[X]_{prompt}$ and use RoBERTa's dynamic MLM task to get the embedding of $[MASK]$ to represent the embedding of $[X]$.

4.1 Generate Embedding Set

$\boldsymbol{KeyGen()} \rightarrow \boldsymbol{M}, \boldsymbol{S}$: *DO* randomly generates a 769×769 invertible matrix M and a symmetric key S.

$\boldsymbol{EmbGen(D)} \rightarrow \boldsymbol{E}$: *DO* uses the prompt-based RoBERTa model to convert each sentence in set D into a 768-dimensional embedding, then adds an additional dimension to store the document ID that the sentence belongs to, resulting in a 769-dimensional vector. It should be noted that before this step, *DO* needs to use the BART model to extract summaries of documents longer than 50% of the average length and replace the original document with its summary.

$\boldsymbol{DocEncry(D, E, M, S)} \rightarrow \boldsymbol{C}, \widetilde{\boldsymbol{E}}$: *DO* uses the key S to encrypt document set D to get ciphertext document set C, and then uses matrix M to encrypt E ($E \times M$) to get $\widetilde{E}$. Finally, *DO* uploads C and $\widetilde{E}$ together to the cloud server.

Inspired by Jiang et al. [16], to accurately capture the semantic information of the sentence during the generation of the embeddings, we use the prompt method to train our model. As shown in Fig. 4, given a sentence $[X]$, we map it to $[X]_{prompt}$ and feed it into RoBERTa. Then, we leverage the embeddings of $[MASK]$ obtained from the Dynamic MLM task of the fine-tuned RoBERTa-base model to represent the embedding of sentence $[X]$. In the model training stage, we use the sentence embeddings generated by the same template as the positive example, combined with supervised negative examples for comparative learning.

4.2 Retrieval for Documents

$\boldsymbol{Trapdoor(Q_i, M, r_i)} \rightarrow \boldsymbol{T_i}$: *DU* sends a query Q_i to *DO*, then *DO* runs *EmbedGen*(Q_i) to obtain E_{Q_i}, and then generates one random number r_i, constructing $T = M^{-1} \times E_{Q_i}^T \cdot r_i$. It should be pointed out that adding a random number r can avoid similar semantic queries (their corresponding sentence vectors are also similar) from producing similar trapdoors so that the cloud server cannot infer the associated document based on a particular query. Adding a random number r will not affect the search result, because we use the cosine distance to calculate the similarity score, and finally select the top-k documents according to the top-k scores.

Algorithm 1. HNSW index construction

Input: Encrypted Embedddings set: $S = \{s_1, \ldots, s_n\}$
Output: Hierarchical graph-based index structure
Preparatory:
$GetNearest(s, en)$: start from en, find the nearest node to s in layer l
$GetNeighbors(s, M, W)$: start from set W, find the M nearest nodes to s in layer l
$AddConnections(W, s)$: add bidirectional connections from W to s
$RandomLayer(s)$: assign layers to s using random function

1: $l_1 \leftarrow RandomLayer(s_1)$
2: Insert s_1 into $hnsw$ at l_1
3: **for** s in $\{S\} - s_1$ **do**
4: $l_s \leftarrow RandomLayer(s)$
5: **for** l in $\{L, \ldots, l_s + 1\}$ **do**
6: $en \leftarrow GetNearest(s, en)$
7: **end for**
8: $W \leftarrow GetNeighbors(s, M, en)$
9: $AddConnections(W, s)$
10: **for** l in $\{l_s - 1, \ldots, 0\}$ **do**
11: $W \leftarrow GetNeighbors(s, M, W)$
12: $AddConnections(W, s)$
13: **end for**
14: **end for**

***Query*($T_i, \widetilde{E}$) $\rightarrow$ D_i:** This algorithm is executed by the cloud server. When this algorithm is executed for the first time, the cloud server will construct the hierarchical graph-based index structure for $\widetilde{E}$, and then insert T_i into the index structure for searching. In the search process, the cosine similarity $score_i$ of T_i and the vector node $\widetilde{s_i}(\widetilde{s_i} \in \widetilde{E_i})$ in the index structure will be calculated as Eq. 1. Then, according to the score from high to low, select top-k sentences with the most similar semantics to query Q_i, and return the corresponding encrypted documents according to the last dimension of their vectors which store the document ID.

$$
\begin{aligned}
score_i &= cosine(\widetilde{s_i} \times T_i) \\
&= cosine((s_i \times M) \times (M^{-1} \times E_{Q_i}^T \cdot r)) \\
&= cosine(s_i \times E_{Q_i}^T \cdot r) \\
&= \frac{s_i \times E_{Q_i}^T \cdot r}{\| s_i \| \cdot \| E_{Q_i}^T \cdot r \|} \\
&= \frac{s_i \times E_{Q_i}^T}{\| s_i \| \cdot \| E_{Q_i}^T \|}
\end{aligned} \tag{1}
$$

In our proposed scheme, the HNSW algorithm [20] is divided into two main processes: index construction and search. Algorithm 1 shows the index construction algorithm. Once the hierarchical graph-based index is constructed, the search process is as illustrated in Fig. 5 (assuming there are 3 levels of indexing):

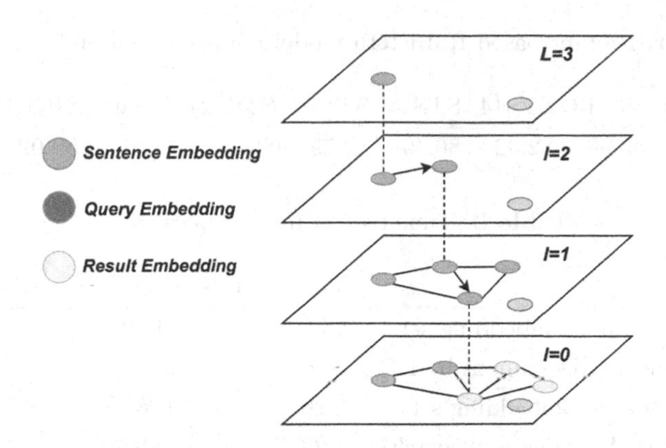

Fig. 5. The retrieval process of HNSW. Layer $l = 0$ contains all the nodes.*RandomLayer*() is used to map nodes randomly to different layers and make the number of nodes decrease with the increase of layer.

The red nodes represent the query vector q, the blue nodes represent the nodes already established in the graph, and the yellow nodes represent the search results. First, in level L, find the nearest node to q, denoted as en, and use it as the input for the next level. In level $l = 2$, starting from en, find the nearest neighbor to q among the neighbors of en, and designate it as the new en and input for the next level. In level $l = 1$, starting from en, find the nearest neighbor to q among the neighbors of en, and designate it as the new en and input for the next level. Finally, in the bottom level, starting from en, search for the top-K nearest nodes to q, which will yield the final results.

Analysis: In our CRE scheme, the cloud server knows only the encrypted document set and the encrypted embedding set stored on it. We encrypt the original document set, and only DO has the secret key. To prevent the cloud server from inferring the association between a specific query and corresponding encrypted documents based on the query history of DU, we added the random number mechanism, so that even if two queries are identical, their query trapdoors are different, the final search results are the same, which can effectively prevent the cloud server from carrying out access mode attacks.

5 Experiment and Performance Analysis

Our experimental platform is Ubuntu 18.04 server with RTX 3050 GPU and xeon 6230 2.1GHz CPU. The dataset in our experiment is *MS MARCO*. In this section, we will give a detailed description of our experimental results.

First, We used the SNLI dataset (contains 314K records) and MNLI dataset (contains 432K records) of the GLUE dataset to train our Prompt-based RoBERTa model. The format of each record is sent1, sent2, sent3. Sent1 and

Table 1. Prompt-based RoBERTa model performs well on STS tasks.

	STS12	STS13	STS14	STS15	STS16	STSBenchmark	SICKRelatedness
score	76.69	85.96	82.32	86.49	82.79	86.03	80.02

Table 2. Time cost of main processes.

	Time cost /s	Proportion /%
Generate Embedding set	66.72	94.89
Encrypt Document by DES	2.06	2.93
Encrypt Embedding set	0.20	0.28
HNSW (index and search)	1.33	1.89

sent2 are positive pairs, and sent3 is a negative example sentence. We use RoBERTa-base model, and set $Batch = 5$, $Epoch = 3$, $Learning\ rate = 5 \times 10^{-5}$.

Complete the training, we choose STS dataset (contains STS12, STS13, STS14, STS15, STS16, STSBenchmark, SICKRelatedness) [3] to evaluate our trained model. As shown in Table 1, The trained model exhibits excellent performance on the STS (Semantic Textual Similarity) task. We use Spearman's rank coefficient of correlation as the evaluation index.

After training the model, we apply it in ciphertext retrieval and test its actual performance. We compared with $SSSW_2$ [19] and $SSRB_2$ [11], which also used the pre-trained language model(respectively Word2vec and BERT) for ciphertext retrieval. Among them, the $SSRB_2$ scheme adopts the STS dataset, whose average length of query and document is close and both are short, which we think is unreasonable. Therefore, we chose a more general $MS\ MARCO$ real dataset for testing, which contains 10,569 records, and the format of each record is Query, Document. Query and Document with an average length of 5.96 and 55.98 respectively. [27] The following are our experiment results:

We set the parameters of the HNSW algorithm as $k = 2$ (number of the nearest elements), $ef = 50$ (control the recall), $space = l2$ (algorithm for calculating distance), $M = 16$ (number of connections for each element), $ef\_construction = 200$ (build speed tradeoff). Using 2000 documents as an example, as shown in Table 2, we tested the CRE scheme's time consumption and its proportion in major processes. We use the DES algorithm to simulate the process of document encryption. It can be seen that most of the time in CRE is spent on generating the corresponding Embedding set.

As shown in Fig. 6(a), we test the change of time required for CRE, $SSSW_2$ and $SSRB_2$ to generate and encrypt the index (embedding set) under different numbers of documents. As we can see, CRE takes much less time to generate the embedding set even though the average document length is longer than that of $SSSW_2$ and $SSRB_2$ schemes. In addition, unlike the other two solutions, CRE

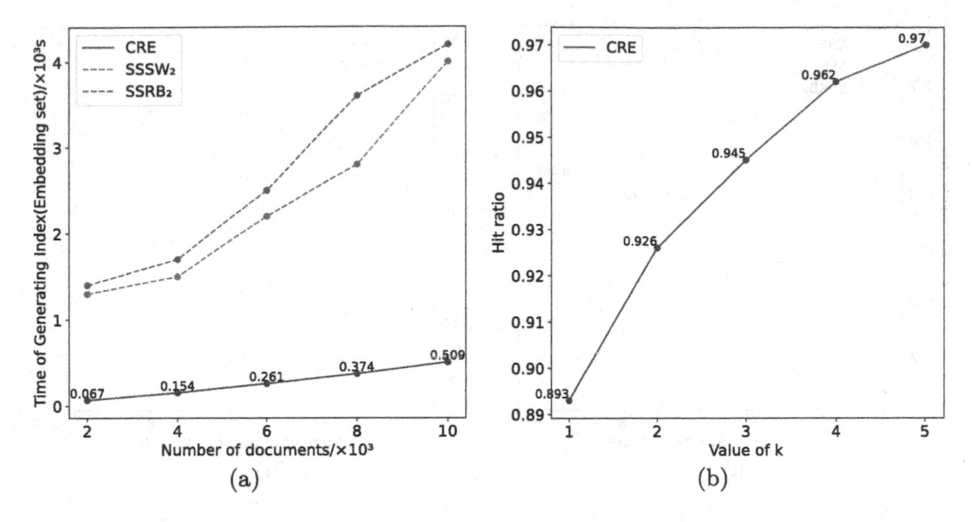

Fig. 6. (a) Time cost of getting index (embedding set). Among them, CRE consumes the shortest time. (b) The hit ratio varies with the value k. When retrieving the top-5 related documents, the hit ratio was as high as 97%.

eliminates the process to extract keywords, which is one of the reasons for our shorter time.

If DO does not modify the document, only the initial embedding set needs to be generated, then the cloud server uses the same encrypted embedding set to match different trapdoors. If DO wants to modify document D_i, it needs to regenerate the embedding set E_i corresponding to D_i and insert it into the index structure.

In HNSW, we need to set the k value to retrieve the top-k sentence embeddings that are most similar to query embedding. We use the hit ratio as the measurement standard of search accuracy. We set the number of queries that successfully retrieve the corresponding document as Q_{hit}, and the number of queries that fail to match the corresponding document as Q_{miss}. The calculation method of hit ratio is shown as Eq. 2.

$$Hit\ ratio = \frac{Q_{hit}}{Q_{hit} + Q_{miss}} \tag{2}$$

From Fig. 6(b), we can see the change in hit ratio with the increase of k. As k increases, the hit ratio increases. With k set to 5, the hit ratio can reach 97%. Neither of the other two schemes gave experimental results on retrieval accuracy or recall rates.

As shown in Fig. 7(a), we tested how the retrieval time of CRE, $SSSW_2$, and $SSRB_2$ changed as the number of documents increased. Parameter k in CRE is set to the default value 1. As the number of documents increases, the number of elements in the Embedding set will also increase dramatically, and the index structure constructed by the HNSW algorithm will also become larger, so the

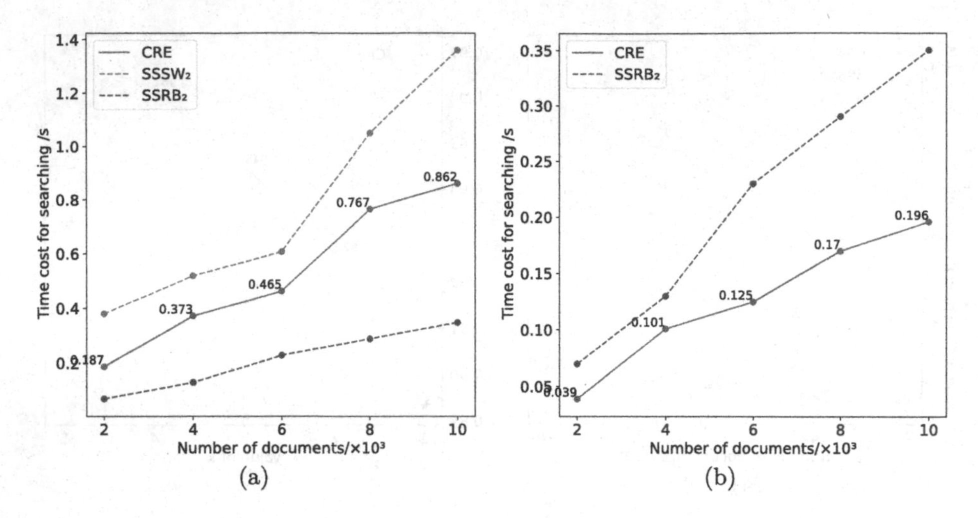

Fig. 7. (a) Time cost for searching. The average document length of the dataset used by CRE is much longer than that of the STS dataset used by $SSRB_2$. (b) Time cost for searching (use STS dataset). When CRE and $SSRB_2$ use the same dataset, the search speed of CRE is faster than that of $SSRB_2$, and the search speed advantage of CRE becomes more obvious as the number of documents increases.

retrieval time will increase. Although the retrieval time of the $SSRB_2$ scheme is lower, this is mainly because the length of documents used in their scheme is much shorter than that of CRE, so we produce much more high-dimensional vectors.

In order to further compare the difference in search speed between the CRE scheme and $SSRB_2$ scheme, we also make the CRE scheme perform retrieval on the STS dataset. As can be seen from Fig. 7(b), when CRE and $SSRB_2$ use the same dataset, the retrieval time of CRE is shorter than that of $SSRB_2$, and the difference in the retrieval time between these two schemes will be larger as the number of documents increases. It also proves that the hierarchical graph-based approximate nearest neighbor search algorithm shows excellent performance when dealing with massive high-dimensional vectors.

6 Conclusion

In this paper, we propose a ciphertext retrieval scheme based on encoder (CRE) for efficiently retrieving outsourced encrypted data. Our approach leverages the RoBERTa model with the prompt-based method to accurately capture the semantic information of both documents and queries. Furthermore, we employ the HNSW algorithm, a hierarchical graph-based approximate nearest neighbor search algorithm, to measure cosine similarity among massive high-dimensional embedding vectors, thereby facilitating accurate and efficient retrieval. Experimental results demonstrate significant improvements in retrieval accuracy and efficiency compared to existing ciphertext retrieval schemes. In future work, we can consider adding more security measures to the CRE solution, such as adding Ethereum smart contract to enable the cloud server to honestly execute search procedures.

References

1. Ahsan, M.M., Chowdhury, F.Z., Sabilah, M., Wahab, A.W.B.A., Idris, M.Y.I.B.: An efficient fuzzy keyword matching technique for searching through encrypted cloud data. In: 2017 International Conference on Research and Innovation in Information Systems (ICRIIS), pp. 1–5. IEEE (2017)
2. Cao, N., Wang, C., Li, M., Ren, K., Lou, W.: Privacy-preserving multi-keyword ranked search over encrypted cloud data. IEEE Trans. Parallel Distrib. Syst. **25**(1), 222–233 (2013)
3. Cer, D., Diab, M., Agirre, E., Lopez-Gazpio, I., Specia, L.: Semantic textual similarity-multilingual and cross-lingual focused evaluation. In: Proceedings of the 2017 SEMVAL International Workshop on Semantic Evaluation (2017). https://doi.org/10.18653/v1/s17-2001
4. Chai, Q., Gong, G.: Verifiable symmetric searchable encryption for semi-honest-but-curious cloud servers. In: 2012 IEEE International Conference on Communications (ICC), pp. 917–922. IEEE (2012)
5. Chen, Z., Wu, A., Li, Y., Xing, Q., Geng, S.: Blockchain-enabled public key encryption with multi-keyword search in cloud computing. Secur. Commun. Netw. **2021**, 1–11 (2021)
6. Cui, J., Sun, Y., Yan, X.U., Tian, M., Zhong, H.: Forward and backward secure searchable encryption with multi-keyword search and result verification. Sci. China Inf. Sci. **65**(5), 159102 (2022)
7. Cui, S., Song, X., Asghar, M.R., Galbraith, S.D., Russello, G.: Privacy-preserving dynamic symmetric searchable encryption with controllable leakage. ACM Trans. Privacy Secur. (TOPS) **24**(3), 1–35 (2021)
8. Dai, H., Dai, X., Yi, X., Yang, G., Huang, H.: Semantic-aware multi-keyword ranked search scheme over encrypted cloud data. J. Netw. Comput. Appl. **147**, 102442 (2019)
9. Dauterman, E., Feng, E., Luo, E., Popa, R.A., Stoica, I.: Dory: an encrypted search system with distributed trust. In: Operating Systems Design and Implementation (2020)
10. Fu, Z., Ren, K., Shu, J., Sun, X., Huang, F.: Enabling personalized search over encrypted outsourced data with efficiency improvement. IEEE Trans. Parallel Distrib. Syst. **27**(9), 2546–2559 (2015)
11. Fu, Z., Wang, Y., Sun, X., Zhang, X.: Semantic and secure search over encrypted outsourcing cloud based on BERT. Front. Comput. Sci. **16**, 1–8 (2022)
12. Fu, Z., Wu, X., Guan, C., Sun, X., Ren, K.: Toward efficient multi-keyword fuzzy search over encrypted outsourced data with accuracy improvement. IEEE Trans. Inf. Forensics Secur. **11**(12), 2706–2716 (2016)
13. Fu, Z., Wu, X., Wang, Q., Ren, K.: Enabling central keyword-based semantic extension search over encrypted outsourced data. IEEE Trans. Inf. Forensics Secur. **12**(12), 2986–2997 (2017)
14. Hu, Z., Dai, H., Yang, G., Yi, X., Sheng, W.: Semantic-based multi-keyword ranked search schemes over encrypted cloud data. Secur. Commun. Netw. **2022** (2022)

15. Hua, Y., Zhang, D., Ge, S.: Research progress in the interpretability of deep learning models. J. Cyber Secur. **5**(3), 1–12 (2020)
16. Jiang, T., et al.: Promptbert: improving BERT sentence embeddings with prompts. arXiv preprint arXiv:2201.04337 (2022)
17. Lee, J., Toutanova, K.: Pre-training of deep bidirectional transformers for language understanding. arXiv preprint arXiv:1810.04805 (2018)
18. Lewis, M., et al.: Bart: denoising sequence-to-sequence pre-training for natural language generation, translation, and comprehension. arXiv preprint arXiv:1910.13461 (2019)
19. Liu, Y., Fu, Z.: Secure search service based on word2vec in the public cloud. Int. J. Comput. Sci. Eng. **18**(3), 305–313 (2019)
20. Malkov, Y.A., Yashunin, D.A.: Efficient and robust approximate nearest neighbor search using hierarchical navigable small world graphs. IEEE Trans. Pattern Anal. Mach. Intell. **42**(4), 824–836 (2018)
21. Moataz, T., Shikfa, A., Cuppens-Boulahia, N., Cuppens, F.: Semantic search over encrypted data. In: ICT 2013, pp. 1–5. IEEE (2013)
22. Nair, M.S., Rajasree, M.S., Thampi, S.M.: Fine-grained, multi-key search control in multi-user searchable encryption. In: International Workshops on Security, Privacy, and Anonymity in Computation, Communication, and Storage; IEEE International Symposium on Ubisafe Computing; IEEE International Workshop on Security in e-Science and e-Research; International Workshop on Trust, SE (2017)
23. Ravichandiran, S.: Getting Started with Google BERT: Build and train state-of-the-art natural language processing models using BERT. Packt Publishing Ltd (2021)
24. Xu, M., Namavari, A., Cash, D., Ristenpart, T.: Searching encrypted data with size-locked indexes. In: USENIX Security (2021)
25. Song, D.X., Wagner, D., Perrig, A.: Practical techniques for searches on encrypted data. In: Proceeding 2000 IEEE Symposium on Security and Privacy. S&P 2000, pp. 44–55. IEEE (2000)
26. Su, J., Zhang, L., Mu, Y.: BA-RMKABSE: blockchain-aided ranked multi-keyword attribute-based searchable encryption with hiding policy for smart health system, vol. 132, pp. 299–309 (2022)
27. Thakur, N., Reimers, N., Rücklé, A., Srivastava, A., Gurevych, I.: BEIR: a heterogenous benchmark for zero-shot evaluation of information retrieval models. arXiv preprint arXiv:2104.08663 (2021)
28. Veretennikov, A.B.: Relevance ranking for proximity full-text search based on additional indexes with multi-component keys (2021)
29. Xia, Z., Zhu, Y., Sun, X., Chen, L.: Secure semantic expansion based search over encrypted cloud data supporting similarity ranking. J. Cloud Comput. **3**, 1–11 (2014)
30. Yu-Jie, X., Lan-Xiang, C., Yi, M.: BTM topic model based searchable symmetric encryption. J. Cryptol. Res. **9**(1), 88–105 (2022)

Sentiment Analysis Based on Pretrained Language Models: Recent Progress

Binxia Yang[1,2], Xudong Luo[1,2(✉)], Kaili Sun[1,2], and Michael Y. Luo[3]

[1] Guangxi Key Lab of Multi-Source Information Mining and Security, Guangxi Normal University, Guilin, China

[2] School of Computer Science and Engineering, Guangxi Normal University, Guilin, China

luoxd@mailbox.gxnu.edu.cn

[3] Emmanuel College, University of Cambridge, Cambridge, UK

Abstract. Pre-trained Language Models (PLMs) can be applied to downstream tasks with only fine-tuning, eliminating the need to train the model from scratch. In particular, PLMs have been utilised for Sentiment Analysis (SA), a process that detects, analyses, and extracts the polarity of text sentiments. To help researchers comprehensively understand the existing research on PLM-based SA, identify gaps, establish context, acknowledge previous work, and learn from methodologies, we present a literature review on the topic in this paper. Specifically, we brief the motivation of each method, offer a concise overview of these methods, compare their pros, cons, and performance, and identify the challenges for future research.

Keywords: Pre-trained language model · Sentiment Analysis · Cross language

1 Introduction

Sentiment Analysis (SA), also known as opinion mining or emotion Artificial Intelligence (AI), is a Natural Language Processing (NLP) technique used to determine the sentiment, emotion, or opinion expressed in a given piece of text, like happiness, sadness, anger, or fear. Its goal is to identify the polarity of the text, which can be positive, negative, or neutral. It has been applied to various text data types, such as reviews, social media posts, comments, and news articles. It helps businesses and researchers understand people's opinions and emotions towards products, services, events, or topics, enabling better decision-making, customer service, and targeted marketing.

There are many methods for SA, one of which is based on Pretrained Language Models (PLMs). PLMs are very useful for SA due to their rich understanding of language, transfer learning capabilities, and ability to handle complex language constructs. Their use in SA can help achieve higher accuracy, reduced development time, and scalability across various languages and domains.

A survey on SA based on PLMs is crucial to understand the current state of the art, identify the pros and cons of existing PLM-based SA approaches,

B. Luo et al. (Eds.): ICONIP 2023, CCIS 1966, pp. 131–148, 2024.
https://doi.org/10.1007/978-981-99-8148-9_11

Table 1. Performance comparison of models based on BERT

Model	Accuracy	Precision	F1	Recall	Dataset
BERT+DEMMT [21]	N/A	N/A	0.5713	N/A	DocRED [55]
DR-BERT [57]	0.7724	N/A	0.7610	N/A	Twitter [18]
DR-BERT [57]	0.9750	0.9780	0.9780	0.9510	TAC 2017 [16]
Pairs-BERT-NLI-M [25]	0.91	N/A	0.90	N/A	Pars-ABSA[a]
LDA-BERT [50]	N/A	N/A	0.81	N/A	SemEval2014 (Restaurant) [43]
LDA-BERT [50]	N/A	N/A	0.74	N/A	SemEval2015 (Restaurants) [42]
LDA-BERT [50]	N/A	N/A	0.75	N/A	SemEval2016 (Restaurants) [41]
BERT+GAT [61]	0.7848	N/A	0.7461	N/A	SemEval2014 (task4-Laptop) [43]
BERT+GAT [61]	0.8488	N/A	0.7819	N/A	SemEval2014 (Restaurant) [43]
DMF-GAT-BERT [59]	0.8038	N/A	N/A	N/A	SemEval2014 (task4-Laptop) [43]
DMF-GAT-BERT [59]	0.8610	N/A	N/A	N/A	SemEval2014 (task4-Restaurant) [43]
DMF-GAT-BERT [59]	0.7622	N/A	N/A	N/A	Twitter [18]
DMF-GAT-BERT [59]	0.8386	N/A	N/A	N/A	MAMS [27]
TAS-BERT [51]	N/A	N/A	0.6611	N/A	SemEval-2015 [42]
TAS-BERT [51]	N/A	N/A	0.7568	N/A	SemEval-2016 [41]
PBERT-MTL [30]	N/A	N/A	0.6892	N/A	SemEval-2015 [42]
PBERT-MTL [30]	N/A	N/A	0.7624	N/A	SemEval-2016 [41]
BERT+BiGRU [3]	N/A	N/A	0.68	N/A	Drug reviews [20]

[a] https://github.com/Titowak/Pars-ABSA.

reveal trends and research gaps, compare methodologies, guide practitioners, promote interdisciplinary research, and help establish benchmark datasets and evaluation metrics for PLM-based SA. Furthermore, it contributes to developing more effective and efficient SA systems and supports the growth and progress of the field. Thus, in this paper, we will survey the recent progress on this topic.

Although there are some surveys on SA and PLMs, our survey is different from theirs. In 2021, Luo *et al.* [37] surveyed Cross-Lingual SA (CLSA) methods. However, theirs focus on PLM-based CLSA but little about monolingual SA with PLMs like ours; and more importantly, this survey includes studies after 2021, while they did not. In 2022, Wang *et al.* [53] reviewed PLMs and their applications, but they mentioned little about PLMs' use in SA. In the same year, Sun, Luo, and Luo [47] also provide a similar survey, but only one subsection briefs PLM-based SA. Moreover, all the studies mentioned therein were conducted before 2022. In contrast, this paper surveys many studies carried out after 2022. In 2023, Cui *et al.* [15] survey SA, but it focuses on the evolution of research methods and topics in last two decades, little about PLM-based SA.

The rest of this paper is organised as follows. Section 2 reviews the application of PLMs in SA. Section 3 summarises SA based on variants of BERT. Section 4 explores SA based on other PLMs. Section 5 discusses the challenges and future research directions for SA with PLMs. Finally, Sect. 6 concludes the paper.

2 Models Based on BERT

This section briefly describes the applications of PLM BERT in SA. Tables 1 and 2 compare their performance, pros, and cons. Figure 1 compares their citations.

Table 2. Pro and con comparison of models based on BERT

Reference	Pro	Con
Han *et al.* (2020) [21]	obtain every reference and type of entity accurately, predicate relationships between entities	its effectiveness depending on the quality of BERT, labeled data required, complex modelling techniques
Zhang *et al.* (2022) [57]	address the limitations of the vanilla pre-trained model in aspect-level SA task, explainable	increase computational complexity, much lower calculation times than BERT-based frameworks, limited generalisability
Jafarian *et al.* (2021) [25]	focus on the Persian language, combine BERT and sentence to improve the accuracy of aspect-level SA, new dataset	fine-tuning requires computational resources, model is compute-intensive limited generalisability
Venugopalan *et al.* (2022) [50]	can distinguish between overlapping themes representing each aspect category, contribute an almost unsupervised model for aspect term extraction	increase computational and time complexity, unlabelled training data not large enough, A certain percentage of aspectual terms are lost in the semantic filter
Zou *et al.* (2022) [61]	integrate graph attention with BERT, capture syntactic information more effectively, no external tools needed to analyse structural features in sentences	limited evaluation, cannot analysis the dependency of review contexts with disordered structures well
Zhou *et al.* (2022) [59]	enable to adaptively adjust the fusion weight of each channel	limited evaluation, resource-intensive fine-tuning, not sensitive to specific labels
Wan *et al.* (2020) [51]	incorporation of joint detection, better understanding of the relationships, can capture the dual dependence of sentiments on both targets and aspects	increase computational and time complexity, limited evaluation, dependency on high-quality PLMs, data construction needs further optimisation
Ke *et al.* (2022) [30]	incorporation of prior knowledge and multi-task learning, alleviate the grossly unbalanced distribution of labels	complexity, limited evaluation, relies on the quality of the pre-trained models
Bensalah *et al.* (2023) [3]	Fewer computing resources and faster speeds, simplicity	limited evaluation, relies on the quality of BERT, lack of interpretability

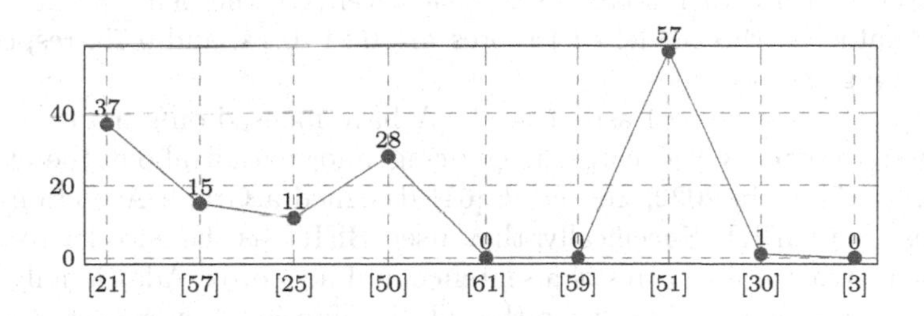

Fig. 1. Citation of the models based on BERT

2.1 Document-Level SA

Document-level SA takes an entire document as an object to determine whether the document expresses positive or negative opinions [40].

For document-level SA on a long text, extracting the relationships between multiple entities in the text is necessary. To address this issue, in 2020, Han *et al.* [21] proposed a one-pass model based on BERT [17] to better predict the relationships between entities by processing the text once. On the DocRED dataset, their model achieved a 6% improvement in F1-score compared to a state-of-the-art model without using any PLM.

2.2 Aspect-Level SA

Aspect-level SA is a fine-grained classification task [41]. Its goal is to identify the aspects involved in a sentence, along with their corresponding sentiment polarity and opinion items. Therefore, it can provide more in-depth information than document-level SA and sentence-level SA.

Although BERT has achieved good results in capturing semantic features, it still faces challenges in handling dynamic semantic changes in aspect-level SA. Thus, in 2022, Zhang *et al.* [57] proposed a method for learning dynamic aspect-oriented semantics for aspect-level SA by incorporating dynamic semantics into Dynamically Reweighted BERT (DR-BERT). They first use the Stack-BERT layer to grasp the overall semantics of a sentence and then integrate it with a lightweight Dynamic Reweight Adapter (DRA) to focus on smaller regions of the sentence to understand perceived emotions better. Their experiments on three benchmark datasets demonstrate their model's high effectiveness.

Although aspect-level SA attracts considerable attention from researchers, there are limited work in Persian. Thus, in 2021, Jafarian *et al.* [25] proposed a BERT-based method for aspect-level SA on texts in Persian. On the Pars-ABSA dataset, using Persian PLM Pars-BERT combined with a natural language inference auxiliary sentence can achieve an aspect-level SA accuracy of up to 91%, meaning a 5.5% improvement over a state-of-the-art baseline.

In aspect-level SA, supervised models have been moderately successful in aspect term extraction, but data annotation is expensive. Thus, in 2022, Venugopalan *et al.* [50] proposed a model for aspect term extraction using an unsupervised method with a BERT-based semantic filter. On the SemEval 2014, 2015, and 2016 datasets, this model's F1 scores are 0.81, 0.74, and 0.75, respectively, outperforming the baselines.

Most previous studies of aspect-level SA have focused on semantic information between contexts and corpora, ignoring information about the structure of sentences. Thus, in 2022, Zou *et al.* [61] designed a Graph Attention (GAT) model based on BERT. Specifically, they used BERT as the encoder to capture the internal structural features of a sentence and its words. Additionally, BERT adopted an attention mechanism to the relationship between aspect items, sentiment polarity words, and context. The output of the attention module is then input to the GAT and classified by the fully connected layer of the GAT and the softmax function. As a result, their model outperforms most baselines on the laptop and restaurant datasets of SemEval 2014.

In 2022, Zhou *et al.* [59] proposed a dynamic multichannel fusion mechanism based on GAT and BERT for aspect-level SA DMF-GAT-BERT. It uses BERT and GAT to capture the text's semantic and syntactic features. It also employs the syntax, syntactic, and syntax-syntax channels to enhance its ability to capture syntactic structure features. Additionally, it has a two-layer dynamic fusion mechanism that can adaptively adjust the fusion weights of the three-channel information when dealing with different datasets to improve the accuracy of the output features. Finally, they use an attentive layer ensemble (ALE) to integrate the syntactic features of the context captured by GAT in different layers to

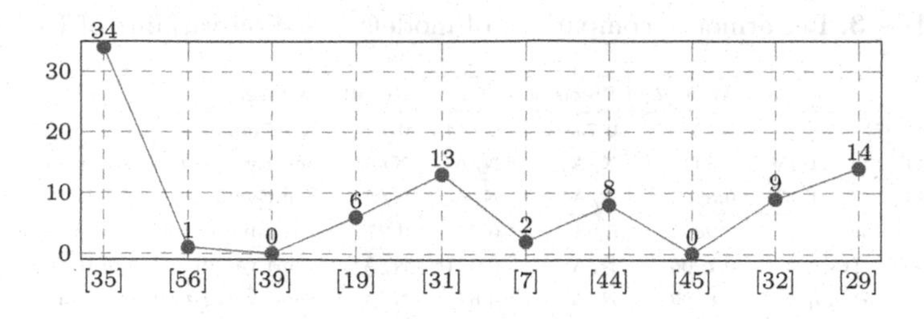

Fig. 2. The citation of the models based on variants of BERT

improve the comprehensiveness of the output features. Their model outperforms other models embedded with BERT on four different datasets.

In 2020, Wan *et al.* [51] proposed a method for aspect-level SA. They used BERT to determine each target in a text and reduce it to a binary text classification problem. The method can also handle implicit targets.

In 2022, Ke *et al.* [30] proposed an aspect-level SA model called Prior-BERT and Multi-Task Learning (PBERT-MTL) for detecting (target, aspect, sentiment) triples from a sentence. They first fine-tuned BERT using the dataset's prior distribution knowledge to address severe label distribution imbalances. Then, they used a multi-layer Bidirectional Long Short-Term Memory (BiLSTM) to handle the long-distance dependency between object and aspect sentiment.

2.3 General SA

Analysing online user opinions and extracting information on drug efficacy and risk factors can improve pharmacovigilance systems in the pharmaceutical industry. To address these issues, in 2023, Bensalah et al. [3] proposed a model that combines static word embeddings with a BERT-based SA language model to optimise the effectiveness of SA for drug reviews. A Bi-directional Gated Recurrent Unit (BiGRU) layer is incorporated on top of the obtained features for further training to enhance the final context vector. Since the proposed SA model utilises BERT and static embedding models, it exhibits promising results on a small training dataset.

3 Models Based on Variants of BERT

This section will explore SA based on variants of BERT, including RoBERTa, mBERT, HAdaBERT, and indicBERT. Tables 3 and 4 compares their performance, pros, and cons. Figure 2 compares their citations.

3.1 Aspect-Level SA Based on RoBERTa

In 2021, Liao *et al.* [35] proposed a multi-task aspect SA model based on the RoBERTa [36], called RACSA. They use the PLM to extract features from text

Table 3. Performance comparison of models based on variants of BERT

Model	Accuracy	Precision	F1	Recall	Dataset
RACSA [35]	N/A	0.748	0.744	0.747	AI Challenger 2018[a]
RoBERTa+SGPT [56]	0.8445	N/A	N/A	N/A	SemEval2014 (Restaurant) [43]
RoBERTa+SGPT [56]	0.8903	N/A	N/A	N/A	SemEval2014 (task4-Laptop) [43]
RoBERTa [39]	0.98	0.98	0.97	0.97	e-commerce order information [39]
mBERT-FC [19]	0.7906	N/A	0.6716	N/A	SemEval2016 (Restaurant) [41]
mBERT-SCP [19]	0.8106	N/A	0.6778	N/A	SemEval2016 (Restaurants) [41]
mBERT-AEN [19]	0.8022	N/A	0.6670	N/A	SemEval2016 (Restaurants) [41]
mBERT [31]	0.8250	0.8135	0.8149	0.8165	UCSA corpus [31]
mBERT+MIX [7]	0.8600	N/A	0.9163	N/A	TripAdvisor reviews [7]
mBERT [44]	0.6617	0.6416	0.6366	0.6599	Persian-English [44]
HAdaBERT [32]	0.675	N/A	N/A	N/A	Yelp-2013 [44]
HAdaBERT [32]	0.751	N/A	N/A	N/A	AAPD [9]
HAdaBERT [32]	0.909	N/A	N/A	N/A	Reuters [9]
IndicBERT+AUX [29]	N/A	N/A	0.6250	N/A	Hostility detection dataset [4]

[a] https://download.csdn.net/download/linxid/11469830.

Table 4. Pro and con comparison of models based on variants of BERT

Reference	Pro	Con
Liao *et al.* (2021) [35]	enhanced context understanding, combining document focus improves the model's ability to fuse features across the tex	less sensitive to long texts without strong sequential patterns, not fully consider dependencies between aspect categories
Yong *et al.* (2023) [56]	capture the relationships between words and phrases, interpretable results	limited evaluation, resource-intensive fine-tuning, poor model generalisation
Mensouri *et al.* (2023) [39]	incorporation of topic modeling, better understanding of customer feedback, high generalisation ability	limited evaluation, lack of interpretability, limited amount of data
Essebbar *et al.* (2021) [19]	robust for the out-of-domain dataset, fill a scarcity of French aspect-level SA	increase computational and time complexity, resource-intensive fine-tuning, limited generalisability, insufficient training data
Khan *et al.* (2022) [31]	deal with complicated and resource-poor languages	dependency on BERT, lack of interpretability
Catelli *et al.* (2022) [7]	represent the Italian text more effectively, cross-lingual transfer learning, interpretable experimental results	small dataset size, multi-label scenarios not fully considered
Sabri *et al.* (2021) [44]	integrate learning with lexicon-based method, multilingual, real application	dependency on pre-existing resources, lack of interpretability
Sarangi *et al.* (2022) [45]	use contextualised fine-tuning of BERT, extract and find opinion tuples effectively	inaccurate boundary detection, misunderstand a sentence with multiple polar expressions, no distinction between themes
Kong *et al.* (2022) [32]	improved classification performance, adaptive fine-tuning, generalisability, efficient capture of local and global information about long documents	lack of detailed analysis, high computational complexity, may not be suitable for low-resource languages, fine-tuning challenges
Kamal *et al.* (2021) [29]	robust and consistent modelling without any ensemblling or complex pre-processing	limited scope, require high-quality labeled data, reliance on external knowledge

and aspect tokens and apply a cross-attention mechanism to guide the model to focus on the features most relevant to a given aspect category. They also integrate 1D-CNN [12] to improve the model's feature extraction ability for phrases. On the AI Challenger 2018 public dataset, their model outperformed other baselines.

Many studies using PLMs for SA have overlooked the significance of sentiment information in models. For this, in 2023, Yong *et al.* [56] based RoBERTa to developed a semantic graph-based pre-training model that leverages semantic graphs with sentiment knowledge for pre-training. Firstly, they explore similar aspects and sentiment words and construct a graph of similar semantic and aspect-sentiment pairs. Next, the aspect-sentiment pairs are utilised to construct the semantic graph. Then, the semantic graph is inputted into the pre-trained language model with sentiment masking. Lastly, the aspect-emotion pairs are jointly optimised for predicting the target and masked language models.

In 2023, to harness the marketing potential of online customer reviews, Mensouri *et al.* [39] used recurrent neural networks (LSTM, GRU, and Bi-LSTM) and RoBERTa for aspect-level SA on the reviews. They are trained to categorise the reviews as either positive or negative. The most effective model is chosen by evaluating each outcome. Ultimately, thematic modelling techniques are used to identify different themes within the data, establish what prompts customers to write reviews, and offer the supplier the appropriate course of action for each situation.

3.2 SA Based on mBERT

Multilingual BERT (mBERT) [17] has been used for aspect-level SA, low resource language SA (referring to the process of performing sentiment analysis on languages for which there is limited annotated data or resources available), and cross-language SA (CLSA) (referring to the process of analysing and determining the sentiment expressed in texts written in mix-codes of different languages). For aspect-level SA, in 2021, Essebbar *et al.* [19] presented a model for French texts. Although there are many SA models of French texts, but few focus on aspect-level SA. Their work aims to address this issue. They also tried to base their model on another two French PLMs: CamemBERT and FlauBERT. They use three methods to fine-tune the PLMs: fully-connected, sentence pair classification, and attention encoder network. On the SemEval 2016 French review dataset, their model outperforms the state-of-the-art baselines.

For low-resource language SA, in 2022, Khan *et al.* [31] used mBERT SA on review texts in Urdu. During training on mBERT, categorical cross-entropy is used as the loss function. Their experiments show that the model using mBERT outperforms all other baseline models. In the same year, Catelli *et al.* [7] presented a study on SA for low-resource languages based on mBERT, by transferring sentiment recognition capabilities from multilingual language models to low-resource languages. They use an Italian dataset explicitly built for this purpose. Their experiments show that mBERT, fine-tuned on a mixed English/Italian dataset, performs much better than the Italian-specific model.

For CLSA, in 2021, Sabri *et al.* [44] used mBERT for SA of Persian-English code-mixed texts. Their experiments show that for CLSA, mBERT outperforms the Bayesian method and the random forest model. In 2022, Sarangi *et al.* [45] integrated mBERT, Stanza vector, and ELMo to build a model for Structured SA (SSA) on cross-language texts called CBERT. First, CBERT uses mBERT as a multilingual encoder to capture global information. Then, it uses the Stanza pre-trained syntax vector and PLM ELMo to get contextualized embeddings and use ELMo's embedding representation are input vectors. Finally, CBERT concatenates contextual information with syntactic information. On SemEval-2022's seven datasets in five languages, CBERT outperforms BiLSTM and VBERT.

3.3 Document-Level SA Based on HAdaBERT

In 2022, Kong *et al.* [32] proposed a hierarchical BERT model, HAdaBERT, with an adaptive fine-tuning strategy for SA. First, its local encoder obtains sentence-level features, and then its global encoder combines these features. During fine-tuning, it takes a portion of the document as a region and divides it into containers. Then, the local encoder adaptively decides which layers of BERT need to be fine-tuned. HAdaBERT for SA outperforms various neural networks and PLMs DocBERT [1], RoBERTa, and ALBERT [33].

3.4 CLSA Based on IndicBERT

The increasing online hostility, particularly during emergencies, can have severe consequences. To address the issue, in 2021, Kamal *et al.* [29] used IndicBERT [28] to identify hatred in Hindi texts. They fine-tuned the PLM on a specific Hindi dataset with adversarial and non-adversarial detection models as auxiliary models. In the CONSTRAINT-2021 Shared Task, they achieved the third place.

Table 5. Performance comparison of models based on other PLMs

Model	Accuracy	Precision	F1	Recall	Dataset
ontology-XLNet [48]	0.9690	0.9420	0.9450	0.9410	Twitter [46]
ontology-XLNet [48]	0.975	0.978	0.978	0.951	WebMD [22]
ontology-XLNet [48]	0.97	0.97	0.96	0.932	n2c2 2018 [23]
XLM-R [24]	N/A	0.931	0.931	0.931	English [9]
XLM-R [24]	N/A	0.610	0.602	0.609	Tamil [9]
XLM-R [24]	N/A	0.859	0.854	0.852	Malayalam [9]
XLM-T [2]	N/A	N/A	0.6935	N/A	SA Twitter dataset [2]
DLAN-XLM-R [52]	0.9223	N/A	N/A	N/A	Websis CLS-10 [60]
SentiX [58]	0.9268	N/A	N/A	N/A	cross-domain sentiment dataset [5]
mBERT+XLM-R [26]	0.7400	N/A	0.7030	N/A	Kannada [22]
mBERT+XLM-R [26]	0.7967	N/A	0.7694	N/A	Tamil [10]
mBERT+XLM-R [26]	0.9700	N/A	0.9676	N/A	Malayalam [8]

Table 6. Pro and con comparison of models based on other PLMs

Ref	Pro	Con
Sweidan *et al.* (2021) [48]	produce a more comprehensive context and enhance the quality of feature extraction, better results with limited data	limited by cost and memory resources when dealing with long sequences, cannot handle multilingual SA
Hossain *et al.* (2021) [24]	represent the mixed-language text more effectively, cross-lingual transfer learning	excessive misclassification, limited evaluation, lack of interpretability
Chen *et al.* (2022) [11]	alleviate the problem of insufficient or missing annotation data, enhance interaction information between parallel data	limited evaluation, dependency on pre-trained models, increase computational and time complexity
Barbieri *et al.* (2022) [2]	large-scale training, suite to handle social media and specifically multilingual SA	limited evaluation, limited explanation of methodology
Wang *et al.* (2022) [52]	can effectively align the representation of source and target languages while keeping the capacity of sentiment predictions	limited evaluation, higher computational cost and longer time, heavily influenced by the availability of language-specific data
Van Thin *et al.* (2023) [49]	capture contextual information more effectively across different languages	the performance of zero-shot learning is not as good as in a monolingual setting
Zhou *et al.* (2020) [58]	reduce overfitting in the source domain, labeled data, and training time	resource-intensive fine-tuning, limited generalisability, up to high-quality data
Leippold (2023) [34]	demonstrates the vulnerability of using dictionaries	potential to undermine the reliability and accuracy of sentiment classification
Jayanthi *et al.* (2021) [26]	identify multilingual offensive language, improve performance of task-adaptive pre-training	limited language scope, lack of ablation study, representation fusion degrades performance in some datasets

4 Models Based on Other PLMs

This section will review SA based other PLMs, including XLNet, XLM-R, XLM-Align, SentiX, GPT, and hybrid PLMs. Tables 5 and 6 compares their performance, pros, and cons. Figure 3 compares their citations.

4.1 Sentence-Level SA Based on XLNet

Since document-level SA takes the entire document as an object, it is challenging to identify the various emotions of different entities within a document. Consequently, sentence-level SA has been proposed to judge whether a sentence in a text reflects a positive, negative, or neutral opinion [38].

In 2021, Sweidan *et al.* [48] presented ontology-XLNet, a sentence-level SA model based on XLNet [54]. It uses a lexicalised ontology to extract indirect relationships between user data and enhances feature extraction by providing a more comprehensive context of connections through XLNet. On a drug reaction dataset, their model achieved an accuracy of 98% and an F1 score of 96.4%.

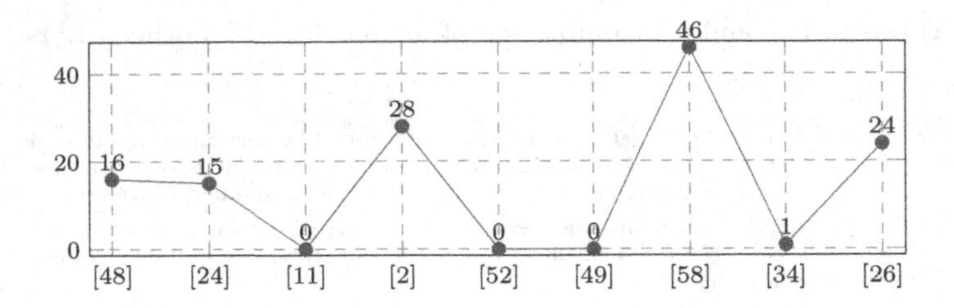

Fig. 3. Citation of the models based on other PLMs

4.2 CLSA Based on XLM-R

In 2021, Hossain, Sharif, and Hoque [24] used XLM-R [14] for CLSA on 28,451, 20,198, and 10,705-word reviews in English, Tamil, and Malayalam. The model outperforms machine learning and deep learning baselines regarding F1-score, precision, and recall.

In 2022, Chen *et al.* [11] proposed an XLM-R based CLSA model. It uses XLM-R as an encoder to handle the interaction information between different languages. Also, the model uses a sequence labelling to enhance structure prediction and uses BiLSTM and multi-layer perceptron for sentence-level sentiment classification. The model ranks the first on the cross-lingual subtask of SemEval-2022 task 10 and the second on the monolingual subtask. Also, in 2022, Barbieri *et al.* [2] presented a multilingual PLM called XLM-T, based on XLM-R and trained on nearly 200 million tweets. They also assembled a set of eight different language Twitter datasets to fine-tune XLM-T for SA.

In response to the lack of a low-resource, linguistically annotated corpus in the CLSA, in 2022, Wang *et al.* [52] proposed Distilled Linguistic Adversarial Network (DLAN) for CLSA. DLAN, based on XLM-R, employs adversarial learning and knowledge distillation to learn language-invariant sentence representations that contain sentiment prediction information for more accurate cross-linguistic transfer. DLAN can effectively adapt the representation of source and target languages while retaining the ability to predict sentiment.

4.3 Aspect-Level SA Based on XLM-Align

To investigate the performance of multilingual models in a zero-shot and joint training aspect-level CLSA task, in 2023, Van Thin *et al.* [49] employed migration learning techniques based on PLMs for zero-shot and joint cross-language training on the aspect-level SA task for five languages. They chose to fine-tune XLM-Align [13] on the source language and make predictions directly in the target language. For the joint training scenario, the models were trained on a combination of multiple source languages. The experimental results indicate that choosing languages other than English as the source language yields promising results in low-resource language scenarios. And the model's performance increases proportionally to the number of data samples in the target language.

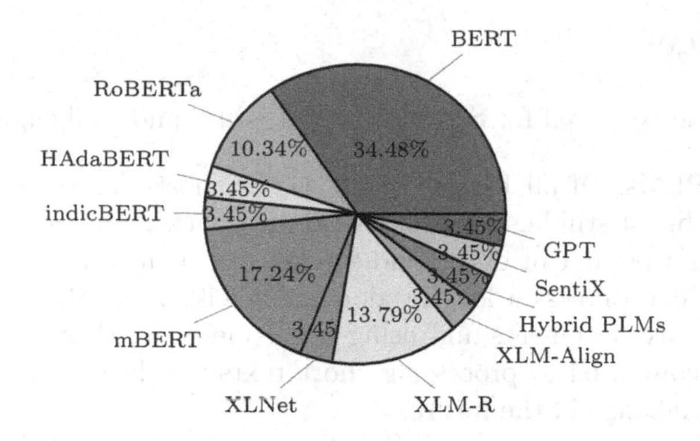

Fig. 4. Percentage of PLMs for SA

4.4 Cross-Domain SA Based on SentiX

In real-life, since users' emotional expressions vary significantly across different domains, fine-tuning a PLM for SA on a source domain may lead to overfitting, resulting in poor SA performance in the target domain. To address this issue, in 2020, Zhou *et al.* [58] pre-trained a sentiment-aware language model (SentiX) via domain-invariant sentiment knowledge from large-scale review datasets. They then used the PLM for cross-domain SA without fine-tuning. Their experiments show that although they trained SentiX using only 1% of samples, it outperformed BERT trained with 90% of samples.

4.5 Sentence-Level SA Based on GPT-3

The dictionary-based SA method extracts negative and affirmative words in each sentence and aggregates them to obtain a sentence-level score for the sentiment. The method remains widespread in financial SA and other financial and economic applications because it appears transparent and explainable. However, in 2023, Leippold [34] used the advantages of GPT-3 [6] as a pure neural decoder network to generate the text required for an adversarial attack. They then demonstrated the vulnerability of using dictionaries by exploiting the eloquence of GPT-3 to generate successful adversarial attacks on dictionary-based methods, achieving a success rate close to 99% for negative sentences in the financial phrase base.

4.6 CLSA Model Based on Hybrid PLMs

In 2021, Jayanthi *et al.* [26] combined the strengths of mBERT and XLM-R to propose a hybrid model for offensive language CLSA. They also introduced a fusion architecture using character-level, sub-word, and word-level embeddings to improve model performance. The model is trained in Malayalam, Tamil, and Kannada. The hybrid model ranked first and second in Malayalam and third in Kannada. The researchers also observed that representation fusion improved performance on some datasets but degraded it on others. However, further evaluations are necessary to determine the usefulness of this fusion approach.

5 Challenges

PLMs have done very well for SA, but several issues and challenges remain:

1. Improving PLMs: Of all PLMs, BERT is the most commonly used for SA (see Fig. 4), but it still has some practical drawbacks. Among them, the most significant is that it can only handle tokens of a maximum length of 512. Although it can process a lengthy document with more than 512 tokens by intercepting text fragments and using other methods, there is still a performance gap compared to processing short texts. Bridging this gap will be a significant challenge in the future.
2. Multilingual and cross-language SA: Current multilingual PLMs, such as mBERT and XLM-R, have been pre-trained in more than 100 languages, but there are about 7,000 languages worldwide. Annotating and training data for many low-resource language categories requires significant human resources, making it expensive and challenging to implement. As a result, existing cross-language PLMs still need to meet the demand. Furthermore, the presentation of different languages varies considerably, making it challenging for PLMs to handle SA in various languages.
3. Multimodal SA: Techniques integrating PLMs with visual, audio, or other modality-specific models are being investigated for SA on multimodal data. However, PLMs need help in processing information from multiple modalities simultaneously. Therefore, using PLMs for SA of individual body expressions in images or videos is complex and requires further research.
4. Domain adaptation: Adapting PLMs to specific domains (*e.g.*, finance, healthcare, and social media) to improve SA performance in these contexts is also an important direction for future research. This can be done by fine-tuning to a specific domain to fit the SA tasks in the domain. Also, the accuracy and effectiveness of SA in the domain can also be improved by introducing domain-specific corpora and building domain-specific sentiment lexicons.
5. Ambiguity understanding: PLMs may struggle to infer the correct sentiment when the text contains double negativity, sarcasm, irony, or innuendo. Additionally, most current research on SA focuses on explicit text classification problems, while detecting and classifying certain implicit words could improve accuracy. In the future, better sentiment classification could be achieved by building a lexicon of implicit sentiment words or using more advanced PLMs (such as GPT-4) to extract semantically relevant information at a deeper level.

Table 7. Summary of various PLMs for various SAs

	Monolingual					Multilingual	
	Document level SA	Sentence level SA	Aspect level SA	General SA	Cross Domain SA	CLSA	low resource language SA
BERT	✓	×	✓	✓	×	×	×
RoBERTa	×	×	✓	×	×	×	×
mBERT	×	×	✓	×	×	✓	✓
HAdaBERT	✓	×	×	×	×	×	×
indicBERT	×	×	×	×	×	✓	×
XLNet	×	✓	×	×	×	×	×
XLM-R	×	×	×	×	×	✓	×
XLM-Align	×	×	✓	×	×	×	×
SentiX	×	×	×	×	✓	×	×
GPT	×	✓	×	×	×	×	×
Hybrid PLMs	×	×	×	×	×	✓	×

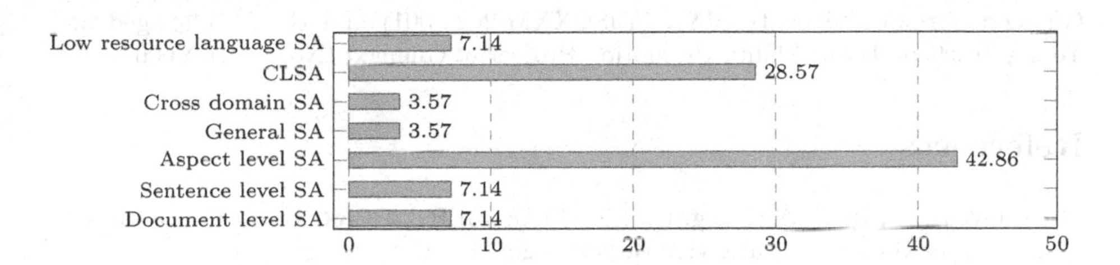

Fig. 5. The number percentage of papers on various topics

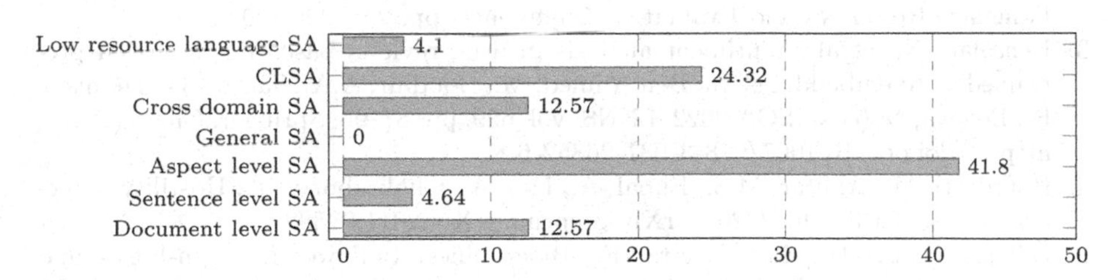

Fig. 6. The citation percentage of papers on various topics

6 Conclusion

SA has significantly progressed recently, largely thanks to the development of PLMs. These models have revolutionised the field by providing a foundation for efficient and effective SA across various languages and domains (see Table 7 for a summary of various PLMs for various SA tasks). Our literature review[1]

[1] See Figs. 5 and 6 for the number and citation percentages of the references on various SA topics.

shows that PLMs have several advantages over traditional SA methods, including their ability to learn from vast amounts of data and generalise to new tasks and domains. Overall, the progress made in SA based on PLMs is an exciting development for natural language processing, with potential applications across a wide range of industries, from marketing to healthcare to finance.

However, there are also challenges to using PLMs for SA, such as the need for large amounts of annotated data and the difficulty of handling multiple modalities simultaneously. Despite these challenges, the future of SA looks promising as new PLMs continue to be developed and researchers explore new ways to improve their performance. The field will continue to evolve rapidly in the coming years as SA is applied to an ever-wider range of applications and new techniques are developed to address the challenges of working with different languages, modalities, and domains.

Acknowledgements. This work was partially supported by a Research Fund of Guangxi Key Lab of Multi-source Information Mining Security (22-A-01-02) and a Graduate Student Innovation Project of School of Computer Science, Engineering, Guangxi Normal University (JXXYYJSCXXM-2021-001) and the Middle-aged and Young Teachers' Basic Ability Promotion Project of Guangxi (No. 2021KY0067).

References

1. Adhikari, A., Ram, A., Tang, R., Lin, J.: DocBERT: BERT for document classification. arXiv preprint arXiv:1904.08398 (2019)
2. Barbieri, F., Anke, L.E., Camacho-Collados, J.: XLM-T: multilingual language models in twitter for sentiment analysis and beyond. In: Proceedings of the 13th Language Resources and Evaluation Conference, pp. 258–266 (2022)
3. Bensalah, N., et al.: Sentiment analysis in drug reviews based on improved pre-trained word embeddings. In: Ben Ahmed, M., Boudhir, A.A., Santos, D., Dionisio, R., Benaya, N. (eds.) SCA 2022. LNNS, vol. 629, pp. 87–96. Springer, Cham (2023). https://doi.org/10.1007/978-3-031-26852-6_8
4. Bhardwaj, M., Akhtar, M.S., Ekbal, A., Das, A., Chakraborty, T.: Hostility detection dataset in Hindi (2020). arXiv preprint arXiv:2011.03588
5. Blitzer, J., Dredze, M., Pereira, F.: Biographies, Bollywood, boom-boxes and blenders: domain adaptation for sentiment classification. In: Proceedings of the 45th Annual Meeting of the Association of Computational Linguistics, pp. 440–447 (2007)
6. Brown, T., et al.: Language models are few-shot learners. In: Advances in Neural Information Processing Systems, vol. 33, pp. 1877–1901 (2020)
7. Catelli, R., et al.: Cross lingual transfer learning for sentiment analysis of Italian tripadvisor reviews. Expert Syst. Appl. **209**, 118246 (2022)
8. Chakravarthi, B.R., Jose, N., Suryawanshi, S., Sherly, E., McCrae, J.P.: A sentiment analysis dataset for code-mixed Malayalam-English. In: LREC 2020 Workshop Language Resources and Evaluation Conference, p. 177 (2020)
9. Chakravarthi, B.R., Muralidaran, V.: Findings of the shared task on hope speech detection for equality, diversity, and inclusion. In: Proceedings of the 1st Workshop on Language Technology for Equality, Diversity and Inclusion, pp. 61–72 (2021)

10. Chakravarthi, B.R., Muralidaran, V., Priyadharshini, R., McCrae, J.P.: Corpus creation for sentiment analysis in code-mixed Tamil-English text. In: LREC 2020 Workshop Language Resources and Evaluation Conference (2020)
11. Chen, C., Chen, J., Liu, C., Yang, F., Wan, G., Xia, J.: MT-speech at SemEval-2022 task 10: Incorporating data augmentation and auxiliary task with cross-lingual pretrained language model for structured sentiment analysis. In: Proceedings of the 16th International Workshop on Semantic Evaluation, pp. 1329–1335 (2022)
12. Chen, Y.: Convolutional neural network for sentence classification. Master's thesis, University of Waterloo (2015)
13. Chi, Z., et al.: Improving pretrained cross-lingual language models via self-labeled word alignment. In: Proceedings of the 59th Annual Meeting of the Association for Computational Linguistics and the 11th International Joint Conference on Natural Language Processing, vol. 1, pp. 3418–3430 (2021)
14. Conneau, A., et al.: Unsupervised cross-lingual representation learning at scale. In: Proceedings of the 58th Annual Meeting of the Association for Computational Linguistics, pp. 8440–8451 (2020)
15. Cui, J., Wang, Z., Ho, S.B., Cambria, E.: Survey on sentiment analysis: evolution of research methods and topics. Artif. Intell. Rev. 1–42 (2023)
16. Demner-Fushman, D., et al.: A dataset of 200 structured product labels annotated for adverse drug reactions. Sci. Data **5**(1), 1–8 (2018)
17. Devlin, J., Chang, M.W., Lee, K., Toutanova, K.: BERT: pre-training of deep bidirectional transformers for language understanding. In: Proceedings of the 17th Annual Conference of the North American Chapter of the Association for Computational Linguistics: Human Language Technologies, pp. 4171–4186 (2019)
18. Dong, L., Wei, F., Tan, C., Tang, D., Zhou, M., Xu, K.: Adaptive recursive neural network for target-dependent Twitter sentiment classification. In: Proceedings of the 52nd Annual Meeting of the Association for Computational Linguistics, vol. 2, pp. 49–54 (2014)
19. Essebbar, A., Kane, B., Guinaudeau, O., Chiesa, V., Quénel, I., Chau, S.: Aspect based sentiment analysis using French pre-trained models. In: Proceedings of the 13th International Conference on Agents and Artificial Intelligence, pp. 519–525 (2021)
20. Gräßer, F., Kallumadi, S., Malberg, H., Zaunseder, S.: Aspect-based sentiment analysis of drug reviews applying cross-domain and cross-data learning. In: Proceedings of the 2018 International Conference on Digital Health, pp. 121–125 (2018)
21. Han, X., Wang, L.: A novel document-level relation extraction method based on BERT and entity information. IEEE Access **8**, 96912–96919 (2020)
22. Hande, A., Priyadharshini, R., Chakravarthi, B.R.: KanCMD: Kannada codemixed dataset for sentiment analysis and offensive language detection. In: Proceedings of the 3rd Workshop on Computational Modeling of People's Opinions, Personality, and Emotion's in Social Media, pp. 54–63 (2020)
23. Henry, S., Buchan, K., Filannino, M., Stubbs, A., Uzuner, O.: 2018 n2c2 shared task on adverse drug events and medication extraction in electronic health records. J. Am. Med. Inform. Assoc. **27**(1), 3–12 (2020)
24. Hossain, E., Sharif, O., Hoque, M.M.: NLP-CUET@ LT-EDI-EACL2021: multilingual code-mixed hope speech detection using cross-lingual representation learner. In: Proceedings of the 1st Workshop on Language Technology for Equality, Diversity and Inclusion, pp. 168–174 (2021)
25. Jafarian, H., Taghavi, A.H., Javaheri, A., Rawassizadeh, R.: Exploiting BERT to improve aspect-based sentiment analysis performance on Persian language. In:

Proceedings of the 2021 7th International Conference on Web Research, pp. 5–8 (2021)
26. Jayanthi, S.M., Gupta, A.: Sj_aj@ dravidianlangtech-eacl2021: task-adaptive pre-training of multilingual BERT models for offensive language identification. In: Proceedings of the 1st Workshop on Speech and Language Technologies for Dravidian Languages, pp. 307–312 (2021)
27. Jiang, Q., Chen, L., Xu, R., Ao, X., Yang, M.: A challenge dataset and effective models for aspect-based sentiment analysis. In: Proceedings of the 2019 Conference on Empirical Methods in Natural Language Processing and the 9th International Joint Conference on Natural Language Processing (EMNLP-IJCNLP), pp. 6280–6285 (2019)
28. Kakwani, D., et al.: IndicNLPSuite: monolingual corpora, evaluation benchmarks and pre-trained multilingual language models for Indian languages. In: Findings of the Association for Computational Linguistics: EMNLP 2020, pp. 4948–4961 (2020)
29. Kamal, O., Kumar, A., Vaidhya, T.: Hostility detection in Hindi leveraging pre-trained language models. In: Chakraborty, T., Shu, K., Bernard, H.R., Liu, H., Akhtar, M.S. (eds.) CONSTRAINT 2021. CCIS, vol. 1402, pp. 213–223. Springer, Cham (2021). https://doi.org/10.1007/978-3-030-73696-5_20
30. Ke, C., Xiong, Q., Wu, C., Liao, Z., Yi, H.: Prior-BERT and multi-task learning for target-aspect-sentiment joint detection. In: 2022 IEEE International Conference on Acoustics, Speech and Signal Processing, ICASSP 2022, pp. 7817–7821 (2022)
31. Khan, L., Amjad, A., Ashraf, N., Chang, H.T.: Multi-class sentiment analysis of Urdu text using multilingual BERT. Sci. Rep. **12**(1), 1–17 (2022)
32. Kong, J., Wang, J., Zhang, X.: Hierarchical BERT with an adaptive fine-tuning strategy for document classification. Knowl.-Based Syst. **238**, 107872 (2022)
33. Lan, Z., Chen, M., Goodman, S., Gimpel, K., Sharma, P., Soricut, R.: ALBERT: a lite BERT for self-supervised learning of language representations. In: Proceedings of the 8th International Conference on Learning Representations (2020)
34. Leippold, M.: Sentiment spin: attacking financial sentiment with GPT-3. Technical report. 23-11, Swiss Finance Institute (2023)
35. Liao, W., Zeng, B., Yin, X., Wei, P.: An improved aspect-category sentiment analysis model for text sentiment analysis based on RoBERTa. Appl. Intell. **51**(6), 3522–3533 (2021)
36. Liu, Y., et al.: RoBERTa: a robustly optimized BERT pretraining approach. arXiv preprint arXiv:1907.11692 (2019)
37. Luo, X., Yin, S., Lin, P.: A survey of cross-lingual sentiment analysis based on pre-trained models. In: Proceedings of the 21st International Conference on Electronic Business, pp. 23–33 (2021)
38. Meena, A., Prabhakar, T.V.: Sentence level sentiment analysis in the presence of conjuncts using linguistic analysis. In: Amati, G., Carpineto, C., Romano, G. (eds.) ECIR 2007. LNCS, vol. 4425, pp. 573–580. Springer, Heidelberg (2007). https://doi.org/10.1007/978-3-540-71496-5_53
39. Mensouri, D., Azmani, A., Azmani, M.: Combining RoBERTa pre-trained language model and NMF topic modeling technique to learn from customer reviews analysis. Int. J. Intell. Syst. Appl. Eng. **11**(1), 39–49 (2023)
40. Moraes, R., Valiati, J.F., Neto, W.P.G.: Document-level sentiment classification: an empirical comparison between SVM and ANN. Expert Syst. Appl. **40**(2), 621–633 (2013)

41. Pontiki, M., et al.: SemEval-2016 task 5: aspect based sentiment analysis. In: Proceedings of the 10th International Workshop on Semantic Evaluation, pp. 19–30 (2016)
42. Pontiki, M., Galanis, D., Papageorgiou, H., Manandhar, S., Androutsopoulos, I.: SemEval-2015 task 12: aspect based sentiment analysis. In: Proceedings of the 9th International Workshop on Semantic Evaluation (SemEval 2015), pp. 486–495 (2015)
43. Pontiki, M., Galanis, D., Pavlopoulos, J., Papageorgiou, H., Androutsopoulos, I., Manandhar, S.: SemEval-2014 task 4: aspect based sentiment analysis. In: Proceedings of the 8th International Workshop on Semantic Evaluation (SemEval 2014), pp. 27–35 (2014)
44. Sabri, N., Edalat, A., Bahrak, B.: Sentiment analysis of Persian-English code-mixed texts. In: Proceedings of 2021 26th International Computer Conference, Computer Society of Iran (CSICC), pp. 1–4 (2021)
45. Sarangi, P., Ganesan, S., Arora, P., Joshi, S.: AMEX AI labs at SemEval-2022 task 10: contextualized fine-tuning of BERT for structured sentiment analysis. In: Proceedings of the 16th International Workshop on Semantic Evaluation, pp. 1296–1304 (2022)
46. Sarker, A., Gonzalez, G.: A corpus for mining drug-related knowledge from Twitter chatter: language models and their utilities. Data Brief **10**, 122–131 (2017)
47. Sun, K., Luo, X., Luo, M.Y.: A survey of pretrained language models. In: Memmi, G., Yang, B., Kong, L., Zhang, T., Qiu, M. (eds.) KSEM 2022. LNCS, vol. 13369, pp. 442–456. Springer, Cham (2022). https://doi.org/10.1007/978-3-031-10986-7_36
48. Sweidan, A.H., El-Bendary, N., Al-Feel, H.: Sentence-level aspect-based sentiment analysis for classifying adverse drug reactions (ADRs) using hybrid ontology-XLNet transfer learning. IEEE Access **9**, 90828–90846 (2021)
49. Van Thin, D., Quoc Ngo, H., Ngoc Hao, D., Luu-Thuy Nguyen, N.: Exploring zero-shot and joint training cross-lingual strategies for aspect-based sentiment analysis based on contextualized multilingual language models. J. Inf. Telecommun. 1–23 (2023)
50. Venugopalan, M., Gupta, D.: An enhanced guided LDA model augmented with BERT based semantic strength for aspect term extraction in sentiment analysis. Knowl.-Based Syst. 108668 (2022)
51. Wan, H., Yang, Y., Du, J., Liu, Y., Qi, K., Pan, J.Z.: Target-aspect-sentiment joint detection for aspect-based sentiment analysis. In: Proceedings of the 34th AAAI Conference on Artificial Intelligence, pp. 9122–9129 (2020)
52. Wang, D., Yang, A., Zhou, Y., Xie, F., Ouyang, Z., Peng, S.: Distillation language adversarial network for cross-lingual sentiment analysis. In: 2022 International Conference on Asian Language Processing (IALP), pp. 45–50 (2022)
53. Wang, H., Li, J., Wu, H., Hovy, E., Sun, Y.: Pre-trained language models and their applications. Engineering (2022)
54. Yang, Z., Dai, Z., Yang, Y., Carbonell, J., Salakhutdinov, R.R., Le, Q.V.: XLNet: generalized autoregressive pretraining for language understanding. In: Advances in Neural Information Processing Systems, vol. 32, pp. 5753–5763 (2019)
55. Yao, Y., et al.: DocRED: a large-scale document-level relation extraction dataset. In: Proceedings of the 57th Annual Meeting of the Association for Computational Linguistics, pp. 764–777 (2019)
56. Yong, Q., Chen, C., Wang, Z., Xiao, R., Tang, H.: SGPT: semantic graphs based pre-training for aspect-based sentiment analysis. World Wide Web 1–14 (2023)

57. Zhang, K., et al.: Incorporating dynamic semantics into pre-trained language model for aspect-based sentiment analysis. In: Findings of the Association for Computational Linguistics, ACL 2022, pp. 3599–3610 (2022)
58. Zhou, J., Tian, J., Wang, R., Wu, Y., Xiao, W., He, L.: SentiX: a sentiment-aware pre-trained model for cross-domain sentiment analysis. In: Proceedings of the 28th International Conference on Computational Linguistics, pp. 568–579 (2020)
59. Zhou, X., Zhang, T., Cheng, C., Song, S.: Dynamic multichannel fusion mechanism based on a graph attention network and BERT for aspect-based sentiment classification. Appl. Intell. 1–14 (2022)
60. Zhou, X., et al.: Attention-based LSTM network for cross-lingual sentiment classification. In: Proceedings of the 2016 Conference on Empirical Methods in Natural Language Processing, pp. 247–256 (2016)
61. Zou, J., et al.: Aspect-level sentiment classification based on graph attention network with BERT. In: Sun, X., Zhang, X., Xia, Z., Bertino, E. (eds.) ICAIS 2022. CCIS, vol. 1586, pp. 231–244. Springer, Cham (2022). https://doi.org/10.1007/978-3-031-06767-9_19

Improving Out-of-Distribution Detection with Margin-Based Prototype Learning

Junzhuo Liu[1], Yuanyuan Ren[2(✉)], Weiwei Li[1], Yuchen Zheng[2], and Chenyang Wang[3]

[1] University of Electronic Science and Technology of China, Chengdu 611700, China
junliu@std.uestc.edu.cn

[2] School of Information Science and Technology, Shihezi University, Shihezi 832000, China
yuanyuanren043@outlook.com

[3] Beijing Institute of Genomics, Chinese Academy of Sciences, Beijing 100000, China
wangchenyang17m@big.ac.cn

Abstract. Deep Neural Networks often make overconfident predictions when encountering out-of-distribution (OOD) data. Previous prototype-based methods significantly improved OOD detection performance by optimizing the representation space. However, practical scenarios present a challenge where OOD samples near class boundaries may overlap with in-distribution samples in the feature space, resulting in misclassification, and few methods have considered the challenge. In this work, we propose a margin-based method that introduces a margin into the common instance-prototype contrastive loss. The margin leads to broader decision boundaries, resulting in better distinguishability of OOD samples. In addition, we leverage learnable prototypes and explicitly maximize prototype dispersion to obtain an improved representation space. We validate the proposed method on several common benchmarks with different scoring functions and architectures. Experiments results show that the proposed method achieves state-of-the-art performance. Code is available at https://github.com/liujunzhuo/MarginOOD.

Keywords: Out-of-Distribution Detection · Prototype learning · Margin-based classification

1 Introduction

Deep neural networks (DNNs) have demonstrated remarkable performance in various vision tasks such as image classification and semantic segmentation. How-

This work was supported by the National Natural Science Foundation of China under Grant 32260388; and The Open Research Fund of National Engineering Research Center for Agro-Ecological Big Data Analysis & Application, Anhui University under Grant AE202203; and Innovation and Cultivation Project for Youth Talents of Shihezi University under Grant CXPY202117; and Startup Project for Advanced Talents of Shihezi University under Grant RCZK2021B21.

J. Liu and Y. Ren—Equal contribution.

B. Luo et al. (Eds.): ICONIP 2023, CCIS 1966, pp. 149–160, 2024.
https://doi.org/10.1007/978-981-99-8148-9_12

ever, in practical scenarios and open-world settings, these models often encounter large amounts of out-of-distribution (OOD) data they have never seen before. Unfortunately, DNNs tend to make overconfident predictions for OOD data [4], significantly undermining their reliability.

Prototype-based methods have shown promising results for OOD detection [12,13,27,31]. Unlike vanilla classification methods that rely on Cross-Entropy loss, prototype-based methods explicitly incorporate the class centers and aim to achieve intra-class compactness and inter-class dispersion. This paradigm shift transforms the training objective from classifying the entire representation space to clustering input features around class prototypes [27]. Combined with distance-based scoring functions [9,17], which measure the distance between input samples and training samples or class centers in the feature space, prototype-based methods show excellent performance in OOD detection. Recently, Ming *et al.* [13] introduced CIDER, a prototype-based method, achieving the state-of-the-art (SOTA) performance by optimizing the hypersphere embedding into a mixture of von Mises-Fisher (vMF) distributions. Restricting the feature and prototype representations to a hypersphere allows for a favorable geometric interpretation, where the angle directly corresponds to the distance in the feature space due to the linearity of angles and radians. As a result, we can jointly optimize the geodesic distance and cosine distance.

Obtaining a feature space that effectively separates OOD and in-distribution (ID) samples is crucial for OOD detection. While the prototype-based methods have shown promising results by improving intra-class compactness and increasing the distance between ID samples and OOD samples, practical scenarios often involve OOD samples that reside near the boundaries of ID sample clusters, leading to significant overlap between OOD and ID samples in the feature space. This problem stems from the neglect of clear decision boundaries. Inspired by margin-based classification methods [2,20], which achieve broader decision boundaries by adding a margin to the classification loss, we propose a margin loss that introduces an angle term into the common instance-prototype contrastive learning loss. The addition of a margin leverages the linear relationship between angles and distances on the hypersphere, increasing the distance between instances and their corresponding prototypes. After the optimization process, a broader decision boundary and improved intra-class compactness are obtained, resulting in better OOD detection performance. Unlike existing methods that employ non-parametric prototypes, our method leverages learnable prototypes, allowing direct updates for prototypes. With the proposed margin loss, learnable prototypes indirectly achieve inter-class dispersion by maximizing the distance between prototypes and their negative samples. Moreover, we explicitly increase the inter-prototype distance during training to fully exploit the potential of learnable prototypes and achieve greater inter-class dispersion. We also employ Exponential Moving Average (EMA) to stabilize the training process.

We conduct comprehensive experiments on the CIFAR-100 and CIFAR-10 benchmarks to validate the performance of our method. Specifically, the proposed method outperforms the previous SOTA method by 4.47% in the average false positive rate at 95% (FPR95) on the CIFAR-100 benchmark and 1.15% on the CIFAR-10 benchmark. We also present detailed ablation experiments to thoroughly analyze the effectiveness of our method. Our contributions can be summarized as follows:

- We propose a novel margin-based OOD detection method, which introduces a margin to the training process. We get the broader boundaries between different classes and improve the intra-class compactness, enhancing OOD detection performance.
- We introduce Learnable Class Prototypes and explicitly increase the inter-prototype distance, promoting better dispersion between different classes in the feature space.
- Extensive experiments show that our method achieves SOTA results on both CIFAR-100 and CIFAR-10 benchmarks, and we further analyze the score distributions of both ID and OOD samples, demonstrating the better representations we have achieved.

2 Related Works

Feature-Based Methods for Out-of-Distribution Detection. Post-hoc methods focus on designing scoring functions to distinguish ID and OOD samples. Some methods directly calculate scores based on the feature embeddings [3,21,30]. For example, in [3,30], feature norm are used for OOD detection. Other methods, known as distance-based methods [28], identify samples in the feature space far away from all known classes as OOD samples. Mahalanobis distance is employed for OOD detection in [9], while some methods [19,24] utilize cosine similarity. Yiyou *et al.* [17] achieves remarkable results using non-parametric KNN.

The success of feature-based scoring methods has inspired researchers to explore training methods for better feature embeddings. Some methods [15,16, 18,22] leverage contrastive learning [6] to achieve better representation space, leading to significant enhancements in OOD detection performance. In addition to sample-based contrastive learning, some methods have incorporated the class prototypes [12,13,27,31]. For instance, Alireza *et al.* [31] enhances the cosine similarity between features and fixed class vectors to obtain a more constraining representation space. Recently, CIDER [13] utilizes the average feature representation of each class as its class prototype, yielding SOTA performance. Unlike previous methods, we combine margin-based methods and use learnable prototypes to achieve broader decision boundaries in the representation space.

Margin-Based Classification. Margin-based classification methods have been extensively utilized in training face recognition (FR) models to enhance their discriminative capabilities, such as CosFace [20] and ArcFace [2]. By incorporating a margin term into the Cross-Entropy loss, these methods address the insufficiency of learned features in distinguishing different classes and further lead to

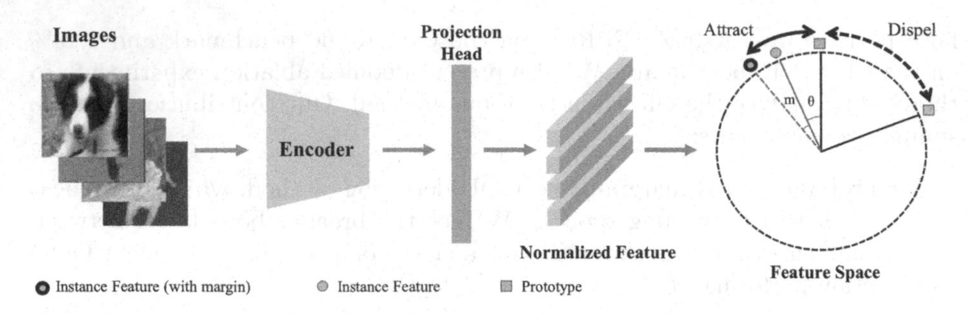

Fig. 1. The pipeline of the proposed method. We optimize the hypersphere feature space in two way: (1) decrease the distance between the prototype and the instance (with margin), and (2) increase the distance between prototypes.

more accurate differentiation between known and unknown individuals in open-set FR scenarios. The open set FR has a similar setting with OOD detection for identifying unknown samples in real world.

3 Preliminaries

Given a labeled dataset $D_{in} = \{(x_i, y_i)\}_{i=1}^{N}$, drawn from the joint distribution $P_{X,Y_{in}}$, where x_i is the input sample and $y_i \in Y_{in}$ is the corresponding label. The multi-class classification task aims to train a model correctly classify input samples into label space Y_{in}. However, during testing, the model may encounter inputs that deviate from the distribution $P_{X,Y_{in}}$. Specifically, these inputs may consist of samples x with labels belonging to Y_{out}, where $Y_{in} \cap Y_{out} = \emptyset$. Such samples are called OOD samples. The task of OOD detection is to identify OOD samples when performing the classification task. Typically, a scoring function $s(x)$ is used, where samples with $s(x) < \lambda$ are considered as OOD. λ is commonly chosen so that a high fraction (e.g., 95%) of ID data is correctly classified.

4 Method

The overall framework is depicted in Fig. 1. In line with the prevalent contrastive learning paradigm [6,13], we apply random data augmentation to the input images twice. The augmented $2N$ images are then encoded into a high dimensional space, resulting in $R = \text{Enc}(X)$, where $R \in \mathbb{R}^{2N \times d_e}$, Enc is the encoder network and d_e is the dimension of the embeddings. Subsequently, the encoded vectors undergo projection using a project head, resulting in $Z = \text{Proj}(R)$, where $Z \in \mathbb{R}^{2N \times d_p}$, and d_p is the dimension of the embeddings. Z can be represented as $(z_1, z_2, \ldots, z_{2N})^\top$.

Learnable prototypes $P \in \mathbb{R}^{C \times d_p}$ are initialized with a normal distribution, where C represents the number of classes. Both Z and P are L2 normalized and thus distributed on the unit hypersphere. In the subsequent sections, we will detail how the proposed method achieves a desirable representation space for OOD detection.

4.1 Intra-class Compactness in Hypersphere

The commonly used loss [7,13] for prototype-based contrastive learning is

$$\mathcal{L}_{\text{pcl}} = -\frac{1}{2N}\sum_{i=1}^{2N}\log\left(\frac{\exp(z_i \cdot p_{y_i}^{\top}/\tau)}{\exp(z_i \cdot p_{y_i}^{\top}/\tau) + \sum_{j\neq y_i}^{C}\exp(z_i \cdot p_j^{\top}/\tau)}\right), \tag{1}$$

where y_i is the label of z_i, τ is a temperature parameter and p_{y_i} is the prototype of the corresponding class y_i. Since both Z and P are L2 normalized, if we define $\theta_{i,k}$ as the angle between z_i and p_k, then $z_i \cdot p_{y_i}^{\top} = \cos\theta_{i,y_i}$. The Eq. 1 can be rewritten as

$$\mathcal{L}_{\text{comp}} = -\frac{1}{2N}\sum_{i=1}^{2N}\log\left(\frac{\exp(\cos\theta_{i,y_i}/\tau)}{\exp(\cos\theta_{i,y_i}/\tau) + \sum_{j\neq y_i}^{C}\exp(\cos\theta_{i,j}/\tau)}\right). \tag{2}$$

The loss encourages samples to move closer to their corresponding prototypes and farther away from negative prototypes, resulting in improved intra-class compactness and inter-class dispersion. Additionally, prototype-based methods provide a more extensive feature space for OOD samples compared to vanilla classification methods, making it less likely for the model to misclassify OOD samples. Motivated by margin-based classification methods, we introduce a margin into the $\mathcal{L}_{\text{comp}}$. Specifically, for each z_i and its corresponding prototype, we introduce an angle m:

$$\mathcal{L}_{\text{margin}} = -\frac{1}{2N}\sum_{i=1}^{2N}\log\left(\frac{\exp(\cos(\theta_{i,y_i}+m)/\tau)}{\exp(\cos(\theta_{i,y_i}+m)/\tau) + \sum_{j\neq y_i}^{C}\exp(\cos\theta_{i,j}/\tau)}\right). \tag{3}$$

As shown in Fig. 1, after adding an angle, the distance between instances and the corresponding prototype becomes larger due to the linear relationship between angle and radian on the hypersphere, resulting in a higher loss value. After the optimization process, we achieve more compact inter-class distances and samples that were distributed around the decision boundaries would diverge from their negative prototypes while move closer to their corresponding prototype, ultimately creating broader decision boundaries. Experiment results confirm that the margin contributes to better intra-class compactness and separation of OOD and ID data in the feature space.

4.2 Inter-class Dispersion

Although $\mathcal{L}_{\text{margin}}$ effectively amplifies the dispersion between training samples and their corresponding negative prototypes, it indirectly increases the distance between class prototypes. To explicitly increase the dispersion between class prototypes, we introduced a dispersion loss

$$\mathcal{L}_{\text{dis}} = \frac{1}{C}\sum_{i=1}^{C}\frac{1}{C-1}\log\sum_{j\neq i}\exp(p_i \cdot p_j^{\top}/\tau), \tag{4}$$

where C denotes the number of classes, p_i represents the learnable prototype vector for class i , and τ is a temperature parameter. The loss aims to maximize the distances between all pairs of negative prototypes, thereby promoting greater inter-class dispersion, which further enables better discrimination of OOD samples.

To ensure a stable training process for learnable prototypes, we use the EMA technique for updating:

$$P_{\text{ema}} = \alpha \cdot P_{\text{ema}} + (1 - \alpha) \cdot P, \tag{5}$$

$$P = \text{sg}(P_{\text{ema}}), \tag{6}$$

where P_{ema} is a non-parametric prototype, α is a hyperparameter controlling the weight of the current and new values, and the stop gradient operation sg$(\cdot)$ is used upon the updated value of P_{ema} to prevent the flow of gradients during training. By applying the EMA, the learnable prototype P is gradually integrated into P_{ema}, which helps stabilize the training process and mitigate the impact of sudden fluctuations in the learnable prototypes.

The final loss function combines $\mathcal{L}_{\text{dis}}$ and $\mathcal{L}_{\text{margin}}$ with a weight factor λ

$$\mathcal{L} = \mathcal{L}_{\text{dis}} + \lambda \cdot \mathcal{L}_{\text{margin}}. \tag{7}$$

5 Experiments

5.1 Implementation Details

Datasets. We use CIFAR-10 and CIFAR-100 [8] datasets as ID datasets, and use five common datasets: Textures [1], SVHN [14], Places365 [32], LSUN [29], and iSUN [26] as OOD datasets following CIDER [13], with the same data preprocessing.

Evaluation Metrics. We use two widely adopted metrics: (1) FPR95, which measures the false positive rate at a true positive rate of 95%. (2) Area Under the Receiver Operating Characteristic Curve (AUROC). Higher AUROC and lower FPR95 indicate better quality in OOD detection.

Training Details. We use ResNet-18 as the backbone for CIFAR-10 and ResNet-34 for CIFAR-100 and a two-layer non-linear projection head with an output dim of 128. To ensure consistency, we adopt the same training settings as CIDER, using stochastic gradient descent with momentum 0.9, weight decay 10^{-4} and the initial learning rate is 0.5 with cosine decay, the batch size is 512, and the temperature τ is 0.1. We train the models for 500 epochs. The default λ is 0.1. We set the margin m to 0.1 for CIFAR-100 and 0.3 for CIFAR-10; the EMA update factor α is 0.8 for CIFAR-100 and 0.9 for CIFAR-10. We use the non-parametric KNN score as our scoring function, where K = 300 for ResNet-34 and K = 100 for ResNet-18 as in CIDER. We also evaluate our method with different scoring functions to verify its applicability. All experiments are conducted on an Nvidia V100 GPU with 32GB VRAM.

Table 1. Performance of OOD detection on CIFAR-100 benchmark.

Method	OOD Dataset										Average	
	SVHN		Places365		LSUN		iSUN		Texture			
	FPR	AUROC	FPR	AUROC	FPR	AUROC	FPR	AUROC	FPR	AUROC	FPR	AUROC
MSP	78.89	79.80	84.38	74.21	83.47	75.28	84.61	74.51	86.51	72.53	83.12	75.27
ODIN	70.16	84.88	82.16	75.19	76.36	80.10	79.54	79.16	85.28	75.23	78.70	79.11
Mahalanobis	87.09	80.62	84.63	73.89	84.15	79.43	83.18	78.83	61.72	84.87	80.15	79.53
Energy	66.91	85.25	81.41	76.37	59.77	86.69	66.52	84.49	79.01	79.96	70.72	82.55
GODIN	74.64	84.03	89.13	68.96	93.33	67.22	94.25	65.26	86.52	69.39	87.57	70.97
LogitNorm	59.60	90.74	80.25	78.58	81.07	82.99	84.19	80.77	86.64	75.60	78.35	81.74
ProxyAnchor	87.21	82.43	**70.10**	79.84	37.19	91.68	70.01	84.96	65.64	84.99	66.03	84.78
CE+SimCLR	24.82	94.45	86.63	71.48	56.40	89.00	66.52	83.82	63.74	82.01	59.62	84.15
CSI	44.53	92.65	79.08	76.27	75.58	83.78	76.62	84.98	61.61	86.47	67.48	84.83
SSD+	31.19	94.19	77.74	**79.9**	79.39	85.18	80.85	84.08	66.63	86.18	67.16	85.90
KNN+	39.23	92.78	80.74	77.58	48.99	89.30	74.99	82.69	57.15	88.35	60.22	86.14
CIDER	23.09	95.16	79.63	73.43	16.16	96.33	71.68	82.98	43.87	90.42	46.89	87.67
Ours	**21.94**	**95.87**	79.85	75.44	**14.73**	**96.86**	**56.41**	**88.37**	**39.17**	**92.01**	**42.42**	**89.71**

Table 2. Performance of OOD detection on CIFAR-10 benchmark.

Method	OOD Dataset										Average	
	SVHN		Places365		LSUN		iSUN		Texture			
	FPR	AUROC	FPR	AUROC	FPR	AUROC	FPR	AUROC	FPR	AUROC	FPR	AUROC
MSP	59.66	91.25	62.46	88.64	45.21	93.80	54.57	92.12	66.45	88.50	57.67	90.86
Energy	54.41	91.22	42.77	91.02	10.19	98.05	27.52	95.59	55.23	89.37	38.02	93.05
ODIN	53.78	91.30	43.40	90.98	10.93	97.93	28.44	95.51	55.59	89.47	38.43	93.04
GODIN	18.72	96.10	55.25	85.50	11.52	97.12	30.02	94.02	33.58	92.20	29.82	92.97
Mahalanobis	9.24	97.80	83.50	69.56	67.73	73.61	**6.02**	**98.63**	23.21	92.91	37.94	86.50
CE+SimCLR	6.98	99.22	54.39	86.70	64.53	85.60	59.62	86.78	16.77	96.56	40.46	90.97
CSI	37.38	94.69	38.31	93.04	10.63	97.93	10.36	98.01	28.85	94.87	25.11	95.71
SSD+	2.47	99.51	**22.05**	95.57	10.56	97.83	28.44	95.67	**9.27**	**98.35**	14.56	97.38
ProxyAnchor	39.27	94.55	43.46	92.06	21.04	97.02	23.53	96.56	42.70	93.16	34.00	94.67
KNN+	2.70	99.61	23.05	94.88	7.89	98.01	24.56	96.21	10.11	97.43	13.66	97.22
CIDER	2.89	**99.72**	23.88	94.09	5.45	99.01	20.21	96.64	12.33	96.85	12.95	97.26
Ours	**2.10**	99.59	25.61	**94.9**	**2.92**	**99.35**	17.73	96.83	10.64	98.21	**11.8**	**97.78**

5.2 Comparison with State-of-the-art Methods

Various methods have been considered in our comparisons. MSP [4], ODIN [10], Energy [11] and Malalanobis [9] are post-hoc methods implemented on pre-trained models; LogitNorm [23] and GODIN [5] modified the loss function; Sim-CLR [25], CSI [18], SSD+ [15], and KNN+ [17] ultized the sample-based contrastive learning; ProxyAnchor [7] and CIDER [13] are prototype-based methods. Methods trained on CIFAR-100 and CIFAR-10 use ResNet-34 and ResNet-18 backbone, respectively.

As shown in Table 1, the proposed method achieves SOTA performance in terms of both FPR95 and AUROC on the CIFAR-100 benchmark, reducing the average FPR95 by 4.47% and gaining 2.04% in AUROC compared to the previous SOTA method CIDER. Similarly, in Table 2, our method consistently outperforms competing methods, yielding the best results in the CIFAR-10 benchmark. These improvements indicate that our method obtains a more desirable representation space on the hypersphere. In the next section, we will thoroughly analyze our method.

5.3 Further Analysis

Ablation Studies. We conduct ablation studies to investigate the contributions of different modules in our method. We use the ResNet18 with a batch size of 128 and train the model for 100 epochs on the CIFAR-100 dataset. The detailed experimental results can be found in Table 3.

Table 3. Ablation Studies on effects of different components.

	L_{comp}	L_{margin}	L_{dis}	EMA	FPR	AUROC
Baseline	✓				56.58	85.78
+EMA	✓			✓	56.76	86.71
$+L_{dis}$ & EMA	✓		✓	✓	48.14	87.51
L_{margin} & EMA		✓		✓	51.24	88.17
Ours		✓	✓	✓	**47.58**	**88.71**

Among the modules studied, $\mathcal{L}_{\text{margin}}$ showed significant improvements in OOD detection performance. The AUROC increased from 85.78% to 88.17%, suggesting that intra-class compactness is essential for discriminating between ID and OOD samples. The L_{dis} also demonstrated positive effects on OOD detection. The combination of these modules yielded the best performance in OOD detection, achieving a low FPR95 of 47.58% and a high AUROC of 88.71%. These results underscore the necessity and effectiveness of these modules.

Hyper-Parameters Sensitivity. We explore the effects of different hyperparameters on the performance of our method in Fig. 2. Specifically, we investigate the impact of the margin angle m and the weight factor λ.

For λ, a value of 0.1 achieved the best performance, while higher values of λ, such as 0.2, led to a notable drop in performance. It indicates that a value around 0.1 effectively balances the importance of different loss components, and the reason behind it is that the scale of $\mathcal{L}_{\text{dis}}$ is relatively small. Regarding m, a margin angle of 0.3 yielded the best results. Finally, a weight factor of 0.1 and a margin angle of 0.3 demonstrated superior performance.

Combination of Different Scoring Functions. We discuss the effectiveness of combining different scoring functions with our proposed method in Table 4. We utilize a widely used distance-based scoring function, the Mahalanobis score [9], and compare it with the latest SOTA method CIDER. Our experiments are

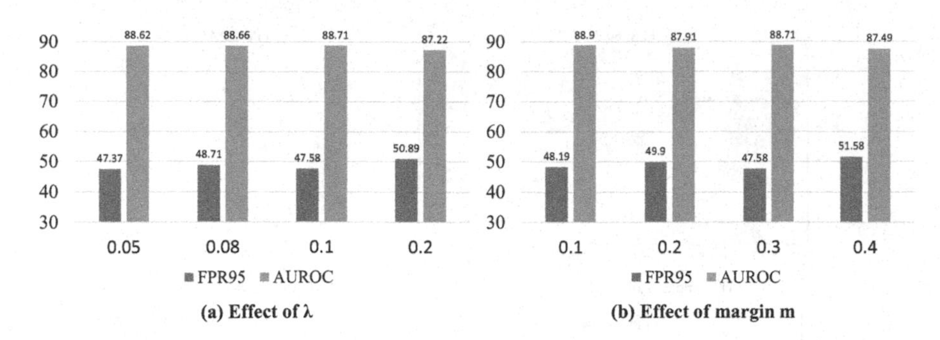

Fig. 2. Hyper-parameters sensitivity analysing.

conducted on CIFAR-100 with ResNet-34. The experiment results demonstrate superior performance over CIDER in all datasets. We have achieved a notable reduction in the average FPR95 by 7.15% and attained a high AUROC above 90%, combined with the Mahalanobis score. The results suggest that our methods can work with other OOD scoring functions.

Table 4. Combination of different Scoring functions

Method	OOD Dataset										Average	
	SVHN		Places365		LSUN		iSUN		Texture			
	FPR	AUROC	FPR	AUROC	FPR	AUROC	FPR	AUROC	FPR	AUROC	FPR	AUROC
CIDER+Mahalanobis	15.67	96.59	79.20	76.52	15.44	97.06	80.01	79.61	47.82	89.46	47.63	87.85
Ours+Mahalanobis	**10.69**	**97.93**	**78.99**	**77.26**	**8.29**	**98.41**	**72.39**	**86.55**	**32.03**	**93.67**	**40.48**	**90.76**
CIDER+KNN	22.19	95.50	80.07	73.28	16.15	96.34	71.64	80.84	43.74	90.42	46.76	87.28
Ours+KNN	**21.94**	**95.87**	**79.85**	**75.44**	**14.73**	**96.86**	**56.41**	**88.37**	**39.17**	**92.01**	**42.42**	**89.71**

Do We Achieve Better Representations? To explore the effectiveness of our method, we plot the KNN score distributions of both ID and OOD samples in Fig. 3. The KNN score measures the cosine distance between input samples and training samples, where higher scores indicate greater similarity. In an ideal scenario, OOD samples should have low scores, while ID samples should have high scores. Train and test sets of CIFAR-100 are used as training and ID datasets, respectively, and we use iSUN as the OOD dataset. In Fig. 3(a), when only the compact loss ($\mathcal{L}_{\text{comp}}$) is employed, we observe that many ID samples have scores that closely resemble those of OOD samples, resulting in a significantly high FPR95. CIDER combines the non-parametric prototype and compact loss, resulting in relatively high scores for ID samples, as shown in Fig. 3(c). However, our method employs parametric prototypes and introduces a margin loss (L_{margin}) to obtain a more compact representation for ID samples. As shown in Fig. 3(b), the density peak of ID samples in the score distribution noticeably

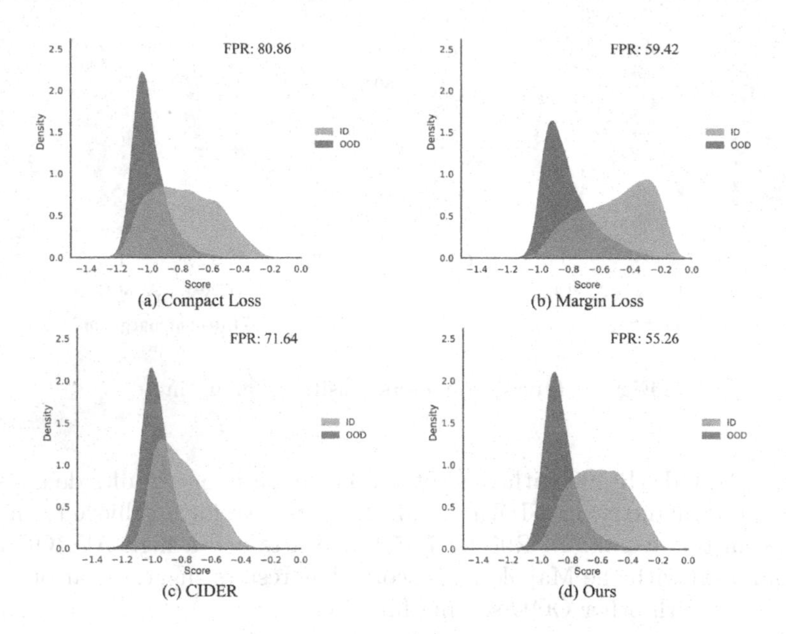

(a) Compact Loss

(b) Margin Loss

(c) CIDER

(d) Ours

Fig. 3. KNN score distribution.

shifts to the right, indicating a closer distance between ID samples and training samples in the feature space. Consequently, the FPR is reduced by 21.44%, demonstrating the enhancement of intra-class compactness in improving OOD detection performance. However, we observe that using the margin loss results in higher scores for OOD samples as well. By incorporating L_{dis}, our method achieves a more desirable score distribution. As shown in Fig. 3(d), the scores of OOD samples decrease while the scores of ID samples become smoother, ultimately yielding an FPR of 55.26%, demonstrating that the enhancement of inter-class dispersion, leads to higher distances between OOD and ID samples in feature space.

6 Conclusion

In this work, we proposed a margin-based prototype learning framework for improving OOD detection performance. By employing margin-based loss and learnable prototypes, we established more distinct decision boundaries, and we further promoted inter-class dispersion, leading to improved discriminability between in-distribution (ID) and OOD samples. Experimental results demonstrated the effectiveness of our framework across various scoring functions. Specifically, the analysis of K-nearest neighbors (KNN) score distributions revealed enhanced intra-class compactness, inter-class dispersion, and improved discrimination between ID and OOD data. We hope that our work could inspire future research in analyzing the representation space of deep neural networks.

References

1. Cimpoi, M., Maji, S., Kokkinos, I., Mohamed, S., Vedaldi, A.: Describing textures in the wild. In: Proceedings of the IEEE Conference on Computer Vision and Pattern Recognition (CVPR) (2014)
2. Deng, J., Guo, J., Xue, N., Zafeiriou, S.: ArcFace: additive angular margin loss for deep face recognition. In: Proceedings of the IEEE/CVF Conference on Computer Vision and Pattern Recognition (CVPR) (2019)
3. Dong, X., Guo, J., Li, A., Ting, W.T., Liu, C., Kung, H.: Neural mean discrepancy for efficient out-of-distribution detection. In: Proceedings of the IEEE/CVF Conference on Computer Vision and Pattern Recognition (CVPR), pp. 19217–19227 (2022)
4. Hendrycks, D., Gimpel, K.: A baseline for detecting misclassified and out-of-distribution examples in neural networks. In: International Conference on Learning Representations (2017)
5. Hsu, Y.C., Shen, Y., Jin, H., Kira, Z.: Generalized ODIN: detecting out-of-distribution image without learning from out-of-distribution data. In: Proceedings of the IEEE/CVF Conference on Computer Vision and Pattern Recognition (CVPR) (2020)
6. Khosla, P., Teterwak, P., Wang, C., et al.: Supervised contrastive learning. In: Advances in Neural Information Processing Systems, vol. 33, pp. 18661–18673 (2020)
7. Kim, S., Kim, D., Cho, M., Kwak, S.: Proxy anchor loss for deep metric learning. In: Proceedings of the IEEE/CVF Conference on Computer Vision and Pattern Recognition (CVPR) (2020)
8. Krizhevsky, A., Hinton, G., et al.: Learning multiple layers of features from tiny images (2009)
9. Lee, K., Lee, K., Lee, H., Shin, J.: A simple unified framework for detecting out-of-distribution samples and adversarial attacks. In: Advances in Neural Information Processing Systems, vol. 31 (2018)
10. Liang, S., Li, Y., Srikant, R.: Enhancing the reliability of out-of-distribution image detection in neural networks. In: International Conference on Learning Representations (2018)
11. Liu, W., Wang, X., Owens, J., Li, Y.: Energy-based out-of-distribution detection. In: Advances in Neural Information Processing Systems, vol. 33, pp. 21464–21475 (2020)
12. Miller, D., Sunderhauf, N., Milford, M., Dayoub, F.: Class anchor clustering: a loss for distance-based open set recognition. In: Proceedings of the IEEE/CVF Winter Conference on Applications of Computer Vision (WACV), pp. 3570–3578 (2021)
13. Ming, Y., Sun, Y., Dia, O., Li, Y.: How to exploit hyperspherical embeddings for out-of-distribution detection? In: The Eleventh International Conference on Learning Representations (2023)
14. Netzer, Y., Wang, T., Coates, A., Bissacco, A., et al.: Reading digits in natural images with unsupervised feature learning (2011)
15. Sehwag, V., Chiang, M., Mittal, P.: SSD: a unified framework for self-supervised outlier detection. In: International Conference on Learning Representations (2021)
16. Sohn, K., Li, C.L., Yoon, J., Jin, M., Pfister, T.: Learning and evaluating representations for deep one-class classification. In: International Conference on Learning Representations (2021)

17. Sun, Y., Ming, Y., Zhu, X., Li, Y.: Out-of-distribution detection with deep nearest neighbors. In: International Conference on Machine Learning, pp. 20827–20840. PMLR (2022)
18. Tack, J., Mo, S., Jeong, J., Shin, J.: CSI: novelty detection via contrastive learning on distributionally shifted instances. In: Advances in Neural Information Processing Systems, vol. 33, pp. 11839–11852 (2020)
19. Techapanurak, E., Suganuma, M., Okatani, T.: Hyperparameter-free out-of-distribution detection using cosine similarity. In: Proceedings of the Asian Conference on Computer Vision (ACCV) (2020)
20. Wang, H., Wang, Y., Zhou, Z., Ji, X., et al.: CosFace: large margin cosine loss for deep face recognition. In: Proceedings of the IEEE Conference on Computer Vision and Pattern Recognition, pp. 5265–5274 (2018)
21. Wang, H., Li, Z., Feng, L., Zhang, W.: ViM: out-of-distribution with virtual-logit matching. In: Proceedings of the IEEE/CVF Conference on Computer Vision and Pattern Recognition (CVPR), pp. 4921–4930 (2022)
22. Wang, H., Zhang, A., Zhu, Y., et al.: Partial and asymmetric contrastive learning for out-of-distribution detection in long-tailed recognition. In: Proceedings of the 39th International Conference on Machine Learning, pp. 23446–23458. PMLR (2022)
23. Wei, H., Xie, R., Cheng, H., et al.: Mitigating neural network overconfidence with logit normalization. In: Proceedings of the 39th International Conference on Machine Learning. Proceedings of Machine Learning Research, vol. 162, pp. 23631–23644. PMLR (2022)
24. Wilson, S., Fischer, T., Sünderhauf, N., Dayoub, F.: Hyperdimensional feature fusion for out-of-distribution detection. In: Proceedings of the IEEE/CVF Winter Conference on Applications of Computer Vision (WACV), pp. 2644–2654 (2023)
25. Winkens, J., et al.: Contrastive training for improved out-of-distribution detection. arXiv preprint arXiv:2007.05566 (2020)
26. Xu, P., Ehinger, K.A., Zhang, Y., Finkelstein, A., Kulkarni, S.R., Xiao, J.: TurkerGaze: crowdsourcing saliency with webcam based eye tracking. arXiv preprint arXiv:1504.06755 (2015)
27. Yang, H.M., Zhang, X.Y., Yin, F., Yang, Q., Liu, C.L.: Convolutional prototype network for open set recognition. IEEE Trans. Pattern Anal. Mach. Intell. **44**(5), 2358–2370 (2022)
28. Yang, J., Zhou, K., Li, Y., Liu, Z.: Generalized out-of-distribution detection: a survey. arXiv preprint arXiv:2110.11334 (2021)
29. Yu, F., Seff, A., Zhang, Y., Song, S., et al.: LSUN: construction of a large-scale image dataset using deep learning with humans in the loop. arXiv preprint arXiv:1506.03365 (2015)
30. Yu, Y., Shin, S., Lee, S., Jun, C., Lee, K.: Block selection method for using feature norm in out-of-distribution detection. In: Proceedings of the IEEE/CVF Conference on Computer Vision and Pattern Recognition (CVPR), pp. 15701–15711 (2023)
31. Zaeemzadeh, A., Bisagno, N., et al.: Out-of-distribution detection using union of 1-dimensional subspaces. In: Proceedings of the IEEE/CVF Conference on Computer Vision and Pattern Recognition (CVPR), pp. 9452–9461 (2021)
32. Zhou, B., Lapedriza, A., Khosla, A., Oliva, A., Torralba, A.: Places: a 10 million image database for scene recognition. IEEE Trans. Pattern Anal. Mach. Intell. **40**(6), 1452–1464 (2018)

Text-to-Image Synthesis with Threshold-Equipped Matching-Aware GAN

Jun Shang[1,4], Wenxin Yu[1](✉), Lu Che[1](✉), Zhiqiang Zhang[1], Hongjie Cai[1], Zhiyu Deng[1], Jun Gong[2], and Peng Chen[3]

[1] Southwest University of Science and Technology, Mianyang, Sichuan, China
yuwenxin@swust.edu.cn

[2] Southwest Automation Research Institute, Chengdu, China

[3] Chengdu Hongchengyun Technology Co. Ltd., Chengdu, China

[4] Instrumentation Technology and Economy Institute, Chengdu, People's Republic of China

Abstract. In this paper, we propose a novel Equipped with Threshold Matching-Aware Generative Adversarial Network (ETMA-GAN) for text-to-image synthesis. By filtering inaccurate negative samples, the discriminator can more accurately determine whether the generator has generated the images correctly according to the descriptions. In addition, to enhance the discriminative model's ability to discriminate and capture key semantic information, a word fine-grained supervisor is constructed, which in turn drives the generative model to achieve high-quality image detail synthesis. Numerous experiments and ablation studies on Caltech-UCSD Birds 200 (CUB) and Microsoft Common Objects in Context (MS COCO) datasets demonstrate the effectiveness and superiority of the proposed method over existing methods. In terms of subjective and objective evaluations, the model presented in this study has more advantages than the recently available state-of-the-art methods, especially regarding synthetic images with a higher degree of realism and better conformity to text descriptions.

Keywords: Text-to-Image Synthesis · Computer Vision · Generative Adversarial Networks · Matching-Aware

1 Introduction

Image generation based on text descriptions is more natural but more challenging than annotation diagrams and sketches. For people, text descriptions are a

This Research is Supported by National Key Research and Development Program from Ministry of Science and Technology of the PRC (No. 2018AAA0101801), (No. 2021ZD0110600), Sichuan Science and Technology Program (No. 2022ZYD0116), Sichuan Provincial M. C. Integration Office Program, And IEDA Laboratory Of SWUST.

B. Luo et al. (Eds.): ICONIP 2023, CCIS 1966, pp. 161–172, 2024.
https://doi.org/10.1007/978-981-99-8148-9_13

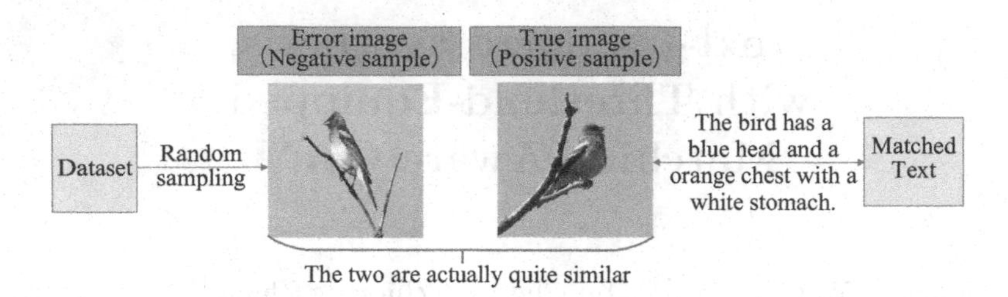

Fig. 1. A randomly selected negative sample in the dataset is similar to a positive sample and in this case, the negative sample is not accurate.

natural and convenient way to express themselves, so it has a broad application prospect in computer-aided structural design and image editing. However, the semantic gap between text and image space is enormous. There are still difficulties in multimodal techniques for synthesizing realistic sense images based on text information. To address this problem, text-to-image synthesis methods based on Generative Adversarial Networks (GANs) [1] have emerged and made some progress. Reed et al. [10] first proposed and used DCGAN [9] to achieve realistic images of 64 × 64 size based on text descriptions. To improve the resolution of the synthesized images, Zhang et al. [18] proposed a staged synthesis method (StackGAN), i.e., the initial image is generated from the noise by first encoding the sentence embedding, and following the low-resolution image is combined with the text description to increase the resolution of the synthesized image to 256 × 256. Its subsequent cascaded generation mode [19] further improves the synthesized authenticity of the images. Subsequently, there are various ways [8,14,17,20] to integrate fine-grained word information into the generative model to guide the generator to improve image detail generation directly. Meanwhile, some studies [3,10] adopt strategies to enhance the discriminative model and thus indirectly guide the generator. For example, in many existing approaches [3,5,10,15], the matching-aware discriminant method is used, i.e., in each training step, in addition to inputting {true image, matched text} and {generated image, matched text} into the discriminator to determine whether the generated images are reasonable or not, the mismatched text-image pairs ({true image, mismatched text} or {error image, matched text}) as negative samples to determine whether the generated image fits the corresponding text. Added mismatched text-image pairs as negative samples are called matching-aware discrimination.

However, as shown in Fig. 1, the specific implementation of the mismatched text-image pairs included in this method has the problem of inaccurate negative samples, i.e., the negative samples may contain much information similar to the positive samples. For example, the error image in {error image, matched text} is randomly sampled so that it may have similar content to the true image in {true image, matched text}. Regretfully, inaccurate negative samples used for training would botch model learning and hinder model performance improvement. To solve this problem, we propose an Equipped with Threshold Matching-Aware

Generative Adversarial Network (ETMA-GAN). In addition, to further improve generation quality, we introduce word-level discrimination in the discriminator. On the one hand, we set the threshold to filter inaccurate negative samples to enhance the matching-aware mechanism and further improve the discriminant model to perceive and judge whether the synthesized image matches the text description; thus, the generative model achieves high-quality semantically consistent image synthesis according to the text. On the other hand, based on the threshold-equipped matching-aware, this study builds a word-image discriminative attention module in the discriminator to strengthen the discriminative model's ability to capture and discriminate critical semantic information, thus effectively reducing the generator to synthesize poor image details to confuse the discriminator.

Overall, our contributions can be summarized as follows:

- An efficient single-stage text-to-image synthesis backbone network is designed; we propose a novel threshold-equipped matching-aware method in the discriminator, and ablation experiments reflect the role of different size thresholds taking on the final synthesis results.
- By constructing a word fine-grained supervision module, which aims to find image regions semantically related to words, the model does not use attention to generate images directly but only employs attention in the discriminator to promote text-image semantic coherence.
- Qualitative and quantitative comparison results on CUB and COCO datasets show our approach's effectiveness and superiority. In addition, we achieve better generation performance while keeping the number of generative model parameters to a minimum.

The rest of the paper is arranged as follows: Sect. 2 presents our approach's network architecture and details, Sect. 3 shows our excellent experimental results, and Sect. 4 summarizes our work.

2 ETMA-GAN for Text-to-Image Generation

2.1 Network Architecture

As shown in Fig. 2, the network architecture of ETMA-GAN has three parts, i.e., it consists of a text encoder for learning text representation, a generator for deepening texts and images fusion and improving the resolution, and a discriminator for determining whether the semantic information of the synthesized images and the given texts match.

For the text encoder, a bi-directional Long Short-Term Memory network (Bi-LSTM) [13] pre-trained in the literature [17] is used to learn text description features. Bi-LSTM encodes the text description as a sentence feature vector $s \in R^D$ and a word feature matrix $w \in R^{D \times L}$, where D represents the feature dimension and L represents the number of words in a sentence description.

Since a multi-stage GAN network architecture results in higher computational effort and a more unstable training process, a single-stage generative model

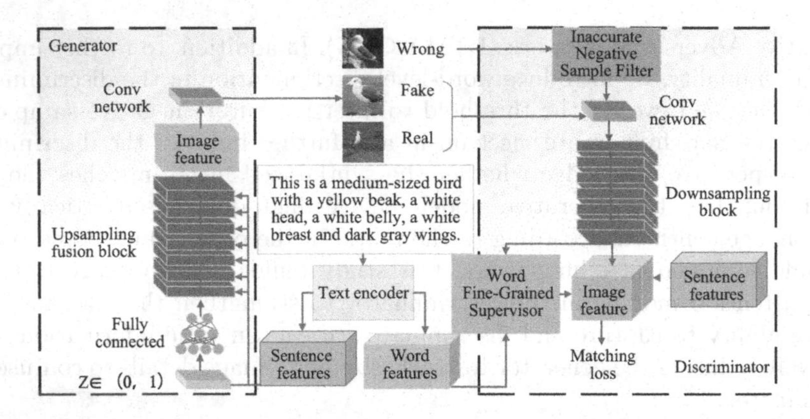

Fig. 2. The network architecture of the proposed ETMA-GAN for text-to-image synthesis. The generator network is on the left, the text encoder is in the middle, and the discriminator network is on the right. ETMA-GAN generates high-resolution images directly through a light-weighted backbone network. Equipped with an inaccurate negative sample filter and a word fine-grained supervisor, our model can synthesize high-quality images more semantically consistent with text descriptions.

with only one generator-discriminator pair was proposed in the literature [15], and this lightweight network structure is also used as the backbone network in this study. Specifically, the generator takes as input the sentence vector s encoded by the text encoder and the noise vector z obeying normal distribution and consists of seven Upsampling fusion blocks, each of which is shown on the left side of Fig. 3 and contains an Upsampling layer, a Residual block and two Deep fusion blocks for synthesizing feature maps of different resolutions. One of the Deep fusion blocks is shown on the right side of Fig. 3 and consists of two layers of Affine transformation and Relu activation function stacking; the Affine layer contains two MLP networks, and the input text feature vector s is used to obtain the channel scaling scale parameter a and the shift parameter b.

$$\begin{aligned} a &= MLP_1(s) \\ b &= MLP_2(s) \end{aligned} \tag{1}$$

Thus, a more profound fusion of image features and text features is achieved, and the Affine transformation process is as follows:

$$affine\left(f_i \mid s\right) = a_i \cdot f_i + b_i \tag{2}$$

where $affine$, f_i, and s denote the Affine transformation, the $i^{th} channel$ of the image feature, and the text feature vector, respectively. The generator synthesizes the image process as follows:

$$\begin{cases} FM_0 = FC(z) \\ FM_i = F_i\left(FM_{i-1}, z, s\right)(i = 1, 2, 3, \cdots, 7) \\ I \quad = Conv\left(FM_7\right) \end{cases} \tag{3}$$

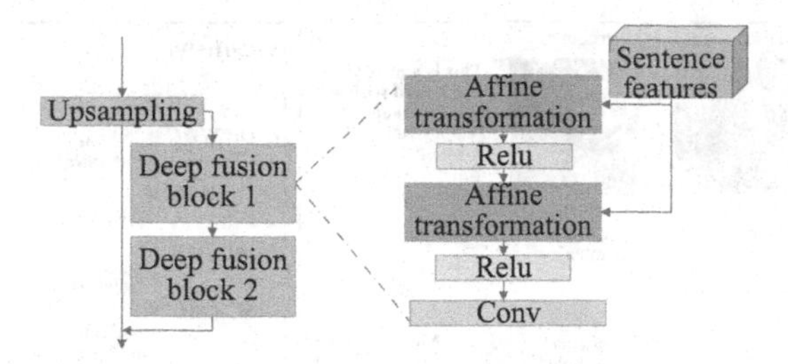

Fig. 3. A single Upsampling fusion block in the generator network. The Upsampling fusion block first upsamples the image features and then deeply fuses the text and image features. This synthesizes feature maps at different resolutions. The Deep fusion block consists of two layers: Affine transformation and ReLU activation function stacking.

where z is the random noise sampled in the standard normal distribution, FC is the Fully connected layer, F_i is the Upsampling fusion block, Conv is the Convolutional network that generates image I, FM_0 is the feature vector after the Fully connected layer, and $FM_1 - FM_7$ are the intermediate representations of the output of the Upsampling fusion block layer.

Our discriminator employs an innovative approach to distinguish synthetic images from real images. First, matching-aware equipped with a threshold is one of the creative design strategies in this study. It uses the threshold to filter inaccurate negative samples to enhance text-image semantic coherence. Then, the convolutional network and six downsampling layers convert the images into image features. Then, the newly introduced word feature vectors can facilitate the identification of image details by the novel word fine-grained supervisor designed in this study. Finally, sentence vectors are spatially replicated and connected with image features to predict a matching loss to determine text-image semantic consistency.

2.2 Inaccurate Negative Sample Filter (INSF)

The discriminator receives three types of inputs: <generated image, matched description>, <true image, matched description>, and <error image, matched description>, which are expressed concretely as shown on the left in Fig. 4, for a negative sample such as <error image, matched description>, where the error image is randomly selected from the dataset. Therefore, for the error image in <error image, matched description>, this so-called negative sample may already contain much of the content of the matched description. However, it is still directly input as a negative sample, which will negatively affect the learning of the matching-aware mechanism in the discriminative model.

The discriminator receives negative samples and passes the signal of image-text non-match to the generator. If the semantic similarity of the error image and matched description in the randomly sampled negative sample(<error image, matched description>) is higher at this time, it will disturb the generative model to synthesize the image according to the above description instead. To address

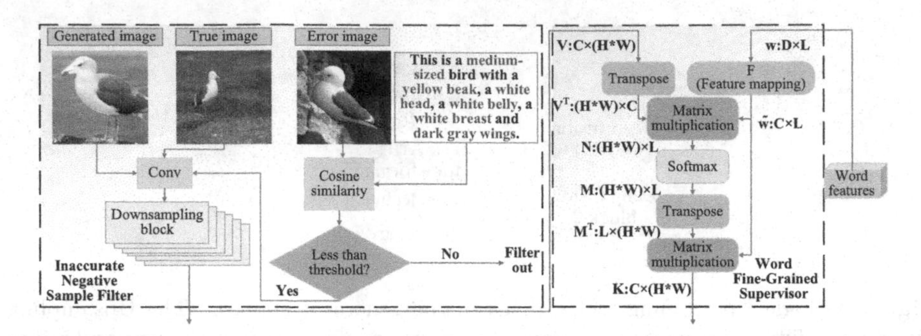

Fig. 4. The discriminator's network structure. At the top are the received image types (generated, true, and error image) and matched text description, and a newly designed inaccurate negative sample filter and word fine-grained supervisor facilitate the generation of images that match the texts by effectively applying negative sample information to image generation as well as enabling supervision of image subregions.

this limitation, we compute the semantic similarity that already exists for the <error image, matched description> pair; for this, to reduce the computational complexity of the different modalities between the error image and the description, it is converted to compute the text semantic similarity between the original text feature $s_{error} \in R^D$ of the corresponding error image and the text feature $s_{real} \in R^D$ of the matched description, expressed as follows:

$$\sigma = \frac{s_{error} \cdot s_{real}}{\|s_{error}\| \, \|s_{real}\|} \tag{4}$$

where σ represents the cosine similarity existing between the error image and matched description in <error image, matched description>, and this study sets different thresholds between $(0, 1)$ to filter the inaccurate negative samples, and then to learn the matching-aware mechanism. Finally, we can find a relatively good threshold to act as a filter to filter out inaccurate negative samples, and instead, good negative samples will be left to enhance the matching-aware mechanism. As a result, it can help the discriminative model to determine better whether the texts match the images. In adversarial networks, the enhanced discriminative model can guide the generative model to generate image content that is more relevant to the text descriptions.

2.3 Word Fine-Grained Supervisor (WFGS)

To make the newly designed threshold-equipped matching-aware discriminator better, we creatively introduce a cross-modal attention mechanism in the discriminator so that the generator can get fine-grained feedback on each visual attribute from the word-level discriminator. This is shown on the right in Fig. 4. The word fine-grained supervisor in this study has two inputs: (1) visual features (true image features, generated image features, and error image features) $V \in R^{C \times (H*W)}$, where H and W denote the height and width of the feature map, respectively; and (2) word features $w \in R^{D \times L}$ encoded by the text encoder. Word features w are first mapped to the same semantic space as the visual features V through the feature mapping layer F.

$$\widetilde{w} = F(w) \tag{5}$$

where $F \in R^{C \times D}$. The similarity weight matrix $N \in R^{(H*W) \times L}$ is then obtained.

$$N = V^T \widetilde{w} \tag{6}$$

The softmax function normalizes the matrix N to obtain the correlation matrix $M \in R^{(H*W) \times L}$, where $M_{i,j}$ represent the ith sub-region of the image with the jth word correlation value of the sentence.

$$M_{i,j} = \frac{\exp(N_{i,j})}{\sum_{l=1}^{L} \exp(N_{i,l})} \tag{7}$$

Finally, the semantically enhanced image features $K \in R^{C \times (H*W)}$, which combine all word information weighted by the word-context normalization matrix M, can be obtained as follows:

$$K = \widetilde{w} M^T \tag{8}$$

By applying the cross-modal attention mechanism, it can be achieved to assign word features to relevant image sub-regions, thus helping the discriminator to determine the detail-matching degree of the image-text pairs.

2.4 Loss Function

The matching loss is designed to measure the degree of semantic matching between the synthetic images and the text descriptions. We use hinge loss to optimize the training of the adversarial network, and the adversarial loss function of the generator is shown as:

$$L_G = -E_{x \sim P_G}[D(x, s)] \tag{9}$$

The adversarial loss function of the corresponding discriminator is shown as:

$$\begin{aligned} L^D_{Adversarial} = &- E_{x \sim P_{data}}[\min(0, -1 + D(x, s))] \\ &- (1/2) E_{x \sim P_G}[\min(0, -1 - D(x, s))] \\ &- (1/2) E_{x \sim P_{error}}[\min(0, -1 - D(x, s))] \end{aligned} \tag{10}$$

where P_{data}, P_G, and P_{error} are the true image, generated image, and error image data distributions, respectively, and s is the sentence feature vector. We introduce MA-GP loss in the discriminator for gradient optimization as in DF-GAN [15] to make the generated image closer to the true image data distribution.

$$L_{MA-GP} = \lambda_1 E_{x \sim P_{data}}[(\|\nabla_x D(x, s)\| + \|\nabla_s D(x, s)\|)^{\lambda_2}] \tag{11}$$

where λ_1 and λ_2 are hyperparameters used to balance the gradient optimization and are set to 6 and 2, respectively. The loss function of the discriminator is:

$$L_D = L^D_{Adversarial} + L_{MA-GP} \tag{12}$$

3 Experiment

In this section, the datasets, training details, and evaluation metrics used in the experiments are first presented. This is followed by quantitative and qualitative evaluations of ETMA-GAN and its variants.

Datasets. We evaluated the proposed model on two challenging datasets, namely CUB [16] and COCO [6]. The CUB dataset contains 8,855 training images and 2,933 test images of birds belonging to 200 species, each with 10 text descriptions. The COCO dataset contains 82,783 training images and 40,504 validation images. Each image in this dataset has 5 sentence descriptions.

Training Details. The text encoder is frozen during network training. We optimized the network using the Adam optimizer, where the generator has a learning rate of 0.0001, and the discriminator has a learning rate of 0.0004. The model was trained on the CUB and COCO datasets using a small batch size of 32 for 1300 and 300 epochs, respectively. It took about five days to train the CUB dataset and two weeks to train the COCO dataset on a single NVIDIA Tesla $V100$ GPU.

Evaluation Metrics. We use the widely used Inception Score (IS) [12] and Fréchet Inception Distance (FID) [2] to evaluate the performance of our proposed network. A higher IS indicates that the obtained image has higher quality. A lower FID score indicates that the synthesized image is closer to the real one. Each model generates 30,000 images from the corresponding dataset to calculate the IS and FID scores. These images are generated from randomly selected text descriptions in the test dataset. Previous works [4,5,15] have demonstrated that IS cannot be evaluated well for synthetic images on the COCO dataset. Therefore, we compare only the FID score for the COCO dataset.

3.1 Quantitative Evaluation

Table 1 shows the results for the CUB and COCO datasets and the Number of Parameters (NoP) used in the generative model. With our ETMA-GAN, the number of parameters of the generator is significantly reduced compared to other dominant models, but the performance is still the best. The method of this study obtained the highest IS (4.96 ± 0.08) and the lowest FID score (11.02) compared to the existing state-of-the-art methods in the CUB dataset. The lowest FID score (14.45) was obtained from the COCO dataset, with a 10% improvement compared to DF-GAN [15]. According to the quantitative results, the newly proposed matching-aware equipped with a threshold strategy in the research demonstrates excellent performance in terms of quality of image synthesis and consistency with text semantics, outperforming the existing state-of-the-art methods.

Table 1. Performance of IS and FID on the CUB and COCO test set for different state-of-the-art methods and our method, as well as the Number of Parameters (NoP) used on their respective generators. The results are taken from the author's own paper. Note that the data reported in DF-GAN† [15] are based on DF-GAN's [15] publicly available code on Github tested under the same experimental conditions as this study. The best result is highlighted in bold.

Model	IS ↑	FID ↓		NoP↓
	CUB	CUB	COCO	
StackGAN [18]	3.70 ± 0.04	–	–	–
StackGAN++ [19]	4.04 ± 0.06	15.30	81.59	–
AttnGAN [17]	4.36 ± 0.03	23.98	35.49	230M
MirrorGAN [8]	4.56 ± 0.05	18.34	34.71	–
DM-GAN [20]	4.75 ± 0.07	16.09	32.64	46M
DF-GAN† [15]	4.88 ± 0.04	12.63	16.07	**10M**
DAE-GAN [11]	4.42 ± 0.04	15.19	28.12	98M
TIME [7]	4.91 ± 0.03	14.30	31.14	120M
ETMA-GAN	**4.96 ± 0.08**	**11.02**	**14.45**	**10M**

Fig. 5. Qualitative comparison of our proposed method with AttnGAN [17], and DF-GAN [15] on the CUB dataset (columns 1–3) and COCO dataset (columns 4–6).

3.2 Qualitative Evaluation

In Fig. 5, the visualization results of AttnGAN [17], DF-GAN [15], and our proposed ETMA-GAN on the CUB and COCO datasets are compared. AttnGAN is a classical multi-stage approach with an attention mechanism, while DF-GAN is a single-stage text-to-image synthesis method. As shown in columns 1–3 in Fig. 5, on the CUB dataset, our model is much clearer in terms of details, such as beaks, claws, and feathers. In addition, the overall effect is closer to the real images in our ETMA-GAN synthesis results. As shown in columns 4–6 in Fig. 5, the advantage of our method is even more significant on the COCO dataset, where it is more realistic than the results synthesized by AttnGAN and DF-GAN and more consistent with the text descriptions.

Table 2. Ablation study of our model on the test set of the CUB dataset by setting different thresholds in the Inaccurate Negative Sample Filter(INSF). This filters the negative samples to varying degrees without adding a Word Fine-Grained Supervisor(WFGS).

ID	Components		FID ↓
	INSF	WFGS	
0	–	–	12.63
1	✓(Threshold = 0.20)	–	11.67
2	**✓(Threshold = 0.35)**	–	**11.06**
3	✓(Threshold = 0.50)	–	11.11
4	✓(Threshold = 0.65)	–	11.43
5	✓(Threshold = 0.80)	–	12.58

Table 3. Ablation study of our model on the test set of the CUB dataset with the Inaccurate Negative Sample Filter(INSF) and Word Fine-Grained Supervisor(WFGS).

ID	Components		FID ↓
	INSF	WFGS	
0	–	–	12.63
1	✓(Threshold = 0.35)	–	11.06
2	✓(Threshold = 0.35)	✓	**11.02**

3.3 Ablation Studies

For the ablation experiments, Tables 2 and 3 give the results of our quantitative evaluation of the FID scores for the model components. This is from the CUB dataset test set. ID_0 represents the baseline DF-GAN [15]. $ID_1 - ID_5$ in Table 2, setting different thresholds to filter negative samples at different levels, respectively, compared with ID_0, we can all see that matching-aware equipped with a threshold has a significant improvement on the model performance.

ID_2 in Table 3 adds cross-modal word-level attention to the discriminator based on a threshold set to 0.35 (taken from the optimal threshold ID_2 in Table 2). The result shows improved performance, indicating that the word fine-grained supervision module can further enhance the matching-aware equipped with a threshold discriminator and overall help the generator synthesize higher quality images.

4 Conclusion and Future Work

In this paper, we use the newly proposed Equipped with Threshold Matching-Aware GAN(ETMA-GAN) to solve the problem of negative samples during text-to-image synthesis. During previous studies, only randomly sampled negative samples to obtain. More advanced, our model can effectively filter out inaccurate negative samples. This will accurately guide the generator to synthesize images that match text semantics. In addition, we fuse the cross-modal word-level attention model into the discriminator to supervise image fine-grained regions and improve semantic consistency between generated images and texts. Experimental results on the CUB and COCO datasets show that our approach can significantly improve the quality of the generated images. In the future, we will continue exploring how to apply the threshold-equipped matching-aware mechanism to other conditional image synthesis tasks to enhance its effectiveness.

References

1. Goodfellow, I., et al.: Generative adversarial nets. In: Proceedings of the 27th International Conference on Neural Information Processing Systems, NIPS 2014, vol. 2, pp. 2672–2680 (2014)
2. Heusel, M., Ramsauer, H., Unterthiner, T., Nessler, B., Hochreiter, S.: GANs trained by a two time-scale update rule converge to a local Nash equilibrium. In: Advances in Neural Information Processing Systems, vol. 30 (2017)
3. Li, B., Qi, X., Lukasiewicz, T., Torr, P.: Controllable text-to-image generation. In: Advances in Neural Information Processing Systems, vol. 32 (2019)
4. Li, W., et al.: Object-driven text-to-image synthesis via adversarial training. In: Proceedings of the IEEE/CVF Conference on Computer Vision and Pattern Recognition, pp. 12174–12182 (2019)
5. Liao, W., Hu, K., Yang, M.Y., Rosenhahn, B.: Text to image generation with semantic-spatial aware GAN. In: Proceedings of the IEEE/CVF Conference on Computer Vision and Pattern Recognition, pp. 18187–18196 (2022)

6. Lin, T.-Y., et al.: Microsoft COCO: common objects in context. In: Fleet, D., Pajdla, T., Schiele, B., Tuytelaars, T. (eds.) ECCV 2014. LNCS, vol. 8693, pp. 740–755. Springer, Cham (2014). https://doi.org/10.1007/978-3-319-10602-1_48
7. Liu, B., Song, K., Zhu, Y., de Melo, G., Elgammal, A.: Time: text and image mutual-translation adversarial networks. arXiv abs/2005.13192 (2020)
8. Qiao, T., Zhang, J., Xu, D., Tao, D.: MirrorGAN: learning text-to-image generation by redescription. In: Proceedings of the IEEE/CVF Conference on Computer Vision and Pattern Recognition, pp. 1505–1514 (2019)
9. Radford, A., Metz, L., Chintala, S.: Unsupervised representation learning with deep convolutional generative adversarial networks. arXiv preprint arXiv:1511.06434 (2015)
10. Reed, S., Akata, Z., Yan, X., Logeswaran, L., Schiele, B., Lee, H.: Generative adversarial text to image synthesis. In: International Conference on Machine Learning, pp. 1060–1069. PMLR (2016)
11. Ruan, S., et al.: DAE-GAN: dynamic aspect-aware GAN for text-to-image synthesis. In: 2021 IEEE/CVF International Conference on Computer Vision (ICCV), pp. 13940–13949 (2021). https://doi.org/10.1109/ICCV48922.2021.01370
12. Salimans, T., Goodfellow, I., Zaremba, W., Cheung, V., Radford, A., Chen, X.: Improved techniques for training GANs. In: Advances in Neural Information Processing Systems, vol. 29 (2016)
13. Schuster, M., Paliwal, K.K.: Bidirectional recurrent neural networks. IEEE Trans. Signal Process. **45**(11), 2673–2681 (1997)
14. Tan, H., Liu, X., Li, X., Zhang, Y., Yin, B.: Semantics-enhanced adversarial nets for text-to-image synthesis. In: 2019 IEEE/CVF International Conference on Computer Vision (ICCV), pp. 10500–10509 (2019). https://doi.org/10.1109/ICCV.2019.01060
15. Tao, M., Tang, H., Wu, F., Jing, X., Bao, B.K., Xu, C.: DF-GAN: a simple and effective baseline for text-to-image synthesis. In: 2022 IEEE/CVF Conference on Computer Vision and Pattern Recognition (CVPR), pp. 16494–16504 (2022). https://doi.org/10.1109/CVPR52688.2022.01602
16. Wah, C., Branson, S., Welinder, P., Perona, P., Belongie, S.: The Caltech-UCSD birds-200-2011 dataset (2011)
17. Xu, T., et al.: AttnGAN: fine-grained text to image generation with attentional generative adversarial networks. In: Proceedings of the IEEE Conference on Computer Vision and Pattern Recognition, pp. 1316–1324 (2018)
18. Zhang, H., et al.: StackGAN: text to photo-realistic image synthesis with stacked generative adversarial networks. In: Proceedings of the IEEE International Conference on Computer Vision, pp. 5907–5915 (2017)
19. Zhang, H., et al.: StackGAN++: realistic image synthesis with stacked generative adversarial networks. IEEE Trans. Pattern Anal. Mach. Intell. **41**(8), 1947–1962 (2018)
20. Zhu, M., Pan, P., Chen, W., Yang, Y.: DM-GAN: dynamic memory generative adversarial networks for text-to-image synthesis. In: Proceedings of the IEEE/CVF Conference on Computer Vision and Pattern Recognition, pp. 5802–5810 (2019)

Joint Regularization Knowledge Distillation

Haifeng Qing[1], Ning Jiang[1,2](✉), Jialiang Tang[3], Xinlei Huang[1,2], and Wengqing Wu[1]

[1] School of Computer Science and Technology, Southwest University of Science and Technology, Mianyang 621000, Sichuan, China
jiangning@swust.edu.cn

[2] Jiangxi Qiushi Academy for Advanced Studies, Nanchang 330036, Jiangxi, China

[3] School of Computer Science and Engineering, Nanjing University of Science and Technology, Nanjing 210094, Jiangsu, China

Abstract. Knowledge distillation is devoted to increasing the similarity between a small student network and an advanced teacher network in order to improve the performance of the student network. However, these methods focus on teacher and student networks that receive supervision from each other independently and do not consider the network as a whole. In this paper, we propose a new knowledge distillation framework called Joint Regularization Knowledge Distillation (JRKD), which aims to reduce network differences through joint training. Specifically, we train teacher and student networks through joint regularization loss to maximize consistency between the two networks. Meanwhile, we develop a confidence-based continuous scheduler method (CBCS), which divides examples into center examples and edge examples based on the example confidence distribution of network output. Prediction differences between networks are reduced when training with a central example. Teacher and student networks will become more similar as a result of joint training. Extensive experimental results on benchmark datasets such as CIFAR-10, CIFAR-100, and Tiny-ImagNet show that JRKD outperforms many advanced distillation methods.

Keywords: Knowledge Distillation · Joint regularization · Continuous scheduler

1 Introduction

Over the past few decades, deep neural networks (DNNs) have enjoyed great success in computer vision fields [20,25], such as real-time semantic segmentation [7], object detection [15]. However, powerful DNNs frequently have larger parameters and require large computational and storage resources, which are undesirable for industrial applications. To address this issue, a number of model compression techniques have been proposed, including model pruning [2,16,30], quantification [9], and knowledge distillation [5], with knowledge distillation proving to be a mature method for improving the performance of small models.

B. Luo et al. (Eds.): ICONIP 2023, CCIS 1966, pp. 173–184, 2024.
https://doi.org/10.1007/978-981-99-8148-9_14

Traditional knowledge distillation (KD) [5] (see Fig. 1(a)) utilizes a soft label from a pretrained teacher to supervise students to obtain similar performance to the teacher, which is a two-stage training process and not flexible. Recently, online knowledge distillation [14,25] proposed a single-stage scheme to encourage networks to train each other and retrain teacher and student networks to improve consistency on different points of view [1,4,23]. For example (see Fig. 1(b)), deep mutual learning (DML) [22] predictions after the classifier of teacher and student. Chung *et al.* [3] introduces the middle layer feature transition between teacher and student, as shown in Fig. 1(c). The existing online knowledge distillation method is a way of teaching and learning collaboratively, and we hope to further enhance this collaboration, bringing students and teachers together as a whole.

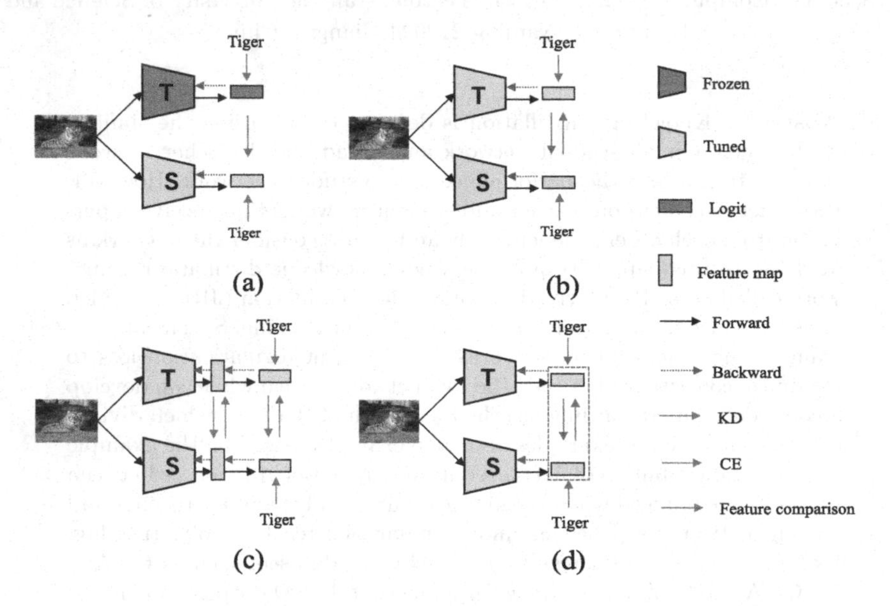

Fig. 1. Illustration of (a) KD, (b) DML, (c) DML with Feature comparison, and (d) Knowledge distillation framework with joint regularization loss.

When the differences between teacher and student models are too great, distillation can adversely affect students [17]. Strengthening connections between teacher and student networks can improve distillation performance, from traditional knowledge distillation to online knowledge distillation (see Sect. 4.2, "Proof"). Recently, Wei *et al.* [29] proposes a robust federated learning method called Jocor to maximize similarity between DNNs by reducing their. Based on this insight, we believe that the teacher and student can obtain consistent joint supervision in predictions, enhancing the integrity of the two classifiers, and thus improving the distillation performance, as shown in Fig. 1(d).

This paper proposes a new knowledge distillation framework called Joint Regularization Knowledge Distillation (JRKD). Specifically, we train teacher and student networks through joint losses to maximize consistency between the two networks. Inspired by "course learning" [21], we propose a confidence-based continuous scheduler method (CBCS), which divides examples into center examples and edge examples based on their-confidence density distributions calculated teacher and student. The central example reduces the prediction error in the joint training of the two networks, promotes their mutual learning, and reduces the accumulation of error streams in the network. The proportion of central examples gradually increases as the training process progresses, ensuring the integrity of the training set. Extensive experiments on three representative benchmarks have shown that our JRKD can effectively train a high-performance student network.

1. We propose a joint regularized knowledge distillation method(JRKD), which can effectively reduce the differences between networks.
2. We used federated regularized loss to normalize teacher and student networks to maximize consistency across networks.
3. We develop a confidence-based continuous scheduling method (CBCS), through which the selection of loss instances can mitigate the negative impact between networks and reduce the difficulty of consistency training.

2 Related Literature

In this section, we will discuss the work related to online knowledge distillation and Disagreement. In both areas, various approaches have been proposed over the past few years. We summarize it below.

2.1 Online Knowledge Distillation

Traditional knowledge distillation is achieved by a network of pre-trained teachers who take their knowledge (extracted logits [5] or intermediate feature forms [11]) and guide students to models during the training process. This method is simple and effective, but it requires a high-performance teacher model. In online knowledge distillation, the teacher and student models update their own parameters at the same time to achieve an end-to-end process. The concept of online distillation was first proposed by Zhang et al. [22] to enable collaborative learning and mutual teaching between students and teachers. In order to address the impact between network mutual learning, SwitKD [25]adaptively calibrates the gap during the training phase through a switching strategy between the two modes of expert mode (pause teacher, keep student learning) and learning mode (restart teacher), so that teachers and students have appropriate distillation gaps when learning from each other. Chung et al. [26] adds a feature-map-based judgment to the original logit-based prediction, and the feature-map-based loss controls the teacher and student to distill each other through the adjudicator.

Online distillation is a single-stage training scheme with efficient parallel computation. The existing online knowledge distillation method is a way of teaching and learning collaboratively, and we hope to further enhance this collaboration, bringing students and teachers together as a whole, rather than individually.

2.2 Disagreement

Weakly supervised learning [27] solves the problem of time-consuming and labor-intensive collection of large and accurate data sets, and the use of online queries and other methods will inevitably be affected by noise labels. In recent years, the "divergence" strategy has been introduced to address such issues. For example, decoupling [19]uses two different networks, and when there is no difference in the predictions of the two networks, the network parameters are not updated, and the network is updated when there is a disagreement. "Divergence" strategy expectations use these examples that produce different predictions to steer the network away from current errors. In 2019, Chen et al [12] combined the "divergence" strategy with Co- teaching [28] in collaborative teaching to provide good performance in terms of DNN's robustness to noise tags. Recently, Wei et al. [29] proposed a robust learning paradigm called JoCoR from different perspectives, which aims to reduce the diversity of training examples of two networks during training, and update the parameters of two networks at the same time by selecting examples with small losses. Under the training of joint loss, the two networks will become more and more similar due to the effects of coregularization.

We hope to be able to use the idea of "divergence" strategy in the field of knowledge distillation, aiming to reduce the differences between teacher and student networks, thereby improving the integrity between networks and improving distillation performance.

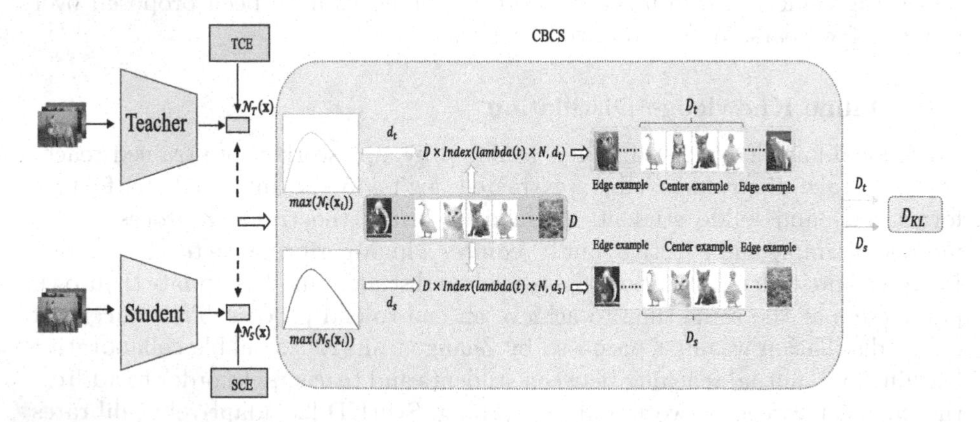

Fig. 2. JRKD flowchart, CBCS selects a central example based on the network output's example confidence, and teachers and students receive joint supervision training.

3 Approach

In this section, we will discuss how CBCS selects central examples (Sect. 3.1) and how joint regularization loss trains the network collaboratively (Sect. 3.2).

3.1 Confidence-Based Continuous Scheduler

According to recent research [29], while networks can improve consistency between them through joint regularization, they are vulnerable to error streams caused by biased selection. To address this issue, we design a confidence-based continuous scheduler (CBCS) that divides the example into center examples and edge examples. Using central example training can better reduce the prediction bias between networks. This is shown in Fig. 2.

Different center examples are chosen by teachers and students; we only show how teachers choose, and students do the same. We use dataset $\mathcal{D} = \{(\boldsymbol{x}_i, y_i)\}_{i=1}^{n}$ as the network input for each batch with n examples. Let the teacher network be $\mathcal{N}_T$, and the prediction probability on the dataset $\mathcal{D}$ be $\mathcal{N}_T(\mathbf{x}_i)_{i=1}^{n}$. For Class m classification tasks, $\max(\mathcal{N}_S(\mathbf{x}_i))$ represents the maximum confidence that the teacher network prediction instance $\boldsymbol{x}_i$ is for one of the classes in class m. The KMeans clustering algorithm is used to obtain the maximum confidence centroid of n examples:

$$M\_p_{\text{target}} = \frac{\sum_{i=1}^{N} \max(\mathcal{N}_S(\mathbf{x}_i))}{N}. \tag{1}$$

M_p_{target} is the centroid of confidence. Calculating the absolute distance from each $\max(\mathcal{N}_S(\mathbf{x}_i))$ to M_p_{target} yields the set $d_t = [d_1, d_2, \ldots, d_n]$. The smaller the value in d_t, the closer to the confidence center.

CBCS controls the central example proportion for each period through a continuous scheduling functions $lambda(t)$. T_{total} is the total training cycle, λ_0 Represents the proportion of the initial central example selection, t stands for epock currently trained:

$$lambda(t) = \min\left(1, \lambda_0 + \frac{1-\lambda_0}{T_{\text{total}}} \cdot t\right). \tag{2}$$

By using d_t as the basis for selecting the central example, the example size is controlled by the $lambda(t)$. Index$(\cdot,\cdot)$ is a function method that returns an index of multiple minimum values. Get the current set of central example D_t, the same process can also obtain D_s:

$$D_t = D \times Index\left(\text{lambda}(t) \times N, d_t\right), \quad D_t \in D, \tag{3}$$

$$D_s = D \times Index\left(\text{lambda}(t) \times N, d_s\right), \quad D_s \in D. \tag{4}$$

3.2 Joint Regularization Knowledge Distillation

For the multi-class classification task for class m. We use two deep neural networks to express the proposed JRKD method. For clarity, we set $p_s = [p_s^1, p_s^2 \ldots, p_s^m]$ and $p_t = [p_t^1, p_t^2, \ldots, p_t^m]$ as the final prediction probabilities of the example $\boldsymbol{x}_i$ by students and teachers, respectively. It is obtained by softening the network output by the $softmax$ function of distillation temperature $T = 3$.

Joint Regularization Loss. We train the two networks together using joint regularization loss, which brings the predictions of each network closer to the peer-to-peer network. Under joint training, networks will become more and more similar to each other. To accomplish this, asymmetric Kullback-Leibler (KL) divergence is used:

$$\mathcal{L}_{con} = D_{\mathrm{KL}}\left(\boldsymbol{p}_s \| \boldsymbol{p}_t\right) + D_{\mathrm{KL}}\left(\boldsymbol{p}_s \| \boldsymbol{p}_t\right). \tag{5}$$

$\mathcal{L}_{con}$ represents the joint regularization loss. CBCS selects different central examples to participate in joint training based on the confidence probability of the examples generated by teachers and students:

$$D_{\mathrm{KL}}\left(\boldsymbol{p}_s \| \boldsymbol{p}_t\right) = \sum_{i=1}^{N}\sum_{m=1}^{M} p_s^m\left(\boldsymbol{x}_i\right)\log\frac{p_s^m\left(\boldsymbol{x}_i\right)}{p_t^m\left(\boldsymbol{x}_i\right)}, x \in D_t, \tag{6}$$

$$D_{\mathrm{KL}}\left(\boldsymbol{p}_t \| \boldsymbol{p}_s\right) = \sum_{i=1}^{N}\sum_{m=1}^{M} p_t^m\left(\boldsymbol{x}_i\right)\log\frac{p_t^m\left(\boldsymbol{x}_i\right)}{p_s^m\left(\boldsymbol{x}_i\right)}, x \in D_s. \tag{7}$$

Total Losses. For JRKD, the joint regularization loss is used to improve the integrity between the networks, and the conventional supervision loss is used to maintain the correctness of the learning. JRKD minimizes the following losses to train the network:

$$\mathcal{L}_T = \mathcal{L}_{TCE} + \mathcal{L}_{con}, \tag{8}$$

$$\mathcal{L}_S = \mathcal{L}_{SCE} + \mathcal{L}_{con}. \tag{9}$$

$\mathcal{L}_{SCE}$ and $\mathcal{L}_{TCE}$ represent conventional supervision loss for students and teachers, respectively. Finally, we give the algorithm flow table of JRKD, as shown in Algorithm 1.

Algorithm 1. JRKD

Input: Network f with $\boldsymbol{\Theta} = \{\boldsymbol{\Theta}_t, \boldsymbol{\Theta}_s\}$, learning rate η, fixed τ, epoch T_k and $T_{\max}$, iteration $I_{\max}$;

1: **for** $t = 1, 2, \ldots, T_{\max}$ **do**
2: Shuffle training set D;
3: **for** for $n = 1, \ldots, I_{\max}$ **do**
4: Fetch mini-batch D_n from D ;
5: $p_s = f_s\left(\boldsymbol{x}, \boldsymbol{\Theta}_s\right), \forall \boldsymbol{x} \in D_n$;
6: $p_t = f_t\left(\boldsymbol{x}, \boldsymbol{\Theta}_t\right), \forall \boldsymbol{x} \in D_n$;
7: Calculate the example size by (2) from $lambda(t)$;
8: Obtain training subset D_s,$D_t$by (3,4)from D_n ;
9: Obtain L_S,L_T by (8, 9) on D_s, D_t ;
10: Update $\boldsymbol{\Theta t} = \boldsymbol{\Theta t} - \eta\nabla L_T, \boldsymbol{\Theta s} = \boldsymbol{\Theta s} - \eta\nabla L_S$;
11: **end for**
12: **end for**

Output: $\boldsymbol{\Theta}_s$ and $\boldsymbol{\Theta}_t$

4 Experiments

In this section, we select three representative image classification tasks for experiments in Sect. 3.1 to evaluate the performance of JRKD. The ablation experiment at Sect. 3.2 confirmed the effectiveness of CBCS and loss of joint regularization. In addition, we analyze the effect of λ_0 initial center example ratio on performance. In Sect. 3.3, visualize the probability distribution of teacher and student network outputs.

Experiment Setup. The configuration of our experiment is to descend SGD with a stochastic gradient and set thc learning rate, weight decay, and momentum to 0.1, 5×10^{-4}, and 0.9, respectively. The dataset uses a standard data augmentation scheme and normalizes [17] the input image using channel means and standard deviations.

4.1 Experiments on Benchmarks

Results on Tiny-ImageNet. It contains 200 categories, each containing 500 training images, 50 validation images, and 50 test images. After using JRKD, the two groups of networks obtained an accuracy of 59.43% and 55.71%, respectively. It can effectively improve the accuracy of the student network. Compare these methods, Our method also achieves good results. The results are shown in Table 1.

Table 1. The accuracy of the comparison method comes from the papers of other authors. JRKD verified accuracy results on the Tiny-ImageNet dataset.

Teacher	ResNet34	WRN40-2
Student	MobileNetV2	ResNet20
DML	55.70	53.98
KDCL	57.79	53.74
SwitOKD [25]	58.79	55.03
JRKD	**59.43**	**55.71**

Results on CIFAR-100. The CIFAR-100 dataset has 100 classes. Each class has 500 sheets as a training set and 100 as a test set. Table 2 shows the experimental results, and JRKD outperforms many other methods on various network architectures. Impressively, JRKD achieves 1.33% (WRN-40-2/WRN-16-2) accuracy improvement to DML on CIFAR-100. Besides, JRKD also shows 0.88% and 0.19% (ResNet32 × 4/ResNet8 × 4) accuracy gain over ReviewKD and DKD, respectively.

Results on CIFAR-10. The CIFAR-10 dataset has a total of 60,000 examples, which are divided into 50,000 training examples and 10,000 test examples. The

Table 2. JRKD verified accuracy results on the CIFAR-100 dataset. W40-2, R32x4, R8x4 and SV1 stand for WRN-40-2, ResNet32 × 4, ResNet8 × 4,ShuffleNetV1. The accuracy of other methods is mainly derived from DKD [22].

Teacher Student	W40-2 W16-2	W40-2 SV1	R32 ×4 R8 ×4	VGG13 VGG8
Teacher	75.61	75.61	79.42	74.64
Student	73.26	70.50	72.50	70.36
KD [5]	74.92	74.83	73.33	72.98
FitNets [8]	73.58	73.73	73.50	71.02
RKD [10]	73.59	72.21	71.90	71.48
CRD [6]	75.48	76.05	75.51	73.94
AT [11]	74.08	73.32	73.44	71.43
CC [13]	75.66	71.38	72.97	70.71
DML [18]	75.33	75.58	74.30	73.64
KDCL [14]	74.25	74.79	74.03	71.26
ReviewKD [19]	76.12	77.14	75.63	N/A
DKD [22]	76.24	76.70	76.32	74.68
JRKD	**76.66**	**77.24**	**76.51**	**74.90**

experimental results are shown in Table 3, using the same experimental configuration as other methods. Our method not only improved student performance, the teacher achieved an accuracy gain of 0.23% and 0.8% over SwithOKD and KDCL, respectively.

Table 3. Ours results are the average over 5 trials. Comparison of performances with powerful distillation techniques using the same 200 training epochs. Performance metrics refer to the original article.

	Backbone	KDCL	SwithOKD	JRKD
Student	WRN-16-1	91.86	92.50	**93.11**
Teacher	WRN-16-8	95.33	94.76	**95.56**

4.2 Ablation Experiments

CIFAR-100 was chosen for the dataset of the ablation experiment. As shown in Table 4, we quantified the gap between teachers and networks using T-S gap, and compared KD and DML, JRKD can effectively reduce the differences between networks and improve distillation performance. The JRKD† compared other distillation methods and showed that joint regularization loss can improve similarity between networks. The comparison of JRKD and JRKD† shows that CBCS is beneficial for online training. In addition, the sensitivity analysis of the λ_0 parameter manually set in the continuous scheduler $lambda(t)$ was performed. As shown in Table 5, The value of λ_0 in the continuous scheduler generally defaults to 0.3, so we only analyze the value around 0.3 and find that the appropriate λ_0 is conducive to distillation.

Table 4. Verify the effectiveness of joint regularization losses and CBCS. The student network is MobileNetV2, the teacher is the VGG13, $KD_{T \to S}$ represents the teacher network to accept student supervision, Top-1 is the classification accuracy of CIFAR-100, T-S gap uses KL to calculate the gap between output logical values between networks. JRKD† refers to the absence of CBCS to select loss instances.

Method	$KD_{T \to S}$	$KD_{S \to T}$	Top-1	T-S gap
KD	✔	✘	67.37	1.12
DML	✔	✔	68.52	0.83
JRKD†	✔	✔	69.12	0.61
JRKD	✔	✔	69.55	0.42

Table 5. The parameter sensitivity experiment of the continuous scheduler of CBCS. The experimental data set uses CIFAR-100, and the experimental accuracy result is averaged 5 times.

	λ_0	0.2	0.3	0.4
Student	WRN-16-2	76.35	**76.66**	76.43
Teacher	WRN-40-2	78.40	**78.89**	78.47

4.3 Visual Analytics

We compare the traditional online knowledge distillation method DML and the JRKD by feeding the same batch of examples into the trained network and visualizing the confidence distribution of the examples by the teacher-student network. As shown in Fig. 3, the confidence distribution of the teacher-student network is more similar in the example output of JRKD, demonstrating that JRKD can improve network similarity.

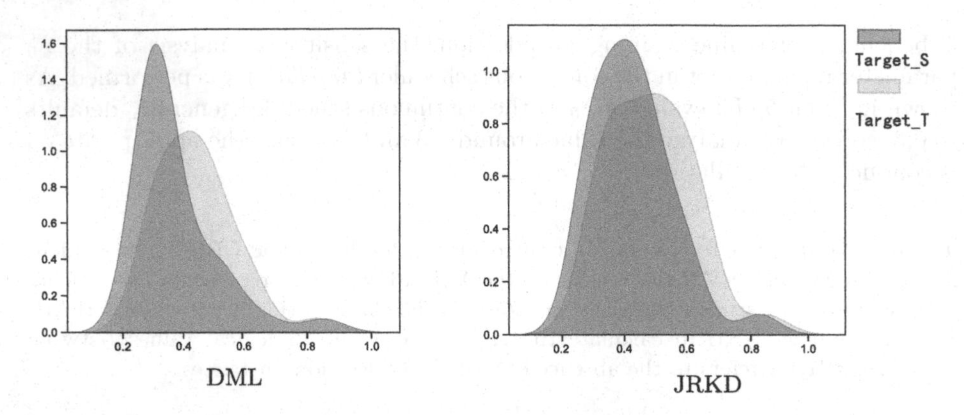

Fig. 3. Two different methods produce confidence profiles.

5 Conclusion

This paper proposes an effective method called JRKD to reduce the differences between networks. The key idea of JRKD is to train the teacher and student networks by jointly regularizing losses to maximize consistency between the two networks. In order to reduce the difficulty of federation, we developed a confidence-based continuous scheduling method (CBCS), which can divide samples into central samples and edge samples according to the sample confidence distribution of network output. In the early stage of joint training, when training with central examples, the prediction difference between networks is reduced, and edge samples are added to the training with the training cycle to ensure the integrity of the training samples. We demonstrated the effectiveness of JEKD with a large number of experiments, and analyzed the joint regularization loss and the training aid of CBCS through ablation experiments. In future work, we will continue to explore the correlation between teacher networks and student networks as a whole training in online knowledge distillation.

Acknowledgement. This research is supported by Sichuan Science and Technology Program (No. 2022YFG0324), SWUST Doctoral Research Foundation under Grant 19zx7102.

References

1. Smith, J., et al.: Always be dreaming: a new approach for data-free class-incremental learning. In: Proceedings of the IEEE/CVF International Conference on Computer Vision (2021)
2. Jiang, Y., et al.: Model pruning enables efficient federated learning on edge devices. IEEE Trans. Neural Netw. Learn. Syst. (2022)
3. Chung, I., Park, S., Kim, J., Kwak, N.: Feature-map-level online adversarial knowledge distillation. In: ICML (2020)

4. Blum, A., Mitchell, T.: Combining labeled and unlabeled data with co-training. In: Proceedings of the Eleventh Annual Conference on Computational Learning Theory, pp. 92–100 (1998)
5. Hinton, G., Vinyals, O., Dean, J.: Distilling the knowledge in a neural network. arXiv preprint arXiv:1503.02531 (2015)
6. Tian, Y., Krishnan, D., Isola, P.: Contrastive representation distillation. In: ICLR (2020)
7. Wang, Y., et al.: LEDNet: a lightweight encoder-decoder network for real-time semantic segmentation. In: 2019 IEEE International Conference on Image Processing (ICIP). IEEE (2019)
8. Romero, A., Ballas, N., Kahou, S.E., Chassang, A., Gatta, C., Bengio, Y.: FitNets: hints for thin deep nets. In: ICLR (2015)
9. Faisant, N., Siepmann, J., Benoit, J.-P.: PLGA-based microparticles: elucidation of mechanisms and a new, simple mathematical model quantifying drug release. Eur. J. Pharm. Sci. **15**(4), 355–366 (2002)
10. Park, W., Lu, Y., Cho, M., Kim, D.: Relational knowledge distillation. In: CVPR (2019)
11. Zagoruyko, S., Komodakis, N.: Paying more attention to attention: improving the performance of convolutional neural networks via attention transfer. In: ICLR (2017)
12. Yu, X., Han, B., Yao, J., Niu, G., Tsang, I.W., Sugiyama, M.: How does disagreement benefit co-teaching? arXiv preprint arXiv:1901.04215 (2019)
13. Peng, B., et al.: Correlation congruence for knowledge distillation. In: ICCV (2019)
14. Guo, Q., et al.: Online knowledge distillation via collaborative learning. In: CVPR (2020)
15. Choi, H., Bajić, I.V.: Latent-space scalability for multi-task collaborative intelligence. In: 2021 IEEE International Conference on Image Processing (ICIP). IEEE (2021)
16. Li, B., Wu, B., Su, J., Wang, G.: EagleEye: fast sub-net evaluation for efficient neural network pruning. In: Vedaldi, A., Bischof, H., Brox, T., Frahm, J.-M. (eds.) ECCV 2020. LNCS, vol. 12347, pp. 639–654. Springer, Cham (2020). https://doi.org/10.1007/978-3-030-58536-5_38
17. Ioffe, S., Szegedy, C.: Batch normalization: accelerating deep network training by reducing internal covariate shift. In: International Conference on Machine Learning. PMLR (2015)
18. Zhang, Y., Xiang, T., Hospedales, T.M., Lu, H.: Deep mutual learning. In: CVPR (2018)
19. Malach, E., Shalev-Shwartz, S.: Decoupling "when to update" from "how to update". In: Advances in Neural Information Processing Systems, pp. 960–970 (2017)
20. Krizhevsky, A., Sutskever, I., Hinton, G.E.: ImageNet classification with deep convolutional neural networks. In: Advances in Neural Information Processing Systems (2012)
21. Wang, X., Chen, Y., Zhu, W.: A survey on curriculum learning. IEEE Trans. Pattern Anal. Mach. Intell. **44**(9), 4555–4576 (2021)
22. Zhao, B., et al.: Decoupled knowledge distillation. In: CVPR (2022)
23. Tanaka, D., Ikami, D., Yamasaki, T., Aizawa, K.: Joint optimization framework for learning with noisy labels. In: Proceedings of the IEEE Conference on Computer Vision and Pattern Recognition, pp. 5552–5560 (2018)

24. Sindhwani, V., Niyogi, P., Belkin, M.: A co-regularization approach to semi-supervised learning with multiple views. In: Proceedings of ICML Workshop on Learning With Multiple Views, pp. 74–79 (2005)
25. Qian, B., Wang, Y., Yin, H., Hong, R., Wang, M.: Switchable online knowledge distillation. In: Avidan, S., Brostow, G., Cissé, M., Farinella, G.M., Hassner, T. (eds.) ECCV 2022. LNCS, vol. 13671, pp. 449–466. Springer, Cham (2022). https://doi.org/10.1007/978-3-031-20083-0_27
26. Chung, I., et al.: Feature-map-level online adversarial knowledge distillation. In: International Conference on Machine Learning. PMLR (2020)
27. Zhou, Z.-H.: A brief introduction to weakly supervised learning. Natl. Sci. Rev. **5**(1), 44–53 (2018)
28. Han, B., et al.: Co-teaching: robust training of deep neural networks with extremely noisy labels. In: Advances in Neural Information Processing Systems, vol. 31 (2018)
29. Wei, H., et al.: Combating noisy labels by agreement: a joint training method with co-regularization. In: Proceedings of the IEEE/CVF Conference on Computer Vision and Pattern Recognition (2020)
30. Tang, J., et al.: Data-free network pruning for model compression. In: 2021 IEEE International Symposium on Circuits and Systems (ISCAS). IEEE (2021)

Dual-Branch Contrastive Learning for Network Representation Learning

Hu Zhang[1(✉)], Junnan Cao[1], Kunrui Li[1], Yujie Wang[1], and Ru Li[1,2]

[1] School of Computer and Information Technology of Shanxi University, Taiyuan, China
zhanghu@sxu.edu.cn

[2] Key Laboratory of Computation Intelligence and Chinese Information Processing, Shanxi, China

Abstract. Graph Contrastive Learning (GCL) is a self-supervised learning algorithm designed for graph data and has received widespread attention in the field of network representation learning. However, existing GCL-based network representation methods mostly use a single-branch contrastive approach, which makes it difficult to learn deeper semantic relationships and is easily affected by noisy connections during the process of obtaining global structural information embedding. Therefore, this paper proposes a network representation learning method based on a dual-branch contrastive approach. Firstly, the clustering idea is introduced into the process of embedding global structural information, and irrelevant nodes are selected and removed based on the clustering results, effectively reducing the noise in the embedding process. Then, a dual-branch contrastive method, similar to ensemble learning, is proposed, in which the two generated views are compared with the original graph separately, and the joint optimization method is used to continuously update the two views, allowing the model to learn more discriminative feature representations. The proposed method was evaluated on three datasets, Cora, Citeseer, and Pubmed, for node classification and dimensionality reduction visualization experiments. The results show that the proposed method achieved better performance compared to existing baseline models.

Keywords: Graph contrastive learning · Network representation learning · self-supervision learning · Clustering · Graph neural network

1 Introduction

Network representation learning [1]aims to generate low-dimensional dense vector representations in Euclidean space for nodes, edges, or subgraphs in a graph. Existing methods based on graph neural networks [2] (GNNs) usually train in a supervised or semi-supervised manner, and their performance on unlabelled data is often poor. In order to address this issue, researchers have introduced self-supervised learning. Graph contrastive learning (GCL) is a typical

Project supported by the National Natural Science Foundation of China (62176145).

B. Luo et al. (Eds.): ICONIP 2023, CCIS 1966, pp. 185–197, 2024.
https://doi.org/10.1007/978-981-99-8148-9_15

self-supervised learning method that typically generates different views of each instance in the dataset through various augmentation techniques, with granularity ranging from node-level, subgraph-level, and global-level perspectives. Therefore, contrastive learning based on mutual information measurement can perform contrast at the same or different granularities. For example, Hafidi [3], Qiu et al. [4] perform contrastive training on views of the same granularity such as global-global and subgraph-subgraph, while Velikovi [5], Hassani [6], Zhu et al. [7] perform contrastive training on views of different granularities such as subgraph-global and node-global, respectively. These contrastive methods usually require obtaining the embedding of the global structural information. However, when obtaining the embedding of global structural information in classification tasks, existing methods typically use aggregation methods or graph pooling methods, which may cause the model to lose unique features in the node embeddings and cannot guarantee that global structural embedding can extract beneficial information from the nodes. Moreover, the presence of noise can also affect the effectiveness of downstream tasks. Therefore, it is necessary to explore more effective methods for embedding global structural information.

In order to effectively address issue of noise in global structural information embedding, Qiu et al. [4] used Graph Convolutional Networks (GCNs) to obtain global structural information embeddings; Hafidi et al. [3] proposed using graph attention mechanisms or graph pooling methods to obtain global structural information embeddings. We propose a simple, effective, and lightweight method for global embedding method. After obtaining the embeddings of each node in the graph, we introduce clustering to filter out and remove irrelevant nodes, obtaining more accurate global structural information embedding. After obtaining the global structural information embedding, training models via the GCL, further learn the feature encoding of the nodes by maximizing the consistency of positive sample pairs and the inconsistency of negative sample pairs.

Currently, many GCL models such as DGI [5], GRACE [7], GCA [8], GMI [9], SimGRACE [10], COSTA [11], ImGCL [12] perform well in maximizing mutual information, but they all use single-branch contrastive learning to train the model. Although this approach has achieved good results, it cannot learn discriminative feature representations and deep semantic relationships. Therefore, we propose a simple and effective Dual-Branch Contrastive Learning for Network Representation Learning (DBCL) method. The method uses an ensemble-learning approach to train an enhanced encoder. After obtaining two views generated by the view generation module from the original graph, we continuously update the generated two views using a joint optimization method. Finally, we train the model using the dual-branch contrastive learning method, enabling the model to continuously acquire discriminative feature representations and deep semantic relationships. The main contributions of this paper are as follows:

(1) Based on the existing single-branch contrast approach, we propose a network representation learning method based on dual-branch contrast learning, which uses dual-branch contrast to construct the contrast loss of the model.

(2) We used the clustering results to effectively remove some of the no-joints when capturing the global structural information embedding.

2 Related Work

2.1 Network Representation Learning

Early network representation learning methods were mainly based on matrix factorization. With the successful application of the Word2Vec [13] model, methods based on random walks have emerged. Later, with the development of neural networks, the superiority of GNN [2] has also been gradually recognized. Compared to traditional methods, GNN has the ability to learn non-linear representations. However, it requires a large amount of labeled data for training. To overcome the problem of data scarcity, researchers began exploring the use of unsupervised representation learning methods. Self-supervised learning is a special case of unsupervised learning, where GCL is a typical method. With the arrival of the big data era, researchers have turned their attention to topology learning [14] in graphs. They utilize the optimized graph structure to generate node representations through message passing and continuously update the iterative graph structure and node representations.

2.2 Graph Contrastive Learning

GCL is a contrastive learning method for graph-structured data that can be used for downstream tasks such as node classification, link prediction, and graph clustering. It has good robustness and performs well in various graph-related tasks. In GCL, mutual information maximization is widely used to establish contrastive objectives between local and global features. DGI [5] first introduced mutual information in GNNs to achieve mutual information maximization between nodes and the global context. GCC [4] is based on subgraph-subgraph contrast patterns and uses information loss to achieve contrastive learning. GRACE [7] uses an enhanced method of deleting edges and feature masking to generate two views and obtains graph representations by maximizing the consistency between the node embeddings of these two views, also based on node-global contrastive learning.

3 Method Overview

In this section, we introduce our model. The general framework diagram of DBCL is shown in Fig. 1. We propose a network representation learning method based on dual-branch contrastive learning, which includes a view generation module and a dual-branch contrastive learning module. In the view generation module, we use the original graph to generate two new views, which are continuously updated using a joint optimization method between the two views. In the dual-branch graph contrastive learning module, we first screen the results of clustering and remove irrelevant nodes to obtain global structural information embedding. Then we use dual-branch contrastive learning to train the model. Finally, the obtained node vectors can be used for downstream tasks.

We propose the dual-branch contrast learning approach that can obtain discriminative feature representations and deep semantic relationships, while the clustering algorithm can obtain global structural information embedding more accurately.

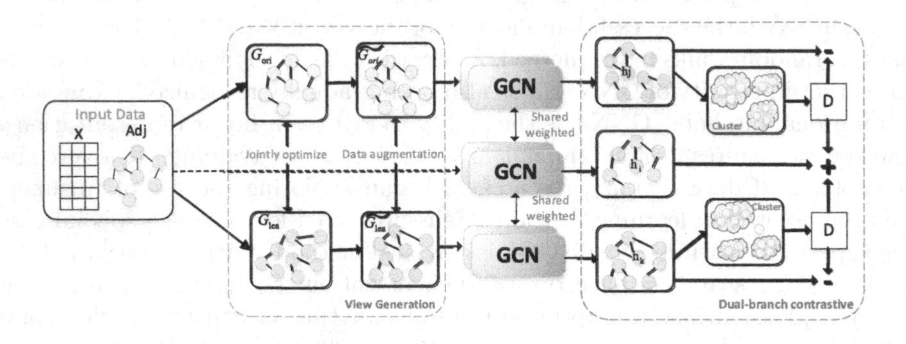

Fig. 1. The Overall Framework of DBCL

3.1 Problem Definition

Given a original graph $G = (V, E)$ with a node set $V = (v_1, v_2, \cdots, v_N)$ of N nodes and an edge set $E = (e_1, e_2, \cdots, e_M)$ of M edges. The feature matrix of the nodes is $X \in R^{N\times F}$,where F represents the dimensionality of the node features, and $x_i \in R^F$ represents the feature vector of each node v_i. The adjacency matrix of nodes is $A \in \{0, 1\}^{N\times N}$. If $(V_i, V_j) \in E$, then $A_{ij} = 1$; otherwise, $A_{ij} = 0$. The focus of the DBCL method is to learn an encoder of graph neural networks, which takes the feature vectors X and adjacency matrix A of the graph as inputs. Through the continuous learning of the encoder, low-dimensional node embedding representations are generated, denoted as $F^{'} \ll F$, where $F^{'}$ represents the vector of generated nodes. The learned node representations are denoted as $H = f(X, A)$, and the embedding representation of node V_i is h_i, and these embedding representations can be applied to downstream tasks.

3.2 View Generation

In this subsection, we will introduce how to generate two new views from the original graph. We define the original graph as G, and the two newly generated views as G_{lea} and G_{ori}. The one view generated by learning the graph topology structural is G_{lea}, which aims to explore potential structural changes at each step. The other view after initialization and normalization is G_{ori}.G_{ori} provides a reference for G_{lea}. G_{lea} and G_{ori} are continuously learned and updated with training.

The generation process of view G_{lea} is shown in Fig. 2. The graph topology learning module first models the connectivity of edges to refine G and generate a sketch adjacency matrix $\tilde{A}$. Then, we model and optimize each parameter using a parameter optimization learning method to obtain G_{lea}, as shown in Eq. (1).

$$\tilde{A} = p_w = \sigma(\Omega) \tag{1}$$

Here, p_w represents the graph topology learning module, $w = \Omega \in R^{N \times N}$ represents the learnable parameter matrix, and $\sigma(\cdot)$ is a non-linear function that aims to improve the robustness of the model. After obtaining $\tilde{A}$, we use a post-processor $q(\cdot)$ to refine $\tilde{A}$ into a sparse and smooth adjacency matrix A. Since the sketch adjacency matrix is usually dense, removing some less closely related edges can improve the robustness of the model and reduce memory usage and computational costs. In the post-processor $q(\cdot)$, we utilize the method of constructing graphs with k-nearest neighbors (KNN) [15] to preserve the top K connection values of each node and set the other edges to 0, and then perform a smoothing operation on the adjacency matrix to ensure that the edge weights are between [0, 1].

The purpose of G_{ori} is to provide stable guidance for G_{lea}, so G_{ori} only initializes and normalizes the nodes based on G. The reason why we scale the node features in G_{ori} to a relatively uniform range is to avoid unstable model training due to large changes in node features, and it also helps to improve the convergence speed and accuracy of the model.

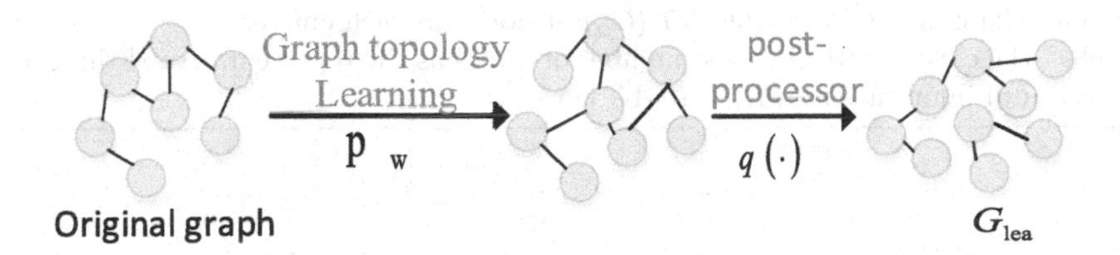

Fig. 2. Graph topology learner to obtain G_{lea}

3.3 Dual-Branch Graph Contrastive Learning Module

After obtaining two new views through the view generation module, then, we use the random walk method to sample a subgraph [16] from the graph as the augmented view for the two views, denoted as $\widetilde{G_{ori}}$ and $\widetilde{G_{lea}}$. Finally, we train the model by maximizing the mutual information between the node-level information and the global structural information on the dual branches.

Global Structural Information. To capture the global structural information embedding (GSE), we combine the idea of clustering to filter nodes, where the clustering algorithm can divide nodes into different clusters, and nodes within each cluster have similar features and attributes.

Based on the above, We apply the K-means clustering algorithm on the target representation H. The algorithms is shown in Algorithm 1. Specifically, we first cluster the nodes in the graph into K clusters, denoted as $G = (\mathrm{G}_1, G_2, \cdots, G_K)$, and $c(h_i) \in \{1, 2, \ldots, K\}$ represents the allocation after clustering, i.e.,$v_i \in G_{c(h_i)}$. Next, we count the number of nodes in each cluster, denoted as $P_a = COUNT(G_a)$, and select the cluster with the most number of nodes as the representative of global semantics, i.e.,$c(h_i) = MAX(P_a)$. We choose this method because it is simple and easy to implement, without the need to consider obtaining vectors for each node's cluster information, saving memory resources and being easily applicable to practical scenarios. At the same time, selecting the cluster with the most number of nodes as the global vector can reflect the features of the graph to a certain extent. Therefore, discarding some of the information does not have a significant impact on the results, but rather helps to obtain better results.

Algorithm 1 Obtain global structural information

Algorithm 1. Obtain global structural information

Require: The original graph G, the feature matrix X of the nodes, the adjacency matrix A

Ensure: Clusters representing global structure information $c(h_i)$

1: The original graph G generates G_{lea} and G_{ori} through the view generation module
2: Obtained by subgraph augmentation $\widetilde{G_{lea}}$ and $\widetilde{G_{ori}}$
3: Obtain the initial node embedding Hh_i via GCN
4: Obtain $G = (\mathrm{G}_1, G_2, \cdots, G_K)$ clusters by k-means clustering
5: Count the number $P_a = COUNT(G_a)$ of nodes in each cluster
6: Select the cluster with the most number of nodes as the representative of the global structural information $c(h_i) = MAX(P_a)$

Dual-Branch Contrastive. Ensemble learning aims to aggregate multiple weak models into a strong model with better performance or lower variance. We propose the dual-branch contrastive method, which is ensemble-learning, serves as an extended training framework that provides a tight way to optimize the model. It focuses on training a coder that relies on maximizing local mutual information to obtain a representation of all global structural informations. To obtain the representation of the vector $\boldsymbol{s}$, by using the GSE method mentioned above, we select the cluster that represents the global structural information and use a readout function $R : R^{N \times F} \rightarrow R^F$ to obtain the embedding of the global structure information, as shown in Eq. (2). We choose the simple mean (avg-readout) readout method to reduce computational complexity and minimize noise interference. Moreover, a simple readout function has better robustness, which can enhance the reliability of the model.

$$(H) = \sigma\left(\frac{1}{N}\sum_{i=1}^{N} \boldsymbol{h}_i\right) \tag{2}$$

We defined a bilinear layer to calculate the similarity scores between nodes in measuring the maximization of local mutual information, which helps the model better understand the relationships and topological structures between nodes and improves the accuracy of downstream tasks. At the same time, the bilinear layer can be used to learn the relationships between nodes, thereby improving the performance and generalization ability of GNNs. By modeling the relationships between nodes, the model can better understand the interactions and dependencies between nodes in the graph, thereby improving the accuracy and robustness of the model. Finally, the bilinear layer can also learn the weights between different nodes, thus weighting and aggregating the hidden states of nodes to improve the expressiveness and generalization ability of the model. Specifically, the node embeddings obtained through the encoder GCN will be used as inputs, and the similarity scores will be calculated through the bilinear layer. We define the aforementioned computation process as a discriminator $D : R^F \times R^F \rightarrow R$. $D(\boldsymbol{h}_i, \boldsymbol{s})$ represents a probability score, where a higher score indicates that the corresponding local representation contains more global structural information, as shown in Eq. (3).

$$(\boldsymbol{h}_i, \boldsymbol{s}) = \sigma\left(\boldsymbol{h}_i{}^T W \boldsymbol{s}\right) \tag{3}$$

We design ensemble-learning dual-branch contrastive learning consists of two branches. One branch G_{ori} generates an augmented view $\widetilde{G_{ori}}$ through subgraph augmentation, which is then contrasted with the original graph G after passing through the GCN encoder and global structural embedding. The goal is to incorporate more global structural information into the node representation. To achieve this, we draw inspiration from the Deep InfoMax (DIM) [17] approach and use binary cross-entropy loss between positive and negative examples to train the model. The loss function $\mathcal{L}_1$ on this branch is given by Eq. (4).

$$\mathcal{L}_1 = \frac{1}{2N}\left(\sum_{i=1}^{N} E_{(X,A)}\left[logD\left(\vec{h_i}, \boldsymbol{s}\right)\right] + \sum_{j=1}^{N} E_{(\tilde{X},\tilde{A})}\left[log\left(1 - D\left(\vec{\tilde{h}_j}, \boldsymbol{s}\right)\right)\right]\right) \tag{4}$$

Where $\tilde{X}$ and $\tilde{A}$ represent the feature matrix and adjacency matrix of $\widetilde{G_{ori}}$, and h_i and h_j represent the node embeddings of G and $\widetilde{G_{ori}}$.The other branch is obtained by applying subgraph augmentation to G_{lea}, resulting in an augmented view $\widetilde{G_{lea}}$. Similar to the first branch, after passing through the GCN encoder and global structural information embedding, $\widetilde{G_{lea}}$ is compared with the original graph. The loss function $\mathcal{L}_2$ for this branch is shown in Eq. (5).

$$\mathcal{L}_2 = \frac{1}{2N}\left(\sum_{i=1}^{N} E_{(X,A)}\left[logD\left(\vec{h_i}, \boldsymbol{s}\right)\right] + \sum_{t=1}^{N} E_{(\tilde{X},\tilde{A})}\left[log\left(1 - D\left(\vec{\tilde{h}_t}, \boldsymbol{s}\right)\right)\right]\right) \tag{5}$$

The variable h_i represents the node embedding representation of $\widetilde{G_{lea}}$. Here we assume that the number of nodes in the augmented graph is the same as that in the original graph, and this loss function can effectively maximize the mutual information between them.

The contrastive process between the two branches constitutes the training process of the entire model. The two views are continually updated through joint optimization, and the model's loss is minimized through a learnable parameter.

Joint Optimization Learn. Inspired by the BYOL [18] algorithm, we incorporate the joint optimization learning method into the process of updating G_{ori} and G_{lea}, which avoids the problem of the graph learner being unable to obtain sufficient information due to the static nature of the views. Specifically, given a learning rate $\tau \in [0, 1]$, view G_{ori} is updated every n rounds using formula Eq. (6). This method can effectively solve the problem of model overfitting and learn new semantic relationships in the process of continuous updating.

$$G_{ori} \leftarrow \tau G_{ori} + (1 - \tau) G_{lea} \tag{6}$$

The above has explained the model's contrast process of the model and the loss functions used in each branch. The overall loss function of the DBCL method can be derived from Eq. (4) and (5) as shown in Eq. (7). The objective is to minimize $\mathcal{L}$ during the training and optimization process of the model, where β is a balancing factor.

$$\mathcal{L} = \beta \mathcal{L}_1 + (1 - \beta) \mathcal{L}_2 \tag{7}$$

4 Experiments

In this section, we have conducted many experiments on accurate benchmarks to verify the proposed method. At the same time, we carried out ablation and dimension reduction visualization experiments on the model to further prove its effectiveness. We conducted experiments on the Cora, Citeseer, and Pubmed citation network datasets.

4.1 Experimental Results and Analysis

Experimental Setup. In this experiment, we use three datasets. Due to their small size, we define the encoder with only one layer of GCN, and the propagation rule of GCN is shown in the formula Eq. (8).

$$\varepsilon(X, A) = \sigma\left(\hat{D}^{-\frac{1}{2}} \hat{A} \hat{D}^{-\frac{1}{2}} X \theta\right) \tag{8}$$

We set the hidden vector dimension of the GCN encoder to 512 and the parameters $\tau = 0.9999$ in the joint optimization learning process.We set the parameters $\beta = 0.5$ for the balancing factor in the loss function and the learning rate of the model is set to $\ell = 0.001$. For subgraph augmentation of the generated two views, the sampling rate is set to 0.2. In capturing the global structure information embedding, we found through experiments that the optimal result was obtained when $K = 3$.

Experimental Results. We evaluate the performance of the DBCL method using node classification experiments and adopt the mean classification accuracy with standard deviation as the evaluation metric. The experimental results are presented in Table 1. The experimental results show that our proposed method outperforms all baseline models, demonstrating the effectiveness of the method described in this paper. We selected two representative categories of baseline models to evaluate the performance of our proposed DBCL method: (1) unsupervised learning methods, including methods using logistic regression classifiers on raw features, classical random walk-based methods, as well as unsupervised contrastive learning methods; (2) supervised learning methods.

The experimental results in Table 1 show that our proposed method outperforms the baseline models in terms of average accuracy for node classification tasks on all three datasets. Compared to the baseline model GRACE, the DBCL method improves the average accuracy by 3.6%, 1.6%, and 0.3% on the three datasets, respectively. Especially compared to DGI, the DBCL method outperforms by 1.3%, 1.5%, and 2.9%, respectively. It is worth noting that although DBCL is based on self-supervised learning, its performance is superior to some supervised learning methods, including GCN, GAT, and GraphSAGE models.

Table 1. Experimental result. The best results are in bold

Model	Input data	Cora	Citeseer	Pubmed
DeepWalk	A	67.2	43.2	63.0
Raw features	X	47.9 ± 0.4	49.3 ± 0.2	69.1 ± 0.2
LP	A,Y	68.0	45.3	63.0
MLP	X,Y	55.1	46.5	71.4
PLANETOID	A,X,Y	75.7	64.7	77.2
Chebyshev	A,X,Y	81.2	69.8	74.4
GraphSAGE	A,X,Y	79.2 ± 0.5	71.2 ± 0.5	73.1 ± 1.4
GCN	A,X,Y	81.5	70.3	79.0
GAT	A,X,Y	83.0 ± 0.7	72.5 ± 0.7	79.0 ± 0.3
Unsup-GraphSAGE	A,X	72.5 ± 1.5	59.4 ± 0.9	70.1 ± 1.4
DGI	A,X	82.3 ± 0.6	71.8 ± 0.7	76.9 ± 0.6
GMI	A,X	82.8 ± 0.3	72.3 ± 0.3	79.8 ± 0.2
GRACE	A,X	80.0 ± 0.4	72.3 ± 0.3	79.5 ± 1.1
GCA	A,X	80.5 ± 0.5	71.3 ± 0.4	78.6 ± 0.6
MVGRL	A,X	82.9 ± 0.7	72.6 ± 0.7	79.4 ± 0.3
DBCL	A,X	$\mathbf{83.7 \pm 0.2}$	$\mathbf{73.4 \pm 0.2}$	$\mathbf{79.8 \pm 0.3}$

Results Analysis. With the above experimental results we can observed that for some baseline models, the experimental results without DBCL are better than those on the Pubmed dataset. This may be due to the relatively small sizes of the Cora and Citeseer datasets, while the Pubmed dataset is larger and has lower dimensional node features. It is difficult to obtain more information from the initial node features alone, and therefore, our proposed ensemble-learning dual-branch contrastive method enables the model to learn more discriminative feature representations and deeper semantic relationships.

After applying the DBCL method to learn representations on the Cora, Citeseer, and Pubmed datasets, we used t-SNE for dimensionality reduction and visualization. We selected a representative baseline model, DGI, for comparison. The experimental results are shown in Fig. 3, 4, and 5, where different colors represent different node categories.

4.2 Ablation Experiments

We introduce the classical unsupervised learning method k-means clustering to capture the global structural information embedding by clustering the nodes in the graph into K clusters. Since the selected datasets have a small size with 3, 6 and 7 categories, we set $K \in \{3, 4, 5\}$. Through experiments, we find that when $K = 3$, the performance of the DBCL method is optimal. The specific experimental results are shown in Table 2.

Meanwhile, to further evaluate the contributions of each module in the proposed model, we conduct the following ablation experiments:

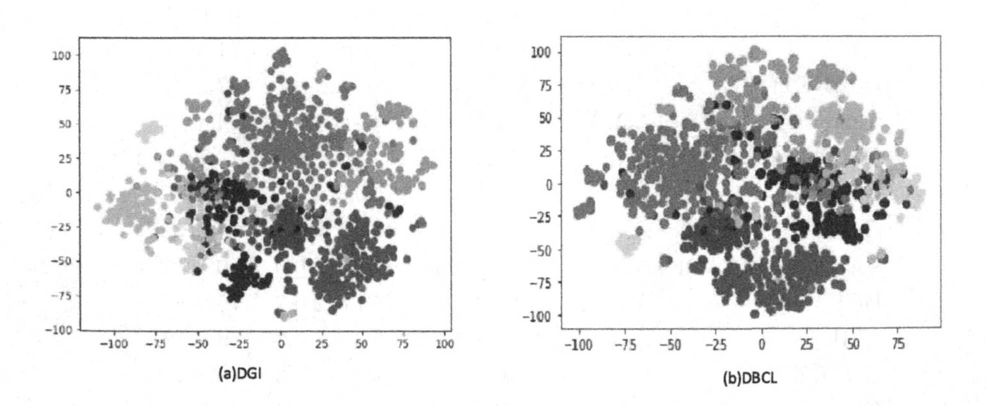

Fig. 3. t-SNE of Cora.

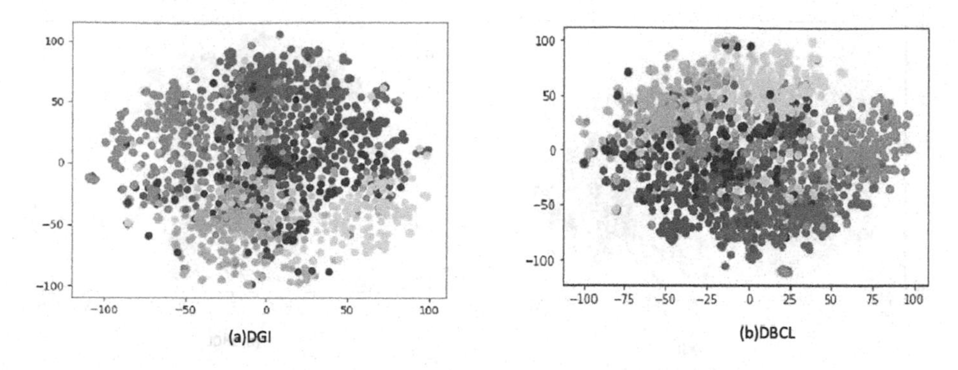

Fig. 4. t-SNE of Citeseer.

(1) -GSE: removing the method of capturing global structural information in the model.
(2) -DBC: removing the dual-branch contrastive learning module and using a single-branch contrastive learning approach instead. We employ two graphs, G_{lea} and G, for single-branch training.

The specific results of the ablation experiments are shown in Table 3.

Table 2. The effect of different K values on experimental results

K	Cora	Citeseer	Pubmed
3	**83.65 ± 0.2**	**73.35 ± 0.2**	**79.79 ± 0.2**
4	83.53 ± 0.2	73.31 ± 0.3	79.74 ± 0.2
5	83.61 ± 0.2	73.28 ± 0.2	79.67 ± 0.1

According to Table 3, we can see that the experimental results on all three datasets show a decrease when we remove the GSE module, which demonstrates the effectiveness of our proposed method in identifying nodes with semantic similarity but not sharing edges, filtering out irrelevant nodes, and reducing the noise problem caused by embedding global structural information. This makes the obtained global structural information more accurate. Similarly, removing the DBC module also leads to a decrease in experimental results, indicating that the proposed dual-branch contrastive learning model can effectively learn the deep semantic and discriminative feature representations of the model.

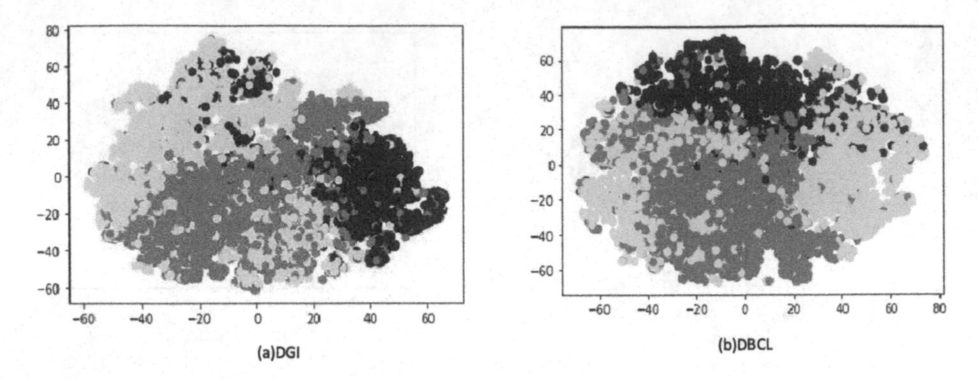

Fig. 5. t-SNE of Pubmed.

Table 3. Ablation results

Model	Cora	Citeseer	Pubmed
DBCL	$\mathbf{83.7 \pm 0.2}$	$\mathbf{73.4 \pm 0.2}$	$\mathbf{79.8 \pm 0.2}$
-GSE	82.4 ± 0.2	72.3 ± 0.1	78.5 ± 0.3
-DBC	82.8 ± 0.1	72.7 ± 0.2	78.2 ± 0.2

5 Conclusion

Our proposed dual-branch contrast learning method (DBCL), which uses a training approach of the ensemble-learning to capture richer input features. By incorporating clustering ideas remove irrelevant nodes to improve the accuracy of global structural embedding. Finally, we use the dual-branch contrastive learning method to train the model and obtain superior embedding representations. The DBCL method effectively combines the diversity of input features, enabling the model to learn deeper semantics and improve the quality of representation. Experiments conduct on three real datasets demonstrate that DBCL outperforms baseline models, indicating that the node embeddings learned by DBCL are better suited for node classification tasks.

Our research mainly focuses on a fully batched scenario and experiments on homogeneous networks. In the next step, we will explore the use of transfer learning or federated learning methods to transfer the trained model of the isomorphic graph to the heterogeneous graph to improve the representation performance of the heterogeneous graph.

References

1. Ju, W., Fang, Z., Gu, Y., et al.: A Comprehensive Survey on Deep Graph Representation Learning. arXiv preprint arXiv:2304.05055 (2023)
2. Zhou, J., Cui, G., Hu, S., Zhang, Z., Yang, C., Liu, Z., et al.: Graph neural networks: a review of methods and applications. AI Open **1**, 57–81 (2020)

3. Hafidi, H., Ghogho, M., Ciblat, P., Swami, A.: Graphcl: contrastive self-supervised learning of graph representations. arXiv 2020. arXiv preprint arXiv:2007.0802 (2020)
4. Qiu, J., et al.: Gcc: graph contrastive coding for graph neural network pre-training. In: Proceedings of the 26th ACM SIGKDD International Conference on Knowledge Discovery & Data Mining, pp. 1150–1160 (2020)
5. Velickovic, P., Fedus, W., Hamilton, W.L., Li, P., Bengio, Y., Hjelm, R.D.: Deep graph infomax. ICLR (Poster) **2**(3), 4 (2019)
6. Hassani, K., Khasahmadi, A.H.: Contrastive multi-view representation learning on graphs. In: International Conference on Machine Learning, pp. 4116–4126. PMLR (2020)
7. Zhu, Y., Xu, Y., Yu, F., Liu, Q., Wu, S., Wang, L.: Deep graph contrastive representation learning. arXiv preprint arXiv:2006.04131 (2020)
8. Zhu, Y., Xu, Y., Yu, F., Liu, Q., Wu, S., Wang, L.: Graph contrastive learning with adaptive augmentation. In: Proceedings of the Web Conference 2021, pp. 2069–2080 (2021)
9. Peng, Z., et al.: Graph representation learning via graphical mutual information maximization. In: Proceedings of the Web Conference 2020, pp. 259–270 (2020)
10. Xia, J., Wu, L., Chen, J., Hu, B., Li, S.Z.: Simgrace: a simple framework for graph contrastive learning without data augmentation. In: Proceedings of the ACM Web Conference 2022, pp. 1070–1079 (2022)
11. Zhang, Y., Zhu, H., Song, Z., Koniusz, P., King, I.: COSTA: covariance-preserving feature augmentation for graph contrastive learning. In: Proceedings of the 28th ACM SIGKDD Conference on Knowledge Discovery and Data Mining, pp. 2524–2534 (2022)
12. Zeng, L., Li, L., Gao, Z., Zhao, P., Li, J.: ImGCL: revisiting graph contrastive learning on imbalanced node classification. arXiv preprint arXiv:2205.11332 (2022)
13. Goldberg, Y., Levy, O.: word2vec explained: deriving Mikolov et al.'s negative-sampling word-embedding method. arXiv preprint arXiv:1402.3722 (2014)
14. Xia, J., Zhu, Y., Du, Y., Li, S.Z.: A survey of pretraining on graphs: taxonomy, methods, and applications. arXiv preprint arXiv:2202.07893 (2022)
15. Lee, N., Lee, J., Park, C.: Augmentation-free self-supervised learning on graphs. In: Proceedings of the AAAI Conference on Artificial Intelligence, vol. 36, no. 7, pp. 7372–7380 (2022)
16. Wang, C., Liu, Z.: Graph representation learning by ensemble aggregating subgraphs via mutual information maximization. arXiv preprint arXiv:2103.13125 (2021)
17. Hjelm, R.D., et al.: Learning deep representations by mutual information estimation and maximization. arXiv preprint arXiv:1808.06670 (2018)
18. Grill, J.B., Strub, F., Altch, F., Tallec, C., Richemond, P., Buchatskaya, E., et al.: Bootstrap your own latent-a new approach to self-supervised learning. Adv. Neural. Inf. Process. Syst. **33**, 21271–21284 (2020)

Multi-granularity Contrastive Siamese Networks for Abstractive Text Summarization

Hu Zhang[1(✉)], Kunrui Li[1], Guangjun Zhang[1], Yong Guan[2], and Ru Li[1,3]

[1] School of Computer and Information Technology of Shanxi University, Taiyuan, China
zhanghu@sxu.edu.cn

[2] Department of Computer Science and Technology of Tsinghua University, Beijing, China

[3] Key Laboratory of Computational Intelligence and Chinese Information Processing, Taiyuan, China

Abstract. Abstractive text summarization is an important task in natural language generation, which aims to compress input documents and generate concise and informative summaries. Sequence-to-Sequence (Seq2 Seq) models have achieved good results in abstractive text summarization in recent years. However, such models are often sensitive to noise information in the training data and exhibit fragility in practical applications. To enhance the denoising ability of the models, we propose a Multi-Granularity Contrastive Siamese Networks for Abstractive Text Summarization. Specifically, we first perform word-level and sentence-level data augmentation on the input text and integrate the noise information of the two granularities into the input text to generate augmented text pairs with diverse noise information. Then, we jointly train the Seq2Seq model using contrastive learning to maximize the consistency between the representations of the augmented text pairs through a Siamese network. We conduct empirical experiments on the CNN/Daily Mail and XSum datasets. Compared to many existing benchmarks, the results validate the effectiveness of our model.

Keywords: Abstractive text summarization · Contrastive learning · Siamese network · Data augmentation

1 Introduction

Text summarization aims to present the core content and main points of the original text in a concise manner while maintaining the information in the original text. It plays a crucial role in the field of natural language processing (NLP), and provides technical support for intelligent reading, information retrieval, knowledge graph construction and other fields. At present, the two main types of text

Project supported by the National Natural Science Foundation of China (62176145).

B. Luo et al. (Eds.): ICONIP 2023, CCIS 1966, pp. 198–210, 2024.
https://doi.org/10.1007/978-981-99-8148-9_16

summarization methods are extractive and abstractive summarization. Extractive summarization methods require the selection of important words, phrases, or sentences from the original text as summaries. This approach can preserve the language style of the original text and make it more readable, but it usually results in redundant information. Abstractive summarization methods are generally based on Seq2Seq models, which treat the original text as a sequence of words and the corresponding summary as another sequence of words. They are more flexible than extractive summarization methods. In recent years, the Seq2Seq network model has made significant progress, especially in the context of large-scale pre-trained language models, such as BART [1] and PEGASUS [2], which are based on the transformer architecture.

Although such models perform well in standard benchmark tests, these deep learning models require large amounts of high-quality input and output data as a basis for training. They are also often fragile when deployed in real systems because the models are not robust to data corruption, distribution shift, or harmful data manipulation [3]. Actually, these models are weakly noise resistant and not robust to noisy information. As a result, directly fine-tuning on the target dataset may not generate satisfactory summaries. Recently, researchers have proposed using contrastive learning methods to enhance text representation, which in turn improves the quality of summary generation. First, a pair of augmented texts with high similarity to the original text are created using data augmentation methods. Then, by pulling similar texts closer and pushing away dissimilar texts to learn an effective representation of the original text, the intrinsic structure and features of the data are mined. Zheng et al. [4] propose the ESACL method, which enhances the robustness of the summary model to global semantic noise using only sentence-level document augmentation methods, but this leads to insufficient robustness to local semantic noise. It also neglects the connection between local and global semantic noises. Moreover, contrastive learning treats two augmented versions of the same instance as positive samples, whereas all other samples in the batch are treated as negative samples. However, there is a sampling bias in randomly sampling the negative samples within the batch, which leads to inconsistencies in the representation space [5] and failure to learn effective representations of text. Finally, owing to the nature of contrastive learning methods, a large number of negative samples and batches, as well as higher computational and memory costs, are required to improve the performance of downstream tasks.

In this paper, we propose a Multi-Granularity Contrastive Siamese Networks for Abstractive Text Summarization (MGCSN). First, we address the issue of insufficient denoising capability in summary models due to the inadequate consideration of local and global noise in the input text by existing contrastive learning methods. We thus perform a multi-granularity data augmentation method that fuses word-level and sentence-level on the input text. Second, since contrastive learning methods generally rely on a large number of negative samples with sampling bias and poor flexibility. We thus integrate Siamese network architecture to perform self-supervised contrastive learning without negative samples. We then construct a transformer-based Seq2Seq joint learning framework with a momentum updating mechanism for optimization. Finally, experimental results and analysis on the CNN/Daily Mail and XSum datasets demonstrate the validity of the model.

Our contributions can be summarized as follows:

(1) We construct a multi-granularity data augmentation method that introduces diverse noise information into the input text through various word-level and sentence-level data augmentation strategies.
(2) We propose MGCSN, which performs contrastive learning in the Seq2Seq model to optimize text representations through a Siamese network, thereby improving the denoising ability of the model and guiding summary generation.
(3) We conduct multiple comparative analysis experiments on the CNN/Daily Mail and XSum datasets, which validate the effectiveness of our model.

2 Related Work

Contrastive Learning. The core idea of contrastive learning is minimizing the distance between different view feature representations of the same image while maximizing the distance between different view feature representations of different images. Chen et al. [6] propose the SimCLR contrastive learning framework, which uses other sample pairs in the same batch as negative samples. He et al. [7] propose the MoCo contrastive learning framework by maintaining a negative sample queue and using the momentum updating for representation updates. They show that using a large number of negative samples and large batch size is essential to learn good representations and achieve better performance. By contrast, the BYOL [8] uses an online network and a target network to form a Siamese network without negative samples. Recently, contrastive learning has been applied to text summarization tasks. Lee et al. [9] propose to mitigate the problem of exposure bias through a contrastive learning framework. Liu et al. [10] focus on applying contrastive learning to bridge the gap between the training objective and evaluation metrics. Sun et al. [11] contrast reference summaries with model-generated summaries of low quality during model inference.

Abstractive Text Summarization Model. Most previous methods focus on modifying the model structure to improve the quality of summary generation. Gu et al. [12] propose the copying mechanism, which predicts the probability distribution of the next word at each step of decoding based on both copying and generating patterns, which has the ability to solve the out-of-vocabulary (OOV) problem. See et al. [13] propose the coverage mechanism, which needs to gradually consider the attention weights of previously generated words in the decoding process to avoid generating the same words repeatedly, it effectively alleviates the problem of generating duplicates. Li et al. [14] propose a new copying mechanism. It not only refers to the attention probability of the current moment but also the copying probability of the historical moment in the decoding process, which improves the coherence and rationality of the copying behavior. With pre-trained language models being widely used to improve the performance of various NLP tasks. Lewis et al. [1] propose the BART model, which effectively improves the denoising ability of the Seq2Seq model. Qi et al. [15] propose the ProphetNet model, which introduces new self-supervised goals: n-gram prediction and n-stream self-attention mechanism, which prevents overfitting of strong local correlations during prediction. These methods have achieved great success in abstractive text summarization.

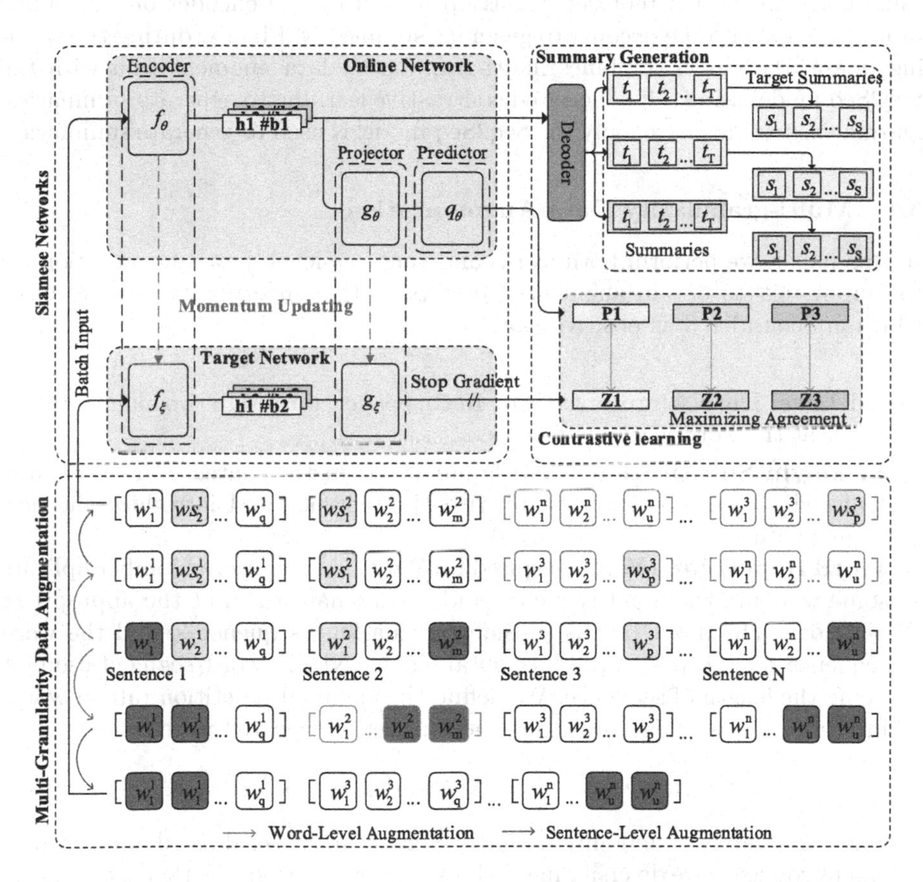

Fig. 1. The overall framework of MGCSN. The light yellow word block indicates synonym substitution, and the orange word block indicates word repetition. The light blue sentence block and red sentence block represent random swap, and the green sentence block represents random deletion. The online network and target network generate vector representations of the augmented text and learn effective representations of the text by maximizing the consistency of latent representations through contrastive learning. Effective text representations are fed into the decoder of the Seq2Seq model for summary generation. The illustration shows multiple pairs of text within a batch.

3 Method

The structure of the MGCSN model proposed in this paper is shown in Fig. 1, which mainly consists of three parts: multi-granularity data augmentation, self-supervised contrastive learning, and summary generation. MGCSN first creates augmented text pairs of the input text through the multi-granularity data augmentation method. MGCSN then performs self-supervised contrastive learning by feeding the augmented text pairs into the online and target branch of Siamese network to obtain vector representations and maximize their similarity. At the same time, the vector representations optimized by the encoder on the online branch are fed into the decoder to generate summaries. Finally, during the training phase, the model combines multi-granularity data augmentation with the Seq2Seq model using self-supervised contrastive learning to generate summaries. During the testing phase, only the Seq2Seq model is used to generate summaries.

3.1 Multi-granularity Data Augmentation

In this paper, we perform both word- and sentence-level data augmentation for the input text to obtain augmented text pairs that incorporate diverse noise, which are considered as positive pairs.

Word-Level Data Augmentation. In this paper, we design two augmentation methods at the word level., as follows:

- **Synonym Substitution (SS):** Some words in the sentence are randomly replaced. We use NLPAug to rewrite the sentence and introduce the new named entity.
- **Word Repetition (WR):** Inspired by Wu et al. [16], we randomly duplicate some words in the input sequence and concatenate them at the appropriate positions. The input text is considered as a long sequence s, and the word sequence $s = \{s_1, s_2, ..., s_N\}$ is obtained by NLTK word segmentation, N being the length of sequence. We define the maximal repetition rate of words in the sequence as *dup_rate*, then the number of repeated words is:

$$dup\_set \in [0, \max(int(dup\_rate * N))] \tag{1}$$

where *dup_len* is a randomly sampled number in the set defined above, which can introduce more diversity in word-level augmentation. We then use uniform distribution to randomly select the *dup_len* words that need to be repeated. We define the set of repeated words as:

$$dup\_len = uniform\,(range = [1, N]\,, num = dup\_len) \tag{2}$$

where the words in the word set are repeated in the sequence.

Specifically, we select n sentences in the input text and use **SS** to generate augmented text, or **WR** for the entire input text. This approach aims to add local noise to the sentences in the input text while ensuring that the semantics of the sentences do not change, where n and *dup_rate* are hyperparameter. The augmentation process is as follows:

$$l_1 = t_1(x) \tag{3}$$

$$l_2 = t_1'(x) \tag{4}$$

where x is the input text, t_1 and t_1' can be different or the same word-level augmentation methods, and l_1 and l_2 are augmented text pairs.

Sentence-Level Data Augmentation. We design three augmentation methods at the sentence level, similar to Qiu et al. [17], and they are as follow:

- **Random Insertion (RI):** Randomly select a sentence from the input text and insert it into a random position in the input text.
- **Random Swap (RS):** Randomly pick two sentences from the input text and swap the positional order.
- **Random Deletion (RD):** Randomly delete a sentence from the input text.

Specifically, we first perform sentence slicing using NLTK and then select one of the methods to perform m times to generate the augmented text. The aim is to increase the global noise while keeping the global structural information of the input text as unchanged as possible, where m is a hyperparameter. The augmentation process is as follows:

$$X_1 = t_2(l) \tag{5}$$

$$X_2 = t_2'(l) \tag{6}$$

where l is the text after word-level augmentation, t_2 and t_2' can be different or the same sentence-level augmentation methods, and X_1 and X_2 are augmented text pairs.

3.2 Self-supervised Contrastive Learning

Self-supervised contrastive learning can encourage the model to identify if two context vectors learned from the encoder represent the same input text by maximizing the similarity between the augmented text pairs. Inspired by this, we employ a Siamese network architecture to build a contrastive learning model consisting of two identical encoders and projectors, with an additional predictor on the online branch.

Encoder. We use two independent encoders with the same initial parameters, i.e., the online encoder f_θ and the target encoder f_ξ. We use the encoder of the pre-trained model BART to encode representations of the augmented text separately:

$$\boldsymbol{h}_1 = f_\theta(X_1) \tag{7}$$

$$\boldsymbol{h}_2 = f_\xi(X_2) \tag{8}$$

where X_1 and X_2 represent the augmented text pairs, which are positive examples of each other. We use the hidden vector corresponding to the start symbol of the last layer of the encoder as the aggregated representation $\boldsymbol{h}$ of the text. The encoder has 6 layers in the base model and 12 layers in the large model.

Projector. The projector uses two independent two-layer MLP with the same initial parameters, i.e., the online projector g_θ and the target projector g_ξ. They project the encoded augmented text vector representation $\boldsymbol{h}$ to the low-dimensional latent space separately:

$$\boldsymbol{z}_1 = g_\theta\left(\boldsymbol{h}_1\right) \tag{9}$$

$$\boldsymbol{z}_2 = g_\xi\left(\boldsymbol{h}_2\right) \tag{10}$$

$$g\left(\boldsymbol{h}_i\right) = w^{(2)}\sigma\left(w^{(1)}\boldsymbol{h}_i\right) \tag{11}$$

where $w^{(1)}$ and $w^{(2)}$ are the two linear layers that can be learned, σ is the ReLU nonlinear activation function, and the batch normalization operation is performed after the first linear layer.

Predictor. The predictor uses the same structure as projector, but it exists only on the online branch i.e., the online predictor q_θ. We apply an additional MLP on top of $\boldsymbol{z}_1$ to obtain a predicted representation $\boldsymbol{p}$:

$$\boldsymbol{p} = q_\theta\left(\boldsymbol{z}_1\right) \tag{12}$$

To be specific, the augmented text pairs X_1 and X_2 are fed into the online branch and target branch to obtain $\boldsymbol{p}$ and $\boldsymbol{z}$, respectively, and maximize the similarity between $\boldsymbol{p}$ and $\boldsymbol{z}$ during the contrastive learning. That is, minimizing the cosine distance between the augmented text pairs incorporating diverse noise information:

$$\mathcal{L}_{\theta,\xi} = -\frac{1}{N}\sum_{i=1}^{N}\frac{\boldsymbol{p}_i \cdot \boldsymbol{z}_i}{\left\|\boldsymbol{p}_i\right\|_2\left\|\boldsymbol{z}_i\right\|_2} \tag{13}$$

At the same time, X_2 and X_1 are fed into the online branch and target branch, respectively, to obtain the symmetry loss $\widetilde{\mathcal{L}}_{\theta,\xi}$. The final loss of self-supervised contrastive learning is as follows:

$$\mathcal{L}_{\theta,\xi}^{MSE} = \mathcal{L}_{\theta,\xi} + \widetilde{\mathcal{L}}_{\theta,\xi} \tag{14}$$

To prevent model collapse, the model optimizes the loss during training and performs back propagation only on the online branch while the gradient is not updated for the target branch. Only the parameter set θ on the online branch is updated, while the parameter set ξ on the target branch is updated by the momentum updating mechanism:

$$\xi = \lambda\xi + (1-\lambda)\,\theta \tag{15}$$

where $\lambda \in [0,1]$ is a hyper-parameter that controls how close ξ remains to θ. At the end of training, we only retain the branch encoder f_θ for summary generation.

3.3 Summary Generation

The purpose of summary generation is to use the vector representation $\boldsymbol{h}$ obtained from the encoder to decode and generate the summary sequence. We use the decoder of the pre-trained model BART for summary generation. During the fine-tuning phase, the loss is computed as the cross-entropy between the generated summary by the model and the reference summary.

$$P_X(y) = \log p(y|X) \tag{16}$$

$$\mathcal{L}_{CE} = \frac{1}{|T|}\sum_{j=1}^{|T|} P_X(T_{i,j}) \tag{17}$$

Where P_X is the conditional language model probability, and T denotes the length of the generated summary.

3.4 Model Training

We calculate the contrastive loss $\mathcal{L}_{\theta,\xi}^{MSE}$ in the self-supervised contrastive learning phase and the cross-entropy loss $\mathcal{L}_{CE}$ in the model fine-tuning phase. How to effectively use contrastive learning in the Seq2Seq framework is the key for enhancing the summary model's denoising ability and deep semantic understanding of the text, ultimately improving the quality of generated summaries. Therefore, to train our model end to end, we define the joint loss, which is the weighted sum of the two parts of the loss. Specifically, the overall objective function is defined as:

$$\mathcal{L} = \alpha\mathcal{L}_{\theta,\xi}^{MSE} + (1-\alpha)\mathcal{L}_{CE} \tag{18}$$

where $\alpha \in [0,1]$ is a balance factor.

4 Experiments

4.1 Datasets

CNN/Daily Mail: [18] The dataset was originally proposed by Hermann et al. for a question-and-answer study using news documents and was subsequently extended to the summary domain by Nallapati et al. The CNN/Daily Mail dataset contains news articles and related summaries from the CNN and Daily Mail websites. We follow the standard dataset consisting of 287,226/13,368/11,490 training/validation/test document-summary pairs.

XSum: [19] The articles in this dataset come from the BBC website and are accompanied by professionally written single sentence summaries, which is a highly abstract summary dataset. We use the original cased version.

4.2 Implementation Details

We use the pre-trained language model BART-large as the backbone which is warmed up by the HuggingFace pre-trained model. We use the AdamW optimizer with a learning rate of $5e-7$. We train 5 epochs with a batch size of 10, including 5 positive example pairs. We empirically choose a relatively large momentum $\lambda = 0.995$. The input document and the corresponding summary are truncated to 1024 and 128, respectively. All experiments are conducted on a single NVIDIA V100 GPU. For CNN/ Daily Mail, we use "facebook/bart-large-cnn", we set $\alpha = 0.6$, which takes about 60 h. We also add the post-processing step after the generation [15],which is common in literature. For XSum, we use "facebook/bart-large-xsum", we set $\alpha = 0.3$, which takes about 30 h.

4.3 Baselines

To verify the validity of the proposed method in Sect. 4.4, we compare the competitive baseline models. Specifically, extractive summarization methods: **Lead3**, **BERTSUMEXT** [20], **GRETEL** [21]; abstractive summarization methods: **BERTSUMEXTABS** [20], **BART** [1], **SKGSUM** [22], **HierGNN** [23]; contrastive learning summarization methods: **ESACL** [4], **CO2Sum** [24], **SeqCo** [25].

4.4 Experimental Results

Results on the CNN/Daily Mail. We use ROUGE F1 scores as the evaluation metric. The word-level data augmentation methods include: 1) **SS**. We randomly select six sentences from the input text for synonym rewriting. 2) **WR**. The maximum word repetition rate is 0.20. The sentence-level data augmentation methods include: 1) **RI**. We randomly select one sentence from the input text and insert it into any position in the input text twice. 2) **RD**. We randomly select one sentence from the input text for deletion twice. Table 1 summarizes our results on the CNN/DailyMail dataset. It can be seen that the MGCSN achieves higher ROUGE scores. Compared with some state-of-the-art models, our model shows strong competitiveness in this research field. The ROUGE-2 results of our model are not the highest, but the ROUGE-1 and ROUGE-L results are the best. The phenomenon is believed to be caused by the word-level augmentation employed by MGCSN, rather than the phrase-level data augmentation.

Table 1. ROUGE F1 results on CNN/DailyMail test set

Model	ROUGE-1	ROUGE-2	ROUGE-L
Lead3	40.34	17.70	36.57
BERTSUMEXT	43.25	20.24	39.63
GRETEL	44.62	20.96	40.69
BERTSUMEXTABS	42.13	19.60	39.18
BART	44.16	21.28	40.90
HierGNN	45.04	**21.82**	41.82
ESACL	44.24	21.06	41.20
CO2Sum	43.51	20.64	40.53
SeqCo	45.02	21.80	41.75
MGCSN	**45.08**	21.57	**41.89**

Table 2. ROUGE F1 results on XSum test set

Model	ROUGE-1	ROUGE-2	ROUGE-L
Lead3	16.30	1.60	11.95
BERTSUMEXTABS	38.81	16.50	31.27
SKGSUM	33.59	13.33	26.53
ESACL	44.64	21.62	36.73
MGCSN	**45.12**	**22.01**	**37.16**

Results on the XSum. Similarly, the word-level data augmentation methods include: 1) **SS**. We randomly select one sentences from the input text for synonym rewriting. 2) **WR**. The maximum word repetition rate is 0.10. The sentence-level data augmentation methods include: 1) **RI**. We randomly select one sentence from the input text and insert it into any position in the input text once. Table 2 summarizes our results on the XSum dataset. The extractive model performs poorly because the summaries in this dataset are highly abstractive. The results show that the MGCSN is superior to the baseline method compared. Overall, our proposed MGCSN outperforms ESACL by 0.48, 0.39, and 0.43 on ROUGE-1, ROUGE-2, and ROUGE-L, respectively.

The experimental results on both datasets and the comparative analysis with ESACL prove the effectiveness of our proposed model. In ESACL [4], the batch size is 16, so we need less batch size and less computational resources in comparison.

4.5 Ablation Study

We further explore the contribution of data augmentation at different granularities to MGCSN alone. We conduct experiments on two datasets. As shown in Table 3, **w/o word-level** is to remove the word-level data augmentation during multi-granularity data augmentation and use the sentence-level data augmentation alone; **w/o sentence-level** is to use the word-level data augmentation alone. The experimental results demonstrate that removing either word-level data augmentation or sentence-level data augmentation on both datasets leads to a performance decline in the model. This indicates that using word-level or sentence-level data augmentation alone is ineffective in improving the model's denoising capability, thus validating the effectiveness of multi-granularity data augmentation.

Table 3. Ablation study on CNN/Daily Mail and XSum datasets

Model	CNN/Daily Mail			XSum		
	R-1	R-2	R-L	R-1	R-2	R-L
MGCSN	**45.08**	**21.57**	**41.89**	**45.12**	**22.01**	**37.16**
w/o word-level	44.95	21.46	41.79	44.98	21.78	36.94
w/o sentence-level	44.95	21.50	41.79	45.07	21.83	37.00

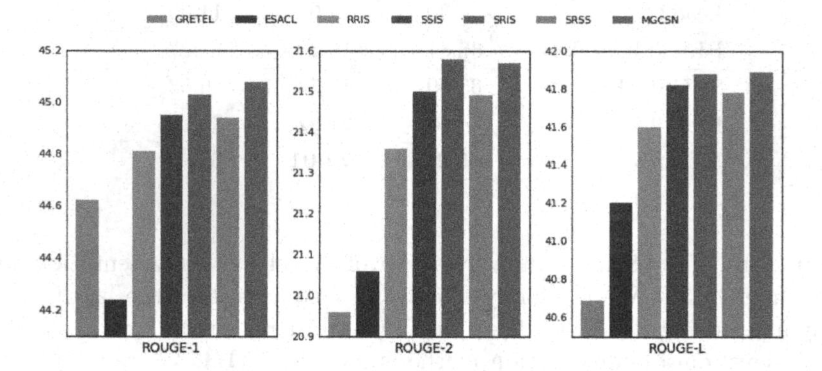

Fig. 2. Comparison results of different data augmentation combinations on the CNN/Daily Mail dataset.

We further explore the impact of different augmentation combinations on the model performance. We conduct several experiments on the CNN/Daily Mail dataset, including four augmentation combinations: 1) **RRIS**. During data augmentation, word-level data augmentation only uses word repetition, whereas sentence-level data augmentation uses random insertion and random swap. 2) **SSIS**. Word-level data augmentation only uses synonym substitution. 3) **SRIS**. Word-level data augmentation uses synonym substitution and word repetition. 4) **SRSS**. Sentence-level data augmentation only uses random swap. Illustrated in Fig. 2, difference in model performance is minimal with different augmentation combinations. Among them, **RRIS** performs the worst, probably due to the significant increase in text length after the word repetition and random insertion strategies, which is truncated during encoding, causing some of the key information to be lost. At the same time, under different multi-granularity data augmentation combinations, the model achieves high ROUGE scores. This outcome demonstrates the flexibility of the proposed multi-granularity contrastive network, which is more robust to augmented hyperparameters and augmentation combinations.

5 Conclusion

In this paper, we propose a multi-granularity contrastive Siamese networks for abstractive text summarization, which guides the summary generation by using contrastive learning to obtain an effective representation of the input text. The proposed model achieves higher quality summaries and ROUGE scores with improved robustness and flexibility of the summary model compared with some existing benchmarks. However, our model lacks richer self-supervised signals to guide the model training encoder during contrastive learning. In future work, we will continue to explore contrastive learning methods to further improve the quality of generated summaries by designing different contrastive learning models and contrastive targets.

References

1. Lewis, M., et al.: Bart: denoising sequence-to-sequence pre-training for natural language generation, translation, and comprehension. arXiv preprint arXiv:1910.13461 (2019)
2. Zhang, J., Zhao, Y., Saleh, M., Liu, P.: Pegasus: pre-training with extracted gap-sentences for abstractive summarization. In: International Conference on Machine Learning, pp. 11328–11339. PMLR. (2020)
3. Jia, R., Liang, P.: Adversarial examples for evaluating reading comprehension systems. arXiv preprint arXiv:1707.07328 (2017)
4. Zheng, C., Zhang, K., Wang, H. J., Fan, L., Wang, Z.: Enhanced Seq2Seq autoencoder via contrastive learning for abstractive text summarization. In: 2021 IEEE International Conference on Big Data (Big Data), pp. 1764–1771. IEEE. (2021)
5. Zhou, K., Zhang, B., Zhao, W. X., Wen, J. R.: Debiased contrastive learning of unsupervised sentence representations. arXiv preprint arXiv:2205.00656 (2022)
6. Chen, T., Kornblith, S., Norouzi, M., Hinton, G.: A simple framework for contrastive learning of visual representations. In: International Conference on Machine Learning, pp. 1597–1607. PMLR. (2020)
7. He, K., Fan, H., Wu, Y., Xie, S., Girshick, R.: Momentum contrast for unsupervised visual representation learning. In: Proceedings of the IEEE/CVF Conference on Computer Vision and Pattern Recognition, pp. 9729–9738 (2020)
8. Grill, J.B., et al.: Bootstrap your own latent-a new approach to self-supervised learning. In: Advances in Neural Information Processing Systems, vol. 33, 21271–21284 (2020)
9. Lee, S., Lee, D.B., Hwang, S.J.: Contrastive learning with adversarial perturbations for conditional text generation. arXiv preprint arXiv:2012.07280 (2020)
10. Liu, Y., Liu, P.: SimCLS: a simple framework for contrastive learning of abstractive summarization. arXiv preprint arXiv:2106.01890 (2021)

11. Sun, S., Li, W.: Alleviating exposure bias via contrastive learning for abstractive text summarization. arXiv preprint arXiv:2108.11846 (2021)
12. Gu, J., Lu, Z., Li, H., Li, V. O.: Incorporating copying mechanism in sequence-to-sequence learning. arXiv preprint arXiv:1603.06393 (2016)
13. See, A., Liu, P.J., Manning, C.D.: Get to the point: summarization with pointer-generator networks. arXiv preprint arXiv:1704.04368 (2017)
14. Li, H., et al.: Learn to copy from the copying history: correlational copy network for abstractive summarization. In: Proceedings of the 2021 Conference on Empirical Methods in Natural Language Processing, pp. 4091–4101 (2021)
15. Qi, W., et al.: Prophetnet: predicting future n-gram for sequence-to-sequence pre-training. arXiv preprint arXiv:2001.04063 (2020)
16. Gao, T., Yao, X., Chen, D.: Simcse: simple contrastive learning of sentence embeddings. arXiv preprint arXiv:2104.08821 (2021)
17. Qiu, S., et al.: Easyaug: an automatic textual data augmentation platform for classification tasks. In: Companion Proceedings of the Web Conference 2020, pp. 249–252 (2020)
18. Hermann, K.M., et al.: Teaching machines to read and comprehend. In: Advances in Neural Information Processing Systems, vol. 28 (2015)
19. Narayan, S., Cohen, S.B., Lapata, M.: Don't give me the details, just the summary! topic-aware convolutional neural networks for extreme summarization. arXiv preprint arXiv:1808.08745 (2018)
20. Liu, Y., Lapata, M.: Text summarization with pretrained encoders. arXiv preprint arXiv:1908.08345 (2019)
21. Xie, Q., Huang, J., Saha, T., Ananiadou, S.: GRETEL: graph contrastive topic enhanced language model for long document extractive summarization. arXiv preprint arXiv:2208.09982 (2022)
22. Ji, X., Zhao, W.: SKGSUM: abstractive document summarization with semantic knowledge graphs. In: 2021 International Joint Conference on Neural Networks (IJCNN), pp. 1–8. IEEE (2021)
23. Qiu, Y., Cohen, S.B.: Abstractive summarization guided by latent hierarchical document structure. arXiv preprint arXiv:2211.09458 (2022)
24. Liu, W., Wu, H., Mu, W., Li, Z., Chen, T., Nie, D.: CO2Sum: contrastive learning for factual-consistent abstractive summarization. arXiv preprint arXiv:2112.01147 (2021)
25. Xu, S., Zhang, X., Wu, Y., Wei, F.: Sequence level contrastive learning for text summarization. In: Proceedings of the AAAI Conference on Artificial Intelligence, vol. 36, no. 10, pp. 11556–11565 (2022)

Joint Entity and Relation Extraction for Legal Documents Based on Table Filling

Hu Zhang[1(✉)], Haonan Qin[1], Guangjun Zhang[1], Yujie Wang[1], and Ru Li[1,2]

[1] School of Computer and Information Technology of Shanxi University, Taiyuan, China
zhanghu@sxu.edu.cn

[2] Key Laboratory of Computation Intelligence and Chinese Information Processing, Beijing, China

Abstract. Joint entity and relation extraction for legal documents is an important research task of judicial intelligence informatization, aiming at extracting structured triplets from rich unstructured legal texts. However, the existing methods for joint entity relation extraction in legal judgment documents often lack domain-specific knowledge, and are difficult to effectively solve the problem of entity overlap in legal texts. To address these issues, we propose a joint entity and relation extraction for legal documents method based on table filling. Firstly, we construct a legal dictionary with knowledge characteristics of the judicial domain based on the characteristics of judicial document data and incorporate it into a text encoding representation using a multi-head attention mechanism; Secondly, we transform the joint extraction task into a table-filling problem by constructing a two-dimensional table that can express the relation between word pairs for each relation separately and designing three table-filling strategies to decode the triples under the corresponding relations. The experimental results on the information extraction dataset in "CAIL2021" show that the proposed method has a significant improvement over the existing baseline model and achieves significant results in addressing the complex entity overlap problem in legal texts.

Keywords: Relation extraction · Joint extraction · Entity overlap · Intelligent justice

1 Introduction

In recent years, with the continuous improvement of the level of information disclosure in the Chinese judicial field, the abundant information in Chinese legal judgment documents has attracted widespread attention. However, these legal judgment documents often have significant variations in format and content, making it difficult for machines to perform large-scale knowledge acquisition and

Project supported by the National Natural Science Foundation of China (62176145).

B. Luo et al. (Eds.): ICONIP 2023, CCIS 1966, pp. 211–222, 2024.
https://doi.org/10.1007/978-981-99-8148-9_17

the construction of judicial knowledge bases. Therefore, extracting key information from the rich legal text has become an important research topic in judicial intelligence and the informatization of judicial operations.

Joint entity and relation extraction, as an important subtask of information extraction, aims to model natural language text and extract meaningful semantic knowledge to obtain all triplets (Subject, Relation, Object) from the text. In legal texts, it primarily involves the precise extraction of key information such as criminal suspects, committed crimes, and relevant locations. The rich information contained in legal judgments can be further represented through structured entity relation triplets, facilitating better understanding and storage by machines. Traditional entity relation extraction methods typically adopt a pipeline approach [1], where named entity recognition (NER) is first used to extract all entities in a sentence, followed by classification of the relations between entities. However, this pipeline approach overlooks the inherent connections and dependencies between the two subtasks, leading to error accumulation and entity redundancy issues. Existing joint learning methods [2] map entity recognition and relation extraction into a unified framework, allowing for simultaneous processing and effective utilization of dependency information between entities and relations to reduce error propagation.

Although joint extraction methods have made remarkable progress in various fields such as finance, news, and healthcare, research focusing on the judicial domain remains relatively limited. In comparison to other domains, legal texts in the judicial field often encompass a plethora of legal terminology and specialized vocabulary. For instance, drug-related legal judgments frequently contain extensive chemical descriptions or colloquial expressions of drug names. However, pre-trained language models are typically trained on general corpora and therefore lack specific domain knowledge. Furthermore, legal judgment documents often involve multiple criminal suspects engaged in various criminal activities, which leads to a significant problem of entity overlap in legal judgment documents. In the drug-related judgment documents dataset provided by CAIL2022 Information Extraction Task, the proportion of entity overlap reaches nearly 70%. As shown in Fig. 1, in drug-related cases, there are primarily two types of entity overlap issues. The first is Single Entity Overlap (SEO), which refers to the presence of at least one entity in a sentence that appears in two or more relational triplets. For example, in Fig. 1, the criminal suspect entity 伍某 appears in two triplets, (伍某, sell_drugs_to, 欧阳 xx) and (伍某, traffic_in, 海洛因) respectively. The second type is Entity Pair Overlap(EPO), which occurs when there are at least two different relational triplets sharing the same entity pair in a sentence. In Fig. 1, the criminal suspect entities "伍某" and "欧阳 xx" have two different relations, "sell_drugs_to" and "provide_shelte_for", which exist in two separate triplets. In order to extract the overlapping relations in sentences, Zeng et al. [3] proposed a CopyRE model, which employs a copying mechanism to solve the overlapping problem. Wei et al. [4] proposed the CasRel model, which converts joint extraction into a sequence labeling problem by first extracting the head entity of the target relation, and then using the subject-to-object mapping function to extract the corresponding object. Although the above methods can solve the entity overlap problem to some extent, these models are not effective in extracting multiple overlapping triples in a sentence in the face of legal document data containing multiple complex overlapping triples.

	文本	关系
SEO	被告人伍某在**区桂花园小区向吸毒人员欧阳xx贩卖50元约0.05克海洛因，后容留欧阳xx在自己家中吸食。	(伍某, sell_drugs_to, 欧阳xx) (伍某, traffic_in, 海洛因)
EPO	被告人伍某在**区桂花园小区向吸毒人员欧阳xx贩卖50元约0.05克海洛因，后容留欧阳xx在自己家中吸食。	(伍某, sell_drugs_to, 欧阳xx) (伍某, provide_shelter_for, 欧阳xx)

Fig. 1. Example of Entity Overlap in Judicial Documents.

To address the aforementioned issues, we propose a joint entity and relation extraction for legal documents method based on table filling. We utilize the drug-related legal judgment documents provided by the CAIL2022 Information Extraction Task as our experimental data, focusing on extracting relational triples from criminal judgment documents related to drug cases. Our approach consists of the following steps. Firstly, we construct a drug name dictionary with judicial domain knowledge characteristics based on the characteristics of drug-related judicial decision documents, and use a multi-headed attention mechanism to incorporate it into a text encoding representation to help the model learn domain-specific knowledge; Secondly, we transform the joint extraction task into a table-filling problem. For each relation, we construct a two-dimensional table that expresses the relations between word pairs. We design three table-filling strategies to decode relational triplets corresponding to the respective relations.

The main contributions of this paper are as follows:

(1) We construct a drug name dictionary as a legal feature specifically for drug-related legal judgment documents. This dictionary helps the model better learn domain knowledge, thereby enhancing the model's entity recognition capability.
(2) We transform the joint extraction task into a table-filling problem and design three table-filling strategies to decode all overlapping relational triplets for specific relations.
(3) We conduct experiments on the dataset provided by the CAIL2022 Information Extraction Task. The results demonstrate that the proposed method achieves significant improvements in handling complex overlapping triplets.

2 Related Work

2.1 Information Extraction in Judicial Field

Judicial information extraction is an important research area in natural language processing for the legal domain, aiming to extract important information from judicial judgment documents for structured storage. Wang et al. [5] proposed a Bi-LSTM model that integrates character, word, and self-attention mechanisms for legal text named entity recognition, effectively improving the efficiency of boundary recognition for legal entities. Gao et al. [6] used improved kernel functions and convolutional neural network methods to extract relational triplets from key paragraphs of legal documents.

2.2 Joint Entity and Relation Extraction

Early research predominantly adopted a pipeline approach, Such as Zhang et al. [7] and Chen et al. [8], They use pipeline methods to extract the triplets, where entity extraction is performed first, followed by relation classification for the extracted entity pairs. The pipeline methods suffer from error propagation and ignore the interactions between the prediction of entities and relations.

To overcome these limitations, joint learning methods have gradually become a research focus. Zheng et al. [9] proposed a new annotation strategy that transformed joint entity and relation extraction into a sequence labeling problem, opening up new avenues for relation extraction. However, this approach couldn't handle entity overlapping in sentences. Zeng et al. [3] addressed the entity overlapping problem by introducing a Seq2seq model with a copying mechanism. They continuously generated triples by copying relevant entities, improving the accuracy of extracting overlapping triples. However, challenges such as incomplete entity recognition remained. Fei et al. [10] treated the complex task of extracting overlapping triples as a multi-prediction problem and used graph attention mechanism to model the relations between entities. Wang et al. [11] regarded the joint extraction task as a table filling problem and designed two different encoders to learn the necessary information for each task. Wang et al. [12] introduced a novel handshaking tagging scheme, which transformed the joint extraction task into a token pair linking problem, enabling single-stage entity relation joint extraction and addressing the issues of exposure bias and entity overlapping.

3 Model

We propose a joint entity and relation extraction for legal documents model based on table filling. The overall structure of the model is shown in Fig. 2, which consists of three main parts: Firstly, we input the judicial text into the encoding layer to get a general text encoding representation; Secondly, we construct a dictionary of drug names with legal features, and use a multi-headed attention mechanism to incorporate it into the text encoding to get a text encoding representation with enhanced legal features; Finally, there is a table-filling module, where we transform the joint extraction task into a table-filling problem, and design three table-filling strategies to decode the relational triples contained in the sentences.

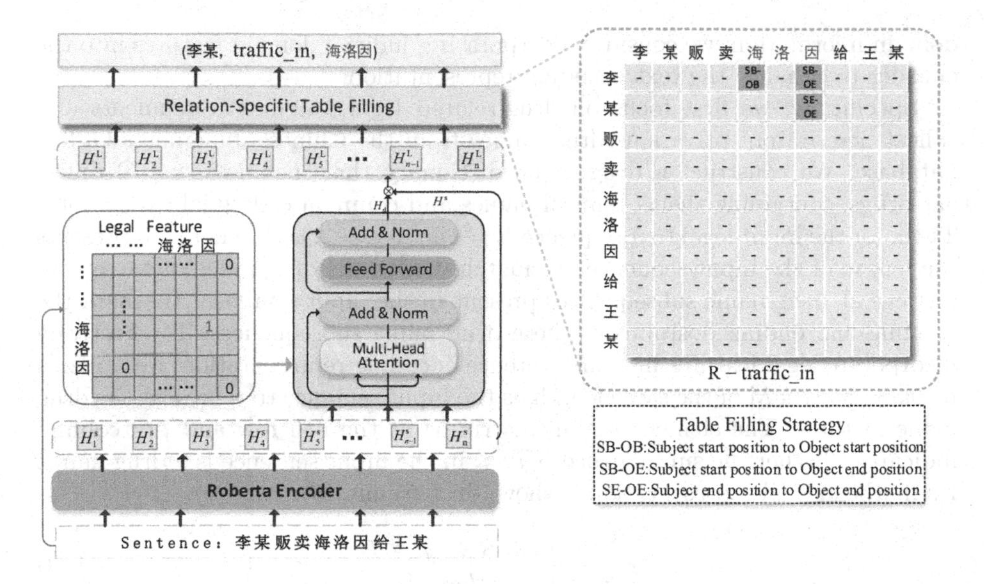

Fig. 2. The overall architecture of the proposed model.

3.1 Encoding Layer

The role of the encoder is to transform the input text sequence into vector representations. In this paper, we employ the pre-trained language model RoBERTa to perform contextual encoding on the input sentences. As the model takes single sentences as input rather than sentence pairs, the input vectors do not include segment embeddings. Instead, they are composed of the sum of word embeddings and position embeddings of the input sentence. We define the Transformer encoding structure as $Trans(x)$ and denote the set of sentences as $S = \{s_1, s_2, \cdots, s_n\}$. The encoding operation for each sentence s_i can be represented by Eq. (1) and Eq. (2):

$$h_0 = W_e + W_p \tag{1}$$

$$h_\alpha = Trans\left(h_{\alpha-1}\right); \alpha \in [1, N] \tag{2}$$

Where W_e is the word embedding matrix, W_p is the positional embedding matrix, and h_α is the hidden state of the input sentence at layer α. The vector representation obtained by encoding the input text through N layers of Transformer is denoted as H^S.

3.2 Legal Feature Enhancement Module

The encoding of pre-trained language models tends to capture general text representations, but lacks the domain-specific knowledge available in specialized domain texts such as judicial documents. To compensate for the lack of legal domain information, we consider incorporating judicial domain features into the encoder to enhance the text encoding representation.

Specifically, we first focus on drug-related legal judgment documents and collect descriptions of various drug names from the Chinese Judgments Online database. We construct a drug name dictionary that includes over 70 different drugs, containing their chemical names and common colloquial expressions. Then, for the input legal text sequence $S = \{w_1, w_2, \cdots, w_n\}$, where n represents the length of the input sequence. we match it with the drug name dictionary to retrieve all drug name subsequences present in the input sentence. We mark the starting and ending positions of these drug name subsequences. To effectively incorporate legal features into the sentence encoding representation, we utilize a masking matrix M of the same length as the input sentence to represent the drug name features. The element m_{ij} in matrix M at the i-th row and j-th column indicates whether the subsequence $S[i:j]$ in the input sentence is a drug name. Its mathematical representation is shown in Formula (3):

$$m_{ij} = \begin{cases} 1 \; ifS[i:j] \in D \\ 0 \quad otherwise \end{cases} \tag{3}$$

where D represents the drug name dictionary, and $m_{ij} = 1$ indicates that the subsequence $S[i:j]$ in the input sentence is a drug name, otherwise it is 0.

Next, to incorporate the drug dictionary features into the sentence encoding representation, we employ a multi-head attention mechanism and a feed-forward neural network layer to perform vectorized calculations on the input sentence and the drug name masking matrix. Firstly, we map the vector representation H^S of the input text, obtained through RoBERTa encoding, into three matrices of dimensions $N \times d_q$, $N \times d_k$, $N \times d_v$, denoted as Q_i, K_i and V_i,respectively. These matrices are used for multi-head attention computation, where N represents the length of the input sentence, and i denotes the i-th head in the multi-head self-attention mechanism. Then, the multi-head self-attention calculation, combined with the drug name masking, is formulated as shown in Eq. (4).

$$Att_i = Softmax\left(\frac{\mathrm{M} \cdot (Q_i K_i^T)}{\sqrt{\mathrm{d_k}}}\right) V_i \tag{4}$$

After the attention computation, we concatenate the outputs of all attention heads and feed the resulting concatenated vector into a feed-forward neural network layer. The final output is denoted as H_d, representing the vector representation that integrates the drug name features. Finally, we perform a weighted sum of the sentence vector representation H^S obtained through RoBERTa encoding and the drug feature representation H_d to obtain a legal feature-enhanced text encoding representation H^L.

3.3 Table Filling Module

Table Construction. To address the issue of entity overlap in legal texts, we transform the joint extraction task into a table filling problem. Specifically, given an input sentence $S = \{w_1, w_2, \cdots, w_n\}$ and a predefined set of relations $R = \{r_1, r_2, \cdots, r_m\}$, we construct a two-dimensional table, denoted as $table_r$, with a size of n x n, where n represents the length of the input sentence. Each entry in the table indicates whether there exists the corresponding labeled relation between token pairs. Next, based on the previously obtained law-enhanced sentence encoding representation $H^L = \{h_1^L, h_2^L, \cdots, h_n^L\}$, we generate an associated feature representation $h_{i,j}^L$ for each token pair (w_i, w_j). This is formulated as shown in Eq. (5).

$$h_{i,j}^L = \sigma\left(W_h \cdot \left[h_i^L; h_j^L\right] + b_h\right) \tag{5}$$

where W_h and b_n are trainable weights and biases, [;] represents the connection operation, and σ is the ReLU activation function.

Finally, we treat the prediction of token pair labels in the table as a multi-label classification task. For a given token pair(w_i, w_j), their associated feature representation $h_{i,j}^L$ is passed through the Softmax function for label prediction, as shown in Eq. (6):

$$P\left(y_{i,j}\right) = Softmax\left(W \cdot h_{i,j}^L + b\right) \tag{6}$$

where W and b are trainable weights and biases.

Table Filling Strategy. To better capture the positional information of each word within an entity in legal texts, we employ the "BE" (Begin, End) symbols to represent entity boundaries. For instance, "SB" (Subject, Begin) signifies the start position of the subject, while "OE" (Object, End) indicates the end position of the object. Guided by the idea that a relational triplet can be determined by aligning the boundary tokens of the subject and object conditioned on a specific relation "r", we have devised three table-filling strategies to assign appropriate labels to each entry in the table. (1) SB-OB: subject start position to object start position. In the table, this is represented by the corresponding row denoting the start token of the subject and the column denoting the start token of the object. For example, in the table on the right side of Fig. 2, the relation "sell_drug_to" exists between the subject "李某" and the object "海洛因" Hence, the token pair ("李", "海") in the table of the corresponding relation is assigned the label "SB-OB". (2) SB-OE: Subject start position to object end position. In the table, this is represented by the corresponding row denoting the start token of the subject and the column denoting the end token of the object. For instance, in the table on the right side of Fig. 2, the token pair ("李", "因") consisting of the start token of the subject "李某" and the end token of the object "海洛因" is assigned the label "SB-OE".(3) SE-OE: Subject end position to object end position. In the table, this is represented by the corresponding row denoting the end token of the subject and the column denoting the end token of the object. For example, in the table on the right side of Fig. 2, the token pair ("某", "因") is assigned the label "SE-OE".

From the table, it can be observed that the three table filling strategies we proposed effectively address the issue of entity overlap in legal texts. Specifically, for the Single Entity Overlap (SEO) problem, if two triplets sharing the same entity have the same relation, the entity pairs will be marked in different sections of the same specific relation table, and then extracted accordingly. Otherwise, they will be marked in different tables based on their respective relations, and subsequently extracted. Additionally, for the Entity Pair Overlap (EPO) problem, identical entity pairs will be marked in different tables based on their respective relations, and then extracted accordingly.

Decoding. The table marks the boundary information of entity pairs and their relations. Therefore, based on the three table filling strategies we designed, the corresponding relational triplet can be decoded by aligning the boundary tokens of the subject and object conditioned on the relation 'r'. In the table on the right side of Fig. 2, with the 'SB-OE' and 'SE-OE' markings, the subject "李某" can be decoded. Additionally, with the 'SB-OB' and 'SB-OE' markings, the object "海洛因" can be decoded. As a result, a triplet (李某, traffic_in, 海洛因) can be decoded.

3.4 Loss Function

The loss function we defined is as follows:

$$\mathcal{L}_{lable} = -\frac{1}{N^2} \sum_{i=1}^{N} \sum_{j=1}^{N} \log P\left(y_{i,j} = g_{i,j}\right) \tag{7}$$

Where N represents the length of the input sentence and $g_{i,j}$ is the true label.

4 Experiment

4.1 Datasets

We conducted experiments on the drug-related legal judgment dataset provided in the 2022 China "Legal AI Cup" Judicial Artificial Intelligence Challenge (CAIL2022) information extraction task. The dataset consists of 1415 annotated text descriptions of drug-related cases. Based on the three most common charges in drug-related judicial judgments documents: illegal drug possession, drug trafficking, and sheltering others for drug use, four predefined relation types were defined: possess (for illegal drug possession), sell_drug_to (for selling drugs to someone), traffic_in (for drug trafficking), and provide_shelter_for (for providing shelter for drug use). To facilitate model training and validation, we split the dataset into training and testing sets in a 4:1 ratio. The distribution of the number of relations in the training and test sets is shown in Table 1.

Table 1. Relation Type Distribution in the Dataset.

Relations	Train	Test	Total
possess	388	90	478
traffic_in	723	184	907
sell_drug_to	784	196	980
provide_shelter_for	683	159	842
Total	2578	629	3270

4.2 Experimental Setup

In our experiment, we utilized the pre-trained language model chinese-roberta-wwm-ext [13] as the encoder. The maximum length of the input sequence was set to 512, and the maximum length of the decoding sequence was set to 10. The hidden layer dimension was set to 768. We employed the Adam optimizer with a learning rate of 1×10^{-5} during the training process. The batch size was set to 32, and the training was conducted for 10 epochs. To prevent overfitting, a dropout rate of 0.3 was applied.

4.3 Experimental Results and Analysis

In order to validate the effectiveness of our proposed model, we conducted comparative experiments with several mainstream joint extraction models, including: NovelTagging [9]: Introduces a novel sequence labeling strategy that incorporates entity and relation information into the labels to predict entities and relations. CopyMTL [14]: Proposes an encoder-decoder based joint extraction framework that uses a copying mechanism to extract entities and relations. PNDec [15]: Utilizes a Bi-LSTM encoder and a decoder based on pointer networks to achieve joint extraction of entities and relations. CasRel [4]: Presents a cascaded extraction framework to address the issue of overlap, where all possible subjects in a sentence are first extracted, followed by the extraction of corresponding objects based on the subjects and relations. GPLinker [16]: Introduces a method based on a global pointer network for joint extraction of entities and relations.

Table 2. Experimental Results of Different Models on Our Dataset.

Model	P	R	F1
NovelTagging	60.9	71.9	65.9
CopyMTL	70.2	74.3	72.2
PNDec	84.4	79.1	81.7
CasRel	84.7	85.2	84.9
GPlinker	85.2	85.1	85.2
Ours-Legal	**85.1**	**87.9**	**86.5**
Ours	**86.1**	**87.8**	**87.0**

Table 2 displays the experimental results of our proposed model compared to the baseline models on the dataset of drug-related legal judgment documents. It can be observed that our model achieves significant improvements over the other baseline models. Compared to the classical tagging model, NovelTagging, our model shows an increase of 21.1% points in F1-score. This improvement is attributed to the fact that NovelTagging adopts a single-layer tagging approach, where each word can only be assigned a single label, which fails to address the issue of entity overlap. Compared to the CopyMTL model, which utilizes a copy mechanism for entity and relation extraction, our model achieves a 14.8% point increase in F1-score. The limitations of the copy mechanism result in lower accuracy. Compared to the PNDec model, our model achieves a 5.3% point increase in F1-score. This improvement is due to the fact that PNDec model extract entities first and then extract the relations between entities, which leads to error accumulation and exposure bias issues. Compared to the classical model CasRel, our model achieved a 2.1% increase in F1 score. This improvement can be attributed to the fact that CasRel, although capable of addressing the overlapping problem, still follows a pipeline extraction pattern, where errors in entity extraction can affect subsequent relation prediction and object extraction. Compared to the above-mentioned models, our proposed model achieves the best performance on the dataset of drug-related legal documents, demonstrating its ability to effectively handle complex entity overlap issues in legal documents and address the issues of error propagation and exposure bias in joint extraction.

Additionally, to verify the impact of the legal feature enhancement module on the performance of our model in extracting triplets, we conducted separate tests by removing the legal dictionary feature enhancement and integrating the legal dictionary features into the model. In Table 2, "Ours-Legal" indicates the model without the legal feature enhancement module. From the experimental results in Table 2, it can be observed that removing the legal feature enhancement module leads to a decrease of 0.5% points in F1-score and a decrease of 1% point in precision. This demonstrates that the proposed legal feature enhancement module contributes to improving the extraction performance of the model. This improvement is particularly evident in precision. The legal feature enhancement module incorporates the drug name dictionary as a legal feature into the sentence encoding representation, enabling the model to more accurately identify complex drug entities in legal documents, thereby enhancing the accuracy of triplet extraction.

4.4 Experiments Based on Different Sentence Types

To verify the ability of our model in handling the overlapping problem and extracting multiple relations, we conducted comparative experiments with different baseline models under different scenarios. The experimental results are shown in Table 3.

From Table 3, it is evident that our model outperforms the other baseline models in sentences with different overlapping patterns. Moreover, as the number of triplets increases, the F1 score shows an upward trend. Even in sentences with

Table 3. F1 score on sentences with different overlapping pattern and different triplet number.

Model	Normal	SEO	EPO	N = 1	N = 2	N = 3	N = 4	N ≥ 5
CopyMTL	78.6	63.4	64.7	79.7	76.3	69.9	67.3	63.4
PNDec	82.5	76.8	78.4	81.9	82.5	80.8	77.4	74.6
CasRel	83.8	85.6	86.0	82.0	84.1	86.6	85.6	84.2
GPlinker	83.5	85.8	86.3	82.2	84.5	86.7	85.9	85.0
Ours	**85.6**	**87.6**	**88.3**	**84.7**	**86.1**	**88.5**	**87.6**	**86.8**

as many as 5 or more triplets, our model achieves the best performance. These results demonstrate the effectiveness of our model in handling complex scenarios where legal texts contain multiple overlapping triples.

5 Conclusion

In this paper, we propose a joint entity and relation extraction for legal documents model based on table filling. The model incorporates judicial domain features into text encoding and transforms the joint extraction task into a table filling problem. We construct a two-dimensional table for each relation to represent the relations between word pairs and design three table filling strategies to decode relational triplets for the corresponding relations. Experimental results on a legal judgment document dataset demonstrate the effectiveness of the proposed model compared to existing models. Although the proposed model effectively solves the problem of entity overlapping in legal texts, it consumes considerable space resources during table construction. In future work, we will further explore more efficient table filling strategies to improve the extraction efficiency of the model and attempt to extend the proposed method to other domains.

References

1. Yang, B., Cardie, C.: Joint inference for fine-grained opinion extraction. In: Proceedings of the 51st Annual Meeting of the Association for Computational Linguistics (Volume 1: Long Papers), pp. 1640–1649 (2013)
2. Katiyar, A., Cardie, C.: Going out on a limb: Joint extraction of entity mentions and relations without dependency trees. In: Proceedings of the 55th Annual Meeting of the Association for Computational Linguistics (Volume 1: Long Papers), pp. 917–928 (2017)
3. Zeng, X., Zeng, D., He, S., Liu, K., Zhao, J.: Extracting relational facts by an end-to-end neural model with copy mechanism. In: Proceedings of the 56th Annual Meeting of the Association for Computational Linguistics (Volume 1: Long Papers), pp. 506–514 (2018)
4. Wei, Z., Su, J., Wang, Y., Tian, Y., Chang, Y.: A novel cascade binary tagging framework for relational triple extraction. arXiv preprint arXiv:1909.03227 (2019)

5. Wang, D.X., Wang, S.G., Pei, W.S., Li, D.Y.: Named entity recognition based on JCWA-DLSTM for legal instruments. J. Chin. Inf. Process. **34**(10), 51–58 (2020)
6. Gao, D., Peng, D.L., Liu, C.: Entity relation extraction based on CNN in large-scale text data. J. Chin. Comput. Syst. **39**(5), 1021–1026 (2018)
7. Zhang, S., Zheng, D., Hu, X., Yang, M.: Bidirectional long short-term memory networks for relation classification. In: Proceedings of the 29th Pacific Asia Conference on Language, Information and Computation, pp. 73–78 (2015)
8. Zhong, Z., Chen, D.: A frustratingly easy approach for entity and relation extraction. arXiv preprint arXiv:2010.12812 (2020)
9. Zheng, S., Wang, F., Bao, H., Hao, Y., Zhou, P., Xu, B.: Joint extraction of entities and relations based on a novel tagging scheme. arXiv preprint arXiv:1706.05075 (2017)
10. Fei, H., Ren, Y., Ji, D.: Boundaries and edges rethinking: an end-to-end neural model for overlapping entity relation extraction. Inf. Process. Manage. **57**(6), 102311 (2020)
11. Wang, J., Lu, W.: Two are better than one: Joint entity and relation extraction with table-sequence encoders. arXiv preprint arXiv:2010.03851 (2020)
12. Wang, Y., Yu, B., Zhang, Y., Liu, T., Zhu, H., Sun, L.: TPLinker: single-stage joint extraction of entities and relations through token pair linking. arXiv preprint arXiv:2010.13415 (2020)
13. Liu, Y., et al.: RoBERTa: a robustly optimized BERT pretraining approach. arXiv preprint arXiv:1907.11692 (2019)
14. Zeng, D., Zhang, H., Liu, Q.: CopyMTL: copy mechanism for joint extraction of entities and relations with multi-task learning. In: Proceedings of the AAAI Conference on Artificial Intelligence, vol. 34, no. 05, pp. 9507–9514 (2020)
15. Nayak, T., Ng, H.T.: Effective modeling of encoder-decoder architecture for joint entity and relation extraction. In: Proceedings of the AAAI conference on artificial intelligence, vol. 34, no. 05, pp. 8528–8535 (2020)
16. Su, J., et al.: Global pointer: novel efficient span-based approach for named entity recognition. arXiv preprint arXiv:2208.03054 (2022)

Dy-KD: Dynamic Knowledge Distillation for Reduced Easy Examples

Cheng Lin[1,3], Ning Jiang[1,3](✉), Jialiang Tang[2], Xinlei Huang[1,3], and Wenqing Wu[1]

[1] School of Computer Science and Technology, Southwest University of Science and Technology, Mianyang 621000, Sichuan, China
jiangning@swust.edu.cn

[2] School of Computer Science and Engineering, Nanjing University of Science and Technology, Nanjing 210094, Jiangsu, China

[3] Jiangxi Qiushi Academy for Advanced Studies, Nanchang 330036, Jiangxi, China

Abstract. Knowledge distillation is usually performed by promoting a small model (student) to mimic the knowledge of a large model (teacher). The current knowledge distillation methods mainly focus on the extraction and transformation of knowledge while ignoring the importance of examples in the dataset and assigning equal weight to each example. Therefore, in this paper, we propose Dynamic Knowledge Distillation (Dy-KD). To alleviate this problem, Dy-KD incorporates a curriculum strategy to selectively discard easy examples during knowledge distillation. Specifically, we estimate the difficulty level of examples by the predictions from the superior teacher network and divide examples in a dataset into easy examples and hard examples. Subsequently, these examples are given various weights to adjust their contributions to the knowledge transfer. We validate our Dy-KD on CIFAR-100 and Tiny-ImageNet; the experimental results show that: (1) Use the curriculum strategy to discard easy examples to prevent the model's fitting ability from being consumed by fitting easy examples. (2) Giving hard and easy examples varied weight so that the model emphasizes learning hard examples, which can boost students' performance. At the same time, our method is easy to build on the existing distillation method.

Keywords: Model Compression · Knowledge Distillation · Curriculum Learning · Dynamic Distillation

1 Introduction

Deep neural networks (DNNs) have made significant progress in computer vision over the recent years, such as image classification [6], object detection [18], and semantic segmentation [30]. However, these advanced DNNs typically have a large number of learnable parameters and are computationally expensive, making them unsuitable for using on intelligence devices with limited resources. To address this issue, many researchers investigate model compression, mainly

B. Luo et al. (Eds.): ICONIP 2023, CCIS 1966, pp. 223–234, 2024.
https://doi.org/10.1007/978-981-99-8148-9_18

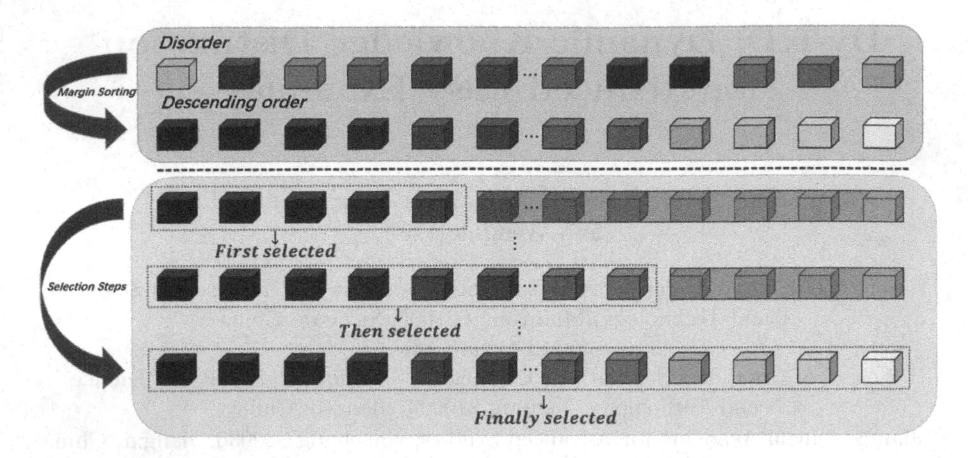

Fig. 1. The first row is the example of randomly shuffling, and the second row is the example in descending order of difficulty. The third row is the example of the model loaded for the first time, and the fourth row is the easy example added. The fifth row is the end of the last iteration, all examples are selected.

including network pruning [5,10,21], network quantization [4,25], lightweight neural networks [8,14], and knowledge distillation [7,17,19]. Among them, knowledge distillation (KD) has recently attracted more attention due to its simplicity and effectiveness in transferring knowledge from a large pre-trained teacher network to a small student network, allowing the student network to achieve comparable performance to the teacher network.

Although simple and effective, existing KD methods usually perform knowledge transfer invariantly. For example, the vanilla KD method [7] aligns the output probabilities of the student model with those of the teacher model. In other words, the knowledge transfer process of KD is always unchanged in these two aspects: (1) KD treats each example equally, which ignores the difficulty levels of examples. (2) KD randomly samples datasets without considering the training status of the student network. There are some examples in the dataset that can be quickly remembered by DNNs during the training [2], which are claimed as easy examples. In contrast, if DNNs have enough representation ability, the dataset's remaining examples (denoted as hard examples) could be learned by repeatedly learning for many iterations. In KD, however, the representational power of the small student network is limited. During the training, the quickly remembered easy examples will continue to consume the student's representation ability, thus resulting in insufficient hard sample learning and ultimately affecting the model's performance.

Content-based selection of training examples has been studied via curriculum learning and anti-curriculum learning approaches [24]. Curriculum learning focuses on training easy examples first and gradually including hard examples in later training stages, where easy examples are typically defined as examples with a large Margin during training and hard examples are typically defined as examples with a low Margin [9,20]. On the other hand, anti-curriculum learning first focuses on hard examples and gradually includes easy examples in later training stages [24]. Due to the limited fitting ability of the model, we follow anti-curriculum learning which discards easy examples.

In this paper, inspired by curriculum learning, we propose a new model compression framework dubbed "Dynamic Knowledge Distillation", which precisely considers hard and easy examples during the training of the student network. Firstly, we use a pre-trained teacher network to rank the dataset according to the size of the Margin value of the teacher network's output probability distribution. Secondly, to fully learn hard examples, we first provide mini-batch data for training the student network, retaining hard examples while discarding easy examples. Then, as the ability of the student model improves, we gradually add easy examples to the model training. Meanwhile, we utilize the difference between the teacher's and student's probability outputs as instance-level weights, which allows the training state of the student model to be properly considered during knowledge distillation. We conduct extensive experiments on the CIFAR-100 and Tiny-ImageNet datasets. Experimental results show that our dynamic knowledge distillation is plug-and-play for the most existing KD methods to improve their performance.

2 Related Work

In this section, we first discuss the history of knowledge distillation and then introduce curriculum learning while pointing out that our proposed method differs from previous work.

2.1 Knowledge Distillation

Knowledge distillation [7] aims to transfer knowledge from a pre-trained teacher model to a compact student model. To improve distillation performance, the existing methods design various forms of knowledge for transfer. It can be roughly divided into three types, logit-based [7,28], feature-based [19,27] and relation-based [13,15,22]. The traditional KD framework is static and always randomly selects a mini-batch of training examples from the dataset and trains all training examples indiscriminately. It is unreasonable to conduct static KD studies as the evolution of a student's model during training. As discussed in [3], different examples should be given different weights during distillation. We are curious whether adaptively tuning the settings of teacher adoption, dataset selection, and supervisory tuning would yield benefits in terms of student's performance and learning efficiency, motivating us to explore a dynamic KD framework.

2.2 Curriculum Learning

Curriculum learning, originally proposed by [2], is a method that can train a network by planning the sequence of learning tasks. This strategy has been widely used in fields such as computer vision and natural language processing. Lesson learning that follows the prescribed sequence and content of educational materials is often used in human learning environment. The model can leverage previously learned concepts and easily learn new and more complex concepts with this guidance. Currently, for content-based training sample selection, two methods of curriculum learning and anti-curriculum learning are studied, and in different cases, both methods can improve the generalization ability of the model more than random curriculum learning [24]. In recent studies, Zhao et al. [29] and Li et al. [12] utilize a teacher network to sort the entire dataset and then train models in the order from easy examples to hard examples. None of them considered the fit ability of the student model. Our proposed method is to discard some easy examples to save the fitting ability of the student model so that the student model has enough fitting ability to learn the hard examples fully.

3 Method

In this section, we first discuss the history of knowledge distillation before introducing the Dy-KD structure and specifics.

3.1 Preliminaries

In knowledge distillation, given a training dataset $\mathcal{D} = \{(\mathbf{x}_i, y_i)\}_{i=1}^{N}$ (N denotes the total number of examples in the dataset), we can transfer knowledge from a pre-trained teacher model $\mathcal{T}$ to a student model $\mathcal{S}$ by following loss term:

$$\mathcal{L}_{transfer} = \mathcal{F}(\sigma(\mathbf{z}^{\mathcal{S}}; \tau), \sigma(\mathbf{z}^{\mathcal{T}}; \tau)), \tag{1}$$

where $\mathbf{z}^{\mathcal{S}}$ and $\mathbf{z}^{\mathcal{T}}$ are the logits from the student and the teacher, respectively. $\sigma(\cdot)$ is the softmax function that yields class probabilities from the logistic, and τ is the temperature hyperparameter to scale the smoothness of the predicted distribution. Specifically, we have $\sigma\left(\mathbf{z}^{\mathcal{T}}; \tau\right) = softmax\left(\mathbf{z}^{\mathcal{T}}/\tau\right)$. $\mathcal{F}(\cdot, \cdot)$ is a loss function that can capture the difference between two categorical distributions, *e.g.*, Kullback-Leibler divergence.

3.2 Dynamic Knowledge Distillation

Knowledge distillation methods usually train students to mimic the teacher's representative knowledge. Under the supervision of this softened knowledge, additional supervised information, such as class correlation, is added to train the student network. However, the previous KD [7,17,19] method assigned the same weight to all examples, which limites the performance of the student model. In fact, the examples are also divided into hard and easy ones, and their influence on the training of the student network should be dynamically adjusted. A reasonable weighting of examples during distillation is more conducive to the fitting of the student model, therefore we design a dynamically example selection method.

In knowledge distillation, the pre-trained teacher network is high-performance which can be utilized to rank the datasets. Specifically, we calculate the gap between target and non-target classes in the teacher's probability outputs to estimate the difficulty level of examples. Intuitively, if the teacher network can accurately recognize an example, the probability of the target is large since the gap value is large. Therefore, we treat the example as an easy example and vice versa. Consequently, we feed all examples into the teacher network to obtain their output probabilities, and the above gap is calculated as follows:

$$Margin = \mathcal{Q}_{top} - \mathcal{Q}_{sec}, \tag{2}$$

$$Scores^T = -Margin, \tag{3}$$

where $\mathcal{Q}_{top}$ is the maximum confidence value of the teacher network's output, and $\mathcal{Q}_{sec}$ is the second largest confidence value of the teacher network's output. $Scores^T$ is the score of each example at the end.

Then, the entire dataset is ranked according to the calculated gap. As shown in Fig. 1, we adopt a training strategy of gradually reducing the number of discarded easy examples. First, we ranked the entire dataset in order of hard to easy. In the first epoch, only half of the more hard examples are used for training, and half of the easy examples are discarded. This allows the model to focus on learning complex patterns at an initial stage. In the second round of the epoch, more easy examples are kept, and only a small number of easy examples are discarded. In the last epoch, we use the entire dataset to train the student model. In this way, it helps the model learn better from hard examples while saving fitting power.

Finally the optimization goal of Dy-KD is to minimize the following loss function:

$$\mathcal{L}_{DyKD}(W_s, W_t) = \mathcal{L}_{Dyce} + \lambda\mathcal{L}(p_s(X_i, W_s), p_t(X_i, W_t)), \tag{4}$$

where the $p_s(X_i, W_s)$ and $p_t(X_i, W_t)$ refer to the corresponding outputs of student and teacher network, respectively. $\mathcal{L}_{Dyce}$ will be introduced in the next section.

3.3 Dynamic Weighted Cross Entropy

In Sect. 3.2, we evaluated the hard and easy examples based on Margin, and then assigned appropriate training samples to the student network based on course learning. In this section, we propose a dynamic cross-entropy loss to dynamically adjust the knowledge transfer from teacher to student.

The existing methods have a flaw in that it does not take into account the learning progress of the student network. If the student network has already fitted some difficult examples in the early stage, these examples should be given smaller weights in the later stage, so we propose dynamic cross-entropy. The difference between the output probabilities of the student network and the teacher network defines the degree of fitting of the student model. In the cross-entropy function, the dynamic weight is used. The formula is as follows:

$$w_i = (\mathbf{z}_i^T - \mathbf{z}_i^S)^2, \tag{5}$$

$$\mathcal{L}_{Dyce} = -\frac{1}{n}\sum_{i=1}^{n} w_i y_i \log\left(z_i^S\right). \tag{6}$$

3.4 Pacing Function

We know the difficulty level of each example through the scoring criteria, but we still need to define how to use the dataset reasonably. We use the concept of pacing function: (1) The first $\alpha\%$ examples are selected for the first iteration. Also discard the remaining easy examples in the dataset. (2) Gradually reduce the number of dropped easy examples during training. (3) Finally, when the iteration is coming to an end, all examples can be selected. Figure 1 shows the selection process, where the f_{pace} formula is as follows:

$$f_{pace}(l) = \begin{cases} \lfloor \alpha \times (1 + \frac{l}{180}) \rfloor, & l \leq 180 \\ \alpha \times 2, & 180 < l \leq 240, \end{cases} \tag{7}$$

where $\lfloor \cdot \rfloor$ means round down and l denotes the training epoch, α is the initial preset value. Our new method Dy-KD is described in Algorithm 1.

Algorithm 1. Dynamic Knowledge Distllation for Reduced Easy Examples

Require: Teacher network N_T, student network N_S, and training dataset, *i.e.*, image data and label data.
$X = \{(x_i, y_i)_{i=1}^N\}$
Ensure: parameters W_S of student model.
Initialization: W_S, W_T and training hyper-parameters
Stage 1: Prepare the teacher network:
1.Repeat:
2.compute $\mathcal{L}(y_{true}, P_T)$.
3.update W_T by gradient back-propagation.
4.Until: $\mathcal{L}(y_{true}, P_T)$ converges.
Stage 2: Training the student via Pacing function:
1.Loading the student model and sort X according to the Pacing function.
2.Repeat:
3.feed subset X_1, ..., $X_i \subseteq X$ in order.
4.compute $\mathcal{L}_{DyKD}(W_s)$ by Eq. 4.
5.Until: $\mathcal{L}_{DyKD}(W_s)$ converges.
6.return (W_s)

Table 1. Top-1 test accuracy of different distillation approaches on CIFAR-100

Teacher Student	WRN-40-2 WRN-16-2	WRN-40-2 WRN-40-1	ResNet32×4 ResNet8×4	VGG13 VGG8
Teacher	76.31%	76.31%	79.42%	74.64%
Student	73.26%	71.98%	72.50%	70.36%
FitNet [19]	73.58%	72.24%	73.50%	71.02%
AT [27]	74.08%	72.77%	73.44%	71.43%
SP [23]	73.83%	72.43%	72.94%	72.68%
PKT [16]	74.54%	73.45%	73.64%	72.88%
CRD [22]	75.64%	74.38%	75.46%	74.29%
KD [7]	74.92%	73.54%	73.33%	72.98%
KCD-KD [11]	75.70%	73.84%	74.05%	73.44%
Dy-KD	75.53%	73.90%	75.66%	73.82%
VID [1]	74.11%	73.30%	73.09%	71.23%
Dy-VID	74.53%	73.74%	74.03%	73.58%
SRRL [26]	75.96%	74.75%	75.92%	74.40%
Dy-SRRL	**75.99%**	**75.04%**	**76.61%**	**74.63%**

4 Experiments

4.1 Datasets and Experiments Configuration

Datasets. We conduct experiments on two benchmark datasets of KD, CIFAR-100 and Tiny-ImageNet. CIFAR-100 contains 50K training images, 500 images per class; 10K test images, 100 images per class, image size is 32×32. Tiny-ImageNet has 200 classes. Each class has 500 training images, 50 validation images and 50 test images. Image size is 64×64.

Implementation Details. We employ a stochastic gradient descent optimizer with a momentum of 0.9, a weight decay of 5×10^{-4}, learning rate initialized to 0.02, and every 30 epochs after the first 150 epochs decay 0.1 until the last 240 epochs. The hyperparameter α is set to 0.5.

4.2 Comparison of Results

Results on CIFAR-100. We compare various representative KD methods, including Vanilla KD [7], Fitnet [19], AT [27], SP [23], VID [1], PKT [16], CRD [22], SRRL [26]. We directly cite quantitative results reported in their papers [22,26]. For the networks of the teacher and student models, we use the WRN-40-2/WRN-40-1 and VGG13/VGG8, *etc.* Surprisingly, the test accuracy of ResNet32 × 4/ResNet8 × 4 on CIFAR-100 has improved by 2.33%. The experimental results on four teacher-student pairs are shown in Table 1. We can see that building the proposed Dynamic Knowledge Distillation (Dy-KD) on KD shows impressive improvements. Furthermore, the proposed Dy-KD constitutes a new learning paradigm, and as a plug-and-play module, we plug our method on top of the VID [1] method and SRRL [26], enabling the advantages of improved accuracy and efficiency.

Results on Tiny-ImageNet. Following common practice, experiments are performed on Tiny-ImageNet using ResNet32×4 (teacher) and ResNet8×4 (student). Table 2 shows the results of Top-1 and Top-5 test accuracy. We can see that building the proposed Dynamic Knowledge Distillation on KD significantly increases the test accuracy.

Table 2. Top-1 test accuracy of different distillation approaches on CIFAR-100

Teacher Student	WRN-40-2 WRN-16-2	WRN-40-2 WRN-40-1	ResNet32×4 ResNet8×4	VGG13 VGG8
Teacher	62.50%	62.50%	65.89%	62.24%
Student	58.31%	56.75%	67.90%	57.40%
KD [7]	61.27%	60.03%	60.54%	61.57%
Dy-KD	**62.32%**	**60.68%**	**62.08%**	**62.62%**

4.3 Ablation

In this section, we verify the effectiveness of the proposed method through different ablation experiments. First, we measure the effect of various ablation components one by one. Table 3 summarizes the accuracy of the results. Then, we did the opposite experiment, first using easy examples, discarding hard examples, and gradually adding hard examples to compare with our proposed method as the model is trained. Table 4 summarizes the accuracy of the results.

Two Components of Dy-KD. On the CIFAR-100 dataset, we use VGG8 as the student and VGG13 as the teacher. The baseline is KD to train.

(1) Only use sorted examples for experiments (Sort only).
(2) Only use dynamic weighted cross entropy for experiments (Dy_ce only).

Table 3. Ablation study of each component on CIFAR-100.

#	Ablation	D_{ce}	P_f	$\mathcal{L}_{ce}$	$\mathcal{L}_{kd}$	VGG8
(1)	KD	-	-	✓	✓	72.98%
(2)	Sort only	-	✓	✓	✓	73.67%
(3)	Dy_ce only	✓	-	✓	✓	73.51%
(4)	Dy-KD	✓	✓	✓	✓	73.82%

From the experimental results we can see that by introducing the component D_{ce} into the system, we observe a significant performance improvement. For example, our model improves accuracy by 0.69% when using the component D_{ce}. By introducing the component P_f into the system, we observe a significant performance boost. For example, our model improves accuracy by 0.53% when using the component P_f. Finally, the best performance is obtained by combining these two components, indicating that the proposed two components are complementary.

From Easy to Hard, Reduced Hard Examples. In order to better demonstrate that hard examples have more information, and easy examples consume the model's fitting ability, we conducted a reverse experiment to further verify this point. Specifically, we design multiple groups of teacher-student pairs to conduct experiments, and gradually introduce hard examples during the training process to observe their impact on model performance.

Table 4. Top-1 test accuracy for baseline, anti-DyKD and Dy-KD on CIFAR-100.

Teacher	WRN-40-2	WRN-40-2	ResNet32×4	VGG13
Student	WRN-16-2	WRN-40-1	ResNet8×4	VGG8
Teacher	76.31%	76.31%	79.42%	74.64%
Student	73.26%	71.98%	72.50%	70.36%
KD [7]	74.92%	73.54%	73.33%	72.98%
Anti-Dy-KD	74.81%	73.48%	74.39%	73.50%
Dy-KD	**75.53%**	**73.90%**	**75.66%**	**73.82%**

Experimental results show that when we use hard examples for training and discard the easy examples in the mini-batch, the model shows significant performance improvement. Hard examples provide more challenging and complex learning scenarios, forcing the model to develop stronger representation capabilities and robustness to effectively deal with more challenging examples in practical applications.

As can be seen from Table 4, when we discard hard examples, the performance of the student model drops dramatically. This is because hard examples are usually samples with high difficulty or extreme features, which provide important challenges and training opportunities for the student model. By processing these hard examples, the student model can learn more complex patterns and regularities.

However, as the model's fitting capability improves, we observed interesting results by retaining hard examples and gradually adding easy examples. While easy examples themselves consume the model's fitting capacity, we found that gradually incorporating easy examples can enhance the model's generalization ability and robustness. This is because easy examples provide a wider range of training instances, which helps the model better understand the data distribution and learn general features, enabling it to perform better on unseen examples.

5 Conclusion

We propose a new learning framework for knowledge distillation that reasonably considers the difficulty of each example to offer suitable data for student network training. The student network can focus on hard examples and reduce the fitting of easy examples to save the model's representation ability. Meanwhile, we assign instance-level dynamic weights to adjust the knowledge transfer from teacher network to student network for feedback-based learning. Extensive experiments demonstrate that our proposed Dy-KD can be used to improve the performance of existing knowledge distillation methods. For future work, we would like to explore strategies for automatically reduced data.

Acknowledgement. This research is supported by Sichuan Science and Technology Program (No. 2022YFG0324), SWUST Doctoral Research Foundation under Grant 19zx7102.

References

1. Ahn, S., Hu, S.X., Damianou, A., Lawrence, N.D., Dai, Z.: Variational information distillation for knowledge transfer. In: Proceedings of the IEEE/CVF Conference on Computer Vision and Pattern Recognition, pp. 9163–9171 (2019)
2. Bengio, Y., Louradour, J., Collobert, R., Weston, J.: Curriculum learning. In: Proceedings of the 26th Annual International Conference on Machine Learning, pp. 41–48 (2009)
3. Chen, G., Choi, W., Yu, X., Han, T., Chandraker, M.: Learning efficient object detection models with knowledge distillation. In: Advances in Neural Information Processing Systems, vol. 30 (2017)
4. Courbariaux, M., Hubara, I., Soudry, D., El-Yaniv, R., Bengio, Y.: Binarized neural networks: training deep neural networks with weights and activations constrained to+ 1 or-1. arXiv preprint arXiv:1602.02830 (2016)
5. Ghosh, S., Srinivasa, S.K., Amon, P., Hutter, A., Kaup, A.: Deep network pruning for object detection. In: 2019 IEEE International Conference on Image Processing (ICIP), pp. 3915–3919. IEEE (2019)

6. He, K., Zhang, X., Ren, S., Sun, J.: Deep residual learning for image recognition. In: Proceedings of the IEEE Conference on Computer Vision and Pattern Recognition, pp. 770–778 (2016)
7. Hinton, G., Vinyals, O., Dean, J.: Distilling the knowledge in a neural network. arXiv preprint arXiv:1503.02531 (2015)
8. Hui, T.W., Tang, X., Loy, C.C.: Liteflownet: a lightweight convolutional neural network for optical flow estimation. In: Proceedings of the IEEE Conference on Computer Vision and Pattern Recognition, pp. 8981–8989 (2018)
9. Kumar, M., Packer, B., Koller, D.: Self-paced learning for latent variable models. In: Advances in Neural Information Processing Systems, vol. 23 (2010)
10. LeCun, Y., Denker, J., Solla, S.: Optimal brain damage. In: Advances in Neural Information Processing Systems, vol. 2 (1989)
11. Li, C., et al.: Knowledge condensation distillation. In: Avidan, S., Brostow, G., Cisse, M., Farinella, G.M., Hassner, T. (eds.) ECCV 2022. LNCS, vol. 13671, pp. 19–35. Springer, Cham (2022). https://doi.org/10.1007/978-3-031-20083-0_2
12. Li, J., Zhou, S., Li, L., Wang, H., Bu, J., Yu, Z.: Dynamic data-free knowledge distillation by easy-to-hard learning strategy. Inf. Sci. **642**, 119202 (2023)
13. Li, L., Jin, Z.: Shadow knowledge distillation: bridging offline and online knowledge transfer. Adv. Neural. Inf. Process. Syst. **35**, 635–649 (2022)
14. Ma, N., Zhang, X., Zheng, H.T., Sun, J.: Shufflenet v2: practical guidelines for efficient CNN architecture design. In: Proceedings of the European Conference on Computer Vision (ECCV), pp. 116–131 (2018)
15. Park, W., Kim, D., Lu, Y., Cho, M.: Relational knowledge distillation. In: Proceedings of the IEEE/CVF Conference on Computer Vision and Pattern Recognition, pp. 3967–3976 (2019)
16. Passalis, N., Tefas, A.: Probabilistic knowledge transfer for deep representation learning. CoRR, abs/1803.10837 **1**(2), 5 (2018)
17. Pintea, S.L., Liu, Y., van Gemert, J.C.: Recurrent knowledge distillation. In: 2018 25th IEEE International Conference on Image Processing (ICIP), pp. 3393–3397. IEEE (2018)
18. Ren, S., He, K., Girshick, R., Sun, J.: Faster R-CNN: towards real-time object detection with region proposal networks. In: Advances in Neural Information Processing Systems, vol. 28 (2015)
19. Romero, A., Ballas, N., Kahou, S.E., Chassang, A., Gatta, C., Bengio, Y.: Fitnets: hints for thin deep nets. arXiv preprint arXiv:1412.6550 (2014)
20. Supancic, J.S., Ramanan, D.: Self-paced learning for long-term tracking. In: Proceedings of the IEEE Conference on Computer Vision and Pattern Recognition, pp. 2379–2386 (2013)
21. Tang, J., Liu, M., Jiang, N., Cai, H., Yu, W., Zhou, J.: Data-free network pruning for model compression. In: 2021 IEEE International Symposium on Circuits and Systems (ISCAS), pp. 1–5. IEEE (2021)
22. Tian, Y., Krishnan, D., Isola, P.: Contrastive representation distillation. arXiv preprint arXiv:1910.10699 (2019)
23. Tung, F., Mori, G.: Similarity-preserving knowledge distillation. In: Proceedings of the IEEE/CVF International Conference on Computer Vision, pp. 1365–1374 (2019)
24. Wu, X., Dyer, E., Neyshabur, B.: When do curricula work? arXiv preprint arXiv:2012.03107 (2020)
25. Yang, J., Martinez, B., Bulat, A., Tzimiropoulos, G.: Knowledge distillation via softmax regression representation learning. In: International Conference on Learning Representations (2020)

26. Yang, J., Martinez, B., Bulat, A., Tzimiropoulos, G., et al.: Knowledge distillation via softmax regression representation learning. In: International Conference on Learning Representations (ICLR) (2021)
27. Zagoruyko, S., Komodakis, N.: Paying more attention to attention: improving the performance of convolutional neural networks via attention transfer. arXiv preprint arXiv:1612.03928 (2016)
28. Zhao, B., Cui, Q., Song, R., Qiu, Y., Liang, J.: Decoupled knowledge distillation. In: Proceedings of the IEEE/CVF Conference on Computer Vision and Pattern Recognition, pp. 11953–11962 (2022)
29. Zhao, H., Sun, X., Dong, J., Dong, Z., Li, Q.: Knowledge distillation via instance-level sequence learning. Knowl.-Based Syst. **233**, 107519 (2021)
30. Zhao, H., Shi, J., Qi, X., Wang, X., Jia, J.: Pyramid scene parsing network. In: Proceedings of the IEEE Conference on Computer Vision and Pattern Recognition, pp. 2881–2890 (2017)

Fooling Downstream Classifiers via Attacking Contrastive Learning Pre-trained Models

Chenggang Li[1], Anjie Peng[1,2](✉), Hui Zeng[1], Kaijun Wu[3], and Wenxin Yu[1]

[1] Southwest University of Science and Technology, Mianyang 621000, China
penganjie200012@163.com

[2] Engineering Research Center of Digital Forensics, Ministry of Education, Nanjing University of Information Science and Technology, Nanjing, China

[3] Science and Technology On Communication Security Laboratory, Chengdu 610022, China

Abstract. Nowadays, downloading a pre-trained contrastive learning (CL) encoder for feature extraction has become an emerging trend in computer vision tasks. However, few works pay attention to the security of downstream tasks when the upstream CL encoder is attacked by adversarial examples. In this paper, we propose an adversarial attack against a pre-trained CL encoder, aiming to fool the downstream classification tasks under black-box cases. To this end, we design a feature similarity loss function and optimize it to enlarge the feature difference between clean images and adversarial examples. Since the adversarial example forces the CL encoder to output distorted features at the last layer, it successfully fools the downstream classifiers which are heavily relied on the encoder's feature output. Experimental results on three typical pre-trained CL models and three downstream classifiers show that our attack has achieved much higher attack success rates than the state-of-the-arts, especially when attacking the linear classifier.

Keywords: Contrastive learning · image classification · adversarial attack

1 Introduction

Deep neural networks (DNNs) with supervised training have achieved remarkable success in various computer vision tasks, including image classification [1] and object detection [2]. However, these well-trained DNNs require large-scale labeled datasets. Unfortunately, it is impractical to collect a large number of labeled images for certain specialized tasks, such as the diagnosis of rare cases. Contrastive learning (CL) [3–6] techniques aim to alleviate these limitations by leveraging unlabeled data to extract useful features and enable downstream tasks to achieve similar results with the supervised learning methods. Therefore, users without sufficient training data or computation resources prefer to employ publicly available pre-trained CL-based encoders to fine-tune or directly extract the feature of CL for their tasks. Because adopting pre-trained CL models for downstream tasks is convenient, cheap yet effective, it becomes more and more popular in practice.

B. Luo et al. (Eds.): ICONIP 2023, CCIS 1966, pp. 235–245, 2024.
https://doi.org/10.1007/978-981-99-8148-9_19

In sharp contrast to the hot contrastive learning, much less attention has been paid to the adversarial robustness of CL pre-trained models [7–11]. To our best knowledge, only a few papers generate adversarial examples targeted at CL pre-trained models. Naseer et al. [13] propose a self-supervised perturbation (SSP) attack for the purpose of training a model to purify adversarial examples. SSP is not specially designed for attacking CL pre-trained models, so it performs not well when attacking some downstream classification tasks. Ban et al. [7] introduce pre-trained adversarial perturbations (PAP) to attack fine-tuning pre-trained models without any knowledge of the downstream tasks. PAP proposed a low-level layer lifting attack (L4A) method to distort the low-level feature, and achieved satisfactory performances against fine-tuning classification tasks. However, we find that PAP frequently fails to attack traditional classifiers, such as linear classifier, and k-nearest neighbor (KNN) classifiers. The reason may be that PAP aims at distorting low-level feature, neglecting to attack high-level feature which is more related to the output label.

In this work, we propose an adversarial attack against CL pre-trained models, aiming at generating adversarial examples to give distorted feature outputs to fool downstream classification tasks, such as linear classifiers, multi-layer perceptions (MLPs), and KNN. These traditional classifiers do not have higher classification accuracies than the hot deep learning-based classifiers, but they are more interpretable and less costly, thus still popular in practical applications. Therefore, we think our work may be of interest to users that concern about the safety of their CL-based applications. To adapt to practical environments, we suppose that we have full knowledge of CL pre-trained models which are usually published on the web, but have no prior knowledge of downstream classification models. In other words, we launch a transferable white-box attack against CL pre-trained model to fool downstream tasks under a black-box situation.

Figure 1 demonstrates the pipeline of our method. As seen, the classifier directly depends on the high-level feature F_x for classification, so we disturb such features to

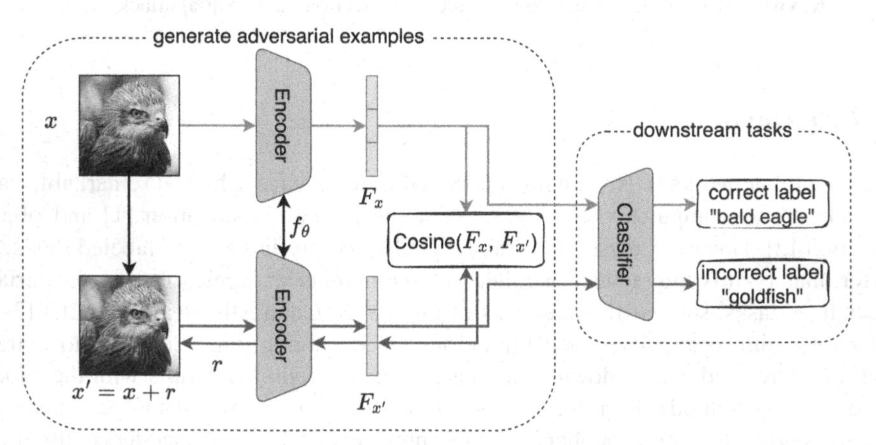

Fig. 1. A pipeline of our method. Green and red line with arrows indicate the inference phase of clean and adversarial image respectively. The adversarial perturbation r is generated by minimizing the loss $consine(F_x, F_{x'})$ with the gradient backward propagation phase denoted by blue line with arrows. (Color figure online)

fool the classifier. The CL algorithms minimize the contrastive loss to learn invariant representations of positive pairs. In contrast, our attack hopes that the clean-adversarial pair are more different on feature maps. To achieve this goal, we design a cosine loss function to measure the similarity between F_x and $F_{x\prime}$, and minimize the loss function via a gradient-based method to generate adversarial examples. After multiple iterations, the features of the adversarial image and the clean image are no longer consistent in the feature space. The red line in Fig. 1 shows that the adversarial example generated from the CL encoder successfully forces the classifier to misclassify "bald eagle" as "goldfish". Experimental results show that our method performs much better than PAP and SSP when attacking different traditional classifiers.

2 Related Works

Contrastive Learning. CL is a self-supervised learning paradigm, which aims to learn an invariant representation of each image in an unlabeled training dataset. Some works [8, 10–12] leverage adversarial training based on CL models to defense against traditional adversarial examples. A well-trained CL model maps images with similar labels to similar locations in the feature space and makes the features of images with different labels as far away as possible. These properties of the CL model are called alignment and uniformity by Wang et al. [14]. Typical CL methods include MoCo [4] and SimCLR [5] which are based on negative sample pairs, BYOL [15] and DINO [25] that based on positive sample pairs, and CLIP [16] that based on images and texts. Among them, the MoCo method is widely used because of its good transferability in downstream tasks. Besides, the DINO method has demonstrated a top-1 accuracy on ImageNet being comparable to supervised learning. These CL models can effectively extract features for downstream tasks, such as image classification, object detection, and image segmentation. In this work, we focus on generating adversarial examples from CL models to fool the downstream classification tasks.

Adversarial Examples. In supervised-based classification, many white-box and black-box adversarial attacks [17–20] have been proposed. With the label of the attacked image, the attacker usually designs a label-based loss to generate adversarial examples. Recently, some feature-level transferable attacks [21, 22] are proposed for supervised classifiers. Although the CL pre-trained model is popular in artificial intelligence tasks, much less effort has been devoted to developing adversarial attacks to concern the security of downstream tasks. Naseer et al. [13] propose an SSP attack method that generates adversarial perturbations by disturbing the features of an intermediate layer. The purpose of SSP is to construct an adversarial example for training a purified model rather than for attacking downstream tasks. Ban et al. [7] propose an L4A method to generate PAP perturbations by lifting the neuron activations of low-level layers of the pre-trained models, supposing that the distorted low-level feature will not change drastically during the fine-tuning procedure. It is worth noting that PAP uses the training dataset to generate adversarial perturbations, which may be sensitive to distributions of images.

3 Methodology

We launch a feature-level attack against contrastive learning models (FACL) to generate adversarial examples to fool the black-box downstream classifiers.

3.1 Notations

Let x and y represent a clean image and its corresponding ground-truth label. f_θ denotes a contrastive learning pre-trained encoder and outputs a feature $F_x \in V$ at the last layer, *i.e.*, $f_\theta(x) = F_x$, where V refers to a high-dimension feature space. Let $C(F_x) \rightarrow y$ denote a downstream classifier, which takes the feature vector F_x as input and outputs a classification label. We generate an adversarial example x' to disturb the feature map of f_θ, expecting $C(F_{x'}) \rightarrow y'$, where $y' \neq y$.

3.2 Our Method

Our goal is to generate an adversarial example from CL pre-trained encoder to fool the black-box downstream classifiers, where we can access the CL encoder, but have no prior knowledge about the classifier. Considering that the downstream classifier depends on the feature extracted from the CL encoder for prediction, we try to contaminate the input feature via an adversarial attack to fool the classifier. In other words, the adversarial example should distort the feature of the corresponding clean image. To this end, we design a feature-level loss and optimize it to make the feature of adversarial example as far away as possible from that of a clean image in the feature embedding, *i.e.*, reducing the similarity between their feature maps. To quantitively measure the similarity of high dimensional features, cosine similarity is utilized. Compared with the L_p norm similarity, the cosine function has a value in $[-1, 1]$ and need not normalization when optimizing the feature-level loss.

$$\underset{x'}{argmin}\ cosine(F_x, F_{x'}), \qquad s.t.\left|\left|x - x'\right|\right|_\infty \leq \varepsilon \tag{1}$$

$$\begin{aligned} BIM :& \phi(g_t) = g_t; \\ MIM :& \phi(g_t) = ug_t + \frac{\nabla_x cosine\left(F_x, F_{x'_t}\right)}{\left|\left|\nabla_x cosine\left(F_x, F_{x'_t}\right)\right|\right|}, g_0 = 0 \end{aligned} \tag{2}$$

Based on the above analysis, we solve the optimization problem (1) to generate the adversarial example x', where the L_∞ constraint controls the attack budget to ensure a human-imperceptible adversarial example. Our attack is feature-based, so the label of the attacked image is not required. The proposed adversarial example is directly generated by minimizing the feature loss. This is the main difference between our attack and the traditional label-based adversarial attack. After getting the loss, inspired by the fast yet effective attack BIM [17] and MIM [18], we utilize an iterative gradient-based method to solve the optimization problem (1). The update direction function ϕ of BIM and MIM are formulated in Eq. (2), where g_t is the gradient at t^{th} step, u is the moment factor, $||\cdot||$ is a L_1 norm. Our FACL attack based on BIM and MIM is called BIM-FACL and MIM-FACL respectively. Of course, our method can also integrate with other optimization algorithms and transferability enhancement, which is our future work. Our method is summarized in Algorithm 1. In the initial step, a Gaussian noised image is taken as an adversarial image to compute the cosine loss.

Algorithm 1 FACL: Feature-level Attack against Contrastive Learning Model

Input: a contrastive learning pre-trained encoder f_θ ; a clean image x; Gaussian noise δ; perturbation budget ε; step size α; iterations T; the function of gradient $\phi(\cdot)$

Output: An adversarial image x'

1: $g_0 = 0$; $x'_0 = x + \delta$

2: Input x to f_θ to obtain the feature F_x

3: for $t = 0$ to T-1 do

4: Input x'_t to f_θ to obtain features $F_{x'_t}$;

5: Compute the gradient $g_t = \nabla_x \, cosine\left(F_x, F_{x'_t}\right)$;

6: Calculate the gradient direction function as in (2);

7: Update x'_{t+1} as:

$$x'_{t+1} = clip(x'_t + \alpha sign(\phi(g_t)), x + \varepsilon, x - \varepsilon)$$

8: end for

9: return x'_T.

Table 1. The average cosine similarity of 2048-D features calculated by 1000 clean images and their adversarial examples generated by the Layer$i$ attack. Layer$i$ (i = 1, 2, 3, 4, 5) means the attack against feature map of the ith block of MoCo pre-trained Resnet-50. The Layer5 attack reduces the most cosine similarity of 2048-D feature.

	Layer1	Layer2	Layer3	Layer4	Layer5(last)
Cosine similarity	0.6472	0.3836	0.2903	0.2484	0.1078

To empirically show that the proposed method disorders the feature output of CL model, we calculate the cosine similarity of feature between clean images and adversarial images at the last layer of MoCo pre-trained Resnet-50. We also attack the feature of different blocks for Resnet-50 to generate adversarial examples. The dimension of feature map for Layer1-Layer5 are $256 \times 56 \times 56$, $512 \times 28 \times 28$, $1024 \times 14 \times 14$, $2048 \times 7 \times 7$ and $2048 \times 1 \times 1$ respectively. Randomly selected 1000 images from ILSVRC 2012 are employed to calculate the cosine similarity of the last features. As can be seen from Table 1, the Layer5 attack significantly reduces the cosine similarity of the feature outputs between clean images and adversarial examples, empirically verifying effectiveness of the proposed attack. The results in Table 1 also show that the Layer5 attack, which directly disturbs feature of MoCo pre-trained Resnet-50, achieves the minimum cosine similarity. In the following, the proposed method targets at the last layer, which will output feature for downstream tasks, to generate adversarial examples.

4 Experiments

4.1 Experimental Setting

Attacked CL Models. MoCo [4], SimCLR [5], and DINO [25] are typical contrastive learning schemes, which utilize Resnet [23] as the backbone for pre-training. We download their pre-trained Resnet-50[1] models to evaluate the performance of attack. Recently, transformer-based methods [24] attract much attention, due to their excellent performance. We also adopt the pre-trained Vit-small[2] model. In particular, the pre-trained Resnet-50 model and Vit-small model have no classification heads. We take their respective 2048-D and 384-D feature outputs as input for the downstream classifiers. More details about CL models can be found on the website[3].

Downstream Classifiers for Evaluation. We generate adversarial examples on CL pre-trained models and test whether these adversarial examples can fool downstream classifiers in the inference phase. Three commonly used traditional classifiers: linear classifier, MLP, and KNN are taken as the downstream tasks for evaluation. The linear classifier is a simple one-layer neural network without activation, which has #feature dimension $\times$ #number of class neurons. For example, when attacking the Vit-small model on 1000-ImageNet, the linear classifier has 384 $\times$ 1000 neurons. MLP with Relu activation function adds a hidden layer using 1000 neurons on the linear classifier and has #feature dimension $\times$ 1000 $\times$ #number of class neurons in total. The KNN classifier employs the cosine function as the similarity metric and votes for the top-20 class for the result. Notice that we have no prior knowledge about these downstream classifiers.

Datasets. Experiments are done on randomly selected 10000 and 8000 images from the validation dataset of ILSVRC 2012 and STL10 test dataset respectively. Before feeding the image into the CL pre-trained model for extracting the feature, we resize the images of ILSVRC 2012 and STL10 to 224 $\times$ 224 $\times$ 3 and normalize them into [0, 1].

Attack Parameters and the Performance Metric. For our method, the iteration number T is set as 20. For compared methods, PAP [7] and SSP [13], the default values according to their papers are used. In order to generate human-imperceptible adversarial examples, the maximal attack budget ε is set to be 8/255. We use the attack success rate (ASR) in (3) to measure the performance of attack, where #attack success indicates the number of adversarial samples that successfully fool the downstream classifier, #prediction success indicates the number of samples that are predicted correctly by the downstream classifier. Table 2 demonstrates the number of correctly predicted images by three downstream classifiers, where the features are extracted from various pre-trained models. As seen, the linear classifier and MLP achieve classification accuracy > 70%. These results verify that the CL model learns some object-aware features.

$$ASR = \frac{\#attack \;\; success}{\#prediction \;\; success} \tag{3}$$

[1] https://dl.fbaipublicfiles.com/moco-v3/r-50-300ep/r-50-300ep.pth.tar.
[2] https://dl.fbaipublicfiles.com/moco-v3/vit-B-300ep/vit-S-300ep.pth.tar.
[3] https://github.com/facebookresearch/moco-v3.

Table 2. The number of correctly predicted images of different downstream classifiers on ILSVRC 2012 and STL10. These images will be used to generate adversarial examples for evaluating *ASR* of attack in the following tables.

Pretrained method	Linear classifier	MLP	KNN
MoCo [4]	7249	7207	6698
SimCLR [5]	7071	7064	6121
DINO [25]	7507	7359	6738
MoCo* [4]	7508	7796	6941

* The classification accuracy of clean image on STL10.

4.2 Comparative Results

In this sub-section, we compare our method with two state-of-the-arts, PAP [7] and SSP [13]. The proposed attacks with two optimization methods, BIM-FACL and MIM-FACL are evaluated. We craft adversarial examples on CL pre-trained Resnet-50 and Vit-small models respectively and evaluate *ASR* on three downstream classification tasks, including the linear classifier, MLP, and KNN.

Table 3. The *ASR* (%) of the attack (in row) on **ILSVRC 2012**. The adversarial examples generated from the CL pre-trained **Resnet50, Vit-small** by **MoCo** [4] to fool downstream classifiers (in columns).

Model	Attack	Linear Classifier	MLP	KNN
Resnet50	PAP [7]	64.0	30.35	24.01
	SSP [13]	80.1	60.59	31.52
	BIM-FACL(ours)	**99.9**	82.35	69.87
	MIM-FACL(ours)	99.85	**96.46**	**94.37**
Vit-small	PAP [7]	81.88	11.99	9.71
	SSP [13]	96.41	49.24	21.93
	BIM-FACL(ours)	97.96	50.04	29.64
	MIM-FACL(ours)	**98.93**	**69.67**	**51.92**

As seen, the proposed method outperforms the compared methods by a large margin when fooling the linear classifier. The results of attacking linear classifiers in Table 3, 4 5 show that, our MIM-FACL obtains nearly 100% of *ASR* on ILSVRC 2012 validation dataset, achieving at least 16.08% and 1.55% higher *ASR* than PAP and SSP respectively. Our method also achieves great improvements in terms of *ASR* when attacking the KNN classifier, as shown in Table 3 and Table 4. For the tests on STL10 database in Table 5, our method achieves similar improvement as that in the ILSVRC 2012 validation database. These promising results of our method owe to that we attack the feature of the last layer of a CL pre-trained model, which directly disturbs the input for the downstream classifiers. PAP and SSP disturb the low-level and middle-level features of the CL model, while these disturbed features may be partly retrieved by the remaining neurons of high layers and thus reducing *ASR*. Due to the strong representation ability of the Vit-small model, all attacks deteriorate the attack performance for attacking KNN and MLP as shown in Table 3. How to effectively fool the downstream classifiers from the transformer-based CL encoder is our future work.

Table 4. The *ASR* (%) of the attack (in row) on **ILSVRC 2012**. The adversarial examples generated from the CL pre-trained **Resnet50** by **SimCLR** [5] and **DINO** [25] to fool downstream classifiers in the column.

CL method	Attack	Linear Classifier	MLP	KNN
SimCLR [5]	SSP[13]	90.92	21.32	21.09
	BIM-FACL(ours)	**99.89**	64.12	40.76
	MIM-FACL(ours)	99.83	**77.23**	**73.49**
DINO [25]	SSP[13]	95.72	32.53	32.08
	BIM-FACL(ours)	99.84	35.07	36.51
	MIM-FACL(ours)	**99.85**	**71.58**	**69.81**

Table 5. The *ASR* (%) of all attacks (in rows) on **STL10**. The adversarial examples generated from the CL pre-trained **Resnet50** by **MoCo** [4] to fool downstream classifiers in the column.

	Linear Classifier	MLP	KNN
PAP [7]	60.74	30.73	60.14
SSP [13]	84.28	83.06	89.08
BIM-FACL(ours)	**91.9**	**94.06**	**91.24**
MIM-FACL(ours)	89.82	93.89	91.04

4.3 Discussion the Influence of Attacking Different Layers

Our method attacks the layer of CL pre-trained model to disrupt the feature outputs for downstream tasks. Therefore, selecting which layer to attack may affect the effectiveness. According to the architecture of Resnet-50 and Vit-small, we attack the feature maps of their 5 blocks, namely layer1, layer2, layer3, layer4 and layer5. To evaluate the attack effectiveness, adversarial examples are generated on the validation dataset of ILSVRC 2012. As shown in Fig. 2, from attacking low-level features at layer1 to high-level features at layer5, the *ASR* gradually increases for CL pre-trained Resnet50 model (blue). Notice that the *ASR* of the proposed method keeps stable for attacking the Vit-small model, which is also reported in the work [7]. The feature of the last layer has the smallest

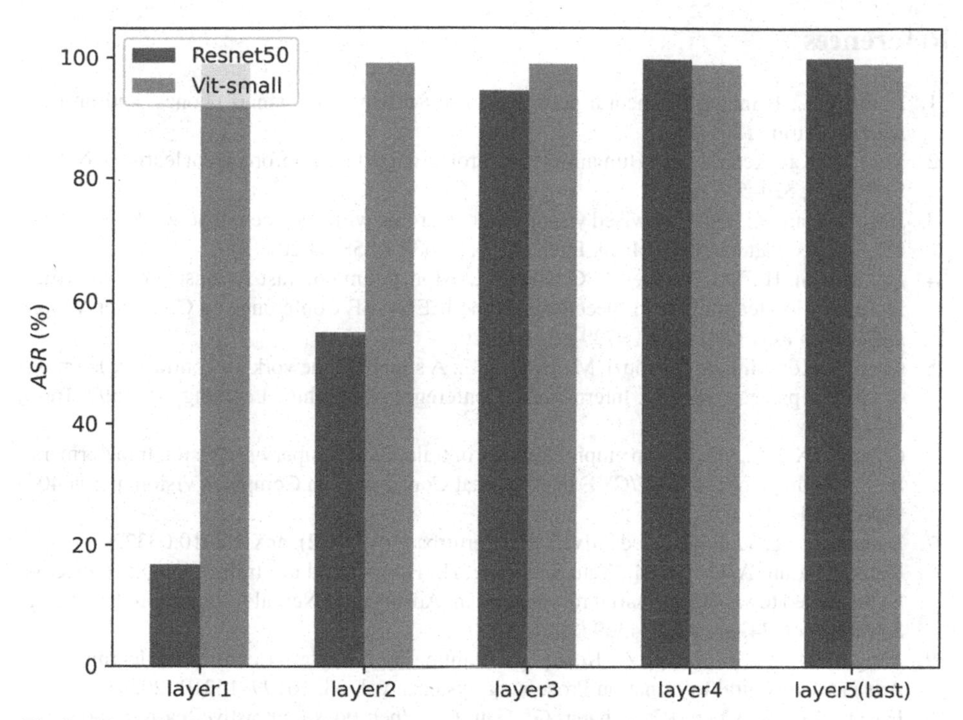

Fig.2. The *ASR* (%) of our BIM-FACL method for attacking different layers of CL pre-trained Resnet50 and Vit-small by MoCo.

dimensions than that of other front layers. Computing the cosine similarity of last layer runs the fastest. So, for a uniform selection of the layer, we choose to attack the last layer of the contrastive learning model.

5 Conclusion

In this paper, we construct adversarial examples from a contrastive learning pre-trained model to blindly attack the downstream tasks under black-box settings. Experimental results validate that our attack against the upstream encoder can fool various downstream classifiers, outperforming state-of-the-arts. Especially, our attack achieves a nearly 100% attack success rate when fooling the linear classifier on the validation dataset of ILSVRC 2012. Our work arises an alert about the security of publicly available contrastive learning pre-trained models.

Acknowledgement. This work was partially supported by NFSC No.62072484, Sichuan Science and Technology Program (No. 2022YFG0321, No. 2022NSFSC0916), the Opening Project of Engineering Research Center of Digital Forensics, Ministry of Education.

References

1. Bourzac K.: Bringing big neural networks to self-driving cars, smart phones, and drones. IEEE Spectrum (2016)
2. Mnih, V., Kavukcuoglu, K.: Human-level control through deep reinforcement learning. Nature **518**(7540), 529–533 (2015)
3. Jing, L., Tian, Y.: Self-supervised visual feature learning with deep neural networks: a survey. IEEE Trans. Pattern Anal. Mach. Intell. **43**(11), 4037–4058 (2020)
4. He, K., Fan, H., Wu, Y., Xie, S., Girshick, R.: Momentum contrast for unsupervised visual representation learning. In: Proceedings of the IEEE/CVF Conference on Computer Vision and Pattern Recognition, pp. 9729–9738 (2020)
5. Chen, T., Kornblith, S., Norouzi, M., Hinton, G.: A simple framework for contrastive learning of visual representations. In: International Conference on Machine Learning, pp. 1597–1607 (2020)
6. Chen, X., Xie, S., He, K.: An empirical study of training self-supervised vision transformers. In: Proceedings of the IEEE/CVF International Conference on Computer Vision, pp. 9640–9649 (2021)
7. Ban, Y., Dong, Y.: Pre-trained Adversarial Perturbations (2022). arXiv:2210.03372
8. Dong, X., Luu, A.T., Lin, M., Yan, S., Zhang, H.: How should pre-trained language models be fine-tuned towards adversarial robustness? In: Advances in Neural Information Processing Systems, vol. 34, pp. 4356–4369 (2021)
9. Jiang, Z., Chen, T., Wang, Z.: Robust pre-training by adversarial contrastive learning. In: Advances in Neural Information Processing Systems, vol. 33, 16199–16210 (2020)
10. Fan, L., Liu, S., Chen, P.Y., Zhang, G., Gan, C.: When does contrastive learning preserve adversarial robustness from pretraining to finetuning? In: Advances in Neural Information Processing Systems, vol. 34, pp. 21480–21492 (2021)
11. Yang, Z., Liu, Y.: On robust prefix-tuning for text classification. In: International Conference on Learning Representations (2022)
12. Ioffe, S., Szegedy, C.: Batch normalization: Accelerating deep network training by reducing internal covariate shift. In: International Conference on Machine Learning, pp. 448–456 (2015)
13. Naseer, M., Khan, S., Hayat, M., Khan, F.S., Porikli, F.: A self-supervised approach for adversarial robustness. In: Proceedings of the IEEE/CVF Conference on Computer Vision and Pattern Recognition, pp. 262–271 (2020)
14. Wang, T., Isola, P.: Understanding contrastive representation learning through alignment and uniformity on the hypersphere. In: International Conference on Machine Learning, pp. 9929–9939 (2020)
15. Grill, J.B., et al.: Bootstrap your own latent-a new approach to self-supervised learning. In: Advances in Neural Information Processing Systems, vol. 33. pp. 21271–21284 (2020)
16. Radford, A., et al.: Learning transferable visual models from natural language supervision. In: International Conference on Machine Learning, pp. 8748–8763 (2021)
17. Kurakin, A., Goodfellow, I.J., Bengio, S.: Adversarial examples in the physical world. In: Artificial Intelligence Safety and Security, pp. 99–112 (2018)
18. Dong, Y., et al.: Boosting adversarial attacks with momentum. In: Proceedings of the IEEE Conference on Computer Vision and Pattern Recognition, pp. 9185–9193 (2018)
19. Pomponi, J., Scardapane, S., Uncini, A.: Pixel: a fast and effective black-box attack based on rearranging pixels. In: 2022 International Joint Conference on Neural Networks (IJCNN), pp. 1–7 (2022)
20. Schwinn, L., Raab, R., Nguyen, A., Zanca, D., Eskofier, B.: Exploring misclassifications of robust neural networks to enhance adversarial attacks.arXiv:2105.10304 (2021)

21. Zhang, J., et al.: Improving adversarial transferability via neuron attribution-based attacks. In: Proceedings of the IEEE/CVF Conference on Computer Vision and Pattern Recognition, pp. 14993–15002 (2022)
22. Wang, Z., Guo, H., Zhang, Z., Liu, W., Qin, Z., Ren, K.: Feature importance-aware transferable adversarial attacks. In: Proceedings of the IEEE/CVF International Conference on Computer Vision, pp. 7639–7648 (2021)
23. He, K., Zhang, X., Ren, S., Sun, J.: Deep residual learning for image recognition. In: Proceedings of the IEEE Conference on Computer Vision and Pattern Recognition, pp. 770–778 (2016)
24. Dosovitskiy, A., et al.: An image is worth 16x16 words: transformers for image recognition at scale.arXiv:2010.11929. (2020)
25. Caron, M., et al.: Emerging properties in self-supervised vision transformers. In: Proceedings of the IEEE/CVF International Conference on Computer Vision, pp. 9650–9660 (2021)

Feature Reconstruction Distillation with Self-attention

Cheng Lin[1,3], Ning Jiang[1,3(✉)], Jialiang Tang[2], and Xinlei Huang[1,3]

[1] School of Computer Science and Technology, Southwest University of Science and Technology, Mianyang 621000, Sichuan, China
jiangning@swust.edu.cn
[2] School of Computer Science and Engineering, Nanjing University of Science and Technology, Nanjing 210094, Jiangsu, China
[3] Jiangxi Qiushi Academy for Advanced Studies, Nanchang 330036, Jiangxi, China

Abstract. A recently proposed knowledge distillation method based on feature map transfer verifies that the intermediate layers of the teacher model can be used as effective targets for training the student model for better generalization. Existing researches mainly focus on how to efficiently transfer knowledge between the intermediate layers of teacher and student models. However, they ignore the increase in the number of channels in the intermediate layer, which will introduce redundant feature information, and there is also a lack of interaction between shallow features and deep features. To alleviate these two problems, we propose a new knowledge distillation method called Feature Reconstruction Knowledge Distillation (FRD). By reconstructing the features of the intermediate layer of the teacher model, the student model can learn more accurate feature information. In addition, to tackle the problem that feature maps have only local information, we use a self-attention mechanism to fuse shallow and deep features, thereby enhancing the model's ability to handle global details. Through extensive experiments on different network architectures involving various teacher and student models, we observe that the proposed method significantly improves performance.

Keywords: Model Compression · Knowledge Distillation · Feature Reconstruction · Self-attention

1 Introduction

Deep neural networks have demonstrated outstanding performance in numerous tasks as a powerful machine learning model. With the development and expansion of models, however, deep neural networks have grown in size and complexity, necessitating substantial computational and storage resources. To handle these issues, researchers have begun investigating model compression techniques [5,6,9,12,14] , among which knowledge distillation [7,13] is a crucial method.

Knowledge distillation [7] is a model compression technique that involves training a small model, commonly referred to as the "student model" based on

B. Luo et al. (Eds.): ICONIP 2023, CCIS 1966, pp. 246–257, 2024.
https://doi.org/10.1007/978-981-99-8148-9_20

a large and complex model, often referred to as the "teacher model" to reduce the model size while maintaining high performance. Traditional knowledge distillation (KD) utilizes the outputs of the teacher model as soft targets during the training of the student model. These soft targets contain the predicted probability information of the teacher model for each sample. By using soft targets, the student model can acquire additional knowledge from the teacher model to enhance its generalization performance. To expand the scope of knowledge, feature distillation [1,4,13,19,21] has been proposed as a variant of traditional knowledge distillation. It differs in that feature distillation aims to transmit the knowledge of the teacher model by using its intermediate layer representations as the target for the student model. Among these techniques, Du et al. [21] utilizes Feature Pyramid Network (FPN) to alleviate the problem of limited geometric information in feature maps of small-sized high-level networks. While shallow networks may contain more geometric information, they lack sufficient semantic features for image classification. Chen et al. [4] addresses the problem of semantic mismatch by leveraging cross-layer interactions to ensure semantic consistency in knowledge transfer.

Traditional feature distillation [13] methods have achieved significant progress in knowledge transfer, but they still have some limitations. Specifically, (1) They ignored the increase in the number of channels in the intermediate layer, which would introduce redundant feature information, some of which may not contribute substantially to the performance of the model. This results in models that are less efficient in terms of feature extraction and representation. (2) Both shallow and deep features fail to capture global information, leading to a lack of multi-scale information.

To overcome these limitations, we propose a Feature Reconstruction Framework (FRD) composed of feature reconstruction (F_R) and feature fusion (F_F) components. The F_R component uses self-attention to calculate the weighted correlation between different channels, effectively improving the expressiveness of features and introducing diversity [8,17]. It acts as a more accurate targeting layer. The F_F component employs self-attention to compute the weighted correlations between different intermediate layers for feature fusion, enabling the student model to learn the fused features from the teacher model and acquire multi-scale information. Extensive experiments are conducted on the STL-10 and CIFAR-100 datasets. The experimental results demonstrate that the proposed Feature Reconstruction Framework outperforms distillation using original features significantly.

In summary, this research makes the following contributions:

- We propose a new feature distillation method, which uses a self-attention mechanism to reconstruct feature maps, improve the expressiveness of features, and introduce diversity. This enables the student model to learn from the target layer with more powerful information.
- Existing feature distillation methods have neglected the capture of global information. We introduce self-attention mechanisms for achieving multi-scale

feature fusion, capturing global contextual information, and better understanding the correlations and semantic information among features.
- Extensive experiments on popular network architectures using STL-10 and CIFAR-100 datasets show that FRD excels in model performance.

2 Related Work

In this section, we introduce the related work in detail. Related works on knowledge distillation and feature distillation are discussed in Sect. 2.1 and Sect. 2.2, respectively.

2.1 Knowledge Distillation

Knowledge distillation [7,10,20] is an effective method to improve the performance of a given student model by exploiting soft targets in a pre-trained teacher model. Compared with one-hot labels, the fine-grained information among different categories provides additional supervision to better optimize the student model. The recently proposed Dynamic Rectification Knowledge Distillation (DR-KD) [2] method involves teachers making wrong predictions during information extraction and then correcting these errors before knowledge distillation occurs. Making DR-KD ensure that teachers never extract incorrect knowledge. Inspired by this, we propose feature map reconstruction, trying to correct the "errors" in the feature map.

2.2 Feature-Map Distillation

Recent approaches not only formalize knowledge into highly abstract forms [1,13,19], such as predictions, but also leverage information from intermediate layers by designing fine-grained knowledge representations. To this end, researchers have proposed a series of techniques. For example, a method for aligning hidden layer responses called hinting was introduced [13]. In addition, there are methods that mimic spatial attention maps [19] and methods that maximize mutual information via variational principles [1]. These methods aim to improve the effectiveness of knowledge distillation and show promising research progress. In contrast, our proposed method utilizes a self-attention mechanism to correct and enhance feature maps, improve feature expressiveness, and introduce diversity. At the same time, the method uses self-attention for feature fusion so that the student model can mimic the global information of teacher features.

3 Method

This section introduces knowledge distillation for feature map reconstruction (FRD). Figure 1 shows an overview of our distillation method. These symbols are in Sect. 3.1, and Sect. 3.2 briefly introduces logit-based distillation. Section 3.3 briefly introduces feature map-based distillation. Section 3.4 and Sect. 3.5 discuss feature map reconstruction and fusion strategies in detail.

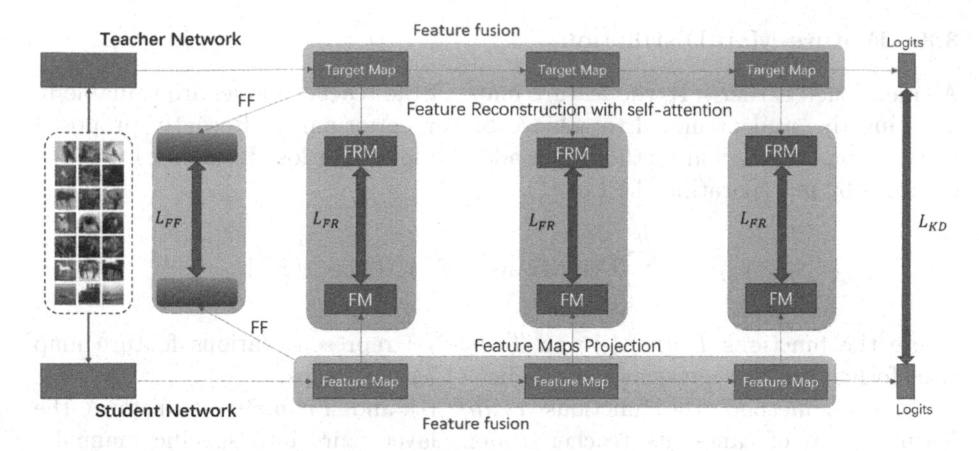

Fig. 1. An overview of the proposed Feature Reconstruction Knowledge Distillation (FRD). Each layer of the teacher performs feature map reconstruction through self-attention to become the target layer. The feature maps of each layer of the student are projected into three separate forms to align with the spatial dimension of each target layer. Multi-layer feature maps of teachers and students generate fused features via self-attention.

3.1 Notation

Given a set of input data $\mathbf{X} = \{x_1, x_2 .., x_k\}$ including $\mathbf{k}$ examples, the label corresponding to each example is denoted as $\mathbf{Y} = \{y_1, y_2 .., y_k\}$. We define a pre-trained and fixed teacher network as $\mathcal{N}_T$ and a student network as $\mathcal{N}_S$. Let $\mathbf{g}^S$ and $\mathbf{g}^T$ denote the logits of the student network and the teacher network, respectively. $\mathbf{F}_j^S$ represents the j-th layer's features of the student network, $j \in \{1, ..., n\}$. $\mathbf{F}_i^T$ represents the i-th layer's features of the teacher network, $i \in \{1, ..., n\}$, where n represents the maximum number of blocks in the network.

3.2 Logit-Based Knowledge Distillation

The purpose of knowledge distillation is to train a student network by mimicking the output of a teacher network. In original knowledge distillation [7,20], the student network is expected to mimic the teacher's soft targets. For image classification tasks, the traditional distillation loss uses the cross-entropy loss and Kullback-Leibler (KL) divergence. The loss function L_{logit} can be expressed as:

$$L_{logit} = (1 - \lambda) L_{ce}(\mathbf{y}, \sigma(\mathbf{g}^S)) + \lambda \tau^2 L_{KL}(\sigma(\mathbf{g}^T/\tau), \sigma(\mathbf{g}^S/\tau)), \quad (1)$$

where $L_{ce}(\cdot,\cdot)$ denotes the cross-entropy loss, $L_{KL}(\cdot,\cdot)$ denotes KL divergence. τ is a hyper-parameter and a higher τ leads to a more considerable softening effect. We set τ to 4 throughout this paper for fair comparison. λ is a balancing hyperparameter.

3.3 Feature-Map Distillation

As mentioned earlier [11], the feature maps of the teacher model are valuable in assisting the student model to achieve better performance. Recently proposed feature map distillation methods can add the following loss term to each batch of size b by incorporating the Eq. (1):

$$\mathcal{L}_{FMD} = \sum_{i=1}^{n} \text{Dist}\left(\text{Trans}^{T}\left(\mathbf{F}_i^T\right), \text{Trans}^{S}\left(\mathbf{F}_i^S\right)\right), \tag{2}$$

where the functions $Trans^T(\cdot)$ and $Trans^S(\cdot)$ represent various feature map transformations, n represents the number of feature maps.

In each method, the functions $Trans^T(\cdot)$ and $Trans^S(\cdot)$ transform the feature maps of candidate teacher-student layer pairs into specific manually designed representations, such as AT [19] or VID [1] leading to the overall loss as:

$$\mathcal{L}_{total} = L_{logit} + \gamma \mathcal{L}_{FMD}, \tag{3}$$

where the hyperparameter γ in the equation is used to balance the two separate loss terms.

3.4 Feature Rectification Distillation

In our FRD method, each teacher layer reconstructs feature maps as target layers using the self-attention mechanism, as shown in Fig. 1. The reconstructed feature maps are then used to train the student model for better regularization. Additionally, FRD can be adapted to different teacher and student models.

First, we use the intermediate feature map of the teacher network as input and obtain Q_i, K_i, and V_i matrices through three Multi-Layer Perceptrons (MLP). The specific formula is as follows:

$$Q_i = MLP_Q\left(\mathbf{F}_i^T\right), K_i = MLP_K\left(\mathbf{F}_i^T\right), V_i = MLP_V\left(\mathbf{F}_i^T\right), \tag{4}$$

where $MLP_Q(\cdot)$, $MLP_K(\cdot)$ and $MLP_V(\cdot)$ represent three mapping functions.

Then, we calculate the attention weights for each feature map. For each pixel m, the relevance between pixel m and other pixels n is measured by computing the similarity between the query vector Q_i and the key vector K_i. This can be done with the dot product operation with the following formula:

$$Attention(Q_i, K_i) = softmax\left(\frac{Q_i * K_i^T}{\sqrt{d_k}}\right), \tag{5}$$

where d_k represents the scaling factor.

Finally, the value vector V_i is weighted and summed using the attention weight matrix *Attention* to obtain the attention output vectors for each pixel. These attention-weighted sums are then added to the input feature maps, and the final reconstructed feature maps are defined as:

$$\mathbf{F}_{z^i}^T = Attention(Q_i, K_i) * V_i + \mathbf{F}_i^T, \tag{6}$$

where $\mathbf{F}_{z^i}^T$ is our reconstructed and repaired feature map.

Each feature map of the student layers is projected into $\mathbf{F}_{iw}^T$ individual forms to align with the spatial dimensions of each target layer for the purpose of performing the following distance calculation:

$$\mathbf{F}_i^S = \mathrm{Proj}\left(\mathbf{F}_i^S \in R^{b \times c_{s_l} \times h_{s_l} \times w_{s_l}}, \mathbf{F}_{iw}^T\right), i \in [1, \ldots, n], \tag{7}$$

$$L_{FR} = \sum_{i=1}^{n} ||\mathbf{F}_i^S - \mathbf{F}_{z^i}^T||_2, \tag{8}$$

where each function Proj($\cdot$, $\cdot$) includes a stack of three layers with 1×1, 3×3, and 1×1 convolutions to meet the demand of capability for transformation. $\mathbf{F}_{iw}^T$ is the dimension of the teacher feature map.

3.5 Feature Fusion with Self-attention

Since layer-by-layer [13] feature distillation will lose global information, we propose a method for feature fusion using a self-attention mechanism. The self-attention mechanism is a technique for establishing the correlation between sequences or features, which can calculate the correlation weight of each feature and its correlation with other features adaptively according to the input features. By leveraging the self-attention mechanism for feature fusion, we are able to better integrate multi-scale features and enable the learned model to understand the input data more comprehensively.

We use multi-layer raw feature maps for self-attention computation to obtain fused features for teacher and student models. Then, we minimize the Mean Square Error (MSE) loss between teacher-fused feature maps (F_F^T) and student-fused feature maps (F_F^S).

First, since the closeness of the pairwise similarity matrix can be regarded as a good measure of intrinsic semantic similarity [16]. These similarity matrices are computed as follows:

$$\mathbf{A}_i^T = R\left(\mathbf{F}_i^T\right) \cdot R\left(\mathbf{F}_i^T\right)^T, \tag{9}$$

where $R(\cdot)$ is the shaping operation and $\mathbf{A}_i^T$ is the b $\times$ b matrix.

Through three MLP networks, the similarity matrix $\mathbf{A}_i^T$ is mapped to query vector Q_i^T, key vector K_i^T, and value vector V_i^T. The specific formula is as follows:

$$Q_i^T = MLP_Q\left(\mathbf{A}_i^T\right), K_i^T = MLP_K\left(\mathbf{A}_i^T\right), V_i^T = MLP_V\left(\mathbf{A}_i^T\right), \tag{10}$$

$$Q_T = Concat(Q_1^T, ..., Q_n^T), K_T = Concat(K_1^T, ..., K_n^T), \tag{11}$$

$$V_T = Concat(V_1^T, ..., V_n^T), \tag{12}$$

where Q_T is the concatenation of Q_i^T generated by all similarity matrices. Likewise, K_T and V_T are the same operations. The $Concat(\cdot)$ function is the concatenation of matrices.

For each feature map similarity matrix, the attention weight matrix *Attention* is calculated by computing the similarity between the query vector Q_T and the key vector K_T. The specific formula is as follows:

$$Attention(Q_T, K_T) = softmax(\frac{Q_T * K_T^T}{\sqrt{d_k}}), \tag{13}$$

where K_T^T is the transpose matrix of K_T.

For each feature map similarity matrix, use the attention weight matrix $Attention(Q_T, K_T)$ to weight and sum the value vector V_T to obtain the fusion feature output $\mathbf{F}_F^T$. The specific formula is as follows:

$$\mathbf{F}_F^T = Attention(Q_T, K_T) * V_T. \tag{14}$$

Similarly, the fusion feature of the student model is $\mathbf{F}_F^S$, and the formula is as follows:

$$\mathbf{F}_F^S = Attention(Q_S, K_S) * V_S, \tag{15}$$

The final loss for feature fusion is as follows:

$$L_{FF} = ||\mathbf{F}_F^T - \mathbf{F}_F^S||_2, \tag{16}$$

The total loss of the proposed FRD algorithm is as follows:

$$L_{FRD} = L_{logit} + \beta(L_{FR} + L_{FF}). \tag{17}$$

where the hyperparameter β in the equation is used to balance the two separate loss terms.

4 Experiments

The training details for all experiments are in Sect. 4.1. The effectiveness of the proposed distillation method has been verified on different datasets. In Sect. 4.2, we provide the results on STL-10 dataset. In Sect. 4.3, we show experimental results of teacher-student networks with the same structures and different structures for distillation on CIFAR-100. To further verify the effectiveness of the method, we provide the results of the comparison of different distillation methods. Finally, the results of ablation experiments are presented in Sect. 4.4.

4.1 Experimental Setup

Dataset. (1) STL-10 contains 13K examples, which have 5K images for training and 8K images for testing. It includes 10 categories and all examples are 96×96 RGB images. (2) CIFAR-100 has 100 classes for image classification with 50K training examples and 10K test examples, where the resolution of each example is 32×32.

Implementation Details. All experiments use the gradient descent algorithm [3]. Horizontal flip and random crop are used for data augmentation in the experiments. The batch size is set to 128 and the weight decay value is $5e^{-4}$ on CIFAR-100 and STL-10 datasets. The temperature hyperparameter is set to 4. For CIFAR-100, all networks are trained for 240 epochs. The initial learning rate for the experiments we set is 0.05. We set the learning rate to drop at 150, 180, and 210 epochs. Our training setup is consistent with CRD [15]. All our results are obtained from training four times. For STL-10, all networks are trained for 100 epochs. During training, the initial learning rate is 0.05 and is decayed at 30, 60, and 90 epochs.

Table 1. Top-1 test accuracy of different distillation approaches on STL-10

Teacher	WRN-40-2	ResNet32×4	ResNet32×4	WRN-40-2
Student	ShuffleNetV1	ShuffleNetV1	ShuffleNetV2	ShuffleNetV2
Teacher	72.00%	69.75%	69.75%	72.00%
Student	60.71%	60.71%	60.40%	60.40%
KD [7]	64.18%	59.46%	69.67%	66.95%
FitNet [13]	68.13%	67.30%	69.77%	70.5%
AT [19]	64.03%	66.78%	68.87%	69.36%
SRRL [18]	64.83%	65.62%	65.81%	65.45%
SemCKD [4]	68.81%	69.22%	69.42%	69.73%
FRD	**70.06%**	**69.48%**	**69.62%**	**71.50%**

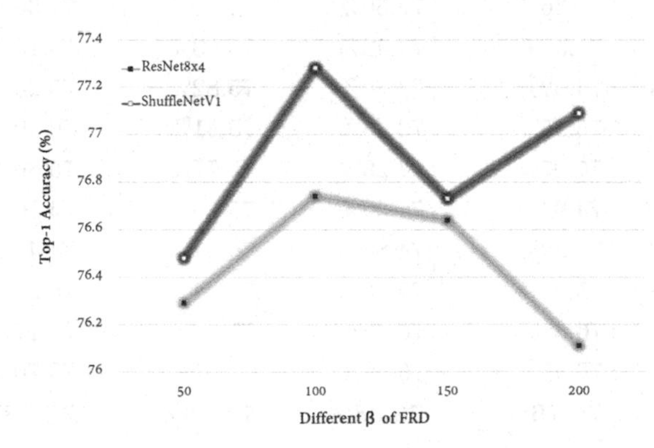

Fig. 2. The influence of the hyperparameter β of ShuffleNetV1 and ResNet8 × 4 on CIFAR-100.

4.2 Experiments of STL-10 Dataset

For STL-10, the combination of WRN model and ResNet32×4 as a teacher and ShuffleNet as a student is evaluated. Table 1 shows the experimental results of our FRD on STL-10. When the teacher is ResNet32×4, and the student is ShuffleV2, the traditional distillation achieves a test accuracy of 70.5% on STL-10. In contrast, using our technique and rectifying the feature maps of ResNet32×4, the test accuracy on the same dataset is 71.50%, a performance improvement of 1% compared to traditional distillation methods.

4.3 Experiments of CIFAR-100 Dataset

Table 2 shows the experimental results of multiple teacher-student architectures on CIFAR-100. We have established five sets of networks with different architectures, such as WRN-40-2/ShuffleV1, ResNet-32 × 4/ResNet-8 × 4. From Table 2, it can be seen that our proposed feature reconstruction distillation method is superior to other distillation methods. Especially, the best example is that in the setting of "ResNet-32 × 4/ResNet-8 × 4", students' performance is even 0.66% higher than DKD.

Table 2. Top-1 test accuracy of different distillation approaches on CIFAR-100

Teacher Student	ResNet32×4 ShuffleNetV1	ResNet32×4 ResNet8×4	ResNet32×4 ShuffleNetV2	WRN-40-2 ShuffleNetV1
Teacher	79.42%	79.42%	79.42%	75.61%
Student	71.36%	72.50%	72.60%	71.36%
KD [7]	71.30%	74.42%	75.60%	75.13%
FitNet [13]	74.52%	73.50%	75.82%	75.28%
AT [19]	75.55%	73.44%	75.41%	76.39%
SP [16]	75.07%	74.29%	75.77%	75.86%
VID [1]	74.04%	74.55%	75.55%	74.10%
HKD [22]	74.32%	74.86%	76.64%	74.43%
CRD [15]	75.43%	75.46%	77.04%	76.28%
SemCKD [4]	76.31%	76.42%	77.63%	77.17%
DKD [20]	76.45%	76.32%	77.07%	76.70%
FRD	**76.76%**	**76.98%**	**77.88%**	**77.28%**

4.4 Ablation Study

In this section, we validated the effectiveness of this method through different ablation experiments. We mainly use the ResNet network as the basic model for validation experiments. First, we visualize the feature reconstruction map to demonstrate that the method can improve feature expressiveness and introduce diversity. Then, we prove their effectiveness by using two components separately. Finally, we evaluate the effect of the hyperparameter β on distillation performance.

Visualization of Feature Map Reconstruction. The experiment explores whether the reconstructed feature map can provide better results than the original feature map after using the self-attention mechanism to reconstruct the feature map in the CIFAR-100 dataset. We choose the ResNet32×4 model for feature extraction and use the self-attention mechanism to reconstruct the feature map to restore the appearance and structure of the original feature map. In order to compare the original and reconstructed feature maps, we visually analyze them using visualization techniques such as heatmaps. As shown in Fig. 3, by observing and comparing the appearance and structure of the two feature maps, we can evaluate whether the reconstructed feature maps present features such as more informative, sharper edges, or stronger responses.

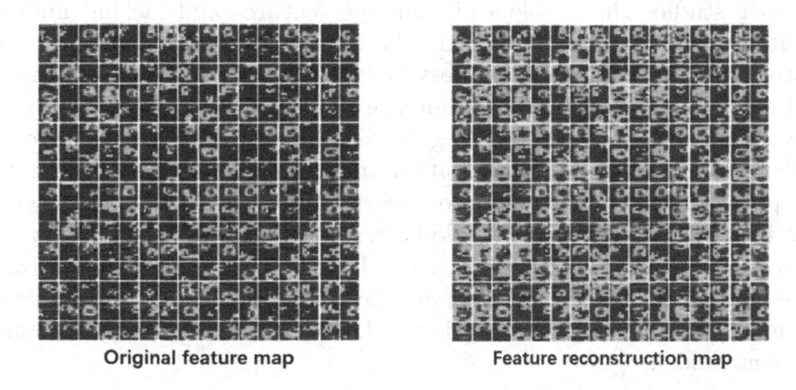

Fig. 3. In the CIFAR-100 dataset, the third-layer original feature map and reconstructed feature map of ResNet32×4

Two Components of FRD. We also conduct thorough experiments to understand the contribution of the proposed feature reconstruction distillation (F_R) and feature fusion distillation (F_F). As mentioned earlier, F_R aims to improve the expressiveness of features and introduce diversity, while F_F aims to learn a global representation to capture long-range dependencies. By covering each of these, the different effects of these two components can be checked. As shown in Table 3, both objectives can significantly improve vanilla students, while (F_R) exhibits more efficacy. Combining the two components achieves the best performance, indicating that the two proposed goals are complementary.

Weight of Loss Hyperparameter. The hyperparameter of the loss function can affect the performance of the network, because the hyperparameter affects

Table 3. Ablation study of each component on CIFAR-100

#	Ablation	F_R	F_F	$\mathcal{L}_{ce}$	$\mathcal{L}_{kd}$	ResNet8×4
(1)	KD	-	-	✓	✓	74.42%
(2)	F_R only	✓	-	✓	✓	76.22%
(3)	F_F only	-	✓	✓	✓	76.34%
(4)	FRD	✓	✓	✓	✓	76.98%

the gradient propagation. Figure 2 shows our experimental results using different hyperparametric values. The hyperparameter β ranges from 50 to 200. The interval between two adjacent values is 50. From Fig. 2 can be seen that a parameter value of 100 is more conducive to network training.

5 Conclusion

This paper studies the problem of "missing features and lacking multi-scale information for targets" in feature distillation. We are different from previous work studying target selection and loss function. We propose to reconstruct the feature map with self-attention to enhance the expressive power of each pixel, thereby compensating for the missing feature information. Since feature reconstruction also enhances local information and lacks multi-scale information, we further propose using self-attention to integrate multi-scale features, allowing the student to mimic the teacher on a global scale, rather than just minimizing the differences between individual components. Extensive experiments show that our proposed method solves the problem of lack of information in feature maps and lack of multi-scale information in objects, thereby improving the performance of the student model.

Acknowledgement. This research is supported by Sichuan Science and Technology Program (No. 2022YFG0324), SWUST Doctoral Research Foundation under Grant 19zx7102.

References

1. Ahn, S., Hu, S.X., Damianou, A., Lawrence, N.D., Dai, Z.: Variational information distillation for knowledge transfer. In: Proceedings of the IEEE/CVF Conference on Computer Vision and Pattern Recognition, pp. 9163–9171 (2019)
2. Amik, F.R., Tasin, A.I., Ahmed, S., Elahi, M., Mohammed, N.: Dynamic rectification knowledge distillation. arXiv preprint arXiv:2201.11319 (2022)
3. Bottou, L.: Stochastic gradient descent tricks. In: Neural Networks: Tricks of the Trade, Second Edition, pp. 421–436. Springer, Berlin (2012)
4. Chen, D., et al.: Cross-layer distillation with semantic calibration. In: Proceedings of the AAAI Conference on Artificial Intelligence, vol. 35, pp. 7028–7036 (2021)

5. Courbariaux, M., Hubara, I., Soudry, D., El-Yaniv, R., Bengio, Y.: Binarized neural networks: training deep neural networks with weights and activations constrained to +1 or -1. arXiv preprint arXiv:1602.02830 (2016)
6. Ghosh, S., Srinivasa, S.K., Amon, P., Hutter, A., Kaup, A.: Deep network pruning for object detection. In: 2019 IEEE International Conference on Image Processing (ICIP), pp. 3915–3919. IEEE (2019)
7. Hinton, G., Vinyals, O., Dean, J.: Distilling the knowledge in a neural network. arXiv preprint arXiv:1503.02531 (2015)
8. Hu, J., Shen, L., Sun, G.: Squeeze-and-excitation networks. In: Proceedings of the IEEE Conference on Computer Vision and Pattern Recognition, pp. 7132–7141 (2018)
9. LeCun, Y., Denker, J., Solla, S.: Optimal brain damage. In: Advances in Neural Information Processing Systems, vol. 2 (1989)
10. Li, L., Jin, Z.: Shadow knowledge distillation: bridging offline and online knowledge transfer. In: Advances in Neural Information Processing Systems, vol. 35, pp. 635–649 (2022)
11. Liu, Z., et al.: Feature distillation: DNN-oriented JPEG compression against adversarial examples. In: 2019 IEEE/CVF Conference on Computer Vision and Pattern Recognition (CVPR), pp. 860–868. IEEE (2019)
12. Ma, N., Zhang, X., Zheng, H.T., Sun, J.: ShuffleNet V2: practical guidelines for efficient CNN architecture design. In: Proceedings of the European conference on computer vision (ECCV), pp. 116–131 (2018)
13. Romero, A., Ballas, N., Kahou, S.E., Chassang, A., Gatta, C., Bengio, Y.: FitNets: hints for thin deep nets. arXiv preprint arXiv:1412.6550 (2014)
14. Tang, J., Liu, M., Jiang, N., Cai, H., Yu, W., Zhou, J.: Data-free network pruning for model compression. In: 2021 IEEE International Symposium on Circuits and Systems (ISCAS), pp. 1–5. IEEE (2021)
15. Tian, Y., Krishnan, D., Isola, P.: Contrastive representation distillation. arXiv preprint arXiv:1910.10699 (2019)
16. Tung, F., Mori, G.: Similarity-preserving knowledge distillation. In: Proceedings of the IEEE/CVF International Conference on Computer Vision, pp. 1365–1374 (2019)
17. Woo, S., Park, J., Lee, J.Y., Kweon, I.S.: CBAM: convolutional block attention module. In: Proceedings of the European Conference on Computer Vision (ECCV), pp. 3–19 (2018)
18. Yang, J., Martinez, B., Bulat, A., Tzimiropoulos, G.: Knowledge distillation via softmax regression representation learning. In: International Conference on Learning Representations (2020)
19. Zagoruyko, S., Komodakis, N.: Paying more attention to attention: improving the performance of convolutional neural networks via attention transfer. arXiv preprint arXiv:1612.03928 (2016)
20. Zhao, B., Cui, Q., Song, R., Qiu, Y., Liang, J.: Decoupled knowledge distillation. In: Proceedings of the IEEE/CVF Conference on Computer Vision and Pattern Recognition, pp. 11953–11962 (2022)
21. Zhixing, D., et al.: Distilling object detectors with feature richness. In: Advances in Neural Information Processing Systems, vol. 34, pp. 5213–5224 (2021)
22. Zhou, S., et al.: Distilling holistic knowledge with graph neural networks. In: Proceedings of the IEEE/CVF International Conference on Computer Vision, pp. 10387–10396 (2021)

DAGAN:Generative Adversarial Network with Dual Attention-Enhanced GRU for Multivariate Time Series Imputation

Hongtao Song, Xiangran Fang, Dan Lu(✉), and Qilong Han

College of Computer Science and Technology,
Harbin Engineering University, Harbin 150000, China
{songhongtao,fxran,ludan,hanqilong}@hrbeu.edu.cn

Abstract. Missing values are common in multivariate time series data, which limits the usability of the data and impedes further analysis. Thus, it is imperative to impute missing values in time series data. However, in handling missing values, existing imputation techniques fail to take full advantage of the time-related data and have limitations in capturing potential correlations between variables. This paper presents a new model for imputing multivariate time series data called DAGAN, which comprises a generator and a discriminator. Specifically, the generator incorporates a Temporal Attention layer, a Relevance Attention layer, and a Feature Aggregation layer. The Temporal Attention layer utilizes an attention mechanism and recurrent neural network to address the RNN's inability to model long-term dependencies in the time series. The Relevance Attention layer employs a self-attention-based network architecture to capture correlations among multiple variables in the time series. The Feature Aggregation layer integrates time information and correlation information using a residual network and a Linear layer for effective imputation of missing data. In the discriminator, we also introduce a temporal cueing matrix to aid in distinguishing between generated and real values. To evaluate the proposed model, we conduct experiments on two real-time series datasets, and the findings indicate that DAGAN outperforms state-of-the-art methods by more than 13%.

Keywords: Time series imputation · Self-Attention · Recurrent Neural Network · Generative Adversarial Network

1 Introduction

Time series data refers to data arranged in chronological order that reflects the state of things as they change over time [25]. Typical time series data includes factory sensor data, stock trading data, climate data for a particular region,

This work was supported by the National Key R&D Program of China under Grant No. 2020YFB1710200 and Heilongjiang Key R&D Program of China under Grant No. GA23A915.

B. Luo et al. (Eds.): ICONIP 2023, CCIS 1966, pp. 258–271, 2024.
https://doi.org/10.1007/978-981-99-8148-9_21

power data, transportation data, and so on. In practical applications, multivariate time series data is more common. However, due to the complexity of real-world conditions, multivariate time series data often contains missing values for various reasons [14,26], such as sensor malfunctions, communication errors, and accidents [13]. At the same time, in reality, due to equipment issues, such as the sampling machine itself not having a fixed sampling rate, even if the dataset is complete, the intervals between the data points are often irregular, which can also be considered a problem of missing data. These missing values undermine the interpretability of the data and can seriously affect the performance of downstream tasks such as classification, prediction, and anomaly detection [12]. To further study this incomplete time series, imputing the missing values is an inevitable step.

In order to solve the problem of missing data, many time series imputation methods have been proposed to infer missing values from observed values. Traditional data imputation methods are based on mathematical and statistical theory and fill in missing values by analyzing the data. However, this method ignores the inherent time correlation imputation in time series data. In contrast, time series imputation methods based on deep learning consider the time information in multivariate time series data and achieve better imputation accuracy. Recurrent neural networks rely on the calculation of hidden states for each unit to capture time dependence, but they still place more emphasis on the outputs of adjacent time steps and have difficulty capturing and utilizing long-term dependencies in time series [9].

In recent years, Generative Adversarial Networks (GANs) have shown outstanding performance in the field of data generation. However, general GANs do not consider the time correlation of time series data and ignore the potential correlation between different features of multivariate time series. Therefore, in this paper, we propose a new multivariate time series imputation model called DAGAN. DAGAN consists of a generator and a discriminator. The generator utilizes a gated recurrent neural network to learn the time information in multivariate time series data. It performs a weighted sum of hidden states in the recurrent neural network using an attention mechanism, which improves the model's ability to learn long-term dependencies in time series while ensuring that the model focuses more on important information. Furthermore, we use a self-attention mechanism to link different variables in multivariate time series, allowing all time steps to participate in each layer to maximize the accuracy of multivariate time series imputation. The task of the discriminator is to distinguish between real and generated values. In this paper, we use a temporal cueing matrix to improve the performance of the discriminator, which contains some missing information from the original time series data. This further forces the generator to learn the real data distribution of the input dataset more accurately.

In summary, compared with existing time series imputation models, the model proposed in this paper has the following contributions:

1) We propose a time series imputation method that combines attention mechanisms and gated recurrent neural networks to enhance the ability of recur-

rent neural networks to utilize the long-term dependencies of time series data, while ensuring that the model focuses on important information and improves the quality of imputation.
2) We use the masked self-attention mechanism to capture relationships between different features in the multivariate time series. By incorporating self-attention at each layer, all time steps participate, maximizing the accuracy of imputation for multivariate time series data.
3) We conduct experiments on two different real datasets, using the root mean square error (RMSE) as the performance indicator. The results show that, in most cases, our model outperforms the baseline method.

2 Related Work

There are two main methods for dealing with missing values in time series analysis: direct deletion and imputation [11]. The main idea behind the direct deletion method is to delete samples or features containing missing values directly. Although this method is simple and easy to implement, it can lead to a reduction in sample size and the loss of important information [10]. The imputation method can be divided into statistical-based, machine-learning-based, and neural network-based methods, based on different imputation techniques and technologies.

Statistical-based imputation methods use simple statistical strategies, such as imputing missing values with mean or median values [11,14]. For example, SimpleMean [7]imputes missing values by calculating the mean value. Although these methods can quickly fill in missing values, they can affect the variance of original data and have poor imputation accuracy.

Common machine-learning-based imputation methods include k-nearest neighbor (KNN) method, expectation-maximization method, and random forest. The basic idea of KNN algorithm is to find the k nearest neighboring data points with known values to a missing data point and then calculate a weighted average by some rule, using the known values of these neighbors, to obtain the imputed value. Random forest is a decision-tree-based imputation method [24]. Although these methods exhibit better imputation performance than the previous two methods, they lack the ability to utilize the time correlation and complex correlation between different variables.

In time series data, variables change over time and are interrelated. Therefore, it is crucial to use the time correlation of time series data to improve imputation accuracy. With the rapid development of neural network technology, researchers have started to apply it to the field of time series data imputation [6,10,18]. To properly handle time series data with missing values, some researchers have proposed using RNN-based methods to process missing values. RNN-based data imputation models can capture the time dependency of time series data [19,22]. Among them, GRUD [3], which is based on gated recurrent neural networks, uses hidden state decay to capture time dependence. BRITS [2] treats missing values as variables and imputes them based on the hidden states of bidirectional RNN. GLIMA [23] uses a global and local neural network model with multi-direction attention to process missing values, which partly overcomes the problem that RNN heavily depends on output from nearby timestamps. The problem with RNN-based models is that after a period of time, the weight of current inputs becomes negligible, but they should not be ignored [2,28]. However, this issue does not exist in GAN-based models. GANs generate data with the same distribution as the original data through the game between generator and discriminator. Examples of GAN-based imputation models include GRUIGAN [15], GAIN [28], BiGAN [8], and SSGAN [17]. These methods take advantage of the benefits of GANs and combine them with RNN to improve their ability to capture time dependence.

3 Preliminary

Before introducing the model proposed by us, we provide a formal definition of multivariate time-series data and explain some symbols used in certain models.

Given a set of multivariate time series $X = (x_1, x_2 \cdots, x_T), \quad \in \mathbb{R}^{T \times D}$ and timestamps $TS = \{t_1, t_2, \ldots, t_T\}$, the t-th observation $\mathbf{x}_t \in \mathbb{R}^D$ consists of D features $\{x_{t1}, x_{t2}, \cdots, x_{tD}\}$ and corresponds to timestamp t_t. We define a mask matrix M, which is used to indicate the positions of missing values in the dataset. When x_{ij} is missing, the corresponding element in the mask matrix is equal to 0, otherwise, it is 1. The mask matrix M has the same size as the input dataset. The specific formula is shown as follows.

$$m_{ij} = \begin{cases} 0 & \text{if } x_{ij} \text{ is missing} \\ 1 & \text{otherwise} \end{cases} \tag{1}$$

We also introduce the time interval matrix δ, which is represented as follows.

$$\delta_{ij} = \begin{cases} \delta_{ij-1} + t_i - t_{i-1}, & \text{if } m_{ij-1} = 0, i > 0 \\ t_i - t_{i-1}, & \text{if } m_{ij-1} = 1, i > 0 \\ 0, & i = 0 \end{cases} \tag{2}$$

The time retention matrix represents the time interval between the current time and the last valid observation. It is also a matrix of the same size as the input data set. Then, we introduce the time decay factor α, which is used to control the influence of past observations. When the δ value is higher, α becomes smaller. This indicates that the further the missing value is from the last true observation value, the more unreliable its value is.

$$\alpha_i = \exp\left(-\max\left(0, W_\alpha \delta_i + b_\alpha\right)\right) \tag{3}$$

4 Model

4.1 The Overall Structure

In this section, we will introduce the overall architecture of the DAGAN model. Figure 2 shows the overall architecture of DAGAN. The input to DAGAN includes time-series data, mask matrix, and time interval matrix. DAGAN consists of a generator and a discriminator. The generator generates imputed data based on the observed values in the time-series data. Its objective is to deceive the discriminator with the generated data. The discriminator takes the estimated time-series matrix and the temporal attention matrix as inputs. It attempts to differentiate between the generated data and the real data. Specifically, we introduce the time-attention matrix to encode a part of the missing information in the time-series data stored in the mask matrix. Below, we will describe the structure of each module in detail (Fig. 1).

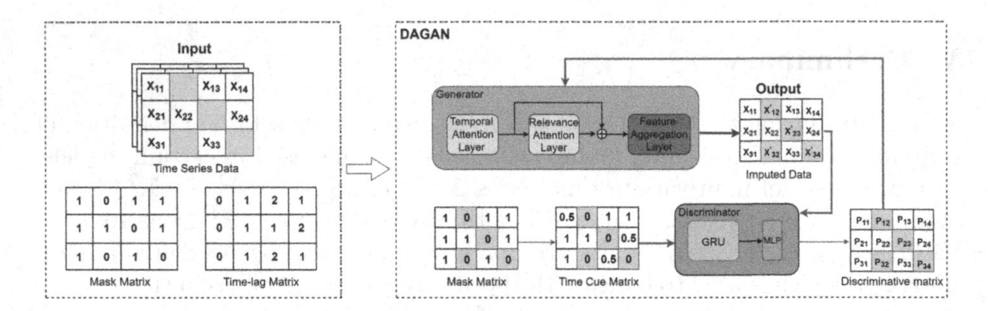

Fig. 1. The overall architecture diagram of DAGAN.

4.2 Generator Network

The goal of the generator is to learn the distribution of multivariate time series data and generate missing values. Our generator includes a Temporal Attention layer, a Relevance Attention layer, and a Feature Aggregation layer. The purpose of the Temporal Attention layer is to capture the temporal dependencies of time series data. Its input is the time series data with missing values, mask vector, and time decay defined in the third section. The Relevance Attention layer captures the correlations between different features. Its input is the output of the Temporal Attention layer. Finally, the time information obtained from the Temporal Attention layer and the Feature Correlation information obtained from the Relevance Attention layer are inputted into the Feature Aggregation layer for aggregation to obtain the final output. The complete dataset obtained in this manner retains true values, replacing missing values with generated values, according to the calculation formula shown below.

$$x_{\text{imputed}} = \mathrm{M} \odot x + (1 - \mathrm{M}) \odot G(x) \tag{4}$$

Here, $x_{imputed}$ is the complete data set interpolated from the input data set x and the mask matrix M, which represents the distribution of missing values in the input data set. G(x) is the output of the generator.

The loss function of the generator includes two parts: adversarial loss and reconstruction loss. The adversarial loss is similar to that of a standard GAN.

The reconstruction loss is used to enhance the consistency between the observed time series and the reconstructed time series. The loss function of the generator is shown below:

$$\mathrm{Loss}_G = \mathrm{Loss}_g + \mathrm{Loss}_r \tag{5}$$

$$\mathrm{Loss}_g = \| x \odot M - x_{\text{imputed}} \odot M \|_2^2 \tag{6}$$

$$\mathrm{Loss}_r = \log\left(1 - D\left(x_{\text{imputed}} \odot (1 - M)\right)\right) \tag{7}$$

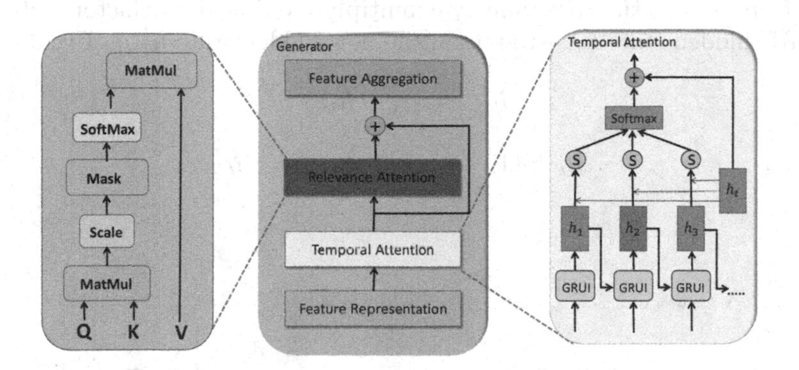

Fig. 2. The structure of DANGAN's Generator.

Improved GRU. A deeper structure is beneficial for a recurrent neural network to better model the time series structure, in order to capture more complex relationships in the time series [21]. Based on this, we designed a feature mapping module to represent the input data in terms of features and map them into potential representations. This can improve the learning ability of RNN without increasing complexity in aggregating at multiple time steps.

$$\boldsymbol{x}_t = \tanh\left(\boldsymbol{W}_m x + \boldsymbol{b}_m\right) \tag{8}$$

where the parameters W_m and b_m are the parameters to be learned, x represents the input data, tanh is a nonlinear activation function.

In our proposed model, we choose a gated recurrent neural network (GRU) to handle the time series input of the generation, which is a network structure adapted from the classical RNN and controls the information transfer in the neural network by adding a gating mechanism (nonlinear activation function, in this paper we use sigmoid activation function). GRU requires fewer parameters to train and converge faster. In a standard GRU model, the input to each GRU unit is the hidden state h_{t-1} of the previous unit's output and the current input x_t. Each GRU has an internal update gate and a reset gate. The data flow inside the GRU can be expressed as follows.

$$\mathbf{h}_t = (1 - \mu_t) \odot \mathbf{h}_{t-1} + \mu_t \odot \tilde{\mathbf{h}}_t \tag{9}$$

$$\tilde{\mathbf{h}}_t = \tanh\left(W_h \mathbf{x}_t + U_h \left(\mathbf{r}_t \odot \mathbf{h}_{t-1}\right) + \mathbf{b}_h\right) \tag{10}$$

$$\mu_t = \sigma\left(W_\mu \mathbf{x}_t + U_z \mathbf{h}_{t-1} + \mathbf{b}_\mu\right) \tag{11}$$

$$\mathbf{r}_t = \sigma\left(W_r \mathbf{x}_t + U_r \mathbf{h}_{t-1} + \mathbf{b}_r\right) \tag{12}$$

where μ_t and $\mathbf{r}_t$ are the update and reset gates of the GRU, respectively. tanh, σ, and $\odot$ denote the tanh activation function, the sigmod function, and the element multiplication.

We integrate the previously introduced Time Retention matrix and time decay factor α into the GRU unit. We multiply the time decay factor alpha with the GRU hidden state $\boldsymbol{h}_{t-1}$ to obtain the new hidden state $\boldsymbol{h}'_{t-1}$ (Fig. 3).

$$\boldsymbol{h}'_{t-1} = \boldsymbol{\alpha}_t \odot \boldsymbol{h}_{t-1} \tag{13}$$

$$\mathbf{h}_t = (1 - \mu_t) \odot \boldsymbol{h}'_{t-1} + \mu_t \odot \tilde{\mathbf{h}}_t \tag{14}$$

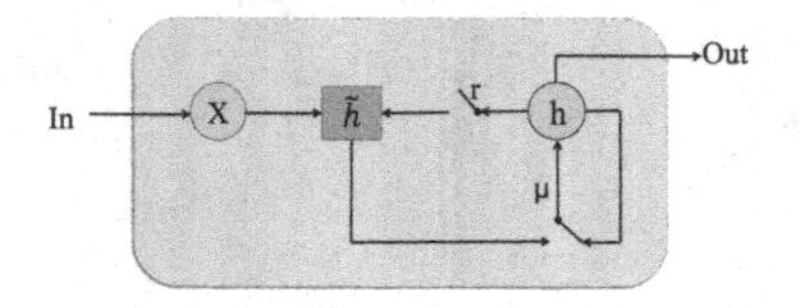

(a) An example of a typical GRU cell.

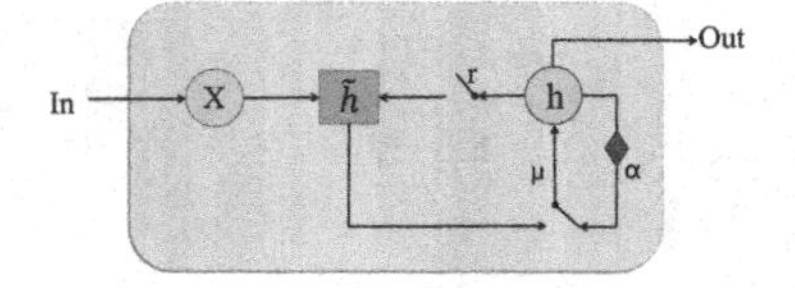

(b) Internal details of our improved GRU cell.

Fig. 3. Standard GRU Cell vs. Our Improved GRU Cell.

Temporal Attention. In order to solve the limitation of memory and excessive attention to adjacent time steps in the recurrent neural network when facing long time sequences, and to enhance the ability of the recurrent neural network to capture important information and long-term dependencies within the time series, we have designed a time recurrent attention mechanism. Our time recurrent attention mechanism weights the hidden states of each time step, and the weighted processing of attention will make the hidden states extracted from each time series contain comprehensive temporal information.

Given a set of hidden states $\mathrm{H} = \{\mathrm{h}_1,\ \mathrm{h}_2,\ \mathrm{h}_3 \ldots, \mathrm{h}_t\}$, calculating the importance score θ of each hidden state and then calculate the weighted sum of the hidden states to obtain the Context Vector v_t. In this way, we can effectively alleviate the disadvantage of GRU's tendency to forget the first few steps in long sequences [1]. The specific calculation formula is shown in Eqs. 15 and 16.

$$\theta_i = \frac{\exp\left(\mathrm{func}\left(\mathbf{h}_t, \mathbf{h}_i\right)\right)}{\sum_{j=1}^{t} \exp\left(\mathrm{func}\left(\mathbf{h}_t, \mathbf{h}_j\right)\right)} \tag{15}$$

$$\boldsymbol{v}_t = \sum_{i=1}^{t} \boldsymbol{\theta}_i \boldsymbol{h}_i \tag{16}$$

where func is the function that calculates the attention score between the current state and the historical hidden states. In this article, we use the hidden state at time t as the query vector and use dot product as the calculation function.

Relevance Attention. In multivariate time series data, there are different variables. By analyzing the correlations between different variables, the ultimate imputation accuracy can be improved. Therefore, in order to utilize the potential correlations between different variables, we designed a correlation attention mechanism based on the self-attention mechanism to capture the potential correlations between different variables. The self-attention mechanism is a mechanism used to calculate the representation of the sequence data. It automatically assigns different weights to each element in the sequence to better capture the relationships between different elements. The self-attention mechanism includes three parts: query, key and value, and calculates the attention representation vector of the sequence through similarity calculation, softmax function and weighted sum. The calculation formula of the self-attention mechanism is shown below.

$$\text{SelfAttention}(Q, K, V) = \text{Softmax}\left(\frac{QK^\top}{\sqrt{d_k}}\right)V \tag{17}$$

In order to enhance the model's interpolation ability, we refer to the self-attention models in DISAN [20], SAITS [4], and XLNet [27]. We apply the diagonal mask matrix to the self-attention mechanism, and set the diagonal items in the attention map to $-\infty$. Therefore, the diagonal attention weight approaches 0 after the softmax function. We use vector $\mathbf{Z} \in \mathbb{R}^{d \times l}$ as inputs, and vector Z is stacked by Context Vector v_t. D is the number of variables in the multivariate time series datasert, and l is the length of v_t. The specific formula for relevance attention is as follows:

$$[\text{Mask}(x)]_{(i,j)} = \begin{cases} -\infty & i = j \\ x_{(i,j)} & i \neq j \end{cases} \tag{18}$$

$$\mathbf{Q} = \mathrm{ZW_q} \quad \mathrm{K} = \mathrm{ZW_k} \quad \mathrm{V} = \mathrm{ZW_v} \tag{19}$$

$$T = \text{Softmax}\left(\text{Mask}\left(\frac{QK^\top}{\sqrt{d_k}}\right)\right) \tag{20}$$

$$\text{MaskSelfAttention}\ (Q, K, V) = TV \tag{21}$$

Q, K, and V respectively refer to the query vector, key vector, and value vector. W_q, W_k, and W_v are trainable parameters. In the calculation, we first use Q and K to calculate the similarity and then use the softmax function to process and obtain similarity scores, reflecting the correlation between different variables. By using a diagonal mask, the input values at time step t cannot 'see' themselves and are not allowed to calculate their own weight in this estimation. Therefore, their estimates only depend on the input values at other time steps. With this setting, we can better capture the relevance between different features and improve the model's interpolation ability using relevance attention.

Feature Aggregation Layer. We refer to the idea of residual connection, where we aggregate Z with temporal information, and $\hat{\mathbf{Z}}$ with correlation information, and finally obtain the final interpolation result after a linear layer.

$$\hat{x} = \text{Linear}(W(\mathbf{Z} + \hat{\mathbf{Z}}) + \text{b}) \tag{22}$$

where W, b are learnable parameters, $\mathbf{Z} \in \mathbb{R}^{d \times l}$ is a Context Vector with temporal information, and $\hat{\mathbf{Z}}$ is a correlation matrix with correlation information. The linear layer produces the final interpolated values.

4.3 Discriminator Network

Following the standard GAN model, we use a discriminator to compete with the generator, which helps the generator to generate more realistic data. Unlike general generative adversarial networks, our discriminator outputs a matrix in which each value indicates the truthfulness of the generated value. To help the discriminator can better distinguish the true values from the generated values, we introduce a temporal cueing matrix inspired by GAIN [28], which contains a portion of missing information. The temporal cueing matrix C is defined as:

$$\mathbf{C} = \mathbf{Y} \odot \mathbf{M} + 0.5(1 - \mathbf{K}) \tag{23}$$

$$\mathbf{Y} = (\mathbf{y}_1, \cdots, \mathbf{y}_i, \cdots, \mathbf{y}_n) \in \{0, 1\}^{d \times n} \tag{24}$$

where each element in y is randomly set to 0 or 1. The discriminator connects the generated time series data and the temporal cueing matrix as input, and the network structure of the discriminator consists of a GRU layer and a linear layer.

The loss function of the discriminator is shown below:

$$\text{Loss}_D = -(\log(D(x_{\text{impute}} \odot M)) + \log(1 - D(x_{\text{impute}} \odot (1 - M)))) \tag{25}$$

where Loss_D refers to the classification loss of the discriminator, we want the discriminator to be able to distinguish the generated values from the true values as much as possible.

5 Experiment

5.1 Datasets

To validate the performance of our proposed model, we conducted experiments on two real datasets: PM2.5 Dataset, Health-care Dataset.

PM2.5 Dataset: This dataset is a public meteorological dataset consisting of air pollutant measurements from meteorological monitoring stations in Chinese cities. The data are collected from 2014-05-01 to 2017-02-28. To evaluate the interpolation performance, we first divide the dataset into a training set and a test set in the ratio of 80% and 20%, and then we randomly remove data from the dataset and use them as missing values for training and testing.

Health-care Dataset: This dataset is from PhysioNet Challenge 2012 [5], and it contains 4000 multivariate clinical time series data, each sample is recorded within the first 48 h after ICU admission. Each time series contains 37 time series variables such as body temperature, heart rate, blood pressure, etc. During training, we will randomly remove data points in the dataset as missing values and then use zeros to fill these missing values.

5.2 Baseline

This section describes the methods and models commonly used in time series interpolation and applies them to the previously mentioned dataset, and finally compares them with the model proposed in this paper.

1) Zero, a simple method of data interpolation, which fills all missing values to zero;
2) Average, where the missing values are replaced by the average of the corresponding features;
3) GRUD [3], a recurrent neural network based data interpolation model which estimates each missing value by the weighted combination of its last observation and the global average and the recurrent component;
4) GRUIGAN [15], an interpolation network combining gated recurrent network and generative adversarial network;
5) BRITS [2], a time series imputation method based on bi-directional RNN;
6) E2GAN [16], an end-to-end time series interpolation model based on generative adversarial networks;
7) SSGAN [17], a semi-supervised generative adversarial network based on bidirectional RNN.

Among the above methods, mean is a simple interpolation method. KNN is a commonly used algorithm for interpolating machine learning data. GRUD and BRITS are both bi-directional RNN-based methods. GRUIGAN, E2GAN, SSGAN are all generative adversarial network based models, we choose these as comparison methods to show the advantages of our model.

5.3 Experimental Setup

On the above dataset, we select 80% of the dataset as the training set and 20% of the dataset as the test set. For all tasks, we normalize the values to ensure stable training. In all the deep learning baseline models, the learning rate is set to 0.001. The number of hidden units in the recurrent network is 100, and the training epoch is set to 30. The dimensionality of random noise in GRUIGAN and the dimensionality of feature vectors in E2GAN are both 64. For SSGAN, we set the cue rate to 0.9 and the label rate to 100%. For the DAGAN model in thiTs paper, the discriminator cue rate is set to 0.9 and the hidden layer cells are set to 100. We apply an early stopping strategy in the model training. We also evaluate the performance of all models with interpolation at different missing rates. The missing rate is the ratio of missing values to the total number of data. It reflects the severity of missingness in the dataset. We randomly remove 10%–70% of the test data points to simulate different degrees of missingness. We use the root mean square error (RMSE) to evaluate the experimental results. A smaller value of RMSE means that the generated value is closer to the true value, and the following is the mathematical definition of the evaluation metric.

$$RMSE = \sqrt{\frac{1}{m}\sum_{i=1}^{m}(\text{ target-estimate })^2} \tag{26}$$

where target and estimate are the true and generated values, respectively, and m is the number of samples.

5.4 Experimental Results and Analysis

Table 1 displays the experimental results of the model with PM2.5 and Healthcare Datasets. Bolded results indicate the best-performing. The first set of experiments compares the performance of various methods at different missing data rates. DAGAN performs well on all datasets, demonstrating good generalization capability. DAGAN outperforms the best available baseline model on average by 13.8%, 12.2%, 8.07%, and 3.9%, respectively, at 10%, 30%, 50%, and 70% missing rates. Its use of temporal attention and relevance attention captures temporal correlation and potential correlation between features to improve interpolation accuracy by leveraging the available information from the time series. Overall, deep learning-based approaches exhibit better interpolation performance compared to statistics-based approaches. Table 1 reveals that the interpolation accuracy of all models declines as missing data increases gradually. This can be attributed to the reduced availability of information for the models to interpolate. Nonetheless, our models still outperform the baseline model.

5.5 Ablation Experiment

To ensure the validity of our model, we conducted ablation experiments, each repeated ten times, and recorded the average root mean square error (RMSE) at a 10% missing rate for the test set. The ablation experiments evaluated the model's performance by removing the temporal attention mechanism, the relevance attention mechanism, and the temporal cueing mechanism, respectively. Our experimental results reveal that the removal of the temporal attention mechanism, the relevance attention mechanism, and the temporal cueing mechanism led to a reduction in performance, indicating that the optimal use of temporal and correlation information between distinct features in the dataset is crucial for achieving high interpolation accuracy. Additionally, the results indicate that our adversarial training benefits from the incorporation of a temporal cueing matrix (Table 2).

Table 1. Performance comparison of time series imputation methods under different missing rates.

Dataset	Missing	Zero	Average	GRUD	GRUIGAN	E2GAN	BRITS	SSGAN	DAGAN
Health-care Dataset	10%	0.793	0.786	0.695	0.674	0.661	0.593	0.586	**0.519**
	30%	0.832	0.847	0.761	0.758	0.725	0.657	0.651	**0.639**
	50%	0.904	0.895	0.793	0.787	0.763	0.757	0.746	**0.722**
	70%	0.934	0.929	0.826	0.804	0.801	0.793	0.776	**0.749**
PM2.5 Dataset	10%	0.779	0.758	0.695	0.685	0.653	0.528	0.429	**0.367**
	30%	0.804	0.799	0.763	0.748	0.668	0.563	0.489	**0.394**
	50%	0.890	0.871	0.793	0.779	0.731	0.574	0.487	**0.423**
	70%	0.921	0.903	0.812	0.786	0.762	0.638	0.601	**0.575**

Table 2. The results of ablation experiments (RMSE).

	Health-care Dataset	PM2.5 Dataset
DAGAN	**0.519**	**0.367**
w/o cue matrix	0.579	0.412
w/o temporal attention	0.683	0.503
w/o multi-head self-attention	0.581	0.393

6 Conclusion

In this paper, we propose a new multivariate time series interpolation model called DAGAN. For time series data, DAGAN consists of two parts: a generator and a discriminator. In the generator, we use a gated recurrent neural network to learn the temporal information in the multivariate time series data and use an attention mechanism to weight the summation of the hidden states in the recurrent neural network, which enhances the ability of the model to learn the long-term dependence of the time series while ensuring that the model can focus more on the important information and compensate for the disadvantages of memory limitation and excessive focus on adjacent time steps of the recurrent neural network, thus improving the interpolation quality. In addition, this study also utilizes a masked self-attentiveness mechanism to correlate different variables in the multivariate time series so that all time steps are involved in each layer through self-attentiveness, thus maximizing the accuracy of multivariate time series interpolation. The generator inputs incomplete data with missing values and outputs the complete interpolation, and the discriminator heaps the generated values and the true values to distinguish them. Experimental results demonstrate that, when compared with other baseline models, our model's imputation performance is better.

References

1. Bahdanau, D., Cho, K., Bengio, Y.: Neural machine translation by jointly learning to align and translate. arXiv preprint arXiv:1409.0473 (2014)
2. Cao, W., Wang, D., Li, J., Zhou, H., Li, L., Li, Y.: Brits: bidirectional recurrent imputation for time series. In: Advances in Neural Information Processing Systems, vol. 31 (2018)
3. Che, Z., Purushotham, S., Cho, K., Sontag, D., Liu, Y.: Recurrent neural networks for multivariate time series with missing values. Sci. Rep. **8**(1), 6085 (2018)
4. Du, W., Côté, D., Liu, Y.: Saits: self-attention-based imputation for time series. Expert Syst. Appl. **219**, 119619 (2023)
5. Feng, F., Chen, H., He, X., Ding, J., Sun, M., Chua, T.S.: Enhancing stock movement prediction with adversarial training. arXiv preprint arXiv:1810.09936 (2018)
6. Fortuin, V., Baranchuk, D., Rätsch, G., Mandt, S.: Gp-vae: deep probabilistic time series imputation. In: International Conference on Artificial Intelligence and Statistics, pp. 1651–1661. PMLR (2020)
7. Fung, D.S.: Methods for the estimation of missing values in time series (2006)
8. Gupta, M., Phan, T.L.T., Bunnell, H.T., Beheshti, R.: Concurrent imputation and prediction on EHR data using bi-directional GANs: Bi-GANs for EHR imputation and prediction. In: Proceedings of the 12th ACM Conference on Bioinformatics, Computational Biology, and Health Informatics, pp. 1–9 (2021)
9. Hochreiter, S., Schmidhuber, J.: Long short-term memory. Neural Comput. **9**(8), 1735–1780 (1997)
10. Hudak, A.T., Crookston, N.L., Evans, J.S., Hall, D.E., Falkowski, M.J.: Nearest neighbor imputation of species-level, plot-scale forest structure attributes from lidar data. Remote Sens. Environ. **112**(5), 2232–2245 (2008)
11. Kaiser, J.: Dealing with missing values in data. J. Syst. Integr. (1804–2724) **5**(1) (2014)
12. Lan, Q., Xu, X., Ma, H., Li, G.: Multivariable data imputation for the analysis of incomplete credit data. Expert Syst. Appl. **141**, 112926 (2020)
13. LIU, S., Li, X., Cong, G., Chen, Y., Jiang, Y.: Multivariate time-series imputation with disentangled temporal representations. In: The Eleventh International Conference on Learning Representations (2023)
14. Liu, X., Wang, M.: Gap filling of missing data for VIIRS global ocean color products using the DINEOF method. IEEE Trans. Geosci. Remote Sens. **56**(8), 4464–4476 (2018)
15. Luo, Y., Cai, X., Zhang, Y., Xu, J., et al.: Multivariate time series imputation with generative adversarial networks. In: Advances in Neural Information Processing Systems, vol. 31 (2018)
16. Luo, Y., Zhang, Y., Cai, X., Yuan, X.: E2gan: end-to-end generative adversarial network for multivariate time series imputation. In: Proceedings of the 28th International Joint Conference on Artificial Intelligence, pp. 3094–3100. AAAI Press (2019)
17. Miao, X., Wu, Y., Wang, J., Gao, Y., Mao, X., Yin, J.: Generative semi-supervised learning for multivariate time series imputation. In: Proceedings of the AAAI Conference on Artificial Intelligence, vol. 35, pp. 8983–8991 (2021)
18. Ni, Q., Cao, X.: MBGAN: an improved generative adversarial network with multi-head self-attention and bidirectional RNN for time series imputation. Eng. Appl. Artif. Intell. **115**, 105232 (2022)

19. Qin, R., Wang, Y.: ImputeGAN: generative adversarial network for multivariate time series imputation. Entropy **25**(1), 137 (2023)
20. Shen, T., Zhou, T., Long, G., Jiang, J., Pan, S., Zhang, C.: Disan: Directional self-attention network for rnn/cnn-free language understanding. In: Proceedings of the AAAI Conference on Artificial Intelligence, vol. 32 (2018)
21. Silva, I., Moody, G., Scott, D.J., Celi, L.A., Mark, R.G.: Predicting in-hospital mortality of ICU patients: the physionet/computing in cardiology challenge 2012. In: 2012 Computing in Cardiology, pp. 245–248. IEEE (2012)
22. Suo, Q., Yao, L., Xun, G., Sun, J., Zhang, A.: Recurrent imputation for multivariate time series with missing values. In: 2019 IEEE International Conference on Healthcare Informatics (ICHI), pp. 1–3. IEEE (2019)
23. Suo, Q., Zhong, W., Xun, G., Sun, J., Chen, C., Zhang, A.: Glima: global and local time series imputation with multi-directional attention learning. In: 2020 IEEE International Conference on Big Data (Big Data), pp. 798–807. IEEE (2020)
24. Tang, F., Ishwaran, H.: Random forest missing data algorithms. Stat. Anal. Data Min.: ASA Data Sci. J. **10**(6), 363–377 (2017)
25. Wang, R., Zhang, Z., Wang, Q., Sun, J.: TLGRU: time and location gated recurrent unit for multivariate time series imputation. EURASIP J. Adv. Signal Process. **2022**(1), 74 (2022)
26. Woodall, P.: The data repurposing challenge: new pressures from data analytics. J. Data Inf. Qual. (JDIQ) **8**(3–4), 1–4 (2017)
27. Yang, Z., Dai, Z., Yang, Y., Carbonell, J., Salakhutdinov, R.R., Le, Q.V.: Xlnet: generalized autoregressive pretraining for language understanding. In: Advances in Neural Information Processing Systems, vol. 32 (2019)
28. Yoon, J., Jordon, J., Schaar, M.: Gain: missing data imputation using generative adversarial nets. In: International Conference on Machine Learning, pp. 5689–5698. PMLR (2018)

Knowledge-Distillation-Warm-Start Training Strategy for Lightweight Super-Resolution Networks

Min Lei[1,2], Kun He[1,2], Hui Xu[1,2,3](✉), Yunfeng Yang[4], and Jie Shao[1,2]

[1] Sichuan Artificial Intelligence Research Institute, Yibin 644000, China
{minlei,kunhe}@std.uestc.edu.cn, {huixu.kim,shaojie}@uestc.edu.cn
[2] University of Electronic Science and Technology of China, Chengdu 611731, China
[3] Intelligent Terminal Key Laboratory of Sichuan Province, Yibin 644000, China
[4] Yibin Great Technology Co., Ltd., Yibin 644000, China
yunfeng.yang@szgreatmobile.com

Abstract. In recent years, studies on lightweight networks have made rapid progress in the field of image Super-Resolution (SR). Although the lightweight SR network is computationally efficient and saves parameters, the simplification of the structure inevitably leads to limitations in its performance. To further enhance the efficacy of lightweight networks, we propose a Knowledge-Distillation-Warm-Start (KDWS) training strategy. This strategy enables further optimization of lightweight networks using dark knowledge from traditional large-scale SR networks during warm-start training and can empirically improve the performance of lightweight models. For experiment, we have chosen several traditional large-scale SR networks and lightweight networks as teacher and student networks, respectively. The student network is initially trained with a conventional warm-start strategy, followed by additional supervision from the teacher network for further warm-start training. The evaluation on common test datasets shows that our proposed training strategy can result in better performance for a lightweight SR network. Furthermore, our proposed approach can also be adopted in any deep learning network training process, not only image SR tasks, as it is not limited by network structure or task type.

Keywords: Training strategy · Knowledge distillation · Super-resolution

1 Introduction

Image Super-Resolution (SR) aims to generate the High-Resolution (HR) image from the corresponding Low-Resolution (LR) image. In terms of model weight and computational cost, there can be two types of models: traditional deep learning methods [7,8,24,37] and lightweight methods [9,10,22,25]. Training much larger and deeper neural networks contributes significantly to the performance

B. Luo et al. (Eds.): ICONIP 2023, CCIS 1966, pp. 272–284, 2024.
https://doi.org/10.1007/978-981-99-8148-9_22

boost of traditional deep learning methods, which means considerable computation costs and thereby makes them difficult to deploy on resource-constrained devices for real-world applications.

As a result, lightweight versions have received more and more attention in the last few years. Lightweight methods are developed to improve the efficiency of deep neural networks via network pruning [4,34], network quantization [30,39], neural architecture search [20,28], knowledge distillation [5,11,17,32], etc. In spite of the fact that the lightweight SR network is computationally efficient and it reduces the number of parameters, it will always have performance flaws.

Advanced training strategies guarantee the performance of lightweight networks. Lots of work [6,12,22,26] prolonged the training. Periodic learning rate scheduler and cosine learning rate scheduler were employed by some researchers, which could help the training to step outside of the local minima [23]. The training of winner solutions in NTIRE 2022 efficient super-resolution challenge [23] typically contain multiple stages, and multi-stage warm-start training strategy has been proven to have the ability to improve model performance in RLFN [22]. However, most current training strategies are designed to maintain training stability and minimize the performance gap between the lightweight network and the large network, so there is still room for further research to improve the performance of the lightweight network.

Knowledge distillation can assist the lightweight network in obtaining knowledge from the complicated network, boosting the lightweight network's performance. However, in practice, it is still difficult for a lightweight network to learn the knowledge of a complicated network with a different structure using the knowledge distillation approach, and the results are not ideal all the time. That is because the teacher network may impede the training of the student network due to the disparity in representation spaces between the two. In this paper, we propose a Knowledge-Distillation-Warm-Start (KDWS) training strategy, which combines the advantages of both warm-start training strategy and knowledge distillation approach. Specifically, knowledge distillation is applied at the end of the warm-start training process. The warm-start approach can utilize the pre-training weights in the previous stage not only to improve the performance of the lightweight network, but also to reduce the negative impact of the teacher network on the lightweight network during knowledge distillation. This allows it to exert its positive effect to its full potential. Experimental results show that the proposed training strategy can enable the lightweight network to progressively optimize in stages and acquire dark knowledge [14] from the large network, which enables the lightweight network to obtain a better solution without increasing the number of parameters.

Our contributions can be summarized as follows:

- We propose a knowledge-distillation-warm-start training strategy, which integrates the benefits of warm-start training strategy and knowledge distillation approach. The strategy enhances the performance of a lightweight network without increasing its parameters.

- We conduct substantial experiments on several typical SR networks and evaluate on several common test datasets. The results demonstrate that the method we proposed can effectively enhance the performance of a lightweight SR network.

2 Related Work

In this section, we will introduce some relevant works about knowledge distillation and the training strategy for neural networks, which are the inspiration for our proposed method.

2.1 Knowledge Distillation

Knowledge Distillation (KD) is a common instance of the transfer learning approach. It entails training a big and powerful teacher network to transfer its knowledge to a small and weak student network in order to improve the performance of this student network. Hinton et al. [17] first proposed the concept of knowledge distortion in the question about image classification. Tung and Mori [32] proposed a new form of knowledge distillation loss that is inspired by the observation that semantically similar inputs tend to elicit similar activation patterns in a trained network. Gao et al. [11] were the first to propose using a teacher-student networking technique to train the SR network, which dramatically enhances the performance of the student network's SR reconstruction without modifying its structure.

Knowledge distillation is split into two groups: homogeneous KD and heterogeneous KD, according to the structural differences between the teacher and the student networks. Homogeneous KD refers to the structure of a teacher and student model that is similar or belongs to the same series, layer-to-layer, or block-to-block one-to-one connection. Common homogeneous KD designs include the ResNet and DenseNet series, among others. Popular techniques for homogeneous KD include similarity-preserving [32], differentiable feature aggregation search [15], and decoupled knowledge distillation [38]. Heterogeneous KD refers to the circumstance in which the network architectures of the teacher and student models are not identical and feature map matching between layers is problematic. In such a case, the intermediate layer must be managed in order to be utilized for homogeneous KD. The most typical works are variational information distillation [2], activation boundaries [16], SimKD [5], and so on. Normally, it is challenging to explicitly match the middle-layer knowledge representation space of teacher-student models with diverse architectures. Homogeneous KD usually gives superior and more stable results, whereas heterogeneous KD might find it challenging to train the student model due to a mismatch in the representation space. At the same time, homogeneous KD is not universal in practice since it requires elaborate construction and is subject to network structure constraints. In this paper, we propose a knowledge-distillation-warm-start training technique, which is a general training strategy unrestricted by network architecture. It utilizes warm-start technology to provide a lightweight network with time to learn independently. After the training of this network has reached maturity, i.e., the correct optimization direction has been determined, the knowledge distillation process will commence, and this network will serve as the student network. This kind of training may aid in reducing heterogeneity's interference, contributing "dark knowledge" from the teacher network to the student network, and gradually enhancing the performance of the student network.

2.2 Training Strategy

Training strategy involves many aspects of training settings, such as data augmentation, learning rate scheduler, and loss function, etc. There have been many studies on these techniques [13,19,29,31,33,35]. Most works extend the training process, which involves multi-stage training mixed with different training settings. A refined multi-stage multi-task training strategy [12] was proposed to improve the performance of online attention-based encoder-decoder models. Mahmud et al. [26] employed a multi-stage training strategy to increase diversity in the feature extraction process, to make accurate recognition of actions by combining varieties of features extracted from diverse perspectives. Recently, Residual Local Feature Network (RLFN) [22] proves that the warm-start training strategy can stabilize the training process and help the model jump out of the local minimum.

Inspired by existing works, we propose a Knowledge-Distillation-Warm-Start (KDWS) training strategy. Specifically, we introduce the supervision of the teacher network into the warm-start training process. This kind of progressive training of the warm-start technology increases the effectiveness of the knowledge distillation technology and boosts the performance of a lightweight network. For a better understanding, consider a lightweight model as a student. Then, data enhancement technology is similar to providing this student with more information to improve its grades. Our proposed KDWS training strategy aims to provide teachers' experience on the basis of this student's self-study, which improves its learning ability, so it can learn better with the same amount of learning materials.

3 Method

3.1 KDWS Training Strategy

Figure 1 illustrates the workflow of our proposed KDWS training strategy. The lightweight SR network RLFN [22] is used as an example in this figure, and the white square represents the randomly initialized RLFN. As depicted by the dashed outlines in Fig. 1, our KDWS training strategy includes three stages. In the first stage, the model is trained from scratch. The square in light blue represents the model after finishing the first stage of training. The second stage

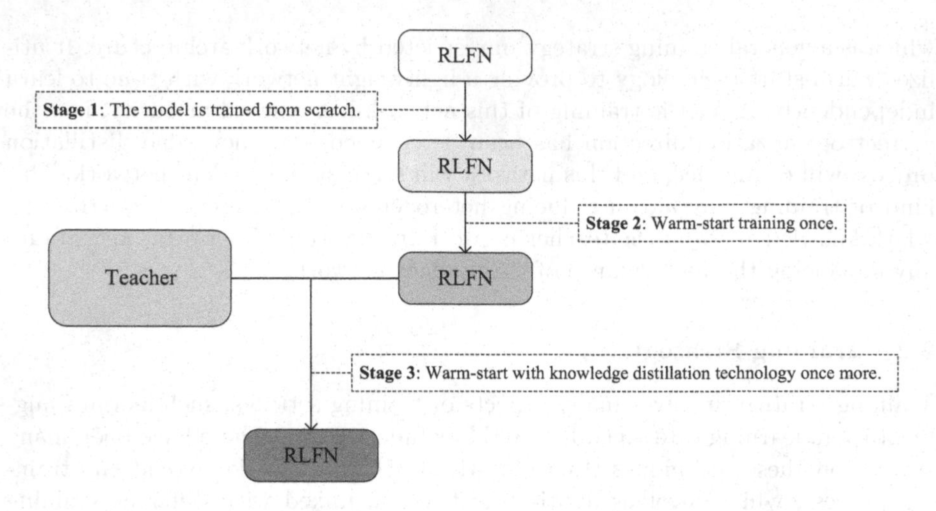

Fig. 1. The workflow of our proposed knowledge-distillation-warm-start training strategy. RLFN [22] is an example of a trainable lightweight network. The white square represents the untrained network. The light blue and dark blue squares represent the pre-trained network from the first and second stages of training, respectively. The green square represents the final model obtained after the third training stage, and the orange square represents the pre-trained teacher network (Color figure online).

loads the pre-trained weights from the previous stage while maintaining the same training settings, namely the warm-start approach. This process can be repeated multiple times, but the results presented in this work are based on only one repetition. The model after the second training stage is shown by the dark blue square. The green square represents the final model after the third training stage. Training at this stage is under the supervision of the pre-trained teacher network. The teacher network can be chosen with flexibility, and in this paper, we only demonstrate the training effects of RCAN [37], SAN [7] and EDSR [24] as teacher networks. From 5×10^{-4} in the first two stages, 1×10^{-4} serves as the initial learning rate in this stage, because it is more challenging to optimize the network after the training in the first two stages, and a smaller learning rate is more likely to produce a better solution. In addition, the objective function of this stage is also changed from L1 loss in the previous two stages to a kind of hybrid loss based on 1NF. More details about the loss function will be provided in the next section.

3.2 Loss Function

During the first two stages of our proposed KDWS training, the model is trained with L1 loss which can be formulated as:

$$L^{l_1}(P) = \frac{1}{N} \sum_{p \epsilon P} |x(p) - y(p)|, \tag{1}$$

where p is the index of the pixel and P is the patch; N is the number of pixels p in the patch. $x(p)$ and $y(p)$ denote the pixel values of the processed patch and the ground-truth, respectively.

In the third stage of training, the supervision of teacher network is added, i.e., the L1 distance between the outputs of teacher network and the network being trained is calculated additionally. The formulation is represented as:

$$L^{KD}(P) = \frac{1}{N}\sum_{p\epsilon P} |x(p) - y_t(p)|, \quad (2)$$

where $y_t(p)$ represents the pixel values of the output patch from teacher network. Then, the comprehensive loss function for the third training stage is formulated as:

$$L(P) = (1-\alpha)L^{l_1}(P) + \alpha L^{KD}(P), \quad (3)$$

where α is a hyperparameter to balance the above two loss functions, and changing α modifies the degree of intervention of teacher network on the network being trained.

4 Experiments

4.1 Setup

Datasets and Evaluation Metrics. For training, we use 800 training images from the DIV2K dataset [1]. The model trained with our proposed KDWS strategy are evaluated on four benchmark datasets: Set5 [3], Set14 [36], BSD100 [27], and Urban100 [18]. We evaluate the PSNR and SSIM on the Y channel of YCbCr space.

Implementation Details. We implement the proposed KDWS training technique using PyTorch framework with one NVIDIA GeForce RTX 3090 GPU. As a baseline, we use the lightweight version of RLFN [22]. Its main idea is using three convolutional layers for residual local feature learning to simplify feature aggregation, which achieves a good trade-off between model performance and inference time and won first place in the runtime track of the NTIRE 2022 efficient super-resolution challenge [23]. As teacher networks, we chose pre-trained classic models RCAN [37], SAN [7] and EDSR [24]. Our implementation code is available at https://github.com/githublei-min/KDWS.

Training Details. With bicubic interpolation, HR images are scaled down to generate LR images. The mini-batch size is set to 16, and we randomly crop HR patches of size 256 × 256 from ground-truth. We augment the training data with flipping and rotation at 90°, 180°, and 270°. The training process has three stages. In the first stage, we train the model from scratch. Then, we employ the warm-start strategy once. In each stage, we adopt the Adam optimizer [21] by setting $\beta_1 = 0.9$, $\beta_2 = 0.999$ and $\epsilon = 10^{-8}$ and minimize the L1 loss following RFDN [25]. The initial learning rate is 5e−4 and is halved at every 200 epochs. The total number of epochs is 1000. In the third stage, we additionally employ knowledge distillation technology. The same training settings as the first two stages are kept to train the model, except that the initial learning rate is replaced by 1e-4 and the loss function is replaced by a hybrid loss with 0.5 of the hyperparameter α in Eq. (3).

Table 1. Comparison of our proposed knowledge-distillation-warm-start training strategy with other training strategies. Baseline refers to training from scratch. WS refers to warm-start training. KDWS-R, KDWS-S, and KDWS-E respectively represent the use of RCAN [37], SAN [7], and EDSR [24] as the teacher network when our proposed knowledge-distillation-warm-start training strategy is implemented. The best ones are marked in red.

Scale	Training Strategy	Set5	Set14	BSD100	Urban100
		PSNR↑/SSIM↑	PSNR↑/SSIM↑	PSNR↑/SSIM↑	PSNR↑/SSIM↑
×2	Baseline	37.9368/0.9603	33.4532/0.9165	32.1226/0.8993	31.8097/0.9253
	WS	37.9944/0.9604	33.4869/0.9170	32.1397/0.8995	31.9665/0.9269
	KDWS-R (ours)	38.0543/0.9608	33.6008/0.9176	32.2052/0.9004	32.1713/0.9289
	KDWS-S (ours)	38.0382/0.9607	33.6085/0.9174	32.1993/0.9003	32.1649/0.9288
	KDWS-E (ours)	38.0341/0.9607	33.6146/0.9179	32.1962/0.9003	32.1649/0.9291
×4	Baseline	31.9958/0.8910	28.4806/0.7779	27.4971/0.7338	25.8803/0.7784
	WS	32.1068/0.8924	28.5179/0.7800	27.5204/0.7349	25.9879/0.7824
	KDWS-R (ours)	32.2362/0.8943	28.6112/0.7819	27.5742/0.7370	26.1326/0.7869
	KDWS-S (ours)	32.2139/0.8941	28.6117/0.7820	27.5776/0.7369	26.1322/0.7869
	KDWS-E (ours)	32.2360/0.8942	28.6182/0.7820	27.5708/0.7367	26.1231/0.7863

4.2 Quantitative Results

Comparison with Other Training Strategies. We compare our proposed strategy with two other commonly used training strategies with scale factors of 2 and 4, which are single-stage training from scratch and warm-start training, respectively. For experiment fairness, the training objects of all methods are unified with the lightweight version of RLFN [22], and the initial training settings are the same (see Sect. 4.1).

The quantitative performance comparison on several benchmark datasets is shown in Table 1. In this table, the baseline refers to training from scratch, and WS refers to warm-start training. The training procedure is as follows:

initially, the model is trained from scratch, and then it is initialized with the pre-trained weights and trained with the same settings as in the previous stage twice more. KDWS-R, KDWS-S, and KDWS-E respectively represent the use of RCAN [37], SAN [7] and EDSR [24] as teacher network. Compared with other training strategies, our proposed KDWS training strategy achieves superior performance in terms of both PSNR and SSIM. In addition, the results here also indicate that the effect of RCAN as a teacher network is generally superior, indicating that the choice of teacher network impacts the degree of performance improvement. For further investigation, we additionally train the model under the supervision of multiple teacher networks and the results and analysis is presented in Sect. 4.3.

KDWS Versus Knowledge Distillation. We compare KDWS and two versions of strategies using knowledge distillation from scratch, KD and multi-KD. Note that KD stands for training with knowledge distillation from scratch. Multi-KD refers to multi-stage knowledge distillation training. The first stage is to train from scratch, and then we train twice loading the weights from the previous stage and using the same training settings. Knowledge distillation is performed at all the three stages. The results in Table 2 show that our proposed KDWS strategy outperforms the traditional knowledge distillation.

Table 2. Comparison of our proposed KDWS with two versions of strategies using knowledge distillation from scratch. KD stands for knowledge distillation training from scratch. Multi-KD refers to multi-stage knowledge distillation training. RCAN [37] is used as teacher network, and scale factor is 4.

Training Strategy	Set5	Set14	BSD100	Urban100
	PSNR↑/SSIM↑	PSNR↑/SSIM↑	PSNR↑/SSIM↑	PSNR↑/SSIM↑
Baseline	31.9958/0.8910	28.4806/0.7779	27.4971/0.7338	25.8803/0.7784
KD	32.0169/0.8916	28.5115/0.7789	27.5061/0.7345	25.8663/0.7783
Multi-KD	32.0428/0.8923	28.5304/0.7820	27.5288/0.7390	26.0177/0.7835
KDWS (ours)	32.2362/0.8943	28.6112/0.7819	27.5742/0.7370	26.1326/0.7869

Table 3. Evaluation of the three stages of our proposed knowledge-distillation-warm-start training strategy. RCAN [37] is used as teacher network, and scale factor is 4.

Stages	Set5	Set14	BSD100	Urban100
	PSNR↑/SSIM↑	PSNR↑/SSIM↑	PSNR↑/SSIM↑	PSNR↑/SSIM↑
Stage 1	31.9958/0.8910	28.4806/0.7779	27.4971/0.7338	25.8803/0.7784
Stage 2	32.0326/0.8919	28.5283/0.7797	27.5200/0.7347	25.9571/0.7813
Stage 3	32.2362/0.8943	28.6112/0.7819	27.5742/0.7370	26.1326/0.7869

4.3 Ablation Study

Effectiveness of Each Stage. Table 3 shows the PSNR and SSIM scores of the models generated during the three stages of our proposed KDWS training strategy on some common test datasets. It can be observed that the PSNR and SSIM scores of the model increase as the training stage progresses. This demonstrates that both warm-start technology and knowledge distillation technology have contributed to the effectiveness of our proposed KDWS approach.

Multiple Teacher Networks. In addition to single-teacher supervision, we also train the model under the supervision of multi-teacher networks in the third stage of our proposed KDWS training strategy. Table 4 shows the comparison based on the teacher network RCAN [37]. In the third stage of KDWS training, let the model be supervised by the following three teacher network schemes: single RCAN, RCAN and SAN, and RCAN and EDSR. The impact of multi-teacher supervision on enhancing model performance is similar to that of single-teacher supervision, as demonstrated in Table 4. Additionally, different configurations of the teacher network will provide various outcomes. A more detailed study of the relationship between teacher network and the final training outcomes is required, which will also be our future research topic. Based on current results, using single-teacher network supervision in the KDWS training process can already lead to a relatively ideal effect in terms of enhancing the performance of lightweight networks.

Table 4. Comparison of single-teacher training and multi-teacher training on scale ×4. The best ones are marked in red.

Teacher	Set5	Set14	BSD100	Urban100
	PSNR↑/SSIM↑	PSNR↑/SSIM↑	PSNR↑/SSIM↑	PSNR↑/SSIM↑
RCAN [37]	32.2362/0.8943	28.6112/0.7819	27.5742/0.7370	26.1326/0.7869
RCAN [37] + SAN [7]	32.2253/0.8942	28.6208/0.7821	27.5779/0.7370	26.1414/0.7870
RCAN [37] + EDSR [24]	32.2372/0.8941	28.6125/0.7820	27.5711/0.7369	26.1202/0.7868

Table 5. Evaluation of FMEN [10] under our proposed KDWS training strategy on scale ×4. The baseline refers to training FMEN from scratch. KDWS-R, KDWS-S, and KDWS-E respectively represent the use of RCAN [37], SAN [7], and EDSR [24] as teacher network. The best ones are marked in red.

Teacher	Set5	Set14	BSD100	Urban100
	PSNR↑/SSIM↑	PSNR↑/SSIM↑	PSNR↑/SSIM↑	PSNR↑/SSIM↑
Baseline	31.8582/0.8894	28.4310/0.7774	27.4665/0.7331	25.7205/0.7729
KDWS-R (ours)	32.2753/0.9210	28.7764/0.8289	27.8011/0.7842	26.8179/0.7962
KDWS-S (ours)	32.1613/0.8990	28.7879/0.8292	27.9067/0.7943	26.8137/0.7961
KDWS-E (ours)	32.1814/0.9111	29.1865/0.8321	28.0026/0.7943	27.0182/0.8063

Generality of KDWS. To demonstrate the generality of our proposed KDWS training strategy, we additionally train other lightweight SR networks with it. Table 5 shows the performance of our method on Fast and Memory-Efficient Network (FMEN) [10], another excellent lightweight SR network that achieves the lowest memory consumption and the second shortest runtime in the NTIRE 2022 challenge on efficient super-resolution [23]. In this table, baseline means training FMEN from scratch. The results show that our proposed KDWS training strategy is still effective for FMEN and improves its performance to a certain extent.

5 Conclusion

In this paper, we propose a knowledge-distillation-warm-start training strategy, where warm-start is used to stabilize the training process and help to jump out of the local minimum during the optimization process. At the same time, knowledge distillation is used to transfer the hidden knowledge of the classic complex SR network to the lightweight one. By combining the above two technologies, a better solution can be found for a lightweight SR network. It has been proven through experiments on several classic SR networks that the proposed KDWS strategy can indeed improve the performance of lightweight networks without increasing the network parameters or complexity. Additionally, although the experiment in this paper is based on the super-resolution network, this strategy is theoretically applicable to any deep learning network training process, not just those involving image SR tasks, as it is not restricted by network structure.

Future research will concentrate on resolving the issue of how to choose from a variety of teacher network structures. We will conduct additional experiments to figure out how various combinations of student networks and teacher networks can be used to optimize the proposed KDWS training strategy, as well as whether additional technologies can be added to KDWS to improve its efficacy.

Acknowledgments. : This work is supported by the Ministry of Science and Technology of China (No. G2022036009L), Open Fund of Intelligent Terminal Key Laboratory of Sichuan Province (No. SCTLAB-2007), Yibin Science and Technology Program (No. 2021CG003) and Science and Technology Program of Yibin Sanjiang New Area (No. 2023SJXQYBKJJH001).

References

1. Agustsson, E., Timofte, R.: NTIRE 2017 challenge on single image super-resolution: dataset and study. In: 2017 IEEE Conference on Computer Vision and Pattern Recognition Workshops, CVPR Workshops 2017, pp. 1122–1131 (2017)
2. Ahn, S., Hu, S.X., Damianou, A.C., Lawrence, N.D., Dai, Z.: Variational information distillation for knowledge transfer. In: IEEE Conference on Computer Vision and Pattern Recognition, CVPR 2019, pp. 9163–9171 (2019)
3. Bevilacqua, M., Roumy, A., Guillemot, C., Alberi-Morel, M.: Low-complexity single-image super-resolution based on nonnegative neighbor embedding. In: British Machine Vision Conference, BMVC 2012, pp. 1–10 (2012)

4. Chang, J., Lu, Y., Xue, P., Xu, Y., Wei, Z.: Global balanced iterative pruning for efficient convolutional neural networks. Neural Comput. Appl. **34**(23), 21119–21138 (2022)
5. Chen, D., Mei, J., Zhang, H., Wang, C., Feng, Y., Chen, C.: Knowledge distillation with the reused teacher classifier. In: IEEE/CVF Conference on Computer Vision and Pattern Recognition, CVPR 2022, pp. 11923–11932 (2022)
6. Clancy, K., Aboutalib, S.S., Mohamed, A.A., Sumkin, J.H., Wu, S.: Deep learning pre-training strategy for mammogram image classification: an evaluation study. J. Digit. Imaging **33**(5), 1257–1265 (2020)
7. Dai, T., Cai, J., Zhang, Y., Xia, S., Zhang, L.: Second-order attention network for single image super-resolution. In: IEEE Conference on Computer Vision and Pattern Recognition, CVPR 2019, pp. 11065–11074 (2019)
8. Dong, C., Loy, C.C., He, K., Tang, X.: Learning a deep convolutional network for image super-resolution. In: Fleet, D., Pajdla, T., Schiele, B., Tuytelaars, T. (eds.) ECCV 2014 Part IV. LNCS, vol. 8692, pp. 184–199. Springer, Cham (2014). https://doi.org/10.1007/978-3-319-10593-2_13
9. Dong, C., Loy, C.C., Tang, X.: Accelerating the super-resolution convolutional neural network. In: Leibe, B., Matas, J., Sebe, N., Welling, M. (eds.) ECCV 2016 Part II. LNCS, vol. 9906, pp. 391–407. Springer, Cham (2016). https://doi.org/10.1007/978-3-319-46475-6_25
10. Du, Z., Liu, D., Liu, J., Tang, J., Wu, G., Fu, L.: Fast and memory-efficient network towards efficient image super-resolution. In: IEEE/CVF Conference on Computer Vision and Pattern Recognition Workshops, CVPR Workshops 2022, pp. 852–861 (2022)
11. Gao, Q., Zhao, Y., Li, G., Tong, T.: Image super-resolution using knowledge distillation. In: Jawahar, C.V., Li, H., Mori, G., Schindler, K. (eds.) ACCV 2018. LNCS, vol. 11362, pp. 527–541. Springer, Cham (2019). https://doi.org/10.1007/978-3-030-20890-5_34
12. Garg, A., Gowda, D., Kumar, A., Kim, K., Kumar, M., Kim, C.: Improved multi-stage training of online attention-based encoder-decoder models. In: IEEE Automatic Speech Recognition and Understanding Workshop, ASRU 2019, pp. 70–77 (2019)
13. Gonzalez, S., Miikkulainen, R.: Improved training speed, accuracy, and data utilization through loss function optimization. In: IEEE Congress on Evolutionary Computation, CEC 2020, pp. 1–8 (2020)
14. Gou, J., Yu, B., Maybank, S.J., Tao, D.: Knowledge distillation: a survey. Int. J. Comput. Vis. **129**(6), 1789–1819 (2021)
15. Guan, Y., et al.: Differentiable feature aggregation search for knowledge distillation. In: Vedaldi, A., Bischof, H., Brox, T., Frahm, J.-M. (eds.) ECCV 2020 Part XVII. LNCS, vol. 12362, pp. 469–484. Springer, Cham (2020). https://doi.org/10.1007/978-3-030-58520-4_28
16. Heo, B., Lee, M., Yun, S., Choi, J.Y.: Knowledge transfer via distillation of activation boundaries formed by hidden neurons. In: The Thirty-Third AAAI Conference on Artificial Intelligence, AAAI 2019, pp. 3779–3787 (2019)
17. Hinton, G.E., Vinyals, O., Dean, J.: Distilling the knowledge in a neural network. CoRR abs/1503.02531 (2015)
18. Huang, J., Singh, A., Ahuja, N.: Single image super-resolution from transformed self-exemplars. In: IEEE Conference on Computer Vision and Pattern Recognition, CVPR 2015, pp. 5197–5206 (2015)

19. Khalifa, N.E.M., Loey, M., Mirjalili, S.: A comprehensive survey of recent trends in deep learning for digital images augmentation. Artif. Intell. Rev. **55**(3), 2351–2377 (2022)
20. Kim, Y., Li, Y., Park, H., Venkatesha, Y., Panda, P.: Neural architecture search for spiking neural networks. In: Avidan, S., Brostow, G., Cissé, M., Farinella, G.M., Hassner, T. (eds.) ECCV 2022 Part XXIV, vol. 13684, pp. 36–56. Springer, Cham (2022)
21. Kingma, D.P., Ba, J.: Adam: a method for stochastic optimization. In: 3rd International Conference on Learning Representations, ICLR 2015, Conference Track Proceedings (2015)
22. Kong, F., et al.: Residual local feature network for efficient super-resolution. In: IEEE/CVF Conference on Computer Vision and Pattern Recognition Workshops, CVPR Workshops 2022, pp. 765–775 (2022)
23. Li, Y., et al.: NTIRE 2022 challenge on efficient super-resolution: Methods and results. In: IEEE/CVF Conference on Computer Vision and Pattern Recognition Workshops, CVPR Workshops 2022, pp. 1061–1101 (2022)
24. Lim, B., Son, S., Kim, H., Nah, S., Lee, K.M.: Enhanced deep residual networks for single image super-resolution. In: 2017 IEEE Conference on Computer Vision and Pattern Recognition Workshops, CVPR Workshops 2017, pp. 1132–1140 (2017)
25. Liu, J., Tang, J., Wu, G.: Residual feature distillation network for lightweight image super-resolution. In: Bartoli, A., Fusiello, A. (eds.) ECCV 2020. LNCS, vol. 12537, pp. 41–55. Springer, Cham (2020). https://doi.org/10.1007/978-3-030-67070-2_2
26. Mahmud, T., Sayyed, A.Q.M.S., Fattah, S.A., Kung, S.: A novel multi-stage training approach for human activity recognition from multimodal wearable sensor data using deep neural network. IEEE Sens. J. **21**(2), 1715–1726 (2021)
27. Martin, D.R., Fowlkes, C.C., Tal, D., Malik, J.: A database of human segmented natural images and its application to evaluating segmentation algorithms and measuring ecological statistics. In: Proceedings of the Eighth International Conference On Computer Vision (ICCV-01), vol. 2, pp. 416–425 (2001)
28. Pham, H., Guan, M.Y., Zoph, B., Le, Q.V., Dean, J.: Efficient neural architecture search via parameter sharing. In: Proceedings of the 35th International Conference on Machine Learning, ICML 2018, pp. 4092–4101 (2018)
29. Raymond, C., Chen, Q., Xue, B., Zhang, M.: Online loss function learning. CoRR abs/2301.13247 (2023)
30. Siddegowda, S., Fournarakis, M., Nagel, M., Blankevoort, T., Patel, C., Khobare, A.: Neural network quantization with AI model efficiency toolkit (AIMET). CoRR abs/2201.08442 (2022)
31. Smith, L.N.: Cyclical learning rates for training neural networks. In: 2017 IEEE Winter Conference on Applications of Computer Vision, WACV 2017, pp. 464–472 (2017)
32. Tung, F., Mori, G.: Similarity-preserving knowledge distillation. In: 2019 IEEE/CVF International Conference on Computer Vision, ICCV 2019, pp. 1365–1374 (2019)
33. Wang, K., Sun, T., Dou, Y.: An adaptive learning rate schedule for SIGNSGD optimizer in neural networks. Neural Process. Lett. **54**(2), 803–816 (2022)
34. Wang, Z., Li, C., Wang, X.: Convolutional neural network pruning with structural redundancy reduction. In: IEEE Conference on Computer Vision and Pattern Recognition, CVPR 2021, pp. 14913–14922 (2021)
35. Xu, M., Yoon, S., Fuentes, A., Park, D.S.: A comprehensive survey of image augmentation techniques for deep learning. Pattern Recognit. **137**, 109347 (2023)

36. Zeyde, R., Elad, M., Protter, M.: On single image scale-up using sparse-representations. In: Curves and Surfaces - 7th International Conference, Revised Selected Papers, pp. 711–730 (2010)
37. Zhang, Y., Li, K., Li, K., Wang, L., Zhong, B., Fu, Y.: Image super-resolution using very deep residual channel attention networks. In: Computer Vision - ECCV 2018–15th European Conference, Proceedings, Part VII, pp. 294–310 (2018)
38. Zhao, B., Cui, Q., Song, R., Qiu, Y., Liang, J.: Decoupled knowledge distillation. In: IEEE/CVF Conference on Computer Vision and Pattern Recognition, CVPR 2022, pp. 11943–11952 (2022)
39. Zhou, A., Yao, A., Guo, Y., Xu, L., Chen, Y.: Incremental network quantization: Towards lossless CNNs with low-precision weights. In: 5th International Conference on Learning Representations, ICLR 2017, Conference Track Proceedings (2017)

SDBC: A Novel and Effective Self-Distillation Backdoor Cleansing Approach

Sheng Ran[1], Baolin Zheng[2](✉), and Mingwei Sun[1](✉)

[1] College of Artificial Intelligence, Nankai University, Tianjin, China
ransheng@mail.nankai.edu.cn, smw_sunmingwei@163.com

[2] School of Computer Science, Wuhan University, Wuhan, China
baolinzheng@whu.edu.cn

Abstract. Deep Neural Networks (DNNs) are vulnerable to backdoor attacks, which only need to poison a small portion of samples to control the behavior of the target model. Moreover, the escalating stealth and power of backdoor attacks present not only significant challenges to backdoor defenses but also enormous potential threats to the widespread adoption of DNNs. In this paper, we propose a novel backdoor defense framework, called Self-Distillation Backdoor Cleansing (SDBC), to remove backdoor triggers from the attacked model. For the practical scenario where only a very small portion of clean data is available, SDBC first introduces self-distillation to clean the backdoor in DNNs. Extensive experiments demonstrate that SDBC can effectively remove backdoor triggers under 6 state-of-the-art backdoor attacks using less than 5% or even less than 1% clean training data without compromising accuracy. Experimental results show that the proposed SDBC outperforms existing state-of-the-art (SOTA) methods, reducing the average ASR from 95.36% to 5.75% and increasing the average ACC by 1.92%.

Keywords: Deep neural networks · Backdoor cleansing · Knowledge Distillation

1 Introduction

In recent years, deep neural networks (DNNs) have been widely adopted in critical applications such as face recognition [20,39] and autonomous driving [9,11]. However, the widespread use of DNNs and their vulnerability to attacks [8,29] have also made DNNs a prime target for adversaries. Among the various attacks [6,7,12,13,25,34–36], backdoor attacks [1,3,4,10,16,23,24,26,27,30,33,41,42] have received increasing attention due to their concealment, harmfulness, and practicality. With a backdoor attack, an attacker only needs to poison a small portion of the training data to establish a strong connection between a backdoor trigger and the predefined label without affecting the performance of

Supported by the National Natural Science Foundation of China (Grant Nos. 62073177, 61973175).

B. Luo et al. (Eds.): ICONIP 2023, CCIS 1966, pp. 285–297, 2024.
https://doi.org/10.1007/978-981-99-8148-9_23

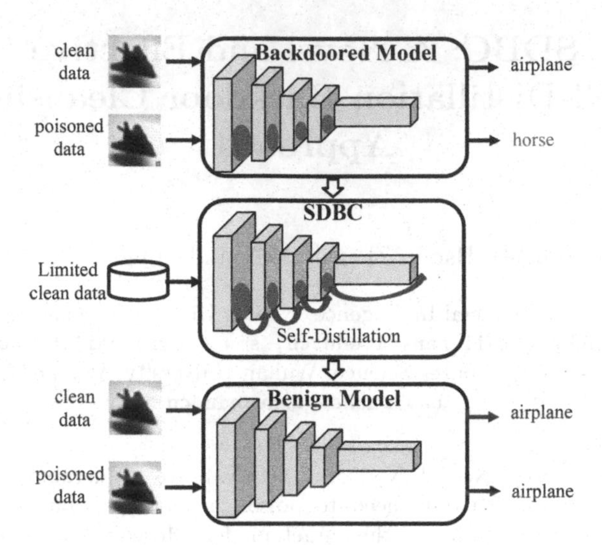

Fig. 1. SDBC utilizes the limited clean data to purify the toxic models with backdoors through self-distillation, eliminating the malicious effects of backdoor triggers.

the model on benign data. Additionally, as the cost of model training increases, the trend toward model sharing and reuse is also on the rise, which poses a potential threat. Therefore, there is an urgent need to explore effective backdoor defense methods.

There are currently two main types of backdoor defense mechanisms. The first is backdoor detection [2,15,32], which is the focus of current research and has achieved great success. Once a backdoor is detected in a model, it needs to be sanitized into a benign model. However, although these methods can detect whether the model is implanted with a backdoor, the backdoor effect remains in the backdoored model. Therefore, the second type is backdoor cleansing [17,19,21,23,28,31,40], which has been explored by few studies so far. In this paper, we focus on backdoor cleansing scenarios. Note that as backdoor attacks become more stealthy and powerful, it is increasingly challenging to completely clean backdoored models. For example, Neural Attention Distillation (NAD) [17], the most effective method for backdoor erasure, still has not achieved satisfactory results against many attacks. Therefore, our research goal is to further reduce the attack success rate (ASR) of backdoored models on toxic data without affecting the classification accuracy (ACC) of clean data.

In this paper, we propose a novel backdoor elimination method, called Self-Distillation Backdoor Cleansing (SDBC), for effective backdoor cleansing in deep-leaning models, as shown in Fig. 1. In our SDBC, Self-distillation [38], which was proposed to improve the model performance by mutual distillation between different layers of the network, is first introduced to purify the backdoor in models with a small amount of clean data. Specifically, SDBC aligns the information of different layers of the network through a clean and non-toxic branch network.

Using mutual distillation can not only continuously reduce the toxicity of the network, but also maintain the performance of the network. Experiments show that the proposed SDBC can not only achieve significantly lower ASR than NAD but also bring ACC to a new high level of accuracy in some cases.

Our major contributions are outlined below:

- We propose a novel and effective backdoor cleansing method called Self-Distillation Backdoor Cleansing (SDBC), which is a powerful distillation-based defense against a variety of attacks.
- For the first time, we introduce self-attention to purify backdoors in deep learning models with a small amount of clean data.
- Extensive experiments demonstrated the proposed SDBC outperforms existing state-of-the-art methods, reducing the average ASR from 95.36% to 5.75% and increasing the average ACC by 1.92%.

2 Related Work

2.1 Backdoor Attack

Backdoor attacks are a category of attack techniques that exploit vulnerabilities in the training pipeline of deep neural networks. Rule-based attacks are commonly employed, where a simple rule is applied to manipulate the target image, such as placing a white square in the lower right corner [10]. Another rule-based approach is to add the trigger image proportionally to the target image [4]. An interesting variation of a rule-based attack utilizes a sine wave signal as the trigger [1]. Recently, a frequency domain-based attack method has been proposed that works in a more stealthy manner by targeting the frequency components of the target [33].

Another popular approach is optimization-based attacks, where triggers are generated by optimization algorithms. A classic method involves identifying the neuron in a deep neural network that has the highest weight on the target class and utilizing that neuron to create optimal triggers [23]. A more recent optimization-based attack method has been developed that generates a unique trigger for each input, making the backdoor triggers more difficult to detect [26]. For a comprehensive overview of backdoor attacks, see [18].

2.2 Backdoor Defense

Backdoor defense can be achieved through two primary approaches [5,22,37]: backdoor detection and backdoor cleansing. Backdoor detection aims to detect the presence of backdoor attacks or filter out suspicious samples [2,15,32]. While detection-based methods are effective in determining whether a model has been compromised by a backdoor, they do not remove the triggers from the model itself. Therefore, a cleansing-based approach is needed to immediately cleanse the poisoned model, to eliminate the harmful effects of the triggers while maintaining the accuracy of the model on clean data.

A simple and direct solution of backdoor cleansing is to fine-tune the model on a limited amount of clean data, which is feasible for defenders [23,28]. However, retraining with a small sample size can suffer from catastrophic forgetting, leading to performance deterioration [14]. To address this problem, fine pruning techniques have been proposed [19], in which less significant nodes are removed before the network is refined. Another approach, known as WILD [21], is proposed to use data augmentation to remove backdoors from deep neural networks. Methods such as mode connectivity repair (MCR) [40] and regularization [31] have also contributed to the development of deletion techniques.

The Neural Attention Distillation (NAD) method [17] utilizes attention features to enhance backdoor elimination by incorporating fine-tuning and distillation operations. However, NAD still suffers from challenges in effectively expressing attention features, resulting in a non-negligible attack success rate (ASR). In this study, we propose a novel distillation-based backdoor cleansing technique that achieves remarkable performance against various well-known and sophisticated backdoor attacks.

3 Our SDBC Approach

In this section, we first describe the defense setup and then introduce the proposed SDBC method.

3.1 Defense Setup

We consider a very general defense setup where the defender has received a backdoored model from an untrusted third party and assumes that the defender has a small portion of clean samples to remove the embedded triggers. The goal of backdoor cleansing is to remove backdoor triggers without compromising the accuracy of the model on clean data.

3.2 Overview

With the self-distillation method and a small number of clean samples, our SDBC method can turn a backdoor model into a benign one. The pipeline of SDBC is shown in Fig. 1. The backdoored model can be manipulated by attackers through poisoned samples before we impose our SDBC defense on the backdoored model. For technical details on the structure of SDBC, refer to Fig. 2. The backbone model is the backdoored model that SDBC needs to purify and the branch models consist of 3 main parts: stacked layers, FC layers, and softmax layers. The reason why the SDBC works so well is due to the knowledge transfer through our designed losses: attention loss, knowledge loss, and Cross-Entropy loss. The attention loss is computed with the middle staked layers' attention map and the map of the last block of the backbone model, which makes the backbone model grab a better attention representation of the correct target. The knowledge loss is to calculate the softmax outputs between the backbone model and branch

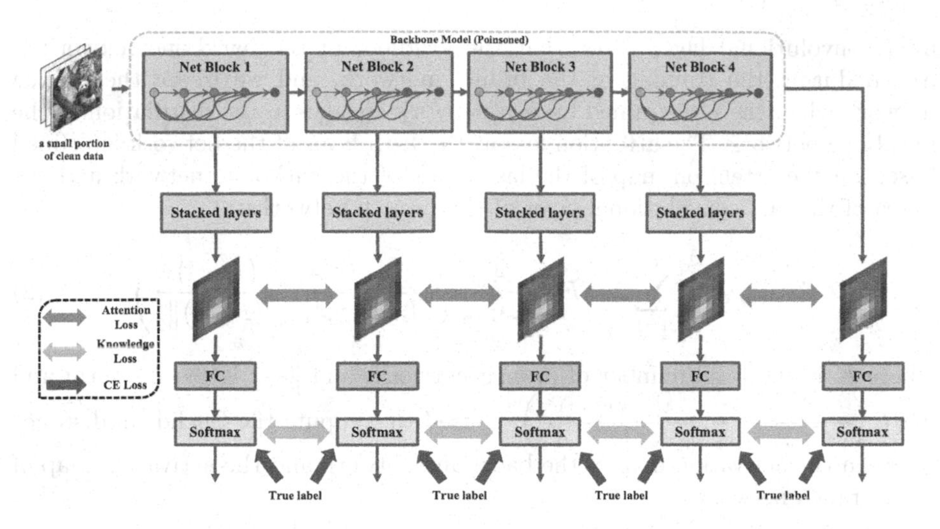

Fig. 2. The overview of the proposed SDBC framework.

models to gain more sufficient information about the classification. Moreover, with the help of Cross-Entropy loss, we can ensure that the purified model will not sacrifice the accuracy of clean data. Overall, relying on the structure and the loss we design, SDBC is capable of achieving remarkably low ASR and high ACC. In the next few subsections, we will formally define the attention representation and soft-label knowledge, as well as the definition of the loss functions which SDBC utilizes in the Self-Distillation Backdoor Cleansing framework.

3.3 Self-Distillation Backdoor Cleansing

Attention Representation. For a deep neural network model F, we define the convolutional activation output at the n-th layer to be $F^n \in \mathbb{R}^{C\times H\times W}$ where C, H, and W are the dimensions of the channel, the height and the width of the activation map, respectively. We define $\mathcal{A} : \mathbb{R}^{C\times H\times W} \rightarrow \mathbb{R}^{H\times W}$ as an operator which maps an activation map to an attention representation:

$$\mathcal{A}^p_{sum}\left(F^n\right) = \sum_{i=1}^{C} |F^n_i|^p \, ; \mathcal{A}^p_{mean}\left(F^n\right) = \frac{1}{C}\sum_{i=1}^{C} |F^n_i|^p \tag{1}$$

where F^n_i is the activation map of the n-th channel. $|*|$ is the absolute value function. $\mathcal{A}^p_{sum}$ mirrors the numerical summation of the activated regions of all channels in the layer and is augmented by the exponent p. $\mathcal{A}^p_{mean}$ is the average of $\mathcal{A}^p_{sum}$ over all channels.

Attention Loss. The framework details of SDBC are shown in Fig. 2. We use one of the attention representations proposed above to compute the attention loss of the backbone network's last convolutional layer and each branch network's final convolutional layer. Note that the gradient of the backbone network is frozen during the training of the branch network, and we freeze the branch network when the well-trained branch network engages in the distillation of the backbone network. The attention loss at the n-th layer of the network is defined based on the attention map of the last block of the backbone network and the maps of the last convolutional layer of the branch networks:

$$\mathcal{L}_{Attention} = \sum_{i=1}^{K} MSE\left(\frac{\mathcal{A}_{sum}^2\left(F_B^n\right)}{\left\|\mathcal{A}_{sum}^2\left(F_B^n\right)\right\|_2}, \frac{\mathcal{A}_{sum}^2\left(F_{bi}^{n-i}\right)}{\left\|\mathcal{A}_{sum}^2\left(F_{bi}^{n-i}\right)\right\|_2}\right) \tag{2}$$

where K is the total number of branch networks and $\|*\|_2$ is the L_2 norm and $MSE\left(\frac{\mathcal{A}_{sum}^2(F_B^n)}{\|\mathcal{A}_{sum}^2(F_B^n)\|_2}, \frac{\mathcal{A}_{sum}^2(F_{bi}^{n-i})}{\|\mathcal{A}_{sum}^2(F_{bi}^{n-i})\|_2}\right)$ is used to compute the Euclidean distance between the activation map of the backbone network and the activation map of each branch network.

Knowledge Loss. Similar to attention loss, knowledge loss is also the information interaction between the backbone network and each branch network. Similarly, during the learning phase of the branch network, the backbone network remains unchanged and provides its soft label information to the branch networks for learning; then, when the branch networks' learning is completed, they will help to improve the performance of the backbone network by transferring their soft label information to the backbone network. The loss based on soft-label knowledge is defined as:

$$\mathcal{L}_{knowledge} = -\sum_{i=1}^{K}\left(\sum_{j}^{N} p_j^T \log\left(q_j^T\right)\right)_i \tag{3}$$

where K is the total number of branch networks; $p_j^T = \frac{\exp(v_j/T)}{\sum_k^N \exp(v_k/T)}$ and $q_j^T = \frac{\exp(z_j/T)}{\sum_k^N \exp(z_k/T)}$; v_j is the logits of the backbone network; z_j are the logits of the branch networks; p_j^T is the softmax output of the backbone network at temperature T on the jth class; q_j^T is the softmax output of the branch networks at temperature T on the jth class; N is the total number of label classes; i represents the serial number of the branch network. The knowledge loss is the sum of the losses of the backbone network and all branch networks, each of which performs Kullback-Leibler divergence at temperature T.

CE Loss. Cross-entropy loss is used for both backbone and branch networks to ensure and improve the performance of the model, which can be formulated as:

$$\mathcal{L}_{ce} = -\sum_{i=1}^{K}\left(\sum_{j}^{N} p_j \log y_j\right) \tag{4}$$

where y is the true label of the clean data, and other parameters represent the same as before. $\mathcal{L}_{ce}$ can measure the classification error of the branch networks and backbone network.

Total Loss. The total loss is the sum of the soft-label-based knowledge loss, the attention loss, and the cross-entropy loss of the backbone and branch networks:

$$\mathcal{L}_{total} = \alpha \cdot \mathcal{L}_{knowledge} + \beta \cdot \mathcal{L}_{attention} + \mathcal{L}_{ce} \tag{5}$$

where α and β are hyperparameters that adjust the strength of knowledge and attention distillation loss separately.

4 Experiments

This section first presents the experimental setting. Then the effectiveness of NAD and SDBC under 6 different state-of-the-art backdoor attacks is evaluated and discussed. In the end, we conduct some research to understand the SDBC approach further.

4.1 Experimental Setting

Backdoor Attacks and Configurations. We consider 6 state-of-the-art backdoor attacks: 1) BadNets [10], 2) Trojan attack [23], 3) Blend attack [4], 4) Sinusoidal signal attack (SIG) [1], 5) Input-aware attack [26], 6) Frequency trojan (FT) [33]. To ensure a valid comparison and evaluation of the results, we follow the same configuration as in the original paper on NAD and backdoor attacks. We verify the performance of all attack and defense strategies using the ResNet-50 model on the CIFAR-10 benchmark.

Defense Configuration. We compare SDBC and NAD under the same conditions of 5% clean training data. For SDBC, we run the self-distillation of the backdoored model on 5% of the available pure data, with about 50 epochs of branch network training and 5 epochs of backbone network training. We employ the Adam optimizer with an initial learning rate of 0.001 when acquiring backdoored model and with a learning rate of 0.01 when carrying self-distillation. Other employment of the Adam optimizer concludes the pair of coefficients for computing running averages of the gradient and its square of (0.9, 0.999), a numerical stability improvement factor of 10^{-8}, and a weight decay factor of 0. The batch size during the whole training is 64, and the data enhancement techniques we adopt include random cropping (padding = 4), horizontal flipping, center cropping, color jitter, and random rotation (degrees = 3). The hyperparameters α and β are set to 1 and 1000 respectively according to the Sect. 4.3, and the temperature T is chosen to be 1000 empirically. Our experiments are conducted on a workstation with an NVIDIA GeForce RTX 3090 Ti and 128G of RAM.

Table 1. Comparison of the effectiveness of 3 backdoor defenses against 6 backdoor attacks in terms of Attack Success Rate (ASR) and Classification Accuracy Rate (ACC). The Deviation row shows the percentage change in performance (ASR and ACC) of the backdoor defense methods concerning the backdoor model. All experiments in the table were performed on CIFAR-10, and the best results are shown in bold.

Backdoor	Before		Finetuning		NAD		SDBC (Ours)	
Attack	ASR	ACC	ASR	ACC	ASR	ACC	ASR	ACC
BadNets	97.10	83.24	58.50	79.45	6.81	78.29	**2.48**	**80.23**
Trojan	98.74	84.16	49.01	**84.54**	11.92	82.91	**5.77**	83.94
Blend	94.89	84.20	38.81	83.86	9.33	82.24	**3.80**	**86.01**
SIG	92.74	82.05	69.37	**84.73**	9.19	80.36	**0.28**	80.84
Input-aware	98.44	73.48	14.56	**73.96**	10.62	72.98	**9.37**	73.25
FT	90.22	70.10	25.77	**85.40**	15.23	84.27	**12.80**	84.47
Average	95.36	79.54	42.67	**81.99**	10.52	80.18	**5.75**	81.46
Deviation	-	-	↓52.69	↑**2.45**	↓84.84	↑0.64	↓**89.61**	↑1.92

Evaluation Metrics. We evaluate the defense algorithms mainly by two performance matrices: the attack success rate (ASR) of outputting attacker-specified labels under toxic samples, and the model accuracy rate (ACC) of outputs under clean samples. The better defense is determined by which method has the higher ACC and lower ASR.

4.2 Effectiveness of SDBC

We verify the effectiveness of our SDBC method through two performance matrices (ASR and ACC) against 6 backdoor attacks. We then compare the performance of SDBC with the NAD defense in Table 1. Our experiment shows that our SDBC approach dramatically reduces the average ASR from 95.36% to 5.75% with an increase in average ACC of 1.92%. On the other hand, the Finetuning and NAD approaches only brought the average ASR reduction to 42.67% and 10.52%, respectively. As evidenced by the most significant reduction in success rate for each attack, our SDBC approach is superior against all 6 attacks. The accuracy rates for the SDBC and Finetuning methods are comparable, with percentages of 81.46% and 81.99%, respectively. The reason for the SDBC's slightly lower average accuracy is due to a trade-off between ASR and ACC.

Performance with Various Proportions of Clean Data. The performance of each defense method under different sizes of clean data is a crucial research point. The detail of the comparison results under varying data sizes is shown in Fig. 3. The curve illustrates the performance of ASR and ACC under Trojan backdoor attacks since the Trojan backdoor is a highly challenging attack to mitigate. The motivation for excluding the cases below 1% in the figures is to

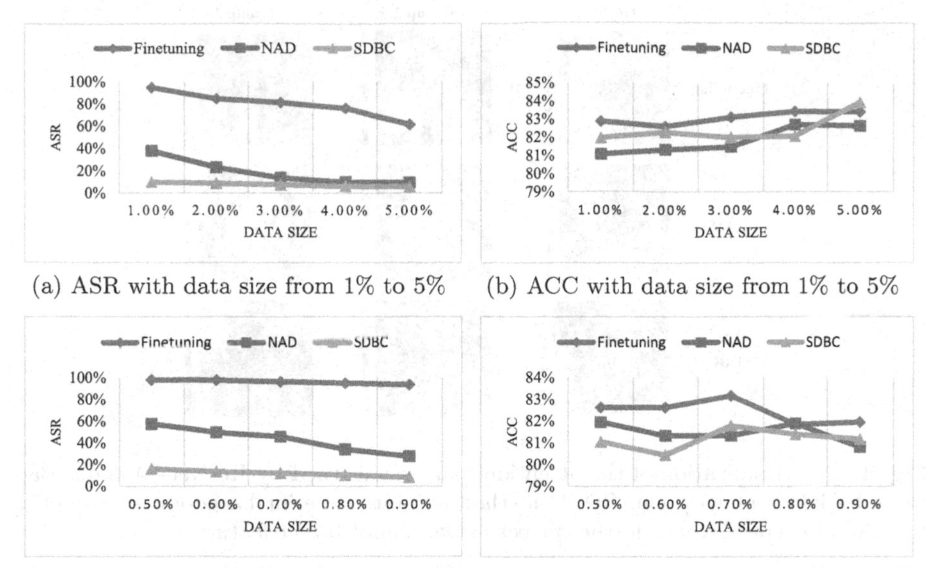

(a) ASR with data size from 1% to 5% (b) ACC with data size from 1% to 5%

(c) ASR with data size from 0.5% to 0.9% (d) ACC with data size from 0.5% to 0.9%

Fig. 3. The results in the figure show that the advantage of our SDBC method is more evident when the amount of data is smaller, at the same time, it successfully maintains the ACC on clean data. For defenses, a smaller ASR indicates a greater effect on backdoor elimination.

highlight the superiority of our SDBC method when there is very limited clean data. The results show that our SDBC method can achieve the lowest ASR and maintain the ACC when the amount of data is small, even when it is less than 1%. As the data volume decreases, We also see that the superiority of SDBC is enhanced. In particular, the ASR after SDBC defense is 81.95% lower than that after Finetuning defense, and 41.45% lower than that after NAD defense at 0.5% of the clean data volume.

4.3 A Comprehensive Understanding of SDBC

In this subsection, we give an intuitive visual representation of the attention map for a better understanding of how the SDBC works and the superiority of the SDBC method. And then, the optimal values of hyperparameters α and β are discussed through experiments.

Understanding Attention Maps. The attention map is derived from the output of the middle layer and allows us to visualize how the backdoor cleansing changes in the different middle layers. We visualize the attention maps of networks under 3 defense methods to show the capabilities and operational details of defense methods. The comparison of the visualization of the attention maps is shown in Fig. 4. We can see that for shallow feature extraction, the

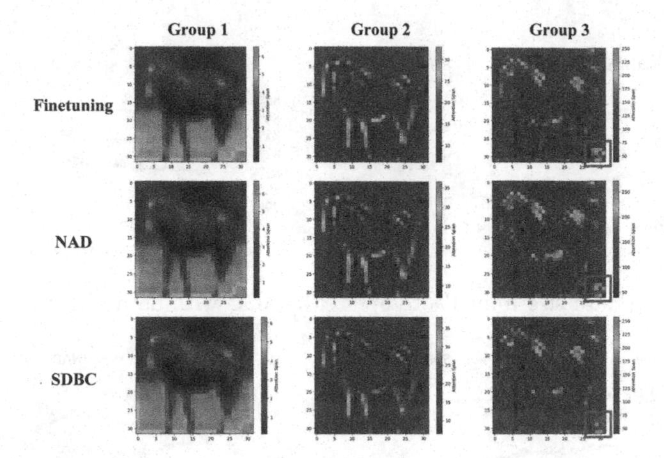

Fig. 4. The visualization of the attention map was learned by ResNet-50 under BadNet's backdoor images. Our SDBC method exhibits superior backdoor erasure effectiveness at deeper layers and the correct extraction of target features.

defenses are empirically similar. However, for deeper feature extraction, SDBC pays less attention to the backdoor trigger in the lower right corner than the other defenses. Furthermore, the feature maps, i.e. the extraction of target features, show a small difference in the accuracy of all methods on clean data, which means that our SDBC method successfully maintains the classification accuracy of clean data.

Effect of Parameters α and β. We study the effect of different values of the hyperparameters α and β on the distillation performance using Trojan as an example, and the results are shown in Fig. 5. The curves in Fig. 5(a) show that ACC is highest when α is set to 1, while the performance of ASR deteriorates with the increase of α, so the optimal α value is around 1. Figure 5(b) reveals that the SDBC approach is robust to variations of the hyperparameter β when it is above 1000, so we generally adopt $\beta = 1000$ in the distillation process for simplicity.

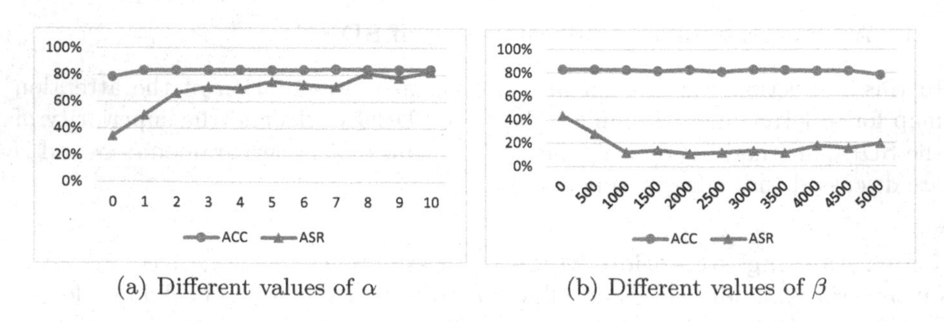

(a) Different values of α

(b) Different values of β

Fig. 5. (a) The SDBC performance on ACC and ASR with different values of α (b) the SDBC performance on ACC and ASR with different values of β. The results show that the ACC of our SDBC is the finest when α is 1 as well as SDBC has robustness for β above 1000.

5 Conclusion

In this paper, we proposed a novel and effective backdoor-cleaning method called Self-Distillation Backdoor Cleansing (SDBC), which can greatly mitigate potential threats from backdoor attacks. We experimentally demonstrated that our method outperforms existing SOTA backdoor defense methods under six state-of-the-art backdoor attacks. The proposed SDBC can reduce the average ASR from 95.36% to 5.75% while increasing the average ACC by 1.92%. Besides, we illustrated an intuitive visual comparison of different methods via attention map visualization. Finally, we investigated the effect of different hyperparameter values on SDBC defense performance, and the results show that our method is robust and we provided optimal hyperparameter benchmark values.

Acknowledgements. This work was supported by the National Natural Science Foundation of China (Grant Nos. 62073177, 61973175).

References

1. Barni, M., Kallas, K., Tondi, B.: A new backdoor attack in CNNs by training set corruption without label poisoning. In: 2019 IEEE International Conference on Image Processing (ICIP), pp. 101–105. IEEE (2019)
2. Chen, H., Fu, C., Zhao, J., Koushanfar, F.: Deepinspect: A black-box trojan detection and mitigation framework for deep neural networks. In: IJCAI. vol. 2, p. 8 (2019)
3. Chen, J., Zheng, H., Su, M., Du, T., Lin, C., Ji, S.: Invisible poisoning: highly stealthy targeted poisoning attack. In: Liu, Z., Yung, M. (eds.) Inscrypt 2019. LNCS, vol. 12020, pp. 173–198. Springer, Cham (2020). https://doi.org/10.1007/978-3-030-42921-8_10
4. Chen, X., Liu, C., Li, B., Lu, K., Song, D.: Targeted backdoor attacks on deep learning systems using data poisoning. arXiv preprint arXiv:1712.05526 (2017)
5. Doan, B.G., Abbasnejad, E., Ranasinghe, D.C.: Februus: input purification defense against trojan attacks on deep neural network systems. In: Annual Computer Security Applications Conference, pp. 897–912 (2020)
6. Duan, R., Ma, X., Wang, Y., Bailey, J., Qin, A.K., Yang, Y.: Adversarial camouflage: Hiding physical-world attacks with natural styles. In: Proceedings of the IEEE/CVF Conference on Computer Vision and Pattern Recognition, pp. 1000–1008 (2020)
7. Fan, Y., et al.: Sparse adversarial attack via perturbation factorization. In: Vedaldi, A., Bischof, H., Brox, T., Frahm, J.-M. (eds.) ECCV 2020 Part XXII. LNCS, vol. 12367, pp. 35–50. Springer, Cham (2020). https://doi.org/10.1007/978-3-030-58542-6_3
8. Goodfellow, I.J., Shlens, J., Szegedy, C.: Explaining and harnessing adversarial examples. arXiv preprint arXiv:1412.6572 (2014)
9. Grigorescu, S., Trasnea, B., Cocias, T., Macesanu, G.: A survey of deep learning techniques for autonomous driving. J. Field Robot. **37**(3), 362–386 (2020)
10. Gu, T., Dolan-Gavitt, B., Garg, S.: BadNets: identifying vulnerabilities in the machine learning model supply chain. arXiv preprint arXiv:1708.06733 (2017)

11. Hu, Y., et al.: Planning-oriented autonomous driving. In: Proceedings of the IEEE/CVF Conference on Computer Vision and Pattern Recognition, pp. 17853–17862 (2023)
12. Huang, L., Wei, S., Gao, C., Liu, N.: Cyclical adversarial attack pierces black-box deep neural networks. Pattern Recogn. **131**, 108831 (2022)
13. Jiang, L., Ma, X., Chen, S., Bailey, J., Jiang, Y.G.: Black-box adversarial attacks on video recognition models. In: Proceedings of the 27th ACM International Conference on Multimedia, pp. 864–872 (2019)
14. Kirkpatrick, J., et al.: Overcoming catastrophic forgetting in neural networks. Proc. Nat. Acad. Sci. **114**(13), 3521–3526 (2017)
15. Kolouri, S., Saha, A., Pirsiavash, H., Hoffmann, H.: Universal litmus patterns: revealing backdoor attacks in cnns. In: Proceedings of the IEEE/CVF Conference on Computer Vision and Pattern Recognition, pp. 301–310 (2020)
16. Li, S., Xue, M., Zhao, B.Z.H., Zhu, H., Zhang, X.: Invisible backdoor attacks on deep neural networks via steganography and regularization. IEEE Trans. Dependable Secure Comput. **18**(5), 2088–2105 (2020)
17. Li, Y., Lyu, X., Koren, N., Lyu, L., Li, B., Ma, X.: Neural attention distillation: erasing backdoor triggers from deep neural networks. arXiv preprint arXiv:2101.05930 (2021)
18. Li, Y., Jiang, Y., Li, Z., Xia, S.T.: Backdoor learning: a survey. IEEE Trans. Neural Netw. Learn. Syst. (2022)
19. Liu, K., Dolan-Gavitt, B., Garg, S.: Fine-pruning: defending against backdooring attacks on deep neural networks. In: Bailey, M., Holz, T., Stamatogiannakis, M., Ioannidis, S. (eds.) RAID 2018. LNCS, vol. 11050, pp. 273–294. Springer, Cham (2018). https://doi.org/10.1007/978-3-030-00470-5_13
20. Liu, W., Wen, Y., Yu, Z., Li, M., Raj, B., Song, L.: Sphereface: deep hypersphere embedding for face recognition. In: Proceedings of the IEEE Conference on Computer Vision and Pattern Recognition, pp. 212–220 (2017)
21. Liu, X., Li, F., Wen, B., Li, Q.: Removing backdoor-based watermarks in neural networks with limited data. In: 2020 25th International Conference on Pattern Recognition (ICPR), pp. 10149–10156. IEEE (2021)
22. Liu, Y., Lee, W.C., Tao, G., Ma, S., Aafer, Y., Zhang, X.: ABS: scanning neural networks for back-doors by artificial brain stimulation. In: Proceedings of the 2019 ACM SIGSAC Conference on Computer and Communications Security, pp. 1265–1282 (2019)
23. Liu, Y., Ma, S., Aafer, Y., Lee, W.C., Zhai, J., Wang, W., Zhang, X.: Trojaning attack on neural networks. In: 25th Annual Network And Distributed System Security Symposium (NDSS 2018). Internet Soc (2018)
24. Liu, Y., Ma, X., Bailey, J., Lu, F.: Reflection backdoor: a natural backdoor attack on deep neural networks. In: Vedaldi, A., Bischof, H., Brox, T., Frahm, J.-M. (eds.) ECCV 2020 Part X. LNCS, vol. 12355, pp. 182–199. Springer, Cham (2020). https://doi.org/10.1007/978-3-030-58607-2_11
25. Ma, X., Niu, Y., Gu, L., Wang, Y., Zhao, Y., Bailey, J., Lu, F.: Understanding adversarial attacks on deep learning based medical image analysis systems. Pattern Recogn. **110**, 107332 (2021)
26. Nguyen, T.A., Tran, A.: Input-aware dynamic backdoor attack. Adv. Neural. Inf. Process. Syst. **33**, 3454–3464 (2020)
27. Ning, R., Li, J., Xin, C., Wu, H.: Invisible poison: a blackbox clean label backdoor attack to deep neural networks. In: IEEE INFOCOM 2021-IEEE Conference on Computer Communications, pp. 1–10. IEEE (2021)

28. Sha, Z., He, X., Berrang, P., Humbert, M., Zhang, Y.: Fine-tuning is all you need to mitigate backdoor attacks. arXiv preprint arXiv:2212.09067 (2022)
29. Szegedy, C., et al.: Intriguing properties of neural networks. arXiv preprint arXiv:1312.6199 (2013)
30. Tran, B., Li, J., Madry, A.: Spectral signatures in backdoor attacks. In: Advances in Neural Information Processing Systems, vol. 31 (2018)
31. Truong, L., et al.: Systematic evaluation of backdoor data poisoning attacks on image classifiers. In: Proceedings of the IEEE/CVF Conference on Computer Vision and Pattern Recognition Workshops, pp. 788–789 (2020)
32. Wang, B., et al.: Neural cleanse: identifying and mitigating backdoor attacks in neural networks. In: 2019 IEEE Symposium on Security and Privacy (SP), pp. 707–723. IEEE (2019)
33. Wang, T., Yao, Y., Xu, F., An, S., Tong, H., Wang, T.: An invisible black-box backdoor attack through frequency domain. In: Avidan, S., Brostow, G., Cissé, M., Farinella, G.M., Hassner, T. (eds.) ECCV 2022 Part XIII. LNCS, vol. 13673, pp. 396–413. Springer, Cham (2022)
34. Wang, Y., Ma, X., Bailey, J., Yi, J., Zhou, B., Gu, Q.: On the convergence and robustness of adversarial training. arXiv preprint arXiv:2112.08304 (2021)
35. Wu, D., Wang, Y., Xia, S.T., Bailey, J., Ma, X.: Skip connections matter: On the transferability of adversarial examples generated with resnets. arXiv preprint arXiv:2002.05990 (2020)
36. Wu, D., Xia, S.T., Wang, Y.: Adversarial weight perturbation helps robust generalization. Adv. Neural. Inf. Process. Syst. **33**, 2958–2969 (2020)
37. Xu, X., Wang, Q., Li, H., Borisov, N., Gunter, C.A., Li, B.: Detecting AI trojans using meta neural analysis. In: 2021 IEEE Symposium on Security and Privacy (SP), pp. 103–120. IEEE (2021)
38. Zhang, L., Song, J., Gao, A., Chen, J., Bao, C., Ma, K.: Be your own teacher: improve the performance of convolutional neural networks via self distillation. In: Proceedings of the IEEE/CVF International Conference on Computer Vision, pp. 3713–3722 (2019)
39. Zhang, X., et al.: P2sgrad: refined gradients for optimizing deep face models. In: CVF Conference on Computer Vision and Pattern Recognition (CVPR). pp. 9898–9906 2019 IEEE (2019)
40. Zhao, P., Chen, P.Y., Das, P., Ramamurthy, K.N., Lin, X.: Bridging mode connectivity in loss landscapes and adversarial robustness. arXiv preprint arXiv:2005.00060 (2020)
41. Zhao, S., Ma, X., Zheng, X., Bailey, J., Chen, J., Jiang, Y.G.: Clean-label backdoor attacks on video recognition models. In: Proceedings of the IEEE/CVF Conference on Computer Vision and Pattern Recognition, pp. 14443–14452 (2020)
42. Zhong, H., Liao, C., Squicciarini, A.C., Zhu, S., Miller, D.: Backdoor embedding in convolutional neural network models via invisible perturbation. In: Proceedings of the Tenth ACM Conference on Data and Application Security and Privacy, pp. 97–108 (2020)

An Alignment and Matching Network with Hierarchical Visual Features for Multimodal Named Entity and Relation Extraction

Yinglong Dai[1], Feng Gao[1], and Daojian Zeng[1,2](✉)

[1] Hunan Provincial Key Laboratory of Intelligent Computing and Language Information Processing, Hunan Normal University, Changsha, China
{daiyl,grident,zengdj}@hunnu.edu.cn

[2] Institute of AI and Targeted International Communication, Hunan Normal University, Changsha, China

Abstract. In this paper, we study the tasks of multimodal named entity recognition (MNER) and multimodal relation extraction (MRE), both of which involve incorporating visual modality to complement text modality. The core issues are how to bridge the modality gap and reduce modality noise. To address the first issue, we introduce an image-text alignment (ITA) module to obtain a better unimodal representation by aligning the inconsistent representations between image and text, which come from different encoders. To tackle the second issue, we propose an image-text matching (ITM) module that constructs hard negatives to improve the model's ability to capture the semantic correspondence between text and image. Besides, we also selectively combine and concatenate the hierarchical visual features obtained from both global and visual objects of Vision Transformer (ViT) as improved visual prefix for modality fusion. We conduct extensive experiments to demonstrate the effectiveness of our method (AMNet) (Code is available in https://github.com/Grident/AMNet) and achieve state-of-the-art performance on three benchmark datasets.

Keywords: Contrastive Learning · Hierarchical Visual Features · Image-Text Matching

1 Introduction

Named entity recognition (NER) and relation extraction (RE) are two crucial tasks for social media analysis. Existing text-based methods suffer from a drastic performance drop when applied to social media data. The reason is that social media texts are short and lack contextual information, which makes it difficult for models to learn robust features from these limited words. Fortunately, **Multimodal NER (MNER)** [1] and **Multimodal MRE** [2] incorporate images in social media as additional input to help with NER and RE.

B. Luo et al. (Eds.): ICONIP 2023, CCIS 1966, pp. 298–310, 2024.
https://doi.org/10.1007/978-981-99-8148-9_24

The essence of MNER and MRE tasks is integrating visual features into textual representation. The Existing models MNER and MRE incorporate visual features in various ways. Early researchers [3, 4] studied how to encode the entire image into raw feature vectors and directly incorporate them into text representation for semantic enhancement. More recent studies [2, 5, 6] focus on encoding visual objects to accomplish MRE and MNER tasks. This includes HVPNeT [8], which proposes hierarchical visual features as visual prefix for fusion to alleviate the error sensitivity of irrelevant object images. Despite the success achieved in MRE and MNER tasks, current methods still face two problems:

Firstly, current methods ignore the modality gap. The textual and visual representations with different feature distributions hinder the model from capturing the cross-modal semantic correlations. For example, in Fig. 1(a), the inconsistent representation hinders the model from capturing the semantic relevance of the textual entity " *Rocky* " and the region of the dog.

Secondly, many methods also ignore the modality noise. Not all image-text pairs are matched. Irrelevant global and object images can mislead the model to make incorrect judgments. For example, in Fig. 1(b), the model may associate the people with "*Disney*" and generate false predictions with irrelevant images.

(a) [Rocky MISC] is ready for snow season.
(b) Coming soon on [Disney MISC] Channel.

Fig. 1. Examples from Twitter dataset for MNER.

In this paper, we propose an alignment and matching network with hierarchical Visual Features (AMNet) for MNER and MRE tasks. To solve the first issue, we propose an image-text (ITM) matching module to predict whether a given image and text match. Besides, we selectively combine and concatenate the hierarchical visual features [8] of the global and object images from ViT as improved visual prefix for fusion, further reducing modality noise. Specifically, to solve the second issue, we propose an image-text alignment (ITA) module to align the modalities before fusion for better single-modality representation and consistency.

Our main contributions are as follows:

- We propose an alignment and matching network with hierarchical visual features for MNER and MRE tasks, effectively reducing the impact of mismatched images on the model and bridging the modality gap.
- We selectively combine and concatenate the hierarchical visual features obtained from the global and visual object features from ViT as improved visual prefix for multimodal fusion, further reducing modality noise.

- We demonstrate the effectiveness of our proposed method through detailed ablation experiments and case studies.

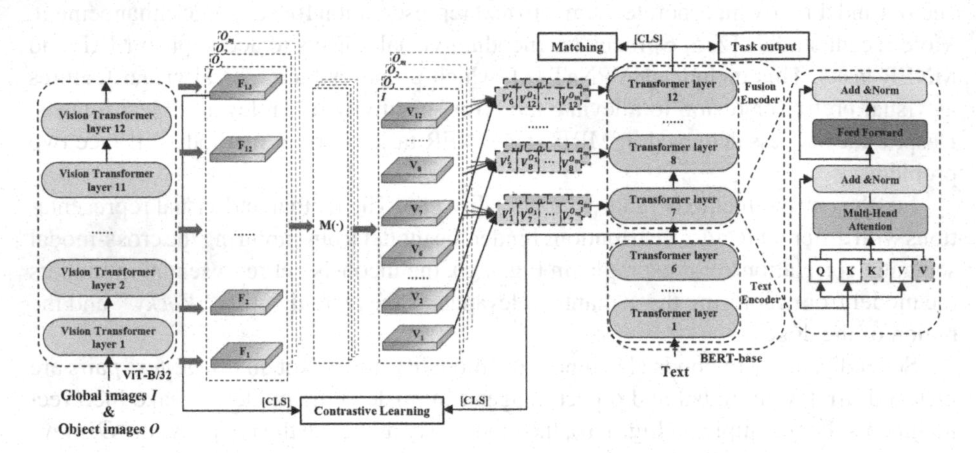

Fig. 2. The overall architecture of AMNet.

2 Task Formulation

In this section, we first define the task we need to solve:

MRE: It is a multi-classification problem. Given a short text $T = (t_1, t_2, \ldots, t_n)$ and its associated image I, the task of MRE is to extract the semantic relationship type $r \in Y$ between two entities E_1 and E_2, exactly $E_1 = (t_i, \ldots, t_{i+|E_1|-1})$ and $E_2 = (t_i, \ldots, t_{i+|E_2|-1})$ [1].

MNER: It is a sequence labeling paradigm. Given a short text $T = (t_1, t_2, \ldots, t_n)$ and associated image I as input, the task of MNER is to classify the named entities containing n tokens into the predefined types $Y = (y_1, y_2, \ldots, y_n)$, where $y_i \in Y$ follows BIO-tagging schema [7].

3 Method

Our alignment and matching network with hierarchical visual features as shown in Fig. 2, consists of five main components: (1) encoding modules convert text and image into a suitable representation; (2) text-image alignment module refers to aligning textual and visual features; (3) text-image matching module predicts whether given image-text pairs match; (4) fusion module adopts improved visual prefix for fusion; (5) prediction module performs the final task predictions.

3.1 Encoding Module

Text Encoder: we use BERT [11] as the text encoder and add [CLS] and [SEP] tokens at the beginning and end of each input sentence which respectively represents as $\mathrm{T}' = (t_0, t_1, t_2, \ldots, t_{n+1})$, where t_0 and t_{n+1} are the inserted [CLS] and [SEP] tokens, and t_1 to t_n represent the textual entity sequence. Inspired by [12], the first 6 and last 6 layers of BERT are initialized as the text encoder and the fusion encoder, respectively. The input text T' is fed into text encoding layers to obtain the output representation as $\mathrm{T} = (\mathrm{t}_0, \mathrm{t}_1, \mathrm{t}_2, \ldots, \mathrm{t}_{n+1})$, where $\mathrm{t}_i \in \mathbb{R}^{\mathrm{d}}$ corresponds to the representation of the i_{th} token.

Image Encoder: images contain both global image features which provide the overall context information of the image, and object features which provide information about specific objects that are aligned to the textual entities. Following HVPNeT [8], we use both granularities of visual information. The multi-scaled visual features obtained by Resnet's blocks are too coarse to match the multi-level language information encoded by BERT [11]. Therefore, we use the vision transformer (ViT) [13, 15] as the visual encoder. ViT learns high-dimensional features by stacking multiple layers of attention, similar to the text encoder BERT. Firstly, we use a visual grounding toolkit [14] to extract m salience objects from the entire image $\boldsymbol{I}$, denoted as $\boldsymbol{O} = \{o_1, o_2, \ldots, o_m\}$. Then the global image $\boldsymbol{I}$ and object image $\boldsymbol{O}$ are resized to 224 × 224 pixels and fed into ViT. They are divided into 49 patches. The ViT produces a list of hierarchical visual features represented as $\{F_0, F_1, \ldots, F_{11}, F_{12}\}$, where $F_0 \in \mathbb{R}^{50\times d}$ corresponds to the initial embedding of the image and $F_l \in \mathbb{R}^{50\times d}$ $(l > 0)$ represents the l-th visual feature. Each visual feature representations as $(v_{cls}, v_1, v_2, \ldots, v_{49})$, where $v_{cls} \in \mathbb{R}^{1\times d}$ is the [CLS] token that represents full-image information at each layer.

3.2 Image-Text Alignment

The textual and visual representations from different encoders have different feature distributions, which hinder the model from capturing the cross-modal semantic correlations. We introduce a contrastive learning method [12] to obtain better unimodal representations before modality fusion to make the modal representations more consistent. The number of negative samples has a certain impact on the effectiveness of contrastive learning. Specifically, within a batch, we have N pairs of matched image-text pairs (V, T) examples (where V refers only to image I). For each text T, V is its positive example, and the remaining (N-1) images are treated as negative examples in the batch. The specific process is as follows:

Firstly, we use MLP layers $g_v(\cdot)$ and $g_w(\cdot)$ to map the [CLS] embeddings $v^I_{cls} \in \mathbb{R}^{1\times d}$ and $w_{cls} \in \mathbb{R}^{1\times d}$ to normalized lower-dimensional(256-d) representations. Transforming [CLS] embeddings through MLP layers can achieve better results compared to simple linear transformations. The [CLS] embeddings come from the last visual feature F^I_{12} and the output T of the text encoder. Then we define the cosine similarity $s(V,T) = g_v(v^I_{cls})^T g_w(w_{cls})$ and $s(T,V) = g_w(w_{cls})^T g_v(v^I_{cls})$ to calculate the similarity between images and texts within a batch [12]. Then, we calculate the in-batch image-to-text and text-to-image similarity as:

$$p^{i2t}(V) = \frac{\exp\left(\frac{s(V,T)}{\tau}\right)}{\sum_{i=1}^{N}\exp\left(\frac{s(V,T_i)}{\tau}\right)}, p^{t2i}(T) = \frac{\exp(s(T,V)/\tau)}{\sum_{i=1}^{N}\exp(s(T,V_i)/\tau)} \tag{1}$$

where τ is a learnable temperature parameter. Let y^{i2t}(I) and y^{t2i}(T) denote the ground-truth one-hot similarity. The contrastive loss is defined as the cross-entropy H between p and y:

$$\mathcal{L}_{ita} = \mathbb{E}_{(I,T)\sim D}[H(y^{i2t}(V), p^{i2t}(V)) + H(y^{t2i}(T), p^{t2i}(T))] \tag{2}$$

3.3 Cross-Modal Fusion

Aggregated Hierarchical Feature

In order to obtain visual hierarchical features that match the level of information in the text, we use the following operations: First, we map the list of hierarchical visual features $\{F_0, F_1, \ldots, F_{11}, F_{12}\}$ with M(·) as follows:

$$G = View([F_0, F_1, \ldots, F_{11}, F_{12}]), Q = MLP(G) \tag{3}$$

where [·, ·] is the concatenation operation. The *view* operation is used to reshape the concatenated hierarchical maps into a tensor $G \in \mathbb{R}^{l\times k}$, where l denotes the number of the visual encoder layers(here is 12 for ViT [21]). The i-th row feature vector of G represents the mixed visual information of F_i and a portion of F_{i+1}. Then we use an *MLP* layer to reduce the feature dimension of G. For the visual feature $Q \in \mathbb{R}^{l\times(2ld)}$, we use a *split* operation to obtain a list of visual representations $\{V_0, V_1, \ldots, V_{11}\}$, where $V_i \in \mathbb{R}^{l\times(2d)}$ represents the i-th visual feature of ViT.

Previous research [15] has shown that ViT learns simple features in shallow layers, while complex object features are learned in deeper layers. In multimodal representations, text and image, entities and visual objects are aligned in semantic concepts. Therefore, we combine the deep visual features of the global image I representing the target objects with the shallow visual features of visual object O that correspond to entities in the text. Then, to obtain the final visual representation $\tilde{V}_i \in \mathbb{R}^{(m+1)l\times(2d)}$ for the i-th modal fusion layer, we use the following concatenation operation [·, ·]:

$$\tilde{V}_i = \left[V_{i+6}^{I}, V_i^{o_1}, V_i^{o_2}, \ldots, V_i^{o_m}\right]. \tag{4}$$

Visual Prefix-Guided Fusion

Compared to cross-attention-based fusion methods, the visual prefix-based attention [8, 10] method has stronger robustness. The i-th hierarchical visual representation $\tilde{V}_i$ will be regarded as visual prefix to prepend to enhance layer-level textual representations. For the i-th visual representation $\tilde{V}_i$, we use a linear transformation $W_i^{\varphi} \in \mathbb{R}^{d\times 2\times d}$ to project $\tilde{V}_i$ into the same embedding space of textual representation, and obtain the visual prompt $\varphi_k^i, \varphi_v^i \in \mathbb{R}^{l(m+1)\times d}$:

$$\{\varphi_k^i, \varphi_v^i\} = \tilde{V}_i W_i^{\varphi} \tag{5}$$

where m denotes the number of visual objects.

Given input sequence X, the context representation of the (i−1)-th layer denoted as $H^{i-1} \in \mathbb{R}^{(n+1)\times d}$, which is projected into query, key, and value vectors as follows:

$$Q^i = H^{i-1}W_i^q, K^i = H^{i-1}W_i^k, V^i = H^{i-1}W_i^v \tag{6}$$

where W_i^q, W_i^k, $W_i^v \in \mathbb{R}^{d\times d_h}$ are the attention mapping parameters. The hidden features at the i-th layer of the fusion encoder based on visual prefix-based attention are as follows:

$$H^i = Attn\left(Q^i, \left[\varphi_k^i; K^i\right], \left[\varphi_v^i; V^i\right]\right). \tag{7}$$

Via Eq. 8, each layer of the fusion encoder captures the semantic relevance between visual objects and textual entities by using an implicit token-object alignment. Through the above visual prefix-based attention operation $U(\cdot)$, we obtain the final text representation $H^L = U(X, \tilde{V}_l)$.

3.4 Image-Text Matching

We introduce the image-text matching task to reduce the effects of irrelevant visual features. The entire process is described as follows:

Firstly, based on the cosine similarity function in Sect. 3.2, we calculate the similarity between the text and image and construct a similarity matrix $S \in \mathbb{R}^{N\times N}$, where N represents the number of input pairs (V, T) as a positive example in the batch.

Secondly, the similarity scores of the matched image-text pairs (V, T) (where V represents the global image I) in the batch are highest and located on the diagonal of the similarity matrix S. $S_{i,j}$ represents the similarity score between text representation of the i-th pairs and image representation of the j-th in the similarity matrix. We sample hard negative images corresponding to the text by Eq. 1. Specifically, we set the diagonal elements of the similarity matrix S to zero to avoid comparing with positive samples. Then, we extract the index corresponding to the highest image-text similarity score from the remaining (N−1) similarity scores of negative examples in the i-th row. Then, we use the index to find the image $I_{negative}$ and its corresponding m visual objects $O_{negative} = \{o_1, o_2, \ldots, o_m\}$ as the hard negative visual concept for each text. We also sample an in-batch hard negative text $T_{negative}$ by Eq. 2 for each visual concept. Then, to obtain the hard negative samples, we use the following concatenation operation $[\cdot, \cdot]$:

$$T' = \left[T, T_{negative}\right], I' = \left[I, I_{negative}\right], O' = \left[O, O_{negative}\right]. \tag{8}$$

The differences between the hard negative samples (T', I', O') and the positive samples (T, I, O) are not significant. The hard negative samples can help the model better distinguish between positive and negative samples, improving the model's generalization and robustness. This enables the model better captures the semantic correlation between text and image.

Finally, we use the same approach to encode and fuse negative samples. Then, the [CLS] token embedding in the last layer of the modal fusion encoder is used to predict the probability of image-text matching. The visual concept V includes image I and objects O. The image-text matching loss is as follows:

$$\mathcal{L}_{itm} = \mathbb{E}_{(V,T)\sim D} H\left(y^{itm}, p^{itm}(V, T)\right). \tag{9}$$

Let y^{itm} denote the ground-truth 2-dimensional one-hot label, where the probability of image-text matching is 1. Minimizing the loss function $\mathcal{L}_{itm}$, the model will better distinguish between positive and negative samples.

3.5 Classifier

For the MNER and MRE tasks, we input the final hidden representation H^L into different classifier layers.

Named Entity Recognition. We feed the final hidden representation H^L into a CRF layer [1, 7], which predicts the probability of the label sequence $y = \{y_1, \ldots, y_n\}$ as follows:

$$p\left(y|H^L\right) = \frac{\prod_{i=1}^{n} S_i(y_{i-1}, y_i, H^L)}{\sum_{y' \in Y} \prod_{i=1}^{n} S_i(y'_{i-1}, y'_i, H^L)} \tag{10}$$

where Y represents the predefined labels (see Sect. 2). The NER loss $\mathcal{L}_{ner}$ is defined as follows:

$$\mathcal{L}_{ner} = -\sum_{i=1}^{M} \log\left(p\left(y^{(i)} | U\left(X^i, \tilde{V}\right)\right)\right). \tag{11}$$

Relation Extraction. The RE task is to predict the relationship r ∈ Y between head entity E_1 and tail entity E_2 of sequence X (see Sect. 2). The corresponding probability distribution over the class Y is:

$$p(r|X, E_1, E_2) = softmax(W[E_1, E_2] + b) \tag{12}$$

where $[\cdot, \cdot]$ is a concatenation operation, W and b are learnable parameters. The MRE loss $\mathcal{L}_{re}$ is defined as follows:

$$\mathcal{L}_{re} = -\log(p(r|X, E_1, E_2)). \tag{13}$$

Overall Objective Function. The final loss function $\mathcal{L}$ based above loss function is as follows:

$$\mathcal{L} = \alpha\mathcal{L}_{ner/re} + \beta\mathcal{L}_{ita} + \gamma\mathcal{L}_{itm} \tag{14}$$

where α, β, and γ are hyperparameters to balance their contributions to the overall loss.

Table 1. Detailed Statistics of Dataset

Dataset	Train	Dev	Test
Twitter-2015	4000	1000	3257
Twitter-2017	3373	723	723
MNRE	12247	1624	1614

4 Experiments

4.1 Datasets

We conducted experiments on three datasets, including two MNER datasets Ttitter2015 and Twitter2017 [1, 7], and one MRE dataset MNRE [2]. Table 1 shows detailed statistics for the three datasets.

4.2 Implementation Details

We implement AMNet with PyTorch 1.7.0 on a single NVIDIA 3090 GPUs. The experimental details of our model are as follows:

Metrics*:* We used Precision (**P**), Recall (**R**), and F1 score(**F1**) to evaluate the performance of our model on the MNER and MRE tasks, which is similar to the evaluation metrics used in the baseline.

Hyperparameters*:* We use CLIP-ViT-Base/32 and $BERT_{base}$ as encoders. The hidden size of representations is 768. We use AdamW [16] for parameter optimization. The decay is 0.01, the batch size N is 16, and the number of objects m is 3. The learning rate is 3e-5. The hyperparameters α, β and γ are 0.8, 0.1, 0.1 for Twitter-2015 and α, β and γ are 0.9, 0.1, 0.1 for Twitter-2017. In the MNRE dataset, α, β and γ are 0.5, 0.4, 0.1.

4.3 Baselines

Text-Based Baselines: We consider a representative set of text-based models to demonstrate the necessity of introducing images. For NER task, we consider the following models: 1) HBiLSTM-CRF [2]; 2) BERT-CRF. The RE baselines contain 1) PCNN [4]; 2) MTB [5] is a RE-oriented pre-training model based on BERT.

Multimodal Baselines: We consider the following approaches for MNRE and MRE tasks: 1) UMGF [6], which is the first method to use multimodal graph fusion for the MNER task; 2) MAF [7], which introduces contrastive learning and a matching module to control image features dynamically; 3) MEGA [2], which is the previous baseline for the MRE task and uses graph alignment to capture correlations between entities and objects; 4) HVPNeT [8], which uses hierarchical visual features as visual prefix to alleviate error sensitivity of irrelevant object images. 5) MKGformer [9], which uses a correlation-aware fusion module to alleviate the noisy information. 6) IFAformer [10] uses a dual transformer architecture and introduces a visual prefix for modal fusion to reduce sensitivity to errors. We refer to the original results of UMGF from Chen et al. [8].

4.4 Main Results

The experimental results of our model and the baselines on the three datasets are shown in Table 2. From the results, our analysis is as follows: Firstly, we find that almost all multimodal models perform better than text-based models. This suggests that using visual features to enhance text semantics can indeed improve the performance of the models on the NER and RE tasks. Secondly, Comparing HVPNeT and MAF results show bridging the modality gap or reducing modality noise can improve limited performance. Finally, our method achieves the best performance on the MNER and MRE tasks, demonstrating the effectiveness of our model.

4.5 Analysis and Discussion

Ablation Study
To investigate the contribution of each module to the task, we conducted an ablation experiment with the same parameters to confirm their effectiveness. Specifically, we remove the ITM module and the ITA module and both two modules in the model, which are represented as "w/o ITA", "w/o ITM" and "w/o ITA + ITM".

As shown in Table 3, removing the ITA or ITM modules degrades the final performance. Without the ITM and ITA module, w/o ITA + ITM drops severely, meaning two modules play a critical role in the model. The effect of the ablation experiment is more significant on MRE than on the MNER task. We think that this is related to the Twitter-2015 dataset containing too much noisy data and the size of Twitter-2017 dataset is small.

Analysis of Hierarchical Feature
To probe how combining hierarchical visual features from the global and object images at different layers impacts performance on two tasks. We conduct this study by sequentially combining hierarchical visual features from the first 6 and last 6 layers of the global image features and the object image features. As shown in Table 3, combining the first 6 layers of the global image features with the last 6 layers of the object image features results the highest scores. This supports our hypothesis that combining deep target object features with shallow visual object features enhances object representation and improves performance.

The 4-layer visual features from ResNet [19] are coarse-grained in HVPNeT. Using dynamic gates to aggregate hierarchical visual features also brings noise. As shown in Table 4, we replace the Resnet encoder in HVPNeT with ViT to create HVPNeT-ViT model. This significantly improved performance, showing that the hierarchical visual features from ViT match the textual representation of each layer better. Then, we remove the dynamic gates to create HVPNeTno_gate-ViT model. This further improved performance, confirming that dynamic gates introduce noise.

4.6 Case Study

To validate our method, we categorize image-text pairs as related, weakly related, and unrelated based on their relevance. For MNER and MRE tasks, we analyze representative cases and compare the results of our method with HVPNeT in Table 5.

Table 2. Over all MNER and MRE results

Methods	Twitter-2015			Twitter-2017			MNRE		
	P	R	F1	P	R	F1	P	R	F1
HBiLSTM-CRF	70.32	68.05	69.17	82.69	78.16	80.37	-	-	-
BERT-CRF	70.32	68.05	71.81	82.69	83.57	83.44	-	-	-
PCNN	-	-	-	-	-	-	62.85	49.69	55.49
MTB	-	-	-	-	-	-	64.46	57.81	60.86
UMGF	74.49	75.21	74.85	86.54	84.50	85.51	64.38	66.23	65.29
MAF	71.86	75.10	73.42	86.13	86.38	86.25	-	-	-
MEGA	70.35	74.58	72.35	84.03	84.75	84.39	64.51	68.44	66.41
IFAformer	-	-	-	-	-	-	82.59	80.78	81.67
MKGformer	-	-	-	86.98	88.01	87.49	82.67	81.25	81.95
HVPNeT	73.87	**76.82**	75.32	85.84	87.93	86.87	83.64	80.78	81.85
Ours	**75.60**	76.42	**76.00**	**87.40**	**88.30**	**87.56**	**85.24**	**83.91**	**84.57**
w/o ITA	74.86	**76.51**	75.68	87.10	87.93	87.51	83.17	81.87	82.52
w/o ITM	75.15	76.32	75.73	87.38	87.64	87.51	83.92	82.34	83.12
w/o ITA + ITM	74.17	76.48	75.30	85.83	86.97	86.40	82.31	81.41	81.85

Table 3. Model performance of combining different hierarchical visual features

Layer		Twitter-2015			Twitter-2017			MNRE		
3154V^I	V^O	P	R	F1	P	R	F1	P	R	F1
1–6	1–6	74.00	**76.86**	75.40	86.98	87.49	87.23	83.78	82.34	83.06
	7–12	74.41	76.67	75.52	**88.89**	86.45	87.52	84.60	83.28	83.94
7–12	**1–6**(Ours)	**75.60**	76.42	**76.00**	87.11	**88.01**	**87.56**	**85.24**	**83.91**	**84.57**
	7–12	75.35	76.40	75.87	86.72	87.49	87.10	84.08	82.50	83.28

Table 4. Study of HVPNeT for MNER and MRE.

Method	Twitter-2015			Twitter-2017			MNRE		
	P	R	F1	P	R	F1	P	R	F1
HVPNeT	73.87	**76.82**	75.32	85.84	**87.93**	86.87	**83.64**	80.78	81.85
HVPNeT-ViT	75.30	76.25	75.36	87.05	87.05	87.05	83.44	82.66	83.05
HVPNeT$_{\text{no-gate}}$-ViT	75.08	76.04	**75.55**	**87.28**	87.34	**87.31**	83.46	**82.81**	**83.14**

Table 5. Three Representative Cases from MNER and MRE and Their Prediction Results on the HVPNeT method and Our method (AMNet, AMNet w/o ITA and AMNet w/o ITM)

	Relevant Pair	Weak Relevant Pair	Irrelevant Pair
MNER	 Downtown Austin with Lady Bird **[Lake LOC]** in the foreground.	 **[Tebow PER]** and company warming up for tonight's MiracleBaseball game vs stluciemets.	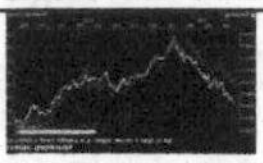 Business: Forget Justin Trudeau, it's oil that's driving this **[loonie O]** rally.
HVPNeT	[Lake PER] ×	[Tebow ORG] ×	[loonie MISC] ×
Ours	[Lake LOC] √	[Tebow PER] √	[loonie O] √
w/o ITA	[Lake LOC] √	[Tebow ORG] ×	[loonie O] √
w/o ITM	[Lake LOC] √	[Tebow PER] √	[loonie O] √
	Relevant Pair	Weak Relevant Pair	Irrelevant Pair
MRE	 Cavs hosted a workout for **Alabama** guard **Collin Sexton** today per @mcten. **/per/org/member_of**	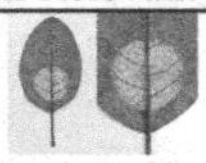 Delicate crocheted/sewn leaf sculptures by **UK** artist **Susanna Bauer** # womensart. **/per/loc/place_of_residence**	 **Montauban les Gilets Jaunes** lancent leur web radio. **/org/loc/locate_at**
HVPNeT	/per/loc/place_of_residence×	/org/org/place_of_birth×	/org/org/subsidiary ×
Ours	/per/org/member_of √	/per/loc/place_of_residence √	/org/loc/locate_at √
w/o ITA	/per/loc/place_of_residence ×	/per/loc/place_of_birth ×	/org/org/subsidiary ×
w/o ITM	/per/loc/place_of_residence ×	/per/loc/place_of_residence √	/org/org/subsidiary ×

For the MNER task, the ITA and ITM modules cannot fully demonstrate the showcase their performance due to dataset limits. Our model could not sufficiently bridge the modality gap and reduce the modality noise. This led to correct predictions without ITA/ITM modules. This also explains the less significant effect of the ablation experiment when presented with weak relevant or irrelevant pairs.

For the MRE task, our model can make correct predictions for irrelevant image-text pairs by reducing modality noise. However, our model cannot bridge the modality gap without ITA, leading to wrong predictions for relevant and weak relevant pairs.

5 Related Work

Multimedia Named Entity Recognition (MNER) and Multimedia Entity Resolution (MRE) are important steps for extracting hidden information from social media posts.

Early researchers [2, 4] studied how to encode the entire image into raw feature vectors and directly incorporate them into text representation for semantic enhancement. These methods cannot establish fine-grained mappings between visual objects and entities. So, Zheng et al. [2] develop a dual graph alignment method to capture the correlations between objects and entities. Xu et al. [17] drop reinforcement learning to train a data discriminator to split the media posts. Yuan et al. [18] propose an edge-enhanced graph alignment network by aligning modes to align objects and entities. Wang et al. [20] propose image-text alignments to align image features into the textual space to utilize textual embeddings better. Chen et al. [9] use a correlation-aware fusion module to alleviate the noisy information. Chen et al. [8] use hierarchical visual features as a visual prefix method to reduce the sensitivity of errors caused by irrelevant objects. However, most previous methods fail to bridge the modality gap and ignore the modality noise. Therefore, in this paper, we propose a method for MNER and MRE tasks that can make the representations between the two modalities more consistent and effectively reduce the impact of irrelevant image pairs and irrelevant objects on the model.

6 Conclusion

In this paper, we propose an alignment and matching Network with Hierarchical Visual Features for MNER and MRE. Specifically, we utilize an ITA module based on contrastive learning to better align the modality representations before fusion. This allows learning single-modality representations to bridge the semantic gap better. We propose an ITM module to make the model better distinguish image-text pairs, reducing the impact of irrelevant images. We combine and concatenate the hierarchical visual features obtained from the global and visual object features from ViT as improved visual prefix for fusion, further reducing modality noise. We demonstrate the effectiveness of our proposed method through detailed ablation experiments and case studies. However, our method also faces some limitations, and it may struggle with datasets that exhibit high noise levels or contain a small number of samples.

Acknowledgements. This work is supported by the National Natural Science Fund of China under Grant No. 62276095 and the National Social Science Fund of China under Grant No. 20&ZD047.

References

1. Moon, S., Neves, L., Carvalho, V.: Multimodal named entity recognition for short social media posts[J]. arXiv preprint arXiv:1802.07862 (2018)
2. Zheng, C., Feng, J., Fu, Z., et al.: Multimodal relation extraction with efficient graph alignment. In: Proceedings of the 29th ACM International Conference on Multimedia, pp. 5298–5306 (2021)
3. Lample, G., Ballesteros, M., Subramanian, S., et al.: Neural architectures for named entity recognition. arXiv preprint arXiv:1603.01360, 2016
4. Distant supervision for relation extraction via piecewise convolutional neural networks
5. Soares, L.B., FitzGerald, N., Ling, J., et al.: Matching the blanks: Distributional similarity for relation learning[J]. arXiv preprint arXiv:1906.03158 (2019)
6. Zhang, D., Wei, S., Li, S., et al.: Multi-modal graph fusion for named entity recognition with targeted visual guidance. In: Proceedings of the AAAI Conference on Artificial Intelligence, vol. 35. no. 16, pp. 14347–14355 (2021)
7. Xu, B., Huang, S., Sha, C., et al.: MAF: a general matching and alignment framework for multimodal named entity recognition. In: Proceedings of the Fifteenth ACM International Conference on Web Search and Data Mining, pp. 1215–1223 (2022)

8. Chen, X., Zhang, N., Li, L., et al.: Good visual guidance makes a better extractor: hierarchical visual prefix for multimodal entity and relation extraction. arXiv preprint arXiv:2205.03521 (2022)
9. Chen, X., Zhang, N., Li, L., et al.: Hybrid transformer with multi-level fusion for multimodal knowledge graph completion. In: Proceedings of the 45th International ACM SIGIR Conference on Research and Development in Information Retrieval, pp. 904–915 (2022)
10. Chen, X., Qiao, S., et al.: On analyzing the role of image for visual-enhanced relation extraction. arXiv preprint arXiv:2211.07504 (2022)
11. Devlin, J., Chang, M.W., Lee, K., et al.: Bert: pre-training of deep bidirectional transformers for language understanding. arXiv preprint arXiv:1810.04805 (2018)
12. Li, J., Selvaraju, R., Gotmare, A., et al.: Align before fuse: vision and language representation learning with momentum distillation. Adv. Neural. Inf. Process. Syst. **34**, 9694–9705 (2021)
13. Dosovitskiy, A., Beyer, L., Kolesnikov, A., et al.: An image is worth 16x16 words: Transformers for image recognition at scale. arXiv preprint arXiv:2010.11929 (2020)
14. Yang, Z., Gong, B., Wang, L., et al.: A fast and accurate one-stage approach to visual grounding. Ion: Proceedings of the IEEE/CVF International Conference on Computer Vision, pp. 4683–493 (2019)
15. Ghiasi, A., Kazemi, H., Borgnia, E., et al.: What do vision transformers learn? a visual exploration. arXiv preprint arXiv:2212.06727 (2022)
16. Hchilov, I., Hutter, F.: Decoupled weight decay regularization. arXiv preprint arXiv:1711.05101 (2017)
17. Xu, B., Huang, S., Du, M., et al.: Different data, different modalities! reinforced data splitting for effective multimodal information extraction from social media posts. In: Proceedings of the 29th International Conference on Computational Linguistics, pp. 1855–1864 (2022)
18. Yuan, L., Cai, Y., Wang, J., et al.: Joint multimodal entity-relation extraction based on edge-enhanced graph alignment network and word-pair relation tagging. arXiv preprint arXiv:2211.15028 (2022)
19. He, K., Zhang, X., Ren, S., et al.: Deep residual learning for image recognition. In: Proceedings of the IEEE Conference on Computer Vision and Pattern Recognition, pp. 770–778 (2016)
20. Wang, X., Gui, M., Jiang, Y., et al.: ITA: image-text alignments for multi-modal named entity recognition. arXiv preprint arXiv:2112.06482 (2021)
21. Li, B., Hu, Y., Nie, X., et al.: DropKey for Vision Transformer. In: Proceedings of the IEEE/CVF Conference on Computer Vision and Pattern Recognition, pp. 22700–22709 (2023)

Multi-view Consistency View Synthesis

Xiaodi Wu[1], Zhiqiang Zhang[1(✉)], Wenxin Yu[1(✉)], Shiyu Chen[1], Yufei Gao[2], Peng Chen[3], and Jun Gong[4]

[1] Southwest University of Science and Technology, Mianyang, China
zzq.zhangzhiqiang2018@gmail.com, yuwenxin@swust.edu.cn
[2] Hosei University, Chiyoda City, Japan
[3] Chengdu Hongchengyun Technology Co., Ltd., Chengdu, China
[4] Southwest Automation Research Institute, San Antonio, USA

Abstract. Novel view synthesis (NVS) aims to synthesize photo-realistic images depicting a scene by utilizing existing source images. The core objective is that the synthesized images are supposed to be as close as possible to the scene content. In recent years, various approaches shift the focus towards the visual effect of images in continuous space or time. While current methods for static scenes treat the rendering of images as isolated processes, neglecting the geometric consistency in static scenes. This usually results in incoherent visual experiences like flicker or artifacts in synthesized image sequences. To address this limitation, we propose Multi-View Consistency View Synthesis (MCVS). MCVS leverages long short-term memory (LSTM) and self-attention mechanism to model the spatial correlation between synthesized images, hence forcing them closer to the ground truth. MCVS not only enhances multi-view consistency but also improves the overall quality of the synthesized images. The proposed method is evaluated on the Tanks and Temples dataset, and the FVS dataset. On average, the Learned Perceptual Image Patch Similarity (LPIPS) is better than state-of-the-art approaches by 0.14 to 0.16%, indicating the superiority of our approach.

Keywords: Novel View Synthesis · Deep Learning · Long Short-Term Memory Mechanism

1 Introduction

Novel view synthesis (NVS) is one of the major research tasks in computer vision and graphics. The goal of NVS is to synthesize images depicting a scene under specified new camera poses, using a set of known images. The core objective of NVS is that the synthesized images must be as realistic as possible. For this objective, many recent methods incorporating deep learning have been able to

This Research is Supported by National Key Research and Development Program from Ministry of Science and Technology of the PRC (No. 2021ZD0110600), Sichuan Science and Technology Program (No. 2022ZYD0116), Sichuan Provincial M. C. Integration Office Program, and IEDA Laboratory of SWUST.

B. Luo et al. (Eds.): ICONIP 2023, CCIS 1966, pp. 311–323, 2024.
https://doi.org/10.1007/978-981-99-8148-9_25

predict high-quality images. In addition to the above, another important objective is the 3D consistency of multiple views, especially in static content. That means the images rendered along a camera track should depict consistent 3D content, avoiding deformations or artifacts that wrongly portray the ground truth. As the field of neural rendering has matured, recent methods that utilize fixed 3D representations [1,7,13,14] can produce relatively reliable views. However, the rendering of a view is only related to the specified camera pose, and the processes of rendering multiple views remained independent of each other. In addition, due to the high dependence on the 2D convolutional networks in the rendering process, although the synthesized views seem sensible, they may still portray the wrong scene content.

Based on the problem above, we propose Multi-view Consistency View Synthesis (MCVS) with the multi-view consistency (MC) module. Different from other NVS methods based on fixed 3D scene representation [1,7,13,14], in the rendering process, the proposed MC module can model the spatial correlation between multiple views. This spatial correlation enables multiple views on a camera track to influence and restrain each other, improving the 3D consistency and hence approaching the real content of the scene, indicating better quantitative results.

The key idea of the proposed MC module is based on the long short-term memory (LSTM) [6] mechanism for sequence modeling. Specifically, we utilize the convolutional LSTM (ConvLSTM) [18] mechanism to transfer the correlations among multiple views, which enables each view to obtain relevance with others. Besides, the MC module takes advantage of the self-attention memory (SAM) [9] mechanism to enhance ConvLSTM's ability to model and convey long-term spatial changes over multiple views. Benefitting from the ConvLSTM mechanism combined with the SAM mechanism, the MC module assigns multi-view consistency to the predicted images while forcing them closer to the ground truth. To sum up, the main contributions of the proposed method are as follows:

- The multi-view consistency (MC) module for novel view synthesis is presented, which successfully models the multi-view spatial correlation of static scenes.
- A new concept to improve the 3D consistency of novel view synthesis is proposed, that is, to optimize the rendering process of the view with long short-term memory and self-attention mechanisms.
- The proposed method is verified on the Tanks and Temples, and the FVS datasets. Experiments show that the quantitative and qualitative results outperform the state-of-the-art methods.

2 Related Work

2.1 Novel View Synthesis

People have been working in the field of New View Synthesis (NVS) since early on, and methods incorporating deep learning have blossomed in recent years. An end-to-end deep learning pipeline based on explicit scene representation is first proposed by Flynn [3] et al. This work designed convolutional neural networks that use plane sweep volume (PSV). In recent work based on fixed 3D representations, Riegler and Koltun [13,14] warp features from selected source images into the target view and blend them using convolutional networks. Solovev [20] et al. represent the scene with a set of front-parallel semitransparent planes and convert them to deformable layers. Suhuai [21] et al. leverage epipolar geometry to extract patches of each reference view, which can predict the color of a target ray directly. Our work is most related to [13,14] but differs from the above, the rendering of multiple views is interrelated.

Methods based on implicit function representations have also achieved significant results in recent years, with Neural radiance fields (NeRF) [10] being a major innovation. NeRF feeds coordinates to a multi-layer perceptron to learn color and body density. After this, the work of Reiser [12] et al. significantly reduced the number of weights of the original NeRF. Garbin [4] et al. compactly cache and subsequently query to calculate the pixel values, which improved the computational speed of NeRF.

2.2 Long Short-Term Memory Mechanism

Long Short-Term Memory (LSTM) [6] is a Recurrent Neural Network (RNN) model for processing time-series data, proposed by German computer scientist Sepp Hochreiter and his colleagues in 1997 to solve the gradient disappearance and gradient explosion problems in standard RNNs, as well as problem of modeling long-term dependencies. LSTM has achieved remarkable results in natural language processing, speech recognition, image recognition, and other fields.

Although LSTM models have achieved remarkable results in sequence modeling, there are still problems in specific application scenarios. Researchers have proposed some improvements and developments to further improve the performance of LSTM models. To enhance the memory capability of LSTM models, researchers have proposed some variants of LSTM models such as convolutional LSTM [18] which is adopted in this paper. Other researchers proposed deeper LSTM models to enhance the expressive power of LSTM models. Wang [22] et al. applied a unified spatiotemporal memory pool based on the ConvLSTM cell. Furthermore, with the advent of the self-attention mechanism, Lin [9] et al. introduced it into LSTM for adaptively selecting and weighing the input information.

3 Method

This section will explain how the multi-view consistency (MC) module optimizes the rendering process in novel view synthesis. Subsection 3.1 gives the specific components of the proposed MC module and corresponding motivations. Subsection 3.2 presents the overall architecture of our method.

3.1 Multi-view Consistency

Convolutional Long Short-Term Mechanism. LSTM [6] is a Recurrent Neural Network (RNN) variant that can model the dependencies of sequence information. The core of LSTM is the cell status C_t which essentially acts as an accumulator of the state information. For multidimensional tensors, because of the extensive spatial information and the strong correlation between each point and its surroundings, it is difficult for the traditional LSTM to portray such spatial features.

Convolutional long short-term memory [18] (ConvLSTM) network is introduced to overcome the shortage of LSTM in processing tensors of multiple dimensions. For each time step t, ConvLSTM replaces part of the join operations in the full-connection LSTM with convolution operations. Denote input to a ConvLSTM cell as X_t, cell outputs as C_t, and hidden state as H_t, the key equations of ConvLSTM are:

$$\begin{aligned}
i_t &= \sigma\left(W_{xi} * X_t + W_{hi} H_{t-1} + W_{ci} \circ C_{t-1} + b_i\right) \\
f_t &= \sigma\left(W_{xf} * X_t + W_{hf} H_{t-1} + W_{cf} \circ C_{t-1} + b_f\right) \\
C_t &= f_t \circ C_{t-1} + i_t \circ \tanh\left(W_{xc} * X_t + W_{hc} * H_{t-1} + b_c\right) \\
o_t &= \sigma\left(W_{xo} * X_t + W_{ho} * H_{t-1} + W_{co} \circ C_t + b_o\right) \\
H_t &= o_t \circ \tanh\left(C_t\right)
\end{aligned} \tag{1}$$

where '$*$' denotes the convolution operator, '$\circ$' denotes the Hadamard product, and σ means the sigmoid activation function.

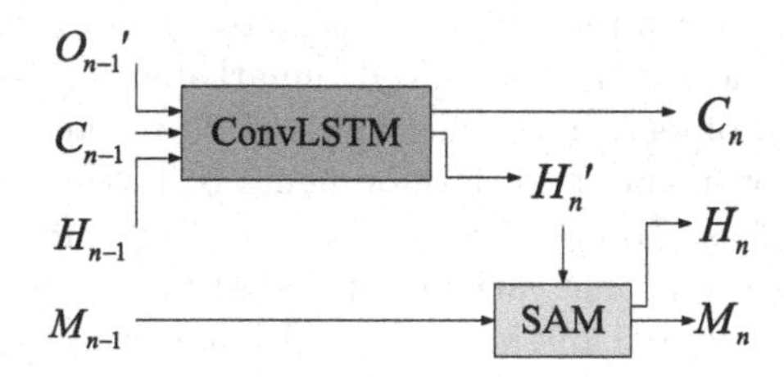

Fig. 1. Work flow between ConvLSTM cell and SAM unit. $O_n{'}$ is the feature map to feed into the module, M_n is the self-attention memory updated by the SAM unit.

Transmission of Spatial Correlations. To model the spatial relevance of multiple views, the basic idea is to use the ConvLSTM [18] mechanism which focuses on handling spatiotemporal correlations. In addition to that, the sequence of views describing a scene can be circumferential. For the ConvLSTM mechanism to hold global spatial information, the self-attention memory [9] (SAM) is adopted here to score the factors of long-term relevance in the spatial and temporal dimensions. Based on the self-attention mechanism, SAM can generate new features by aggregating the features of each position of the input feature itself and memory features, which allows the LSTM mechanism to learn and transmit long-term feature changes. Figure 1 shows how a SAM [9] unit works together with a ConvLSTM [18] cell.

Below is the key equation of similarity score e in the SAM unit, where Q_h^T is a feature space mapped from the hidden state ($H_n{}'$ in Fig. 1), K_m is a feature space mapped from the memory of the last time step (M_{n-1} in Fig. 1), and N equals the multiplication of feature width and height.

$$\mathrm{e} = \mathrm{Q}_h^T \mathrm{K}_m \in \mathbb{R}^{N \times N} \tag{2}$$

This is exactly the main difference between the SAM and the standard self-attention model: the similarity score is not only derived from a single feature map but also counts in the memory information of the last time step.

In the following text, the combination of the ConvLSTM cell and the SAM unit will be referred to as the SA-ConvLSTM cell. So far the calculation of rendering interrelated target features can be expressed by:

$$\begin{aligned} H_n{}', C_n &= \mathrm{ConvLSTM}(O_n{}', H_{n-1}, C_{n-1}) \\ H_n, M_n &= \mathrm{SAM}\left(H_n{}', M_{n-1}\right) \end{aligned} \tag{3}$$

where H_n is the target feature to be rendered to calculate the loss with ground truth.

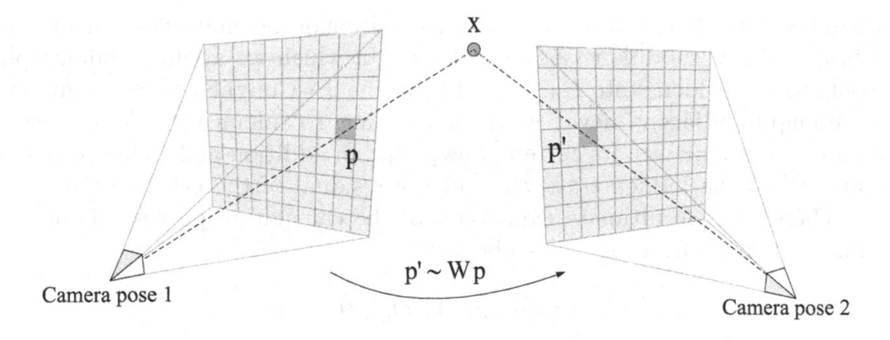

Fig. 2. Homography warping. '∼' denotes the projective equality, and 'x' is a surface point in 3D space.

Reflection of Spatial Correlations. The model mentioned above is theoretically sufficient to establish the correlation between multiple views but is limited to the two-dimensional level. In NVS, the changes in camera poses are three-dimensional. To address the above, we use homography warping (illustrated in Fig. 2) to reflect the 3D correlations between multiple views. The homograph is used to describe the position mapping relationship between objects in the world coordinate system and pixel coordinate system. It is commonly adopted in image correction, image stitching, camera pose estimation, and other fields.

For a certain pixel p in a view, we map its pixel position p' ~ Wp to another view using the homograph matrix W. Denote camera intrinsics as K, rotation matrix as R, and translation vector as t. The affine transformation matrix W (here is the homograph matrix) of two views (indexed with $n-1$ and n) can be calculated as:

$$\mathrm{W} = \mathrm{K}_{n-1} \cdot [\mathrm{R}_{n-1}, \mathrm{t}_{n-1}] \cdot [\mathrm{R}_n, \mathrm{t}_n]^{-1} \cdot \mathrm{K}_n^{-1} \tag{4}$$

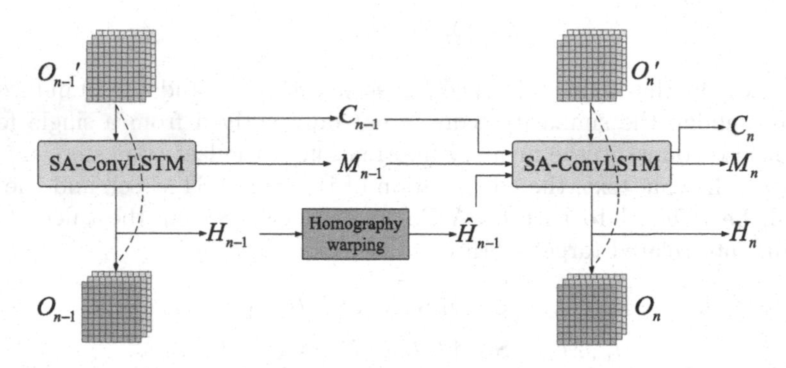

Fig. 3. Flow of multi-view consistency module. O_n is the output of the MC module, combining the cell hidden state H_{n-1} and the residual of O_n'.

Summary. Given the transmission and acquisition of the multi-view correlation, the proposed MC module is shown in Fig. 3. In which we applied homography warping to the hidden state H_n derived from the SA-ConvLSTM cell to map its feature map to adjacent views. At the same time, to enhance the dominance of the input feature map in its current viewpoint, an addition and Relu operations are applied to the hidden state H_n and the residual of the cell input (namely O_n'). Therefore, the ultimate expression of the calculation process of the MC module is changed from Eq. (1) to below:

$$\begin{aligned} H_n', C_n &= \mathrm{ConvLSTM}(O_n', \hat{H}_{n-1}, C_{n-1}) \\ H_n, M_n &= \mathrm{SAM}\left(H_n', M_{n-1}\right) \\ \hat{H}_{n-1} &= \mathrm{warping}(H_{n-1}, \mathrm{W}) \\ O_n &= \mathrm{Re}\,lu(H_n + O_n') \end{aligned} \tag{5}$$

where O_n is the output of the MC module, also the feature map to be further processed into the final target image.

3.2 Overall Architecture

In this paper, the proposed MC module is deployed based on the pipeline of Stable View Synthesis (SVS) [14], an NVS method that leverages geometric representation and deep learning approaches. See Fig. 4, the input is a sequence of source images $\{I_n\}_{n=1}^{N}$. The goal is to predict the target images O given a sequence of target camera poses: rotation matrix with translation vector (R_t, t_t) and camera intrinsics K_t.

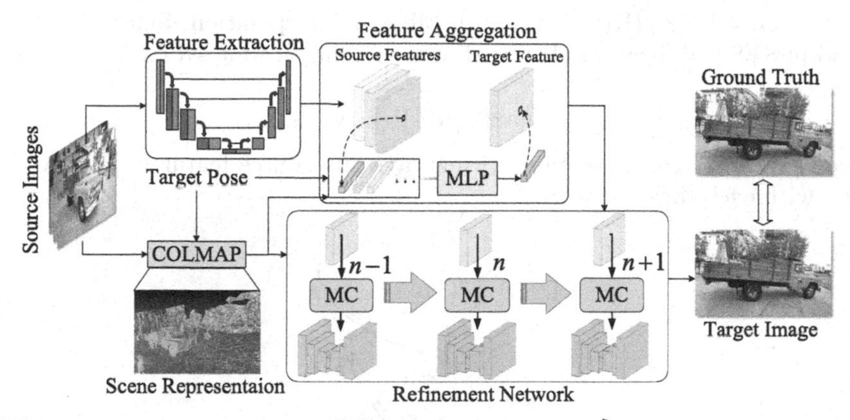

Fig. 4. Overview of Multi-view Consistency View Synthesis. (a) Extract feature maps from source images and construct the geometric scene using structure-from-motion, multi-view stereo, and surface reconstruction. (b) Based on the geometric information, the associated feature vectors for each target pixel are aggregated and assembled into a target feature. (c) Render the final novel view with the multi-view consistency module in the refinement network.

Pre-process. There are two pre-processes on the source images. I. Use an ImageNet-pre-trained ResNet18 [5] to encode each I_n to obtain the feature tensor F_n. Noting this network as ϕ_{enc}, we obtain the encoded feature tensor by $F_n = \phi_{enc}(I_n)$. II. Use COLMAP [16,17] to compute the point cloud and the camera poses of each source image, including the rotation matrix $\{R_n\}_{n=1}^{N}$, the camera intrinsic $\{K_n\}_{n=1}^{N}$ and the translation vectors $\{t_n\}_{n=1}^{N}$. The final scene geometric Γ is the surface mesh derived from the point cloud through Delaunay-based 3D surface reconstruction.

Feature Selection and Aggregation. After preprocessing, the source feature vectors for each pixel in the target view O are selected and aggregated based on certain rules.

As illustrated in Fig. 5, for a certain point $x \in \Gamma \subset \mathbb{R}^3$, we select only the visible source feature vectors $f_k(x)$ for aggregation. Given a newly specified camera K_t, R_t and t_t, the center of each pixel in O combined with $x \in \Gamma \subset \mathbb{R}^3$ is backprojected to obtain the set of 3D points corresponding to each pixel $\{x_{w,h}\}_{w,h=1,1}^{W \times H}$, from which the source feature vectors $\{f_k(x)\}_{k=1}^{K}$ visible for each x are derived. At this point, a multi-layer perceptron (MLP) is applied to obtain each target feature vector corresponding to $\{f_k(x)\}_{k=1}^{K}$ and to the associated view rays $\{v_k\}_{k=1}^{k}$, and u. Denote a target feature vector as g, the network used for aggregation as ϕ_{aggr}, g is computed by

$$g(v, u) = \phi_{aggr}(u, \{(v_k, f_k(x))\}_{k=1}^{K}), \tag{6}$$

where $f_k(x) = F_k(K_k(R_k x + t_t))$ with bilinear interpolation. Since accuracy is proved in SVS [14], here an MLP with max pooling is adopted for feature aggregation:

$$\phi_{aggr} = \max_{k=1}^{K} \mathrm{MLP}([u, v_k, f_k(x)]), \tag{7}$$

where $[\cdot]$ means a concatenation of view rays and source features that are associated with each target feature.

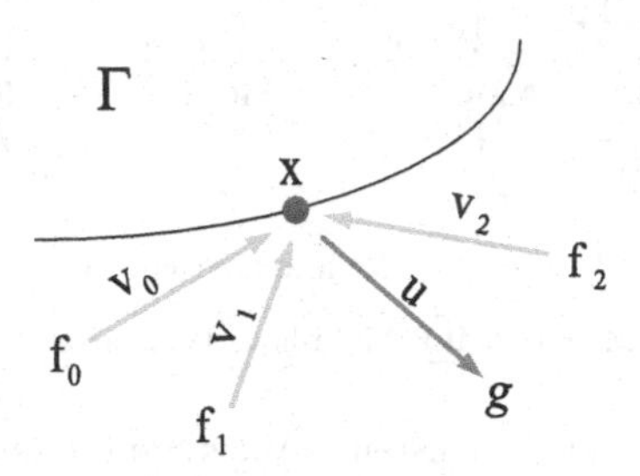

Fig. 5. Feature Selection and Aggregation. With associated (the 3D point $x \in \Gamma \subset \mathbb{R}^3$ is visible to which) feature vectors f_k, source view rays v_k and target view ray u, the target feature vector for each x is able to be computed.

Refinement with MC Module. Note that the source features are selected and aggregated on a pixel-by-pixel basis, to obtain the final photo-realistic view, the feature vectors $\{g_{w,h}\}_{w,h=1,1}^{W\times H}$ obtained by ϕ_{aggr} are assembled into a tensor $O' = \{g_{w,h}\}_{w,h=1,1}^{W\times H}$, which is the input of the refinement stage.

Based on the observation of the performance in experiments, we use the sequence consisting of nine U-Nets [15] and combined it with the MC module to build ϕ_{ref} in an end-to-end way (shown in Fig. 4), where each U-Net learns the residual of its input to mitigate the degradation in deep networks. Denote the MC module enhanced refinement network as ϕ_{ref}, the final output of the proposed method is $O = \phi_{ref}(O')$.

Loss Function. Inspired by Chen and Koltun [2], the basic idea of the loss function is to match activations in a visual perception network, applying to the synthesized image and separately to the reference image. Therefore, the loss to minimize is:

$$L(O, I_n) = \|O - I_n\|_1 + \sum_l \lambda_l \|\phi_l(O) - \phi_l(I_n)\|_1 \tag{8}$$

where ϕ is a pre-trained VGG-19 [19] network, the hyperparameter λ_l is to adjust the weight of each layer l. For layers ϕ_l $(l \geqslant 1)$ we actually adopt 'conv1_2', 'conv2_2','conv3_2', 'conv4_2', and 'conv5_2'. This design allows the evaluation to match lower-layer and higher-layer activations in the perception network ϕ and guides the synthesis network to learn fine details and global layouts.

4 Experiments

We applied PyTorch [11] to build the network. The three networks: ϕ_{enc} for feature extraction, ϕ_{aggr} for feature aggregation, and ϕ_{ref} for refinement are trained using Adam optimizer setting $\beta_1 = 0.9$, $\beta_2 = 0.9999$, learning rate $= 10^{-4}$. We use the FVS dataset [13] and the Tanks and Temples [8] dataset for training (images for evaluation are not included). The proposed model is trained on a single V100 GPU with a batch size of 1. For the fairness of the comparison, all results listed are derived after 600,000 iterations of the entire network without fine-tuning. For the Tanks and Temples dataset, we use LPIPS [25], PSNR, and SSIM [23] as accuracy metrics, and LPIPS and SSIM for the FVS dataset. The following subsections provide a comparison with SOTA NVS methods, as well as the ablation study.

Table 1. Compare with State-to-art on Tanks and Temples dataset

Methods	Accuracy per Scene											
	Truck			M60			Playground			Train		
	↓*LPIPS%*	↑*PSNR*	↑*SSIM*	↓*LPIPS%*	↑*PSNR*	↑*SSIM*	↓*LPIPS%*	↑*PSNR*	↑*SSIM*	↓*LPIPS%*	↑*PSNR*	↑*SSIM*
NeRF [10]	50.74	20.85	0.738	60.89	16.86	0.701	52.19	21.55	0.759	64.64	16.64	0.627
NeRF++ [24]	30.04	22.77	0.814	43.06	18.49	0.747	38.70	22.93	0.806	47.75	17.77	0.681
FVS [13]	13.06	22.93	0.873	30.70	16.83	0.783	19.47	22.28	0.846	24.74	18.09	0.773
SVS [14]	12.41	23.07	0.893	23.70	19.41	0.827	17.38	23.59	0.876	19.42	18.42	0.809
NMBC [7]	16.84	**24.03**	0.888	**23.15**	19.54	0.815	22.72	23.59	0.87	24.17	17.78	0.799
Ours	**11.99**	23.17	**0.894**	23.64	**19.67**	**0.831**	**17.29**	**23.66**	**0.881**	**19.35**	**18.54**	**0.81**

Numbers in bold are the best. On average, our method outperforms the SOTA on LPIPS by 0.16%, and PSNR by 0.03 dB.

Table 2. Compare with State-to-art on FVS dataset

Methods	Accuracy per Scene									
	Bike		Flowers		Pirate		Digger		Sandbox	
	↓*LPIPS%*	↑*SSIM*	↓*LPIPS%*	↑*SSIM*	↓*LPIPS%*	↑*SSIM*	↓*LPIPS%*	↑*SSIM*	↓*LPIPS%*	↑*SSIM*
NeRF++ [24]	27.01	0.715	30.3	0.816	41.56	0.712	34.69	0.657	23.08	0.852
FVS [13]	27.83	0.592	26.07	0.778	35.89	0.685	23.27	0.668	30.2	0.77
SVS [14]	21.1	0.752	20.66	**0.848**	29.21	0.76	**17.3**	0.779	21.55	0.852
Ours	**20.88**	**0.763**	**20.61**	0.845	**28.8**	**0.787**	17.32	**0.788**	**21.49**	**0.862**

Numbers in bold are the best. On average, our method outperforms the SOTA on LPIPS by 0.14%, and SSIM by 0.01.

Quantitative Results. The quantitative results on the Tanks and Temples dataset [8] and FVS dataset [13] are listed in Table 1 and Table 2 respectively. The results of SVS [14] are derived from the provided pre-trained model. Results of other SOTA methods are from Riegler et al. [13,14] and Jena et al. [7] using published models or network parameters. On average, the proposed method is 0.14 to 0.16% better than SOTA on LPIPS, and 0.03 dB higher on PSNR. While improving above, our results still hold the performance on SSIM. It is hence proved that with the MC module, our method utilizes the multi-view consistency to force synthesized images more realistic.

Fig. 6. Qualitative results on the scene Horse of the Tanks and Temples dataset.

Qualitative Results. Figure 6 displays samples of our qualitative results. To demonstrate the multi-view consistency, the main contribution of our method, two adjacent views (the actual synthesized views are much more) of one scene are listed and marked with different camera poses. The image details prove that the MC module reduced the blurring and artifacts of the synthesized images and forces them closer to the ground truth. Meanwhile, the content of different views (of different camera poses) has become more consistent. Benefiting from the MC module, the two aspects above complement each other and bring better qualitative results.

Table 3. Average Accuracy of Ablation study

Datasets Structures	T&T			FVS	
	↓*LPIPS*%	↑*PSNR*	↑*SSIM*	↑*SSIM*	↓*LPIPS*%
w SAM, w/o Residual	18.17	21.13	**0.851**	0.806	21.95
w/o SAM, w/o Residual	18.21	21.21	0.845	0.794	22.05
Full Method	**18.06**	**23.23**	**0.851**	**0.809**	**21.82**

Numbers in bold are the best.

Ablation Study. In Table 3 we performed ablation experiments on the proposed MC module. The first line indicates that the experiment uses an MC module without residual connections between the input and output. Likewise, the second row indicates that the experiment uses an MC module without a SAM [9] unit and skip connections, which is equivalent to using only the ConvLSTM [18] mechanism. On average, rendering an image without the SAM unit takes around 1 s, and 1.2 s for the full method on the same device.

Recall the original purpose of including the SAM [9] mechanism, to aid the ConvLSTM [18] mechanism model global spatial correlations over multiple views. The results in the second row proved that the proposed method does benefit from the SAM mechanism, bringing in 3D consistency and delivering more realistic images.

5 Conclusion

This paper proposes Multi-view Consistent View Synthesis (MCVS), which combines long short-term memory mechanisms and the self-attention memory mechanism to model the spatial correlation between synthetic images and make the synthetic images closer to the ground truth. It provides a new idea to improve the 3D consistency of images in the field of new view synthesis. Experiments proved that our method outperforms the state-of-the-art view synthesis method. On average, the LPIPS is 0.14 to 0.16% higher than SOTA and the PSNR is 0.03 dB higher. Qualitative results showed our method significantly reduces the artifacts and blurring in synthesized images.

References

1. Aliev, K.-A., Sevastopolsky, A., Kolos, M., Ulyanov, D., Lempitsky, V.: Neural point-based graphics. In: Vedaldi, A., Bischof, H., Brox, T., Frahm, J.-M. (eds.) ECCV 2020, Part XXII. LNCS, vol. 12367, pp. 696–712. Springer, Cham (2020). https://doi.org/10.1007/978-3-030-58542-6_42
2. Chen, Q., Koltun, V.: Photographic image synthesis with cascaded refinement networks. In: Proceedings of the IEEE International Conference on Computer Vision, pp. 1511–1520 (2017)
3. Flynn, J., Neulander, I., Philbin, J., Snavely, N.: Deepstereo: learning to predict new views from the world's imagery. In: Proceedings of the IEEE Conference on Computer Vision and Pattern Recognition, pp. 5515–5524 (2016)
4. Garbin, S.J., Kowalski, M., Johnson, M., Shotton, J., Valentin, J.: FastNeRF: high-fidelity neural rendering at 200FPS. In: Proceedings of the IEEE/CVF International Conference on Computer Vision, pp. 14346–14355 (2021)
5. He, K., Zhang, X., Ren, S., Sun, J.: Deep residual learning for image recognition. In: Proceedings of the IEEE Conference on Computer Vision and Pattern Recognition, pp. 770–778 (2016)
6. Hochreiter, S., Schmidhuber, J.: Long short-term memory. Neural Comput. **9**(8), 1735–1780 (1997)

7. Jena, S., Multon, F., Boukhayma, A.: Neural mesh-based graphics. In: Karlinsky, L., Michaeli, T., Nishino, K. (eds.) ECCV 2022, Part III. LNCS, vol. 13803, pp. 739–757. Springer, Cham (2023). https://doi.org/10.1007/978-3-031-25066-8_45
8. Knapitsch, A., Park, J., Zhou, Q.Y., Koltun, V.: Tanks and temples: benchmarking large-scale scene reconstruction. ACM Trans. Graph. (ToG) **36**(4), 1–13 (2017)
9. Lin, Z., Li, M., Zheng, Z., Cheng, Y., Yuan, C.: Self-attention ConvLSTM for spatiotemporal prediction. In: Proceedings of the AAAI Conference on Artificial Intelligence, vol. 34, pp. 11531–11538 (2020)
10. Mildenhall, B., Srinivasan, P.P., Tancik, M., Barron, J.T., Ramamoorthi, R., Ng, R.: NeRF: representing scenes as neural radiance fields for view synthesis. Commun. ACM **65**(1), 99–106 (2021)
11. Paszke, A., et al.: PyTorch: an imperative style, high-performance deep learning library. In: Advances in Neural Information Processing Systems, vol. 32 (2019)
12. Reiser, C., Peng, S., Liao, Y., Geiger, A.: KiloNeRF: speeding up neural radiance fields with thousands of tiny MLPS. In: Proceedings of the IEEE/CVF International Conference on Computer Vision, pp. 14335–14345 (2021)
13. Riegler, G., Koltun, V.: Free view synthesis. In: Vedaldi, A., Bischof, H., Brox, T., Frahm, J.-M. (eds.) ECCV 2020, Part XIX. LNCS, vol. 12364, pp. 623–640. Springer, Cham (2020). https://doi.org/10.1007/978-3-030-58529-7_37
14. Riegler, G., Koltun, V.: Stable view synthesis. In: Proceedings of the IEEE/CVF Conference on Computer Vision and Pattern Recognition (CVPR), pp. 12216–12225 (2021)
15. Ronneberger, O., Fischer, P., Brox, T.: U-net: convolutional networks for biomedical image segmentation. In: Navab, N., Hornegger, J., Wells, W.M., Frangi, A.F. (eds.) MICCAI 2015, Part III. LNCS, vol. 9351, pp. 234–241. Springer, Cham (2015). https://doi.org/10.1007/978-3-319-24574-4_28
16. Schonberger, J.L., Frahm, J.M.: Structure-from-motion revisited. In: Proceedings of the IEEE Conference on Computer Vision and Pattern Recognition, pp. 4104–4113 (2016)
17. Schönberger, J.L., Zheng, E., Frahm, J.-M., Pollefeys, M.: Pixelwise view selection for unstructured multi-view stereo. In: Leibe, B., Matas, J., Sebe, N., Welling, M. (eds.) ECCV 2016, Part III. LNCS, vol. 9907, pp. 501–518. Springer, Cham (2016). https://doi.org/10.1007/978-3-319-46487-9_31
18. Shi, X., Chen, Z., Wang, H., Yeung, D.Y., Wong, W.K., Woo, W.C.: Convolutional LSTM network: a machine learning approach for precipitation nowcasting. In: Advances in Neural Information Processing Systems, vol. 28 (2015)
19. Simonyan, K., Zisserman, A.: Very deep convolutional networks for large-scale image recognition. arXiv preprint arXiv:1409.1556 (2014)
20. Solovev, P., Khakhulin, T., Korzhenkov, D.: Self-improving multiplane-to-layer images for novel view synthesis. In: Proceedings of the IEEE/CVF Winter Conference on Applications of Computer Vision, pp. 4309–4318 (2023)
21. Suhail, M., Esteves, C., Sigal, L., Makadia, A.: Generalizable patch-based neural rendering. In: Avidan, S., Brostow, G., Cissé, M., Farinella, G.M., Hassner, T. (eds.) ECCV 2022. LNCS, vol. 13692, pp. 156–174. Springer, Cham (2022). https://doi.org/10.1007/978-3-031-19824-3_10
22. Wang, Y., Long, M., Wang, J., Gao, Z., Yu, P.S.: PredRNN: recurrent neural networks for predictive learning using spatiotemporal LSTMs. In: Advances in Neural Information Processing Systems, vol. 30 (2017)
23. Wang, Z., Bovik, A.C., Sheikh, H.R., Simoncelli, E.P.: Image quality assessment: from error visibility to structural similarity. IEEE Trans. Image Process. **13**(4), 600–612 (2004)

24. Zhang, K., Riegler, G., Snavely, N., Koltun, V.: NeRF++: analyzing and improving neural radiance fields. arXiv preprint arXiv:2010.07492 (2020)
25. Zhang, R., Isola, P., Efros, A.A., Shechtman, E., Wang, O.: The unreasonable effectiveness of deep features as a perceptual metric. In: Proceedings of the IEEE Conference on Computer Vision and Pattern Recognition, pp. 586–595 (2018)

A Reinforcement Learning-Based Controller Designed for Intersection Signal Suffering from Information Attack

Longsheng Ye[1], Kai Gao[1,2(✉)], Shuo Huang[1], Hao Huang[1], and Ronghua Du[1,2]

[1] College of Automotive and Mechanical Engineering, Changsha University of Science and Technology, Changsha 410114, China
kai_g@csust.edu.cn

[2] Hunan Key Laboratory of Smart Roadway and Cooperative Vehicle-Infrastructure Systems, Changsha University of Science and Technology, Changsha 410114, China

Abstract. With the rapid development of smart technology and wireless communication technology, Intelligent Transportation System (ITS) is considered as an effective way to solve the traffic congestion problem. ITS is able to collect real-time road vehicle information through sensors such as networked vehicles (CV) and cameras, and through real-time interaction of information, signals can more intelligently implement adaptive signal adjustment, which can effectively reduce vehicle delays and traffic congestion. However, this connectivity also poses new challenges in terms of being affected by malicious attacks that affect traffic safety and efficiency. Reinforcement learning is considered as the future trend of control algorithms for intelligent transportation systems. In this paper, we design reinforcement learning intelligent control algorithms to control the intersection signal imposed by malicious attacks. The results show that the reinforcement learning-based signal control model can reduce vehicle delay and queue length by 22% and 23% relative to timing control. Meanwhile, the intensity learning is a model-free control method, which makes it impossible for attackers to target flaws in specific control logic and evaluate the impact of information attacks more effectively. Designing a coordinated state tampering attack between different lanes, the results show that the impact is greatest when the attacked states are in the same phase.

Keywords: intelligent transportation system · reinforcement learning · information attack

1 Introduction

With the high-speed development of society and the increasing number of cars, traffic congestion has increasingly become an important factor limiting urban development. And an excellent signal control system is of great significance for improving traffic efficiency [1]. A key point of traffic signal optimization is traffic status sensing, and in the

This work was supported in part by the Natural Science Foundation of Hunan under Grant 2021JJ40575, in part by The Open Fund of Hunan Key Laboratory of Smart Roadway and Cooperative Vehicle-Infrastructure Systems (Changsha University of Science & Technology) under grant kfj190701.

B. Luo et al. (Eds.): ICONIP 2023, CCIS 1966, pp. 324–337, 2024.
https://doi.org/10.1007/978-981-99-8148-9_26

past, intersections were used to obtain real-time traffic status through induction coils and cameras, but these devices have the disadvantages of high installation and maintenance costs and short service life [2, 3]. Thanks to the development of vehicle networking and Vehicle-to-Everything (V2X) communication, a Connected Vehicle (CV) driving on the road can act as a powerful motion detector to provide high-quality and inexpensive traffic status input to signal intersections [4–6]. With the further development of communication and artificial intelligence technologies, Intelligent Transportation System (ITS), which combines the concepts of vehicle networking and intelligent control, is gradually becoming the future direction of transportation systems, making it possible to improve traffic efficiency at urban intersections on a large scale and in all directions [7, 8].

Combined with a data-driven control approach, the efficiency of traffic junction passage is effectively improved. Reinforcement learning (RL) is an unsupervised machine learning algorithm. Unlike supervised learning approaches that require pre-labeling, reinforcement learning models learn entirely through interaction with the environment until the best policy is learned to guide the action, which is a very advanced adaptive control method [9]. Data-driven reinforcement learning control enables intersection signal control to be fully integrated with the real-time characteristics of traffic flow, which is important for improving traffic flow efficiency [10]. Traditional reinforcement learning is suitable for simple models such as segmented constant tables and linear regression, leading to limited scalability or optimality in practice. Deep Neural Network (DNN), which has the ability to learn complex tasks [11, 12], combined with reinforcement learning solves this problem and brings a new direction to intersection adaptive control. To improve the stability of the algorithm, empirical replay and target network mechanisms are used. Simulation results show that the algorithm reduces vehicle delays by 47% and 86%, respectively, compared to two other popular traffic signal control algorithms, the longest queue priority algorithm and the fixed time control algorithm [12]. Liang et al. use a rasterized state input representation to quantify the traffic flow information by dividing the intersection into occupancy raster maps. Multiple states such as vehicle position and speed are fused and different green light times are set as action spaces. A convolutional network is used to approximate the Q-network and mechanisms such as empirical replay and target networks are employed to stabilize the Q-network [13].

This paper focuses on designing a reinforcement learning based control model for intersection signal controller suffering from information attack. There are two main reasons, firstly to evaluate how much improvement the reinforcement learning model can bring to the efficiency of the intersection communication. Secondly due to the model-free nature of reinforcement learning, the evaluation of the impact of being attacked is more convincing, this is because the attacker cannot set a specific targeting strategy based on the flaws of the model.

The main contributions of this paper are summarized as follows:

1. A reinforcement learning-based signal control model is designed that uses a deep neural network to approximate the maximum cumulative return and an empirical replay mechanism to stabilize the network.

2. Performing a signaling attack on the reinforcement learning control model. The interference of the information attacks is removed by using the model-free property of reinforcement learning.
3. The state of the controller is tampered with to explore the best attack strategy.

2 Problem Description

2.1 Application Scenarios

To solve traffic control problems, intelligent transportation systems (ITS) have been created by integrating advanced technologies such as edge, cloud computing, V2X, and IoT into traffic monitoring and management [14]. V2X technology connects vehicle and road infrastructure, making adaptive real-time signal control at intersections possible.

On Board Unit (OBU) is the main application of vehicle side, each CV has an OBU, which mainly realizes the collection of vehicle status information. Road Side Unit (RSU) is the main application product on the road side, deployed at each intersection, mainly detecting the status information of itself or the surrounding environment, such as ETC and other non-contact sensing devices, and receiving and feeding back communication data through microwave. The controller mainly implements specifically adapted traffic control algorithms. For example, at a queuing lane, the OBU transmits CV position, speed, heading and acceleration information to the RSU in the form of Basic Safety Messages (BSMs), and consecutive BSMs represent the vehicle trajectory. At the same time, the RSU and the controller broadcast the signal timing plan to the traffic monitoring center for area-wide coordination [15].

In the ITS environment, the crossed vehicle states can be obtained in real time by the track information transmitted by the networked vehicles or by the cameras. The state volume should try to use real-time information from all vehicles as much as possible, and in general, using more information to describe the state can help the network to determine the value of the state more precisely. One type of state description is raster occupancy [16, 17]. The working principle is shown in Fig. 1:

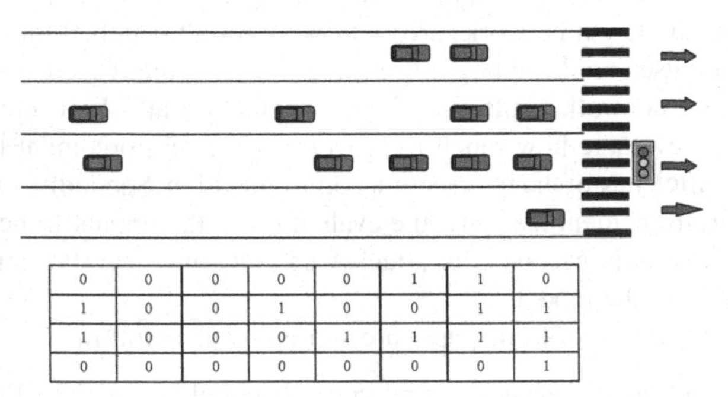

Fig. 1. Environment perception of raster occupancy, for each lane approaching an intersection in the diagram, divide it into multiple cells. If a cell is occupied by one car, that cell will return 1, indicating that there is a car in that cell, otherwise it will return 0.

This state description method provides a large amount of information to the controller, almost similar to image data input. Therefore, this method is expected to be perfectly integrated with the image recognition vehicle detector.

In an intelligent transportation system, it is assumed that an attacker is able to tamper with the status information obtained in the RSU. There are two main ways to do this attack, firstly, the attacker is able to access the internal network of V2X and the attacker tampers with the state input in the RSU. Secondly, the status can also falsify the CV trajectory at the end of the queue to create the illusion that a vehicle is parked at the end of the queue.

A state table attack is defined as a direct modification of the state table so that all lanes of an intersection are filled with networked cars, where the state of all lanes is tampered with, and the map of the occupied grid will change from 0 to 1. The state table attack is a direct attack that must invade the controller's interior, and this attack is more difficult to implement. There are few quantitative evaluations of information attacks, and the specific impact such an attack would have on the traffic control system is something that needs to be evaluated, such as how the attacker would attack to maximize overall traffic congestion.

3 RL Signal Controller for Intersection Suffering from Information Attack

3.1 Reinforcement Learning Model Controller

Intersection Scene Description. In this paper, a typical urban road intersection controlled by traffic signals is considered.

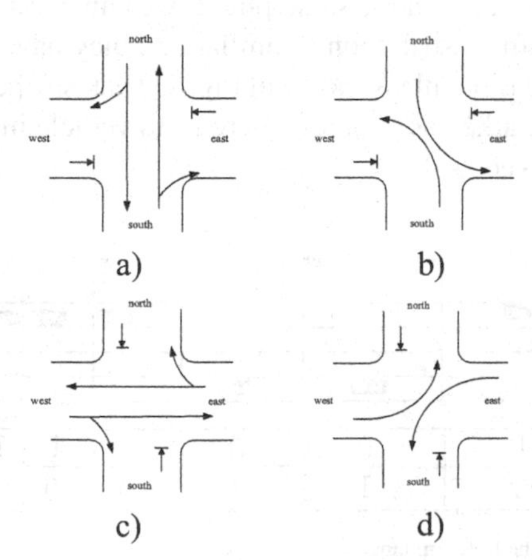

Fig. 2. Cross-port phase setting. There are four traffic signal phases available, and after executing a certain phase, the corresponding priority time interval is assigned to a set of non-conflicting vehicle movements.

As shown in Fig. 2, there are four traffic signal phases available, and after executing a certain phase, the corresponding priority time interval is assigned to a set of non-conflicting vehicle movements. The lanes available for each of these phases are marked in green. This control process can be formulated as a reinforcement learning (RL) problem. The environment is a Markov process and reinforcement learning model learn by interacting with the environment. In this case, our environment is an isolated signal intersection.

Single Intersection Reinforcement Learning Model. In RL, there are three important elements (S, A, R) in the training process, where is the state space, $(s \in S)$, A is the action space, $(a \in A)$, R is the reward function. In the case of traffic signal control, we define these elements as follows:

State: The input data is a description of the current state. In general, using more information to describe the state can help the network to determine the value of the state more precisely. One of the state descriptions uses raster occupancy, which is described in Sect. 2.2.

For the traditional grid division, it is usually divided by equal distance, but the intersection is usually several hundred meters, and the equal distance division will lead to too large feature state set, which affects the calculation. This subsection makes some improvements to this state input, taking into account the arrival distribution of vehicles in the traffic flow, and divides the grid by distance. Specifically, the grid is divided in front of the frequently formed queue according to the length of the vehicle, and the length of the grid increases as the distance increases. The intervals are [0, 7], [7, 14, 21], [21, 28], [28, 40], [40, 60], [60, 100], [100, 160], [160, 400], [400, 750], which ensure that the focus is on the vehicle information in the queue, but also to ensure that we do not lose the information of the vehicles that are moving at high speed.

Similarly, for lanes that are in the same phase, we can combine them into the same state table since their traffic distribution is similar and they have the same travel route.

The detailed working principle is shown in Fig. 3. That is, whenever there is a vehicle in one lane, this grid is marked as 1, unless there is no vehicle in the straight lane of the same grid, then it is marked as 0.

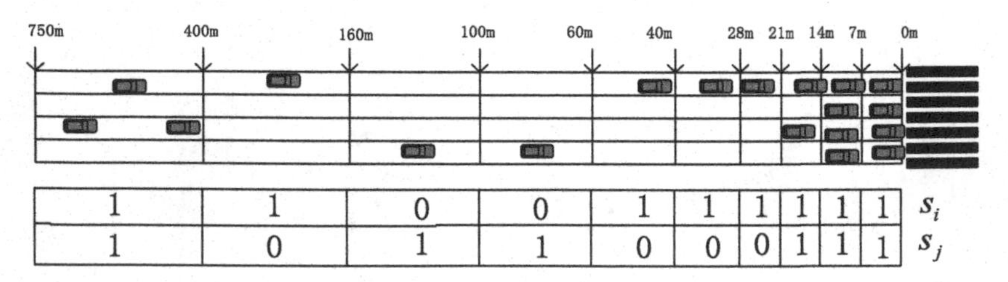

Fig. 3. Intersection state awareness. That is, whenever there is a vehicle in one lane, this grid is marked as 1, unless there is no vehicle in the straight lane of the same grid, then it is marked as 0.

For each entrance lane, the state can be defined as s_i, s_j is an array of 10 elements recording whether a vehicle occupies or not, and i represents its lane number. Then the set of states of each lane constitutes the whole set of states S.

$$S = \{s_1, \cdots, s_i, \cdots, s_8\} \ (i = 1, 2, \cdots, 8) \tag{1}$$

Action: The action here means selecting a signal phase. The possible signal phase configuration settings are shown in Fig. 3. The smart body picks up one of these four phases every second. The set of actions is as follows:

$$A = \{1, 2, 3, 4\} \tag{2}$$

Because the green light possesses a time interval limit, if the intelligence changes the current phase. Conversely, if the intelligent body keeps the current phase, it can choose the same signal phase action.

Reward: the reward is a guide that allows an intelligent body to perform an action in the right direction. In this case, we choose the difference between the cumulative vehicle wait times of two adjacent actions as the reward. The vehicle waiting time is defined as the time spent waiting in the vehicles in the import lane since the vehicles emerged from the environment.

$$W_t = \sum\nolimits_{i=1}^{16} w_t^i \tag{3}$$

W_t indicates the delay time for vehicles in all lanes, and w_t^i denotes the cumulative waiting time for all vehicles in the lane i at the time t.

Since the intelligence is engaged in a continuous decision process, for each state action, an immediate reward is generated, called R_t. Here we use the difference between the cumulative vehicle wait time W_{t-1} for the previous action a_t. And the cumulative vehicle wait time W_t for the current action a_t as R_t. The following is shown:

$$R_t = W_{t-1} - W_t \tag{4}$$

It means that R_t will encourage the intelligence to make the current action towards a smaller cumulative vehicle wait time than past actions, so as to ensure a larger reward and thus reduce the overall cumulative vehicle wait time.

Q-Network Design. Unlike model-based control methods that require full information about the environment, RL intelligences can learn optimal strategies for traffic signal control by interacting with the traffic environment. We use Q-networks in RL approach. It's a well-designed neural network is used as a Q-function approximator.

The goal of the Q-network is to train traffic signals to adapt to the phase and phase duration of intersections based on their real-time traffic patterns. This can be achieved by selecting an action in each training step ($a_t \in A$) that maximizes the expectation of future cumulative rewards.

$$Q^*(s_t, a_t) = \max E[R_t + \gamma R_{t+1} + \gamma^2 R_{t+2} + \ldots |\pi] \tag{5}$$

where $\gamma \in (0,1)$ is a discount factor that represents the trade-off between future and immediate payoffs. The current policy π is defined as the probability of taking action a_t

in state s_t. According to dynamic programming theory, the optimal Q-function in Eq. (5) can be rewritten in the form of Bellman's equation, as shown in the following equation:

$$Q^*(s_t, a_t) = E[R_t + \gamma \max_{a'} Q^*(s_{t+1}, a')] \tag{6}$$

The traditional Q-learning algorithm solves the Bellman equation in an iterative manner, which relies on the concept of Q-table, which is a discrete table of Q-values. However, in our traffic signal control formulation, traditional Q-learning suffers from the curse of dimensionality because the state-behavior space becomes huge due to the existence of the definition of the state space[18].

Therefore, in this paper, we adopt the recent prominent idea of Q-network, which is to use neural network to approximate the Q-function:

$$Q(s, a : \theta) \approx Q^*(s, a) \tag{7}$$

where θ denotes the parameter of the neural network.

Our Q-network for traffic signal control is shown in Fig. 4. At the beginning of each training step, state information is collected as the input layer of the neural network, and then feedforward is performed to estimate the Q-value at the output layer of the neural network.

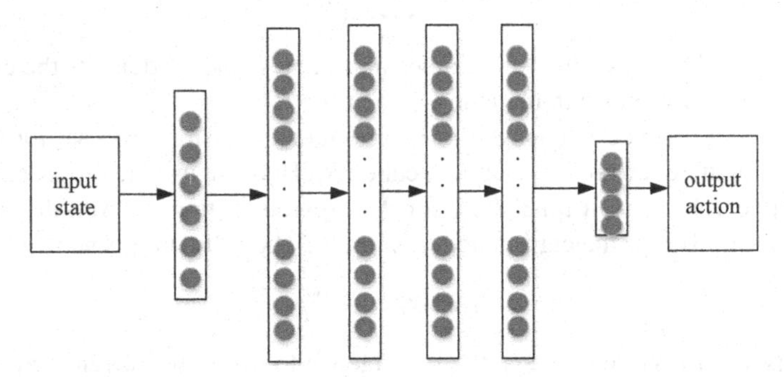

Fig. 4. Q network structure diagram. It contains state inputs, six layers of neurons and action outputs.

To fit the function $Q^*(s, a)$, the deep neural network shown in Fig. 4 is used, and the network structure consists of six layers of neurons, with the first layer receiving the state S of the intersection, containing 80 neurons, and the middle four layers with 400 neurons. The last layer is the output layer and possesses four neurons that fit the action value function for each of the four phase actions. The middle neuron uses relu as the activation function, which has a good performance in fitting the nonlinear function. A linear function is used as the activation function for the final output. The maximum cumulative return is approximated effectively.

Algorithm Process. In the reinforcement learning model designed in this paper, first the model observes the state of the environment, and then the Q-network is used to fit the action value function and output the decided action, the environment generates a new state at the same time, and so on and so forth. Until the best strategy is learned, which eventually makes the vehicle traffic flow at the traffic junction more efficient and achieves the purpose of reducing the delay.

In signal control at intersections, since the traffic flow state and phase settings are directly related, then there is a strong correlation between the samples (s, a, r, s') generated by the reinforcement learning model, which must be disrupted and redistributed in order to prevent overfitting of the deep neural network. To improve the stability of Q-network training, the mechanism of experience replay is used in our reinforcement learning. Specifically, experience replay is the storage of historical samples (s_t, a_t, r_t, s_{t+1}) in a memory m that records historical samples. When the Q-network is trained, a certain number of historical samples are randomly selected as training data in each training process.

The exact process is shown in Algorithm 1:

Algorithm 1: Reinforcement Learning Signal Control Algorithm

1.Initialize Q network parameters and set parameter θ
2.Initialization
3.For epoch=1 to N do
  4.Initialize intersection state , action
  5.For 1 to T do
    6.Select actions a based on strategy π
    7.Execute the action a , observe the reward r_t and the next state s_{t+1}
    8.If (memory) $m>M$
      9.Delete the oldest data in memory
    10.End if
    11.Store (s_t, a_t, r_t, s_{t+1}) in M
    12.Update θ
    13.From s_t to s_{t+1}
  14.End for
15.End for

3.2 Signaling Attack Model Based on State Table

Assume that an attacker can directly manipulate the controller's state table input $S = \{s_1, \cdots s_i, \cdots s_8\}$. By directly modifying the state vector S, make the controller incorrectly believe that there are cars in the free lane.

Modify a phase lane state to s_a, so that each grid of this lane becomes 1:

$$s_a = \{1, 1, 1, 1, 1, 1, 1, 1, 1, 1\} \tag{8}$$

The complete state is generally collected by multiple RSUs, and we assume that the attacker has limited ability to attack and can only interfere with the state perception of certain lanes. To explore the impact of state attacks, a state table attack is designed to attack different lanes.

Attacking a single lane of traffic:

$$s_i = s_a \tag{9}$$

Attacking two lanes of traffic:

$$\begin{cases} s_i = s_a \\ s_j = s_a \end{cases} (\{i, j\} \in \{1, 2, 3, 4, 5, 6, 7, 8\} \& i \neq j) \tag{10}$$

When i and j are in different relative positions, there are at least three different scenarios, i.e., (1) lane i and lane j are lanes in different phases in the same entrance direction, (2) lane i and lane j are lanes in different entrance directions and in different phases, and (3) lane i and lane j are lanes in the same phase.

The lane state is modified to measure the extent to which the controller is affected by the signal attack and to confirm whether this causes other congestion. Revealing the degree of impact of a information attacks can be an indication of the necessity of defense against potential information attacks.

4 Simulation Result

4.1 Simulation Settings

The simulation environment uses SUMO, a microscopic traffic simulation platform, to model intersections, and TraCI, a secondary development interface, using SUMO software and python language for secondary development, and TensorFlow framework for modeling and training of Q-networks. A joint SUMO-TensorFlow simulation platform was built to test and evaluate the signal control effects of the reinforcement learning model.

The intersection road network is built using SUMO's own road network editor, netedit, to create a two-way with 8-lane intersection, where the leftmost lane is a left-turn lane. The middle two lanes are straight lanes, the rightmost lane is a straight and right-turn lane, and the road length is set to 750 m.

Traffic Distribution: For each simulation instance, the initial state is a traffic network with no vehicles, and then vehicles are randomly inserted at the destination and the corresponding path. Each simulation lasts for 1.5 h (5400 s). As we seek to propose a method that can be generalized to any case, the traffic distribution is Poisson to simulate the general case. That is, the traffic flow is 4500 veh/h. The left-turn and right-turn lanes are allocated 25% of the traffic flow.

Signal Timing Scheme: Since this scheme is a timing scheme without fixed phase sequence, the phase settings are the same as in Sect. 3.2, which is worth noting. After performing the control of the reinforcement learning model, the adjustment of the phase is completely determined by the reward, and it is up to the intelligence to decide the action to be taken for the next phase.

4.2 Reinforcement Learning Q-Network Training

First, we examine the simulation data, and the results show that our algorithm does learn good action strategies that are effective in reducing vehicle dwell time, thereby reducing vehicle delays and traffic congestion, and that our algorithm is stable in terms of control decisions. The controller does not oscillate between good and bad action strategies, or even favor bad action strategies.

Since we set the reward as the difference of vehicle waiting time between two adjacent actions, it is impossible to guarantee that the next action will always have less cumulative vehicle waiting time than the previous action since the traffic flow arrives randomly. In order to better evaluate the training process, only rewards with negative reward values are collected and summed in this paper, and the change of cumulative negative rewards reflects the whole process of training. As shown in Fig. 5a):

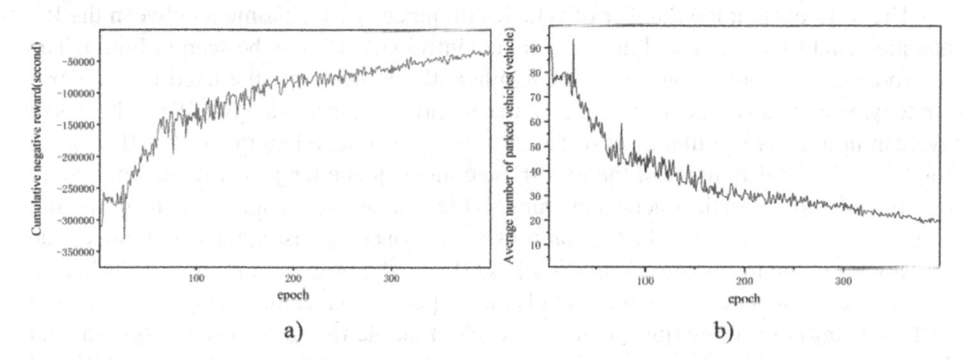

Fig. 5. a): Cumulative negative reward. b): Average number of parked vehicles.

As the number of times the model is trained increases, the cumulative negative reward value of the vehicles rapidly decreases the reward reaches convergence at around 370 turns. This indicates that the model does learn good action strategies from training. The same trend can be seen in Fig. 5b), where the average number of parked vehicles also remains stable at smaller values after 370 rounds, with a significant decrease in the number of vehicles in the queue. This indicates that the algorithm in this paper converges on a good action strategy and algorithmic stabilization mechanism, and that the empirical replay mechanism used by the model is effective.

4.3 Reinforcement Learning Control Model Results Analysis

The RL model does not need to follow a specific phase cycle, but runs acyclically. Each action is chosen to maximize the reward. In contrast to fixed-time control, which does not exhibit this goal-directed behavior and instead performs a relatively poor primitive logic. To evaluate the performance of our proposed reinforcement learning model, a fixed-time control strategy is used for comparison. The fixed-time control involves pre-calculating the signal phase duration based on the proportion of vehicle arrivals and fixing it for the duration of the run.

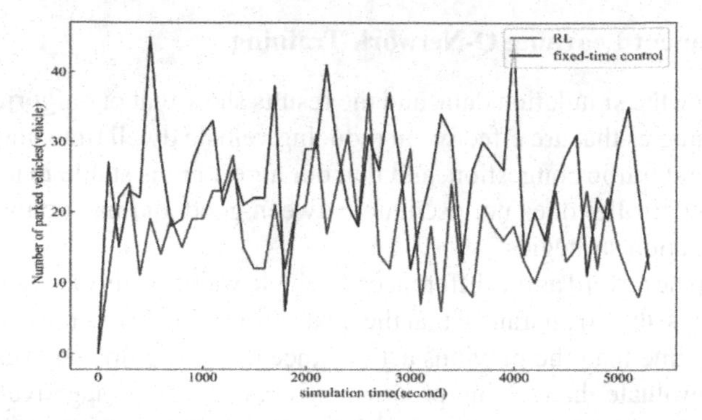

Fig. 6. Comparison of the average number of stops of the two controllers.

The difference in the number of vehicles in queue at each moment between the RL adaptive control strategy and the fixed-time control strategy can be seen in Fig. 6. The reinforcement learning control model significantly outperforms the fixed-time control strategy. Specifically, due to the effect of the adaptive control strategy of RL. The naked eye can intuitively see that most of the instantaneous queue lengths of the RL control model are lower than those of the fixed-time control queue lengths, with the maximum number of vehicles in the queue remaining stable at about 40 vehicles and the minimum at about 5 vehicles when in the RL control strategy. In comparison, the maximum queue of vehicles in the fixed-time control reaches 60, and the minimum is about 10 vehicles.

In order to better evaluate the overall control performance, the average queue length and total parking waiting time are used as evaluation metrics, as shown in Figs. 7a) and Figs. 7b), the number of queuing vehicles and total parking time are reduced by 23% and 22%, respectively, which proves the effectiveness of our reinforcement learning control model.

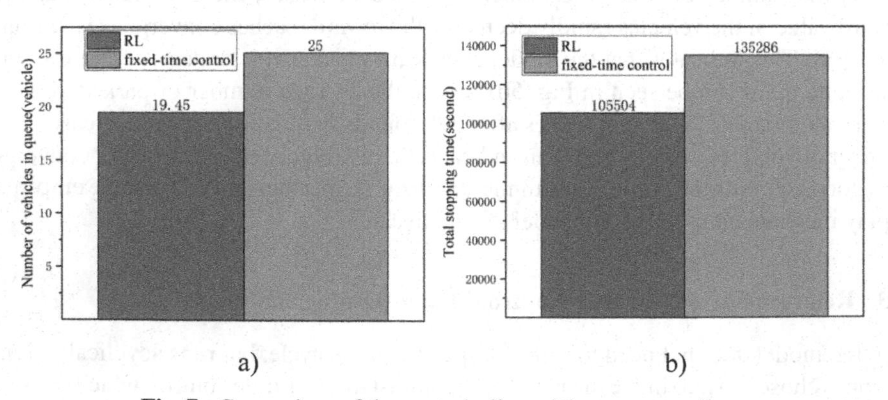

Fig. 7. Comparison of the control effect of the two controllers.

4.4 Analysis of Signal Attack Results Based on State Table

On the basis of the established reinforcement learning model, the state is modified with corresponding attacks according to different scenarios. Similar to the previous section, we also used the average number of vehicles in queue and the total parking time as evaluation metrics. The specific results are shown in Table 1.

From the data in Table 1, it can be seen that, unlike what is imagined, the total average queue length improves by 7% if only one lane is attacked for state awareness, indicating that the attack (Type1) does not have a significant impact on the controller. However, when attacking the perception of two lanes and both lanes are in the same phase (Type2), the average queue length and total stopping time are significantly increased by 253%, which shows the significant interference of signal attack on the intersection controller. The results of attacking two lanes in different phases in the same entrance direction (Type3) and two lanes in different phases in different entrance directions (Type4) are improved by 60% and 19.6%, respectively. Both scenarios are in different phases, but there is also a big difference, which is because the same direction contains left-turn lanes, and the RL model assigns fewer actions to the left-turn lanes in order to release more vehicles to pass, when the left-turn lane falsifies the attack, making a large difference in the results.

Table 1. The degree of influence of the controller by the signal attack.

Attack Type	Total stopping time	Percentage increase in stopping time	The number of vehicles in queue	Percentage increase in vehicles in queue
No attack	105004	–	19.45	–
Type1	112237	7%	20.8	7%
Type2	371291	253%	68.75	253%
Type3	168843	60%	31.2672	60%
Type5	125639	19.6%	23.3	19.6%

The above results show that the RL model is robust and resistant to interference, and can still make suboptimal decisions even when a portion of state perception in the same phase is compromised by an attack. However, when the attacker attacks the overall perception of this phase, the impact on the system is particularly large, which confirms the great impact of network attacks on the signal control system. Fortunately, such direct attacks are easily identified by cyber security systems, but strengthening signal control system cyber security is also an urgent issue.

5 Conclusion

V2X, 5G and other communication technologies that connect vehicles, roads and infrastructure have undoubtedly brought new solutions to traffic signal control [19]. But at the same time, this extensive connectivity makes traffic signal systems very vulnerable to information attacks [20]. The first step in how to protect against such attacks is to analyze how and to what extent such attacks affect signal controllers. In this paper, we evaluate the impact of the attack more objectively by building a model-free controller with reinforcement learning. We also argue that reinforcement learning not only has excellent control performance, but also has some robustness, which coincides with the literature that mentions reinforcement learning as the next generation of signal controllers.

For intersection signal controllers, the impact is huge if they are attacked. Conversely, if only some of the lane states in the same phase are attacked, the impact is limited and within an acceptable range, and a smart attacker will surely take advantage of this flaw [21, 22].

In the future it will be necessary to develop a series of defensive measures that will not only work on the level of network security to prevent malicious attackers from penetrating the internal network. It is also important to focus on the way attackers use personal vehicles to send fake tracks and establish reasonable screening methods. Finally, there is also a need for some degree of verification using data from other sensors, and a certain amount of signal redundancy is necessary.

References

1. Wu, M., Xiong, N.N., Tan, L.: An intelligent adaptive algorithm for environment parameter estimation in smart cities. IEEE Access **6**, 23325–23337 (2018)
2. Xing, Y., Lv, C., Wang, H., et al.: Driver lane change intention inference for intelligent vehicles: framework, survey, and challenges. IEEE Trans. Veh. Technol. **68**(5), 4377–4390 (2019)
3. Florin, R., Olariu, S.: A survey of vehicular communications for traffic signal optimization. Veh. Commun. **2**(2), 70–79 (2015)
4. Li, W., Xu, H., Li, H., et al.: Complexity and algorithms for superposed data uploading problem in networks with smart devices. IEEE Internet Things J. **7**(7), 5882–5891 (2019)
5. Lee, J., Park, B.: Development and evaluation of a cooperative vehicle intersection control algorithm under the connected vehicles environment. IEEE Trans. Intell. Transp. Syst. **13**(1), 81–90 (2012)
6. Du, R., Qiu, G., Gao, K., et al.: Abnormal road surface recognition based on smartphone acceleration sensor. Sensors **20**(2), 451 (2020)
7. Lu, N., Cheng, N., Zhang, N., et al.: Connected vehicles: solutions and challenges. IEEE Internet Things J. **1**(4), 289–299 (2014)
8. Chen, Z., Luo, Z., Duan, X., et al.: Terminal handover in software-defined WLANs. EURASIP J. Wirel. Commun. Netw. **2020**(1), 1–13 (2020)
9. Zheng, G., Xiong, Y., Zang, X., et al.: Learning phase competition for traffic signal control. In: Proceedings of the 28th ACM International Conference on Information and Knowledge Management, pp. 1963–1972 (2019)
10. Zhao, T., Wang, P., Li, S.: Traffic signal control with deep reinforcement learning. In: 2019 International Conference on Intelligent Computing, Automation and Systems (ICICAS), pp. 763–767. IEEE (2019)
11. Chu, T., Wang, J., Codecà, L., et al.: Multi-agent deep reinforcement learning for large-scale traffic signal control. IEEE Trans. Intell. Transp. Syst. **21**(3), 1086–1095 (2019)
12. Gao, J., Shen, Y., Liu, J., et al.: Adaptive traffic signal control: Deep reinforcement learning algorithm with experience replay and target network. arXiv preprint arXiv:170502755 (2017)

13. Liang, X., Du, X., Wang, G., et al.: A deep reinforcement learning network for traffic light cycle control. IEEE Trans. Veh. Technol. **68**(2), 1243–1253 (2019)
14. Sumalee, A., Ho, H.W.: Smarter and more connected: future intelligent transportation system. IATSS Res. **42**(2), 67–71 (2018)
15. Alonso, B., Ibeas, Á., Musolino, G., et al.: Effects of traffic control regulation on network macroscopic fundamental diagram: a statistical analysis of real data. Transp. Res. Part A: Policy Pract. **126**, 136–151 (2019)
16. Marcianò, F.A., Musolino, G., Vitetta, A.: Signal setting optimization on urban road transport networks: the case of emergency evacuation. Saf. Sci. **72**, 209–220 (2015)
17. Filocamo, B., Ruiz, J.A., Sotelo, M.A.: Efficient management of road intersections for automated vehicles—the FRFP system applied to the various types of intersections and roundabouts. Appl. Sci. **10**(1), 316 (2019)
18. Lin, Y., Wang, P., Ma, M.: Intelligent transportation system (ITS): concept, challenge and opportunity. In: 2017 IEEE 3rd International Conference on Big Data Security on Cloud (bigdatasecurity), IEEE International Conference on High Performance and Smart Computing (hpsc), and IEEE International Conference on Intelligent Data and Security (IDS), pp. 167–172. IEEE (2017)
19. Liu, W., Zhou, Q., Li, W., et al.: Path optimization of carpool vehicles for commuting trips under major public health events. J. Changsha Univ. Sci. Tech. **1**, 112–120 (2023)
20. Zhang, Z.Y., Liu, Z., Jiang, L.: Follow-along model considering lateral spacing in vehicle networking environment. J. Changsha Univ. Sci. Tech. **3**, 62–68 (2021)
21. Balint, K., Tamas, T., Tamas, B.: Deep reinforcement learning based approach for traffic signal control. Transp. Res. Procedia **62**, 278–285 (2022)
22. Liu, W., Wang, Z., Liu, X., et al.: A survey of deep neural network architectures and their applications. Neurocomputing **234**, 11–26 (2017)

Dual-Enhancement Model of Entity Pronouns and Evidence Sentence for Document-Level Relation Extraction

Yurui Zhang, Boda Feng, Hui Gao, Peng Zhang(✉), Wenmin Deng, and Jing Zhang

Tianjin University, Tianjin 300000, China
{rachelz6164,pzhang}@tju.edu.cn

Abstract. Document-level relation extraction (DocRE) aims to identify all relations between entities in different sentences within a document. Most works are committed to achieving more accurate relation prediction by optimizing model structure. However, the usage of entity pronoun information and extracting evidence sentences are limited by incomplete manual annotation data. In this paper, we propose a Dual-enhancement model of entity pronouns and EvideNce senTences (DeepENT), which efficiently leverages pronoun information and effectively extracts evidence sentences to improve DocRE. First, we design an Entity Pronouns Enhancement Module, which achieves co-reference resolution and automatic data fusion to enhance the completeness of entity information. Then, we define two types of evidence sentences and design heuristic rules to extract them, used in obtaining sentence-aware context embedding. In this way, we can logically utilize complete and accurate evidence sentence information. Experimental results reveal that our approach performs excellently on the Re-DocRED benchmark, especially in predicting inter-sentence expression relations.

Keywords: Document-level relation extraction · Co-reference resolution · Evidence extraction

1 Introduction

Relation extraction (RE) is one of the most fundamental tasks in Natural Language Processing. It aims to identify semantic relations between entities from text. Many previous works focused on sentence-level relation extraction [17,20,21], which is to predict the relations between entities within a sentence. However, in reality, many relations need to be expressed by multiple sentences (document), and the document-level relation extraction (DocRE) task is gradually attracting the attention of researchers [1,13,22].

Different from Sentence-level RE, DocRE has longer text length, more entities with different mentions, and more relations expressed by multiple sentences.

B. Luo et al. (Eds.): ICONIP 2023, CCIS 1966, pp. 338–349, 2024.
https://doi.org/10.1007/978-981-99-8148-9_27

Document : Nisei (The X-Files)
[1] Nisei is the ninth episode of the third season of the American science fiction television series The X - Files . [2] It premiered on the Fox network on . [3] It was directed by David Nutter , and written by Chris Carter , Frank Spotnitz and Howard Gordon . ... [8] The show centers on FBI special agents Fox Mulder (David Duchovny) and Dana Scully (Gillian Anderson) who work on cases linked to the paranormal , called X - Files . ...

Example 1 :
Subject : Nisei **Object** : Chris Carter **Relation** : Screenwriter **Evidence** : 1,3

Example 2 :
Subject : Chris Carter **Object** : Fox Mulder **Relation** : Creator **Evidence** : 1,3,8

Fig. 1. A sample document and two example relation triples in Re-DocRED. Different colors mark different entities in the document. Mentions of the same entity are highlighted in the same color. Underline marks unlabeled pronouns in the dataset.

Therefore, the DocRE requires the model to have the ability to capture information from long texts, which means that the model needs to pay more attention to sentences relevant to relation prediction and less to irrelevant contexts.

Current works focus on the research to enhance the model's ability in semantic understanding. Some works [9,16,19] based on graph neural networks capture complex interactions between multiple elements, such as mentions and entities, by constructing graphs to obtain relevant information; Some Transformer-based models [12,13,22] utilize the rich contextual information brought by pre-trained language models to learn relevant features.

However, these methods neglect two important feature information. Firstly, at the entity-level, pronouns are ignored. As shown in Example 1 of Fig. 1, humans can easily obtain the relations between "Nisei" and "Chris Carter" through sentence 3, since we know that the pronoun "it" in sentence 3 is a mention of entity "Nisei". Current models neglect the presence of pronoun mentions, resulting in incomplete learning information about entities. Secondly, at the sentence-level, evidence sentences have not been fully explored. In Example 2 of Fig. 1, we can only base on sentence 1, sentence 3 and sentence 8 to predict the relations between "Chris Carter" and "Fox Mulder" while ignoring other sentences. The existing works learning evidence sentence information through neural networks are limited by the incomplete annotation of the dataset.

In order to solve the problem, we propose a Dual-enhancement model of entity pronouns and EvideNce senTences (DeepENT), which efficiently leverages pronouns information and effectively extracts evidence sentences to improve DocRE. On the one hand, we utilize entity pronouns to enhance the completeness of entity-level information and propose an *Entity Pronouns Enhancement Module* based on co-reference resolution. In this module, we use the pre-trained large model to annotate documents and design a set of algorithms to achieve automatic data fusion. The pre-trained model can avoid co-reference resolution errors propagating downwards, while data fusion algorithms can avoid manual annotation; On the other hand, we propose a *Sentence-aware Context Embed-*

ding , where define two types of evidence sentences and design heuristic rules to focus on the relation-related sentence information. This way avoids the tedious process of manually annotating evidence sentences and conforms to actual logic to achieve relationship inference. Based on the attention mechanism, we obtain two different levels of feature embeddings: document-aware and sentence-aware. Finally, we propose an adaptive fusion module to filter and fuse the features.

We conduct experiments on Re-DocRED [10] and results show that DeepENT achieves excellent performance. Notably, the improvement of DeepENT is most significant in inter-sentence entity pairs, suggesting that the leverage of evidence sentences in DeepENT is effective in reasoning across sentences.

In short, the contributions of this work are: (1) Designing entity pronouns enhancement method to improve the completeness of entity-level information. (2) Designing evidence sentence extraction method to improve the accuracy of sentence-level information. (3) Proposing DeepENT based on the above and proving its validity for DocRE on the Re-DocRED dataset.

2 Related Work

Transformer-Based DocRE. Modeling DocRE with a Transformer-based system is a more popular and effective method . These works model relations by implicitly capturing the long-distance token dependencies by PLM encoder. Zhou *et al.* [22] use a self-attention module to extract context information related to the entity pair. Zhang *et al.* [18] view DocRE as a semantic segmentation task and utilizes a U-net to capture the dependencies between entities. Xu *et al.* [14] incorporate dependencies between entities within the self-attention mechanism and encoding stage to guide relation prediction. Jiang *et al.* [7] propose a key instance classifier to identify key mention pairs. However, these models ignore the pronouns referring to entities in the document, leading to incomplete learning of entity information and biased learning of relation information. In contrast, we provide a process of co-reference resolution, focusing on the pronouns' presence to make the entities' information more complete.

Evidence Extraction in DocRE. Huang *et al.* [6] first point out that more than 95% of triples require no more than three sentences as supporting evidence. Afterward, some studies began to focus on the extraction and utilization of evidence sentences. Huang *et al.* [5] propose predicting evidence sentences to guide DocRE. Xie *et al.* [13] jointly train a RE model with an evidence extraction model. Xiao *et al.* [12] design an evidence retrieval module to guide prediction. These works train the evidence sentence extraction with relation extraction and rely too much on the annotations of evidence sentences in the dataset. We perform evidence sentence extraction by heuristic rules which will not be limited by the human annotations.

3 Methodology

The process of our method is shown in Fig. 2, and we will discuss the details.

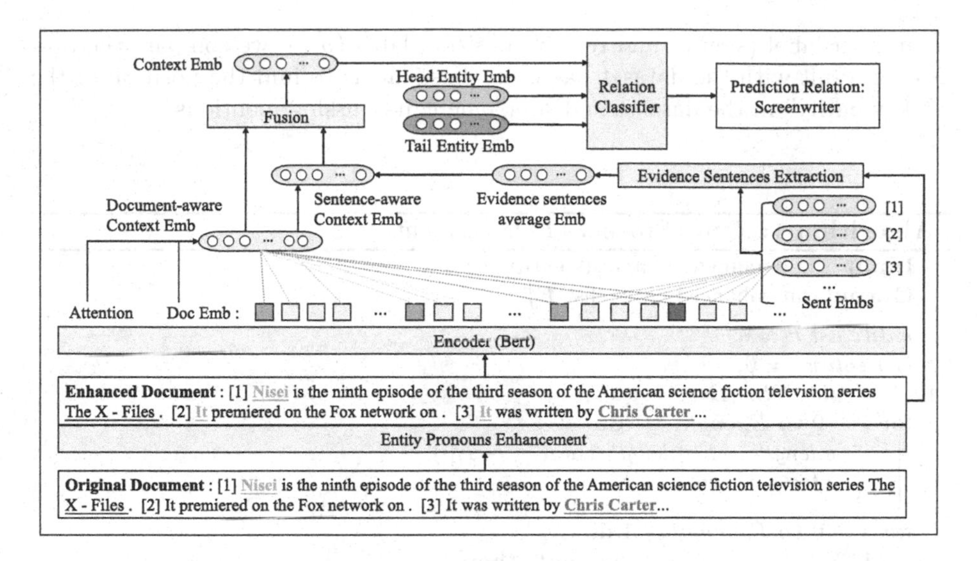

Fig. 2. Diagram of DeepENT framework.

3.1 Problem Formulation

Given a document D containing N sentences $\{s_n\}_{n=1}^{N}$ with L tokens $\{w_l\}_{l=1}^{L}$ and E entities $\{e_i\}_{i=1}^{E}$. Each entity $\{e_i\}$ may be expressed multiple times as different mentions $\{m_j^i\}_{j=1}$ in the document. DocRE aims to predict all possible relations between each entity pair (e_h, e_t) from a pre-defined relation set $R \cup \{NA\}$,where e_h and e_t refer to the head entity and tail entity respectively. A relation r exists between the entity pair (e_h, e_t) if it is expressed by any pair of their mentions, and we define the corresponding evidence sentences $K_{h,t} = \{s_k\}_{k=1}^{K}$ as the subset of sentences in the document which are essential and necessary to infer the relations.

3.2 Entity Pronouns Enhancement Module

In order to ensure the accuracy and completeness of co-reference resolution, avoiding errors propagating downwards caused by insufficient training data, we apply the AllenNLP [3] to achieve co-reference resolution. In practice, the result of co-reference resolution has a different format from the dataset but contains some duplicate information. We design an algorithm to detect and fill in incomplete mentions in the dataset automatically instead of human annotating:

- **Step1. Choosing:** Since a document has only one topic, a document usually has a crucial entity that is mentioned most often. We select the mention list with the most extended length from the co-reference resolution result as the "key entity" which will be compared with the mention list of the corresponding entity in the dataset.
- **Step2. Matching and Filling:** Since the mention lists from both the process result and dataset contain the mention's position information: sentence

number label ($sent_{id}$) and relative position label (pos), we compare the process result with the dataset based on these labels to find the position of the "key entity" in the dataset and supplement its missing mentions.

Algorithm 1: Entity Pronouns Enhancement

Input : a document D and its entity list E_d
Output: an enhanced entity list E_d^+

```
1  entity list E = ∅;
2  key entity e = ∅;
3  E ← AllenNLP(D);
4  for i ← 0 to E.length − 1 do
5      if e.length < E[i].length then e ← E[i]
6      else continue
7  for i ← 0 to E_d.length − 1 do
8      if E_d[i] and e is the same entity then
9          tmp ← 0;
10         while E_d[i].length < e.length do
11             if e[tmp] not in E_d[i] then
12                 E_d[i].append(e[tmp]);
13             tmp ← tmp + 1;
14     else
15         continue ;
16 E_d^+ ← E_d;
```

3.3 Text Encoding Module

In this section, we use the pre-trained language model BERT [2] to obtain the text embedding at different levels.

Given a document D, we insert a special token, "*" , at the start and end of each mention. Then, we feed the document into BERT to obtain the embedding:

$$H = [h_1, h_2, \cdots, h_l] = BERT([w_1, w_2, \cdots, w_l]) \tag{1}$$

where $H \in \mathbb{R}^{l \times d}$ is the document embedding, l is the number of tokens and d is the encoding dimension.

For the j-th mention of the entity e_i, we use the embedding of the start symbol "*" as its embedding m_j^i. Then, we obtain the entity embedding $e_i \in \mathbb{R}^d$ by adopting LogSumExp Pooling over the embedding of all its mentions:

$$e_i = \log \sum_j \exp(m_j^i) \tag{2}$$

For a sentence, we use the Average Pooling over the embedding of all its tokens to get the sentence embedding $s_n \in \mathbb{R}^d$:

$$s_n = Average(h_t, \cdots, h_{t+s_{len}-1}) \tag{3}$$

3.4 Document-Aware Context Embedding

In this section, for a given entity pair (e_h, e_t), we use the entity attention matrix from the encoder to locate relevant contextual information from the entire document, thereby obtaining a document-aware context embedding.

From the encoder in Sect. 3.3, we can obtain the pre-trained multi-head attention matrix $A \in \mathbb{R}^{head \times l \times l}$, where A_{ijk} represents attention score from token j to token k in the i-th attention head. We take the attention from the "*" symbol before the mentions as the mention-level attention and average the attention overall mentions of the entity to obtain the entity-level attention $A_i^E \in \mathbb{R}^{head \times l}$ which denotes attention from the i-th entity to all tokens. For the given entity pair (e_h, e_t), the importance of each token is computed as:

$$A^{(h,t)} = A_h^E \cdot A_t^E \tag{4}$$

$$q^{(h,t)} = \sum_{i=1}^{head} A_i^{(h,t)} \tag{5}$$

$$a^{(h,t)} = q^{(h,t)} / 1^T q^{(h,t)} \tag{6}$$

where $a^{(h,t)} \in \mathbb{R}^l$ is a distribution that reveals the importance of each token to the entity pair (e_h, e_t). Then we compute the entity pair's document-aware context embedding by interacting with the document embedding $H \in \mathbb{R}^{l \times d}$:

$$c^{(h,t)} = H^T a^{(h,t)} \tag{7}$$

where $c^{(h,t)} \in \mathbb{R}^d$ is the document-aware context embedding.

3.5 Sentence-Aware Context Embedding

In this section, we further utilize the representation of evidence sentences to enhance important local information from it. We extract all evidence sentences and use these representations to obtain the sentence-aware context embedding, which is local and sentence-based.

Due to the incomplete annotation of evidence sentences in the dataset, we cannot directly use them to train the model. Designing a set of heuristic rules to extract evidence sentences from the document is a desirable approach:

- **Explicit Co-occur:** If two entities appear in the same sentence, we consider all sentences where they appear together as explicit co-occur evidence sentences.

- **Implicit Co-occur:** Suppose two entities appear in the same sentence with the third entity respectively (the third entity referred to as "the bridge entity"). In this case, we consider these sentences as implicit co-occur evidence sentences between the two entities.

To extract evidence sentences based on these heuristic rules, we construct an entity-level co-occur matrix M, where M_{ij} stores all evidence sentence ids that determine the relationship between entity e_i and e_j.

Record explicit co-occur evidence sentences in M_{ij} at first. Then, compare the elements in the i-th and j-th rows one by one. If the elements in the k-th column are not empty for the two rows, entity e_i and e_j have explicit co-occurrence sentences with entity e_k. Entity e_k is "the bridge entity". Then, the sentence ids from M_{ik} and M_{jk} are added to the M_{ij}. At this point, the set in M_{ij} contains both explicit and implicit co-occurrence sentences, a complete list of evidence sentences we extracted. Taking the entity pair (The X-Files, Chris Carter) in Fig. 1 as an example, the process of extracting their evidence sentences is shown in Fig. 3. We first record explicit co-occur sentences between entities in the document. Due to the entity pronouns enhancement, we can get the complete explicit co-occur. Then, to predict the relation between "The X-Files" and "Chris Carter", compare the elements of these two lines one by one and find that they both co-occur with the entity "Nisei" in different sentences. Here, "Nisei" is "the bridge entity". We append these two elements to their evidence sentences list.

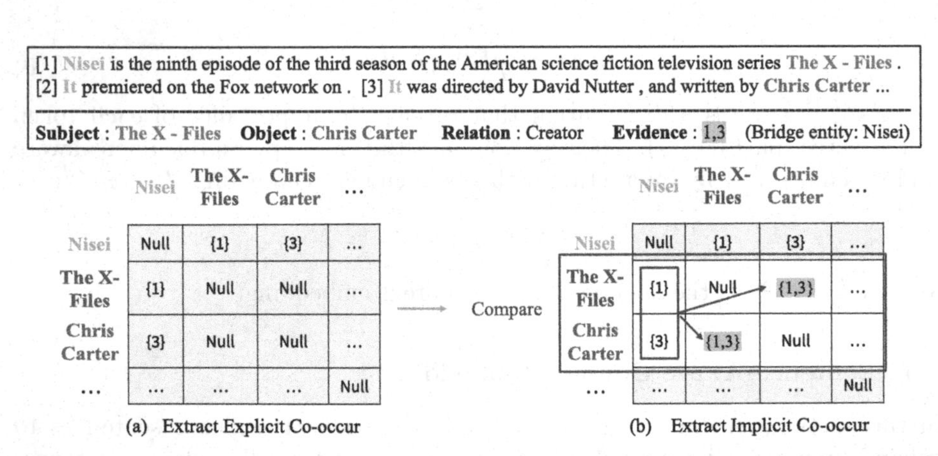

Fig. 3. The process of extracting evidence sentences. In (a), the matrix records the explicit co-occur sentences between entity pair according to the labels in the enhanced document. In (b), the matrix appends the implicit co-occur sentences according to our defined rules.

So far, for the entity pair (e_h, e_t), we obtain all their evidence sentences from the co-occur matrix M_{ht}. Then we average the embedding of these sentences to obtain an evidence sentence representation and interact with the document-aware context embedding $c^{(h,t)}$ to get a sentence-aware context embedding $s^{(h,t)}$:

$$s^{(h,t)} = Average([s_{evi_1}, s_{evi_2}, \cdots, s_{evi_n}]) \odot c^{(h,t)} \tag{8}$$

where $[s_{evi_1}, s_{evi_2}, \cdots, s_{evi_n}]$ is the list of evidence sentence embedding according to the M_{ht}, which is extracted by the heuristic rules mentioned above. $\odot$ denotes element-wise product operation and $s^{(h,t)} \in \mathbb{R}^d$.

3.6 Fusion and Classification Module

The adaptive fusion module retains relevant, useful information from two different levels and filters out redundant and repetitive information to avoid introducing noise. Fusing the document-aware context embedding and the sentence-aware context embedding, we obtain a context vector:

$$g = \sigma(W(c^{(h,t)} \oplus s^{(h,t)})) \tag{9}$$

$$v^{(h,t)} = gs^{(h,t)} + (1-g)c^{(h,t)} \tag{10}$$

where W is a trainable parameter, $\sigma(\cdot)$ is a sigmoid activation function, $\oplus$ denotes concatenation operation and $v^{(h,t)} \in \mathbb{R}^d$. We add the context vector $v^{(h,t)}$ to the entity embedding obtained by Eq. (2) to get enhanced entity representation based on entity pairs:

$$z_h^{(h,t)} = \tanh\left(W_s e_h + W_{c_1} v^{(h,t)}\right) \tag{11}$$

$$z_t^{(h,t)} = \tanh\left(W_s e_t + W_{c_2} v^{(h,t)}\right) \tag{12}$$

where $W_s, W_c \in \mathbb{R}^{d \times d}$ are trainable parameters and the final entity representation $z_h^{(h,t)}$ and $z_t^{(h,t)} \in \mathbb{R}^d$ .To predict the relations for a given entity pair (e_h, e_t), we follow the practice of ATLOP [22]. Using a group bi-linear to split the embedding dimensions into k equal-sized groups and applying bi-linear within the groups to calculate the probability for each relation:

$$[z_h^1; \ldots; z_h^k] = z_h \tag{13}$$

$$[z_t^1; \ldots; z_t^k] = z_t \tag{14}$$

$$P(r|e_h, e_t) = \sigma\Big(\sum_{i=1}^{k} {z_h^i}^T W_r^i z_t^i + b_r\Big) \tag{15}$$

where $W_r^i \in \mathbb{R}^{d/k \times d/k}$ for $i = 1 \cdots k$ are model parameters and $P(r|e_h, e_t)$ is the probability of whether relation r is the correct relation for (e_h, e_t).

3.7 Loss Function

In order to address the problem of multi-label and sample imbalance, we adopt an Adaptive Thresholding Loss as the classification loss to train model [22]. To be specific, a special relation category TH is introduced. This threshold class learns an entities-dependent threshold value which pushes positive classes to have higher probabilities than TH and the negative classes to have lower probabilities than TH:

$$\mathcal{L}_1 = -\sum_{r \in P_T} \log\left(\frac{\exp(logit_r)}{\sum_{r' \in P_T \cup \{TH\}} \exp(logit_{r'})}\right) \tag{16}$$

$$\mathcal{L}_2 = -\log\left(\frac{\exp(logit_{TH})}{\sum_{r' \in N_T \cup \{TH\}} \exp(logit_{r'})}\right) \tag{17}$$

$$\mathcal{L} = \mathcal{L}_1 + \mathcal{L}_2 \tag{18}$$

The first part $\mathcal{L}_1$ involves positive classes and the TH class and pushes the logits of all positive classes to be higher than the TH class. The second part $\mathcal{L}_2$ involves the negative classes and TH class and pushes the logits of all negative classes to be lower than the TH class. The final loss is the sum of the two.

4 Experiments

4.1 Setting

Dataset. Our model is evaluated on Re-DocRED [10], a revised version of a classic dataset DocRED [15] in Document-level Relation Extraction, which corrects the false negative examples and made significant re-annotate to it. Specifically, Re-DocRED's train/dev/test set contains 3053/500/500 documents, with an average entity length of 19.4/19.4/19.6, an average number of triples of 28.1/34.6/34.9, and an average word length of 7.9/8.2/7.9.

Configuration. Our model is implemented based on PyTorch and Huggingface's Transformers [11]. We use cased BERT$_{base}$ [2] as the base encoder and use mixed-precision training based on the Nvidia Apex library. We optimize our model using AdamW [8] with learning rate of 5e−5 and adopt a linear warmup [4] for the first 6% steps followed by a linear decay to 0. The batch size(number of documents per batch) is set to 4, and we clip the gradients of model parameters to a max norm of 1.0. We perform early stopping based on the F1 score on the development set, with a maximum of 30 epochs. We train and evaluate our model on a single Tesla K80 12 GB GPU.

4.2 Main Results

We compare our DeepNET with seven different methods on the dataset Re-DocRED. Following previous works, we use **F1** and **Ign-F1** as the primary evaluation metrics for DocRE, where **Ign-F1** measures the **F1** score excluding the relations facts shared by the training and dev/test sets. Table 1 lists the performance of DeepNET and existing models.

Table 1. Experimental results on the dev and test set of Re-DocRED. Results with * are reported by Jiang *et al.* [7], and others are based on our implementation.

Model	Dev		Test	
	Ign-F1	F1	Ign-F1	F1
CNN*	53.95	55.60	52.80	54.88
LSTM*	56.40	58.30	56.31	57.83
BiLSTM*	58.20	60.04	57.84	59.93
SSAN [14]	73.15	73.91	59.56	60.07
Eider [13]	73.21	74.01	**73.37**	**74.37**
ATLOP [22]	73.29	74.08	73.12	73.88
DeepNET(ours)	**73.38**	**74.22**	72.32	73.53

We simultaneously tested and compared our model with three representative transformer-based models on four metrics proposed by Tan *et al.* [10]. (1) **Freq-F1**: only considering the ten most common relation types in the training set . (2) **LT-F1**: only considering the long-tail relation types. (3) **Intra-F1**: evaluating on relation triples that appear in the same sentence. (4) **Inter-F1**: evaluating on relation triples that appear inter-sentence. We show the results in Table 2.

Table 2. Comprehensive evaluation results on the dev set of Re-DocRED.

Model	Freq-F1	LT-F1	Intra-F1	Inter-F1
SSAN [14]	75.46	68.04	75.99	70.17
Eider [13]	77.43	**69.03**	**77.08**	71.3
ATLOP [22]	77.71	68.83	76.66	71.82
DeepNET(ours)	**78.00**	68.85	76.66	**72.11**

Table 1 and Table 2 show that DeepNET outperforms other methods on the Re-DocRED dev set. In particular, DeepNET performs better than ATLOP, by 0.09/0.14 points on **Ign-F1/F1**. Notably, DeepNET significantly improves the inter-sentence relations, by 0.29 points on **Inter-F1**. We hypothesize that the bottleneck of inter-sentence pairs is locating the relevant context in the document. The process of Entity Pronouns Enhancement helps the model notice the presence of pronoun information in the sentence, and the two types of evidence sentences extracted in the Sentence-aware Context Embedding Module constitute a multi-hop connection path between entities, enabling DeepNET to complete the relation inference of inter-sentence entity pairs better.

Table 3. Ablation studies evaluated on the Re-DocRED dev set.

Setting	Ign-F1	F1	Freq-F1	LT-F1	Inter-F1
DeepNET(our)	**73.38**	**74.22**	**78.00**	**68.85**	**72.11**
w/o Pronouns	72.64(−0.74)	73.50(−0.72)	77.59(−0.41)	67.42(−1.43)	71.21(−0.90)
w/o Sentence-aware	73.31(−0.07)	74.19(−0.03)	77.83(−0.17)	68.76(−0.09)	71.71(−0.40)

4.3 Ablation Studies

Table 3 shows the ablation studies that analyze each module's utility in DeepNET. We can see that when we remove Entity Pronouns Enhancement Module, the model's overall performance significantly deteriorates, with **Ign-F1/F1** decreasing by 0.74/0.72 respectively. It indicates that improving the integrity of entity information in the document and emphasizing the mention of entity pronouns is valuable. Meanwhile, without Sentence-aware Context Embedding Module, the model's performance was also decreasing by 0.07/0.03 on **Ign-F1/F1** respectively, indicating that the extraction and utilization of evidence in this model is meaningful. In addition, the reduction of other indicators (**Freq-F1/LT-F1/Inter-F1**) can also demonstrate the effectiveness of two modules.

5 Conclusion

In this work, we propose DeepNET, which improves DocRE by efficiently leveraging pronoun information and effectively extracting evidence sentences. In model, the Entity Pronouns Enhancement Module updates the dataset by co-reference resolution and a data fusion algorithm. The Sentence-aware Context Embedding designs heuristic rules to extract evidence sentences witch conform to actual logic, and this embedding focuses on more critical tokens for predicting the relations. In the future, we will optimize the inference process in DocRE, and use the Chain-of-Though (COT) to improve the application performance of RE.

References

1. Christopoulou, F., Miwa, M., Ananiadou, S.: Connecting the dots: document-level neural relation extraction with edge-oriented graphs. arXiv preprint arXiv:1909.00228 (2019)
2. Devlin, J., Chang, M.W., Lee, K., Toutanova, K.: BERT: pre-training of deep bidirectional transformers for language understanding. ArXiv (2018)
3. Gardner, M., et al.: AllenNLP: a deep semantic natural language processing platform. arXiv preprint arXiv:1803.07640 (2018)
4. Goyal, P., et al.: Accurate, large minibatch SGD: training imagenet in 1 hour. arXiv preprint arXiv:1706.02677 (2017)
5. Huang, K., Qi, P., Wang, G., Ma, T., Huang, J.: Entity and evidence guided document-level relation extraction. In: Proceedings of the 6th Workshop on Representation Learning for NLP (RepL4NLP-2021), pp. 307–315 (2021)

6. Huang, Q., Zhu, S., Feng, Y., Ye, Y., Lai, Y., Zhao, D.: Three sentences are all you need: Local path enhanced document relation extraction. arXiv preprint arXiv:2106.01793 (2021)
7. Jiang, F., Niu, J., Mo, S., Fan, S.: Key mention pairs guided document-level relation extraction. In: Proceedings of the 29th International Conference on Computational Linguistics, pp. 1904–1914 (2022)
8. Loshchilov, I., Hutter, F.: Decoupled weight decay regularization. arXiv preprint arXiv:1711.05101 (2017)
9. Nan, G., Guo, Z., Sekulić, I., Lu, W.: Reasoning with latent structure refinement for document-level relation extraction. arXiv preprint arXiv:2005.06312 (2020)
10. Tan, Q., Xu, L., Bing, L., Ng, H.T.: Revisiting DocRED-addressing the overlooked false negative problem in relation extraction. ArXiv (2022)
11. Wolf, T., et al.: Huggingface's transformers: state-of-the-art natural language processing. arXiv preprint arXiv:1910.03771 (2019)
12. Xiao, Y., Zhang, Z., Mao, Y., Yang, C., Han, J.: SAIS: supervising and augmenting intermediate steps for document-level relation extraction. arXiv preprint arXiv:2109.12093 (2021)
13. Xie, Y., Shen, J., Li, S., Mao, Y., Han, J.: Eider: empowering document-level relation extraction with efficient evidence extraction and inference-stage fusion. In: Findings of the Association for Computational Linguistics: ACL 2022, pp. 257–268 (2022)
14. Xu, B., Wang, Q., Lyu, Y., Zhu, Y., Mao, Z.: Entity structure within and throughout: modeling mention dependencies for document-level relation extraction. In: Proceedings of the AAAI conference on artificial intelligence, vol. 35, pp. 14149–14157 (2021)
15. Yao, Y., et al.: DocRED: a large-scale document-level relation extraction dataset. arXiv preprint arXiv:1906.06127 (2019)
16. Zeng, S., Xu, R., Chang, B., Li, L.: Double graph based reasoning for document-level relation extraction. arXiv preprint arXiv:2009.13752 (2020)
17. Zeng, X., Zeng, D., He, S., Liu, K., Zhao, J.: Extracting relational facts by an end-to-end neural model with copy mechanism. In: Proceedings of the 56th Annual Meeting of the Association for Computational Linguistics, vol. 1: Long Papers, pp. 506–514 (2018)
18. Zhang, N., et al.: Document-level relation extraction as semantic segmentation. arXiv preprint arXiv:2106.03618 (2021)
19. Zhao, C., Zeng, D., Xu, L., Dai, J.: Document-level relation extraction with context guided mention integration and inter-pair reasoning. ArXiv (2022)
20. Zheng, S., Wang, F., Bao, H., Hao, Y., Zhou, P., Xu, B.: Joint extraction of entities and relations based on a novel tagging scheme. ArXiv (2017)
21. Zhou, W., Chen, M.: An improved baseline for sentence-level relation extraction. In: AACL (2021)
22. Zhou, W., Huang, K., Ma, T., Huang, J.: Document-level relation extraction with adaptive thresholding and localized context pooling. ArXiv (2020)

Nearest Memory Augmented Feature Reconstruction for Unified Anomaly Detection

Honghao Wu[1], Chao Wang[3], Zihao Jian[1], Yongxuan Lai[1,2](✉), Liang Song[1], and Fan Yang[4]

[1] Shenzhen Research Institute/School of Informatics, Xiamen University, Xiamen, China
wuhonghao@stu.xmu.edu.cn, laiyx@xmu.edu.cn
[2] Mathematics and Information Engineering School, Longyan University, Longyan, China
[3] School of Physics and Electronic Information, Dezhou University, Dezhou, China
[4] School of Aerospace Engineering, Xiamen University, Xiamen, China
yang@xmu.edu.cn

Abstract. Reconstruction-based anomaly detection methods expect to reconstruct normality well but fail for abnormality. Memory modules have been exploited to avoid reconstructing anomalies, but they may overgeneralize by using memory in a weighted manner. Additionally, existing methods often require separate models for different objects. In this work, we propose nearest memory augmented feature reconstruction for unified anomaly detection. Specifically, the novel nearest memory addressing (NMA) module enables memory items to record normal prototypical patterns individually. In this way, the risk of over-generalization is mitigated while the capacity of the memory item is fully exploited. To overcome the constraint of training caused by NMA that has no real gradient defined, we perform end-to-end training with straight-through gradient estimation and exponential moving average. Moreover, we introduce the feature reconstruction paradigm to avoid the reconstruction challenge in the image space caused by information loss of the memory mechanism. As a result, our method can unify anomaly detection for multiple categories. Extensive experiments show that our method achieves state-of-the-art performance on MVTecAD dataset under the unified setting. Remarkably, it achieves comparable or better performance than other algorithms under the separate setting.

Keywords: Anomaly detection · Memory mechanism · Feature reconstruction

1 Introduction

Anomaly detection has a wide range of application scenarios [1,4] with the goal of identifying samples that differ from normal. In reality, anomalies are often

B. Luo et al. (Eds.): ICONIP 2023, CCIS 1966, pp. 350–361, 2024.
https://doi.org/10.1007/978-981-99-8148-9_28

rare or even absent and have various forms, making it difficult to define them in advance. Although humans can easily identify anomalies, manual detection is not always feasible due to physical and mental limitations. As a result, many works focusing on automated machine vision inspection have emerged. Due to the absence of anomalies, anomaly detection is typically defined as an unsupervised learning task, where only normal samples are available during training, and the model needs to distinguish between normal and abnormal during testing.

Existing works based on reconstruction have demonstrated their effectiveness in anomaly detection [10,23,25]. The reconstruction network, which typically consists of an encoder and a decoder, minimizes the reconstruction error between the input and output to learn the distribution of the training data. Such methods rely on the hypothesis that the reconstruction model trained solely on normal data is incapable of reconstructing anomalies, which results in a large reconstruction error that serves as a useful indicator for detecting anomalies. However, there is a potential pitfall known as the "identity shortcut" [23], where the model simply outputs a copy of the input, potentially leading to the reconstruction of anomalies. Recently, memory-augmented reconstruction methods [9,10,14] have been proposed to tackle the problem. They maintain a memory bank consisting of several memory items to record normal prototypical patterns during the training phase. In particular, the input to the decoder is not the embedding from the encoder, but the combination of memory items to which the embedding is mapped. Specifically, the feature vector at each location of the embedding is considered as a query and is mapped to memory items by memory addressing. Intuitively, the memory mechanism cuts off the identity shortcut so that the result reconstructed from memory items will have a large gap with the abnormal input. However, previous memory addressing modules map the query to the weighting of all memory items. This is likely to synthesize anomalous features by weighting and unintentionally enhance generalizability. In addition, most of the existing works require training a model for each category (separate setting), which increases the training and inference overhead. Hence, methods that can unify anomaly detection for multiple categories are highly required.

In this paper, we propose nearest memory augmented feature reconstruction for unified anomaly detection. Specifically, we propose nearest memory addressing (NMA) module, where the query is mapped to the closest memory item. NMA forces each memory item to record different normal patterns individually and avoids the possibility of synthesizing anomalous feature due to weighting. As the *argmin* operator of NMA interrupts the backpropagation of gradient, we utilize the straight-through gradient estimation [3] and exponential moving average (EMA) inspired by VQVAE [21] to enable end-to-end training, as shown in Fig. 1. Furthermore, we follow the feature reconstruction paradigm [19], which can be seen as the relaxed version of image reconstruction. Unlike image reconstruction, the object of feature reconstruction is the feature of data extracted by a pre-trained model. It contributes to reduce the difficulty of reconstruction due to information loss caused by memory mechanism. Both feature reconstruction and memory mechanism focus on feature space. Combining the two fully exploits the capability of representation learning. As a result, our method is able to learn the distributions of different normality simultaneously and use only one model to make decisions of anomaly for different categories (unified setting).

In summary, the main contributions of this paper are as follows:

- We propose a memory mechanism with nearest memory addressing to mitigate the identity shortcut phenomenon and control the generalizability. Besides, we also propose to train the model end-to-end by straight-through gradient estimation and exponential moving average.
- We combine the memory mechanism with the feature reconstruction paradigm to unify anomaly detection for various categories.
- We provide extensive experiments to demonstrate the effectiveness of the proposed method. Experimental results on the MVTecAD [4] dataset show that our method achieves state-of-the-art performance under the unified setting while matching or even surpassing the performance of other algorithms under the separate setting.

2 Related Work

The works related to anomaly detection can be classified into two groups of methods, reconstruction-based and representation-based. In the following, We give an introduction to them.

2.1 Reconstruction-Based Methods

Such methods assume that the reconstruction model trained on normality produces a large reconstruction error for the anomaly, which is adopted as anomaly score to detect anomaly. In addition to image reconstruction error, some methods [2,18] have exploited feature differences in the hidden space to identify anomaly. However, the fundamental hypothesis of reconstruction-based methods is not always satisfied, i.e., anomalies can also be reconstructed. Subsequent methods have worked to alleviate this problem. Perera et al. [15] ensure that the normal representations uniformly distribute in a bounded latent space to constrain the latent space to represent the normality. Zavrtanik et al. [25] propose DRAEM to recover pseudo-anomaly to normality and then predict their difference by discriminator to achieve nice localization performance. Furthermore, Gong et al. [9] propose a memory-augmented autoencoder to record the normal prototypical patterns in memory items. It assumes that the results reconstructed by memory items will be close to normal. Hou et al. [10] further modulate the generalizability of the model by varying the granularity of feature map division. The memory addressing modules of existing methods map the query to the weighting of all memory items, while our NMA maps the query to the closest one, which avoids unintended synthesis of anomalous features. In recent years, researchers have proposed the feature reconstruction paradigm [19] to detect anomaly based on the reconstruction error of features extracted by a pre-trained model. You et al. [23] propose a unified anomaly detection framework UniAD upon this basis, which further avoids identity shortcut with the help of neighbor masked encoder, layer-wise query decoder, and feature jittering strategy. Our method also follows the feature reconstruction paradigm but uses a different feature extraction way.

2.2 Representation-Based Methods

Learning discriminative features is the concern of representation-based methods. Hence, deep learning networks and traditional one-class classification methods are naturally combined for anomaly detection [17]. Some authors apply proxy tasks to learn discriminative features. For example, synthesizing pseudo-anomalies to construct a classification task [12] and aligning images [11]. Furthermore, several works [16,24] estimate the probability density of normal features with the help of normalizing flow, and then the estimated probability is taken as the anomaly score. Recently, knowledge distillation has been applied to anomaly detection [5]. Such methods assume that the student model receives only the normal information from the teacher. Then the difference in the representations of the teacher and student models can be considered as the anomaly score. Also, methods [6,7] that directly utilize pre-trained model without fine-tuning have achieved favorable performance. These methods leverage the difference in representation between normal and abnormal data, as learned from the model trained on large external datasets.

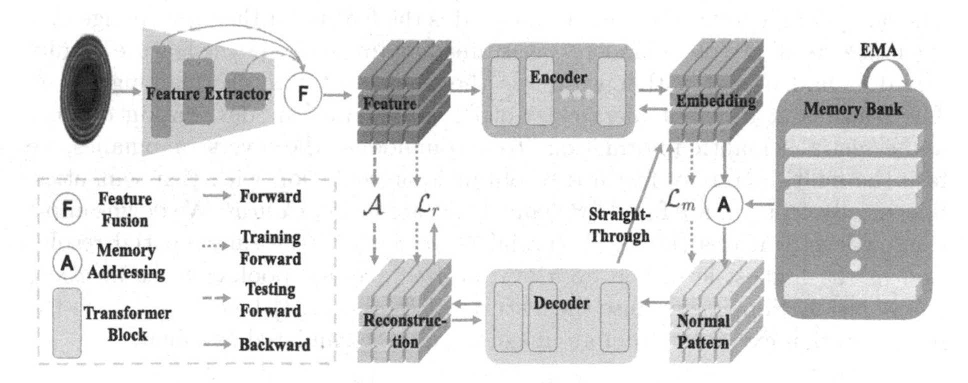

Fig. 1. The framework of the proposed method. $\mathcal{L}_m, \mathcal{L}_r, \mathcal{A}$ denote memory loss, reconstruction loss, and anomaly score map, respectively.

3 Method

3.1 Problem Definition and Basic Methodology

Given a normal dataset $X = \{x_i | i = 1, 2, ..., N_x\}$, anomaly detection can be considered as learning the mapping ($\Phi : x \rightarrow \mathbb{R}$) from the input to the anomaly score based on the normal dataset. The input is regarded as anomalous if its anomaly score is greater than the threshold during testing. Our method follows the feature reconstruction paradigm. Unlike image reconstruction which uses image reconstruction error as the anomaly score, we use feature reconstruction error instead. Figure 1 describes the framework of the proposed method. We first extract and fusion the multi-hierarchy features of input x with a pre-trained model. The fused feature $\mathcal{F} \in \mathbb{R}^{C \times H \times W}$ plays the role of reconstruction object. Furthermore, since the Visual Transformer [8] can model the global context which is beneficial for anomaly detection based on global information, we use multiple transformer blocks to form the encoder and decoder. The encoder maps $\mathcal{F}$ to the embedding $\mathcal{Z} \in \mathbb{R}^{C_e \times H \times W}$ of equal spatial dimension. The memory mechanism connects the encoder and decoder. Feature vector at each location of $\mathcal{Z}$ is considered as a query and will be mapped into the corresponding memory item to form $\hat{\mathcal{Z}} \in \mathbb{R}^{C_e \times H \times W}$ according to memory addressing. Then, the decoder receives $\hat{\mathcal{Z}}$ to get the reconstruction $\hat{\mathcal{F}} \in \mathbb{R}^{C \times H \times W}$. The memory mechanism has two main modules, a memory bank and nearest memory addressing, which can record normal prototypical patterns. For convenience, we use the terms $f_i, \hat{f}_i, z_i$, and $\hat{z}_i$ to denote the feature vectors of $\mathcal{F}, \hat{\mathcal{F}}, \mathcal{Z}$, and $\hat{\mathcal{Z}}$ in spatial dimension i, respectively, where $i = 1, 2, ..., HW$. In the absence of ambiguity, we will ignore the subscript i.

3.2 Feature Extraction

The object of reconstruction in our method is the feature of the input image, not the image itself. We use a model ϕ pre-trained on ImageNet as the feature extractor. It should be noted that we freeze the weights of ϕ in the training phase. The features at different hierarchies of the pre-trained model contain distinct and essential semantic information. To accommodate the diverse anomalies, we fuse the multi-hierarchy features to obtain representation with rich semantics. Specifically, features are fused by channel concatenation $Concat$. We downsample the low-level features (i.e., high spatial resolution) to the same spatial resolution as the highest-level one by a overlap-free average pooling layer $avgpool_h$ before fusion. Suppose we fuse the features of $H = \{1, 2, 3\}$ hierarchies, then the representation extracted from an image x can be formulated as follows:

$$\mathcal{F} = F(x) = Concat(\{avgpool_h(\phi_h(x)) | h \in H\}) \tag{1}$$

where $\phi_h(x)$ denotes the final output of the spatial resolution blocks at the h-th level.

3.3 Nearest Memory Addressing

We define a learnable memory bank $\mathcal{M} = \mathbb{R}^{N \times C_e}$, which is represented as a matrix containing N vectors with dimension C_e. Each item in the memory bank $m_i \in \mathbb{R}^{C_e}, i \in \{1, 2, 3, ..., N_m\}$ records the normal prototypical pattern during training. Unlike the previous works that obtain the addressing result by fusing each item of the memory bank by similarity in a one-to-many format, our nearest memory addressing module maps the feature vector z at each location of the embedding, i.e., the query, to a memory item in a one-to-one format. This is achieved by selecting the memory item that is closest to the query z:

$$\hat{z} = m_k, \quad k = \arg\min_i D(z, m_i) \tag{2}$$

where D denotes the distance between two points in the space. Therefore, each memory item records different normal patterns individually, which directly corresponds to each location of the feature map input to the decoder, avoiding the synthesis of abnormal features. However, nearest memory addressing has no real gradient defined, resulting in the encoder being unable to obtain the gradient from reconstruction loss. Inspired by VQ-VAE [21], straight-through gradient estimation of mapping from z to $\hat{z}$ is applied, and the memory items are optimized by exponential moving average.

$$\hat{z}^{sg} = z + sg[\hat{z} - z] \tag{3}$$

Equation 3 allows the backpropagation to skip computing the gradient of equation $\hat{z} - z$ by stop-gradient operator sg, while the gradient of the reconstruction loss can propagate to the encoder through the query z. This simple gradient estimation approach works efficiently and is the basis for the one-to-one mapping of the query to the memory item. Moreover, since the stop-gradient operator prevents the memory items from receiving the gradient, we additionally optimize the memory items with exponential moving average. By applying EMA, the memory items can record normal prototypical patterns steadily. Suppose the set of queries that map to memory item m_k in the t-th iteration is $\{z_{k,i} | i = 1, 2 ..., n_k^t\}$, then m_k is updated in the t-th iteration by the following formula.

$$N_k^t = \gamma N_k^{t-1} + (1 - \gamma) n_k^t \tag{4}$$

$$M_k^t = \gamma M_k^{t-1} + (1 - \gamma) \sum_{i=1}^{n_k^t} z_{k,i} \tag{5}$$

$$m_k^t = \frac{M_k^t}{N_k^t} \tag{6}$$

where $\gamma \in [0, 1]$ is the hyperparameter of EMA, and N_k^t and M_k^t denote the exponential moving average of the number and sum of queries mapping to m_k in the t-th iteration, respectively.

3.4 Loss Function

The overall loss function of our method is defined by the Eq. 7. It contains the reconstruction loss $\mathcal{L}_r$ and the memory loss $\mathcal{L}_m$. The α and β are weighting parameters controlling the contribution of the two losses. We adopt mean square error and cosine similarity as the reconstruction loss $\mathcal{L}_r$. The encoder and decoder will receive the gradient from it. The memory loss $\mathcal{L}_m$ trains the encoder to bring the queries close to the memory items, while the memory items will be optimized to approach the queries by exponential moving average to record normal patterns.

$$\mathcal{L} = \alpha\mathcal{L}_r + \beta\mathcal{L}_m \tag{7}$$

$$\mathcal{L}_r = \frac{1}{H \cdot W}\sum_{i=1}^{H \cdot W}\left[||f_i - \hat{f}_i||_2^2 + \left(1 - \frac{f_i\hat{f}_i}{||f_i|| \cdot ||\hat{f}_i||}\right)\right] \tag{8}$$

$$\mathcal{L}_m = \frac{1}{H \cdot W}\sum_{i=1}^{H \cdot W}||z_i - \hat{z}_i^{sg}||_2^2 \tag{9}$$

3.5 Anomaly Score

We use the $L1$ norm of the feature reconstruction difference as the anomaly score map $\mathcal{A}$, as shown in Eq. 10. For pixel-level anomaly detection (anomaly localization), $\mathcal{A}$ is upsampled to the image resolution with bilinear interpolation to obtain the localization result $\mathcal{A}_p$, where the value of each pixel represents the anomaly score at the corresponding location. For image-level anomaly detection, we follow the literature [23] and take the maximum value of $\mathcal{A}_p$ after applying overlap-free average pooling with kernel size k as the anomaly score $\mathcal{A}_{img}$.

$$\mathcal{A} = ||\mathcal{F} - \hat{\mathcal{F}}||_1 \in \mathbb{R}^{H \times W} \tag{10}$$

$$\mathcal{A}_{img} = \max(avgpool(\mathcal{A}_p, k)) \tag{11}$$

4 Experiments

4.1 Experimental Setup

Dataset and Evaluation Metric. We validate the performance of our method on the challenging industrial anomaly detection dataset MVTecAD [4]. It contains 15 categories with various anomalies per category, covering object and texture types. The dataset has a total of 5354 samples, of which 1725 are test samples. Furthermore, we adopt the Area Under the curve of the Receiver Operating Characteristics (AUROC) as the evaluation metric, consistent with previous works [7,10,23,25]. The image-level AUROC is used to evaluate anomaly detection performance, while the pixel-level AUROC is for anomaly localization.

Implementation Details. All our experiments follow the unified anomaly detection setup. EfficientNet-B4 [20] is adopted as the feature extractor. Before feature extraction, all images are downsampled to 224 × 224 resolution, and no data augmentation is applied. The features from the first four hierarchies of EfficientNet-B4 are fused to generate the feature map with channel $C = 272$. The channel and number of memory items in the memory bank are set to $C_e = 256$ and $N_m = 30$ respectively. Both the encoder and decoder consist of four transformer blocks. The encoder first reduces the input channel C to C_e by linear projection, while the decoder recovers the channel to C before output. We use the cosine distance as the metric D of nearest memory addressing. In order to fit the exponential moving average with the cosine distance, the query is normalized before addressing, and all memory items are similarly normalized after each update iteration. The hyperparameter of EMA is set to $\gamma = 0.99$. Weight parameters for balancing different optimization objectives are chosen as $\alpha = 1, \beta = 5$. The model is trained for 1000 epochs by the AdamW optimizer [13] with a learning rate of 4×10^{-4} and weight decay of 1×10^{-3}. The kernel size k for the average pooling in Eq. 11 is set to 16.

Table 1. Anomaly detection and localization performance on MVTecAD dataset (Image-level AUROC, Pixel-level AUROC). The best results are marked in bold.

Category		Methods							
		MemAE [9]	DAAD [10]	US [5]	PSVDD [22]	PaDiM [7]	DRAEM [25]	UniAD [23]	Ours
Texture	carpet	45.4, –	86.6, –	91.6, 93.5	92.9, 92.6	99.8, **99.0**	97.0, 95.5	99.8, 98.5	**99.9**, **99.0**
	grid	94.6, –	95.7, –	81.0, 89.9	94.6, 96.2	96.7, 97.1	**99.9**, **99.7**	98.2, 96.5	**99.9**, 97.7
	leather	61.1, –	86.2, –	88.2, 97.8	90.9, 97.4	**100**, **99.0**	**100**, 98.6	**100**, 98.8	**100**, 98.9
	tile	63.0, –	88.2, –	99.1, 92.5	97.8, 91.4	98.1, 94.1	99.6, **99.2**	99.3, 91.8	**99.8**, 93.2
	wood	96.7, –	98.2, –	97.7, 92.1	96.5, 90.8	**99.2**, 94.1	99.1, **96.4**	98.6, 93.2	99.0, 93.8
Object	bottle	95.4, –	97.6, –	99.0, 97.8	98.6, 98.1	99.9, 98.2	99.2, **99.1**	99.7, 98.1	**100**, 98.5
	cable	69.4,	84.4, –	86.2, 91.9	90.3, 96.8	92.7, 96.7	91.8, 94.7	95.2, 97.3	**99.3**, **98.2**
	capsule	83.1, –	76.7,	86.1, 96.8	76.7, 95.8	91.3, 98.6	**98.5**, 94.3	86.9, 98.5	91.1, **98.0**
	hazelnut	89.1, –	92.1, –	93.1, 98.2	92.0, 97.5	92.0, 98.1	**100**, **99.7**	99.8, 98.1	99.8, 98.5
	metal nut	53.7, –	75.8, –	82.0, 97.2	94.0, 98.0	98.7, 97.3	98.7, **99.5**	99.2, 94.8	**99.8**, 96.2
	pill	88.3, -	90.0, –	87.9, 96.5	86.1, 95.1	93.3, 95.7	**98.9**, **97.6**	93.7, 95.0	97.0, 96.5
	screw	99.2, –	98.7, –	54.9, 97.4	81.3, 95.7	85.8, 98.4	**93.9**, 97.6	87.5, 98.3	92.2, **98.5**
	toothbrush	97.2, –	99.2, –	95.3, 97.9	**100**, 98.1	96.1, **98.8**	**100**, 98.1	94.2, 98.4	94.7, 98.7
	transistor	79.3, –	87.6, –	81.8, 73.7	91.5, 97.0	97.4, 97.6	93.1, 90.9	99.8, 97.9	**100**, **98.9**
	zipper	87.1, –	85.9, –	91.9, 95.6	97.9, 95.1	90.3, 98.4	**100**, **98.8**	95.8, 96.8	97.3, 97.4
Average		80.2, –	89.5, –	87.7, 93.9	92.1, 95.7	95.5, 97.4	**98.0**, 97.3	96.5, 96.8	**98.0**, **97.5**

4.2 Results

Quantitative Results. Table 1 compares the quantitative performance of the proposed method with some baselines for anomaly detection and localization on the MVTecAD dataset. We can observe that our method achieves 98.0 AUROC in anomaly detection parallel to the strong competitor DRAEM [25] based on reconstruction, and surpasses it in anomaly localization reaching 97.5 AUROC. It is worth noting that the baselines are experimented under the separate setting except for UniAD. The performance of DRAEM under the unified setting is substantially reduced with respect to the separate setting [23]. Compared with UniAD experimented under the same unified setting, we outperform 1.0 and 1.5 AUROC in anomaly detection and localization, respectively. Remarkably, our method achieves 18.2% and 8.7% AUROC gain in anomaly detection compared to the previous memory-augmented methods MemAE [9] and DAAD [10], respectively.

Qualitative Results. Figure 2 shows the qualitative results of the proposed method on the MVTecAD dataset. We visualize the results of anomaly localization for representative samples in the test set as heatmaps. It can be observed that our method gives accurate determination for object and texture anomalies. To verify the ability of the memory mechanism in recording normal patterns, we additionally train a separate decoder to reconstruct images from features, and the reconstructions are shown in Fig. 2 (Recon). It is necessary to note that the decoder is only used for visualization. It can be observed that the reconstructed features obtained via the memory mechanism return to normal images after decoding, which indicates that the normal patterns are effectively recorded in the memory items.

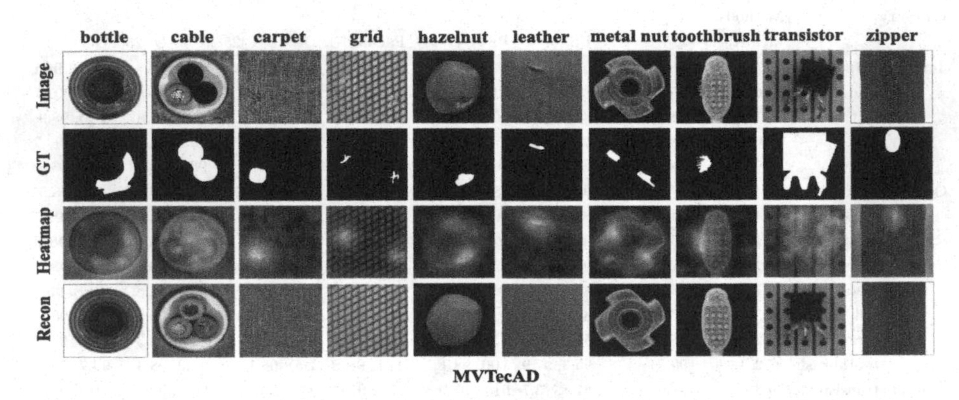

Fig. 2. Qualitative results on the MVTecAD dataset. From top to bottom: input images, ground truths, anomaly localization heatmaps, and reconstructions obtained by the additional decoder.

4.3 Ablation Study

Generalization of Memory Addressing. To verify that the nearest memory addressing is effective in controlling generalizability compared to weighted memory addressing (WMA) [10], we conduct experiments replacing NMA with WMA under the same other settings. The decoding results of the reconstructed features by the two addressing methods are shown in Fig. 3. The reconstructions with WMA produce anomalous pixels in the case of bottle (Fig. 3a) and metal nut (Fig. 3b). In the case of transistor (Fig. 3c), the method with WMA does not reconstruct the metal part but the background instead. In contrast, our approach restores all anomalies to normal, demonstrating that NMA has better ability to control generalizability and assist in recording normal prototypical patterns. Moreover, we can obviously observe that our reconstruction is closer to normal compared to WMA from Fig. 3d.

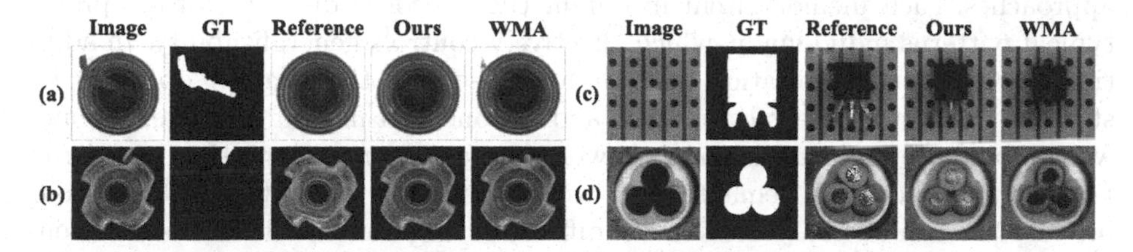

Fig. 3. Visualization of reconstructions for four representative images. From left to right: input images, ground truths, reference normals, reconstructions of our method, and reconstructions with the weighted memory addressing (WMA).

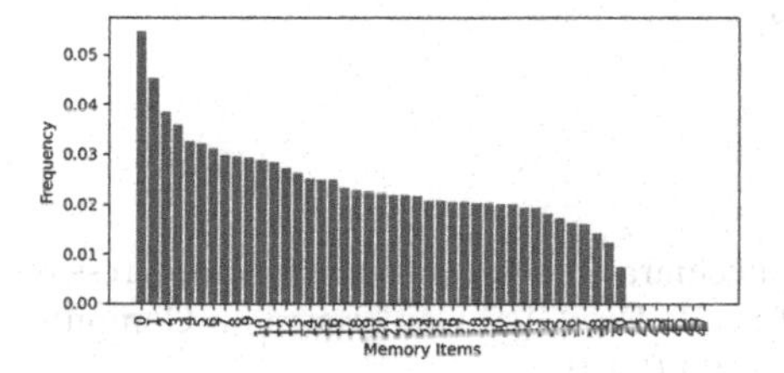

Fig. 4. The number of queries mapped to each memory item in the MVTecAD training set when $N_m = 48$. (Normalized)

Table 2. Ablation study of memory bank scale with Image-level AUROC. The best results are marked in bold.

Channel	No. of memory items				
	0	24	30	36	48
128	88.1	96.7	96.7	96.5	96.5
256	90.3	97.7	**98.0**	97.9	**98.0**
384	90.8	**98.0**	97.8	97.7	97.9

Memory Bank Scale. We investigate the impact of the memory bank scale in terms of the number of memory items N_m and memory item channel C_e. Table 2 with $N_m = 0$ indicates feature reconstruction without memory mechanism, i.e., autoencoder. As a baseline, our method surpasses a large gap compared to it. In addition, the memory item does not have enough capacity to record normal patterns causing performance degradation when the channel is small. We achieve the maximum at $N_m = 30, C_e = 256$ and $N_m = 48, C_e = 256$ and $N_m = 24, C_e = 384$, the first setting of which is lightest and our optimal choice. Besides, the fluctuation of AUROC is small as the number of memory items increases. We visualize the number of all queries mapped to each memory item in the MVTecAD training set when $N_m = 48$ as shown in Fig. 4. Some of the memory items are not used, indicating that our method is less sensitive to the increasing number of memory items and automatically discards the redundant ones.

5 Conclusion

In this paper, we propose nearest memory augmented feature reconstruction for unified anomaly detection. First, we propose the nearest memory addressing to record normal patterns. Compared with the previous memory addressing approaches, each memory item in our method records distinct normal prototypical patterns individually, which effectively controls generalizability. In addition, we avoid the limitation of training caused by the *argmin* operator by straight-through gradient estimation and exponential moving average. Second, we combine the memory mechanism with the feature reconstruction paradigm, both of which focus on semantic context and reinforce each other, allowing our method to learn anomaly decisions of different categories simultaneously. Extensive experiments demonstrate our excellent anomaly detection and localization performance in the more challenging unified setting.

Acknowledgements. This work is supported in part by the Natural Science Foundation of Fujian under Grant 2023J01351; in part by the Natural Science Foundation of Guandong under Grant 2021A1515011578.

References

1. Abati, D., Porrello, A., Calderara, S., Cucchiara, R.: Latent space autoregression for novelty detection. In: Proceedings of the IEEE/CVF Conference on Computer Vision and Pattern Recognition, pp. 481–490 (2019)
2. Akcay, S., Atapour-Abarghouei, A., Breckon, T.P.: GANomaly: semi-supervised anomaly detection via adversarial training. In: Jawahar, C.V., Li, H., Mori, G., Schindler, K. (eds.) ACCV 2018. LNCS, vol. 11363, pp. 622–637. Springer, Cham (2019). https://doi.org/10.1007/978-3-030-20893-6_39
3. Bengio, Y., Léonard, N., Courville, A.: Estimating or propagating gradients through stochastic neurons for conditional computation. arXiv preprint arXiv:1308.3432 (2013)
4. Bergmann, P., Fauser, M., Sattlegger, D., Steger, C.: MVTec AD-a comprehensive real-world dataset for unsupervised anomaly detection. In: Proceedings of the IEEE/CVF Conference on Computer Vision and Pattern Recognition, pp. 9592–9600 (2019)
5. Bergmann, P., Fauser, M., Sattlegger, D., Steger, C.: Uninformed students: student-teacher anomaly detection with discriminative latent embeddings. In: Proceedings of the IEEE/CVF Conference on Computer Vision and Pattern Recognition, pp. 4183–4192 (2020)
6. Cohen, N., Hoshen, Y.: Sub-image anomaly detection with deep pyramid correspondences. arXiv preprint arXiv:2005.02357 (2020)
7. Defard, T., Setkov, A., Loesch, A., Audigier, R.: PaDiM: a patch distribution modeling framework for anomaly detection and localization. In: Del Bimbo, A., et al. (eds.) ICPR 2021. LNCS, vol. 12664, pp. 475–489. Springer, Cham (2021). https://doi.org/10.1007/978-3-030-68799-1_35
8. Dosovitskiy, A., et al.: An image is worth 16×16 words: transformers for image recognition at scale. arXiv preprint arXiv:2010.11929 (2020)

9. Gong, D., et al.: Memorizing normality to detect anomaly: memory-augmented deep autoencoder for unsupervised anomaly detection. In: Proceedings of the IEEE/CVF International Conference on Computer Vision, pp. 1705–1714 (2019)
10. Hou, J., Zhang, Y., Zhong, Q., Xie, D., Pu, S., Zhou, H.: Divide-and-assemble: learning block-wise memory for unsupervised anomaly detection. In: Proceedings of the IEEE/CVF International Conference on Computer Vision, pp. 8791–8800 (2021)
11. Huang, C., Guan, H., Jiang, A., Zhang, Y., Spratling, M., Wang, Y.F.: Registration based few-shot anomaly detection. In: Computer Vision-ECCV 2022: 17th European Conference, Tel Aviv, Israel, 23–27 October 2022, Proceedings, Part XXIV, pp. 303–319. Springer, Heidelberg (2022). https://doi.org/10.1007/978-3-031-20053-3_18
12. Li, C.L., Sohn, K., Yoon, J., Pfister, T.: Cutpaste: self-supervised learning for anomaly detection and localization. In: Proceedings of the IEEE/CVF Conference on Computer Vision and Pattern Recognition, pp. 9664–9674 (2021)
13. Loshchilov, I., Hutter, F.: Decoupled weight decay regularization. arXiv preprint arXiv:1711.05101 (2017)
14. Park, H., Noh, J., Ham, B.: Learning memory-guided normality for anomaly detection. In: Proceedings of the IEEE/CVF Conference on Computer Vision and Pattern Recognition, pp. 14372–14381 (2020)
15. Perera, P., Nallapati, R., Xiang, B.: OCGAN: one-class novelty detection using GANs with constrained latent representations. In: Proceedings of the IEEE/CVF Conference on Computer Vision and Pattern Recognition, pp. 2898–2906 (2019)
16. Rudolph, M., Wandt, B., Rosenhahn, B.: Same same but differnet: semi-supervised defect detection with normalizing flows. In: Proceedings of the IEEE/CVF Winter Conference on Applications of Computer Vision, pp. 1907–1916 (2021)
17. Ruff, L., et al.: Deep one-class classification. In: International Conference on Machine Learning, pp. 4393–4402. PMLR (2018)
18. Schlegl, T., Seeböck, P., Waldstein, S.M., Schmidt-Erfurth, U., Langs, G.: Unsupervised anomaly detection with generative adversarial networks to guide marker discovery. In: Niethammer, M., et al. (eds.) IPMI 2017. LNCS, vol. 10265, pp. 146–157. Springer, Cham (2017). https://doi.org/10.1007/978-3-319-59050-9_12
19. Shi, Y., Yang, J., Qi, Z.: Unsupervised anomaly segmentation via deep feature reconstruction. Neurocomputing **424**, 9–22 (2021)
20. Tan, M., Le, Q.: EfficientNet: rethinking model scaling for convolutional neural networks. In: International Conference on Machine Learning, pp. 6105–6114. PMLR (2019)
21. Van Den Oord, A., Vinyals, O., et al.: Neural discrete representation learning. Adv. Neural Inf. Process. Syst. **30**, 1–10 (2017)
22. Yi, J., Yoon, S.: Patch SVDD: patch-level SVDD for anomaly detection and segmentation. In: Proceedings of the Asian Conference on Computer Vision (2020)
23. You, Z., et al.: A unified model for multi-class anomaly detection. In: Neural Information Processing Systems (2022)
24. Yu, J., et al.: Fastflow: unsupervised anomaly detection and localization via 2D normalizing flows. arXiv preprint arXiv:2111.07677 (2021)
25. Zavrtanik, V., Kristan, M., Skočaj, D.: Draem-a discriminatively trained reconstruction embedding for surface anomaly detection. In: Proceedings of the IEEE/CVF International Conference on Computer Vision, pp. 8330–8339 (2021)

Deep Learning Based Personalized Stock Recommender System

Narada Wijerathne[1], Jamini Samarathunge[1], Krishalika Rathnayake[1](✉), Supuni Jayasinghe[1], Sapumal Ahangama[1], Indika Perera[1], Vinura Dhananjaya[2], and Lushanthan Sivaneasharajah[2]

[1] Department of Computer Science and Engineering, University of Moratuwa, Moratuwa, Sri Lanka
{narada.18,jamini.18,krishalika.18,supuni.18,sapumal, indika}@cse.mrt.ac.lk

[2] IronOne Technologies, Colombo, Sri Lanka
{vinurad,lushanthans}@irononetech.com

Abstract. This research paper introduces a personalized recommender system tailored specifically for the stock market. With the increasing complexity and variety of investment options, individual investors face significant challenges in making informed decisions. Traditional stock market recommendations often offer generic advice that fails to account for investors' unique preferences and risk appetites. We propose a personalized recommender system that utilizes deep learning techniques to provide customized stock recommendations. Our approach combines collaborative filtering (CF) and content-based (CB) filtering methodologies which form a hybrid system capable of generating personalized recommendations. Collaborative filtering utilizes the behaviour of similar investors to identify stocks that are likely to appeal to the user, while content-based filtering matches stock characteristics with the user's preferences and investment history. Experimental evaluations demonstrate that the proposed personalized recommender system outperforms existing algorithms and approaches trained on user interaction data taken from the stock market domain, providing investors with tailored stock recommendations that align with their personal needs and preferences.

Keywords: Stock Market · Hybrid Recommender System · Collaborative Filtering · Content-Based Approach · Implicit Feedback

1 Introduction

A significant part of our lives is spent online as a consequence of the Web 2.0 Revolution [1]. This has led to many user activities and preferences being stored electronically and to be studied. Meanwhile, online services such as streaming services and e-commerce sites offer an overwhelming number of options for users. A recommender system is a type of filtering mechanism designed to provide users with personalised recommendations by narrowing down the enormous number

B. Luo et al. (Eds.): ICONIP 2023, CCIS 1966, pp. 362–374, 2024.
https://doi.org/10.1007/978-981-99-8148-9_29

of options to a manageable number of options by turning the user data and preferences into predictions of the user's future possible likes and interests. Collaborative filtering (CF) and content-based (CB) approaches are well-known methods of recommendation [2].

The stock market is where investors buy and sell stocks (instruments) of companies. Investors in stock markets have different styles of investing and risk appetites combined with an enormous number of available stocks. By most estimates, approximately 630,000 companies are now traded publicly throughout the world [3]. Since the stock market is highly volatile [4], making the right decisions in stock investment is critical for an investor. Unlike in e-commerce and entertainment services, investors typically do not give explicit ratings for stocks they interact with which makes it difficult to extract investors' exact preferences. Existing recommender systems for the stock market domain focus on recommending items based on profitability alone which results in recommending the same set of items for every user irrespective of their investment portfolios [5–7]. The proposed system will provide recommendations for the users to select the stocks by themselves without getting support from a brokerage signifying that, this system will help to establish democratised stock investment.

The main contributions of this work are as follows.

1. We propose a hybrid personalized stock recommendation system combining the latest machine learning-based collaborative and content-based recommendation methods to identify different investment styles and risk appetites of the investors and recommend investment strategies based on the investment profiles.
   - We use a deep learning approach for collaborative filtering using the Bilateral Variational Autoencoder (BiVAE) model to generate personalized stock recommendations.
   - A Variational Autoencoder (VAE) is used in the content-based recommendation model to learn the latent feature representations of instruments.
2. We use stock order flow data to capture investors' preferences instead of user portfolios.

The rest of the paper is organized as follows. In Sect. 2, the related work is discussed. The methodology of the proposed model is presented in Sect. 3. Section 4 presents the experiments conducted on the proposed system. Finally, concluding remarks are given in Sect. 5.

2 Related Work

Researchers have been working hard over the past few years to investigate various approaches to improve recommender systems and the quality of recommendations. Among these approaches, CF, CB methods, and hybrid recommendation approaches have gained significant attention. In this study, we combined deep learning techniques with these models. Deep learning techniques help to enhance the quality of recommendations [8,9].

Limited research has been conducted on stock recommendations, despite the fact that many financial sectors, including banking, stock, equity, loan, and insurance, may benefit from recommendation principles [2]. When recommending useful stocks for investors to buy, recommender systems are critical. Taghavi, Bakhtiyari, and Scavino [10] developed a hybrid approach to the stock recommendation. They used a CB approach to explore investor preferences, macroeconomic variables, and stock portfolios and CF to make recommendations by assessing the investing practices of similar investors. A recommendation system for a diversified stock portfolio has been developed by Shen, Liu, Liu, Xu, Li, and Wang [11] utilizing reinforcement learning from the S&P 500 index in order to capture complex connections between stocks and sectors, the system creates a stock relation graph. Broman [12] constructed a stock recommender system based on investor portfolios by assessing different approaches using various data splitting and filtering techniques. According to the findings of these researchers' study, the LightGCN model performed better across all evaluation metrics than traditional approaches. Swezey and Charron [13] integrated CF with Modern Portfolio Theory, putting investment portfolio management under consideration in order to produce stock recommendations that are in line with individual preferences and risk aversion. In order to produce stock recommendations that are in line with individual preferences and risk aversion, putting investment portfolio management under consideration.

CF is a method to predict users' preferences by identifying the similarity between users and items [14]. Truong, Salah, and Lauw [15] have proposed CF based on a BiVAE. In the model, they have combined dyadic data with neural network parameterized item-based and user-based inference models. Following that, in this work, we used CF utilized with BiVAE to learn user-item interactions. Liang, Krishnan, Hoffman, and Jebara [14] improved their application to incorporate implicit feedback input, by utilizing VAE for CF. They contended that this nonlinear probabilistic model allows them to surpass the linear factor models which triumphed over CF research. Since this is a stock recommender system, implicit data must have to be used in our work as well. Another state-of-the-art approach in CF is the LightGCN model developed by He, Deng, Wang, Li, Zhang, and Wang [16]. That study facilitates a graph convolution model to predict more suitable recommendations. And the neural network-based CF (NCF) framework developed by He, Liao, Zhang, Nie, Hu, and Chua [17] offers a general solution for CF. That framework enriches NCF with nonlinearities by increasing the capabilities of matrix factorization and employing multi-layer perceptrons.

In this system, except for CF, we used a CB approach as well. In the CB approach, similarities between items are considered to recommend similar items to the past preferences of users [18]. Lops, Gemmis, and Semeraro [19] studied the state-of-art systems utilized in numerous application areas by deeply discussing traditional and innovative techniques for representing items and user profiles. In our study, we used a CB approach with VAE. VAE can be used to obtain latent representation for input data [20].

Chikhaoui, Chiazzaro, and Wang [21] used a hybrid approach to create a recommender system for dynamically forecasting ratings. Methodology on the combination of CF, CB and demographic approaches by assigning a level of confidence to each approach has been discussed in that study. The number of user reviews available was used to establish the confidence level. We followed the findings of this study to implement the hybrid approach that combines the results of CF and CB approaches. Additionally, Oyebode and Orji [22] studied a dynamic weighting technique to combine the results of different recommender systems.

3 Methodology

3.1 Preprocessing

In the implementation [1], market data and order details datasets are processed together to capture user-item interactions from order details data. First, the buy orders are filtered out from the order details dataset since we only need to generate recommendations for buy orders. The mean market price for the executed month is calculated for each stock available in the market data dataset. Then the two datasets are combined in such a way that each buy order contains the order price as well as the mean market price for the stock.

$$OrderPrice = MarketPrice * (1 - \frac{neighborhood}{100}) \quad (1)$$

If the order price falls in the range defined above around the mean market price, that order is considered to be Genuine. All user-item pairs in genuine buy orders are used to create the user-item interaction matrix.

3.2 Collaborative Filtering Approach

CF approach considers user-item interaction learned through a BiVAE. BiVAE is a state-of-art deep learning model for collaborative filtering. In comparison to the traditional one-sided VAE, the BiVAE is capable of capturing uncertainty from both perspectives of dyadic data, which refers to user-item interactions. This capability enhances the model's robustness and performance, particularly when dealing with sparse preference data [15]. The BiVAE model consists of a pair of inference models containing two VAEs for user and item data separately. Prepared user embeddings and item embeddings from the user-item interaction matrix were fed into each user and item encoder (Fig. 1). Each encoder was developed such that it returns the mean and covariance matrices. The latent variable of each VAE was taken by reparameterizing these two matrices with epsilon. These latent variables were passed into the decoders to reconstruct the original user and item data points. The predictions were calculated by using the means of latent representations.

[1] For the source code, visit the Git repository: https://github.com/Stock-Recommender-System-FYP/stock-recommender-system.

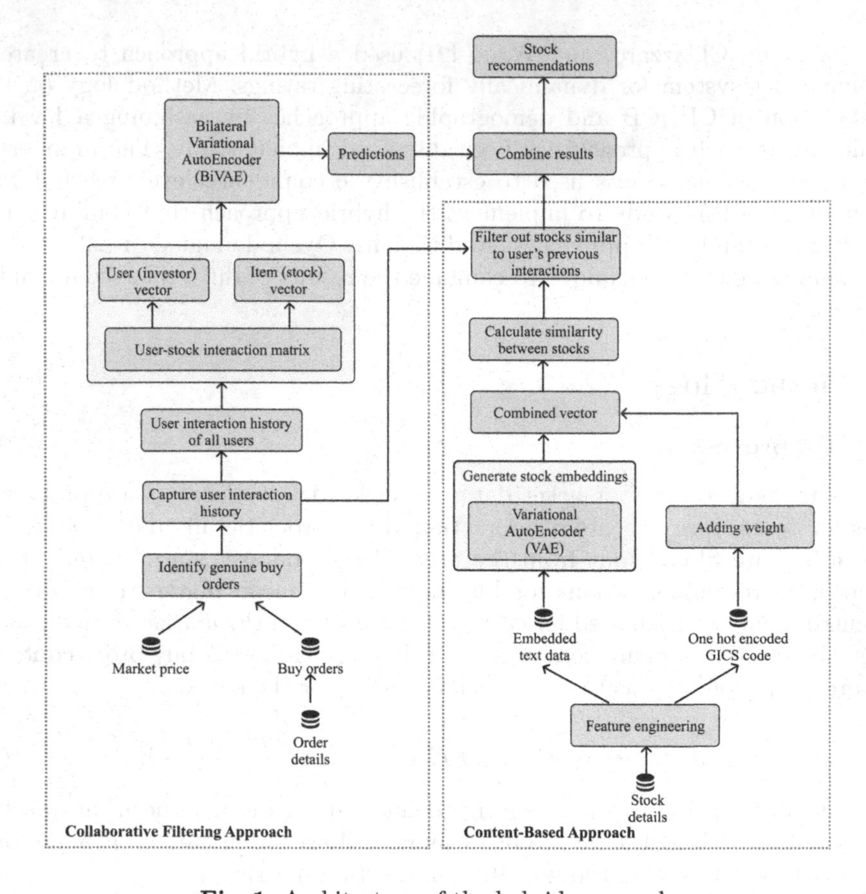

Fig. 1. Architecture of the hybrid approach

From the preprocessed order flow dataset, experiments were conducted for the CF model by changing minimum items and minimum interaction counts. Here "minimum item count" refers to the count that determines how many times each stock should be interacted with in the dataset and "minimum interaction count" refers to the minimum count of interactions each user interacted with.

3.3 Content Based Approach

The main objective of the CB approach is to recommend comparable stocks to a user by considering their previous interactions.

- Ex: If a user buys 'JETS' (JETWING SYMPHONY PLC), then 'MRH' (MAHAWELI REACH HOTELS PLC) is a possible recommendation.

CB approach part of Fig. 1 displays the flow of the CB approach. The feature engineering was mostly conducted to convert the text data to vectors. Hence, one hot encoding was used for the GICS code, and for the business name and business summary, the hashing vectorizer with a total of 5000 columns was used. For each text data (business name, business summary) a vector with 5000 was obtained. This leads to the final dimension of the combined embedded vector to 10000. VAE [14] was applied to reduce the dimension of the embedded text data and create the item latent vector of size 50. In this scenario, we avoided combining one hot encoded GICS code with embedded text data before applying VAE, because they are more important when representing stocks. Therefore they were used without reducing dimensions and we added a weight to give them more significance. Combined vectors of stocks were used to calculate the similarity scores between each other. Ultimately, the similarity matrix and the user interaction history dataset were used to find similar stocks to users' previous interactions.

3.4 Hybrid Approach

The final prediction scores in this study were obtained through a hybrid recommender system which is a combination of the above-mentioned CF and CB approaches. The integration of these approaches enhanced the quality and relevance of the recommendations.

When integrating CF and CB recommendations a dynamic weight was assigned to each recommendation based on each user's user-item interaction count. This weight determines the contribution of both approaches to the final prediction scores. The final prediction score was calculated using Eq. 2.

$$r_{u,i} = \alpha CF_{u,i} + (1 - \alpha) CB_{u,i} \tag{2}$$

where the $CF_{u,i}$ denotes the prediction score generated by CF approach and the $CB_{u,i}$ denotes the prediction score generated by the CB approach.

$$\alpha = \frac{1}{1 + e^{-\beta n}} \tag{3}$$

The sigmoid function was employed to calculate the weight α assigned to each recommendation as it can provide more weight to users with a higher number of interactions than the users with less number of interactions [21]. Equation 3 was used to calculate α. According to the NDCG scores, the highest weight was given to the CF approach.

The minimum value for n was 15 as this study considered the 3,15 user-item interaction split. The CB approach could achieve its highest weight when n equalled 15. Equation 4 was used to calculate the highest weight for the CB approach which considered the NDCG score of both CF and CB approaches. β in Eq. 3 was selected as $\frac{1}{5}$ based on the obtained maximum weight for the CB approach.

$$CB\_max\_weight = \frac{CB\_NDCG\_score}{CB\_NDCG\_score + CF\_NDCG\_score} \tag{4}$$

This dynamic weighted hybrid recommender system adapts to varying user behaviours and interests. Therefore, it ensures personalised recommendations for each user.

3.5 Datasets and Evaluation Metrics

Datasets. We have 3 main datasets which are stock details, order flow data and market data.

Stock Details. This dataset contains the details related to a stock listed in the Colombo stock exchange [23]. Stock symbol, Business name, Business summary and GICS (Global Industry Classification Standard) code are the main features in this dataset. By the time we used this dataset, it contained 294 companies (This count can be changed) and 38 GICS groups.

Order Flow Data. Order flow dataset contains details about all buy orders placed through the stock trading system from August to December 2022. Masked user ID, Security, Order Price, Executed price and Trade date were taken from this dataset.

Market Data. Market dataset contains details about all the successful trades completed through the trading system. Security, Trade date and Executed price from this dataset were acquired from this dataset.

Evaluation Metrics. Hit Ratio (HR) and Normalised Discounted Cumulative Gain (NDCG) [24] were used to measure the performance of the obtained stock recommendation list. The existence of the test item on the top-L list is estimated by the Hit Ratio. NDCG takes into account the position of a hit and assigns higher scores to hits that appear at the top ranks. These evaluation metrics were important to measure the both relevance of results and their ordering.

4 Experiments

The primary objective of this research is to implement and evaluate a personalized stock recommendation system to be used in the stock market domain. The proposed approach employs deep learning techniques to generate recommendations. Our aim is to capture each user's unique risk appetite and investment strategy by leveraging buy order flow data. This research also looks to evaluate generated recommendations and evaluate them with the user's historical preferences to measure effectiveness.

Table 1. Proposed hybrid recommender system performance against existing approaches

Method	Top_k = 5		Top_k = 10		Top_k = 15		Top_k = 20	
	NDCG	Hit ratio %	NDCG	Hit ratio %	NDCG	Hit ratio %	NDCG	Hit ratio %
BPR	0.2956	20.90	0.3171	31.33	0.3475	38.77	0.3740	45.90
LightGCN	0.3216	22.74	0.3435	34.68	0.3773	42.73	0.4049	49.71
VAE	0.3232	21.88	0.3454	33.79	0.3789	44.10	0.4068	51.43
CF (BiVAE)	**0.3408**	**24.54**	**0.3676**	**36.48**	**0.4018**	**45.93**	**0.4290**	**52.52**
Hybrid	**0.3432**	**24.89**	**0.3714**	**37.23**	**0.4050**	**45.68**	**0.4324**	**51.83**

The results from this research disclose multiple insights about the effectiveness of a personalized recommendation system in a stock brokering application. Through extensive evaluation, it was observed that the system achieves an NDCG score of 0.3714 and a Hit Ratio of 37.23% for top 10 recommendation results (Table 1). This result implies that the system is capable of capturing user preference and recommending new stocks according to the investor's personal taste.

Collaborative Filtering. Implemented CF-based recommender system shows higher NDCG scores when the minimum interaction count is set to a higher value compared to lower values (Table 3). This implies that the BiVAE learns a better latent representation of the user. When the minimum interaction count is higher, the original user vector taken from the utility matrix is less sparse which means that the user encoder of BiVAE is fed more data. However, when the minimum number of interactions goes above 20, NDCG starts to decline again. This indicates that the system does not have enough data points to learn from even though the vectors are less sparse.

Table 2. Total interactions with respect to the minimum item and minimum interaction values

Data split		Users	Stocks	Interaction count
Minimum item count	Minimum interaction count			
1	4	3285	272	41242
3	10	1486	263	30457
5	10	1486	258	30446
3	15	873	263	23320
5	15	873	258	23309
3	20	546	263	17843
5	20	546	258	17833

Table 3. Recommender system performance with respect to different min item and min interactions constraint values

Minimum item count	Minimum interaction count	Top_k = 5		Top_k = 10		Top_k = 15		Top_k = 20	
		NDCG	Hit ratio %	NDCG	Hit ratio %	NDCG	Hit ratio %	NDCG	Hit ratio %
1	4	0.2882	31.16	0.3305	43.41	0.3578	51.35	0.3776	57.59
3	10	0.3139	26.51	0.3530	38.90	0.3863	47.01	0.4112	54.04
5	10	0.3117	26.75	0.3513	39.30	0.3841	47.59	0.4092	53.87
3	15	0.3408	24.54	0.3676	36.48	0.4018	45.93	0.4290	52.52
5	15	0.3407	24.68	0.3674	36.37	0.4010	45.70	0.4288	52.23
3	20	0.3510	23.12	0.3571	35.39	0.3904	43.13	0.4192	48.95
5	20	0.3590	20.92	0.3643	31.96	0.3954	41.76	0.4242	48.12

Contrary to the above observation, the Hit ratio of CF recommendations decreases with the threshold we set for the minimum number of interactions (Table 3). When this threshold is lowered, it increases the number of users added to the training dataset which allows the model to learn from more user-item interactions (Table 2). Even though the quality of the order decreases, the recommender system learns to generate hits more often.

Since our dataset has a massive difference between the number of users and the number of items, we decided to test different encoder structures for users and items in our BiVAE model. We find that it only makes a marginal difference in NDCG and Hit Ratio scores of generated recommendations.

Content Based Approach. Compared to CF approaches, CG filtering does not provide good NDCG and Hit Ratio scores (Table 4). Initially, the GICS code was concatenated to the text embeddings which were fed to the VAE for dimensionality reduction. As a step to enhance the recommendations, one hot-encoded GICS code was added after reducing the dimension of the text embeddings through the VAE. This improved the performance of the CB model to some extent. This implies that data about the GICS code of a specific stock contains valuable information and needs more weight in vector embedding compared to text data.

Table 4. Hit Ratio@k% and NDCG@k at k=5, 10, 15, 20 for content-based approach

	k = 5	k = 10	k = 15	k = 20
NDCG	0.0143	0.0235	0.0330	0.0417
Hit Ratio%	1.406	3.389	5.649	7.817

In order to convert text data into numerical vector format, Count vectorizer, TF-IDF and Hashing vectorizer were tested. Changing vectorizers did not show any significant improvement in the NDCG or Hit Ratio scores. Hashing vectorizer was selected based on its low total loss (Fig. 2) during the dimensionality reduction process through the VAE.

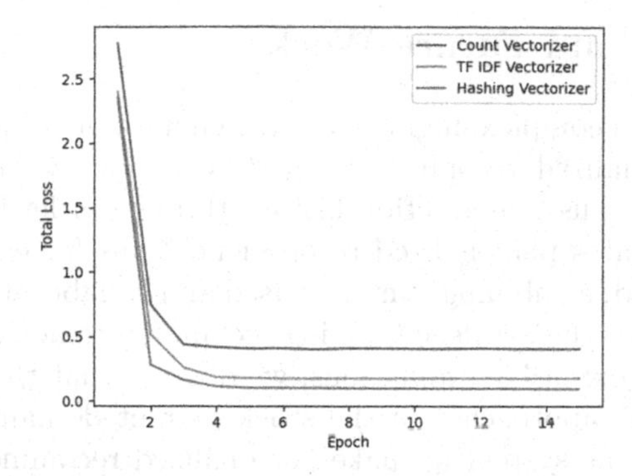

Fig. 2. Total Loss of Each Epoch for Text Vectorizers

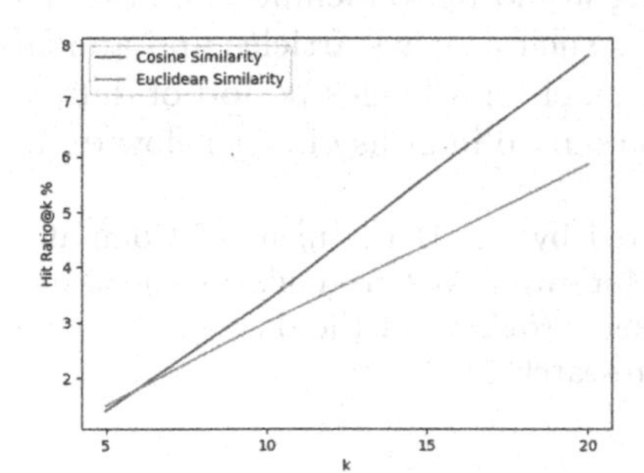

Fig. 3. Hit Ratio for Similarity Metrics

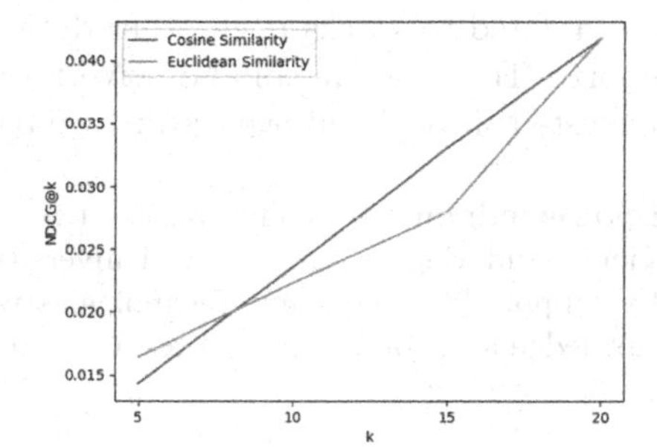

Fig. 4. NDCG for Similarity Metrics

Cosine similarity was chosen as the similarity measure to calculate the semantic similarity between the final latent vectors of stocks [25]. After evaluating the performance against Euclidean distance, cosine similarity showed consistency when the number of recommendations(K) generated was increased. Euclidean distance performed better at the lower K values but was not consistent with higher K values (Fig. 3 & Fig. 4).

Text content for CB approach only included 2 textual features extracted from stock market itself.

Hybrid Approach. Combining the CF and CB approaches through a sigmoid function slightly improved the overall NDCG and Hit Ratio values (Table 1). Above all it allows the system to generate recommendations for new users and items which couldn't make the minimum interaction and minimum item occurrence limits we set for collaborative filtering.

5 Conclusion and Future Work

The stock market is a complex, highly dynamic environment. This research paper introduces a personalized recommender system explicitly made for the stock market. By capturing user interaction history through order flow data, the proposed system generates personalized recommendations for each user. A hybrid approach is proposed, combining content-based and collaborative filtering implemented using state-of-the-art deep learning techniques to achieve maximum performance. Through extensive evaluations, we observe that the proposed system outperforms existing approaches in the stock market domain which highlights the effectiveness of the system to make personalized recommendations and the potential to democratize investing for everyone.

An improvement for content-based filtering would be to include more textual data related to stocks from other data sources such as news articles and annual reports. Training the collaborative filtering model on a longer period of data is suggested since the above results are based on only 6 months of order flow data.

Acknowledgements. This research is conducted by the Department of Computer Science and Engineering of the University of Moratuwa. We gratefully acknowledge the support from IronOne Technologies with their provision of the datasets, domain knowledge and computational resources in this research.

Disclaimer. It is advisable to interpret the results of experiments involving stock market transactions with caution due to the personalized behaviour considered.

References

1. Kwon, J., Kim, S.: Friend recommendation method using physical and social context. Int. J. Comput. Sci. Netw. Secur. **10**(11), 116–120 (2010)
2. Sharaf, M., Hemdan, E.E.-D., El-Sayed, A., El-Bahnasawy, N.A.: A survey on recommendation systems for financial services. Multimed. Tools Appl. **81**(12), 16761–16781 (2022). https://doi.org/10.1007/s11042-022-12564-1
3. Stock exchanges around the world, Investopedia. https://www.investopedia.com/financial-edge/1212/stock-exchanges-around-the-world.aspx. Accessed 30 May 2023
4. Chikwira, C., Mohammed, J.I.: The impact of the stock market on liquidity and economic growth: evidence of volatile market. In: Proceedings of the 25th International Conference on Economics, vol. 11, no. 6, pp. 155 (2023). https://doi.org/10.3390/economies11060155
5. Nair, B.B., Mohandas, V.P.: An intelligent recommender system for stock trading. Intell. Decis. Technol. **9**(3), 243–269 (2015). https://doi.org/10.3233/IDT-140220
6. Gottschlich, J., Hinz, O.: A decision support system for stock investment recommendations using collective wisdom. Decis. Support Syst. **59**(1), 52–62 (2014). https://doi.org/10.1016/j.dss.2013.10.005
7. Paranjape-Voditel, P., Deshpande, U.: A stock market portfolio recommender system based on association rule mining. Appl. Soft Comput. J. **13**(2), 1055–1063 (2013). https://doi.org/10.1016/j.asoc.2012.09.012

8. Bobadilla, J., Alonso, S., Hernando, A.: Deep learning architecture for collaborative filtering recommender systems. Appl. Sci. **10**(7), 2441 (2020). https://doi.org/10.3390/app10072441
9. Batmaz, Z., Yurekli, A., Bilge, A., Kaleli, C.: A review on deep learning for recommender systems: challenges and remedies. Artif. Intell. Rev. **52**(1), 1–37 (2019). https://doi.org/10.3390/app122111256
10. Taghavi, M., Bakhtiyari, K., Scavino, E.: Agent-based computational investing recommender system. In: Proceedings of the 7th ACM Conference on Recommender Systems, pp. 455–458. Association for Computing Machinery, Hong Kong (2013). https://doi.org/10.1145/2507157.2508072
11. Shen, Y., Liu, T., Liu, W., Xu, R., Li, Z., Wang, J.: Deep reinforcement learning for stock recommendation. J. Phys: Conf. Ser. **2050**(1), 12012 (2021). https://doi.org/10.1088/1742-6596/2050/1/012012
12. Broman, N.: Comparison of recommender systems for stock inspiration (2021)
13. Swezey, R.M.E., Charron, B.: Large-scale recommendation for portfolio optimization. In: Proceedings of the 12th ACM Conference on Recommender Systems, pp. 382–386. Association for Computing Machinery, Canada (2018). https://doi.org/10.1145/3240323.3240386
14. Liang, D., Krishnan, R.G., Hoffman, M.D., Jebara, T.: Variational autoencoders for collaborative filtering. In: Proceedings of the 2018 World Wide Web Conference, pp. 689–698. International World Wide Web Conferences Steering Committee, Lyon (2018). https://doi.org/10.1145/3178876.3186150
15. Truong, Q.-T., Salah, A., Lauw, H.W.: Bilateral variational autoencoder for collaborative filtering. In: Proceedings of the 14th ACM International Conference on Web Search and Data Mining, pp. 292–300 (2021). https://doi.org/10.1145/3437963.3441759
16. He, X., Deng, K., Wang, X., Li, Y., Zhang, Y., Wang, M.: LightGCN: simplifying and powering graph convolution network for recommendation. In: Proceedings of the 43rd International ACM SIGIR Conference on Research and Development in Information Retrieval, pp. 639–648 (2020). https://doi.org/10.1145/3397271.3401063
17. He, X., Liao, L., Zhang, H., Nie, L., Hu, X., Chua, T.-S.: Neural collaborative filtering. In: Proceedings of the 26th International Conference on World Wide Web, pp. 173–182 (2017). https://doi.org/10.1145/3038912.3052569
18. Rajaraman, A., Ullman, J.D.: Mining of Massive Datasets, pp. 319–353. Cambridge University Press (2011)
19. Lops, P., De Gemmis, M., Semeraro, G.: Content-based recommender systems: state of the art and trends. Recomm. Syst. Handb. 73–105 (2011). https://doi.org/10.1007/978-0-387-85820-3_3
20. Li, X., She, J.: Collaborative variational autoencoder for recommender systems. In: Proceedings of the 23rd ACM SIGKDD International Conference on Knowledge Discovery and Data Mining, pp. 305–314 (2017). https://doi.org/10.1145/3097983.3098077
21. Chikhaoui, B., Chiazzaro, M., Wang, S.: An improved hybrid recommender system by combining predictions. In: Proceedings of the 2011 IEEE Workshops of International Conference on Advanced Information Networking and Applications, pp. 644–649 (2011). https://doi.org/10.1109/WAINA.2011.12
22. Oyebode, O., Orji, R.: A hybrid recommender system for product sales in a banking environment. J. Bank. Financ. Technol. **4**, 15–25 (2020). https://doi.org/10.1007/s42786-019-00014-w

23. CSE Homepage. https://www.cse.lk/. Accessed 30 May 2023
24. He, X., Chen, T., Kan, M.-Y., Chen, X.: TriRank: review-aware explainable recommendation by modeling aspects. In: CIKM, pp. 1661–1670 (2015). https://doi.org/10.1145/2806416.2806504
25. Jurafsky, D., Martin, J.H.: Speech and Language Processing. Pearson (2019)

Feature-Fusion-Based Haze Recognition in Endoscopic Images

Zhe Yu[1,2], Xiao-Hu Zhou[1,2(✉)], Xiao-Liang Xie[1,2], Shi-Qi Liu[1,2], Zhen-Qiu Feng[1,2], and Zeng-Guang Hou[1,2,3,4(✉)]

[1] The State Key Laboratory of Multimodal Artificial Intelligence Systems, Institute of Automation, Chinese Academy of Sciences, Beijing 100190, China
{yuzhe2021,xiaohu.zhou,zengguang.hou}@ia.ac.cn

[2] The School of Artificial Intelligence, University of Chinese Academy of Sciences, Beijing 100049, China

[3] The CAS Center for Excellence in Brain Science and Intelligence Technology, Beijing 100190, China

[4] The Joint Laboratory of Intelligence Science and Technology, Institute of Systems Engineering, Macau University of Science and Technology, Taipa, Macau

Abstract. Haze generated during endoscopic surgeries significantly obstructs the surgeon's field of view, leading to inaccurate clinical judgments and elevated surgical risks. Identifying whether endoscopic images contain haze is essential for dehazing. However, existing haze image classification approaches usually concentrate on natural images, showing inferior performance when applied to endoscopic images. To address this issue, an effective haze recognition method specifically designed for endoscopic images is proposed. This paper innovatively employs three kinds of features (i.e., color, edge, and dark channel), which are selected based on the unique characteristics of endoscopic haze images. These features are then fused and inputted into a Support Vector Machine (SVM) classifier. Evaluated on clinical endoscopic images, our method demonstrates superior performance: (Accuracy: 98.67%, Precision: 98.03%, and Recall: 99.33%), outperforming existing methods. The proposed method is expected to enhance the performance of future dehazing algorithms in endoscopic images, potentially improving surgical accuracy and reducing surgical risks.

Keywords: Endoscopic image · Haze recognition · Feature fusion

1 Introduction

Recent technological advancements in endoscopy have enabled surgeons to visualize surgical fields clearly during surgeries. Endoscopic images provide a view that is nearly equivalent to the surgeon's own eyes. However, some haze may be generated during surgical operations, which can obscure visibility and significantly affect a surgeon's ability to operate. Therefore, dehazing task for endoscopic images is necessary.

B. Luo et al. (Eds.): ICONIP 2023, CCIS 1966, pp. 375–387, 2024.
https://doi.org/10.1007/978-981-99-8148-9_30

Although many studies have focused on the dehazing task, existing image dehazing methods often ignore whether an image contains haze and are applied to the image directly. However, for endoscopic images, it is necessary to distinguish whether an image needs dehazing. This is important for two reasons: first, if the dehazing algorithm is used directly without judging whether the image contains haze, the resulting image may have lower visibility than the original image. Second, using the dehazing algorithm is time-consuming and can affect the real-time performance of endoscopic surgery images. Selectively using the dehazing algorithm can reduce the time required. Therefore, a haze image classification model is needed to improve the performance of dehazing algorithms.

Many studies have used traditional Machine Learning (ML)-based approaches to classify haze images. Yu *et al.* [16] extracted three features(image visibility, intensity of dark channel, and image contrast) and combined them with Support Vector Machine (SVM) to distinguish between haze and non-haze images. Zhang *et al.* [17] used the variance of HSI images as a feather and computed the angular deviation of different haze images compared to clear day images as another feature, they also used SVM for classification. Pal *et al.* [10] proposed a framework for image classification based on nine features and K-Nearest Neighbor (KNN) method. In [13], nineteen classification techniques were applied using five training parameters: area, mean, minimum intensity, maximum intensity, and standard deviation. Additionally, Wan *et al.* [15] used Gaussian Mixture Model to learn the probability density of three situations: non-haze, haze, and dense haze. The model's parameters were learned using the expectation-maximization algorithm.

Apart from traditional ML-based approaches, some methods are proposed based on Convolutional Neural Networks (CNN). Guo *et al.* [7] combined transfer learning and proposed a haze image classification method based on the Alexnet network transfer model. Pei *et al.* [11] proposed an end-to-end Consistency Guided Network, for classifying haze images. In [12], the authors proposed a deep neural network to identify haze images. Chincholkar *et al.* [5] presented the implementation of CNN to detect and classify the images into haze and non-haze based on the factors such as image brightness, luminance, intensity, and variance.

Existing methods usually focus on natural images, showing inferior performance when applied to endoscopic images. Therefore, this paper proposes a method to judge whether endoscopic images are haze or non-haze. The method implementation process is as follows: first, images are captured from different endoscopic surgery videos to form the training and test set. After extensive experimental analyses, three kinds of features including the color feature, the edge feature, and the dark channel feature are extracted from images in the training set and then fused together. An SVM model is trained to identify whether an image is haze or non-haze. The trained model is tested using features extracted from the test set and demonstrates high classification results.

The main contributions of this paper can be summarised as follows:

- Based on the unique characteristics of endoscopic haze images, three kinds of features (i.e., color, edge, and dark channel) are extracted from images and fused to distinguish between haze and non-haze endoscopic images.
- Various machine learning algorithms and fusion strategies based on the three kinds of features are experimented with the goal of optimizing the classification results.
- Extensive experiments demonstrate that our method can achieve a classification accuracy of 98.67% on endoscopic images, surpassing existing approaches.

The remainder of this paper is structured as follows: Sect. 2 depicts the proposed method. The experimental results are presented in Sect. 3. Section 4 discusses the advantages of our method. Finally, the conclusion of this paper is provided in Sect. 5.

2 Method

2.1 Features Extraction

In order to accurately distinguish between haze images and non-haze images, extracting appropriate features is the key point. This section introduces the three kinds of features extracted from images, which are the color feature, the edge feature, and the dark channel feature. All the features are fused to form the final feature.

Color Feature. Generally, haze can cause color changes in images, so the first kind of feature chosen is the color feature. There are two commonly used categories of color models: the hardware oriented and the color processing application oriented color model. The RGB color model is the most widely used hardware oriented color model owing to easily storing and displaying images on computers. However, the non-linearity with visual perception makes it unsuitable for this task. Therefore, the HSI color model, which is widely used in image processing applications, is chosen instead.

The HSI color model is based on hue, saturation, and intensity, which conforms to how people describe and interpret colors. Hue (H) represents the main pure color perceived by an observer with a range of angles between 0 and 2π. Red is represented by both 0 and 2π, green is represented by $2/3\pi$, and blue is represented by $4/3\pi$. Saturation (S), ranging from 0 to 1, refers to the relative purity of a color, or the amount of white light mixed with a color. Intensity (I) refers to the brightness of light, also ranging from 0 to 1, where black is represented by 0, and white is represented by 1. In contrast to the RGB color model, the HSI color model separates brightness and color information, making it more advantageous for endoscopic haze classification tasks. HSI values can be computed from RGB values as follows, where R, G, and B represent values of the three color channels in the image:

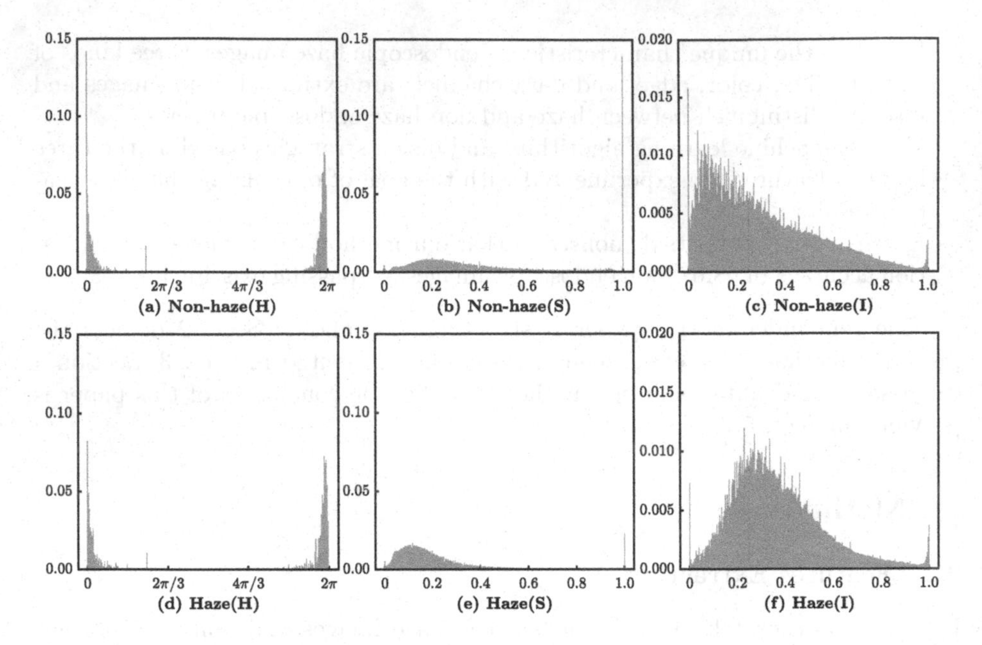

Fig. 1. The average HSI histograms of haze images and non-haze images.

$$\begin{cases} H = \begin{cases} \theta, & B \le G \\ 2\pi - \theta, & B > G \end{cases} \\ S = 1 - \frac{3}{(R+G+B)} \left[\min(R, G, B) \right] \\ I = \frac{1}{3} (R + G + B) \end{cases} \tag{1}$$

$$\theta = \arccos \left\{ \frac{\frac{1}{2} \left[(R - G) + (R - B) \right]}{\left[(R - G)^2 + (R - B)(G - B) \right]^{\frac{1}{2}}} \right\} \tag{2}$$

Once the HSI color model is obtained, separate histograms are drawn for H, S, and I. Histograms can directly reflect the distribution of variables. Figure 1 shows the average H, S, and I histograms of 230 haze images and 270 non-haze images. Figure 1(a) and (b) show that haze changes the hue distribution. Non-haze images mainly contain a red hue owing to their origin in endoscopic surgery videos. In contrast, haze images exhibit a decreased ratio of pure red and a more dispersed hue distribution. Additionally, comparing Fig. 1(b) and (e), Fig. 1(c) and (f), it is evident that the haze also reduces saturation and increases intensity by adding more white light to images.

Edge Feature. One distinct feature for effectively distinguishing between haze and non-haze images is the image edge feature. Image edges contain rich information, the more details the edges of the image contain, the higher sharpness the image has. Moreover, haze reduces image sharpness and blurs image edges.

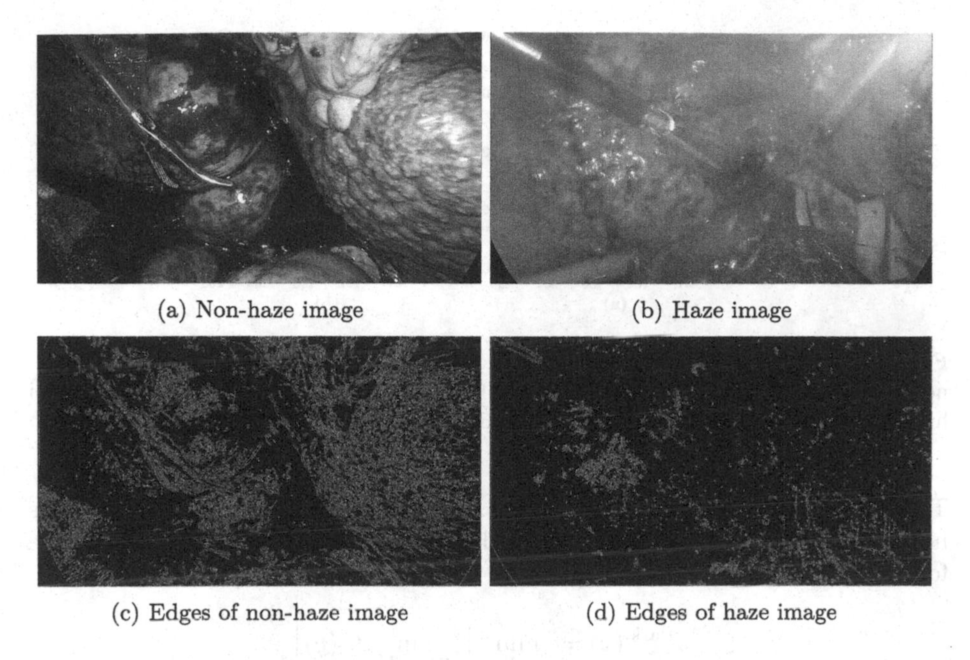

(a) Non-haze image (b) Haze image

(c) Edges of non-haze image (d) Edges of haze image

Fig. 2. The edges of haze image and non-haze image.

To obtain the edge information of an image, it is necessary to choose a suitable method. There are many common methods for edge detection, such as Sobel operator, Kirsch operator, Laplacian operator, and Canny operator. Among them, the Canny operator is chosen. Canny edge detection algorithm is a multi-level edge detection algorithm developed by John F. Canny in 1986 [2]. This algorithm is considered by many to be the best edge detection algorithm. Compared with other edge detection algorithms, its accuracy in recognizing image edges is much higher and it has more accurate positioning of edge points. These advantages enable the algorithm to process blurry images, such as haze images. To use the Canny edge detection algorithm, the original image is first converted to a grayscale image. The algorithm consists of four steps: noise reduction, gradient calculation, non-maximum suppression, and double threshold edge tracking by hysteresis. After these steps, a binary edge image is obtained from the original image. Figure 2 shows the edges of a haze image and a non-haze image obtained by the Canny operator, clearly demonstrating that haze reduces image edges. Therefore, the edge feature of images can be used as another classification basis.

Dark Channel Feature. He *et al.* [8] proposed that in a non-haze image divided into multiple sub-blocks, at least one color channel of some pixels in each sub-block will have a very low value (nearly zero) in the non-sky region. This law is called the prior law of dark channel and the channel with the lowest value in the three color channels is called the dark channel. Conversely, in a haze area, the intensity of the dark channel is higher because of the influence of haze.

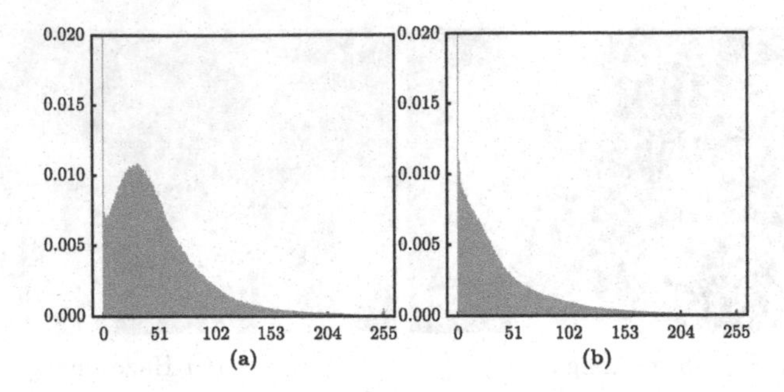

Fig. 3. Dark channel histograms of haze images and non-haze images. (a) Dark channel histogram of haze images and probability of 0 value is 0.2100 (b) Dark channel histogram of non-haze images and probability of 0 value is 0.5711

Therefore, this prior law can be used to distinguish whether the image contains haze or not. To obtain the dark channel of an image, perform according to the following formula:

$$J^{\text{dark}}(x) = \min_{y \in \Omega(x)} \left[\min_{c \in \{r,g,b\}} J^c(y) \right] \tag{3}$$

J^{dark} means dark channel value, $\Omega(x)$ means a local patch centered at x and J^c means value of each color channel. After conducting numerous experiments, a patch size of 15×15 is selected. Figure 3 displays the average histograms of the dark channel for 230 haze images and 270 non-haze images separately. There is a clear contrast between the dark channel histograms of haze and non-haze images, it can be seen that haze makes the distribution of the dark channel shift backward and reduces the proportion of pixels having zero dark channel value. Therefore, distinguishing between haze and non-haze images can be achieved by analyzing the dark channel histogram.

2.2 Classification Using SVM

SVM is a frequently used classifier in supervised learning. SVM developments started from Boser *et al.* [1] in 1992, this kind of SVM is known as the hard margin SVM. And in 1995, Cortes and Vapnik [6] improved SVM to the soft margin SVM in order to reduce the influence of noise and outliers by introducing slack variables. The soft margin SVM is the most popular and widely used SVM, which is also used in our classification task, given its strong performance with high dimensional data and small sample data. SVM algorithm aims to find the optimal hyperplane, which can not only separate two classes of samples correctly but also maximize the width of the gap between them. The widest gap is called optimal margin and the borderline instances are called Support Vectors [4]. For binary classification tasks, it is supposed that the training set comprises points of the form (x_i, y_i), $y_i \in \{-1, +1\}$, for $i = 1, 2, \ldots, m$, where $x_i \in \mathbb{R}^n$, n represents the number of features and m represents the number of points in the training set. Then the hyperplane is set to be $\omega^T x + b = 0$. The hard margin SVM solves the following optimization problem to obtain the parameters ω, b:

$$\begin{aligned} &\min_{\omega,b} \tfrac{1}{2}\|\omega\|^2 \\ s.t.\quad & y_i\left(\omega^T x_i + b\right) \geq 1, i = 1, 2, \ldots, m \end{aligned} \tag{4}$$

Formula 4 represents the prime form of the hard margin SVM which can be solved using Lagrangian Multipliers. The dual form is as follows, where α is vector of dual variables α_i:

$$\begin{aligned} &\max_\alpha -\tfrac{1}{2}\textstyle\sum_{i=1}^m \sum_{j=1}^m \alpha_i\alpha_j y_i y_j x_i^T x_j + \sum_{i=1}^m \alpha_i \\ s.t.\quad & \textstyle\sum_{i=1}^m \alpha_i y_i = 0, \alpha_i \geq 0, i = 1, 2, \ldots, m \end{aligned} \tag{5}$$

As for the soft margin SVM, to reduce the impact of noise and outliers, slack variables are introduced as $\xi_i = \max[0, 1 - y_i(\omega^T x_i + b)]$. This modification changes the prime form to:

$$\begin{aligned} &\min_{\omega,b} \tfrac{1}{2}\|\omega\|^2 + C\textstyle\sum_{i=1}^m \xi_i \\ s.t.\quad & y_i\left(\omega^T x_i + b\right) \geq 1 - \xi_i, i = 1, 2, \ldots, m \end{aligned} \tag{6}$$

C is the penalty parameter to control the punishment of outliers, and it is a hyper-parameter that needs to be set in advance and adjusted during application. Additionally, the dual form of the problem is obtained:

$$\begin{aligned} &\max_\alpha -\tfrac{1}{2}\textstyle\sum_{i=1}^m \sum_{j=1}^m \alpha_i\alpha_j y_i y_j x_i^T x_j + \sum_{i=1}^m \alpha_i \\ s.t.\quad & \textstyle\sum_{i=1}^m \alpha_i y_i = 0, 0 \leq \alpha_i \leq C, i = 1, 2, \ldots, m \end{aligned} \tag{7}$$

For non-linear separable data, the soft margin SVM may not perform well, so kernel functions are utilized. Kernel functions can transform the low-dimensional space to high-dimensional space where the feature is linearly separable. SVM with kernel function doesn't need to calculate the mapping of feature to high dimensional space; instead, it only needs to calculate the inner product of the mapping. Therefore, the cost of using SVM with kernel functions is lower than that of other non-linear classifiers.

To mitigate the impact of noise and outliers, the soft margin SVM is chosen to be used. Given the small scale of our training set, reducing the dimension of the feature vector is crucial for improving classification results. Thus, for the color feature and the dark channel feature, 256-dimensional histograms are simplified to 16-dimensional histograms. Additionally, since the edge feature histogram reflects a binary image with only two values, the 0 value of the histogram is obtained. After that, these features are integrated together as a 65-dimention final feature which is utilized in SVM.

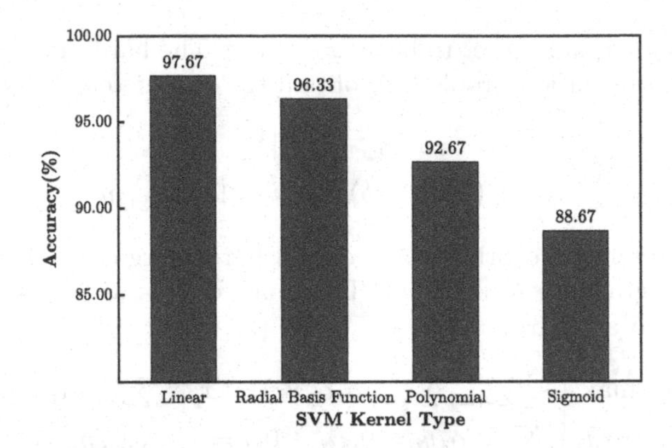

Fig. 4. Performance of SVM using different kernel.

3 Experiments

3.1 Dataset and Experimental Settings

The training set comprises 500 endoscopic images, including 230 haze images and 270 non-haze images, which is obtained from an endoscopic surgery video. Additionally, a test set of 300 images including 150 haze images and 150 non-haze images is obtained from another video. The SVM model used is C-SVM in LIBSVM [3], and the SVM classifier is trained using the feature set and labels obtained from the training set. The parameters are chosen through 10-fold cross-validation. Subsequently, the accuracy rate is assessed by inputting the feature set of the test set into the SVM model and comparing the classification results with the labels. All experiments are conducted on a PC with a CPU (Intel i7 12700H, 2.30 GHz). The input endoscopic images, which are manually labeled, have a resolution of 1920×1080.

3.2 Classification Results with Different Kernel and Dimension

The classification effect of different kernel functions, including linear function, polynomial function, radial basis function, and sigmoid function is compared to select the appropriate kernel function for SVM. The selected kernel function is the linear function, as it achieves the highest classification accuracy, as shown in Fig. 4. Next, various dimensions of histograms are contrasted. Figure 5 shows how changes in histogram dimensions from 8 to 256 affect classification performance. Dimensions of 8, 16, 32, 64, 128, and 256 are chosen to ensure that all the modified columns consist of an integral number of columns. The results indicate that the dimension of the histogram have little impact on classification accuracy after each histogram's dimensions are set to 16. Consequently, to achieve higher accuracy and lower dimension, 16 dimensions is selected for each histogram.

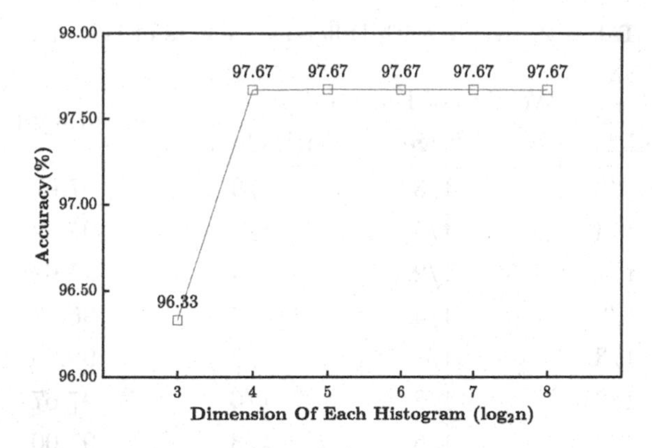

Fig. 5. Performance of features with different dimensions.

Table 1. Result with Different Fusions of Features

Features			Accuracy (%)
Color	Edge	Dark Channel	
✓			92.33
	✓		73.67
		✓	96.33
✓	✓		93.33
✓		✓	95.33
	✓	✓	97.00
✓	✓	✓	97.67

3.3 Classification Results with Different Weights of Features

Since the final feature is fused with three kinds of features, the importance of each kind of feature is verified. In Table 1, the classification accuracy with fusion of different features is given. The results indicate that fusing all three features result in better performance than using a single or two features. The weight of each feature is also varied to search for the optimal fusion strategy, and classification results are compared as shown in Table 2, with the sum of weights set to 1 and the weights of the three kinds of features changed separately. It is found when the weights of the color, edge and dark channel features are assigned to 1/6, 1/2 and 1/3, the highest accuracy 98.67% is achieved on the test set.

4 Discussion

The main challenge of classifying endoscopic haze images is selecting features that can best distinguish between haze and non-haze images. Therefore, three

Table 2. Results with Different Weights of Features

Weights of Features			Accuracy (%)
Color	Edge	Dark Channel	
1/3	1/3	1/3	97.67
1/4	1/4	1/2	98.33
1/4	1/2	1/4	97.67
1/2	1/4	1/4	96.67
1/3	1/6	1/2	98.00
1/3	1/2	1/6	97.67
1/2	1/6	1/3	97.00
1/2	1/3	1/6	96.33
1/6	1/3	1/2	98.33
1/6	1/2	1/3	98.67

types of image features are chosen based on the characteristics of endoscopic haze images. First, the physical knowledge that haze can reduce the saturation of images and increase the brightness of the images is considered. Hence, the HSI color model which can represent the saturation and brightness of the image is chosen as the first kind of feature. The second kind of feature is the edge information of images, as haze tends to blur edges. Finally, the dark channel feature is chosen as the third kind of feature, which is commonly used in dehazing tasks and can tell the difference between haze and non-haze images well. Experiments are conducted to verify the effectiveness of these features.

Figure 6 and Fig. 7 show the confusion matrices for different weights. Figure 6 indicates that using a single edge feature results in the worst classification effect, while a single dark channel feature leads to the best classification effect. Fusing all three features achieves a better result than using one or two features alone, demonstrating the importance of all three features. From Fig. 7, it can be inferred that assigning high weights to the dark channel and edge features and low weights to the color feature improves the classification effect. This may be owing to the low dimensionality of the edge feature, necessitating higher weights to avoid ignoring it. Conversely, the color feature has a high dimensionality, requiring lower weights. Additionally, because of the good performance of the dark channel feature in classification, its weight should be high.

To prove feasibility and effectiveness, our method is compared with other methods. Both traditional ML-based methods and CNN-based methods are employed for the classification task. In CNN based methods, ResNet50 [9] and VGG16 [14] are chosen. Owing to the small scale of our training set, the ResNet50 model which is pre-trained on the ImageNet dataset is fine-tuned by changing the fully connected layer's parameters. The same was done for VGG16. The results, shown in Table 3, indicate an accuracy of 97.00% with ResNet50 and 93.67%

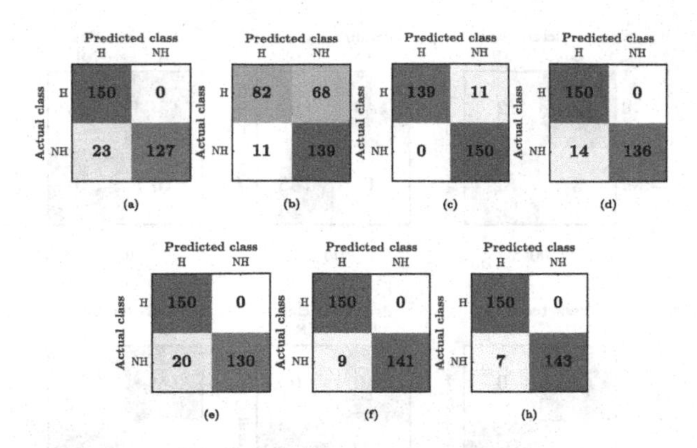

Fig. 6. Confusion matrices of fusions with different features. (a) color feature (b) edge feature (c) dark channel feature (d) color feature and dark channel feature (e) color feature and edge feature (f) edge feature and dark channel feature (g) color feature, edge feature, and dark channel feature

Table 3. Results with different methods

Methods	Accuracy (%)	Precision (%)	Recall (%)
KNN	94.67	90.36	100.0
RF	92.67	87.21	100.0
VGG16	94.67	92.45	98.00
ResNet50	97.00	94.90	99.33
Ours	98.67	98.03	99.33

with VGG16 on the test set, both lower than our SVM method. In addition, our method outperforms them in precision and recall values. This may be because our training set is small and the feature extracted is approximately linearly separable. Other traditional ML-based methods like KNN and Random Forest (RF) are also used, using the same features as SVM. The results, presented in Table 3, show that KNN achieves an accuracy of 94.67% and RF achieves an accuracy of 92.67%, both lower than our SVM method. However, these two methods have higher recall values than SVM. That may be because these two methods pay more attention to outliers and noise, which result in higher recall and lower precision. In summary, our SVM method is proved to be effective and feasible.

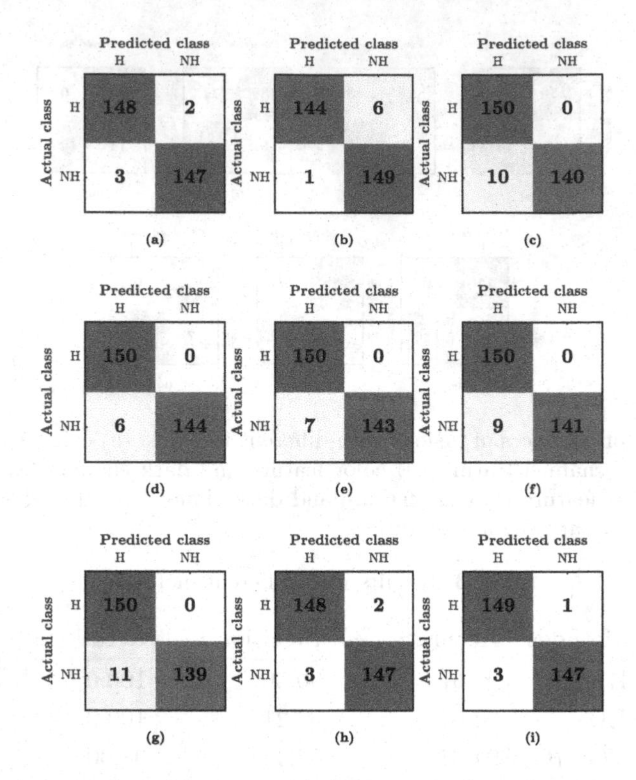

Fig. 7. Confusion matrices of features with different weights. The proportion of weights of color feature, edge feature, and dark channel feature (w_c: color feature, w_e: edge feature, w_d: dark channel feature): (a) w_c: w_e: $w_d = 1:1:2$ (b) w_c: w_e: $w_d = 1:2:1$ (c) w_c: w_e: $w_d = 2:1:1$ (d) w_c: w_e: $w_d = 2:1:3$ (e) w_c: w_e: $w_d = 2:3:1$ (f) w_c: w_e: $w_d = 3:1:2$ (g) w_c: w_e: $w_d = 3:2:1$ (h) w_c: w_e: $w_d = 1:2:3$ (i) w_c: w_e: $w_d = 1:3:2$

5 Conclusion

This paper proposes a novel SVM-based approach for effectively distinguishing between haze and non-haze endoscopic images. Our method selects three kinds of features (i.e., color, edge, and dark channel) based on the unique characteristics of endoscopic haze images. These features can be fused with a suitable strategy to recognize the haze in endoscopic images effectively, achieving superior performance than existing methods. This method can be applied before dehazing algorithms to improve their efficacy on endoscopic images, potentially improving surgical accuracy and reducing surgical risks.

Acknowledgements. This work was supported in part by the National Natural Science Foundation of China under Grant 62003343, Grant 62222316, Grant U1913601, Grant 62073325, Grant U20A20224, and Grant U1913210; in part by the Beijing Natural Science Foundation under Grant M22008; in part by the Youth Innovation Promotion Association of Chinese Academy of Sciences (CAS) under Grant 2020140; in part by the CIE-Tencent Robotics X Rhino-Bird Focused Research Program.

References

1. Boser, B.E., Guyon, I.M., Vapnik, V.N.: A training algorithm for optimal margin classifiers. In: Proceedings of the Fifth Annual Workshop on Computational Learning Theory, pp. 144–152 (1992)
2. Canny, J.: A computational approach to edge detection. IEEE Trans. Pattern Anal. Mach. Intell. **6**, 679–698 (1986)
3. Chang, C.C., Lin, C.J.: LIBSVM: a library for support vector machines. ACM Trans. Intell. Syst. Technol. (TIST) **2**(3), 1–27 (2011)
4. Chauhan, V.K., Dahiya, K., Sharma, A.: Problem formulations and solvers in linear SVM: a review. Artif. Intell. Rev. **52**(2), 803–855 (2019)
5. Chincholkar, S., Rajapandy, M.: Fog image classification and visibility detection using CNN. In: Pandian, A.P., Ntalianis, K., Palanisamy, R. (eds.) ICICCS 2019. AISC, vol. 1039, pp. 249–257. Springer, Cham (2020). https://doi.org/10.1007/978-3-030-30465-2_28
6. Cortes, C., Vapnik, V.: Support-vector networks. Mach. Learn. **20**, 273–297 (1995)
7. Guo, L., et al.: Haze image classification method based on AlexNet network transfer model. J. Phys.: Conf. Ser. **1176**, 032011 (2019)
8. He, K., Sun, J., Tang, X.: Single image haze removal using dark channel prior. IEEE Trans. Pattern Anal. Mach. Intell. **33**(12), 2341–2353 (2010)
9. He, K., Zhang, X., Ren, S., Sun, J.: Deep residual learning for image recognition. In: Proceedings of the IEEE Conference on Computer Vision and Pattern Recognition, pp. 770–778 (2016)
10. Pal, T., Halder, M., Barua, S.: Multi-feature based hazy image classification for vision enhancement. Procedia Comput. Sci. **218**, 2653–2665 (2023)
11. Pei, Y., Huang, Y., Zhang, X.: Consistency guided network for degraded image classification. IEEE Trans. Circuits Syst. Video Technol. **31**(6), 2231–2246 (2020)
12. Satrasupalli, S., Daniel, E., Guntur, S.R., Shehanaz, S.: End to end system for hazy image classification and reconstruction based on mean channel prior using deep learning network. IET Image Process. **14**(17), 4736–4743 (2020)
13. Shrivastava, S., Thakur, R.K., Tokas, P.: Classification of hazy and non-hazy images. In: Proceeding of 2017 International Conference on Recent Innovations in Signal processing and Embedded Systems (RISE), pp. 148–152 (2017)
14. Simonyan, K., Zisserman, A.: Very deep convolutional networks for large-scale image recognition. arXiv preprint arXiv:1409.1556 (2014)
15. Wan, J., Qiu, Z., Gao, H., Jie, F., Peng, Q.: Classification of fog situations based on gaussian mixture model. In: Proceeding of 2017 36th Chinese Control Conference (CCC), pp. 10902–10906 (2017)
16. Yu, X., Xiao, C., Deng, M., Peng, L.: A classification algorithm to distinguish image as haze or non-haze. In: Proceeding of 2011 Sixth International Conference on Image and Graphics, pp. 286–289 (2011)
17. Zhang, Y., Sun, G., Ren, Q., Zhao, D.: Foggy images classification based on features extraction and SVM. In: Proceeding of 2013 International Conference on Software Engineering and Computer Science, pp. 142–145 (2013)

Retinex Meets Transformer: Bridging Illumination and Reflectance Maps for Low-Light Image Enhancement

Yilong Cheng[1], Zhijian Wu[2], Jun Li[1](✉), and Jianhua Xu[1]

[1] School of Computer and Electronic Information, Nanjing Normal University, Nanjing 210023, China
{cyl,lijuncst,xujianhua}@njnu.edu.cn

[2] School of Data Science and Engineering, East China Normal University, Shanghai 200062, China
zjwu_97@stu.ecnu.edu.cn

Abstract. Low-light image enhancement, which is also known as LLIE for short, aims to reconstruct the original normal image from its low-illumination counterpart. Recently, it has received increasingly attention in image restoration. In particular, with the success of deep convolutional neural network (CNN), Retinex-based approaches have emerged as a promising line of research in LLIE, since they can well transfer adequate prior knowledge from an image captured under sufficient illumination to its low-light version for image enhancement. However, existing Retinex-based approaches usually overlook the correlation between Illumination and Reflectance maps which are both derived from the same feature extractor, leading to sub-optimal reconstructed image quality. In this study, we propose a novel Transformer architecture for LLIE, termed Bridging Illumination and Reflectance maps Transformer which is shortly BIRT. It aims to estimate the correlation between Illumination and Reflectance maps derived from Retinex decomposition within a Transformer architecture via the Multi-Head Self-Attention mechanism. In terms of model structure, the proposed BIRT comprises Retinex-based and Transformer-based sub-networks, which allow our model to elevate the image quality by learning cross-feature dependencies and long-range details between Illumination and Reflectance maps. Experimental results demonstrate that the proposed BIRT model achieves competitive performance on par with the state-of-the-arts on the public benchmarking datasets for LLIE.

Keywords: Low-light Image Enhancement · Retinex Decomposition · Illumination and Reflectance Maps · Multi-Head Self-Attention · Multi-scale Feature Fusion

Supported by the National Natural Science Foundation of China under Grant 62173186 and 62076134.

B. Luo et al. (Eds.): ICONIP 2023, CCIS 1966, pp. 388–402, 2024.
https://doi.org/10.1007/978-981-99-8148-9_31

1 Introduction

As an important research area in image restoration, Low-Light Image Enhancement (LLIE) aims to improve the quality of images taken under suboptimal illumination conditions [11]. In real-world scenarios, massive amount of low-light images are captured in various undesirable circumstances, such as insufficient light sources and short exposure time. Consequently, the resulting degraded images provide much less visual clues with significant information loss, which can adversely impact high-level tasks such as object detection [19,23] and autonomous driving [20].

Early representative LLIE algorithms like Histogram Equalization [18] and Gamma Correction [7] often introduce artifacts during restoration process, since both of the methods tend to overlook crucial illumination factors. In addition, the traditional approaches function by simply learning a mapping function between paired normal and low-light images, which is inappropriate and often results in blurry images [28]. Unlike the previous methods, Retinex-based [10] approaches are more consistent with the perception mechanism of human vision system, and thus become the dominant line of research in LLIE particularly with the help of the successful deep convolutional neural networks (CNNs) [25]. They usually decompose paired normal-light and low-light images into respective Illumination and Reflectance maps, optimize them with hand-crafted priors subsequently, and concatenate two components ultimately. In the conventional Retinex-based methods, however, the training process of multiple sub-networks are separate and time-consuming. Although some deep models directly concatenate enhanced Illumination and Reflectance components [25,33], we assume this straightforward fusion method fails to explore the intrinsic correlation between them and is likely to produce biased recovered results. On the other hand, the Retinex-based approaches using CNN architecture, such as RetinexNet [25], are incapable of capturing global long-range dependency and learning cross-feature similarity. Thus, the potential properties of low-light images are still not fully explored.

Recently, the Transformer architecture has demonstrated its superiority in capturing global long-range dependencies [6]. In this paper, to address the aforementioned problems, we propose a novel architecture for LLIE termed Bridging Illumination and Reflectance maps Transformer (BIRT). In particular, we utilize self-attention mechanism derived from Transformer architecture to encode cross-feature dependencies between Illumination and Reflectance maps, yielding robust features with sufficient representation capability. Motivated by the observation that both the two components are derived from the same Retinex Decomposition Network, more specifically, our focus is delving deeper into their correlations for improved recovered results. In previous research, most work regarding attention mechanisms primarily focuses on re-weighting feature maps along either the channel or spatial dimension, whereas we calculate cross-feature similarity for estimating the correlation between texture and color feature, leading to improved recovered results. To sum up, the contributions of this paper are threefold as follows:

- We propose a novel LLIE model termed BIRT for Bridging the Illumination and Reflectance maps. In contrast to other attention-based LLIE methods, exploring cross-feature correlations enables our network to capture more relevant information and provide more guidance for Reflectance maps.
- We utilize a multi-scale fusion strategy within our BIRT by mapping features into different scales, which can obtain texture features at multiple scales and capture long-range dependencies effectively.
- Extensive experiments on two public benchmarking datasets demonstrate that our proposed BIRT model achieves promising performance which is competitive with the state-of-the-arts.

The remainder of this paper is organized as follows. After reviewing the related work in Sect. 2, we will introduce the proposed BIRT method in details in Sect. 3. In Sect. 4, we present extensive experimental evaluations and discussions before the paper is concluded in Sect. 5.

2 Related Works

In general, LLIE methods can be classified into two groups, namely the conventional methods and learning-based approaches. In this section, we will briefly review the related work of the two groups.

2.1 Conventional Method

Histogram Equalization [18] and Gamma Correction [7] are classic LLIE methods. They achieve normal-light image recovery by simply amplifying the contrast of pairwise images. However, barely taking illumination factors into consideration makes the enhanced images perceptually unaligned with normal scenes [2].

2.2 Learning-Based Method

With the success of deep convolutional neural networks (CNNs), learning-based methods have emerged as the dominant line of research in LLIE. Earlier methods are based on CNN architecture, e.g., LLNet [16] is an end-to-end network with stacked encoders to enhance and denoise images jointly. Later, RetinexNet [25] fuses Retinex theory into deep learning methods, and a series of RetinexNet variants have been proposed recently [14,17,26]. In addition, KinD [33] replaces noise suppression with another CNN and introduces a hyperparameter to control the illumination level. However, most of CNN methods handle Illumination(I) and Reflectance(R) components separately, making it difficult to fuse color and texture feature. Besides, CNN-based approaches are not capable of capturing long-range feature dependencies.

Recently, Transformer [1,15,30] has revealed its superior performance in different vision tasks, for example, object detection [34] and image restoration [27,31]. Different from previous Transformer-based LLIE methods, we calculate cross-feature similarity to estimate the correlations between I and R inspired by their mutual derivation from the same Decomposition network. Furthermore, the Illumination component encodes more color appearances while Reflectance maps indicate more visual properties as texture features, which implies that exploring the mutual correlation between them greatly profits the enhanced feature representation capability.

3 The Proposed Method

In this section, we firstly introduce the overall framework of our proposed BIRT model followed by detailed description of the crucial modules within our network. Finally, the loss function for model training is also discussed.

3.1 The Model Framework

Figure 1 illustrates the overall framework of our BIRT model which is composed of two sub-networks: Retinex-based and Transformer-based architectures. As shown in the blue area indicated as the former sub-network, there are two primary modules: Decomposition and Adjustment. I_{low} and R_{low} denote Illumination and Reflectance maps of input low-light image, respectively. In this part, we apply Retinex theory to decompose the input of a low-light image into Illumination and Reflectance components, and optimize them respectively. In the yellow area demonstrating the Transformer-based architecture, we leverage $\hat{I}_{low}$ and $\hat{R}_{low}$ as input for performing image enhancement. More specifically, Feature Extractor aims at generating color and texture features from $\hat{I}_{low}$ and $\hat{R}_{low}$. VGG-like network is employed to learn mutual features between the two components, which allows to generate features representation of different scales via max-pooling layers. Subsequently, we calculate correlation matrices via the Multi-Head Self-Attention mechanism within Transformer architecture. Concatenated with $\hat{R}_{low}$ subsequently, the enhanced image can be obtained by Feature Fusion module. Next, we will elaborate on the critical modules within our BIRT model.

3.2 Retinex-Based Network

According to Retinex theory [10], an image is defined as $S \in \mathbb{R}^{H \times W \times 3}$. After Decomposition, $I \in \mathbb{R}^{H \times W \times 1}$ and $R \in \mathbb{R}^{H \times W \times 3}$ will be obtained by

$$S = I \odot R, \tag{1}$$

where $\odot$ represents element-wise multiplication.

The Decomposition network is constructed by stacked consecutive *conv* 3×3 layers, while the Adjustment counterpart is built with an encoder-decoder structure through successive max-pooling 2×2 and upsampling operations. We also introduce skip-connections here into its symmetrical block, which can transform residuals across different scales. Terminally, a *conv* 3×3 is applied to generate

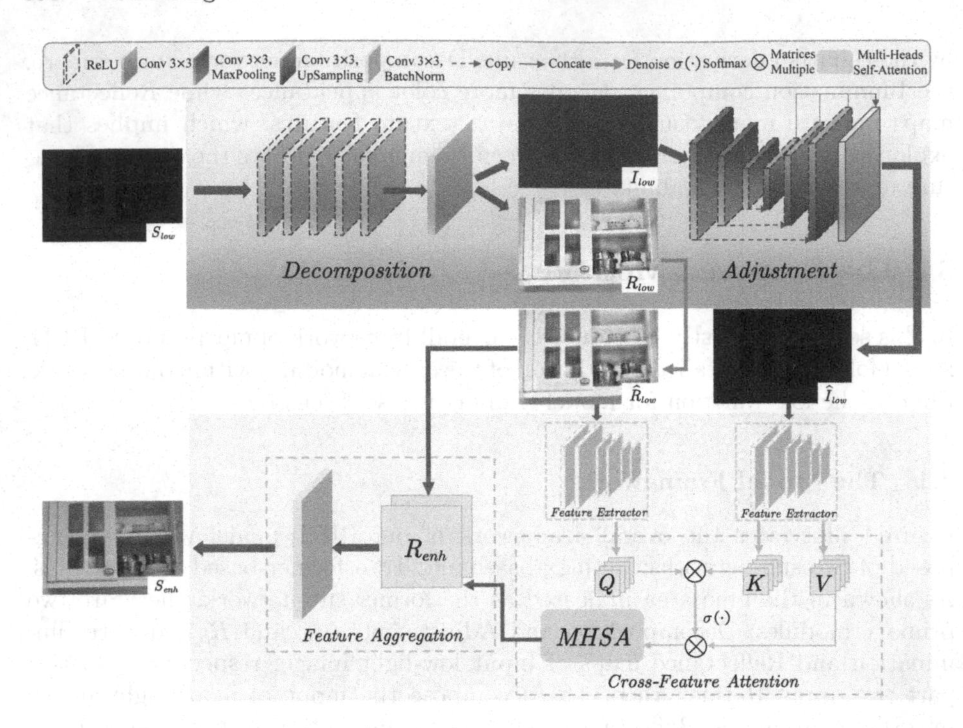

Fig. 1. The architecture of our proposed BIRT model. The above blue area is Retinex-based Network which decomposes input images into Illumination (I) and Reflectance (R) components. The below area is Transformer-based Network which aims at estimating cross-feature correlation between I and R components with incorporated multi-scale strategy. (Color figure online)

$\hat{I}_{low}$. Mathematically, the results $\hat{I}_{low}$ and $\hat{R}_{low}$ are formulated as intermediate responses for next-step enhancement:

$$\hat{I}_{low} = F_{Adj}(I_{low}), \ \hat{R}_{low} = D(R_{low}), \tag{2}$$

where $F_{Adj}(\cdot)$ is Adjustment output and $D(\cdot)$ denotes BM3D [4] denoising. I_{low} and R_{low} are the Decomposition results.

3.3 Transformer-Based Network

We pre-process the inputs through bicubic interpolation, which is consistent with following attention mechanism.

$$\mathbf{I} = Upscale\left(\hat{I}_{low}\right)_{\times 4}, \ \widetilde{\mathbf{I}} = Downscale\left(\mathbf{I}\right)_{\times 4}, \ \mathbf{R} = \hat{R}_{low}, \tag{3}$$

Feature Extractor. We utilize VGG-like [21] architecture to generate color and texture feature representation, as VGG-like network normally encourages learning mutual information between features. The difference between two extractors is individual number of input channels, where $\mathbf{I}$ and $\widetilde{\mathbf{I}}$ is 1 while $\mathbf{R}$ is 3. We set three blocks in Feature Extractor, each of which consists of three components, *conv* 3×3 layer, BatchNorm layer and ReLU. Max-pooling 2×2, stride=2 is utilized to gain features at different spatial scales in the beginning of the second and the third block. Q, K, V are defined as:

$$Q = f_{c=3}(\mathbf{R}),\ K = f_{c=1}(\mathbf{I}),\ V = f_{c=1}(\widetilde{\mathbf{I}}), \tag{4}$$

where $f(\cdot)$ is Feature Extractor and c is the number of input channels.

Cross-Feature Attention. Existing methods use element-wise multiplication to fuse the texture and color maps and a convolution is adopted to refine the resulting map. We assume that this approach does not effectively utilize the cross-feature space dependency between texture and color maps, and the long-range details of image during the textural-colored mixing process. Although the convolution operation can properly fuse texture and color features to some extent, it fails to adjust each pixel value according to the global spatial details of the feature map, which is not conducive to accurate low-light enhancement. To achieve more effective fusion, we propose a novel cross-feature attention to model the relational interaction of texture and color features. Similar to the self-attention mechanism, we use fully connected layers to map the reflectance map to the query Q and project from the illumination map to obtain K and V. After computing scores between different input vectors with $S = Q \cdot K^{\mathrm{T}}$, we normalize the scores for the stability of gradient with $S = S/\sqrt{d}$. Then, we translate the scores into probabilities with softmax function $P = softmax(S)$, and obtain the weighted value matrix with $Z = V \cdot P$. The whole processing is formulated as follow:

$$Attention\,(Q, K, V) = softmax\left(\frac{\mathbf{Q} \cdot \mathbf{K}^{\mathrm{T}}}{\sqrt{d}}\right) \cdot \mathbf{V}, \tag{5}$$

where d is feature dimension. For further performance improvement, we follow [6] to adopt multi-head self-attention mechanisms.

$$MultiHead\,(\mathbf{Q}', \mathbf{K}', \mathbf{V}') = Concat\,(head_1, ..., head_h)\,\mathbf{W}^o, \tag{6}$$

where $head_i = Attention\,(Q_i, K_i, V_i)$, $\mathbf{Q}'$ (similarly $\mathbf{K}'$ and $\mathbf{V}'$) is the matrix of input $\mathbf{Q}$, i is the index of h and $\mathbf{W}^{\mathbf{o}}$ is the projection weight.

Feature Aggregation. After effectively fusing the color and texture maps, we introduce a feature aggregation strategy to obtain finer results. The fused images are fed to multiple multi-head self-attention blocks for further feature refinement. Next, the texture maps are concatenated and a convolutional aggregation with a BatchNorm layer is used to synthesize the final output. Mathematically, it can be formulated as:

$$R_{enh} = Cat(F,\ \hat{R}_{low}), \tag{7}$$

where Cat is concatenation operation along channel dimension while F represents the result of Cross-Feature Attention module. Next, the aggregated feature is obtained by:

$$\mathbf{F} = BN(Conv(R_{enh})), \tag{8}$$

$\mathbf{F}$ denotes the output of Transformer-based network. BN denotes BatchNorm layer.

Multi-scale Strategy. We also incorporate a multi-scale strategy into our BIRT by down-scaling $\hat{R}_{low}$ to different spatial scales. We define quarter spatial scale as $\times 1 \uparrow$, half spatial scale as $\times 2 \uparrow$ and $\hat{R}_{low}$ original scale as $\times 4 \uparrow$. Three scales of Transformer blocks will receive its corresponding scale of $\mathbf{I}$ and $\mathbf{R}$ as inputs. The outputs of Transformer blocks ($\mathbf{F}^{\times 4\uparrow}, \mathbf{F}^{\times 2\uparrow}$ and $\mathbf{F}^{\times 1\uparrow}$) are concatenated by upscaling into $\hat{R}_{low}$ size, yielding $\mathbf{F}_{enh}$ denoted as the recovered image.

3.4 Loss Functions

Here, we will introduce the loss function used for training our model. Overall, the loss function consists of two parts, namely reconstruction loss and VGG perceptual loss.

Reconstruction Loss. In our model, ℓ_1 loss is unanimously used for our reconstruction loss:

$$\mathcal{L}_{recon} = \begin{cases} \|R_{low} - R_{nor}\|_1, & Decom \\ \|R_{low} \circ \hat{I}_{low} - S_{nor}\|_1, & Adjust \\ \frac{1}{CHW}\|R_{nor} - \hat{R}_{low}\|_1, & Attention \end{cases} \tag{9}$$

where S_{nor} and R_{nor} are the input normal-light image and its Reflectance map after Decomposition.

VGG Perceptual Loss. The VGG perceptual loss has demonstrated its effectiveness in super-resolution tasks [9]. We leverage it for enhancing the correlation between reference image and our recovered result, which is formulated as:

$$\mathcal{L}_{vgg} = \frac{1}{CWH}\|f_i^{vgg}(\mathbf{F}_{ref}) - f_i^{vgg}(\mathbf{F}_{enh})\|_2, \tag{10}$$

where $f_i^{vgg}(\cdot)$ represents i^{th} feature map of VGG19 [21], $\mathbf{F}_{ref}$ is the reference image and $\mathbf{F}_{enh}$ is our recovered image. C, H and W are the shape of ith feature map layer.

The overall loss functions in our BIRT is defined as follows:

$$\mathcal{L}_{all} = \mathcal{L}_{recon} + \lambda \mathcal{L}_{vgg}, \tag{11}$$

where λ is the trade-off coefficient of $\mathcal{L}_{vgg}$. Empirically, we set $\lambda = 0.1$ in this paper.

4 Experiments

4.1 Datasets

We have evaluated our BIRT network on LOL (V1 [25] and V2 [29]) datasets, which are publicly available and widely used for LLIE task. LOL-V1 has a total of 500 low-light and normal-light image pairs of which 485 pairs are used for training while the remaining for testing. Each pair consists of a low-light image captured in weak illumination condition and a corresponding reference image of well-exposure. It is the most popular dataset for LLIE task. As a further expansion version of LOL-V1, LOL-V2 is divided into synthetic and real subsets. LOL-V2 real subset has 689 image pairs for training and 100 for testing, while LOL-V2 synthetic subset is obtained by processing raw images manually.

4.2 Experimental Settings

Our model is trained with Adam optimizer for 10^5 iterations, while learning rate is set as 1×10^{-4}. In addition, the patch size is set to 160×160, and batch size is 12. Our proposed BIRT is configured as follows: the number of heads in MHSA is set to 4. Transformer-based blocks under three different scales are utilized. For performance measure, we adopt the peak signal-to-noise (PSNR) and structural similarity (SSIM) as metrics. In implementation, all the experiments are conducted on a server with one NVIDIA 3090 GPU using PyTorch framework.

4.3 Quantitative Results

We compare our method with a wide range of LLIE algorithms including classic and state-of-the-art methods. Table 1 reports the results of different competing methods using the two metrics.

It is illustrated that our BIRT approach reports the highest 21.337 dB PSNR on LOL-V1 dataset, and significantly outperforms the other Retinex-based learning methods including DeepUPE [22], RetinexNet [25], RUAS [14], KinD [33], KinD++ [32]). For example, our model consistently surpasses the KinD++ algorithm, achieving dramatic improvements of over 3.5 dB. In terms of SSIM metric, our approach still reports competitive performance comparable to the other competitors. On LOL-V2, similar performance advantages are also achieved by our model.

In addition to the above comparisons, we also compare our BIRT with other competitors including ZeroDCE [5], EnGAN [8], ZeroDCE++ [12] and RUAS [14]. The comparison results demonstrate that our BIRT provides substantial improvements of 6.476, 5.987, 3.854, 3.111 dB on LOL-V1. Compared with Transformer-based learning methods including IPT [3] and Uformer [24], our BIRT reports performance gains of 5.067, 4.977 dB and 1.753, 1.733 dB on both LOL-V1 and LOL-V2 real. All the above comparative studies show the advantage of combining Retinex theory and Transformer architecture.

Table 1. Quantitative Results on LOL (V1 [25] & V2-real [13]). The highest scores are highlighted **in bold** while the second best results are underlined. Our BIRT method significantly outperforms the other competing algorithms by reporting the highest PSNR scores, while reports the competitive performance with SSIM metric.

Method	LOL-V1		LOL-V2	
	PSNR	SSIM	PSNR	SSIM
DeepUPE [22]	14.380	0.446	13.270	0.452
RetinexNet [25]	16.774	0.462	16.097	0.567
ZeroDCE [5]	14.861	0.589	14.321	0.511
ZeroDCE++ [12]	15.350	0.570	18.490	0.580
IPT [3]	16.270	0.504	19.800	<u>0.813</u>
Uformer [24]	16.360	0.771	19.820	0.771
EnGAN [8]	17.483	0.667	18.230	0.671
RUAS [14]	18.226	0.717	18.370	0.723
KinD++ [32]	17.752	0.766	17.661	0.769
KinD [33]	17.648	0.779	19.740	0.761
URetinex-Net [26]	<u>21.328</u>	**0.834**	<u>21.221</u>	**0.859**
BIRT(Ours)	**21.337**	<u>0.780</u>	**21.553**	0.780

4.4 Qualitative Results

The visual comparison with several other methods is demonstrated in Fig. 2. From the whole view, we can observe that classical Retinex-based models usually introduce more noise and color distortion, leading to the results which are inconsistent with reference image.

Although KinD [33] focuses on setting a hyperparameter to tweak the illumination level, the colorful quality is still undesirable with blurred results produced. Moreover, some methods like RetinexNet [25], LLNet [16] remain unfriendly exposure restoration or fail to suppress noise well. Besides, EnGAN [8] brings more obvious artifacts. The best visual effect is provided by URetinex-Net [26] and BIRT. In terms of overall performance, BIRT excels in color restoration and demonstrates better results in matching the reference image. On the other hand, the results of URetinex-Net exhibit a whiter style (the glass of the showcase and the wallpaper on the background), leading to a loss of color information. In contrast, our BIRT can effectively preserve the color information for the low-light image, showing the importance of estimating the correlation between Illumination and Reflectance maps and fusing features of color and texture.

4.5 Ablation Study

In this section, we conduct extensive ablation studies and analyze the effect of individual component on the model performance.

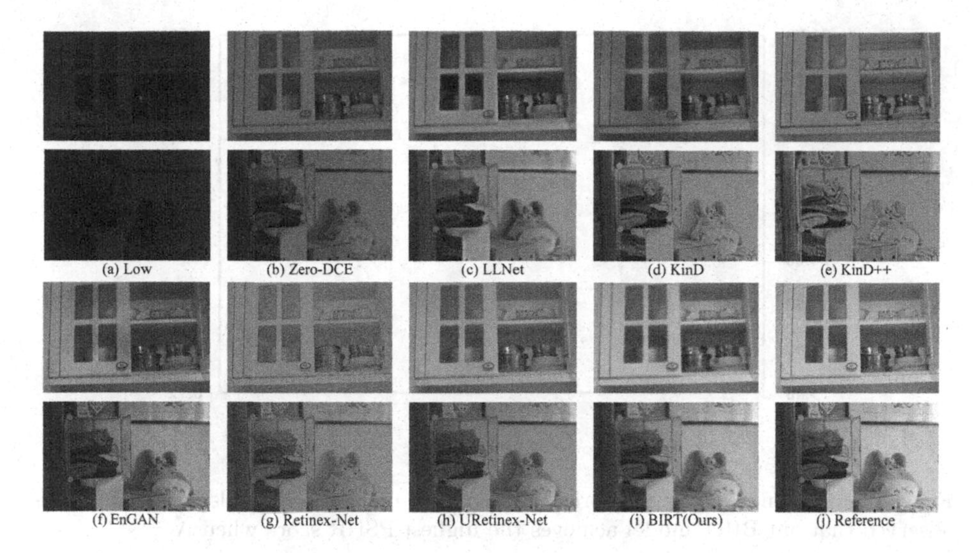

Fig. 2. Visual Results on LOL-V1 test. The competing methods include Zero-DCE [5], LLNet [16], KinD [33], KinD++ [32], EnGAN [8], RetinexNet [25] and URetinex-Net [26]. Our BIRT is superior to other models in recovering color and texture information.

Multi-Head Self-Attention (MHSA) Mechanism. We first demonstrate the effectiveness of Multi-Head Self-Attention mechanism on LLIE, and the result is shown on Table 2. The performance of RetinexNet is used as the baseline result. For the Transformer-based method, we set $N = 0$ and $N = 4$ for MHSA as comparison, where N represents the number of heads. Our BIRT dramatically improves the baseline on LOL-V1 dataset, with 2.714 dB for PSNR and 0.272 for SSIM when $N = 0$ suggesting MHSA is excluded from our BIRT. Furthermore, further performance gains of 1.849 dB PSNR and 0.046 SSIM can be observed when utilizing MHSA module ($N = 4$). These experiments fully validate the efficacy of our BIRT in capturing long-range dependency and cross-feature interaction.

Table 2. Ablation study results on Multi-Head Self-Attention mechanism (MHSA) for LOL-V1 dataset.

Module	PSNR ↑	SSIM ↑
RetinexNet [25]	16.774	0.462
BIRT w/o MHSA	19.488	0.734
BIRT	21.337	0.780

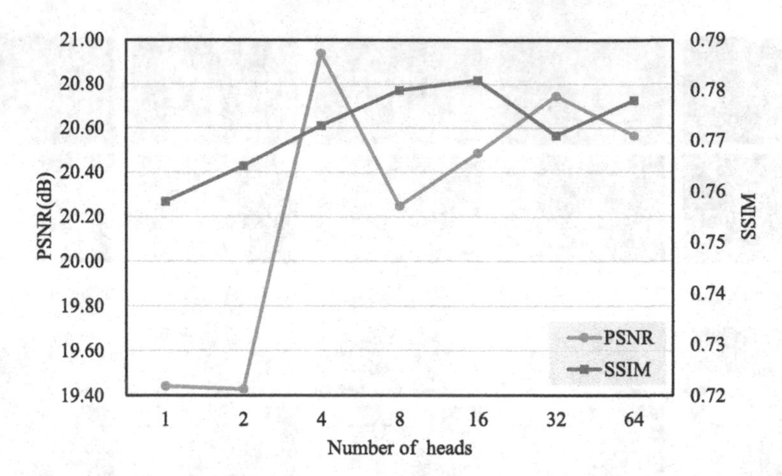

Fig. 3. Performance of our network using varying numbers of heads in MHSA. It is observed that our BIRT model achieves the highest PSNR score when $N = 4$.

Number of Heads in MHSA. Figure 3 shows the performance of our BIRT with the number of heads varying from 1 to 64. It should be noted that VGG-loss is not involved in this part. With the increasing number of heads, it is shown that our BIRT exhibits a significant improvement before $N = 4$ followed by a subsequent decline in the evaluation metrics. The performance drop can be observed with further increase in N. Consequently, we empirically set $N = 4$ in our method.

Multi-scale Strategy. We conduct an ablation study on the effect of multi-scale scheme. The result is reported in Table 3. We incorporate different scales of **R** features into our network, $\hat{R}_{low}\times1\uparrow$, $\hat{R}_{low}\times2\uparrow$ and $\hat{R}_{low}\times4\uparrow$. As can be seen from the table, Reflectance maps at multiple scales can improve the model performance significantly. Furthermore, it is observed that the best result of our BIRT comes from fusing all the three scales, reporting 20.939 dB PSNR and 0.773 SSIM respectively.

Table 3. Ablation study on multi-scale strategy (w/o VGG-loss).

×1	×2	×4	PSNR ↑	SSIM ↑
√			19.852	0.744
	√		19.763	0.746
		√	19.780	0.751
√	√		19.813	0.746
√		√	20.574	0.763
	√	√	20.882	0.771
√	√	√	**20.939**	**0.773**

Analysis of VGG Perceptual Loss. In this part, we introduce VE-LOL [13] as a benchmark to evaluate the effectiveness of VGG-loss. It is a large-scale dataset consisting of 2500 paired images, demonstrating a greater diversity of scenes and visual contents. We use one of its low-level subset, VE-LOL-L-Cap, which contains 400 real-captured pairwise images. The training strategy is consistent with LOL-V1 and LOL-V2.

Table 4. Ablation results on VGG-loss for LOL-V1 [25], LOL-V2 [29] and VE-LOL [13] datasets. The first and the second row indicate the results achieved w/o and with VGG-loss. It is shown that VGG-loss provides consistent performance improvements on both metrics.

Dataset	PSNR ↑	SSIM ↑
LOL-V1	20.939	0.773
	21.337 (+0.398)	0.780 (+0.007)
LOL-V2	21.163	0.777
	21.553 (+0.390)	0.780 (+0.003)
VE-LOL	19.720	0.672
	20.863 (+1.143)	0.776 (+0.102)

Table 4 shows the comparison of results achieved w/o and w/ the VGG loss on three datasets. Combined with the standard ℓ_1 loss, VGG-loss further improves by 0.398, 0.390 and 1.143 dB PSNR, along with 0.007, 0.003, and 0.102 SSIM.

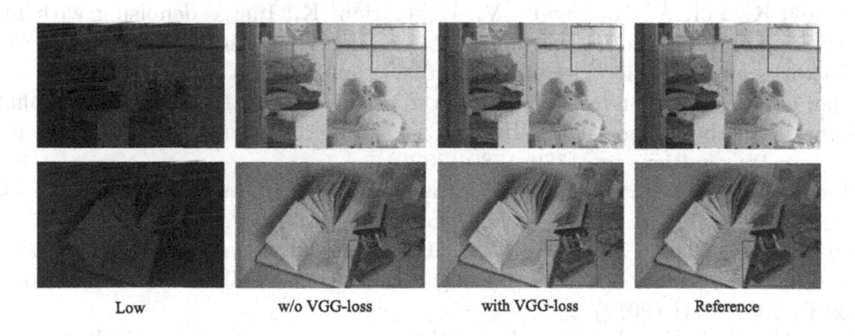

Fig. 4. Qualitative results on LOL-V1 test using VGG-loss. Objects in red bounding boxes illustrate the beneficial effect of VGG-loss, suggesting their color is more consistent with reference image. (Color figure online)

Figure 4 illustrates the effectiveness of VGG-loss intuitively. We can easily observe that VGG-loss is capable of preserving color information and details well. For example, the frame and stapler in red bounding boxes resemble their counterparts in reference images. This demonstrates that VGG-loss helps to preserve important color details, textures, and overall image appearance during the restoration process.

5 Conclusion

In this paper, we have proposed a novel architecture for LLIE, namely BIRT. We further estimate the correlation between Illumination maps and Reflectance maps through a Multi-Head Self-Attention module, which aims at capturing long-range and cross-feature dependencies. Different from classic methods, we view the whole R and I as Q, K and V, which is inspired by their derived network. A multi-scale strategy is also utilized in our BIRT, in terms of learning more powerful textural feature representation at different scales. Adequate experiments are conducted on public benchmarking datasets to demonstrate the efficacy of our proposed model both quantitatively and qualitatively.

In future research, we primarily aim to further improve the robustness of our method, with less processing steps and lower computational complexity. For denoising operation, we need more adaptive methods instead of classic method.

References

1. Bandara, W.G.C., Patel, V.M.: Hypertransformer: a textural and spectral feature fusion transformer for pansharpening. In: Proceedings of the IEEE/CVF Conference on Computer Vision and Pattern Recognition, pp. 1767–1777 (2022)
2. Cai, Y., Bian, H., Lin, J., Wang, H., Timofte, R., Zhang, Y.: Retinexformer: one-stage retinex-based transformer for low-light image enhancement. arXiv preprint arXiv:2303.06705 (2023)
3. Chen, H., et al.: Pre-trained image processing transformer. In: Proceedings of the IEEE/CVF Conference on Computer Vision and Pattern Recognition, pp. 12299–12310 (2021)
4. Dabov, K., Foi, A., Katkovnik, V., Egiazarian, K.: Image denoising with block-matching and 3D filtering. In: Image Processing: Algorithms and Systems, Neural Networks, and Machine Learning, vol. 6064, pp. 354–365. SPIE (2006)
5. Guo, C., et al.: Zero-reference deep curve estimation for low-light image enhancement. In: Proceedings of the IEEE/CVF Conference on Computer Vision and Pattern Recognition, pp. 1780–1789 (2020)
6. Han, K., et al.: A survey on vision transformer. IEEE Trans. Pattern Anal. Mach. Intell. **45**(1), 87–110 (2023). https://doi.org/10.1109/TPAMI.2022.3152247
7. Huang, S.C., Cheng, F.C., Chiu, Y.S.: Efficient contrast enhancement using adaptive gamma correction with weighting distribution. IEEE Trans. Image Process. **22**(3), 1032–1041 (2012)
8. Jiang, Y., et al.: Enlightengan: deep light enhancement without paired supervision. IEEE Trans. Image Process. **30**, 2340–2349 (2021)
9. Johnson, J., Alahi, A., Fei-Fei, L.: Perceptual losses for real-time style transfer and super-resolution. In: Leibe, B., Matas, J., Sebe, N., Welling, M. (eds.) ECCV 2016. LNCS, vol. 9906, pp. 694–711. Springer, Cham (2016). https://doi.org/10.1007/978-3-319-46475-6_43
10. Land, E.H.: The retinex theory of color vision. Sci. Am. **237**(6), 108–129 (1977)
11. Li, C., et al.: Low-light image and video enhancement using deep learning: a survey. IEEE Trans. Pattern Anal. Mach. Intell. **44**(12), 9396–9416 (2021)
12. Li, C., Guo, C., Loy, C.C.: Learning to enhance low-light image via zero-reference deep curve estimation. IEEE Trans. Pattern Anal. Mach. Intell. **44**(8), 4225–4238 (2021)

13. Liu, J., Xu, D., Yang, W., Fan, M., Huang, H.: Benchmarking low-light image enhancement and beyond. Int. J. Comput. Vision **129**, 1153–1184 (2021)
14. Liu, R., Ma, L., Zhang, J., Fan, X., Luo, Z.: Retinex-inspired unrolling with cooperative prior architecture search for low-light image enhancement. In: Proceedings of the IEEE/CVF Conference on Computer Vision and Pattern Recognition, pp. 10561–10570 (2021)
15. Liu, Z., et al.: Video swin transformer. In: Proceedings of the IEEE/CVF Conference on Computer Vision and Pattern Recognition, pp. 3202–3211 (2022)
16. Lore, K.G., Akintayo, A., Sarkar, S.: Llnet: a deep autoencoder approach to natural low-light image enhancement. Pattern Recogn. **61**, 650–662 (2017)
17. Ma, L., Liu, R., Wang, Y., Fan, X., Luo, Z.: Low-light image enhancement via self-reinforced retinex projection model. IEEE Trans. Multimedia (2022)
18. Pizer, S.M., et al.: Adaptive histogram equalization and its variations. Comput. Vis. Graph. Image Process. **39**(3), 355–368 (1987)
19. Ren, S., He, K., Girshick, R., Sun, J.: Faster R-CNN: towards real-time object detection with region proposal networks. In: Advances in Neural Information Processing Systems, vol. 28 (2015)
20. Shao, H., Wang, L., Chen, R., Li, H., Liu, Y.: Safety-enhanced autonomous driving using interpretable sensor fusion transformer. In: Conference on Robot Learning, pp. 726–737. PMLR (2023)
21. Simonyan, K., Zisserman, A.: Very deep convolutional networks for large-scale image recognition. arXiv preprint arXiv:1409.1556 (2014)
22. Wang, R., Zhang, Q., Fu, C.W., Shen, X., Zheng, W.S., Jia, J.: Underexposed photo enhancement using deep illumination estimation. In: Proceedings of the IEEE/CVF Conference on Computer Vision and Pattern Recognition, pp. 6849–6857 (2019)
23. Wang, W., et al.: Internimage: exploring large-scale vision foundation models with deformable convolutions. In: Proceedings of the IEEE/CVF Conference on Computer Vision and Pattern Recognition, pp. 14408–14419 (2023)
24. Wang, Z., Cun, X., Bao, J., Zhou, W., Liu, J., Li, H.: Uformer: a general U-shaped transformer for image restoration. In: Proceedings of the IEEE/CVF Conference on Computer Vision and Pattern Recognition, pp. 17683–17693 (2022)
25. Wei, C., Wang, W., Yang, W., Liu, J.: Deep retinex decomposition for low-light enhancement. arXiv preprint arXiv:1808.04560 (2018)
26. Wu, W., Weng, J., Zhang, P., Wang, X., Yang, W., Jiang, J.: Uretinex-net: retinex-based deep unfolding network for low-light image enhancement. In: Proceedings of the IEEE/CVF Conference on Computer Vision and Pattern Recognition, pp. 5901–5910 (2022)
27. Xu, X., Wang, R., Fu, C., Jia, J.: SNR-aware low-light image enhancement. In: 2022 IEEE/CVF Conference on Computer Vision and Pattern Recognition (CVPR), pp. 17693–17703 (2022)
28. Yang, S., Zhou, D., Cao, J., Guo, Y.: Lightingnet: an integrated learning method for low-light image enhancement. IEEE Trans. Comput. Imaging **9**, 29–42 (2023)
29. Yang, W., Wang, W., Huang, H., Wang, S., Liu, J.: Sparse gradient regularized deep retinex network for robust low-light image enhancement. IEEE Trans. Image Process. **30**, 2072–2086 (2021)
30. Yuan, L., et al.: Tokens-to-token VIT: training vision transformers from scratch on imagenet. In: Proceedings of the IEEE/CVF International Conference on Computer Vision, pp. 558–567 (2021)

31. Zamir, S.W., Arora, A., Khan, S., Hayat, M., Khan, F.S., Yang, M.H.: Restormer: efficient transformer for high-resolution image restoration. In: Proceedings of the IEEE/CVF Conference on Computer Vision and Pattern Recognition, pp. 5728–5739 (2022)
32. Zhang, Y., Guo, X., Ma, J., Liu, W., Zhang, J.: Beyond brightening low-light images. Int. J. Comput. Vision **129**, 1013–1037 (2021)
33. Zhang, Y., Zhang, J., Guo, X.: Kindling the darkness: a practical low-light image enhancer. In: Proceedings of the 27th ACM International Conference on MultiMedia, pp. 1632–1640 (2019)
34. Zhu, X., Su, W., Lu, L., Li, B., Wang, X., Dai, J.: Deformable DETR: deformable transformers for end-to-end object detection. arXiv preprint arXiv:2010.04159 (2020)

Make Spoken Document Readable: Leveraging Graph Attention Networks for Chinese Document-Level Spoken-to-Written Simplification

Yunlong Zhao[1,2], Haoran Wu[1,2], Shuang Xu[1], and Bo Xu[1,2(✉)]

[1] Institute of Automation, Chinese Academy of Sciences, Beijing, China
{zhaoyunlong2020,wuhaoran2018,shuang.xu,xubo}@ia.ac.cn
[2] School of Artificial Intelligence, University of Chinese Academy of Sciences, Beijing, China

Abstract. As people use language differently when speaking compared to writing, transcriptions generated by automatic speech recognition systems can be difficult to read. While techniques exist to simplify spoken language into written language at the sentence level, research on simplifying spoken language has various spoken language issues at the document level is limited. Document-level spoken-to-written simplification faces challenges posed by cross-sentence transformations and the long dependencies of spoken documents. This paper proposes a new method called G-DSWS (**G**raph attention networks for **D**ocument-level **S**poken-to-**W**ritten **S**implification) using graph attention networks to model the structure of a document explicitly. G-DSWS utilizes structural information from the document to improve the document modeling capability of the encoder-decoder architecture. Experiments on the internal and publicly available datasets demonstrate the effectiveness of the proposed model. And the human evaluation and case study show that G-DSWS indeed improves spoken Chinese documents' readability.

Keywords: Spoken language processing · Document modeling · Graph attention network · Document-level spoken-to-written simplification

1 Introduction

Text readability is a critical topic in spoken language understanding. Thanks to advanced automatic speech recognition (ASR) technology, the ASR systems provide people, including hard-of-hearing people, another way to access information in speech. However, people use language differently when they speak than when they write [1]. Spoken-style text generated by the ASR system is difficult to understand due to disfluency [2], syntactic and grammatical errors [4], and other noises [12]. Therefore, simplifying spoken-style text into written-style text is very important for spoken document access. Previous studies [2,5,6,9] focus on

B. Luo et al. (Eds.): ICONIP 2023, CCIS 1966, pp. 403–416, 2024.
https://doi.org/10.1007/978-981-99-8148-9_32

sentence-level tasks, and only on singular types of spoken language issues. There has been limited research on document-level spoken-to-written simplification, which is more valuable in realistic scenarios.

Why is it valuable to study document-level spoken-to-written simplification? Figure 1 shows an example of Chinese document-level spoken-to-written simplification. In the real world, the ASR system transcribes a long speech, and the final system output is a long spoken document which is punctuated using a off-the-shelf punctuation prediction model. The transcribed spoken document is likely to have multiple spoken language issues, such as disfluencies, syntactic errors, punctuation errors, which can significantly hamper its readability. Making spoken documents readable has broader application demand and is more valuable. It is meaningful for human user reading and many downstream tasks. For instance, we can clean a large corpus of noisy spoken language documents obtained through the ASR system to obtain clean written language documents, which can be used to train the large language model.

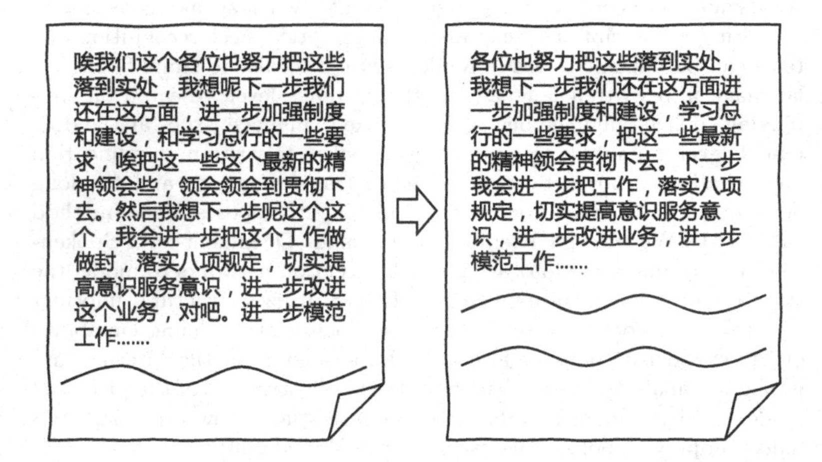

Fig. 1. An example of Chinese document-level spoken-to-written simplification. We can see that the spoken document is lengthy and poorly organized, and spoken language issues occur across sentence boundaries.

Because of long and complex input, document-level spoken-to-written simplification becomes more complicated and faces these challenges: Firstly, spoken language issues occur across sentence boundaries. Due to pauses in speech and errors in predicting punctuation, the sentence boundaries in spoken documents are not always accurate. Cross-sentence transformations are essential for document-level spoken text simplification. Thus, simplifying sentences in a document one by one is not proper because they may be influenced by preceding or succeeding sentences. Secondly, the spoken document is lengthy and not well organized. More than removing the spoken language issues, document-level spoken-to-written simplification aims to simplify and reorganize spoken documents to make them more concise. Thus, the spoken document structure is essential for this high-level simplification.

This paper proposes a novel document-level spoken-to-written simplification method called G-DSWS. To leverage document structure in sequence text data, we extend the standard sequence encoder with a graph encoder. Our proposed graph encoder consists of two graph propagation components: hierarchical graph attention and document graph attention, which utilize graph attention networks. These two graph propagation components model cross-sentence interaction and document-sentence interaction, respectively. The graph encoder enhances the model's document modeling ability, improving document simplification. To evaluate the effectiveness of our method, we conduct experiments on our internal Chinese document-level spoken-to-written simplification dataset. Additionally, we conduct experiments on a similar document-level text simplification task [18] that has a publicly available dataset to verify the effectiveness of our method on other document-level tasks.

We highlight our contributions:

1. To the best of our knowledge, we are the first who focus on Chinese document-level spoken-to-written simplification.
2. We propose a novel method, G-DSWS, which outperforms previous strong baselines and demonstrates advantages in document-level simplification.
3. Qualitative and quantitative experimental results demonstrate that our method improves the readability of spoken Chinese documents.

2 Related Work

Spoken-to-written simplification has been studied widely, many research focus on specific spoken-to-written simplification task, e.g. punctuation restoration [21] disfluency detection [2,24], inverse text normalization [17]. However, few works dedicate to spoken-to-written simplification because of limited data resources. [9] propose a parallel corpus for Japanese spoken-to-written simplification. Some works explore pointer network and pretraining methods for spoken-to-written simplification [5,7,8]. A joint modeling method is proposed for simultaneously executing multiple simplification [6] without preparing matched datasets. All methods focus on sentence-level task setting. To the best of our knowledge, no previous study explores document-level spoken-to-written simplification.

In recent times, there has been a surge in research interest towards document-level generation tasks, including tasks like document-level machine translation [25] and document-level text simplification [18]. Document-level text simplification is similar to our spoken-text simplification task. It also aims to make the document more readable and focus on reducing linguistic complexity. However, document simplification is also a new task, and the research on this task is also limited [18]. Graph-based modeling method has been explored in document understanding tasks [3,27], and we extend it to the document-level simplification task.

3 Preliminary

3.1 Problem Formulation

Spoken-to-written simplification is a monolingual translation task aiming to convert spoken text into written text. The document-level spoken-to-written simplification task can be defined as follows. Given a source spoken-style document D_s which has multiple sentences $\{S_1, \cdots, S_n\}$, we aim to transfer it to a target written-style document D_t which has multiple sentences $\{T_1, \cdots, T_m\}$. Typically, the number of target sentences m is smaller than the number of source sentences n, as the target document is usually more concise than the source.

3.2 Graph Attention Network

Graph attention network [23] is an attention-based graph neural network. It takes a set of nodes features, $\mathbf{h} = \{\boldsymbol{h}_1, \boldsymbol{h}_2, \ldots, \boldsymbol{h}_N\}, \boldsymbol{h}_i \in \mathbb{R}^F$, as its input and produces a new set of node features, $\mathbf{h}' = \{\boldsymbol{h}'_1, \boldsymbol{h}'_2, \ldots, \boldsymbol{h}'_N\}, \boldsymbol{h}'_i \in \mathbb{R}^{F'}$ as its output. $\mathcal{G}$ is the graph. We denote the GAT layer works as following:

$$\mathbf{h}' = \mathrm{GAT}(\mathbf{h}; \mathcal{G}) \tag{1}$$

4 The Proposed Model

This section introduces our proposed model G-DSWS (graph-based document-level spoke-to-written simplification). As Fig. 2 shows, G-DSWS is an encoder-decoder architecture with a graph encoder. The details are as follows.

4.1 The Encoder-Decoder Architecture

The Encoder-Decoder model [20] is a conventional architecture for the sequence-to-sequence task. The most popular encoder-decoder model is Transformer [22], which is based solely on the attention mechanism. Transformer achieves superior results in many sequence generation tasks, and we adopt it as our baseline model. We use a shared word embedding layer for the encoder and the decoder.

4.2 Graph Encoder

Graph Construction. In this paper, we adopt a simple and intuitive document graph structure. Our graph contains token-level nodes, sentence-level nodes, and one document-level node. Given a document D_s, which has multiple sentences $\{S_1, \cdots, S_n\}$, every sentence S_i will be represented as one single sentence-level node and has an edge connected to the document node corresponding to D_s. As for particular sentence s_i which has multiple tokens $\{w_1, \cdots, w_l\}$, every token w_k will be represented as one single token-level node and has an edge connected to S_i node. We denote the document graph as $\mathcal{G}_{doc}$, and it is shown in Fig. 2.

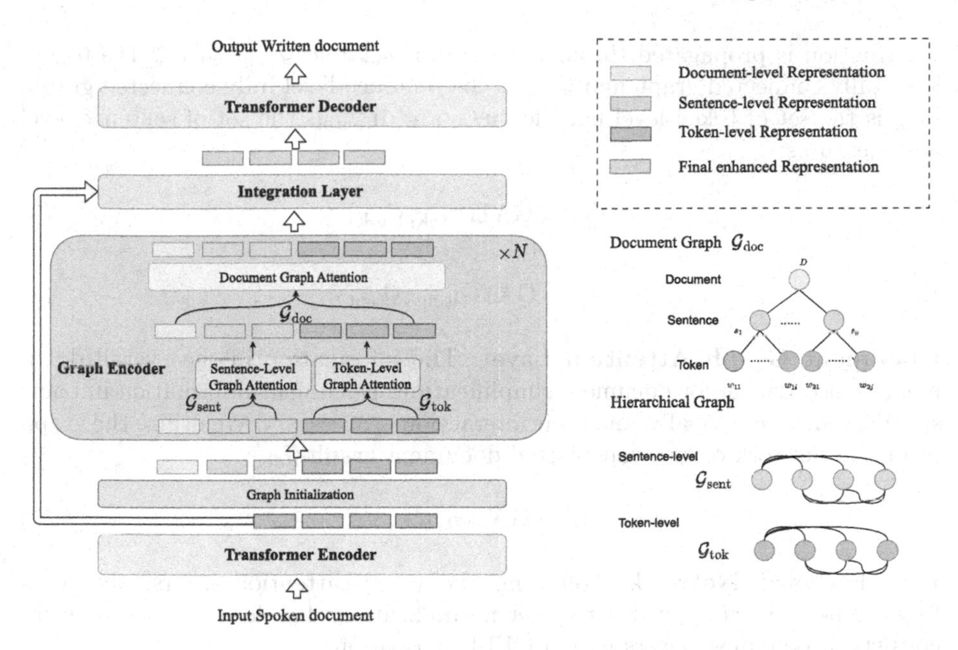

Fig. 2. Overview of our proposed G-DSWS. G-DSWS is an encoder-decoder architecture with a graph encoder. The graph encoder, composed of graph attention layers, aims to capture document structure ignored in the sequence transformer encoder. The integration layer integrates graph representation and original sequence representation for decoder decoding.

Graph Initialization. Due to the encoder can only produce token-level representations, we use a bottom-up average pooling strategy to initialize sentence-level and document-level nodes. We add learnable node embedding to represent positional information of high-level (sentence or document) nodes. Suppose $l_i \in \{0, 1, 2\}$ indicates the level of node i (token, sentence, document), N_i indicates the i neighborhood nodes set:

$$\boldsymbol{h}_i = \underset{j \in \mathcal{N}_i, l_i = l_{j+1}}{\text{average}} \{\boldsymbol{h}_j\} + \boldsymbol{b}_i \tag{2}$$

where $\boldsymbol{b}_i$ represents node embedding.

Graph Propagation. Graph propagation is the critical part of the model. We use graph attention networks (GAT) to perform message passing over the graph. Graph propagation consists of the hierarchical graph attention layer and the document graph attention layer.

Hierarchical Graph Attention Layer: Cross-sentence information transfer is essential for this task. To make interactions among same-level nodes, including cross-sentences interactions, we utilize the hierarchical graph attention layer, which consists of sentence-level graph attention and token-level graph attention.

Information is propagated through the graph $\mathcal{G}_{\text{tok}}$ and $\mathcal{G}_{\text{sent}}$. $\mathcal{G}_{\text{tok}}$ is the token-level fully connected graph and $\mathcal{G}_{\text{sent}}$ is the sentence-level fully connected graph. $\mathbf{h}_{\text{tok}}$ is the set of token-level node features and $\mathbf{h}_{\text{sent}}$ is the set of sentence-level node features.

$$\mathbf{h}'_{\text{tok}} = \text{GAT}(\mathbf{h}_{\text{tok}}; \mathcal{G}_{\text{tok}}) \tag{3}$$

$$\mathbf{h}'_{\text{sent}} = \text{GAT}(\mathbf{h}_{\text{sent}}; \mathcal{G}_{\text{sent}}) \tag{4}$$

Document Graph Attention Layer: The information in the overall document is also crucial for document simplification. Document information impacts specific sentences globally, and their interaction is necessary. We utilize the graph attention network on our constructed document graph $\mathcal{G}_{doc}$:

$$\mathbf{h}' = \text{GAT}(\mathbf{h}; \mathcal{G}_{doc}) \tag{5}$$

Feed-Forward Network: Following the graph attention layers, the feed-forward network is applied with layer normalization. The feed-forward network consists of two linear layers with a GELU activation.

4.3 Integration of Graph Representation

We suppose graph representation as $\mathbf{G}$ and origin encoder representation as $\mathbf{H}$. We use a gated attention mechanism for integrating $\mathbf{G}$ into $\mathbf{H}$. The gated attention is calculated as follows:

$$\hat{\mathbf{G}} = \alpha \cdot \mathbf{G} \tag{6}$$

$$\alpha = \text{Softmax}\left(\text{Relu}\left(\mathbf{W_G G}\right) \cdot \text{Relu}\left(\mathbf{W_H H}\right)^T\right) \tag{7}$$

$\hat{\mathbf{G}}$ is the new graph representation that attends to the origin encoder representation. We further integrate attentive graph representation and encoder representation. The formula is as follows:

$$\hat{\mathbf{H}} = \mathbf{W_p}\left[\sigma\left(\mathbf{W}_s\left[\mathbf{H}; \hat{\mathbf{G}}\right]\right) \cdot \text{Tanh}\left(\mathbf{W}_t\left[\mathbf{H}; \hat{\mathbf{G}}\right]\right)\right] \tag{8}$$

$\hat{\mathbf{H}}$ is the new enhanced representation with the same sequence length as the original encoder representation. The decoder consumes $\hat{\mathbf{H}}$ and generates target sequence tokens.

5 Experiments

5.1 Datasets

Document-Level Spoken-to-Written Simplification. We use our internal Chinese dataset to conduct document-level spoken-to-written simplification experiments. The original spoken documents in the dataset are transcribed from long audio such as live conferences and TV interviews. The dataset with spoken-style and written-style document pairs is built using crowdsourcing. Every spoken-style document has one corresponding written-style document reference. Because some documents are too long, we split the documents over 1000 tokens (chars) into short documents. The dataset has 13k document pairs, and We divide the dataset into a training set with 11k document pairs, a validation set with 1k document pairs, and a test set with 1k document pairs. Figure 1 shows an example. Table 1 shows the statistical analysis of the dataset. We can see that spoken-level documents have more characters and sentences than written-style documents. It illustrates that removing spoken language issues in the spoken-style document meanwhile reorganizing and simplifying it is the focus of the simplification task.

Table 1. The statistical analysis of the dataset. "Avg. #Char" means average character number, and "Avg. #Sent" means average sentence number.

	Avg. #Char	Avg. #Sent
Spoken-style document	511	20
Written-style document	423	16

Document-Level Text Simplification. We use a publicly available document-level text simplification dataset D-Wikipedia [18] further to validate the performance of our method on document-level tasks. D-Wikipedia is built based on English Wikipedia and Simple English Wikipedia. One article has mostly 1000 words. We use byte pair encoding to segment words into sub-words in preprocessing.

5.2 Experimental Settings

Configuration. Our code is based on Fairseq [15] toolkit. Our baseline encoder-decoder model is a vanilla transformer. We use 6 transformer layers for both the encoder and decoder. Each transformer layer comprises 512 hidden units, 8 attention heads, and 2048 feed-forward hidden units. The graph attention network layer uses the same hyperparameters as the above setting of the Transformer.

We conduct all experiments in the same experimental setting. We set the learning rate to 1e-4 and the batch size to 16. We use an adam optimizer and label smoothing with a value of 0.1. We use characters as token level. We use a beam search decoding strategy during inference with a beam size of 5.

Baseline Models. We select four strong baselines for document-level spoken-to-written simplification:

- **Transformer** Vanilla Transformer in document-level.
- **Sentence-Transformer** We use heuristic rules to align and segment documents and construct a sentence-level spoken-to-written simplification dataset. We train a Sentence-Transformer in the sentence-level dataset. We convert each sentence in test spoken documents one by one and evaluate them in a document-level setting.
- **BertEncoder** We implement a Transformer using the pre-trained model Bert [10] as the encoder. Due to Bert's input length limitation, we do not use Bert's absolute position embedding. We use the same vocabulary as the baseline Transformer and load Bert's word embedding for initialization.
- **BART** BART [11] is a pre-trained model which achieves state-of-the-art results on many sequence generation tasks. Because our dataset is Chinese, We use mBART (multilingual version BART).

The baselines for the document-level text simplification task are followed [18], and these models are similar to the above baselines. SUC [19] is a sentence-level simplification model using contextual information. BertSumextabs [14] and BART [11] are pre-trained models achieving excellent results.

Evaluation Metrics. To evaluate the performance of our method, we select some automatic evaluation metrics:

BLEU [16], and **ROUGE** [13] are well-known evaluation metrics and have been used in many natural language generation tasks, such as machine translation. We use BLEU-4 and ROUGE-L as our evaluation metrics. SARI [26] is a commonly used evaluation metric in simplification tasks. It computes average F1 scores of the added, kept, and deleted n-grams considering the system output, references, and input. **D-SARI** [18] is based on the SARI metric and changes some details to adapt to the document-level task. D-SARI is more suitable for monolingual document-level simplification tasks, so we choose D-SARI as one of the most critical metrics for automatic evaluation.

We also conducted a human evaluation on 20 randomly sampled documents. We invite three evaluators to evaluate system outputs from two perspectives: readability (Is the document easy to read or access?) and content consistency (Is the document's content the same as the source spoken-style document?). Each property is assessed with a score from 1 (worst) to 5 (best).

5.3 Main Results and Analysis

Automatic Results. Table 2 shows the results of document-level spoken-to-written simplification. Sentence-Transformer gets the worst results compared with document-level methods. It shows that document-level simplification is very different from sentence-level simplification, and directly converting sentences in a document one by one is not proper. Document-level spoken-to-written simplification has its challenges. Pre-trained models BART and BertEncoder are strong baselines and achieve improvements over baseline Transformer. G-DSWS, which has two graph encoder layers, achieves the best result. It improves the D-SARI by 3.03 points and BLEU by 1.52 points compared with the baseline Transformer. G-DSWS also outperforms pre-trained models in D-SARI and BLEU with a slight drop in ROUGE. It indicates that document structure information benefits performance more than a good language model initialization.

Table 2. Results of automatic evaluation. G-DSWS achieves the best result in two main metrics, D-SARI and BLEU.

Model	D-SARI	BLEU	ROUGE
Sentence-Transformer	44.31	76.36	92.63
Transformer	51.29	83.43	97.92
BART	52.33	84.21	**98.19**
BertEncoder	53.15	84.36	97.90
G-DSWS	**54.31**	**84.95**	97.80

Human Evaluation. Table 3 shows human evaluation results. We can see that our G-DSWS achieves the highest readability score with a slightly lower content consistency score. It shows that G-DSWS could improve the readability of the document meanwhile maintaining content consistency of the source spoken-style document.

Table 3. Results of human evaluation. Readability and Content Consistency are scored between 0 and 5, with higher scores representing better performance.

Model	Readability	Content Consistency
Sentence-Transformer	3.48	4.12
Transformer	3.77	**4.30**
G-DSWS	**4.23**	4.27

Ablation Study. We investigate the effect of each component in our graph encoder. Table 4 shows the results of the ablation study. We can see that the performance will decrease without any components of the graph encoder, especially components in graph propagation. It shows that each component is essential for our proposed method.

Table 4. Ablation study. We remove key components in the graph encoder layer, respectively. The performance drop shows that each individual component is essential for our proposed model G-DSWS.

Model	D-SARI	BLEU	ROUGE
G-DSWS	54.31	84.95	97.80
- Edge Embedding	51.15	83.65	97.91
- Hierarchical Graph Attention	49.49	82.28	96.96
- Document Graph Attention	49.00	82.35	97.55

We also investigate the influence of graph encoder layer numbers, illustrated in Table 5. 0-layer is the vanilla Transformer. When we increase the number of layers from 0 to 2, we can see a significant improvement in the performance of our models. Then the model performance declines with the number of network layers increasing. We attribute it to the fact that the enhancement of the document structure will affect the original sequence structure.

Table 5. Influences of graph layer numbers

Model	D-SARI	BLEU	ROUGE
0-layer	51.29	83.43	97.92
1-layer	51.51	83.77	97.94
2-layer	54.31	84.95	97.80
3-layer	51.72	83.64	97.96

Case Study. Furthermore, we manually analyze G-DSWS, and several baselines' system outputs to properly evaluate the proposed model's performance. Table 6 shows cases of several system outputs. Because the documents are too long, we show a few characters of the documents' heads and tails. G-DSWS's system outputs are more written style while keeping the original document's meaning. We notice that G-DSWS performs better than the vanilla Transformer. Utilizing the document information, G-DSWS better simplify and reorganize spoken documents making them more concise. For example, G-DSWS merges the first two sentences because they are divided by the spoken language issues. And G-DSWS deletes the meaningless conjunction '然后' because the context shows that it is a spoken language issue in Chinese.

5.4 Results on D-Wikipedia

To further verify our method's effectiveness on other document-level tasks, we conduct experiments on a publicly available document-level text simplification

Table 6. Case Study. For clarity, we translate the Chinese cases into English. Text marked in red is a spoken term that is not processed. Text marked in blue is a good case in which a spoken language issues that occurs across sentence boundaries is processed.

Original Spoken document	嗯，嗯呃，写还是写的蛮嗯。还是写的蛮清清晰的。对然后那个昨天三十六客，也对我们做了一个报道，......，还有一个是线下的这个校园大使，嗯。嗯校园活动就说我们通过那个线下的呃活动推广，吸引我们的粉丝，嗯也就是大学生。 Hmm, um, writing it is quite, um, good. It's written quite clearly. And then, yesterday "36Kr" also did a report on us... And there's also the campus ambassador program, um. Campus activities. Through this offline activity, we promote and attract our fans, um, which are mainly college students.
Sentence-Transformer	写还是写的蛮。还是写的蛮清晰。昨天三十六客，也对我们做了一个报道，......，还有一个是线下的校园大使。校园活动就说我们通过线下的活动推广，吸引我们的粉丝，也就是大学生 The writing is quite, umm. It's written quite clearly. Yesterday "36Kr" also did a report on us... And there's also the campus ambassador program. Through this campus activities, umm, offline activity, we promote and attract our fans, which are mainly college students.
Transformer	写还是写的蛮。还是写的蛮清晰的。然后昨天三十六客，也对我们做了一个报道，......，还有一个是线下的校园大使，校园活动就说我们通过线下的活动推广，吸引我们的粉丝，也就是大学生。 The writing is quite, umm. It's written quite clearly. And then, yesterday "36Kr" also did a report on us... And there's also the campus ambassador program. Through this campus activities, umm, offline activity, we promote and attract our fans, which are mainly college students.
G-DSWS	写还是写的蛮清晰的。昨天三十六客，也对我们做了一个报道，......，还有一个是线下的校园大使，校园活动我们通过线下的活动推广，吸引我们的粉丝，也就是大学生。 The writing is quite clearly. Yesterday "36Kr" also did a report on us... And there's also the campus ambassador program. Through this offline campus activity, we promote and attract our fans, which are mainly college students.

dataset. Table 7 shows the results of document-level text simplification. The first to third rows are the results of the baselines. The fourth row is the result of our reimplementation of the transformer model. The fifth row is the proposed model, with one graph encoder layer. Compared with the baseline transformer model, the proposed model improves the D-SARI by 2.83 points and SARI by 0.59 points. As for the main metric, D-SARI, the proposed model outperforms previous methods by 0.44 points. Moreover, the proposed model outperforms previous methods by 0.59 points for text simplification metrics SARI. The results on document-level text simplification show that our proposed model works well in other document-level generation tasks.

Table 7. Results of document-level text simplification, * denotes the result of the transformer baseline that we have re-implemented.

Model	D-SARI	SARI
SUC	12.97	34.13
Transformer	37.38	44.46
BART	37.24	48.34
BertSumextabs	39.88	47.39
Transformer*	37.49	48.22
G-DSWS	**40.32**	**48.81**

6 Conclusion and Future Work

This paper focuses on document-level spoken-to-written simplification and proposes a novel model, G-DSWS. Our key innovation is using a graph encoder with graph attention networks to enhance the document modeling capabilities of the encoder-decoder model. We evaluate our proposed model on the internal and publicly available datasets. The experiment results demonstrate the effectiveness of the proposed method. G-DSWS's enhanced document modeling capability enables it to tackle the challenges posed by cross-sentence transformations and long dependencies in spoken documents. As shown by the human evaluation and case study, the readability of spoken Chinese documents is significantly improved as a result.

In the future work, we plan to further explore G-DSWS with more well-designed document graph structures. We also plan to develop more suitable evaluation metrics for document-level spoken-to-written simplification, which will enable us to better assess model performance and promote further development in this field.

Acknowledgements. This work was supported by the National Key R&D Program of China under Grant No. 2020AAA0108600 and the Strategic Priority Research Program of the Chinese Academy of Sciences under Grant No. XDC08020100.

References

1. Chafe, W., Tannen, D.: The relation between written and spoken language. Annu. Rev. Anthropol. **16**(1), 383–407 (1987)
2. Dong, Q., Wang, F., Yang, Z., Chen, W., Xu, S., Xu, B.: Adapting translation models for transcript disfluency detection. In: Proceedings of the AAAI Conference on Artificial Intelligence, vol. 33, pp. 6351–6358 (2019)
3. Fang, Y., Sun, S., Gan, Z., Pillai, R., Wang, S., Liu, J.: Hierarchical graph network for multi-hop question answering. In: Proceedings of the 2020 Conference on Empirical Methods in Natural Language Processing (EMNLP), pp. 8823–8838 (2020)
4. Hrinchuk, O., Popova, M., Ginsburg, B.: Correction of automatic speech recognition with transformer sequence-to-sequence model. In: ICASSP 2020-2020 IEEE International Conference on Acoustics, Speech and Signal Processing (ICASSP), pp. 7074–7078. IEEE (2020)
5. Ihori, M., Makishima, N., Tanaka, T., Takashima, A., Orihashi, S., Masumura, R.: MAPGN: masked pointer-generator network for sequence-to-sequence pre-training. In: ICASSP 2021-2021 IEEE International Conference on Acoustics, Speech and Signal Processing (ICASSP), pp. 7563–7567. IEEE (2021)
6. Ihori, M., Makishima, N., Tanaka, T., Takashima, A., Orihashi, S., Masumura, R.: Zero-shot joint modeling of multiple spoken-text-style conversion tasks using switching tokens. arXiv preprint arXiv:2106.12131 (2021)
7. Ihori, M., Masumura, R., Makishima, N., Tanaka, T., Takashima, A., Orihashi, S.: Memory attentive fusion: external language model integration for transformer-based sequence-to-sequence model. In: Proceedings of the 13th International Conference on Natural Language Generation, pp. 1–6 (2020)
8. Ihori, M., Takashima, A., Masumura, R.: Large-context pointer-generator networks for spoken-to-written style conversion. In: ICASSP 2020-2020 IEEE International Conference on Acoustics, Speech and Signal Processing (ICASSP), pp. 8189–8193. IEEE (2020)
9. Ihori, M., Takashima, A., Masumura, R.: Parallel corpus for Japanese spoken-to-written style conversion. In: Proceedings of the 12th Language Resources and Evaluation Conference, pp. 6346–6353 (2020)
10. Kenton, J.D.M.W.C., Toutanova, L.K.: Bert: pre-training of deep bidirectional transformers for language understanding. In: Proceedings of NAACL-HLT, pp. 4171–4186 (2019)
11. Lewis, M., et al.: Bart: denoising sequence-to-sequence pre-training for natural language generation, translation, and comprehension. In: Proceedings of the 58th Annual Meeting of the Association for Computational Linguistics, pp. 7871–7880 (2020)
12. Liao, J., et al.: Improving readability for automatic speech recognition transcription. arXiv preprint arXiv:2004.04438 (2020)
13. Lin, C.Y.: Rouge: a package for automatic evaluation of summaries. In: Text Summarization Branches Out, pp. 74–81 (2004)
14. Liu, Y., Lapata, M.: Text summarization with pretrained encoders. In: Proceedings of the 2019 Conference on Empirical Methods in Natural Language Processing and the 9th International Joint Conference on Natural Language Processing (EMNLP-IJCNLP), pp. 3730–3740 (2019)
15. Ott, M., et al.: fairseq: a fast, extensible toolkit for sequence modeling. In: Proceedings of the 2019 Conference of the North American Chapter of the Association for Computational Linguistics (Demonstrations), pp. 48–53 (2019)

16. Papineni, K., Roukos, S., Ward, T., Zhu, W.J.: Bleu: a method for automatic evaluation of machine translation. In: Proceedings of the 40th Annual Meeting of the Association for Computational Linguistics, pp. 311–318 (2002)
17. Pusateri, E., Ambati, B.R., Brooks, E., Platek, O., McAllaster, D., Nagesha, V.: A mostly data-driven approach to inverse text normalization. In: INTERSPEECH, Stockholm, pp. 2784–2788 (2017)
18. Sun, R., Jin, H., Wan, X.: Document-level text simplification: dataset, criteria and baseline. In: Proceedings of the 2021 Conference on Empirical Methods in Natural Language Processing, pp. 7997–8013 (2021)
19. Sun, R., Lin, Z., Wan, X.: On the helpfulness of document context to sentence simplification. In: Proceedings of the 28th International Conference on Computational Linguistics, pp. 1411–1423 (2020)
20. Sutskever, I., Vinyals, O., Le, Q.V.: Sequence to sequence learning with neural networks. In: Advances in Neural Information Processing Systems, vol. 27 (2014)
21. Tilk, O., Alumäe, T.: Bidirectional recurrent neural network with attention mechanism for punctuation restoration. In: Interspeech, pp. 3047–3051 (2016)
22. Vaswani, A., et al.: Attention is all you need. In: Advances in Neural Information Processing Systems, vol. 30 (2017)
23. Veličković, P., Cucurull, G., Casanova, A., Romero, A., Lio, P., Bengio, Y.: Graph attention networks. arXiv preprint arXiv:1710.10903 (2017)
24. Wang, S., Che, W., Liu, Q., Qin, P., Liu, T., Wang, W.Y.: Multi-task self-supervised learning for disfluency detection. In: Proceedings of the AAAI Conference on Artificial Intelligence, vol. 34, pp. 9193–9200 (2020)
25. Xu, M., Li, L., Wong, D.F., Liu, Q., Chao, L.S.: Document graph for neural machine translation. In: Proceedings of the 2021 Conference on Empirical Methods in Natural Language Processing, pp. 8435–8448 (2021)
26. Xu, W., Napoles, C., Pavlick, E., Chen, Q., Callison-Burch, C.: Optimizing statistical machine translation for text simplification. Trans. Assoc. Comput. Linguist. **4**, 401–415 (2016)
27. Zheng, B., et al.: Document modeling with graph attention networks for multi-grained machine reading comprehension. In: ACL (2020)

MemFlowNet: A Network for Detecting Subtle Surface Anomalies with Memory Bank and Normalizing Flow

Le Huang[1,2,3(✉)], Fen Li[1,2,3], Dongxiao Li[1,2,3(✉)], and Ming Zhang[1,2]

[1] College of Information Science and Electronic Engineering, Zhejiang University, Hangzhou 310027, China
{lehuang,lidx}@zju.edu.cn
[2] Zhejiang Provincial Key Laboratory of Information Processing, Communication and Networking, Hangzhou 310027, China
[3] Jinhua Institute of Zhejiang University, Jinhua 321299, China

Abstract. Detection of subtle surface anomalies in the presence of strong noise is a challenging vision task. This paper presents a new neural network called **MemFlowNet** for detecting subtle surface anomalies by combining the advantages of the memory bank and normalizing flow. The proposed method consists of two stages. The first stage achieves pixel-level segmentation of anomalies using noise-insensitive average features in the memory bank and Nearest Neighbor search strategy, and the second stage performs image-level detection using normalizing flows and multi-scale score fusion. A new dataset called **INSCup** has been developed to assist this research by acquiring inner surface images of stainless steel insulated cups with ultra-wide lenses. The performance of MemFlowNet has been validated on INSCup dataset by surpassing other mainstream methods. In addition, MemFlowNet achieves the best performance with an image-level AUROC of 99.57% in anomaly detection of MVTec-AD benchmark. It shows a great potential to apply MemFlowNet to automated visual inspection of surface anomalies.

Keywords: Anomaly detection · Memory bank · Normalizing flow

1 Introduction

In recent years, with the widespread application of machine learning and deep learning, machine vision technology has received increasing attention in industrial production. Among them, anomaly detection based on computer vision is very popular. Typical defect detection objects can be classified into object class and texture class, which have apparent features or regular textures. The surfaces of polished steel products usually lack apparent features, and it is not easy to distinguish inconspicuous anomalies from noise of various kinds.

Detecting subtle inner-surface defects has been a significant problem that has not yet been well solved. Due to the lack of corresponding public datasets,

B. Luo et al. (Eds.): ICONIP 2023, CCIS 1966, pp. 417–429, 2024.
https://doi.org/10.1007/978-981-99-8148-9_33

supervised methods require a lot of images and cannot be conveniently applied to real scenarios. Meanwhile, existing unsupervised or semi-supervised methods do not work well for non-obvious anomalies in the presence of noise interference.

NEU surface defect database [17] and Severstal steel defect dataset [6] are based on flat raw steel, and the defects are apparent. For magnetic tile surface defect dataset [9] and Kolektor Surface-Defect Dataset [18], the objects are non-planar, and the sample contains more irrelevant noisy background interference. Still, the goal is to detect obvious structural anomalies such as crack and fray, and no attention is paid to surface defects.

In this work, we developed a new defect detection dataset called INSCup, comprising inner surface images of stainless steel cups, which represents a more challenging type of task, with the following characteristics: 1. The anomalies are inconspicuous, and the processing objects are texture-free surfaces of polished stainless steel products. 2. The surface to be inspected is cylindrical or spherical, not planar. 3. The surface to be inspected is the inner surface of an object rather than the outer surface, which is not convenient for acquiring high-quality images.

To address the challenges mentioned above, this paper presents a new method called MemFlowNet for detecting subtle anomalies by combining the advantages of memory banks and normalizing flows. Existing methods tend to overestimate the abnormality of noise, but an average feature memory can solve this problem well. Due to noise interference, traditional statistical methods cannot achieve satisfactory results in the detection task based on the segmentation results. Normalizing flows [4,13] map the features to a new distribution, thus significantly improving the detection performance.

Our method has tremendous potential for applications in other industrial scenarios. We demonstrate the reliability of our method on the MVTec-AD dataset [2], outperforming mainstream methods. The average AUROC score for image-level detection is 99.57%, while 98.5% for pixel-level segmentation.

We make the following contributions in this paper.

- We propose a one-class classification method called MemFlowNet, which achieved good results in subtle anomaly detection and segmentation tasks.
- We propose a new dataset called INSCup, captured by a specific camera system to facilitate the research of detecting subtle defects on inner surfaces.
- Combining the advantages of memory banks and normalizing flow. Averaging features and Nearest Neighbor search strategies suppress noise in segmentation while the normalizing flow further improves anomaly detection.
- MemFlowNet achieves better results in the defect detection task on INSCup and MVTec-AD datasets than current mainstream methods.

2 Related Work

Using pre-trained networks as feature extractors is a common technique, as proposed in [3,7,11,15,19,20], which use ResNet [8], Wide-ResNet [21], or other models to extract features with different scales. Regarding one-class classification methods, where only normal samples are available for training, and there may be significant differences in feature representation for abnormal samples in

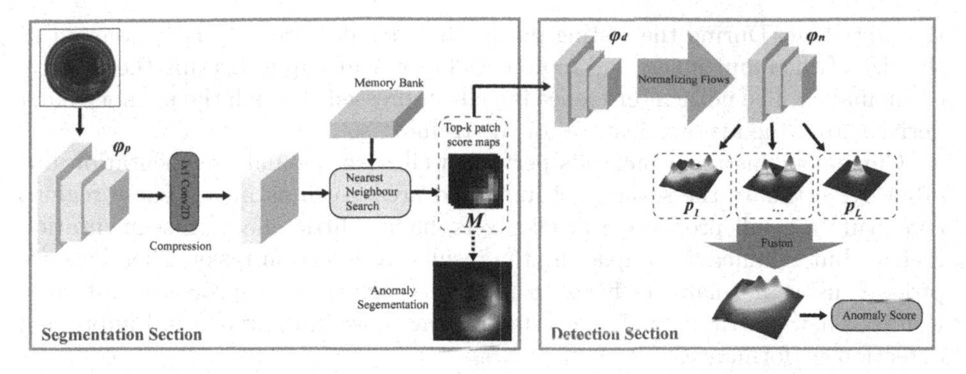

Fig. 1. Overview of the proposed method. Averaging features in the memory bank and NN search strategy suppress noise in segmentation. Normalizing flow and score fusion further improves performance in detection.

the test set, contrastive learning methods are well suited for such tasks. Memory bank can also be seen as a contrastive strategy that compares the features of training and testing sets [3,11,15,19]. When facing a large number of features obtained from extensive training data, feature compression is also a necessary process. PaDiM [3] learns Gaussian parameters from the training embeddings, PatchCore [15] employs k-means clustering, while CFA [11] uses a 1×1 CoordConv layer to compress features but introduces position redundancy.

Normalizing Flows (NF) [4,5] is a technique that enables complex probability distribution manipulation. It transforms a simple probability distribution, such as a Gaussian distribution, into a more complex one using reversible transformation equations. Recently, NF has been applied in the field of anomaly detection. CFLOW [7] adds position encoding in multi-scale features. FastFlow [20] can be used as a plugin module and has high efficiency. AltUB [10] converts normal images into an arbitrary normal distribution, improving its stability.

In addition, some methods apply data augmentation. DRÆM [22] generates random anomalous regions using Perlin noise and adds external texture information to form anomaly images. CutPaste [12] utilizes its own or patterned patches for splicing. MemSeg [19] uses binarized foreground segmentation to superimpose abnormal regions onto the foreground.

Although all these methods have achieved promising results in various domains, most have limitations when detecting subtle defects with noise interference. Our proposed method aims to break through these limitations and improve the accuracy of anomaly detection.

3 Proposed Methods

3.1 Overview

We propose a combined method using segmentation and detection techniques to show inner-surface anomalies of cups. Figure 1 shows an overview of the proposed method. An anomaly segmentation network is trained using defect-free images in the first stage. During the testing phase, the output score map is upsampled to the size of the input image and represented as a heatmap indicating the severity of anomalies. The noise interference is well suppressed through the memory bank derived from the average features of the training set.

Current mainstream methods perform well with anomaly segmentation but often overestimate the severity of unrelated areas such as noise. These regions constitute a small proportion of all pixels, having little effect on segmentation metrics but significantly impacting the results in detection tasks. Therefore, we propose using normalizing flows to map the segmentation-processed information to a new distribution, further suppressing noise interference and improving detection performance.

3.2 Memory Bank and Feature Domain Adaptation

Our method is based on the belief that identifying regions of interest first can simplify anomaly detection in images. To realize this, we use a memory-augmented convolutional neural network to segment abnormal regions.

Given a dataset X, we denote normal samples as X_N ,$\forall x \in X_N : y_x = 0$ and test samples as X_T, $\forall x \in X_T : y_x \in \{0, 1\}$. The network is trained on defect-free images and uses two well-known networks, ResNet18 [8] and Wide ResNet50-2 [21], as feature extractors. The feature extraction network is denoted as φ_P, where $\varphi_l \in \mathbb{R}^{c_l \times h_l \times w_l}$ represents the l-th feature layer ($l \in \{1, 2, 3, 4\}$). For each $(i, j) \in (h_l, w_l)$, the vector $\varphi_l(i, j) \in \mathbb{R}^{c_l}$ describes the patch of the input image. As the number of layers l increases, the patch size represented by vector φ_l increases.

The memory bank contains the average feature representation of the dataset as a reference for extracting features from newly segmented images. The memory bank B needs to be initialized before training. We utilize a pre-trained network to extract features from the trainset, compress and average them, to obtain the average normal features. This process can be completed without training. And it helps to effectively control the computational costs and memory usage required for inferring a single sample.

Since φ_p preserves the relative positional relationship between patches, unlike coordconv used in CFA [11], we reduce the redundancy of positional information and use a simple 1×1 Conv2D module to compress features. The memory bank $B = avg_{x_i \in X_N}\{C(\varphi_p(x_i))\}$, where $C(\cdot)$ represents the compression process.

During training, we compare the compressed features with those stored in the memory bank to determine the most similar pairs. We use kNN (k-nearest neighbor) to search for the k nearest pairs of features in the memory bank. For each feature $p_m \in C(\varphi_p(x))$ representing a patch, $p_m^k \in B$ represents the k-th nearest neighbor of p_m in B. Using Euclidean distance, we calculate the distance between the compressed feature and its nearest feature in the memory bank. Thus, $d(k) = \|p_m^k - p_m\|_2$.

Deep SVDD has been applied to domain adaptation for one-class classification [16]. Similarly, our feature domain adaptation uses the K nearest neighbors of p_m to calculate the loss to find a suitable hypersphere radius r. The first $K/2$

are used to incorporate suitable features into the hypersphere, the latter $K/2$ are used to prevent r from increasing indefinitely, and μ is the weight coefficient of the former. The loss function can be expressed as:

$$L_{seg} = \mu \sum_{k=1}^{K/2} max\{0, d(k) - r^2\} + (1-\mu) \sum_{k=K/2+1}^{K} max\{0, r^2 - d(k) - \theta\}, \quad (1)$$

where θ is the balancing parameter and helps to stabilize r during training.

The patch map score $M = d(k)$ is calculated using the distance obtained by k-NN search, where $k \in \{1, 3\}$. In the segmentation stage, $k = 1$, the patch map score is upsampled to the size of the original image using bilinear interpolation and Gaussian smoothing to obtain abnormal segmentation results. When $k = 3$, more information can be retained for subsequent detection tasks.

3.3 Flow-Based Detection

Due to the lack of clear boundaries between subtle defects and noise, traditional image processing and statistical methods are unsuitable for detection tasks. Therefore, we introduce a mapping that applies normalizing flows to the features, which maps the features to a new distribution.

After generating patch score maps, we extract features using ResNet18 to mainly utilize its pyramid-shaped feature structure to describe patches of different sizes. Applying a bijection function, a simple prior distribution $Z \sim N(0, I)$ is mapped to a complex posterior distribution M, which refers to the patch score map output by the first-stage network. This process can be achieved through a reversible transformation that can be represented as:

$$M \underset{f_1^{-1}}{\overset{f_1}{\rightleftharpoons}} H_1 \underset{f_2^{-1}}{\overset{f_2}{\rightleftharpoons}} H_2 \underset{f_3^{-1}}{\overset{f_3}{\rightleftharpoons}} \cdots \underset{f_K^{-1}}{\overset{f_K}{\rightleftharpoons}} Z.$$

With the transformation $f : M \to Z$, the log-likelihoods of patch distribution of patch score map M can be represented as:

$$\log p_M(m) = \log p_Z(f(m)) + \log \left| \det \frac{\partial f(m)}{\partial m} \right|, \quad (2)$$

where patch score map feature $m \in p_M(m)$, hidden variable $z \in p_Z(z)$, and mapping function $f = f_K \circ f_{K-1} \circ \cdots \circ f_1(M)$.

We employ affine coupling layers in each block, as shown in Fig. 2. For more details about affine coupling layers, please refer to Real NVP [5]. Input features are mapped to outputs of the same scale through multiple affine coupling layers. The input is channel-permuted and split into two parts, denoted as m_1 and m_2. After passing through the affine coupling layers, the resulting features are represented as z_1 and z_2.

Their relationship can be expressed as:

$$\begin{cases} z_1 = m_1 \\ z_2 = m_2 \odot s(m_1) + t(m_1), \end{cases} \quad (3)$$

where s and t are computed by networks containing 3×3 Conv2D layers.

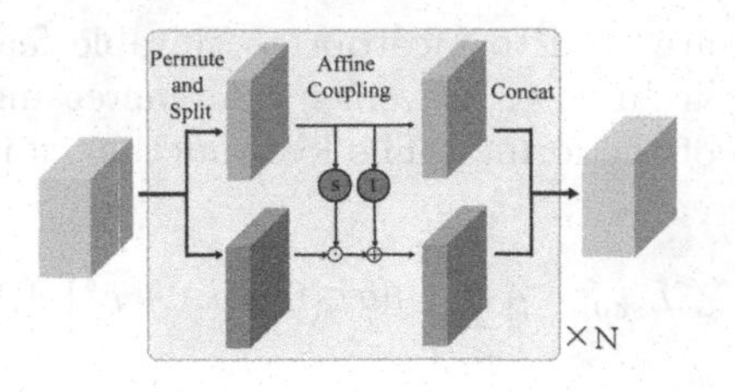

Fig. 2. Affine coupling layers in Normalizing Flows.

The training goal of the detection stage is to maximize the Formula (2). The loss function is defined as:

$$L_{det} = \frac{1}{L}\sum_{l=1}^{L}[-log(p_l) - log|detJ_l|], \quad (4)$$

J_l is the Jacobian matrix of the mapping function f in the l-th layer. p_l is the average probabilities of layer l in φ_n, calculated by log-likelihoods as [7], where $(l \in \{1, 2, \cdots, L\})$, the number of the height and width of each layer are not the same, they are recorded as h_l and w_l respectively. For every $(i, j) \in (h_l, w_l)$, vector $\varphi_n(l, i, j) \in \mathbb{R}^{c_l}$ describes the input samples of different sizes of patches. In particular, when $l = 1, 2$ *and* 3, they correspond to the patches of 4×4, 8×8, and 16×16 in the original figure, respectively.

Anomalies tend to cluster together, while noise tends to be more dispersed. The detection of image anomalies is related to the size of the patch. If the patch is too small, the anomaly may be severed. If the patch is too large, there will be an increase in noise interference. We employ the segmentation results for further feature extraction and normalized flow models to learn the mapping from normal patch score to known distribution.

We average p_l to obtain abnormal probability distribution maps of different scales. To fuse abnormal probability distribution maps of different scales, upsample p_l by bilinear interpolation $b(\cdot)$ to the same size as p_1, and then add them together. The detection score is $\mathbf{S} = max(\sum_{l=1}^{L} b(p_l))$.

The maps of multiple layers are superimposed, with the abnormal regions being highlighted. At this point, the maximum value of the entire map fuses the anomaly information of different scales, which can more effectively represent the degree of anomalies in the image.

4 INSCup Dataset and Augmentation

Most existing datasets for metal materials are based on flat and outer surfaces with obvious defects [6,9,17,18]. There is no suitable dataset for detecting non-planar, non-obvious inner-surface defects.

To facilitate research on these issues, we construct the INSCup dataset, which consists of inside images of stainless steel insulated cups acquired by industrial cameras equipped with ultra-wide-angle lenses. Figure 3 shows an example image and the segmentation results of a defect.

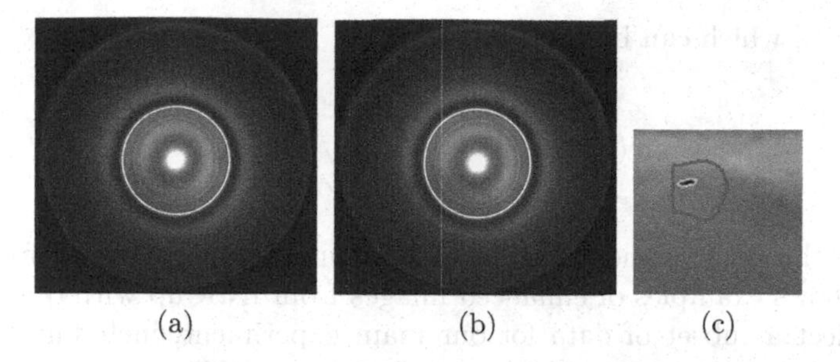

Fig. 3. Example of original collected images and segmentation results using our method. (a) Original image. (b) Segmentation result. (c) Zoomed-in result in the flattened image. Green: label, red: segmentation results. (Color figure online)

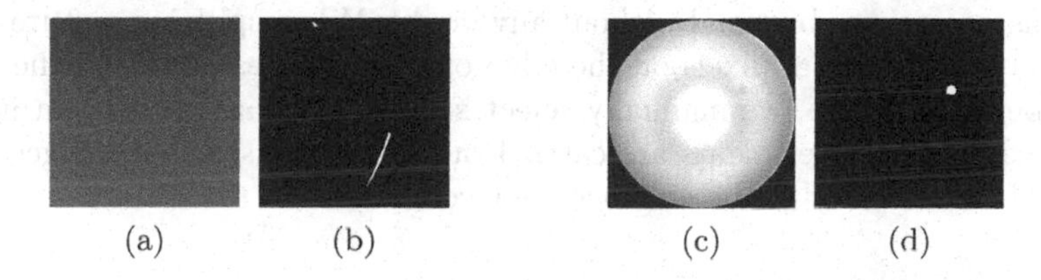

Fig. 4. Example of images and masks from INSCup dataset. (a)Side-wall image with the anomaly. (b) Side mask. (c) Bottom image with the anomaly. (d) Bottom mask.

The dataset includes images of the inner side walls and bottoms. The side is a cylindrical surface, and the bottom is a spherical crown surface. We employ an ultra-wide lens facing to the bottom of the cups at specified working distances to capture images. This imaging technique offers a unique perspective and captures comprehensive coverage of the inner wall of the cup. We utilize Hough circle detection to identify the cup's bottom and segment the side wall area. Next, we expand the ring form image of the cylindrical inwall to a rectangular shape and apply Automatic Color Equalization(ACE) [14] algorithm to facilitate the manual labeling of the abnormal areas. The images are cropped to 256×256 and augmented to facilitate storage and reading. There are 3,927 normal images and 957 defective images for the inner side walls, and 27 normal images and 135 defective images for the bottom of the inner surfaces. All defective images have pixel-level mask labels provided. Figure 4 shows examples of images and masks from the INSCup dataset.

ACE algorithm can be simply described as follows:

$$R(p) = \sum_{j \in Subset, p \neq j} \frac{r(I(p) - I(j))}{d(p, j)}, \; O = \frac{R - min(R)}{max(R) - min(R)}, \tag{5}$$

where I is the input image, O is the output image, $I(p)$ represents any pixel in the image, and $I(j)$ is other pixel points in the window. $d(\cdot)$ represents the distance between two points, and $r(\cdot)$ is the contrast adjustment function with range $[-1, 1]$, which can be expressed as:

$$r(x) = \begin{cases} -1, x \in (-\infty, -1/\alpha) \\ \alpha x, \ x \in [-1/\alpha, 1/\alpha] \\ 1, \quad x \in (1/\alpha, +\infty) \end{cases}, \tag{6}$$

where α is the slope of the linear part of the function, which is set from 3 to 15. Figure 5 shows examples of enhanced images from INSCup with $\alpha = 6$.

We select a subset of data for our main experiment, including 200 normal images in the training set and a total of 87 images (55 normal images and 32 abnormal images) in the test set containing defects such as scratches, stains, dust, spot, and blur, as shown in Fig. 5. The images in the original dataset are directly cropped, while the selected images have clear anomaly types and can be judged as normal or abnormal without any doubt. When splitting pictures, there may be a few abnormal pixels on the edge of some images, affecting the image-level labels. Therefore, we manually select some images and put them into the correct category. Experiments are carried out using the selected images except for the ablation study of the enhancement coefficient.

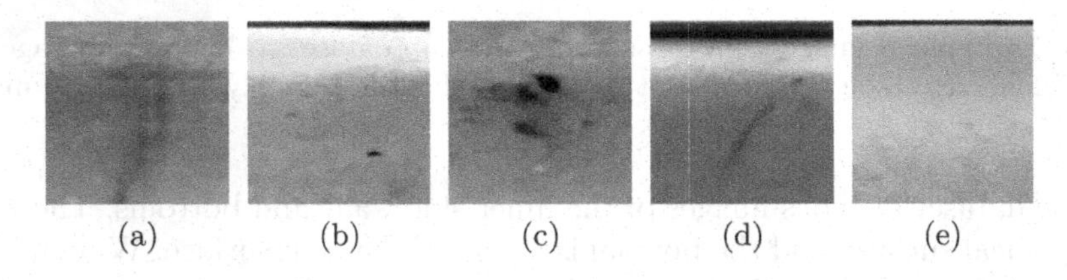

Fig. 5. Example of enhanced images from INSCup dataset. (a) Blur. (b) Spot. (c) Dust. (d) Scratch. (e) Good.

5 Experiments

5.1 Setup

MVTec-AD dataset [2] is a mainstream dataset for anomaly detection focused on industrial inspection. It contains high-resolution images of 15 different object categories. Each category contains images with different types of anomalies and detailed pixel-level masks are provided. To evaluate the performance of our proposed method, we conduct experiments on the INSCup and MVTec-AD datasets and compare our results with other mainstream approaches. Our experiments were run on a server with an Nvidia RTX A5000 GPU, 128 GB RAM, and Intel(R) Xeon(R) Silver 4210R CPU.

In the segmentation phase, we use ResNet18 and Wide ResNet50-2 pre-trained on ImageNet as feature extractors, selecting layer1-3 features. As the MVTec-AD images are large, they are resized to 256×256 as input. We employ AdamW optimizer, with the initial learning rate of 1e$-$3, the weight decay of

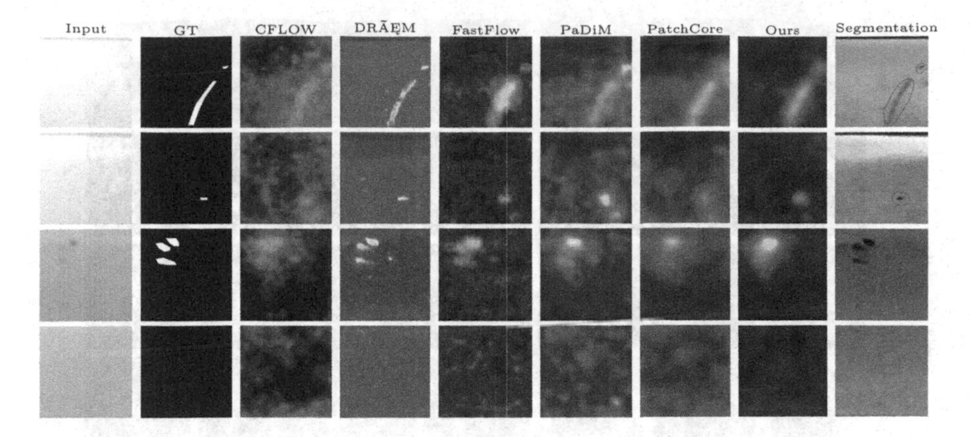

Fig. 6. Comparison of the heatmaps of MemFlowNet and mainstream methods using INSCup dataset. The input images for all methods are the same and are enhanced. Our method dramatically suppresses noise, and segmentation results are added to the enhanced images in the segmentation column.

5e−4, and the batch size of 8, trained for 80 epochs. In the detection phase, we employ AdamW optimizer with the initial learning rate of 1e−3 and the weight decay of 1e−5, and trained for 500 epochs with the batch size of 32. We apply two standard evaluation metrics for both image- and pixel level anomaly detection to compare the performance of our approach with other methods: image AUROC and pixel AUROC. The implementation of other methods refers to anomalib[1] [1]. We use this implementation to train and test the models on INSCup.

5.2 Results

As we can see from Table 1, our proposed method MemFlowNet gets the highest image-level AUROC of 96.31% and pixel-level AUROC of 97.71% based on ResNet18. For WRN50-2, we get image-level AUROC of 98.98%, which is higher than all other methods.

Some of the segmentation results images are shown in Fig. 6. The normal area in the heat map is black, and the abnormal area is red. MemFlowNet can highlight the abnormal area in the INSCup dataset, while the normal area can be stably displayed in the dark. However, the results of CFLOW, Padim, and PatchCore also have obvious red areas in the normal area, and the results of DRÆM and FastFlow overly highlight the noise area.

We further explore the application of MemFlowNet to other industrial scenarios, such as other objects or textures in MVTec-AD, as shown in Table 2 and 3. The subclass results of the mainstream methods are all from their papers. Our method obtains the highest average detection results and achieves the best detection results on nine categories, including Cable, Carpet, Grid, Hazelnut

[1] Open source implementation is available at github.com/openvinotoolkit/anomalib.

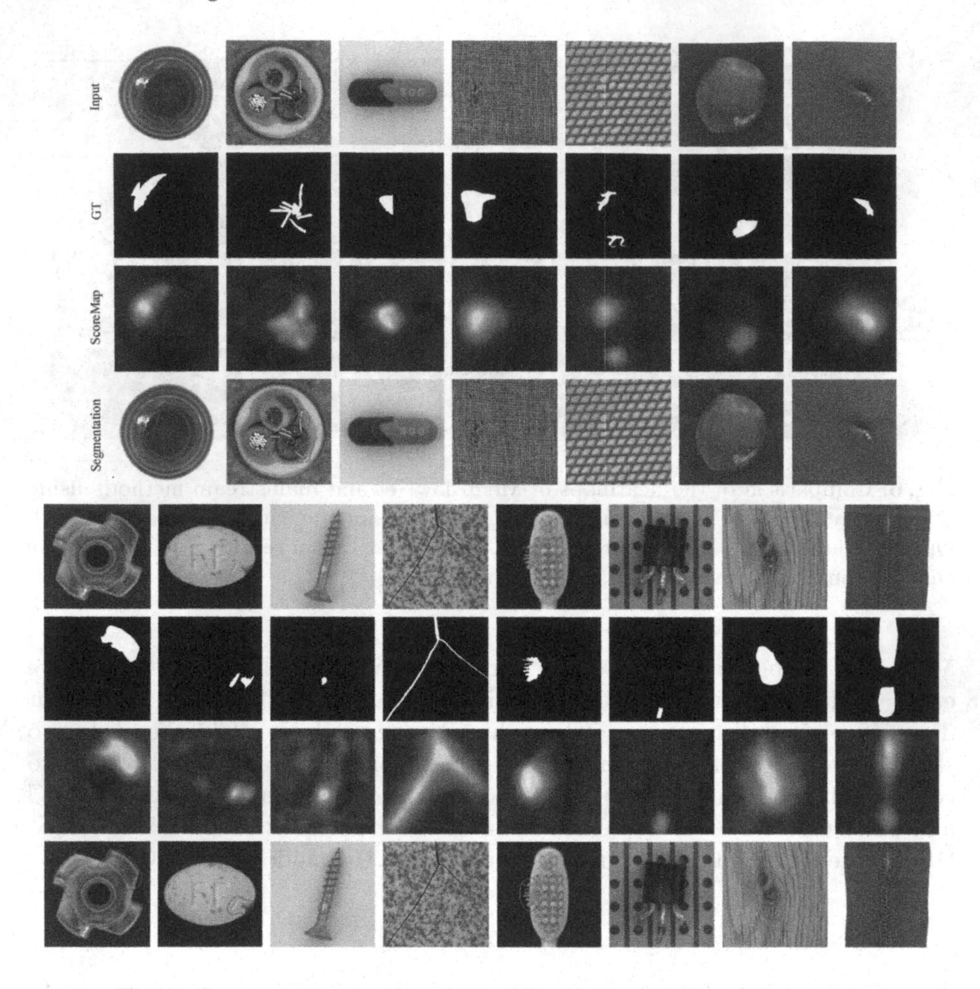

Fig. 7. Segmentation results of MemFlowNet on MVTec-AD dataset.

Leather, Tile, Toothbrush, Transistor, and Wood. Moreover, our segmentation results are comparable to the best method. Partial visualization of the segmentation results is shown in Fig. 7. While ensuring accurate anomaly segmentation, the mis-segmentation of noise is significantly reduced.

5.3 Ablation Studies

Effect of Enhancement Coefficient. The influence of parameter α in the ACE preprocessing is shown in Table 4. Note that unfiltered datasets are used for experimentation. As $\alpha = 6$ has the highest AUROC, it is used in all other experiments on INSCup.

Table 1. Results on INSCup using (image AUROC, pixel AUROC)%.

backbone	CFLOW [7]	FastFlow [20]	padim [3]	PatchCore [15]	CFA [11]	Ours
ResNet18	(95.97, 96.05)	(91.59, 97.43)	(85.91, 94.36)	(81.36, 97.12)	(94.89, 96.77)	(**96.31**, **97.71**)
WRN50-2	(90.28, **98.90**)	(88.64, 98.46)	(92.78, 97.68)	(82.33, 98.73)	(98.64, 98.75)	(**98.98**, 98.69)

Table 2. Image-level anomaly detection results on MVTec-AD using AUROC%.

Method	Avg	Bottle	Cable	Capsule	Carpet	Grid	Hazeln.	Leather	Metal.	Pill	Screw	Tile	Toothb.	Trans.	Wood	Zipper
DRAEM [22]	98.0	99.2	91.8	98.5	97.0	99.9	**100.0**	**100.0**	98.7	98.9	93.9	99.6	**100.0**	93.1	99.1	**100.0**
PatchCore [15]	99.1	**100.0**	99.5	98.1	98.7	98.2	**100.0**	**100.0**	**100.0**	96.6	**98.1**	98.7	**100.0**	**100.0**	99.2	99.4
CFLOW [7]	98.3	**100.0**	97.6	97.7	98.7	99.6	**100.0**	**100.0**	99.3	96.8	91.9	99.9	99.7	95.2	99.1	98.5
MemSeg [19]	99.56	**100.0**	99.3	**100.0**	99.6	**100.0**	**100.0**	**100.0**	**100.0**	**99.0**	97.8	**100.0**	**100.0**	99.2	99.6	**100.0**
Ours	**99.57**	99.8	**100.0**	98.9	**99.8**	**100.0**	**100.0**	**100.0**	99.5	98.8	97.1	**100.0**	**100.0**	**100.0**	**99.7**	99.9

For shallower networks like ResNet18, ACE processing have noticeable effects. Since the WRN50-2 network is deeper and can essentially perform similar functions to ACE, the effect of ACE on the WRN50-2 network is not as evident as on the ResNet18 network.

Effect of Top-k Score and Score Fusion. The detection effect of an NF model with different numbers of input channels is compared on INSCup, as shown in Table 5. Top1 represents taking the maximum patch score from the segmentation stage network output as the NF input. In contrast, top3 represents combining the three largest scores from each patch score candidate value in the segmentation stage into three maps as the NF input. It can be seen that the top3 performs better than the top1 based on WRN50-2 and NF layer fusion, so the top3 is used as the NF input in the MVTec-AD experiments. The impact of NF layer fusion on MVTec is shown in Table 6.

NF layers represent information from patches of different sizes. The size of abnormal areas in the samples determines the differences in the metrics of different layers. NF layer fusion achieves the best results in most categories. Still, NF

Table 3. Pixel-level anomaly localization results on MVTec-AD using AUROC%.

Method	Avg	Bottle	Cable	Capsule	Carpet	Grid	Hazeln.	Leather	Metal.	Pill	Screw	Tile	Toothb.	Trans.	Wood	Zipper
DRAEM [22]	97.3	99.1	94.7	94.3	95.5	99.7	99.7	98.6	99.5	97.6	97.6	99.2	98.1	90.9	96.4	98.8
PatchCore [15]	98.1	98.6	98.4	98.8	99.0	98.7	98.7	99.3	98.4	97.4	99.4	95.6	98.7	96.3	95.0	98.8
CFLOW [7]	98.6	99.0	97.6	99.0	99.3	99.0	98.9	99.7	98.6	99.0	98.9	98.0	98.9	98.0	96.7	99.1
MemSeg [19]	98.8	99.3	97.4	99.3	99.2	99.3	98.8	99.7	99.5	99.5	98.0	99.5	99.4	97.3	98.0	98.8
Ours	98.5	98.9	99.0	99.1	99.2	98.1	99.0	99.4	99.1	99.0	98.9	96.4	99.1	98.3	94.8	99.0

Table 4. Comparison of enhancement coefficient α on INSCup using (image AUROC, pixel AUROC)% metrics.

α	3	6	9	12	15	Without Enhancement
ResNet18	(90.5, 96.7)	(91.6, 95.9)	(89.4, 96.5)	(89.0, 96.0)	(89.5, 95.2)	(87.2, 93.4)
WRN50-2	(90.8, 97.9)	(**93.3**, **98.2**)	(90.7, 98.0)	(88.9, 98.1)	(89.9, 97.5)	(92.6, **98.2**)

Table 5. Detection performance of NF layers on INSCup (Image-level AUROC%). φ_P is pre-trained net in segmentation section, M(as NF input) is the patch score map(s) from the segmentation section, p_l is the average probabilities of the l-th NF layer.

φ_P	Topk Scores	M	p_1	p_2	p_3	Fusion
ResNet18	Top1	93.58	95.85	91.93	95.97	96.82
	Top3	72.73	96.59	93.81	93.47	96.31
WRN50-2	Top1	96.70	98.64	98.07	93.92	98.75
	Top3	96.70	98.58	98.69	94.55	**98.98**

Table 6. Detection performance of NF layers on MVTec-AD (Image-level AUROC%).

Method	Avg	Bottle	Cable	Capsule	Carpet	Grid	Hazeln.	Leather	Metal.	Pill	Screw	Tile	Toothb.	Trans.	Wood	Zipper
M	98.76	**100.00**	99.70	98.92	99.60	99.16	**100.00**	**100.00**	**100.00**	98.42	92.60	**100.00**	95.00	**100.00**	99.30	98.69
p_1	98.88	99.76	99.93	99.12	99.76	**100.00**	99.71	**100.00**	99.51	98.25	95.02	**100.00**	96.39	96.71	99.12	99.95
p_2	98.71	99.44	99.46	98.44	99.76	**100.00**	99.89	**100.00**	98.24	97.49	94.06	**100.00**	94.17	99.75	**100.00**	99.89
p_3	99.20	**100.00**	99.63	**99.56**	99.36	99.92	**100.00**	**100.00**	97.31	97.41	95.57	**100.00**	99.44	**100.00**	99.82	**99.92**
Fusion	**99.57**	99.84	**99.96**	98.88	**99.80**	**100.00**	**100.00**	**100.00**	99.46	**98.83**	**97.11**	**100.00**	**100.00**	**100.00**	99.74	**99.92**

layer3 achieves the best results in some categories, possibly because there are more anomalies in large areas and NF layer3 have better perception ability for large anomalies. In contrast, NF layer1 has better perception ability for small anomalies.

6 Conclusion

We propose MemFlowNet for the segmentation and detection of subtle inner-surface defects along with a new dataset called INSCup. We achieve a 98.98% image-level AUROC and 98.69% pixel-level AUROC on the INSCup dataset, outperforming mainstream methods. The reliability of our approach is also demonstrated on the MVTec-AD dataset, achieving 99.57% image-level AUROC and 98.5% pixel-level AUROC.

Our work can be utilized in automated product inspection to reduce manual labor costs in the manufacturing industry. Additionally, further research could explore the application of our approach in other fields, such as the inspection of agricultural products and medical images.

References

1. Akcay, S., Ameln, D., Vaidya, A., Lakshmanan, B., Ahuja, N., Genc, U.: Anomalib: a deep learning library for anomaly detection (2022)
2. Bergmann, P., Fauser, M., Sattlegger, D., Steger, C.: MVTec AD-a comprehensive real-world dataset for unsupervised anomaly detection. In: Proceedings of the IEEE/CVF Conference on Computer Vision and Pattern Recognition, pp. 9592–9600 (2019)

3. Defard, T., Setkov, A., Loesch, A., Audigier, R.: PaDiM: a patch distribution modeling framework for anomaly detection and localization. In: Del Bimbo, A., et al. (eds.) ICPR 2021. LNCS, vol. 12664, pp. 475–489. Springer, Cham (2021). https://doi.org/10.1007/978-3-030-68799-1_35
4. Dinh, L., Krueger, D., Bengio, Y.: NICE: non-linear independent components estimation. arXiv preprint arXiv:1410.8516 (2014)
5. Dinh, L., Sohl-Dickstein, J., Bengio, S.: Density estimation using real NVP. arXiv preprint arXiv:1605.08803 (2016)
6. Grishin, A., BorisV: Severstal: steel defect detection (2019). https://kaggle.com/competitions/severstal-steel-defect-detection
7. Gudovskiy, D., Ishizaka, S., Kozuka, K.: CFLOW-AD: real-time unsupervised anomaly detection with localization via conditional normalizing flows. In: Proceedings of the IEEE/CVF Winter Conference on Applications of Computer Vision, pp. 98–107 (2022)
8. He, K., Zhang, X., Ren, S., Sun, J.: Deep residual learning for image recognition. In: Proceedings of the IEEE Conference on Computer Vision and Pattern Recognition, pp. 770–778 (2016)
9. Huang, Y., Qiu, C., Yuan, K.: Surface defect saliency of magnetic tile. Vis. Comput. **36**, 85–96 (2020)
10. Kim, Y., Jang, H., Lee, D., Choi, H.J.: AltUB: alternating training method to update base distribution of normalizing flow for anomaly detection. arXiv preprint arXiv:2210.14913 (2022)
11. Lee, S., Lee, S., Song, B.C.: CFA: coupled-hypersphere-based feature adaptation for target-oriented anomaly localization. IEEE Access **10**, 78446–78454 (2022)
12. Li, C.L., Sohn, K., Yoon, J., Pfister, T.: CutPaste: self-supervised learning for anomaly detection and localization. In: Proceedings of the IEEE/CVF Conference on Computer Vision and Pattern Recognition, pp. 9664–9674 (2021)
13. Loshchilov, I., Hutter, F.: Decoupled weight decay regularization. arXiv preprint arXiv:1711.05101 (2017)
14. Rizzi, A., Gatta, C., Marini, D.: A new algorithm for unsupervised global and local color correction. Pattern Recogn. Lett. **24**(11), 1663–1677 (2003)
15. Roth, K., Pemula, L., Zepeda, J., Schölkopf, B., Brox, T., Gehler, P.: Towards total recall in industrial anomaly detection. In: Proceedings of the IEEE/CVF Conference on Computer Vision and Pattern Recognition, pp. 14318–14328 (2022)
16. Ruff, L., Vandermeulen, R., Goernitz, N., Deecke, L., Siddiqui, S.A., Binder, A., Müller, E., Kloft, M.: Deep one-class classification. In: International Conference on Machine Learning, pp. 4393–4402. PMLR (2018)
17. Song, K., Yan, Y.: A noise robust method based on completed local binary patterns for hot-rolled steel strip surface defects. Appl. Surf. Sci. **285**, 858–864 (2013)
18. Tabernik, D., Šela, S., Skvarč, J., Skočaj, D.: Segmentation-based deep-learning approach for surface-defect detection. J. Intell. Manuf. **31**(3), 759–776 (2020)
19. Yang, M., Wu, P., Feng, H.: MemSeg: a semi-supervised method for image surface defect detection using differences and commonalities. Eng. Appl. Artif. Intell. **119**, 105835 (2023)
20. Yu, J., et al.: Fastflow: unsupervised anomaly detection and localization via 2D normalizing flows. arXiv preprint arXiv:2111.07677 (2021)
21. Zagoruyko, S., Komodakis, N.: Wide residual networks. arXiv preprint arXiv:1605.07146 (2016)
22. Zavrtanik, V., Kristan, M., Skočaj, D.: DRAEM-a discriminatively trained reconstruction embedding for surface anomaly detection. In: Proceedings of the IEEE/CVF International Conference on Computer Vision, pp. 8330–8339 (2021)

LUT-LIC: Look-Up Table-Assisted Learned Image Compression

SeungEun Yu and Jong-Seok Lee(✉)

Yonsei University, Seoul, Republic of Korea
{seungeun.yu,jong-seok.lee}@yonsei.ac.kr

Abstract. Image compression is indispensable in many visual applications. Recently, learned image compression (LIC) using deep learning has surpassed traditional image codecs such as JPEG in terms of compression efficiency but at the cost of increased complexity. Thus, employing LIC in resource-limited environments is challenging. In this paper, we propose an LIC model using a look-up table (LUT) to effectively reduce the complexity. Specifically, we design an LUT replacing the entropy decoder by analyzing its input characteristics and accordingly developing a dynamic sampling method for determining the indices of the LUT. Experimental results show that the proposed method achieves better compression efficiency than traditional codecs with faster runtime than existing LIC models.

Keywords: Learned Image Compression · Deep Learning · Look-up Table

1 Introduction

Image compression plays a crucial role in the transmission and storage of visual data. In recent years, learned image compression (LIC) methods using deep learning models [2,5–7,9,10,13,18,22–24] have demonstrated significant advancements in terms of compression efficiency and image quality, outperforming traditional image compression codecs such as JPEG [27], JPEG 2000 [25], and WebP [8] which follow a typical pipeline of transformation, quantization, and entropy coding.

With the proliferation of the Internet of Things (IoT) industry, there is a growing demand for collecting, storing, and transmitting image data from various end-devices. However, the practical application of LIC models is often hindered by their device-dependency, large size, and high computational requirements. Many end-devices such as automobiles, home appliances, and other IoT devices utilize micro-controllers (MCUs) or micro-processors (MPUs), which are

This work was supported in part by the Ministry of Trade, Industry and Energy (MOTIE) under Grant P0014268, and in part by the NRF grant funded by the Korea government (MSIT) (2021R1A2C2011474).

B. Luo et al. (Eds.): ICONIP 2023, CCIS 1966, pp. 430–441, 2024.
https://doi.org/10.1007/978-981-99-8148-9_34

slower compared to graphics processing units (GPUs) commonly used to run deep learning models. These MCUs/MPUs lack the processing power to handle complex calculation that deep learning models require, rendering them inefficient for LIC-based image compression tasks.

To address this challenge, we propose a novel LIC model which utilizes a look-up table (LUT) to reduce computational complexity and model size. By leveraging the one-to-one mapping between input and output values, the LUT enables efficient storage and retrieval of pre-computed values. This approach eliminates the need for extensive calculations during inference for encoding and decoding, reducing the reliance on low-performance hardware and alleviating computational overhead caused by redundant computations for repeatedly appearing input values. Additionally, by employing appropriate sampling techniques, we can maintain a sufficiently small LUT size without experiencing significant performance degradation. Therefore, incorporating an LUT in the LIC model can be a practical solution which makes LIC utilized in devices with limited computational capability.

The use of LUTs as an alternative to deep learning models has been explored previously in the domain of single-image super-resolution (SR). The SR using LUT model [16] is the first to introduce an LUT for replacing the whole SR network. This is extended in [20] by employing multiple LUTs to further enhance the performance.

However, it is not straightforward to apply LUTs to LIC as in SR. First, an LIC model is typically composed of multiple modules, for each of which a dedicated LUT needs to be designed. Above all, the image encoder and bitstream decoder work separately for converting the given image into a bitstream and restoring the given bitstream into an image, respectively. In addition, the encoder part consists of the transform encoder and hyper encoder, and the decoder part includes the transform decoder and hyper decoder (see Fig. 1). The characteristics of each of these four modules need to be analyzed and the LUTs need to be designed accordingly. Second, the input value ranges of the modules are not fixed, especially in the cases of hyper encoder/decoder and transform decoder. Since the input values are used as the indices of an LUT, this issue makes it difficult to build an LUT with pre-determined indices. This is a new challenge in LIC, whereas in the case of SR, the input to the model is always an image whose range is fixed between 0 and 255.

Our proposed model, called LUT-assisted LIC (LUT-LIC), focuses on utilizing an LUT to substitute the hyper decoder, enabling practical usage with enhanced speed and reduced memory requirements without significant performance degradation. In order to deal with the aforementioned issue of the unbounded input range, we propose a sampling method to determine optimal indices of the LUT. In addition, we address the issue of determining the receptive field (RF) of the hyper encoder/decoder. Here, the RF means the input region (in both spatial and channel dimensions) by which a particular output value is influenced. Thus, the size of an LUT exponentially increases with the RF size. We carefully design the hyper encoder and decoder modules with a limited RF size in order to keep the size of the LUT manageable and minimize performance degradation.

The contribution of this paper lies in the exploration of LUT-based solutions to improve the usability of deep learning-based image compression models on resource-constrained devices. By specifically targeting the hyper decoder, we aim to address the limitations posed by the computational requirements of deep learning models. Through our proposed approach, we strive to enable efficient and effective image compression on resource-constrained end-devices, thereby facilitating the seamless integration of LIC techniques into the IoT ecosystem. To the best of our knowledge, we are the first to integrate LUTs with LIC networks.

The rest of the paper is organized as follows. Section 2 briefly reviews related works. The proposed method is presented in Sect. 3. Experimental validation is shown in Sect. 4. Finally, Sect. 5 concludes the paper.

2 Related Works

2.1 Learned Image Compression

Traditional image codecs such as JPEG [27], JPEG2000 [25], and WebP [8] typically employ lossless transformations such as discrete cosine transformation or wavelet transformation, drop insignificant information, and apply lossless entropy coding. In contrast, LIC [2,5,9,10,18,22,23] introduces a learning-based pipeline, which commonly consists of a lossy transformation model and quantization, followed by a joint backward-and-forward adaptive lossless entropy model.

However, incorporating a backward-adaptive entropy model slows down the process, particularly when an autoregressive context model with serial decoding in the spatial direction is included in the entropy model [5,9,10,18,22,23]. To address this, He et al. [10] propose the checkerboard context model, which parallelizes the serial autoregressive process to some extent. Similarly, Minnen et al. [23] suggest a method to replace the serial spatial decoding with channel-oriented decoding. He et al. [9] introduce a multi-dimensional context model by leveraging the characteristics of both context models.

In general, the goal of an LIC model is to minimize the expected length of the compressed bitstream (i.e., rate) and, at the same time, to reduce the distortion of the reconstructed image. Thus, the loss function to be minimized during training can be written as

$$\text{Loss} = \text{Rate} + \lambda \cdot \text{Distortion} \tag{1}$$

where λ is a balancing parameter adjusting the trade-off between the rate and the distortion. The rate term can be estimated using the cross-entropy between the distribution of the quantized latent representation and the learned entropy model. For the distortion term, the mean squared error (MSE) or multi-scale structural similarity (MS-SSIM) [29] can be used.

2.2 Model Compression Techniques

Various model compression methods, such as pruning [21,28], knowledge distillation [12], quantization [14], and LUTs [16,20], have been extensively researched

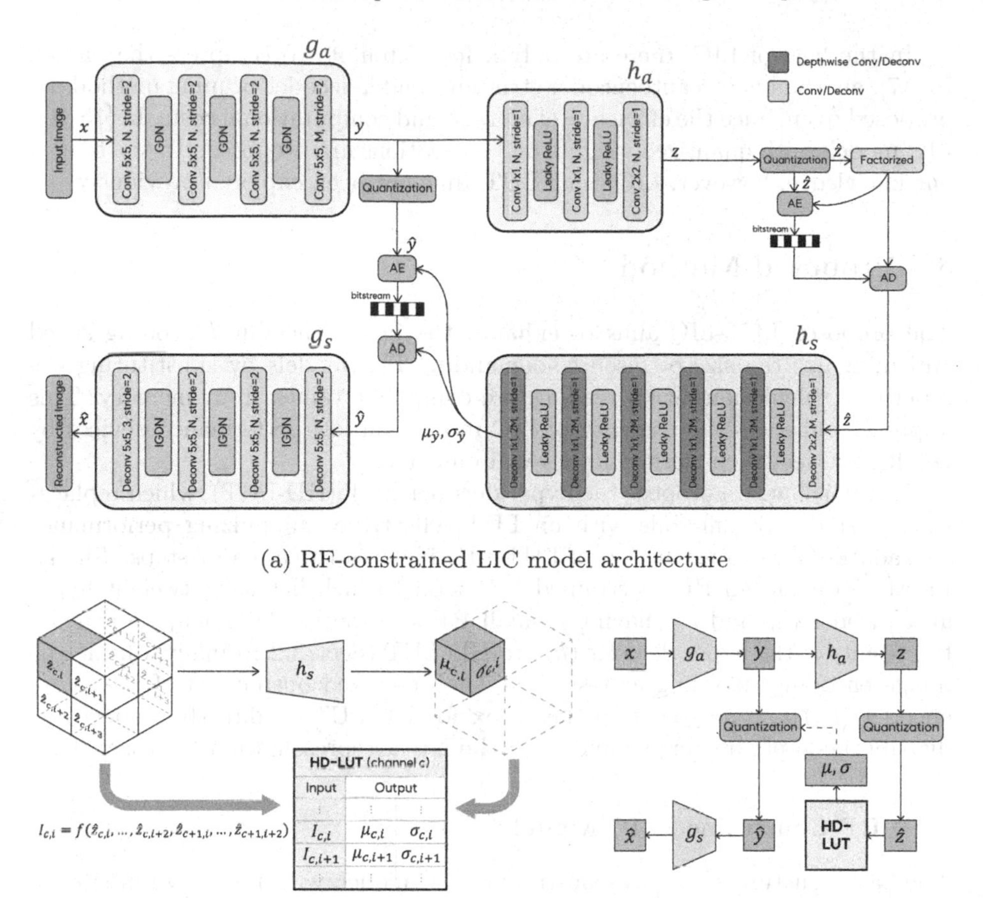

(b) Transferring

(c) Inference

Fig. 1. Overview of the proposed method. (a) LIC model architecture which limits the RF size in the hyperprior autoencoder (h_a and h_s). The target module to transfer to an LUT (h_s) is colored in yellow. AE and AD mean the arithmetic encoder and decoder, respectively. GDN is the generalized divisive normalization. (b) Process to transfer the hyper decoder to HD-LUT. The function $f(.)$ calculates index $I_{c,i}$ using the quantized hyper latents ($\hat{z}$) within the RF. (c) Final LUT-LIC model at inference time. The hyper decoder is replaced with HD-LUT.

for implementing lightweight deep learning models. These techniques aim to reduce the model size, computational complexity, or memory requirements while preventing or minimizing the degradation in model performance. Pruning removes redundant or less important connections from the network. Quantization reduces the precision of model parameters, typically from floating-point to fixed-point representation. LUTs leverage pre-computed values to reduce the reliance on complex network operations.

In the field of LIC, there are only a few attempts to compress the model. In [17], an asymmetric autoencoder structure and a decoder pruning method are proposed to enhance the efficiency of storage and computational costs. In [15,26], LIC models with quantized weights and activations are proposed. To the best of our knowledge, however, LIC using LUTs has not been explored previously.

3 Proposed Method

The proposed LUT-LIC aims to enhance the image encoding/decoding speed and minimize the size of resource-demanding LIC models by substituting the repetitive calculations with reading pre-computed values from memory. This approach enhances the usability of LIC models on low-performance devices by helping to meet their performance requirements.

In particular, we propose the hyper decoder LUT (HD-LUT), which replaces the hyper decoder module with an LUT, effectively minimizing performance degradation. The construction of LUT-LIC involves three major steps (Fig. 1). Firstly, we train an RF-constrained LIC model which has a lightweight hyper autoencoder (h_a and h_s) having a small RF size (Sect. 3.1). Then, we transfer the results of the hyper decoder (h_s) to HD-LUT (Sect. 3.2). Finally, during the image encoding/decoding at test time, the quantized output ($\hat{z}$) of the hyper encoder (h_a) is converted into the index for HD-LUT to directly retrieve the distributional parameters (μ and σ) of the latent representation ($\hat{y}$) (Sect. 3.3).

3.1 RF-Constrained LIC Model

The base structure of our RF-constrained LIC model, which is shown in Fig. 1a, is inspired by the architecture in [22]. It consists of four modules: transform encoder (g_a), hyper encoder (h_a), transform decoder (g_s), and hyper decoder (h_s). The transform encoder converts the input image (x) to the compact latent representation (y). The quantized latent representation ($\hat{y}$) is converted to a bit-stream by entropy coding. The latent representation also undergoes the hyper encoder (h_a) to produce the hyper latent (z), which is also converted to a bit-stream after quantization and entropy coding using the fixed factorized entropy model. The quantized hyper latent ($\hat{z}$) is used to obtain the Gaussian parameters ($\mu_{\hat{y}}$ and $\sigma_{\hat{y}}$) of $\hat{y}$ through the hyper decoder (h_s). They are used for arithmetic encoding/decoding of $\hat{y}$. Finally, the transform decoder (g_s) outputs the reconstructed image ($\hat{x}$) from $\hat{y}$.

As the RF size of the hyper decoder increases, the LUT size grows exponentially. Thus, it is necessary to keep the RF size small. However, a small RF size causes performance degradation. As a balance between this trade-off, the convolutional layers in the hyper autoencoder use 1×1 and 2×2 kernels. Furthermore, the hyper decoder employs depthwise convolution to constrain the RF size along the channel dimension. However, we empirically observed that when all layers of the hyper decoder use depthwise convolution, the rate-distortion performance significantly drops because exploiting the cross-channel information through convolution across channels is critical. Thus, at the first layer of the hyper decoder, convolution operates on a pair of adjacent channels by setting the kernel size in the channel dimension to 2 (i.e., RF of 2 × 2 × 2 instead of 2 × 2 × 1).

The model is trained to minimize the loss function given by

$$L = \underbrace{\mathbb{E}_{x \sim p_x}[-\log_2 p_{\hat{x}|\hat{z}}(\hat{x}|\hat{z}) - \log_2 p_{\hat{z}}(\hat{z})]}_{\text{Rate}} + \lambda \cdot \underbrace{\mathbb{E}_{x \sim p_x}[d(x, \hat{x})]}_{\text{Distortion}} \tag{2}$$

The rate term includes the rates for both $\hat{y}$ and $\hat{z}$. We use MSE for the distortion term. Detailed training parameters are given in Sect. 4.1.

3.2 Transferring Hyper Decoder to LUT

After training the RF-constrained LIC model, the hyper decoder is transferred to an LUT, called HD-LUT. It basically lists the input ($\hat{z}$) as the index and the corresponding output (μ and σ pair).

To design an effective process to obtain HD-LUT, we first analyze the characteristics of the quantized hyper latent $\hat{z}$ having integer values. First, the value of $\hat{z}$ is theoretically unbounded since $\hat{z}$ is a feature containing the information about the correlation in the latent $\hat{y}$, which makes it difficult to determine the range of the index of HD-LUT in advance. Second, different channels have different ranges of $\hat{z}$ empirically. A few channels show values of $\hat{z}$ over wide ranges, while in the majority of channels the values of $\hat{z}$ are concentrated within small ranges. Third, the distributions of $\hat{z}$ are typically bell-shaped, with the maximum frequency appearing almost at the center of the range.

Based on these observations, we develop a dynamic sampling method to sample the values of $\hat{z}$ used for indices of HD-LUT as follows. Considering the discrepancy in the distribution of $\hat{z}$ across channels, we build a channel-specific LUT. For a certain channel, the minimum and maximum values of $\hat{z}$ for the training data are recorded. The interval between them is set to the sample size as long as it is smaller than a pre-defined value k_{max}, i.e.,

$$k_c = \min\{\max(\hat{z}_c) - \min(\hat{z}_c) + 1, k_{max}\} \tag{3}$$

where c refers to a specific channel. Considering the bell-shaped distribution of $\hat{z}$, the values of $\hat{z}$ are sampled around the center of the distribution. For this, we find the starting value of the sampling range, s_c, as

$$s_c = \begin{cases} \min(\hat{z}_c), & \text{if } k_c \neq k_{max} \\ \lfloor \{\max(\hat{z}_c) + \min(\hat{z}_c) - k_c\}/2 \rceil, & \text{if } k_c = k_{max} \end{cases} \tag{4}$$

The values of $\hat{z}$ are normalized using s_c as

$$\hat{z}_c^{'} = \hat{z}_c - s_c \tag{5}$$

and stored after clipping if a value is outside the range determined by k_c as follows:

$$\hat{z}_c^{''} = \begin{cases} 0, & \text{if } \hat{z}_c^{'} < 0 \\ \hat{z}_c^{'}, & \text{if } 0 \leq \hat{z}_c^{'} < k_c \\ k_c, & \text{if } k_c \leq \hat{z}_c^{'} \end{cases} \tag{6}$$

Finally, the index I to retrieve (μ, σ) for the ith spatial location in the cth channel from HD-LUT is calculated as

$$I_{c,i} = \sum_{c^{'}=c}^{c+1} \sum_{j=i}^{i+RF-1} \hat{z}_{c^{'},j}^{''} \times (k_{c^{'}})^{(i+RF-1)-j} \tag{7}$$

where RF means the total number of values of $\hat{z}$ under the RF (i.e., $2{\times}2{\times}2 = 8$). The values of s_c and k_c are separately stored along with HD-LUT.

3.3 Inference Using LUT-LIC

After constructing HD-LUT, the hyper decoding process can now be performed solely using HD-LUT (Fig. 1c). When $\hat{z}$ is given, the index can be computed using Eqs. 5 to 7, where the stored s_c and k_c are also used. Then, the index $I_{c,i}$ is used to look up HD-LUT and read the corresponding $\mu_{c,i}$ and $\sigma_{c,i}$.

4 Experiments

4.1 Settings

Following the previous sutdies [2,5,18,22,23], we train our model using the Vimeo90K dataset [30] which is a large-scale high-quality video dataset consisting of approximately 1,850K images, and for evaluation we use the Kodak dataset [19]. The overall training settings are based on the literature [2,18,22]. We use the CompressAI pytorch library [3] for implementation.

We set λ = {0.0035, 0.0067, 0.0130, 0.0250, 0.0483, 0.0932, 0.1800}, which is the basic setting in CompressAI. Our training process uses a fixed learning rate of 10^{-4}, a batch size of 8, and 200 training epochs. We empirically set the hyperparameters as $k_{max} = 25$, $N = 192$, and $M = 192$ for the three smallest λs and $M = 384$ for the other four. All evaluations are conducted on a desktop environment using Intel Core i5-9400F CPU and NVIDIA GeForce RTX 2060.

4.2 Comparison

To evaluate the performance of our method, we employ three conventional image codecs widely used on edge devices, including JPEG [27], JPEG2000 [25], and WebP [8], and three representative LIC methods, including Minnen2018 [22],

Table 1. Comparison of different compression methods including the proposed LUT-LIC methods, conventional codecs, and LICs. LUT-LIC is used as the anchor to compute the BD-Rate and BD-SNR. A positive BD-Rate means that the rate-distortion performance of the method in terms of rate is worse than LUT-LIC, and vice versa. And, a negative BD-SNR means that the rate-distortion performance of the method in terms of image quality is worse than LUT-LIC, and vice versa. The runtime of LIC and LUT-LIC models is the average value from 30 attempts.

	Method	BD-Rate (%) ↓	BD-SNR (dB) ↑	Size (MB) ↓	Runtime on CPU (s) ↓	
					Enc.	Dec.
Proposed	LUT-LIC	0	0	24.11	2.382	3.589
Conventional	JPEG [27]	+94.04	−3.82	-	0.032	0.001
	JPEG2000 [25]	+36.42	−0.20	-	0.547	0.491
	WebP [8]	+17.09	−0.50	-	0.185	0.029
LIC	Minnen2018 [22]	−25.04	+2.11	56.52	28.403	49.752
	He2021 [10]	−24.58	+1.48	56.52	2.663	4.110
	He2022 [9]	−60.99	+5.10	163.06	8.855	10.512

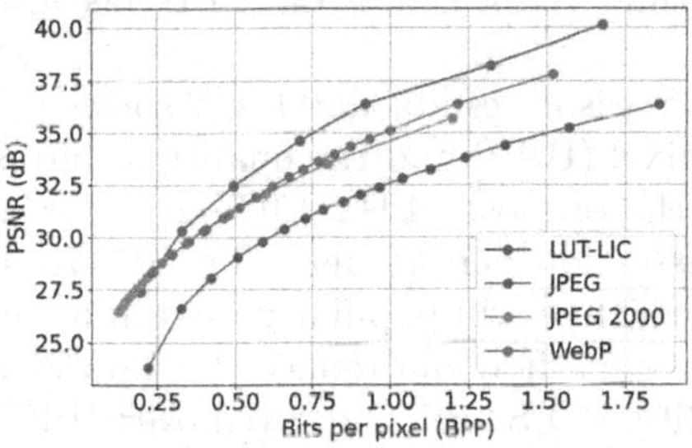

(a) Comparison with codecs (PSNR)

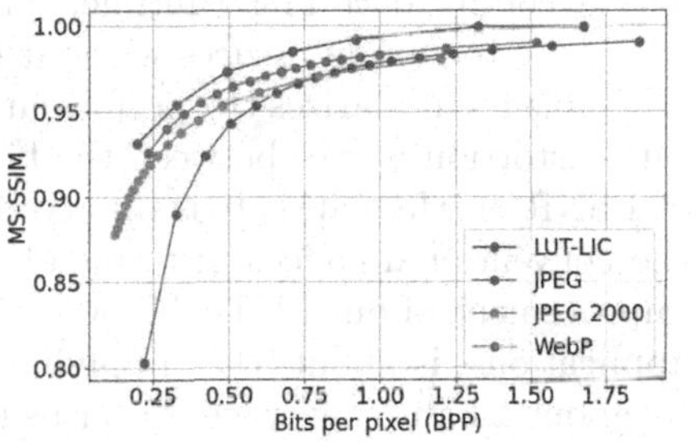

(b) Comparison with codecs (MS-SSIM)

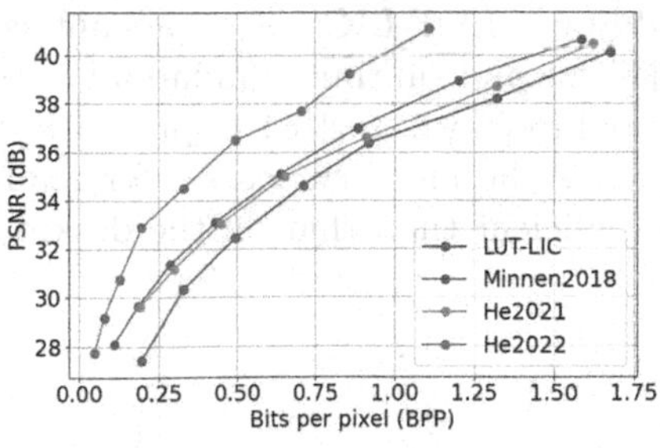

(c) Comparison with LICs (PSNR)

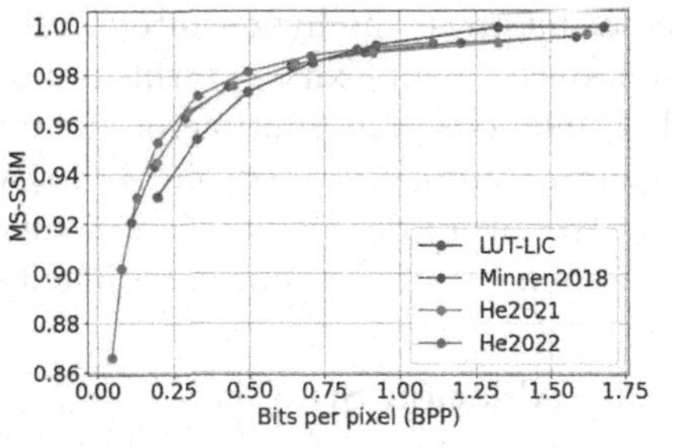

(d) Comparison with LICs (MS-SSIM)

Fig. 2. Rate-distortion curves for comparing coding performance using the Kodak dataset [19].

Table 2. Effects of the kernel size and depthwise convolution in the first layer of the hyper decoder in our RF-constrained LIC model in terms of the rate-distortion performance (after transferring the hyper decoder to HD-LUT). The first row, which is the anchor to compute the BD-Rate and BD-SNR, corresponds to our proposed LUT-LIC.

Kernel Size	Depthwise	BD-Rate (%) ↓	BD-SNR (dB) ↑
2×2		0	0
2×2	✓	+100.98	−3.19
3×3	✓	+210.81	−4.94

He2021 [10], and He2022 [9]. Note that He2021 and He2022 particularly aim at efficient operations.

The performance measures include the Bjöntegaard delta bitrate (BD-Rate) and Bjöntegaard delta peak signal-to-noise ratio (PSNR) (BD-SNR) [4] summarizing the rate-distortion performance (or compression efficiency), model size, and encoding/decoding runtime. The runtime is measured on CPU because LUTs are to run on devices without GPU.

Table 1 summarizes the performance comparison results, and Fig. 2 shows the rate-distortion curves between the bit-per-pixel (BPP) and the quality in terms of PSNR or MS-SSIM. It is observed that the proposed LUT-LIC outperforms the conventional codecs in terms of rate-distortion performance. The BD-Rate improvement of our LUT-LIC over JPEG, which is still popularly used in many applications, is about 94%, implying that our method can reduce the data rate by about a half on average. In terms of quality, BD-SNR is improved over JPEG by 3.82 dB, which is also significant. Note that the lowered rate-distortion performance of our method compared to the LIC methods is mostly due to the restricted RF size using the reduced kernel sizes (2×2 and 1×1) and depthwise convolution in the hyper autoencoder. In addition, LUT-LIC has an advantage of reduced complexity in terms of both model size and runtime compared to the LIC methods. The runtime of our method would be even further improved if it is actually implemented on MPUs/MCUs, since the memory access (for reading HD-LUT) on such devices is much more efficient than that on the desktop environment used in our experiment [1,11].

4.3 Ablation Studies

Cross-Channel Correlation. In our RF-constrained LIC model (Fig. 1a), the first layer of the hyper decoder does not employ depthwise convolution to allow extraction of correlation information across channels. To examine the importance of this cross-channel correlation, Table 2 shows the rate-distortion performance (BD-Rate and BD-SNR) after transferring the hyper decoder to HD-LUT when different RF sizes are used in the first layer of the hyper decoder. The first row of the table corresponds to our LUT-LIC. The second row, which corresponds to the case where the first layer of the hyper decoder also uses depthwise convolution, shows significantly deteriorated performance, indicating that exploiting

Table 3. Rate-distortion performance of our LUT-LIC in terms of BD-Rate and BD-SNR when a context model (either the autoregressive model in Minnen2018 [22] or the checkerboard model in He2021 [10]) is additionally used in the RF-constrained LIC model. The RF size of the context model is set to 2×2 with depthwise convolution.

Context Model	BD-Rate (%) ↓	BD-SNR (dB) ↑
None	0	0
Minnen2018 [22]	+224.01	−83.72
He2021 [10]	+134.94	−6.54

the cross-channel information is crucial. We also test the case with an increased kernel size to 3×3 in the last row of the table, which shows that an increase of the RF size in the spatial dimension cannot compensate for the performance drop due to the depthwise convolution operation. In this case, since the RF size is increased to 9, the LUT size should be increased significantly; due to an excessive increase of the LUT size, however, we use an LUT with a moderately increased size, which does not sample $\hat{z}$ sufficiently and thus the performance is even further lowered.

Context Modeling. Context modeling is often used in existing LIC methods [10,22] to improve the rate-distortion performance. A context model typically uses $\hat{y}$ as input to produce certain features used to determine the distributional parameters (such as μ and σ) together with the output of the hyper decoder. We test the effect of inclusion of context modeling in our RF-constrained LIC model. When a context model is used in our model, the parameters μ and σ become to depend on not only $\hat{z}$ but also the output of the context model. Then, multiple cases with different values of μ and σ may occur for the same value of $\hat{z}$, among which we need to choose one case to build the LUT. In this ablation study, we choose the most frequently occurring pair of (μ, σ). Table 3 shows the BD-Rate and BD-SNR of our LUT-LIC when the context model in Minnen2018 [22] or He2021 [10] is used in our RF-constrained LIC model. We observe that inclusion of context modeling deteriorates the performance, indicating that the ambiguity issue is significantly disadvantageous.

5 Conclusion

We proposed a method to employ an LUT in LIC towards improved usability of LIC in resource-limited devices. Our method replaced the hyper decoder to an LUT (HD-LUT), for which we analyzed the characteristics of the input to the hyper decoder and developed a dynamic sampling technique. The proposed LUT-LIC model achieved better rate-distortion performance than conventional codecs and, at the same time, exhibited reduced complexity (model size and runtime) compared to the LIC models. In our future work, we will investigate ways to improve the rate-distortion performance of our method and explore the feasibility of performing LIC with end-to-end LUTs.

References

1. Aljumah, A., Ahmed, M.A.: Design of high speed data transfer direct memory access controller for system on chip based embedded products. J. Appl. Sci. **15**(3), 576–581 (2015)
2. Ballé, J., Minnen, D., Singh, S., Hwang, S.J., Johnston, N.: Variational image compression with a scale hyperprior. In: Proceedings of the International Conference on Learning Representations (ICLR) (2018)
3. Bégaint, J., Racapé, F., Feltman, S., Pushparaja, A.: CompressAI: a PyTorch library and evaluation platform for end-to-end compression research. arXiv preprint arXiv:2011.03029 (2020)
4. Bjøntegaard, G.: Calculation of average PSNR differences between RD-curves. ITU SG16 Doc. VCEG-M33 (2001)
5. Cheng, Z., Sun, H., Takeuchi, M., Katto, J.: Learned image compression with discretized gaussian mixture likelihoods and attention modules. In: Proceedings of the IEEE Conference on Computer Vision and Pattern Recognition (CVPR) (2020)
6. Cui, Z., Wang, J., Gao, S., Guo, T., Feng, Y., Bai, B.: Asymmetric gained deep image compression with continuous rate adaptation. In: Proceedings of the IEEE Conference on Computer Vision and Pattern Recognition (CVPR) (2021)
7. Gao, G., et al.: Neural image compression via attentional multi-scale back projection and frequency decomposition. In: Proceedings of the IEEE International Conference on Computer Vision (ICCV) (2021)
8. Google: WebP. https://developers.google.com/speed/webp/docs/compression
9. He, D., Yang, Z., Peng, W., Ma, R., Qin, H., Wang, Y.: ELIC: efficient learned image compression with unevenly grouped space-channel contextual adaptive coding. In: Proceedings of the IEEE/CVF Conference on Computer Vision and Pattern Recognition (CVPR), pp. 5718–5727 (2022)
10. He, D., Zheng, Y., Sun, B., Wang, Y., Qin, H.: Checkerboard context model for efficient learned image compression. In: Proceedings of the IEEE Conference on Computer Vision and Pattern Recognition (CVPR) (2021)
11. Hestness, J., Keckler, S.W., Wood, D.A.: A comparative analysis of microarchitecture effects on CPU and GPU memory system behavior. In: Proceedings of the IEEE International Symposium on Workload Characterization (IISWC), pp. 150–160. IEEE (2014)
12. Hinton, G., Vinyals, O., Dean, J.: Distilling the knowledge in a neural network. In: Proceedings of the NIPS Deep Learning and Representation Learning Workshop (2015)
13. Hu, Y., Yang, W., Liu, J.: Coarse-to-fine hyper-prior modeling for learned image compression. In: Proceedings of the AAAI Conference on Artificial Intelligence (AAAI) (2020)
14. Jacob, B., et al.: Quantization and training of neural networks for efficient integer-arithmetic-only inference. In: Proceedings of the IEEE Conference on Computer Vision and Pattern Recognition, pp. 2704–2713 (2018)
15. Jeon, G.W., Yu, S., Lee, J.S.: Integer quantized learned image compression. In: Proceedings of the IEEE International Conference on Image Processing (ICIP) (2023)
16. Jo, Y., Joo Kim, S.: Practical single-image super-resolution using look-up table. In: Proceedings of the IEEE Conference on Computer Vision and Pattern Recognition (CVPR) (2021)

17. Kim, J.H., Choi, J.H., Chang, J., Lee, J.S.: Efficient deep learning-based lossy image compression via asymmetric autoencoder and pruning. In: Proceedings of the IEEE International Conference on Acoustics, Speech and Signal Processing (ICASSP), pp. 2063–2067 (2020)
18. Kim, J.H., Heo, B., Lee, J.S.: Joint global and local hierarchical priors for learned image compression. In: Proceedings of the IEEE Conference on Computer Vision and Pattern Recognition (CVPR) (2022)
19. The Kodak PhotoCD dataset. http://r0k.us/graphics/kodak/
20. Li, J., Chen, C., Cheng, Z., Xiong, Z.: MuLUT: cooperating multiple look-up tables for efficient image super-resolution. In: Avidan, S., Brostow, G., Cissé, M., Farinella, G.M., Hassner, T. (eds.) ECCV 2022 Part XVIII, vol. 13678, pp. 238–256. Springer, Cham (2022)
21. Liu, Z., Sun, M., Zhou, T., Huang, G., Darrell, T.: Rethinking the value of network pruning. In: Proceedings of the International Conference on Learning Representations (ICLR) (2019)
22. Minnen, D., Ballé, J., Toderici, G.: Joint autoregressive and hierarchical priors for learned image compression. In: Proceedings of the Advances in Neural Information Processing Systems (NeurIPS) (2018)
23. Minnen, D., Singh, S.: Channel-wise autoregressive entropy models for learned image compression. In: Proceedings of the IEEE International Conference on Image Processing (ICIP) (2020)
24. Qian, Y., et al.: Learning accurate entropy model with global reference for image compression. In: Proceedings of the International Conference on Learning Representations (ICLR) (2021)
25. Rabbani, M.: JPEG2000: image compression fundamentals, standards and practice. J. Electron. Imaging **11**(2), 286 (2002)
26. Sun, H., Yu, L., Katto, J.: Q-LIC: quantizing learned image compression with channel splitting. IEEE Trans. Circ. Syst. Video Technol. 1–1 (2022). https://doi.org/10.1109/TCSVT.2022.3231789
27. Wallace, G.K.: The JPEG still picture compression standard. IEEE Trans. Consum. Electron. **38**(1), xviii-xxxiv (1992)
28. Wang, X., Zheng, Z., He, Y., Yan, F., Zeng, Z., Yang, Y.: Progressive local filter pruning for image retrieval acceleration. IEEE Trans. Multimedia (2023). https://doi.org/10.1109/TMM.2023.3256092
29. Wang, Z., Simoncelli, E.P., Bovik, A.C.: Multiscale structural similarity for image quality assessment. In: Proceedings of the Asilomar Conference on Signals, Systems & Computers, 2003, vol. 2, pp. 1398–1402 (2003)
30. Xue, T., Chen, B., Wu, J., Wei, D., Freeman, W.T.: Video enhancement with task-oriented flow. Int. J. Comput. Vis. (IJCV) **127**(8), 1106–1125 (2019)

Oil and Gas Automatic Infrastructure Mapping: Leveraging High-Resolution Satellite Imagery Through Fine-Tuning of Object Detection Models

Jade Eva Guisiano[1,2,3](✉), Éric Moulines[1], Thomas Lauvaux[4], and Jérémie Sublime[2]

[1] École Polytechnique, Palaiseau, France
jade-guisiano@outlook.fr
[2] ISEP School of Engineering, Paris, France
[3] United Nations Environment Program, Paris, France
[4] Université de Reims, Reims, France

Abstract. The oil and gas sector is the second largest anthropogenic emitter of methane, which is responsible for at least 25% of current global warming. To curb methane's contribution to climate change, emissions behavior from oil and gas infrastructure must be determined by an automated monitoring across the globe. This requires, as first step, an efficient solution to automatically detect and identify these infrastructures. In this extended study, we focus on automated identification of oil and gas infrastructure by using and comparing two types of advanced supervised object detection algorithms: Region-based Object Detector (YOLO and FASTER-RCNN) and Transformer-based Object Detector (DETR) with fine-tuning on our customized high-resolution satellite image database (Permian Basin U.S). The pre-training effect of each of these algorithms on detection results is studied and compared with non-pre-trained algorithms. The performed experiments demonstrate the general effectiveness of pre-trained YOLO v8 model with a Mean Average Precision over 90. The non-pre-trained model of this last one also over perform compare to FASTER-RCNN and DETR.

Keywords: Object detection · Remote sensing · Deep Learning · Computer vision · Oil and gas

1 Introduction

Methane, an exceptionally potent greenhouse gas, has a much higher global warming potential than carbon dioxide, exacerbating the current climate crisis. Reducing methane emissions is an effective strategy to significantly slow the pace of global warming and its associated environmental impacts. The oil and gas industry (O&G) is a particular contributor to methane emissions, as it is the

B. Luo et al. (Eds.): ICONIP 2023, CCIS 1966, pp. 442–458, 2024.
https://doi.org/10.1007/978-981-99-8148-9_35

Fig. 1. Example of a detected methane plume associated with infrastructure at its source. In 3 automatic steps: detection of methane plume, detection of infrastructure, association of each plume and infrastructure. *Source: @Google Earth*

second largest anthropogenic source [1]. Methane is unintentionally released at various stages of the industry's supply chain. To effectively reduce these emissions in the O&G sector, a comprehensive understanding of the emissions profiles of individual operators, specific sites, and associated infrastructure is needed. This knowledge would inform the formulation and refinement of regulatory measures and potential penalties to ensure they are appropriately tailored and thus optimally effective. The United Nations Environment Program has launched the Methane Alert and Response System (MARS) program to link detected methane plumes to their specific sources. This is intended to provide operators with timely warning of discovered leaks. However, to enable near-continuous monitoring of the world's oil and gas resources, it is essential that this process be supported by an automated detection and attribution system. Certain methods such as OGNET [2] and METER-ML [3] use deep neural networks to identify specific types of oil and gas site. This study focuses on the topic of automated detection of oil and gas infrastructures a topic that has not been explored in depth in the existing literature. Sites in the oil and gas industry that contain wells, storage tanks, or compressor infrastructures are considered significant contributors to fugitive emissions and therefore form the targets we seek to automatically identify. Existing approaches to oil and gas infrastructure detection typically do not allow for the simultaneous detection of multiple infrastructures. With the goal of enabling the automatic detection of compressors, tanks, and well infrastructures simultaneously, this paper focuses on supervised object detection methods, specifically using and comparing the YOLO, FASTER-RCNN, and DETR algorithms. These algorithms, initially trained on the COCO database, are **fine-tuned** using the Oil and Gas (OG) database. The database OG, which was developed specifically for this study, contains aerial photographs with high spatial resolution (less than 1 m). The images in the OG database are extracted from the Permian Basin, the most substantial oil and gas basin in the world, located in the states of New Mexico and Texas (US).

In the first part, the state of the art of object detection algorithms will be presented, along with their applications in the O&G sector. The OG database and its characteristics will then be detailed. Next, the YOLO and FASTER-RCNN algorithms and their parameters will be presented. Finally, the results section details the performance of each **pre-trained and non-pre-trained** algorithm (Fig. 1).

2 State of the Art

Object recognition algorithms, a subset of computer vision techniques, facilitate the automatic identification and location of multiple instances of a given class of objects in images or videos. These algorithms predominantly use either neural network-based methods or non-neural techniques. The non-neural strategies typically integrate SIFT [4] or HOG [5] (for feature extraction) with a classification algorithm such as Support Vector Machines (SVM). Despite their usefulness, recent studies suggest that neural-based object recognition methods generally outperform their non-neural counterparts [6]. Neural approaches to object recognition can be divided into three categories depending on the degree of supervision in their learning process: supervised, semi-supervised, and self-supervised models [7]. In this paper, we will mainly focus on different supervised methods. Supervised object recognition models require an annotated image database for effective training. In this context, annotating an image involves identifying objects of interest by enclosing them in a bounding box and labeling them appropriately. During training, a supervised object recognition algorithm learns to locate and subsequently recognize the targeted objects. This complicated process can be executed over two primaries architectural frameworks [8]:

- **Two-stage detector:** is based on two main models, firstly Region Proposal Network (RPN) which is a fully convolutional network used to extract regions of objects, and secondly an extra model is used to classify and further refine the localization of each region proposal. RCNN [9] architecture is based on a selective search algorithm to propose regions of interest and then applies a CNN to each region to classify it as an object or background.
  As this method is particularly slow, the authors proposed Fast-RCNN [10], an optimized approach to RCNN by sharing computation across all regions proposed in an image. Finally, FASTER-RCNN [11], based on the architecture of Fast-RCNN, replaces the selective search algorithm with a RPN, which is trained to directly predict regions of interest. This latest version reduces computation time and improves the detection accuracy;
- **one stage detector:** Contrary to one stage detector, one stage detector don't need to integrate RPN to generate a region proposal, it can directly obtain the classification accuracy of the object and its coordinate position. These algorithms have the advantage of being faster than two-step algorithms. In this category we find YOLO [12] and its different versions [13], SSD [14] and RetinaNet [15]. Review studies compares the latter 3 methods, for example [16] for the pill identification task, showing that YOLO v3 offers the best performance in terms of execution time but the lowest accuracy. Another study [17] focuses on the comparison of SSD, RetinaNet, YOLO V4 and FASTER-RCNN for Tethered Balloon detection. It was show that YOLO v4 achieved the best trade-off between speed and accuracy with a precision of 90.3%. [18] also concludes that YOLO has better accuracy (increasing with version) via a broad comparison of RCNN and YOLO models and their variants;

- **Others:** There are also object detection methods based on approaches other than the one-two stage approaches detailed above. For example, DETR [19] is a transformer based detector with a 3 parts architecture constitute of a CNN, encoder-decoder transformer and a feed-forward network (FFN).

Object Detection O&G Applications. Remote sensing object detection can be applied to a variety of problems, various studies [20–23] summarizes object recognition algorithms applied to various remote sensing topics. For example, [24] summarizes the performance of FASTER-RCNN, SSD, and YOLO V3 algorithms for agricultural greenhouse detection based on high-resolution satellite imagery. [25] proposes automatic detection of earthquake-induced ground failure effects by using FASTER-RCNN. Others [26–28] focus on comparing one and two-stage object detection algorithms on satellite and aerial images. [29] uses DETR for object detection with enhanced multi-spectral feature extraction. In particular, object detection algorithms are also used for problems in the oil and gas sector. For example, in the work [30,31] YOLO V4 is used to detect oil spills with Sentinel-1 SAR images. Some studies are also looking at oil and gas infrastructure detection:

- **Oil Tanks**: [32] proposes a recognition algorithm that harnesses deep environmental features, using the convolutional neural network (CNN) model and SVM classifier for oil tank recognition. Another study employs FASTER-RCNN for the same objective;
- **Oil Wells**: [33] introduces an enhanced version of YOLO v4 for detection using high-resolution images, similar to [34], where the authors utilize the faster R-CNN. [35] presents a database, dubbed Northeast Petroleum University-Oil Well Object Detection Version 1.0 (NEPU-OWOD V1.0), which includes the geographical locations of oil wells. This database was constructed via the application and comparison of nine object detection algorithms;
- **Pipelines**: In the context of pipelines, [36] utilizes a deep learning approach for object detection in underwater pipeline images, employing various YOLO configurations;
- **Oil & Gas Sites**: On a broader scale encompassing entire infrastructures, [37] employs high-resolution satellite images and YOLO V2 for automatic recognition of oil industry facilities, with a particular emphasis on well-sites.

In the field of object detection, a significant portion of existing methods are dedicated to the identification of specific infrastructures. While this focused approach proves beneficial in studies examining a single infrastructure, it may not be entirely sufficient when examining methane emissions in the oil and gas (O&G) sector. This is because such emissions can come from a variety of infrastructures. Recognising this multi-faceted challenge, this study broadens its scope to include three types of infrastructure that are essential to the O&G sector: Wells, Tanks, and Compressors. This more comprehensive approach provides a broader perspective and leads to a better understanding of the various sources of methane emissions in the sector. In addition, three different object detection algorithms are comparatively analysed in this study: YOLO v8, which follows a single-stage detection paradigm; FASTER-RCNN, a two-stage method; and DETR, an encoder-decoder based detection approach. Each of these algorithms features a unique recognition strategy, providing a broad understanding of object recognition methods in the context of O&G infrastructures. This study also examines the impact of pre-training these models, in particular, how the pre-training phase influences detection outcomes and performance is investigated. The detailed findings from this study are presented and discussed in detail in this paper to further our understanding of the nuanced role that pre-training plays in object recognition.

3 An Oil and Gas Infrastructure Database

Algorithms employed in supervised object recognition necessitate a learning phase involving substantial interaction with a large repository of images. These images must be labelled with the target object in order to enable practical training. In the specific context of identifying wells, tanks, and compressors, this database must be replete with a variety of aerial photographs in which each of these objects or infrastructures is unambiguously identifiable. The procurement of such specialized labelled images, due to the lack of public availability, demanded the development of a dedicated database specifically designed for this purpose. In this study, we chose to extract high-resolution satellite images only from the Permian Basin region (over the states of New Mexico and Texas in the USA), which is the largest O&G basin in the world. 930 Google Earth images of sites with O&G infrastructures were extracted, with resolutions ranging from 15cm to 1m. Each of these images was then manually annotated by drawing bounding boxes around each well, compressor or tank present, as shown in Fig. 2. Each of these boxes is associated with 1 of our 3 objects (label). In total, out of the 930 images, 1951 objects were annotated: Compressor 706 objects, Well 630 objects and Tank 615 objects All the images are in a 640×640 size format, each featuring between one and multiple instances of key infrastructure such as wells, tanks, and compressors. Another aspect worth mentioning is the special consideration given to wells in our database. Given the limited resolution of satellite imagery, it is often difficult to discern the structural details of wells.

Fig. 2. Example of images and annotated objects from OG database: tank (red), compressor (purple) and well (blue) *source: @Google Earth* (Color figure online)

Therefore, the recognizable shadows of wells that are present even at lower resolutions are included in the bounding boxes (as shown in the right column of Fig. 2). This database is hosted on the open-source Roboflow platform and can be accessed via the following link: https://universe.roboflow.com/thesis-ffaad/og-otgc5/dataset/6. Following the requirements of a rigorous study design, we have divided our dataset into different subsets for training, validation, and testing. Of the total images, 80% (744 images) are used for training, 13% (or 120 images) for validation, and the remaining 7% (66 images) for testing.

4 Object Detection Algorithms Presentation

The structural organization of object detection algorithms is usually defined by three main components:

- **Backbone**: This refers to a deep learning architecture, usually a convolutional neural network (CNN), that is tasked with the essential function of feature extraction. Through this process, the backbone identifies and abstracts the salient features from the input data;
- **Neck**: Serving as an intermediary between the backbone and the head the neck performs a fusion of the features extracted from the different layers of the backbone model. This synthesized information forms the basis for the subsequent predictions performed by the head;
- **Head**: The head forms the final component of the object recognition model and is responsible for predicting the classes and bounding box regions. These predictions form the final output of the object recognition model. In particular, the head can produce a number of outputs, typically configured to detect objects of different sizes in an image.

Backbone Pre-training. The majority of object recognition models, including but not limited to YOLO, FASTER-RCNN, and DETR, provide an option for a pre-trained version of the backbone. This pre-training generally helps to improve recognition performance. The pre-training of these algorithms is done using extensive databases of thousands of image categories, ranging from everyday objects such as airplanes and dogs to more specific objects such as apples and chairs. Prominent among these databases are ImageNet [38], which contains 200 classes and about half a million annotated objects, and COCO [39], which contains 80 classes and nearly 1 million annotated objects. In addition, the Pascal database VOC [40] includes about 20 classes with about 63,000 annotated objects. Most modern object recognition algorithms are pre-trained on the COCO dataset. A main advantage of pre-training backbones is the significant reduction in the custom dataset training phase. Pre-trained backbones that have already learned to recognize general features and patterns from large databases can transfer this knowledge to the object recognition task at hand. This not only minimizes training time, but also enables the use of smaller datasets. Prebuilt models also play a critical role in mitigating the problem of over-fitting, which occurs when the model over-learns from the training data, compromising its ability to generalize to new data. Considering these factors, the architecture of recognition algorithms can be classified into three different families: single-stage, two-stage, and other algorithms. In this study, a representative algorithm from each of these families is evaluated: YOLO v8, FASTER-RCNN, and DETR (Fig. 3).

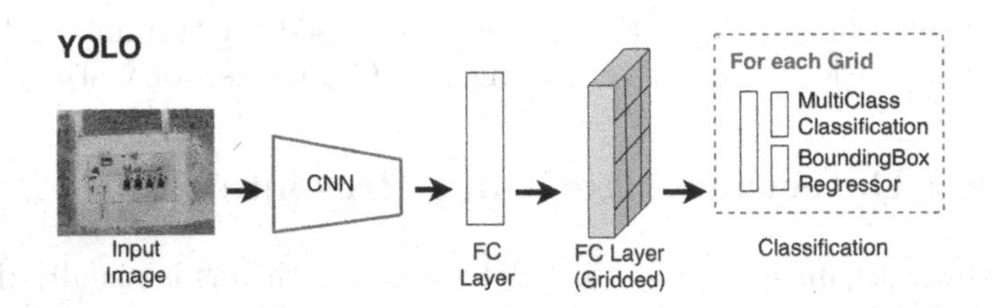

Fig. 3. YOLO architecture: Convolutional Neural Network (CNN), Fully Connected (FC) layer, Girdded FC layer

4.1 One Stage Object Detector: YOLO

You Only Look Once (YOLO) [12] v8 is one of the most recent versions which outperforms previous versions in term of precision as illustrated in Figure 4. YOLO v8 architecture is based on the ResNet-50 (CNN) backbone which has been pre-trained on the ImageNet dataset. The ResNet-50 backbone is then fine-tuned on the COCO dataset to learn to detect objects in 80 different categories. YOLO v8 has a declination of 5 pre-trained models (n, s, m, l, x) trained on COCO 2017 dataset. These models vary according to the number of parameters they hold directly influencing the level of precision, thus, the more parameters a model has, the better its accuracy (cf Fig. 4). The 3 last pre-trained models (m, l, x) with the highest number of parameters were chosen and fine tuned with the OG database (image size 640×640), with 100 epochs, 16 batches, learning rate 0.001 (Fig. 5).

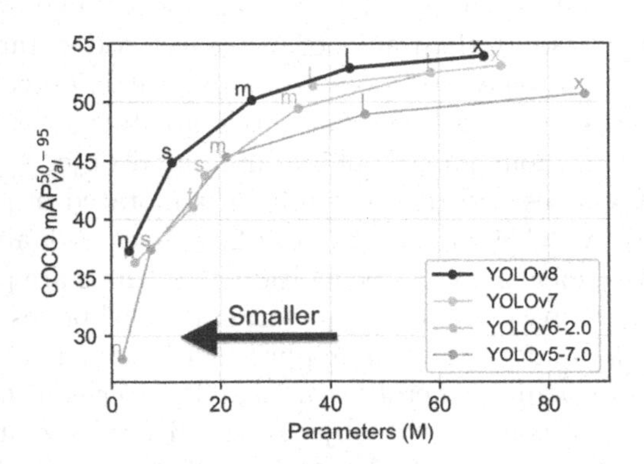

Fig. 4. YOLO Mean Average Precision (mAP) for COCO object detection by versions and models. *source:* https://github.com/ultralytics/ultralytics

4.2 Two Stage Object Detector: FASTER-RCNN

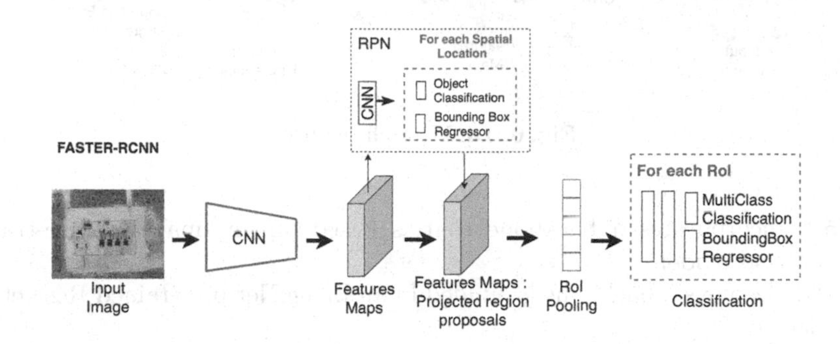

Fig. 5. FASTER-RCNN architecture

FASTER-RCNN [11] process in 2 main steps, first it uses use a Region Proposal Network (RPN) to generate regions of interests and secondly it send the region proposals down the pipeline for object classification and bounding-box regression. FASTER-RCNN architecture is based on 3 principal components: the backbone (CNN type varies according to chosen model), the RPN, and the ROI heads (classification and regression). FASTER RCNN provide 3 backbones architectures pre-trained on COCO 2017 base (train2017 and val2017):

- **Feature Pyramid Network (FPN)**: Use a ResNet+FPN backbone with standard conv and FC heads for mask and box prediction;
- **C4**: Use a ResNet conv4 backbone with conv5 head which correspond to the original baseline in the Faster R-CNN paper;
- **Dilated-C5 (DC5)**: Use a ResNet conv5 backbone with dilations in conv5, and standard conv and FC heads for mask and box prediction, respectively.

We have fine-tuned 2 FPN model with ResNet50 and ResNet101, but also a DC5 model based on ResNet101. Epochs were fixed to 100, batches to 64 and learning rate to 0,001 (Fig. 6).

4.3 Encoder-Decoder Object Detector: DETR

Unlike one-stage and two-stage detectors, DETR [19] is designed as a direct set prediction problem encompassing a unified architecture. DETR employs a backbone (with varying architecture contingent on the selected model), a transformer encoder-decoder architecture, and a bipartite matching between predicted and ground-truth objects. By uniting the backbone and transformer, DETR successfully simplifies the architecture by eliminating specific components to one and two-stage approaches such as anchor generation and non-maximum suppression (NMS). The following pre-trained backbone models are available, all of which have been pre-trained on the COCO 2017 database:

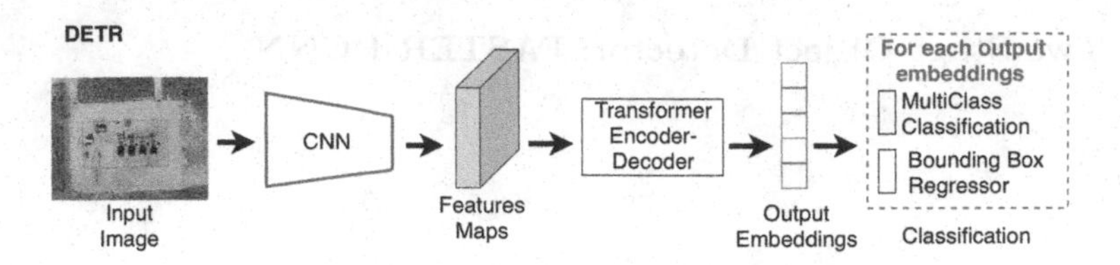

Fig. 6. DETR architecture

- **R50**: Incorporates a backbone that is based on an ImageNet pre-trained ResNet-50 model.
- **R101**: Deploys a backbone grounded in an ImageNet pre-trained ResNet-101 model.
- **R50-DC5**: Increases the feature resolution by employing dilation in the final stage of the backbone. The backbone in this model is based on ResNet-50.
- **R101-DC5**: Implements a similar process to R50-DC5 but relies on a backbone built on ResNet-101.

Pre-trained model R50, R101 and R101-DC5 were selected for test with epochs fixed to 100, batches to 2 and learning rate to 0,001.

4.4 Model Evaluation

Average precision (AP) is a widely used metric for evaluating the efficiency of object recognition tasks.

The AP combines the precision and recognition curves into a single scalar quantity. The AP value ranges from 0 to 1 and tends toward 1 when both precision and recall are high, while it tends toward zero when either metric is low over a spectrum of confidence thresholds. AP is computed by calculating the difference between the current and subsequent recalls and multiplying that difference by the current precision:

$$AP = \sum_{k=0}^{k=n-1} [\text{Recalls}(k) - \text{Recalls}(k+1)] \times \text{Precisions}(k)$$

where k is the number of object and n is the number of threshold. In addition, the mean average precision (mAP) is often used. It represents the average of AP calculated over all classes:

$$mAP = \frac{1}{n} \sum_{k=n}^{k=1} AP_k$$

where AP_k is the AP of the class k and n the number of classes.

5 Results

5.1 Algorithms and Models Performance Comparisons

For each algorithms (YOLO v8, FASTER-RCNN, and DETR), 3 models with different parameters and architectures were selected and compared. For each models, the output corresponds to the AP by class (Compressor, tank and well) and the mAP for the general model. The experiments were conducted with the use of a GPU NVIDIA GeForce RTX 3090 with 24 GO of memory. The experiments required the use of 3 distinct environments for the 3 algorithms with the following packages. YOLO v8: ultralytics (Python 3.8 environment with PyTorch 1.8). FASTER-RCNN: Detectron2 with torch 1.5 and torchvision 0.6 and DETR with PyTorch 1.7 and torchvision 0.7.

Table 1. Pre-trained Algorithms Average Precision (AP) results in % on OG database. **The number of parameters is expressed in millions*

Model	Parameters*	Average Precision (AP)			
		Compressor	Tank	Well	Total
YOLO v8					
8m	25.9	**99.5**	**98.8**	79.4	**92.6**
8l	43.7	**99.5**	88.1	**80.3**	89.3
8x	68.2	98.8	90.9	73.6	87.8
FASTER-RCNN					
R50-FPN	41.7	51.6	51.1	40.5	47.7
R101-FPN	60.6	**53.2**	**57.8**	35.4	**48.8**
R101-DC5	184.5	52.1	47.1	**42.9**	47.4
DETR					
R50	41	94.9	75.1	72.7	80.9
R101	60	**100**	**80.4**	**77.1**	**85.8**
R101-DC5	60	91.9	69.9	67.9	76.3

The Table 1 show that the ensemble of mAP result (Total) of YOLO v8 model over-perform compare to those from FASTER-RCNN. DETR mAP results for each model are lower than those from YOLO v8 but are not very far. Surprisingly, YOLO v8 model (8m) with lower number of parameters performs better than the other with a higher number of parameters (which is the contrary for the COCO database as illustrated of Fig. 4). It appears that for FASTER-RCNN and DETR models based on a simple Resnet101 architecture present a higher mAP compare to others and lower results with a DC5 architecture.

Compressor. YOLO v8 and FASTER-RCNN present AP over 90% for compressor recognition, especially the FASTER-RCNN R101 model with has an AP of 100%. YOLO v8 8m and 8l models are also very close this last one with a 99,5% AP. On average and for compressor, YOLO v8's 3 models offer an AP of 99.3%, compared with 95.6% for DETR. On models average, YOLO v8 and FASTER-RCNN has better AP (with respectively 99,2% and 95,6%) than FASTER-RCNN (52,3%) for compressor recognition. FASTER-RCNN with lowest AP and mAP, has also on average the best results for compressor recognition.

Tank. The best AP performance is obtained thanks to the YOLO v8 8m model with a score of 98,8%. On models average, YOLO has an AP of 92,6% against 75,1% for DETR and 52% for FASTER-RCNN. For tank recognition YOLO v8 largely over perform compare to others.

Well. The best AP is still maintain by YOLO v8 8m model with a score of 92,6%. However, on models average for well recognition, YOLO v8 AP (89.9%) get closer to that of the DETR (81%). Indeed lose almost 10% of AP from compressor to well recognition. Concerning FASTER-RCNN, on models average, obtain its lower AP (39,6%).

5.2 Algorithms Pre-training Effect

In general, pre-trained models offer numerous advantages over non-pre-trained models, including the need for less data (fine-tuning) and an improvement in accuracy. The models selected for this study were trained using the COCO dataset, which consists of annotated everyday objects. However, the COCO dataset does not contain any objects from industry that could resemble the objects in the OG database. This discrepancy raises the question of the extent to which pre-trained models, originally trained on objects that are significantly different from our target objects, can still outperform the predictive accuracy of non-pre-trained models. To investigate this, the algorithms and models discussed previously were run without the weights from the pre-training phase.

According to the results delineated in Table 2, it is observed that the mean Average Precision (mAP) for the non-pre-trained YOLO v8 models is marginally lower than that for the pretrained models. Indeed, the average mAP of the YOLO v8 models is 89.4%, while that of the non-pre-trained models is 88.6%, indicating an overall decrease in mAP of 0.8%. While the difference may seem negligible, this result substantiates the assertion that pre-training YOLO v8 contributes to enhanced performance.

In compliance with the results delineated in Table 2, we observe that the mean average precision (mAP) for the non-pre-trained YOLO v8 models is marginally lower than that of the pre-trained models. In fact, the average mAP of the YOLO v8 models is 89.4%, while that of the non-pre-trained models is 88.6%, which represents an overall decrease in mAP of 0.8%. Even though the difference seems negligible, this result supports the claim that YOLO v8 pre-training contributes to improved performance.

Table 2. Non pre-trained algorithms Average Precision (AP) results in % on OG database. The empty spaces translate the non-convergence of the models and then the absence of results.

Model	Parameters*	Average Precision (AP)			
		Compressor	Tank	Well	Total
YOLO v8					
8m	25.9	97.2	**90.6**	**81.4**	**89.7**
8l	43.7	97.7	90.4	78.0	88.7
8x	68.2	**98.1**	87.2	77.2	87.5
FASTER-RCNN					
R50-FPN	41.7	25.4	2.3	7.2	11.6
R101-FPN	60.6	—	—	—	—
R101-DC5	184.5	—	—	—	—
DETR					
R50	41	—	—	—	—
R101	60	—	—	—	—
R101-DC5	60	—	—	—	—

In terms of average precision (AP) by class, the results mirror those of the pre-trained YOLO v8 models; AP remains higher for the compressor class and lower for the well class. An interesting observation is the comparative analysis between the non-pre-trained YOLO v8 model and the pre-trained FASTER-RCNN and DETR models. The non-pre-trained YOLO v8 model outperforms all pre-trained FASTER-RCNN and DETR models in terms of mAP. This remarkable result demonstrates the superior efficacy of YOLO v8.

As regards of FASTER-RCNN, the pre-trained R50-FPN model shows significantly low AP and mAP. For the other models of FASTER-RCNN and all DETR models, convergence proved difficult even after increasing the number of iterations 20-fold and decreasing the learning rate by a factor of 1000. Non-pretrained models are notorious for their difficulty in achieving convergence, especially when dealing with smaller databases. The OG database is comparatively small, which may explain the observed lack of convergence, especially when compared to the larger COCO database.

5.3 Applications

To facilitate visual inspection of the previous results, the pre-trained model with the highest average precision (mAP) was selected for each algorithm and tested against the Oil and Gas (OG) database test data. Figure 7 illustrates the recognition performance of each algorithm model when applied to four different images from the test data.

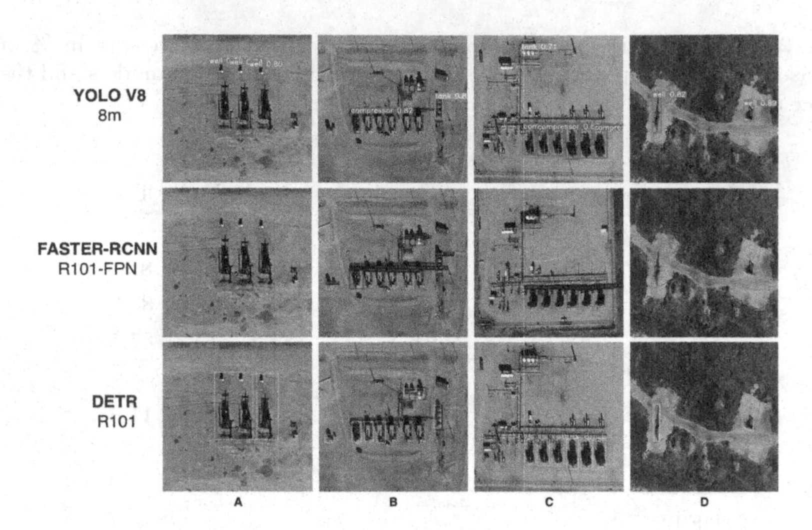

Fig. 7. Visual object detection results from pre-trained YOLO v8, FASTER-RCNN and DETR on 4 test images from OG database (images sources @Google Earth)

Case A: This scenario showcases a view of three wells that are closely spaced. YOLO v8 is able to detect and correctly discriminate each well. In contrast, DETR detects the presence of wells but combines the first two into a single object. FASTER-RCNN, which has significantly low average precision (AP) for the well class (as shown in Table 1), is not able to detect a well in this particular scenario.

Case B: This case represents an unusual circumstance where the appearance of compressors is underrepresented in the Permian Basin and consequently in the learning database OG (the most common representation is shown in Case B). YOLO v8 can only detect one of the five compressors and one of the three tank units. DETR shows a slight improvement and detects three of the five compressors and all three tank units. Interestingly, FASTER-RCNN shows superior performance accurately recognizing all infrastructures without error.

Case C: This scenario presents a view of a typical compressor type found in the Permian Basin. In this specific instance, all algorithms correctly identify the six compressors and a single tank unit.

Case D: This case showcases a view of two sites each with a well. Unlike Case A, the image resolution in this case is lower and the wells are more widely spaced. YOLO v8 and DETR successfully recognize the two wells, while FASTER-RCNN fails to recognize either.

As highlighted in Case B, the visual representation of compressors in the Permian Basin is variable. The OG database contains a few cases where compressors are protected by a roof (as shown in the Case B images). To evaluate the detection capabilities of the algorithms in these particular circumstances, tests were extended to three additional images from the test database showing covered compressors (see Fig. 8 results).

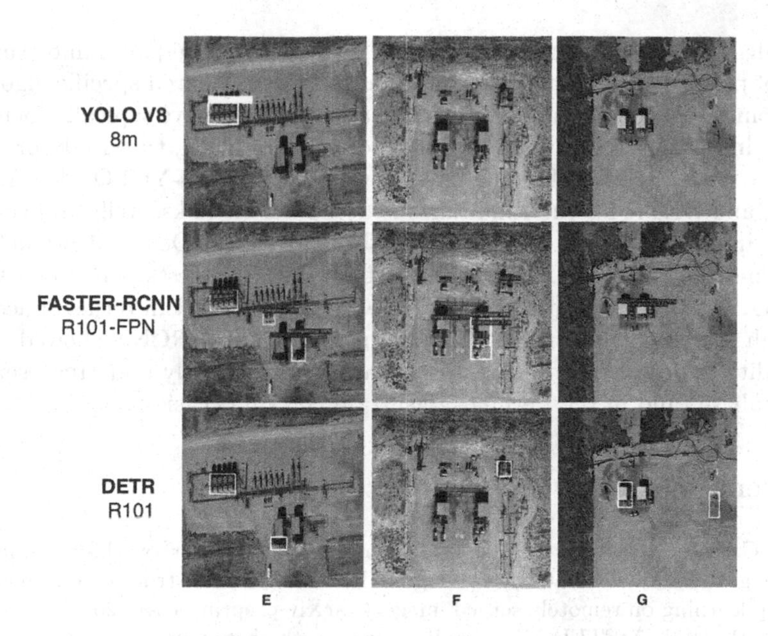

Fig. 8. Comparison of pre-trained YOLO v8, FASTER-RCNN and DETR on a special (less representative) compressor architecture in the Permian Basin @Google Earth

Case E: This scenario involves two unusual compressors along with a tank unit. FASTER-RCNN manages to identify all infrastructures, but it also mistakenly recognizes an additional compressor. DETR delivers an intriguing result by identifying a compressor through only a small segment protruding from the roof, as well as the tank unit. As for YOLO v8, it only manages to recognize the tank unit.

Case F: This scene provides a view of two unusual compressors and a tank unit. Both YOLO v8 and DETR fail to recognize the compressors, with DETR only acknowledging the tank unit. Yet again, FASTER-RCNN successfully identifies all the infrastructures as expected.

Case G: This scene presents a view of two unusual compressors exclusively. YOLO v8 is unable to detect either of them, while DETR correctly identifies one of the two compressors. It also detects an additional one, which does not correspond to a compressor but rather a small piece of infrastructure. Finally, FASTER-RCNN is also able to correctly recognize both compressors in this case.

6 Conclusion

To effectively mitigate methane emissions in the oil and gas (O&G) sector, a comprehensive emissions profile of each O&G infrastructure is essential which allows for an in-depth understanding of individual emission trends. This requires, as first step, an efficient solution to automatically detect and identify these infrastructures, a task that can be well addressed by detection algorithms. However, these algorithms come in myriad forms, each with a unique architecture and range of performance. While previous studies have compared specific algorithms for automatically detecting O&G infrastructures, these have typically focused on a single infrastructure type. In response to this limitation, this study presents a comparative analysis of three main supervised algorithms-YOLO v8, FASTER-RCNN, and DETR-for the simultaneous detection of tanks, wells and compressors. A unique database of various aerial snapshots of O&G infrastructure in the Permian Basin, USA, was used for the study. The tests performed showed that YOLO v8 outperformed FASTER-RCNN and DETR in terms of accuracy, although it was not pre-trained. Interestingly, FASTER-RCNN showed a superior ability to detect compressor styles that are significantly underrepresented in our database (and in the Permian Basin) on certain occasions.

References

1. IPCC. https://www.ipcc.ch/report/sixth-assessment-report-working-group-3
2. Sheng, H., et al.: OGNet: towards a global oil and gas infrastructure database using deep learning on remotely sensed imagery. arXiv preprint arXiv:2011.07227 (2020)
3. Zhu, B., et al.: METER-ML: a multi-sensor earth observation benchmark for automated methane source mapping. arXiv preprint arXiv:2207.11166, 2022
4. Lindeberg, T.: Scale Invariant feature transform **7**, 05 (2012)
5. Dalal, N., Triggs, B.: Histograms of oriented gradients for human detection. In: 2005 IEEE Computer Society Conference on Computer Vision and Pattern Recognition (CVPR 2005), vol. 1, pp. 886–893 (2005)
6. Wang, P., Fan, E., Wang, P.: Comparative analysis of image classification algorithms based on traditional machine learning and deep learning. Pattern Recogn. Lett. **141**, 61–67 (2021)
7. Huang, G., Laradji, I., Vazquez, D., Lacoste-Julien, S., Rodriguez, P.: A survey of self-supervised and few-shot object detection. IEEE Trans. Pattern Anal. Mach. Intell. **45**(4), 4071–4089 (2022)
8. Xiao, Y., et al.: A review of object detection based on deep learning. Multimedia Tools Appl. **79**, 23729–23791 (2020)
9. Girshick, R., Donahue, J., Darrell, T., Malik, J.: Rich feature hierarchies for accurate object detection and semantic segmentation. In Proceedings of the IEEE Conference on Computer Vision and Pattern Recognition, pp. 580–587 (2014)
10. Girshick, R.: Fast R-CNN. In: Proceedings of the IEEE International Conference on Computer Vision, pp. 1440–1448 (2015)
11. Ren, S., He, K., Girshick, R., Sun, J.: Faster R-CNN: towards real-time object detection with region proposal networks. In: Advances in Neural Information Processing Systems, vol. 28 (2015)
12. Redmon, J., Divvala, S., Girshick, R., Farhadi, A.: You only look once: unified, real-time object detection, pp. 779–788 (2016)
13. Terven, J., Cordova-Esparza, D.-M.: A comprehensive review of yolo: from yolov1 to yolov8 and beyond (2023)
14. Liu, W., et al.: SSD: single shot multibox detector. In: Leibe, B., Matas, J., Sebe, N., Welling, M. (eds.) ECCV 2016. LNCS, vol. 9905, pp. 21–37. Springer, Cham (2016). https://doi.org/10.1007/978-3-319-46448-0_2

15. Lin, T. Y., Goyal, P., Girshick, R., He, K., Dollár, P.: Focal loss for dense object detection. In: Proceedings of the IEEE International Conference on Computer Vision, pp. 2980–2988 (2017)
16. Tan, L., Huangfu, T., Liyao, W., Chen, W.: Comparison of RetinaNet, SSD, and YOLO v3 for real-time pill identification. BMC Med. Inform. Decis. Mak. **21**, 11 (2021)
17. Dos Santos, D.F., Françani, A.O., Maximo, M.R., Ferreira, A.S.: Performance comparison of convolutional neural network models for object detection in tethered balloon imagery. In: 2021 Latin American Robotics Symposium (LARS), 2021 Brazilian Symposium on Robotics (SBR), and 2021 Workshop on Robotics in Education (WRE), pp. 246–251 (2021)
18. Jakubec, M., Lieskovská, E., Bučko, B., Zábovská, K.: Comparison of CNN-based models for pothole detection in real-world adverse conditions: overview and evaluation. Appl. Sci. **13**(9), 5810 (2023)
19. Carion, N., Massa, F., Synnaeve, G., Usunier, N., Kirillov, A., Zagoruyko, S.: End-to-end object detection with transformers. In: Vedaldi, A., Bischof, H., Brox, T., Frahm, J.-M. (eds.) ECCV 2020. LNCS, vol. 12346, pp. 213–229. Springer, Cham (2020). https://doi.org/10.1007/978-3-030-58452-8_13
20. Bhil, K., et al.: Recent progress in object detection in satellite imagery: a review. In: Aurelia, S., Hiremath, S.S., Subramanian, K., Biswas, S.K. (eds.) Sustainable Advanced Computing. LNEE, vol. 840, pp. 209–218. Springer, Singapore (2022). https://doi.org/10.1007/978-981-16-9012-9_18
21. Wang, Y., Bashir, S.M.A., Khan, M., Ullah, Q., Wang, R., Song, Y., Guo, Z., Niu, Y.: Remote sensing image super-resolution and object detection: benchmark and state of the art. Expert Syst. Appl. **197**, 116793 (2022)
22. Li, Z., et al.: Deep learning-based object detection techniques for remote sensing images: a survey. Remote Sens. **14**(10), 2385 (2022)
23. Kang, J., Tariq, S., Han, O., Woo, S.S.: A survey of deep learning-based object detection methods and datasets for overhead imagery. IEEE Access **10**, 20118–20134 (2022)
24. Li, M., Zhang, Z., Lei, L., Wang, X., Guo, X.: Agricultural greenhouses detection in high-resolution satellite images based on convolutional neural networks: Comparison of faster R-CNN, YOLO v3 and SSD. Sensors **20**(17), 4938 (2020)
25. Hacıefendioğlu, K., Başağa, H.B., Demir, G.: Automatic detection of earthquake-induced ground failure effects through faster R-CNN deep learning-based object detection using satellite images. Nat. Hazards **105**, 383–403 (2021)
26. Demidov, D., Grandhe, R., Almarri, S.: Object detection in aerial imagery (2022)
27. Li, Q., Chen, Y., Zeng, Y.: Transformer with transfer CNN for remote-sensing-image object detection. Remote Sens. **14**(4), 984 (2022)
28. Tahir, A.: Automatic target detection from satellite imagery using machine learning. Sensors **22**(3), 1147 (2022)
29. Zhu, J., Chen, X., Zhang, H., Tan, Z., Wang, S., Ma, H.: Transformer based remote sensing object detection with enhanced multispectral feature extraction. IEEE Geosci. Remote Sens. Lett. 1–1 (2023)
30. Yang, Y.-J., Singha, S., Mayerle, R.: A deep learning based oil spill detector using sentinel-1 SAR imagery. Int. J. Remote Sens. **43**(11), 4287–4314 (2022)
31. Yang, Y.J., Singha, S., Goldman, R.: An automatic oil spill detection and early warning system in the Southeastern Mediterranean Sea. In: EGU General Assembly Conference Abstracts, EGU General Assembly Conference Abstracts, pp. EGU22-8408 (2022)

32. Zhang, L., Shi, Z., Jun, W.: A hierarchical oil tank detector with deep surrounding features for high-resolution optical satellite imagery. IEEE J. Sel. Top. Appl. Earth Observations Remote Sens. **8**(10), 4895–4909 (2015)
33. Shi, P., et al.: Oil well detection via large-scale and high-resolution remote sensing images based on improved YOLO v4. Remote Sens. **13**(16), 3243 (2021)
34. Song, G., Wang, Z., Bai, L., Zhang, J., Chen, L.: Detection of oil wells based on faster R-CNN in optical satellite remote sensing images. In: Image and Signal Processing for Remote Sensing XXVI, vol. 11533, pp. 14–121. SPIE (2020)
35. Zhibao Wang, L., et al.: An oil well dataset derived from satellite-based remote sensing. Remote Sens. **13**(6), 1132 (2021)
36. Gašparović, B., Lerga, J., Mauša, G., Ivašić-Kos, M.: Deep learning approach for objects detection in underwater pipeline images. Appl. Artif. Intell. **36**(1), 2146853 (2022)
37. Zhang, N.: et al.: Automatic recognition of oil industry facilities based on deep learning. In: IGARSS 2018–2018 IEEE International Geoscience and Remote Sensing Symposium, pp. 2519–2522 (2018)
38. Deng, J., Dong, W., Socher, R., Li, L. J., Li, K., Fei-Fei, L.: Imagenet: a large-scale hierarchical image database. pp. 248–255. IEEE (2009)
39. Lin, T.-Y., et al.: Microsoft COCO: common objects in context. In: Fleet, D., Pajdla, T., Schiele, B., Tuytelaars, T. (eds.) ECCV 2014. LNCS, vol. 8693, pp. 740–755. Springer, Cham (2014). https://doi.org/10.1007/978-3-319-10602-1_48
40. Everingham, M., Gool, L., Williams, C.K., Winn, J., Zisserman, A.: The Pascal visual object classes (VOC) challenge. Int. J. Comput. Vis. **88**(2), 303–338 (2010)

AttnOD: An Attention-Based OD Prediction Model with Adaptive Graph Convolution

Wancong Zhang[1], Gang Wang[2,3](✉), Xu Liu[2], and Tongyu Zhu[1]

[1] SKLSDE, Beihang University, Beijing, China
{wancong,zhutongyu}@buaa.edu.cn
[2] Highway Monitoring and Emergency Response Center, Ministry of Transport of the P.R.C., Beijing, China
[3] School of Vehicle and Mobility, Tsinghua University, Beijing, China
wang.gang@hmrc.net.cn

Abstract. In recent years, with the continuous growth of traffic scale, the prediction of passenger demand has become an important problem. However, many of the previous methods only considered the passenger flow in a region or at one point, which cannot effectively model the detailed demands from origins to destinations. Differently, this paper focuses on a challenging yet worthwhile task called Origin-Destination (OD) prediction, which aims to predict the traffic demand between each pair of regions in the future. In this regard, an Attention-based OD prediction model with adaptive graph convolution (AttnOD) is designed. Specifically, the model follows an Encoder-Decoder structure, which aims to encode historical input as hidden states and decode them into future prediction. Among each block in the encoder and the decoder, adaptive graph convolution is used to capture spatial dependencies, and self-attention mechanism is used to capture temporal dependencies. In addition, a cross attention module is designed to reduce cumulative propagation error for prediction. Through comparative experiments on the Beijing subway and New York taxi datasets, it is proved that the AttnOD model can obtain better performance than the baselines under most evaluation indicators. Furthermore, through the ablation experiments, the effect of each module is also verified.

Keywords: Traffic Prediction · OD Prediction · Attention · Encoder-Decoder · Self-Adaptive Graph Convolution

1 Introduction

In recent years, with the continuous growth of traffic scale, the prediction of traffic demand, volume, velocity and other parameters has become an important problem. For example, the prediction of demand for taxis can help the dispatching platform to allocate drivers to areas with higher demand.

B. Luo et al. (Eds.): ICONIP 2023, CCIS 1966, pp. 459–470, 2024.
https://doi.org/10.1007/978-981-99-8148-9_36

However, many of the previous methods only considered the traffic demand at one point or region, but ignored the modeling demands between origins and destinations [1–3]. The predicting based only on one site cannot meet the need of some applications. For example, if we can only observe the inflow and outflow of each subway station, it is hard to tell which section is overloaded since the interval information is missed, but it is an essential information for traffic management and urban planning.

This paper aims to solve a challenging but worthwhile task called Origin-Destination (OD) prediction, which aims to predict the traffic demand between each pair of regions. In this paper, a spatial-temporal fusion prediction method AttnOD is designed based on self-attention and adaptive graph convolution for OD prediction. The contributions includes:

Contribution 1: A self-adaptive graph is first introduced for OD demand prediction. And the OD matrix and the DO matrix are symmetrically handled to capture the bilateral correlation.

Contribution 2: Self-attention-based temporal module is designed to capture the temporal correlations, and a cross-attention module is proposed to reduce cumulative propagation error.

Contribution 3: A STVec is introduced to embed the extra temporal and spatial information.

2 Related Works

2.1 Temporal Dependencies

To solve the traffic prediction problem, there are many models developed to model the temporal dependencies. In the early stages, statistic-based methods, such as ARIMA [4] and Marcov algorithm [5] are proposed. However, as the observation window increases, these methods show severe limitations. To overcome the problem, some deep learning methods are proposed, such as Recursive Neural Network (RNN), Gated Recurrent Unit (GRU) and Long-Short Time Memory (LSTM) [6]. But as the prediction length increases, these models would encounter the problem of cumulative error propagation. Zheng et al. proposed an encoder-decoder model named GMAN [2], which uses spatial attention to capture the spatial dependencies between nodes and temporal attention to capture historical time effects. However, it is designed for traffic flow and thus cannot be easily transferred into OD prediction due to the relation computing for each OD pair of time series.

2.2 Spatial Dependencies

To model the complex spatial dependencies of OD data, various deep neural network models have been proposed. Liu et al. and Shi et al. used grid division to deal with the OD prediction problem [7,8]. However, there are often non-Euclidean correlations between traffic nodes, such as subway stations. Therefore, graph convolution networks (GCNs) were recently used to efficiently capture spatial information from non-Euclidean structured data [3]. Wang et al. proposed GEML [9] to aggregate spatial dependencies by defining geographic and semantic neighbors.

The above prediction methods are based on a predefined static graph structure to capture dependencies between regions, which ignored dependencies that may change over time. Currently, there have been literature attempting to capture dynamic graph patterns from adjacency matrices. For example, Zhang et al. proposed DNEAT, which developed a new neural layer named k-hop temporal node-edge attention layer (k-TNEAT) to capture the dynamic node structures in dynamic OD graphs instead of the predefined relationships among regions [10]. But the above models still failed to simultaneously capture the dynamic and bidirectional properties of OD diagrams with static structural information.

In summary, current models are not sufficient to adequately capture the dynamic spatial-temporal dependencies of the OD matrices. Therefore, we proposed a self-adaptive prediction method based on self-attention and adaptive graph convolution, named AttnOD, which aims to fully capture the spatial-temporal dependencies of OD data.

3 Methodology

3.1 Problem Definition

A transportation network with n stations, such as subway network and road network as shown in Fig. 1a, is represented as $\mathcal{G} = \{\mathbb{V}, \mathbb{E}\,\mathbf{A}\}$, where $\mathbb{V} = \{v_1, v_2, \cdots, v_n\}$ is the set of stations, $\mathbb{E}$ is the set of relations between stations and $\mathbf{A}$ is the adjacency weight matrix. If $(v_i, v_j) \in \mathbb{E}$, $\mathbf{A}_{i,j}$ will be calculated by the distance between v_i and v_j. And if $(v_i, v_j) \notin \mathbb{E}$, $\mathbf{A}_{i,j}$ will be 0.

For the OD matrix $\mathbf{M}^t \in \mathbb{R}^{n \times n}$, we define the i-th row of the OD matrix to denote departures from site i and the j-th column to denote arrivals at site j at time slot t. The i, j-entry of the OD matrix denotes the crowd flow from site i to site j. As shown in Fig. 1b:

Given P OD matrices of the history $\mathbf{M}^{(t-P)}, \mathbf{M}^{(t-P+1)}, ..., \mathbf{M}^{(t-1)}$, we need to predict an OD matrix $\mathbf{M}^{(t)}$ in the next time step:

$$\left\{\mathbf{M}^{(t-P)}, \mathbf{M}^{(t-P+1)}, ..., \mathbf{M}^{(t-1)}\right\} \xrightarrow{F} \mathbf{M}^{(t)}, \tag{1}$$

where F is the prediction model. As shown in Fig. 2:

3.2 The Attention-Based OD Prediction Model (AttnOD)

As shown in Fig. 3, our attention based OD prediction model (AttnOD) consists of an encoder and a decoder, both of which contain L STFusion blocks with residual connections [11]. Each STFusion block consists of an adaptive graph convolution layer and a temporal attention module. Between the encoder

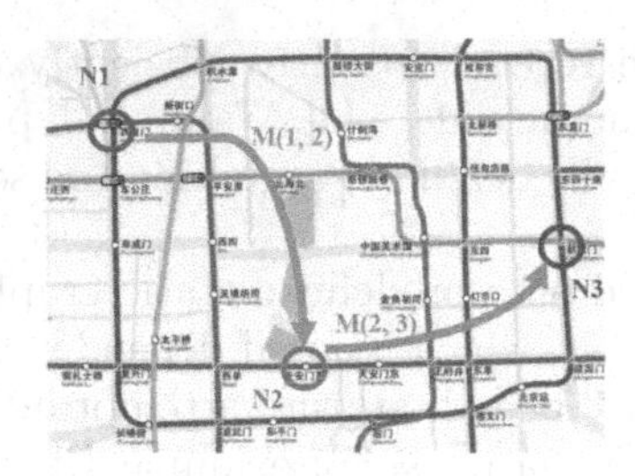

(a) Stations in Beijing Subway. $M(i, j)$ is the OD flow from i to j.

	1	2	...	j	n
1	8	13	...	...	13
2	22	4	...	...	16
...	...	...	...	...	...
i	...	...	...	$M_{i,j}$	...
n	8	9	...	...	56

(b) An OD Matrix.

Fig. 1. The Definition of OD Demand.

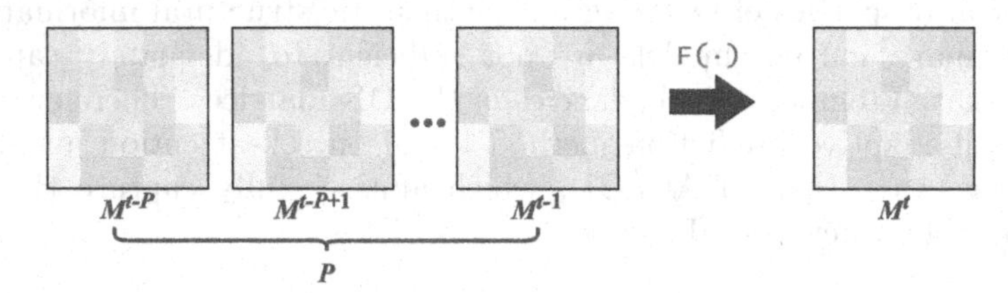

Fig. 2. OD Matrix Prediction.

and the decoder, a cross-attention module is added in the network to combine the encoded spatial-temporal features with the decoder. Furthermore, we setup the residual connections in the encoders and decoders. All layers produce D-dimensional outputs. The STVec module is used for extra spatial and temporal feature embedding. Through the Node2Vec algorithm [12], the adjacency matrix of traffic nodes is used to produce spatial vector $\mathbf{V}^s \in \mathbb{R}^{N \times D}$. For temporal features, we use one-hot coding to encode Day of Week and Hour of Day in the observation window into $\mathbb{R}^{P \times 7}$ and $\mathbb{R}^{P \times 24}$, and concatenate them into $\mathbb{R}^{P \times 31}$. Through a full connection network, we produced the temporal vector $\mathbf{V}^t \in \mathbb{R}^{P \times D}$. Finally, we define the matrix $\mathbf{V}_{\text{ext}} = \mathbf{V}^s \oplus \mathbf{V}^t$, which element v_{n_i,t_j} is defined as $v^s_{n_i} + v^t_{t_j}$. The $\mathbf{V}_{\text{ext}} \in \mathbb{R}^{N \times P \times D}$ will be used in the temporal and cross attention module.

3.3 Adaptive Graph Convolution

Graph convolution is a fundamental operation to extract node features given structural information. A first-order approximation to the Chebyshev spectral filter was proposed by Kipf et al. [13]. Let $\mathbf{A} \in \mathbb{R}^{N \times N}$ represents the adjacency matrix including self-loops, $\mathbf{H} \in \mathbb{R}^{N \times D}$ denotes the input vector, $\mathbf{Z} \in \mathbb{R}^{N \times D}$ denotes the output, $\mathbf{W} \in \mathbb{R}^{D \times D}$ represents the learnable parameters. The graph convolution layer is defined as:

$$\mathbf{Z} = \mathbf{AHW}. \tag{2}$$

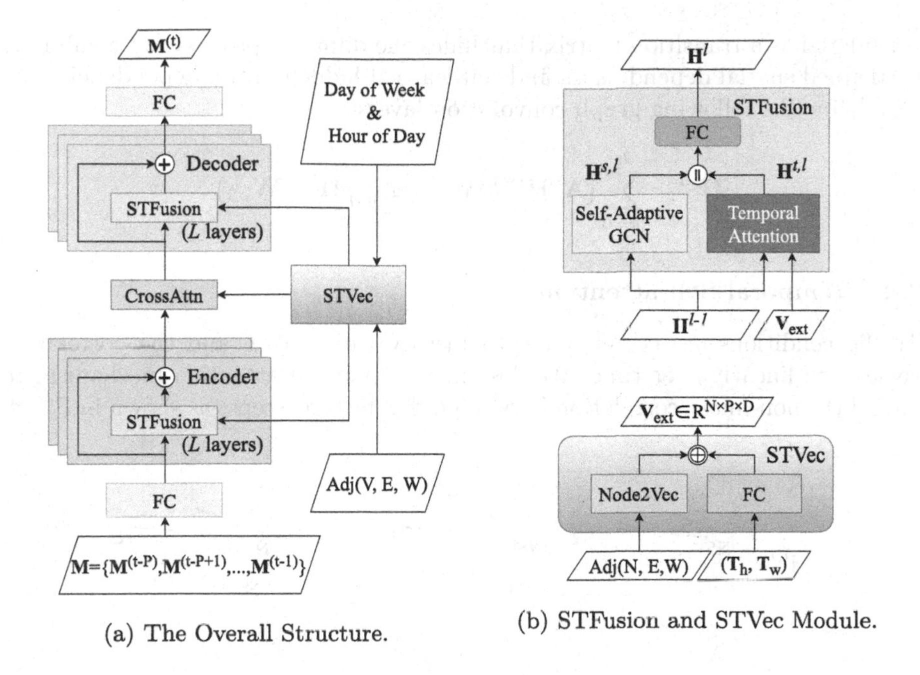

(a) The Overall Structure.

(b) STFusion and STVec Module.

Fig. 3. AttnOD structure.

Li et al. proposed a dilated convolution layer, which has been shown effective in spatial-temporal modeling [14]. They simulated the graph signal diffusion process by K steps. We generalize its dilated convolution layer in the form of Eq. 2, which results in:

$$\mathbf{Z} = \sum_{k=0}^{K} \mathbf{A}^k \mathbf{H} \mathbf{W}_k, \tag{3}$$

where $\mathbf{A}^k$ represents the power series of the transition matrix.

Meanwhile, considering that the spatial relation between traffic nodes may be latent and can be directly designed by hand, we further apply the adaptive adjacency matrix $\mathbf{A}_{adp}$ proposed to the Self-Adaptive GCN in our STFusion module. The adaptive adjacency matrix does not require any prior knowledge and is learned via random gradient descent. Therefore, the model can discover hidden spatial dependencies on its own. We randomly initialize two learnable matrices as $\mathbf{E}_1, \mathbf{E}_2 \in \mathbb{R}^{N \times C}$. The adaptive adjacency matrix is defined as:

$$\mathbf{A}_{adp} = \text{Softmax}(\text{ReLU}(\mathbf{E}_1 \mathbf{E}_2^T)), \tag{4}$$

where $\mathbf{E}_1$ acts as the source node embedding, and $\mathbf{E}_2$ acts as the target node embedding. By multiplying $\mathbf{E}_1$ and $\mathbf{E}_2$, we get the spatial dependent weights between source and target nodes. We use the ReLU activation function to eliminate weak connections. The Softmax function is used to normalize the adaptive adjacency matrix. Therefore, the normalized adaptive adjacency matrix can be considered as a transition matrix that hides the diffusion process. By combining predefined spatial dependencies and self-learned hidden graph dependencies, we can define the following graph convolution layers:

$$\mathbf{H}^{s,l} = \sum_{k=0}^{K} \left(\mathbf{A}^k \mathbf{H}^{\mathbf{l}-1} \mathbf{W}_{k1} + \mathbf{A}^k_{adp} \mathbf{H}^{\mathbf{l}-1} \mathbf{W}_{k2} \right). \tag{5}$$

3.4 Temporal Self-attention

Traffic conditions are correlated with history time series, and the correlation varies non-linearly over time. We designed a temporal attention mechanism to model the non-linear correlation between different time steps, as shown in Fig. 4.

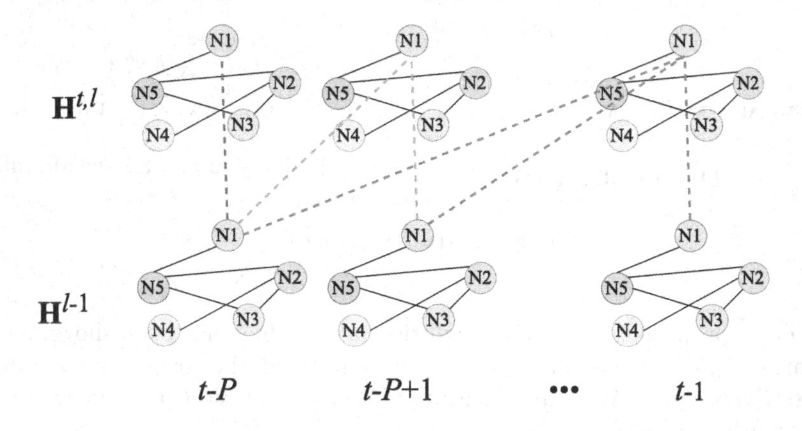

Fig. 4. Temporal Self Attention.

Note that temporal correlation is not only affected by historical series, but also related to time information. For example, congestion usually occurs during the morning or evening rush hour on weekdays. Therefore, we concatenate hidden states with spatial-temporal embedding $\mathbf{V}_{\text{ext}}$ to compute attention scores. For traffic node n_i, the attention score between time steps t_j and t is defined as:

$$\begin{aligned} u^{(k)}_{t_j,t} &= \frac{\left\langle \mathcal{Q}^{(k)}_t (\mathbf{H}^{l-1}_{n_i,t_j} \| v_{n_i,t_j}), \mathcal{K}^{(k)}_t \left(\mathbf{H}^{l-1}_{n_i,t} \| v_{n_i,t}\right) \right\rangle}{\sqrt{D}}, \\ \tau^{(k)}_{t_j,t} &= \frac{\exp\left(u^{(k)}_{t_j,t}\right)}{\sum_{t_r \in \mathcal{N}_{t_j}} \exp\left(u^{(k)}_{t_j,t_r}\right)} = \text{Softmax}\left(u^{(k)}_{t_j,t}\right), \end{aligned} \tag{6}$$

where $\|$ is the concatenating operation. $u^{(k)}_{t_j,t}$ represents the correlation between time steps t_j and t. $\tau^{(k)}_{t_j,t}$ is the attention score of the k-th head, which represents importance of t to t_j. $\mathcal{Q}^{(k)}_t(X) = W^{\mathcal{Q}}_t X$, $\mathcal{K}^{(k)}_t(X) = W^{\mathcal{K}}_t X$ are two different learnable matrices. $\mathcal{N}_{t_j}$ covers the time steps in the observation window before t_j. Once we get the attention score, the temporal hidden layer of node v_i at time step t_j will update as follows:

$$\mathbf{H}_{v_i,t_j}^{t,l} = \|_{k=1}^{K} \left\{ \sum_{t \in \mathcal{N}_{t_j}} \tau_{t_j,t}^{(k)} \cdot \mathcal{V}_t^{(k)}(\mathbf{H}_{v_i,t}^{l-1}) \right\},$$
$$\mathbf{H}^{t,l} = \text{Softmax}\left(\frac{\mathcal{Q}_t(\mathbf{H}^{l-1}||v) \cdot \mathcal{K}_t^T(\mathbf{H}^{l-1}||v)}{\sqrt{D}} \right) \mathcal{V}_t(\mathbf{H}^{l-1}), \tag{7}$$

where $\mathcal{V}_t^{(k)}(\cdot)$ represents a learnable matrix.

After we got the $\mathbf{H}^{t,l}$, we concatenate it with the output of self-adaptive graph convolution module $\mathbf{H}^{s,l}$. Then we got $\mathbf{H}^l$ through a full connection layer and the residual connection:

$$\mathbf{H}^l = \text{FC}(\mathbf{H}^{s,l}||\mathbf{H}^{t,l}) + \mathbf{H}^{l-1}. \tag{8}$$

3.5 Cross Attention

To reduce the error propagation between different prediction time steps in a long time, we add a cross-attention layer between the encoder and decoder. It captures the direct relationship between the future time step and each historical time step. As shown in Fig. 5:

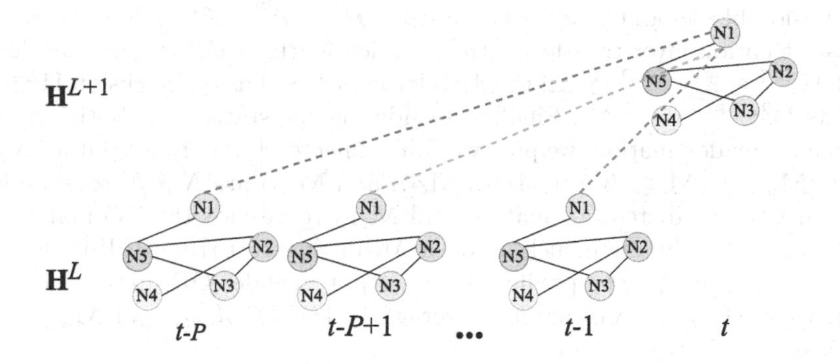

Fig. 5. Cross Self Attention.

For node n_i, the correlation between future time step t and historical time steps $t-1, ..., t-P$ is measured by spatial-temporal embedding of output and input:

$$\lambda_{t,t_j}^{(k)} = \frac{\left\langle \mathcal{Q}_c^{(k)}(v_{n_i,t}), \mathcal{K}_c^{(k)}(v_{n_i,t_j}) \right\rangle}{\sqrt{D}},$$
$$\gamma_{t,t_j}^{(k)} = \frac{\exp(\lambda_{t,t_j}^{(k)})}{\sum_{t_r=t-1}^{t-P} \exp(\lambda_{t,t_r}^{(k)})} = \text{Softmax}(\lambda_{t,t_j}^{(k)}), \tag{9}$$

The cross attention score $\gamma_{t_j,t}^{(k)}$ shows the attention score between history spatial-temporal embedding and future spatial-temporal embedding. We use this score to produce the decoder input:

$$\mathbf{H}_{v_i,t}^{L+1} = \|_{k=1}^{K} \left\{ \sum_{t_j=t-1}^{t-P} \gamma_{t,t_j}^{(k)} \cdot \mathcal{V}_c^{(k)}(\mathbf{H}_{v_i,t_j}^{L}) \right\},$$
$$\mathbf{H}^{L+1} = \text{Softmax}\left(\frac{\mathcal{Q}_c(v_t) \cdot \mathcal{K}_c^T(v_P)}{\sqrt{D}} \right) \mathcal{V}_c(\mathbf{H}^L), \tag{10}$$

where v_t is future spatial-temporal embedding, and v_P is history embedding.

3.6 Encoder-Decoder Structure

As shown in Fig. 3, our AttnOD is an encoder-decoder architecture. Note that our input $\mathbf{M} \in \mathbb{R}^{P\times N\times N}$ is an OD matrix sequence, where P is the observation window length, and N is the number of traffic nodes. As the traffic node can be both the origin and the destination, in order to capture the spatial characteristics symmetrically, we concatenate OD matrix $\mathbf{M}$ and its transpose $\mathbf{M}^T$. Then we fuse them by full connected layer to generate a new matrix sequence $\mathbf{M}' \in \mathbb{R}^{P\times N\times N}$:

$$\mathbf{M}' = \text{FC}(\mathbf{M}\|\mathbf{M}^T). \tag{11}$$

Then, a full connected layer is used to transform $\mathbf{M}' \in \mathbb{R}^{P\times N\times N}$ into $\mathbf{H}^0 \in \mathbb{R}^{P\times N\times D}$ before entering the encoder. $\mathbf{H}^0$ is fed into an encoder with L STFusion blocks and produces an output $\mathbf{H}^L \in \mathbb{R}^{P\times N\times D}$. After the encoder, a cross attention layer transforms the encoded features $\mathbf{H}^L$ to generate decoder input $\mathbf{H}^{L+1} \in \mathbb{R}^{1\times N\times D}$. Next, the decoder uses L STFusion blocks on $\mathbf{H}^{L+1}$ and outputs $\mathbf{H}^{2L+1} \in \mathbb{R}^{1\times N\times D}$. Finally, considering the station can both act as the origin and the destination, we put two full connected layers to output a $1N \times 2N$ matrix $\mathbf{M}_{out} = (\mathbf{M}_{OD}, \mathbf{M}_{DO})$. Both $\mathbf{M}_{OD}$ and $\mathbf{M}_{DO}$ are $N \times N$ tensors. $\mathbf{M}_{OD}$ represents the predicted OD matrix, and $\mathbf{M}_{DO}$ represents the DO matrix.

We use a loss function such as Root Mean Square Error (RMSE) for back-propagation training. The prediction we output includes OD matrix $\mathbf{M}_{out1}$ and DO matrix $\mathbf{M}_{out2}$, so we take the average RMSE of $\mathbf{M}_{out1}$ and $\mathbf{M}_{out2}^T$ as the final loss function:

$$\mathcal{L} = \frac{\text{RMSE}(\mathbf{M}_{OD}, \mathbf{M}_{real}) + \text{RMSE}(\mathbf{M}_{DO}^T, \mathbf{M}_{real})}{2}. \tag{12}$$

4 Experiments

4.1 Datasets

- **BJSubway** includes passenger data of 268 stations. The dataset contains the transaction records from 6:00 to 22:00 in 56 days (2017.10.8–2017.12.2).
- **NYTaxi** includes taxi order data of 63 areas in New York city. The data set contains the daily taxi order in 250 days (2016.1.1–2016.9.6) (Table 1).

Table 1. Statistics of Datasets.

Dataset	BJSubway	NYTaxi
Number of Stations or Regions	268	63
Number of Days	56	250
Time of Everyday	6:00-22:00	0:00-24:00
Time slot	30 min	60 min
Number of Rows	1792	6000
Train:Valid:Test	8:1:1	8:1:1

4.2 Baselines

We compare our model with following models:

- **HA**: History Average, predicting the next step using the average value of the observation window.
- **LAST**: Predicting the next step using the value of last time step.
- **LR**: Linear Regression, predicting the next step using Least Square Method.
- **XGBoost** [15]: XGBoost regression is an extensible Boosting algorithm, which is an improved version of GBDT (Gradient Boosting Decision Tree).
- **GRU** [16]: Gated Recurrent Unit is the simple version of Long-Short Term Memory (LSTM), which can avoid gradient vanishing or explosion in RNN and has less computing complexity than LSTM.
- **GEML** [9]: GEML uses multi-task learning which focuses on modeling temporal attributes and capturing several objectives of the problem.
- **DNEAT** [10]: DNEAT developed a k-hop temporal node-edge attention layer (k-TNEAT) to capture the temporal evolution of node topologies in dynamic OD graphs instead of the pre-defined adjacency.
- **MPGCN** [17]: MPGCN uses LSTM to capture temporal dependencies and uses 2D graph convolution network to capture spatial dependencies.
- **ODformer** [18]: ODFormer is a Transformer-based model proposed in 2022, which uses 2DGCN proposed in MPGCN to capture spatial correlation, and proposed a new method to compute attention scores between node and node.

4.3 Experimental Setups and Computation Cost

Our experiments are conducted under a computer environment with Intel(R) Xeon(R) Gold 5218 CPU @ 2.30GHz and one NVIDIA RTX 3090 GPU card which has 24 GB of graphic memory. We use 8 encoders and 8 decoders. We train our model using Adam optimizer with an initial learning rate of 0.001. Dropout with rate 0.3 is applied to the outputs of the adaptive graph convolution layer. Observation length P is 16 time steps (8 h). The loss function for training is root mean squared error (RMSE). The evaluation metrics we choose are Mean Absolute Error (MAE), Root Mean Squared Error (RMSE), Pearson Correlation Coefficient (PCC) and Mean Absolute Percentage Error (MAPE).

The training process spent 2 h and used 5 GB of graphic memory on BJSubway, and spent 5 h and used 18 GB of graphic memory on NYTaxi.

4.4 Results

The results are shown in Table 2. We put the best under each metric in bold, and the second-best underlined.

It can be observed that the effects of classic machine learning algorithms such as LR and XGBoost are generally better than statistical models such as HA and LAST; while deep neural network methods are generally better than classic machine learning models. AttnOD achieved the best results except the MAPE of NYTaxi. The MAE and MAPE of DNEAT is very close to our model, but the RMSE is much larger than our AttnOD. On the NYTaxi dataset, MPGCN's MAPE achieved the best results. Note that the other metrics of MPGCN is also very close to AttnOD. One possible reason is that MPGCN adopts the two-dimensional graph convolution (2DGCN) method. Considering that OD prediction is a bilateral prediction, two-dimensional graph convolution may be more suitable than the classic one-dimensional graph convolution. In addition, in the MAPE on NYTaxi dataset, LAST surpasses most deep neural network models. This may be caused by small fluctuation of the demand in NYTaxi.

Table 2. Results of comparative experiment with 9 baselines.

Model	BJSubway				NYTaxi			
	MAE	RMSE	PCC	MAPE	MAE	RMSE	PCC	MAPE
HA	2.177	7.029	0.500	0.602	0.733	1.593	0.805	0.374
LAST	1.918	5.751	0.748	0.549	0.717	1.548	0.830	0.327
LR	1.876	5.175	0.744	0.692	0.694	1.365	0.858	0.410
XGBoost	1.782	5.592	0.692	0.652	0.710	1.390	0.853	0.425
GRU	1.523	4.343	0.842	0.561	0.632	1.286	0.860	0.345
GEML	1.854	4.901	0.800	0.655	0.658	1.293	0.874	0.387
MPGCN	1.552	4.096	<u>0.868</u>	0.558	0.628	<u>1.243</u>	<u>0.883</u>	**0.293**
DNEAT	<u>1.475</u>	5.417	0.750	<u>0.538</u>	<u>0.624</u>	1.452	0.840	0.328
ODFormer	1.493	<u>4.031</u>	0.865	0.542	0.643	1.318	0.879	0.339
AttnOD	**1.467**	**3.973**	**0.874**	**0.533**	**0.618**	**1.230**	**0.886**	<u>0.309</u>

4.5 Ablation Experiments

In order to explore the effect of each module in the AttnOD model on the prediction accuracy, we designed an ablation experiment. The three baselines of the ablation experiment are:

1. **Spatial-Attn**: In each Encoder and Decoder, only the adaptive graph convolution is reserved, that is, only the spatial relationship is considered.
2. **Temporal-Attn**: In each Encoder and Decoder, only the temporal self-attention is reserved, that is, only the temporal relationship is considered.
3. **Non-Cross-Attn**: Adaptive graph convolution and temporal self-attention are reserved, while the cross self-attention is removed.

Table 3 shows the results of the ablation experiments. Under most indicators, the AttnOD model achieved the best results. It can be seen that the gap between Temporal-Attn and AttnOD is smaller than that of Spatial-Attn and AttnOD, indicating that the contribution of temporal self-attention to the model is higher than that of adaptive graph convolution. On NYTaxi, PCC of Temporal-Attn even beats the complete AttnOD model. In addition, the gap between Non-Cross Attn and AttnOD proves the effect of the cross attention module.

Table 3. Ablation experiment results.

Dataset	BJSubway				NYTaxi			
Model	MAE	RMSE	PCC	MAPE	MAE	RMSE	PCC	MAPE
Spatial-Attn	1.538	4.223	0.857	0.574	0.632	1.254	0.882	0.358
Temporal-Attn	1.494	4.081	0.871	0.549	0.624	1.236	**0.887**	0.334
Non-Cross-Attn	1.486	4.035	0.870	0.545	0.627	1.232	0.885	0.349
AttnOD	**1.467**	**3.973**	**0.874**	**0.533**	**0.618**	**1.230**	0.886	**0.309**

5 Conclusion

In this paper, aiming at the origin-destination (OD) prediction problem, an Attention-based AttnOD model is proposed. AttnOD uses adaptive graph convolution to capture spatial correlation and self-attention to capture temporal dependencies. In addition, the model also sets up a cross attention module to reduce the error propagation in a long observation window. Through comparative experiments and ablation experiments on the Beijing subway and New York taxi datasets, it is proved that AttnOD achieved better results than the baselines.

Acknowledgment. This work was supported by the National Natural Science Foundation of China (No. 62272023) and the Fundamental Research Funds for the Central Universities (No. YWF-23-L-1203).

References

1. Geng, X., et al.: Spatiotemporal multi-graph convolution network for ride-hailing demand forecasting. In: Proceedings of the AAAI Conference on Artificial Intelligence, vol. 33, pp. 3656–3663 (2019)
2. Zheng, C., Fan, X., Wang, C., Qi, J.: GMAN: a graph multi-attention network for traffic prediction. In: Proceedings of the AAAI Conference on Artificial Intelligence, vol. 34, pp. 1234–1241 (2020)
3. Yu, B., Yin, H., Zhu, Z.: Spatio-temporal graph convolutional networks: a deep learning framework for traffic forecasting. arXiv preprint arXiv:1709.04875 (2017)
4. Deng, Z., Ji, M.: Spatiotemporal structure of taxi services in Shanghai: using exploratory spatial data analysis. In: International Conference on Geoinformatics (2011)
5. Kai, Z., Khryashchev, D., Freire, J., Silva, C., Vo, H.: Predicting taxi demand at high spatial resolution: approaching the limit of predictability. In: IEEE International Conference on Big Data (2017)
6. Lai, G., Chang, W.C., Yang, Y., Liu, H.: Modeling long- and short-term temporal patterns with deep neural networks. ACM (2018)
7. Liu, L., Qiu, Z., Li, G., Wang, Q., Ouyang, W., Lin, L.: Contextualized spatial-temporal network for taxi origin-destination demand prediction. IEEE Trans. Intell. Transp. Syst. **20**(10), 3875–3887 (2019)
8. Shi, X., Chen, Z., Wang, H., Yeung, D.Y., Wong, W.K., Woo, W.C.: Convolutional LSTM network: a machine learning approach for precipitation nowcasting. In: Advances in Neural Information Processing Systems, vol. 28 (2015)
9. Wang, Y., Yin, H., Chen, H., Wo, T., Xu, J., Zheng, K.: Origin-destination matrix prediction via graph convolution: a new perspective of passenger demand modeling. In: Proceedings of the 25th ACM SIGKDD International Conference on Knowledge Discovery & Data Mining, pp. 1227–1235 (2019)
10. Zhang, D., Xiao, F., Shen, M., Zhong, S.: Dneat: a novel dynamic node-edge attention network for origin-destination demand prediction. Transp. Res. Part C Emerg. Technol. **122**, 102851 (2021)
11. He, K., Zhang, X., Ren, S., Sun, J.: Deep residual learning for image recognition. In: Proceedings of the CVPR, pp. 770–778 (2016)
12. Grover, A., Leskovec, J.: node2vec: scalable feature learning for networks. In: Proceedings of the 22nd ACM SIGKDD International Conference on Knowledge Discovery and Data Mining, pp. 855–864 (2016)
13. Kipf, T.N., Welling, M.: Semi-supervised classification with graph convolutional networks. arXiv preprint arXiv:1609.02907 (2016)
14. Li, Y., Yu, R., Shahabi, C., Liu, Y.: Diffusion convolutional recurrent neural network: data-driven traffic forecasting. arXiv preprint arXiv:1707.01926 (2017)
15. Chen, T., Guestrin, C.: Xgboost: a scalable tree boosting system. In: Proceedings of the 22nd ACM SIGKDD, pp. 785–794 (2016)
16. Chung, J., Gulcehre, C., Cho, K., Bengio, Y.: Empirical evaluation of gated recurrent neural networks on sequence modeling. arXiv preprint arXiv:1412.3555 (2014)
17. Shi, H., Yao, Q., Guo, Q., Li, Y., Zhang, L.: Predicting origin-destination flow via multi-perspective graph convolutional network. In: 2020 IEEE 36th International Conference on Data Engineering (ICDE), pp. 1818–1821. IEEE (2020)
18. Huang, B., Ruan, K., Yu, W., Xiao, J., Xie, R., Huang, J.: Odformer: spatial-temporal transformers for long sequence origin-destination matrix forecasting against cross application scenario. Expert Syst. Appl. **222**, 119835 (2023)

CMMix: Cross-Modal Mix Augmentation Between Images and Texts for Visual Grounding

Tao Hong[1], Ya Wang[2], Xingwu Sun[2], Xiaoqing Li[3], and Jinwen Ma[1](✉)

[1] School of Mathematical Sciences, Peking University, Beijing, China
paul.ht@pku.edu.cn, jwma@math.pku.edu.cn
[2] Tencent Inc., Shenzhen, China
connorywang@tencent.com, sunxingwu01@gmail.com
[3] Capital University of Economics and Business, Beijing, China
xqli@cueb.edu.cn

Abstract. Visual grounding (VG) is a representative multi-modal task that has recently gained increasing attention. Nevertheless, existing works still face challenges leading to under-performance due to insufficient training data. To address this, some researchers have attempted to generate new samples by integrating each two (image, text) pairs, inspired by the success of uni-modal CutMix series data augmentation. However, these methods mix images and texts separately and neglect their contextual correspondence. To overcome this limitation, we propose a novel data augmentation method for visual grounding task, called Cross-Modal Mix (CMMix). Our approach employs a fine-grained mix paradigm, where sentence-structure analysis is used to locate the central noun parts in texts, and their corresponding image patches are drafted through noun-specific bounding boxes in VG. In this way, CMMix maintains matching correspondence during mix operation, thereby retaining the coherent relationship between images and texts and resulting in richer and more meaningful mixed samples. Furthermore, we employ a filtering-sample-by-loss strategy to enhance the effectiveness of our method. Through experiments on four VG benchmarks: ReferItGame, RefCOCO, RefCOCO+, and RefCOCOg, the superiority of our method is fully verified.

Keywords: Data augmentation · Mixed sample · Visual grounding · Cross-modal

1 Introduction

Recently, Deep Neural Networks (DNNs) have shown remarkable performance in various fields, particularly in Computer Vision (CV) [8,12] and Natural Language Processing (NLP) [9]. CV and NLP concentrate on the modality of image (vision) and text (language), respectively. Apart from a single modality, cross-modal tasks are hot topics nowadays which attend to utilize the connection

Supported by the Natural Science Foundation of China under grant 62071171.

B. Luo et al. (Eds.): ICONIP 2023, CCIS 1966, pp. 471–482, 2024.
https://doi.org/10.1007/978-981-99-8148-9_37

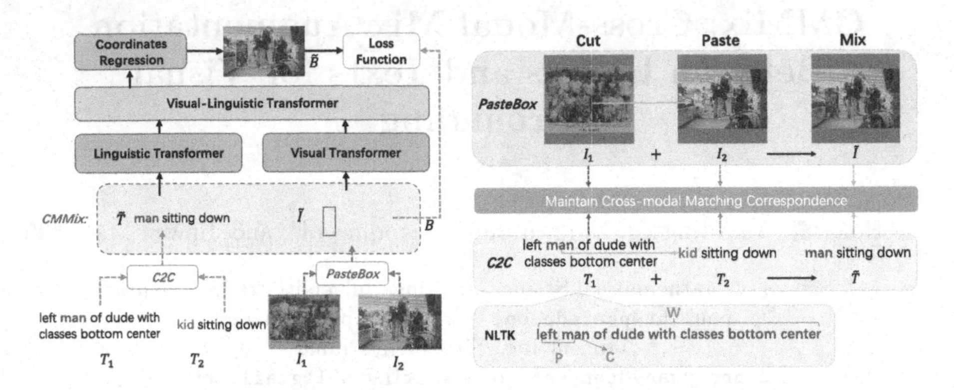

Fig. 1. The overview of CMMix on VG task: overall process including neural network module (left) and specific CMMix operation (right), DA is usually conducted as the input stage of neural network training. The displayed mode is *PasteBox+C2C*, where the central word is located with the NLTK package.

between different modalities to fulfill more precise representation, including but not limited to image caption [18], visual grounding (VG) [10] and so on.

Visual grounding is a task that aims at predicting the location of an image region denoted by a specific linguistic expression. The codebase awesome-grounding [15] provides a comprehensive repository for VG. Its solutions can be grouped into three categories based on the type of information available as marks: (image, text, box) for fully-supervised, (image, text) for weakly-supervised, and neither for unsupervised. Fully-supervised degenerates into weakly-supervised when object-phrase boxes are absent, while weakly-supervised becomes unsupervised if (image, text) pairs are not available. The progress of fully-supervised VG can be divided into one-stage and two-stage methods, somewhat similar to the categorization of object detection tasks. With the emergence of TransVG [3], the Transformer-based [17] models with image-text feature fusion can be categorized into a third category.

Due to manual labeling being expensive, most benchmarks in the VG area are limited by the shortage of samples required to train DNNs, therefore additional countermeasure strategies are necessary. One potential solution is to utilize existing data augmentation (DA) techniques, including random flipping, rotation, and scaling. Among these techniques, CutMix [21] proposes combining two samples to create a new one. It does so by cropping a random region from an image or text and pasting it onto another sample. It has been established as one of the state-of-the-art approaches in enhancing the diversity of uni-modal training sets, such as SaliencyMix [16] in CV and SenMixup [6] in NLP. Despite the success of CutMix in uni-modal tasks, its exploration in VG is still limited. Two existing cross-modal mix DA works are MixGen [7] and VLMixer [19]. However, both of them apply mix DA in the pre-training stage, which mainly benefits from the availability of a large corpus of data. Moreover, MixGen simply splices two texts, neglecting more precise expressions when referring to corresponding objects. VLMixer uses the token embeddings of Transformer [17] to mix and randomly replace the image token with a text token, which is difficult to implement, and is very demanding for pre-trained encoder representation.

To address these issues, we propose **CMMix**, a **C**ross-**M**odal **Mix** DA method for visual grounding task. Instead of pre-training, it is directly conducted in the fine-tuning stage to facilitate the lacking training data, which is more targeted and more efficient in VG. As shown in Fig. 1, we establish the mix operation while ensuring the semantic consistency of both image and text patches to generate more intricate and meaningful (image, text) pairings. We introduce various mix modes that correspond to diverse data characteristics, in particular, the structural connections of texts. Additionally, we apply a filtering-sample-by-loss method to strengthen the power of mix DA. The superiority of our method is demonstrated on four VG benchmarks: ReferItGame [11], RefCOCO [20], RefCOCO+ [20], and RefCOCOg [13]. It is worth noting that this study primarily focuses on the mix DA of fully-supervised VG, and the adopted state-of-the-art model is TransVG (TransVG++ is clarified better, but it's a pre-print version and has not been open-source yet). The primary contributions of our work are summarized below.

- We propose a simple yet effective multi-modal mix DA, named CMMix, which is the first work to directly conduct mix DA for downstream multi-modal tasks.
- We provide a fine-grained paradigm for (image, text) mixing, which can maintain the contextual correspondence and semantic relationship (ignored by previous works) between image and text modalities, resulting in richer and more meaningful samples.
- Our research suggests a filtering-sample-by-loss strategy as a form of regularization to enhance VG performance further.

2 Proposed Approach

In this section, we present the detailed aspects of our CMMix approach, whose overview is illustrated in Fig. 1. Since VG learns about two modalities, we introduce the mix strategies of them separately to display how real-matched (image, text) pairs are generated.

2.1 Mix Image

We use $(\boldsymbol{I}, \boldsymbol{T}, \boldsymbol{B})$ to refer to the three components (image, text, box) of VG. Images have dimensions of $\boldsymbol{I} \in \mathbb{R}^{H \times W \times C}$, where H, W, and C denote height, width, and channel, respectively. Similarly, tokenized texts have dimensions of $\boldsymbol{T} \in \mathbb{R}^{S \times V}$, where S denotes the number of tokens and V denotes the embedding dimension. We denote the 4-dimensional regression boxes as either a center plus height and width $(c_h, c_w, \Delta H, \Delta W)$ or upper left and lower right coordinates (h_1, w_1, h_2, w_2). The VG task involves learning a function $f : (\boldsymbol{I}, \boldsymbol{T}) \rightarrow \boldsymbol{B}$ that maps an $(\boldsymbol{I}, \boldsymbol{T})$ pair to $\boldsymbol{B}$.

To generate a new sample $(\widetilde{\boldsymbol{I}}, \widetilde{\boldsymbol{T}}, \widetilde{\boldsymbol{B}})$ through mix DA, we consider two samples denoted as $(\boldsymbol{I}_1, \boldsymbol{T}_1, \boldsymbol{B}_1)$ and $(\boldsymbol{I}_2, \boldsymbol{T}_2, \boldsymbol{B}_2)$. We follow the paradigm of CutMix [21] and ignore the text modality to express the mixed image formulation as follows:

$$\widetilde{\boldsymbol{I}} = \boldsymbol{M} \odot \boldsymbol{I}_1 + (\mathbf{1} - \boldsymbol{M}) \odot \boldsymbol{I}_2 \tag{1}$$

where $\boldsymbol{M} \in \{0, 1\}^{H \times W}$ denotes a binary mask matrix indicating where to drop out and fill in from two images, $\mathbf{1}$ is a matrix filled with ones, and $\odot$ is element-wise multiplication. The mask $\boldsymbol{M}$ is taken as a rectangular region and it fits very well with the regression goal of VG: bounding box $\boldsymbol{B}$.

We propose three modes for cutting a patch from $\boldsymbol{I}_1$ to paste on $\boldsymbol{I}_2$: *CutBox*, *PasteBox*, and *RandBox*. The cut region of $\boldsymbol{I}_1$ in all modes is labeled box $\boldsymbol{B}_1$, and differences arise from the pasted regions. As indicated literally, *CutBox* involves pasting the cut $\boldsymbol{B}_1$ region of $\boldsymbol{I}_1$ onto the same location of $\boldsymbol{I}_2$, while *PasteBox* resizes the cut region as the size of $\boldsymbol{B}_2$ and then pastes it on the $\boldsymbol{B}_2$ region of $\boldsymbol{I}_2$. RandBox means the pasted region is randomly selected. For the mix process, we express *CutBox* and *PasteBox* as:

$$\widetilde{\boldsymbol{I}}_{CutBox} = \boldsymbol{B}_1 \odot \boldsymbol{I}_1 + (\mathbf{1} - \boldsymbol{B}_1) \odot \boldsymbol{I}_2 \tag{2}$$

$$\widetilde{\boldsymbol{I}}_{PasteBox} = \Phi_{\boldsymbol{B}_2}(\boldsymbol{B}_1) \odot \boldsymbol{I}_1 + (\mathbf{1} - \boldsymbol{B}_2) \odot \boldsymbol{I}_2 \tag{3}$$

respectively, where $\Phi_{\boldsymbol{B}_2}(\boldsymbol{B}_1)$ denotes resizing cut $\boldsymbol{B}_1$ to the shape of $\boldsymbol{B}_2$ and then shifting it to $\boldsymbol{B}_2$. The proposed *CutBox* implementation is straightforward and produces satisfactory results. Meanwhile, some VG datasets contain words indicating interactions with the surrounding environment (e.g., directional words like "left" or "right"). In such cases, *PasteBox* may yield better results when collaborating with a suitable text mix strategy.

2.2 Mix Text

Texts face more challenging obstacles than images due to their semantic structure, logical relationships, and irregular data forms. Upon analyzing the VG task expressions, the noun in a sentence is considered the core word, while the other words serve as pre- or post-modifying components. An example of this is text $\boldsymbol{T}_1$: "left **man** of dude with classes bottom center." Here, we refer to the central noun "man" as *C*, the noun with a prefix modifier "left man" as *P*, and the entire sentence as *W*. Then the remaining question is, which part shall we choose to mix?

Despite variations in text features across different VG datasets (as demonstrated in the experimental section), they generally follow the sentence structure we have established. The central noun *C* serves as the strongest link between the image and text. Therefore, we use the abbreviation *C2C* to indicate the replacement of at least the central noun of pasted text $\boldsymbol{T}_2$ with that of cut text $\boldsymbol{T}_1$. Given the text $\boldsymbol{T}_2$ as "kid sitting down", we generate the *C2C*-mixed text "man sitting down" by replacing "kid" with the central noun "man" from $\boldsymbol{T}_1$. In addition to *C2C*, we exploit modes of *C2W*, *P2W* and *W2W*, *etc.*

For the *CutBox*-mixed image, we typically adopt **2W*-mixed text where * always serves as a placeholder for W. In datasets like RefCOCOg, where many words interact with the central nouns, * may be taken as either C or P to reduce the noise of redundant words in mixed texts. However, when dealing with directional words in datasets such as RefCOCO, we avoid using *W2W* as it produces many unmatched pairs of (image, text). For instance, when pasting the "left horse" from $\boldsymbol{T}_1$ onto the "right car" from $\boldsymbol{T}_2$, *PasteBox* mode results in an image with a horse on the right while the *W2W*-mixed text remains the original: "left horse", resulting in an error. Using the *C2C* mode solves this issue more effectively (in this case, replacing the "car" from $\boldsymbol{T}_2$ with the "horse" from $\boldsymbol{T}_1$ to obtain the mixed text: "right horse"), leading to more abundant and meaningful pairs of (image, text).

After giving a general understanding of mix DA for $(\boldsymbol{I}, \boldsymbol{T}, \boldsymbol{B})$ samples, the question remains: How do we identify the central noun of a sentence? Fortunately, VG datasets have a set of attributes for referring expressions, which is represented by a 7-tuple $R = \{r_1, r_2, r_3, r_4, r_5, r_6, r_7\}$ [11], where r_1 is an entry-level category attribute and is the central noun we are seeking. One might argue that the demand for attribute annotations restricts the applicability of our CMMix approach. However, this is no cause for concern as we can easily replace attribute information using third-party word splitting tools. For our experiments, we used Natural Language ToolKit (NLTK) [1] to extract central nouns from expressions. NLTK assigns tags to different words, such as Nouns ("NN"), Adjectives ("JJ"), and Adverbs ("RB"). By locating the first "NN" tag in a given expression using NLTK, we can accurately identify the central word.

2.3 Filter Samples by Loss

During training, the model learns samples of varying levels of difficulty that can be determined by the training loss. Based on this insight, we propose a filtering technique to remove certain samples based on their losses. This technique sorts samples in a batch by their losses, and then filters the top-k samples with the minimum or maximum losses, referred to as *FilterMin* and *FilterMax*, respectively.

The *FilterMin* strategy directs the model to pay greater attention to difficult samples, thereby resulting in better VG performance. The resulting set of samples after *FilterMin* can be expressed as

$$\mathcal{X} - \{\boldsymbol{x} \in \mathcal{X} | \boldsymbol{x} \ s.t. \ \min_{\text{top-}k} \ell(\boldsymbol{x})\} \tag{4}$$

where $\mathcal{X}$ represents the training samples in a batch, and ℓ denotes the loss function. It is worth mentioning that this strategy can be applied not only to VG tasks but also to other deep learning tasks.

To summarize, we present the CMMix technique using the *PasteBox+C2C* mode in Algorithm 1. It is worth noting that we do not differentiate strictly between single samples and batch samples to simplify the representation.

Algorithm 1. CMMix algorithm (*PasteBox+C2C* mode).

Require: Training set $\mathcal{X}$: (image $\boldsymbol{I}$, text $\boldsymbol{T}$, box $\boldsymbol{B}$); model f; loss function ℓ
1: **for** $(\boldsymbol{I}, \boldsymbol{T}, \boldsymbol{B}) \in \mathcal{X}$ **do**
2: $(\boldsymbol{I}_1, \boldsymbol{T}_1, \boldsymbol{B}_1) = (\boldsymbol{I}, \boldsymbol{T}, \boldsymbol{B})$
3: $(\boldsymbol{I}_2, \boldsymbol{T}_2, \boldsymbol{B}_2) = \text{Shuffle}(\boldsymbol{I}, \boldsymbol{T}, \boldsymbol{B})$
4: $\widetilde{\boldsymbol{I}} = \Phi_{\boldsymbol{B}_2}(\boldsymbol{B}_1) \odot \boldsymbol{I}_1 + (\mathbf{1} - \boldsymbol{B}_2) \odot \boldsymbol{I}_2$ ▷ Mix image via *PasteBox*, refer to Eq. (3)
5: $\widetilde{\boldsymbol{B}} = \boldsymbol{B}_2$
6: $\widetilde{\boldsymbol{T}} \leftarrow \text{NLTK}(\boldsymbol{T}_1, \boldsymbol{T}_2)$ ▷ After NLTK, mix text via *C2C*
7: $(\widetilde{\boldsymbol{I}}, \widetilde{\boldsymbol{T}}, \widetilde{\boldsymbol{B}}) \leftarrow \text{FilterMin}(\widetilde{\boldsymbol{I}}, \widetilde{\boldsymbol{T}}, \widetilde{\boldsymbol{B}})$ ▷ Filter samples via loss ℓ, refer to Eq. (4)
8: $f : (\widetilde{\boldsymbol{I}}, \widetilde{\boldsymbol{T}}) \mapsto \widetilde{\boldsymbol{B}}$ ▷ Regular training
9: **end for**

3 Experimental Results

In this section, we first demonstrate the comprehensive superiority of our CMMix technique over several VG datasets. Next, we focus on ablating implementation details on the RefCOCO dataset, including mix probability and filtering samples based on loss. Finally, we analyze the outcomes visually.

3.1 Datasets and Implementation Details

We conducted experiments on four representative VG datasets: ReferItGame [11], RefCOCO [20], RefCOCO+ [20], and RefCOCOg [13]. ReferItGame consists of 20,000 images from the SAIAPR-12 dataset [5], wherein each image has one or multiple regions with corresponding referring expressions. We follow the common division: a training set with 54,127 referring expressions, a validation set with 5,842 referring expressions and a test set with 60,103 referring expressions. RefCOCO includes 19,994 images with 50,000 referred objects. Each object has more than one referring expression. The number of total expressions is 142,210 and the division of (train, validation, testA, testB) is (120,624; 10,834; 5,657; 5,095). Similarly, RefCOCO+ contains 19,992 images with 49,856 referred objects and 141,564 referring expressions. Its division of (train, validation, testA, testB) is (120,191; 10,758; 5,726; 4,889). RefCOCOg has 25,799 images with 49,856 referred objects and expressions. We adopt RefCOCOg-google (val-g) [13] as the split protocol (the other protocol is RefCOCOg-umd [14]).

We develop CMMix by building upon the codebase of TransVG model, using its parameters without modification. The visual branch is the backbone and encoder of DETR model [2], and the linguistic branch is the basic BERT model [4]. The initial learning rate of both the visual and linguistic branches is set to 10^{-5}, while that of the V-L module and prediction head is set to 10^{-4}. We employ the AdamW optimizer with a weight decay of 10^{-4} and set the batch size to 64 (distributed across 4 GPUs). Except for RefCOCO+ (total 180 epochs), all the default training epochs are set to 90 and the learning rate is decayed by 0.1 every 60 epochs. It's worth noting that during our experiments, we find that some training process is not completely converged, so we extend the total

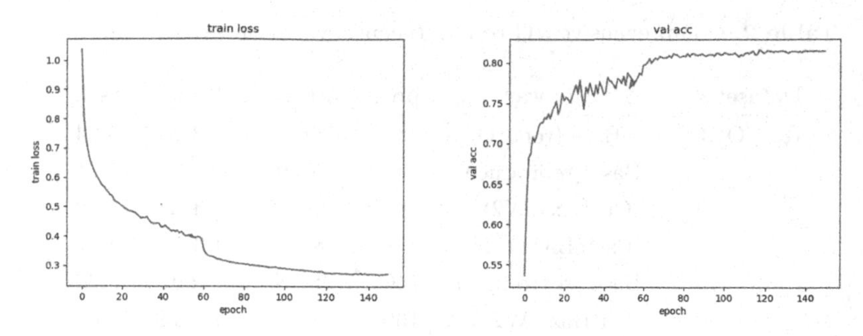

Fig. 2. The curves of training loss (left) and validation accuracy (right) on RefCOCO.

Table 1. VG results (accuracy, %) on RefCOCO. The best and second-place results for each column are bold and underlined. (report) and (reproduce) represent the reported and our reproduced results of baseline model, respectively (the same as below). *MixupI+SpliceT* is a contrast experiment with MixGen [7]. Our CMMix belongs to the last 5 rows, the mix probability is 0.05 and the filtering-sample-by-loss strategy is not adopted yet.

Mix way \ Train epoch	90			150		
	Val	TestA	TestB	Val	TestA	TestB
Base (report)	80.49	83.28	75.24	-	-	-
Base (reproduce)	79.96	81.93	75.71	80.81	82.92	76.66
MixupI+SpliceT	80.82	82.85	76.04	81.20	83.20	76.03
RandBox+P2W	**81.11**	83.35	76.25	81.18	83.45	77.08
CutBox+W2W	80.83	82.55	75.86	81.54	<u>84.02</u>	**77.19**
CutBox+P2W	<u>81.05</u>	<u>83.36</u>	**76.84**	**81.70**	84.00	76.90
PasteBox+P2W	80.69	82.25	<u>76.70</u>	80.76	82.59	<u>77.15</u>
PasteBox+C2C	80.59	**83.37**	76.13	<u>81.61</u>	**84.04**	76.33

epoch to 150 to compensate for convergence issues. We display a training curve in Fig. 2, from which we can see the specific convergence performance. The input image size is 640×640 and traditional DA on images includes random cropping, scaling, and translating. As for the maximum length of expression's token, RefCOCOg's is 40 and others are 20. The evaluation metric is the top-1 accuracy (%). All the experiments are conducted with PyTorch framework on NVIDIA A100 GPUs.

3.2 Visual Grounding Results

Table 1 illustrates the VG results of RefCOCO, with a more detailed exploration presented in Sect. 3.3. In our experiments, we compared with MixupI+SpliceT, wherein we Mixup [22] two images and splice two texts directly, as the imitation

Table 2. Comprehensive VG results (accuracy, %) on several datasets.

Dataset	Mix way	Epoch	Val/Val-g	TestA	TestB
RefCOCO	Base (report)	90	80.49	83.28	75.24
	Base (reproduce)	90	79.96	81.93	75.71
	CutBox+W2W	90	80.87	**83.91**	75.88
	PasteBox+C2C	90	**81.41**	**83.91**	**76.58**
	Base (reproduce)	150	80.81	82.92	76.66
	CutBox+W2W	150	81.16	**84.18**	76.64
	PasteBox+C2C	150	**81.86**	84.04	**77.23**
ReferItGame	Base (report)	90	69.76	-	-
	Base (reproduce)	90	71.70	-	-
	CutBox+W2W	90	72.24	-	-
	PasteBox+C2C	90	**72.42**	-	-
RefCOCOg	Base (report)	90	66.35	-	-
	Base (reproduce)	150	67.15	-	-
	CutBox+P2W	150	**68.50**	-	-
RefCOCO+	Base (report)	180	66.39	70.55	57.66
	Base (reproduce)	180	66.17	70.51	**57.87**
	CutBox+W2W	180	**66.41**	**71.31**	57.46

Table 3. VG results (accuracy, %) of optimization for mix probability on RefCOCO. The training epoch is 90.

Mix way	CutBox+W2W			PasteBox+C2C		
Mix probability	Val	TestA	TestB	Val	TestA	TestB
0 (report)	80.49	**83.28**	75.24	80.49	83.28	75.24
0 (reproduce)	79.96	81.93	75.71	79.96	81.93	75.71
0.02	80.04	81.95	75.92	**80.59**	**83.37**	**76.13**
0.05	80.83	82.55	75.86	**80.59**	82.45	75.43
0.1	**81.01**	83.24	**76.64**	80.07	82.16	75.25
0.25	80.79	82.34	75.72	79.72	81.28	75.86
0.5	79.04	81.47	74.80	79.10	81.10	74.78

comparison of MixGen [7]. Our results exhibit that CMMix outperforms both the baseline and MixGen. Specifically, we recommend using *CutBox+W2W* and *PasteBox+C2C* as the mix ways. The former only mixes the image modality while keeping one sentence matched with a cut-mixed image (**Box+W2W* can be considered equivalent to only mixing the image modality), which is easier to implement and yields satisfactory results. In contrast, *PasteBox+C2C* generates more abundant samples with rich meanings and achieves superior VG

Table 4. VG results (accuracy, %) on RefCOCO with the strategy of filtering top-k (1/4 proportion) samples with maximum or minimal losses.

Train Epoch		90			150		
Mix way	Filter	Val	TestA	TestB	Val	TestA	TestB
Base	-	79.96	81.93	75.71	80.81	82.92	76.66
	Min	80.44	82.41	75.76	80.79	82.64	75.98
CutBox+W2W	-	80.83	82.55	75.86	81.54	84.02	77.19
	Min	80.87	**83.91**	75.88	81.16	**84.18**	76.64
PasteBox+C2C	-	80.59	83.37	76.13	81.61	84.04	76.33
	Min	**81.41**	**83.91**	**76.58**	**81.86**	84.04	**77.23**
	Max	73.36	74.56	71.89	74.33	75.53	72.23

results. Note that the filtering-sample strategy demonstrates a more remarkable outperformance of *PasteBox+C2C*, as evidenced in Table 4. Table 2 presents comprehensive VG results across 4 representative datasets, which demonstrate significant improvements and offer different combination strategies of mix DA for varying dataset characteristics.

3.3 Exploration Study

This subsection focuses on RefCOCO as a representative dataset to elaborate on the exploration details of our CMMix.

Mix Probability. Mix probability p is a crucial parameter in the mix series DA approach, representing the proportion of mixed samples compared to the total number of samples. In image mix DA, a very small p results in suboptimal usage of data augmentation, while an excessively large p may decrease the learning effect on the raw samples. Empirically, $p = 0.5$ is a common mix probability. However, VG mix DA has a stricter demand on mix probability, as demonstrated in Table 3. A large $p = 0.5$ has a negative effect compared to the baseline. For optimal mix DA benefits, a relatively smaller p is required. We set p to 0.05 in our CMMix without otherwise specified.

Filter Samples by Loss. We have discussed the filtering-sample strategy in Sect. 2.3, and its experimental exploration is presented in Table 4. Our findings indicate that applying the *FilterMin* strategy further boosts VG accuracy. By enforcing networks to pay more attention to samples with higher losses, which are more challenging to represent, *FilterMin* serves as a type of regularization. On the other hand, *FilterMax* exacerbates the overfitting of networks to simpler samples, resulting in lower accuracy. We believe this filtering strategy has broader applicability beyond VG tasks. Supportively, the baseline model plus *FilterMin* yields better performance, as shown in the 4th row (Min) of Base compared to

Fig. 3. Visualization examples of *PasteBox+C2C* CMMix on RefCOCO. The three columns from left to right are the cut, pasted, and mixed (image, text) pairs, respectively.

the 3rd row (-) in Table 4. For fairness, we applied the filtering operation to only 5% of the samples, similar to our mix probability $p = 0.05$ for CMMix.

More Design Details. We present additional design details of CMMix in this section. In the *CutBox* mode, the mixed image still contains two patches that correspond to the output bounding box, *i.e.*, $\boldsymbol{B}_1$ and $\boldsymbol{B}_2$. The *W2W* mode only matches $\boldsymbol{B}_1$, ignoring the second expression $\boldsymbol{T}_2$. To reduce the impact of unlabeled $\boldsymbol{B}_2$, we employ a cut-out strategy where we set the pixels in $\boldsymbol{B}_2$ to 0. However this improvement does not bring significant changes. We also try to cut more area than the exact area of $\boldsymbol{B}_1$ to make the boundary of the regressed box smoother with an overlapping strategy, which also doesn't take much effect.

As for the text modality, we utilize NLTK to locate the first noun as the central word. However, due to reasons like data annotation errors, the location of the noun may not always be accurate. For instance, some RefCOCO expressions include only a directional word like "left", resulting in "left" being located as the central noun by NLTK. This systematic bias issue can be addressed by further refining the data annotation process.

In addition, we discuss the mixed level of texts, which can be input-level or embedding-level. MixGen [7] reports that "both input-level and embedding-level MixGen perform better than the baseline, and input-level performs consistently better than embedding-level". Our experimental findings also support this analysis, and mixing at the input-level is more straightforward to implement.

Last but not least, we present an illustration of the time efficiency of CMMix. DA is efficient enough compared to the training of baseline model. Word splicing of texts and cutting and pasting of images can be completed rapidly using only the CPU, of which the time consumption is negligible compared to neural network training on the GPU. Training 150 epochs on RefCOCOg using 4 NVIDIA A100 GPUs, we observed that the training time of baseline model is 11 h 5 min, while the training time is 11 h 10 min for MixGen and 11 h 12 min for CMMix, respectively.

3.4 Visual Analysis

We display several visualization examples of VG in Fig. 3, utilizing the *PasteBox+C2C* mode of CMMix. The *C2C* mode enriches the context to enhance the effect of DA, while *C2C* matched with *PasteBox* is more resistant to modified words, especially directional words like "right" and "top". RefCOCO and ReferItGame data sets have numerous directional words, making *PasteBox+C2C* a suitable option for them. However, RefCOCO+ expressions significantly weaken interactions with surrounding objects, so **2W* mode is more suitable for them. In the *PasteBox* mode, cut patches need to be resized, and if the size difference between $\boldsymbol{B}_1$ and $\boldsymbol{B}_2$ is substantial, the *CutBox* mode produces better visual results.

4 Conclusion

We introduce CMMix, a mix data augmentation method for visual grounding in downstream cross-modal tasks, which is the first work to address the contextual correspondence problem in cross-modal DA. By simultaneously mixing images and texts, we generate a more extensive set of meaningful (image, text) pairs. In addition, we recommend two mix modes, *CutBox+W2W* and *PasteBox+C2C*. Furthermore, we propose a universal strategy to filter samples by loss to improve the training process. The effectiveness of our CMMix approach is demonstrated through quantitative experiments. In future work, it's worth applying mix data augmentation to other cross-modal tasks, such as visual entailment, and developing more general paradigms.

References

1. Bird, S., Klein, E., Loper, E.: Natural Language Processing with Python: Analyzing Text with the Natural Language Toolkit. O'Reilly Media Inc., Sebastopol (2009)
2. Carion, N., Massa, F., Synnaeve, G., Usunier, N., Kirillov, A., Zagoruyko, S.: End-to-end object detection with transformers. In: Vedaldi, A., Bischof, H., Brox, T., Frahm, J.-M. (eds.) ECCV 2020. LNCS, vol. 12346, pp. 213–229. Springer, Cham (2020). https://doi.org/10.1007/978-3-030-58452-8_13
3. Deng, J., Yang, Z., Chen, T., Zhou, W., Li, H.: Transvg: end-to-end visual grounding with transformers. In: IEEE/CVF International Conference on Computer Vision, pp. 1769–1779 (2021)

4. Devlin, J., Chang, M.W., Lee, K., Toutanova, K.: Bert: pre-training of deep bidirectional transformers for language understanding. In: Conference of the North American Chapter of the Association for Computational Linguistics: Human Language Technologies, pp. 4171–4186 (2019)
5. Escalante, H.J., et al.: The segmented and annotated IAPR TC-12 benchmark. Comput. Vis. Image Underst. **114**(4), 419–428 (2010)
6. Guo, H., Mao, Y., Zhang, R.: Augmenting data with mixup for sentence classification: an empirical study. arXiv preprint arXiv:1905.08941 (2019)
7. Hao, X., et al.: Mixgen: a new multi-modal data augmentation. In: IEEE/CVF Winter Conference on Applications of Computer Vision, pp. 379–389 (2023)
8. He, K., Zhang, X., Ren, S., Jian, S.: Deep residual learning for image recognition. In: IEEE Conference on Computer Vision and Pattern Recognition (2016)
9. Hermann, K.M., et al.: Teaching machines to read and comprehend. In: Advances in Neural Information Processing Systems, vol. 28 (2015)
10. Karpathy, A., Joulin, A., Fei-Fei, L.F.: Deep fragment embeddings for bidirectional image sentence mapping. In: Advances in Neural Information Processing Systems, vol. 27 (2014)
11. Kazemzadeh, S., Ordonez, V., Matten, M., Berg, T.: Referitgame: referring to objects in photographs of natural scenes. In: Conference on Empirical Methods in Natural Language Processing, pp. 787–798 (2014)
12. Li, X., Zhang, X., Cai, Z., Ma, J.: On wine label image data augmentation through viewpoint based transformation. J. Signal Process. **38**(1), 1–8 (2022)
13. Mao, J., Huang, J., Toshev, A., Camburu, O., Yuille, A.L., Murphy, K.: Generation and comprehension of unambiguous object descriptions. In: IEEE Conference on Computer Vision and Pattern Recognition, pp. 11–20 (2016)
14. Nagaraja, V.K., Morariu, V.I., Davis, L.S.: Modeling context between objects for referring expression understanding. In: Leibe, B., Matas, J., Sebe, N., Welling, M. (eds.) ECCV 2016. LNCS, vol. 9908, pp. 792–807. Springer, Cham (2016). https://doi.org/10.1007/978-3-319-46493-0_48
15. Sadhu, A.: Awesome visual grounding (2022). https://github.com/TheShadow29/awesome-grounding
16. Uddin, A.S., Monira, M.S., Shin, W., Chung, T., Bae, S.H.: Saliencymix: a saliency guided data augmentation strategy for better regularization. In: International Conference on Learning Representations (2020)
17. Vaswani, A., et al.: Attention is all you need. In: Advances in Neural Information Processing Systems, vol. 30 (2017)
18. Vinyals, O., Toshev, A., Bengio, S., Erhan, D.: Show and tell: a neural image caption generator. In: IEEE Conference on Computer Vision and Pattern Recognition, pp. 3156–3164 (2015)
19. Wang, T., et al.: Vlmixer: unpaired vision-language pre-training via cross-modal cutmix. In: International Conference on Machine Learning, pp. 22680–22690. PMLR (2022)
20. Yu, L., Poirson, P., Yang, S., Berg, A.C., Berg, T.L.: Modeling context in referring expressions. In: Leibe, B., Matas, J., Sebe, N., Welling, M. (eds.) ECCV 2016. LNCS, vol. 9906, pp. 69–85. Springer, Cham (2016). https://doi.org/10.1007/978-3-319-46475-6_5
21. Yun, S., Han, D., Oh, S.J., Chun, S., Choe, J., Yoo, Y.: Cutmix: regularization strategy to train strong classifiers with localizable features. In: IEEE International Conference on Computer Vision, pp. 6023–6032 (2019)
22. Zhang, H., Cisse, M., Dauphin, Y.N., Lopez-Paz, D.: Mixup: beyond empirical risk minimization. In: International Conference on Learning Representations (2018)

A Relation-Oriented Approach for Complex Entity Relation Extraction

Xinliang Liu[1,2,3] and Mengqi Zhang[1(✉)]

[1] College of Information and Electrical Engineering, China Agricultural University, Beijing 100083, China
771093379@qq.com

[2] Key Laboratory of Agricultural Informatization Standardization, Ministry of Agriculture and Rural Affairs, Beijing 100083, China

[3] National Engineering Research Center for Agri-Product Quality Traceability, Beijing Technology and Business University, Beijing 100048, China

Abstract. Entity relation extraction targets the extraction of structured triples from unstructured text, and is the start of the entire knowledge graph lifecycle. In recent advances, Machine Reading Comprehension (MRC) based approaches provide new paradigms for entity relationship extraction and achieve the state-of-the-art performance. Aiming at the features of nested entities, overlapping relationships and distant entities in recipe texts, this study proposes a relation-oriented approach for complex entity relation extraction. This approach addresses the entity redundancy problem caused by the traditional pipeline models which are entity-first methods. By predicting the starting and ending of entities, it solves the problem of nested entities that cannot be identified by traditional sequence labeling methods. Finally, the error propagation issue is mitigated by the triple determination module. We conduct extensive experiments on multi-datasets in both English and Chinese, and the experimental results show that our method significantly outperforms the baseline model in terms of both precision, recall and micro-F1 score.

Keywords: Machine Reading Comprehension, · Named Entity Recognition · Relation Extraction · Pipeline Approach · Question Answering

1 Introduction

Entity-relation extraction has received wide attention in both academia and industry [1–3], which refers to extract structured relational triples information from natural language text. A relational triple describe entities and the correspondence between them, formally represented as (head-entity type, relation, tail-entity type). In Fig. 1, given a sentence '干煸四季豆是一道特色传统菜肴, 属于川菜菜系, 主要制作材料有四季豆, 猪肉、红干辣椒等' (Dry-fried string beans is a dish belonging to the Sichuan cuisine, the main ingredients are string beans, pork, red dried chili and so on.), entity and relation extraction aim to obtain triples (干煸四季豆, 属于, 川菜) (Dry-fried string beans, Belong_To,

B. Luo et al. (Eds.): ICONIP 2023, CCIS 1966, pp. 483–498, 2024.
https://doi.org/10.1007/978-981-99-8148-9_38

Sichuan cuisine), (干煸四季豆, 原料是, 四季豆) (Dry-fried string beans, Ingredient_Is, string beans), (干煸四季豆, 原料是, 猪肉) (Dry-fried string beans, Ingredient_Is, pork), (干煸四季豆, 原料是, 红干辣椒) (Dry-fried string beans, Ingredient_Is, red dried chili).

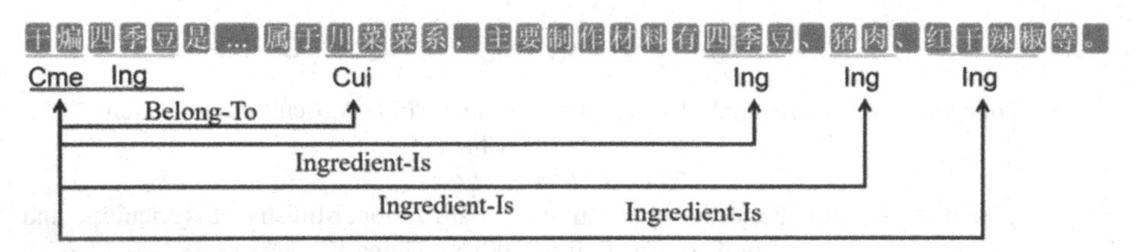

Fig. 1. An illustration of entity-relation extraction. The Cme, Ing, and Cui are short for cooking method, ingredient, and cuisine, separately.

Named entity recognition [4] and relation extraction [5] are subtasks of entity-relation extraction. Named entity recognition, also known as entity extraction, and refers to identify a text span such as "干煸四季豆" and the entity type it belongs to. The type is predefined based on the text corpus features and extraction goals. Relation extraction aim to discern the relation between the head-entity and tail-entity. Traditional methods are to annotate the text manually, which are time-consuming and laborious [6]. With its strong representation of data through multi-layer processors, deep learning is a milestone both theoretically and technologically [7]. Up to now, the transformative feature of deep learning techniques have led to significant improvements in the efficiency of entity-relation extraction. The mainstream methods are broadly categorized as pipelined models and joint learning models. Pipelined models first identifying entities by tagging models, and then identifying the relation between each pair of head-entity and tail-entity. Pipelined models can flexibly integrate entity extraction models and relation extraction models. However, due to the errors in the former model cannot be corrected in the latter model, such models suffer from the problem of error propagation. To alleviate the error propagation problem of pipelined models, joint learning models use end-to-end approach to make extraction in a unified structural framework by constraints or parameters sharing.

As shown in Fig. 1, there are three distinctive features: (1) the tail-entity 四季豆 is part of the head-entity 干煸四季豆. We define it as nested entity. (2) In the text, (干煸四季豆, 属于, 川菜) is a pair and (干煸四季豆, 原料是, 四季豆) is a pair, and those relation spans have overlaps. We define it as relation overlapping. (3) The entity 干煸四季豆 and 红干辣椒 are too far away. The entity-relation extraction becomes complex due to these feature and it is hard for the model to capture all the lexical, semantic and syntactic cues.

In this paper, we propose a new architecture for entity relation extraction using a new extraction paradigm which is based on a machine reading comprehension (MRC) framework. The core idea of MRC-based entity-relation extraction model is to transform the task into a multi-turn question answering (QA) task by constructing a fixed query for each entity and relation type in the dataset. For example, a query "find the process of making a certain dish in the text" is constructed for the entity type "process".

Then, the query and the text to be recognized are used as input to the MRC model, and the output answer string is the entity to be recognized in the text.

The main contributions of this paper are summarized as follows.

- We propose a MRC-IE model, which transform entity-relation extraction task into multi-turn QA task. It fully considers the importance of entity's category and triple pattern and encode the important prior knowledge in the question sentences through the question and answer format.
- A relation-first pipeline model address the entity redundancy problem caused by traditional entity-first extraction order. Using two binary classifiers to identify the starting and ending positions of entities respectively to tackle the entity nested problems. A triples determination module alleviate the error accumulation problem inherent in the pipelined model to a certain extent.
- In this paper, we construct an extraction dataset named Recipes, and the training data are in the form of <question, sentence>. Our MRC-IE model achieves optimal performance on the Recipes dataset.

The rest of this article is organized as follows. Section 2 introduces the related works on entity-relation extraction. Section 3 describes the system model in detail. We conduct an extensive experiment in Sect. 4 to elaborate our proposed model. Section 5 concludes this paper.

2 Related Work

Traditional approaches to entity relation extraction are rule-based [8, 9], feature-based [10] and unsupervised learning [11]. Deep Learning-based (DL-based) approaches have achieved state-of-the-art (SOTA) results, which is beneficial in discovering hidden features. In this work, we review three lines of DL-based to entity relation extractions: flat, nested and MRC-based.

2.1 Entity Relation Extraction

Majority approaches tackle entity relation extraction as a sequence labeling task using conditional random fields (CRF) with begin, inside, outside (BIO) scheme or its variant as a backbone. Santos and Guimaraes [12] used character embedding to augmented CNN-CRF structure which is first presented by Collobert et al. [13]. Ma et al. [14] introduced a LSTM-CNN-CRF architecture without extra feature engineering or data processing. Zheng et al. [15] proposed an end-to-end model with tagging scheme for joint extraction of entity and relation. Cao et al. [16] adopted a Bi-LSTM-CRF model to extract information from biomedical context based on the BIOES (Begin, Inside, Outside, End, Single) scheme with an additional M tag. With the development of large-scale language model pre-training methods, this task has yielded SOTA performance. Wei et al. [17] presented a novel cascade binary tagging model based on a Transformer encoder. Sui et al. [18] formulated entity relation extraction as a set prediction problem, which used the BERT model as the encoder.

2.2 Nested Entity Relation Extraction

The traditional sequence labeling methods assign only one tag to each token, which cannot solve the nested entity recognition problem. To address this issue, some previous work investigates span-based methods. Inspired by coreference resolution [19], Dixit et al. [20] first used span-based method to model all span to address nested entity relation extraction. DyGIE [21] captured the interaction of spans by LSTM encoding, while DyGIE++ [22] captured the semantic information with BERT-based representations. SpERT [23] was a simple but effective model for this task as it is a light-weight reasoning on BERT embeddings. More recently, SpIE [24] was proposed for complex information extraction, which considered the impact of entity type to relation classification.

2.3 MRC-Based Entity Relation Extraction

Recently, MRC-based approaches have been proposed to address NLP (Natural Language Processing) tasks. MRC-based approaches are to extract the span from a specific text given queries. Further, the span extraction includes predicting the starting and ending. Levy et al. [25] first reduced relation extraction to answering reading comprehension questions and generalized to unseen relations. Li et al. [26] used the MRC-based approach to cast entity relation extraction as a multi-turn question answering task, and further generalized this approach to a unified framework for named entity recognition [27]. The above approaches attempt to extract the answer spans through effective questions, while ignoring entity redundancy problem and error propagation issue. Thus our proposed model to address these shortcomings.

3 Method

In this paper, a novel model is proposed to deal with the entity-relation task in the field of recipes. Different with traditional pipeline approach of entity recognition followed by relation classification, we proposed a MRC-based framework, which fully considers the importance of the predefined type information by defining questioning templates to first fix the entity type and relation type, and then extract the head and tail entities among them. In addition, unlike using a multi-label classifier to map the contextual representation to the label space, this study transforms the information extraction task into a binary classification task based on machine reading comprehension. The method effectively alleviates the problems of entity redundancy and inefficient extraction of overlapping entities and triples containing multiple relations in previous research work. The method mainly contains three main modules, which are the relation recognition module, the entity recognition module, and the triples determination module. The overall framework of the model is shown in Fig. 2.

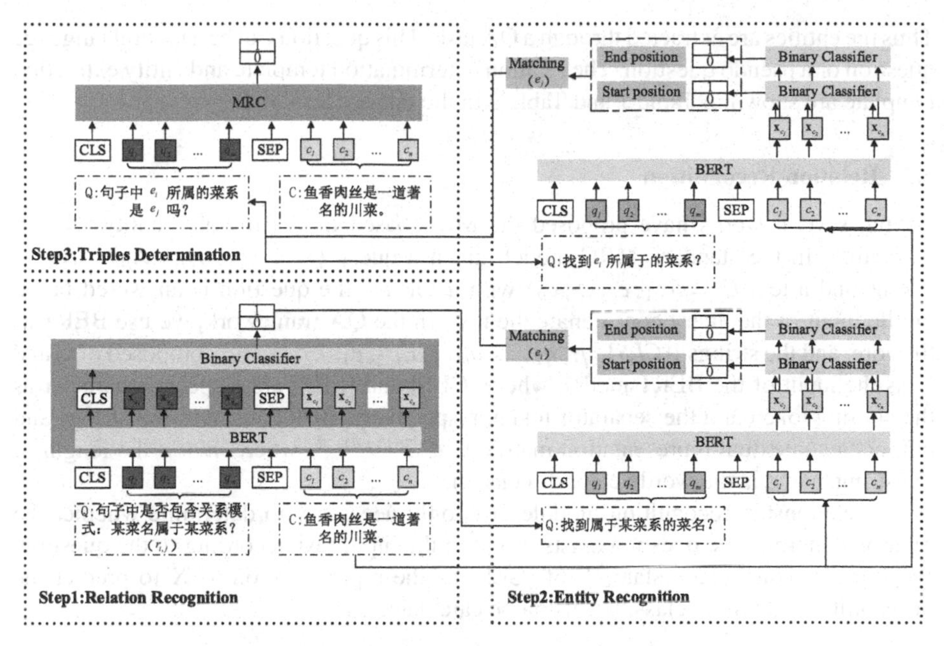

Fig. 2. The overall architecture of the MRC-IE model

3.1 Problem Definition

The purpose of this paper is to perform information extraction in the recipe domain. Given a natural language input $C = \{c_1, c_2, ..., c_n\}$ with n tokens, we aim to find an entities set $E = \{e_1, e_2, ..., e_n\}$ and assign the corresponding entity label set $Y = \{y_1, y_2, ..., y_n\}$ to E, where Y is predefined. And then determine the relation $r_{i,j}$ between each entity pair $\{e_i, e_j\}$.

3.2 Query Generation

In our model, firstly, the pre-defined relation type are transformed into fixed relation patterns by answering questions such as "Does the sentence contain the relation pattern: does the name of a dish belong to a certain cuisine?" The presence of the pre-defined relation pattern is determined by a binary classifier. Then, the head entities e_i and e_j are obtained by asking questions with the relation $r_{i,j}$ and the entity type as parameters.

Define Y and R as predefined sets of entity types and relations, respectively. For each given triple (head-entity, relation, tail-entity) there is a corresponding pattern (head-entity type, relation, tail-entity type), and define this corresponding pattern as $p = (et_i, r_{i,j}, et_j)$, where $et_i, et_j \in \varepsilon, r_{i,j} \in \mathcal{R}, p \in P$, ε is a set of predefined entity categories,P is the set of p. The corresponding schema exists in every natural language sentence containing a triple that undergoes information extraction.

During information extraction, each pattern **h** can generate a fixed question template to determine whether there is a triple or not. In addition, by combining the relation-specific templates with the extracted entity types, a fixed question template is generated.

Thus the entities are extracted through a QA task. This question can be a natural language question or a pseudo question. The relation determination template and entity extraction template are shown in Table 1 and Table 2 in the dataset Sect. 4.1.

3.3 Relation Recognition

By far, several works have proposed the MRC-based model to solve entity-relation extraction. In the standard MRC model, given a query $Q = (q_1, q_2, ..., q_m)$ with m tokens and a text $C = (c_1, c_2, ..., c_n)$ with n tokens, the question is answered based on the input of the query concatenate the text. In the QA framework, we use BERT as the core, and the strings $\{[CLS], q_1, q_2, ..., q_m, [SEP], c_1, c_2, ..., c_n\}$ composed of C and Q as the input of the BERT model, where [CLS] and [SEP] are a special character as the starting token and the separator token, respectively. BERT receives the strings and outputs a contextual representation matrix $X \in \mathbb{R}^{(m+n)\times d}$, where $m + n$ the length of the string, and d is the word vector dimension.

In relationship recognition module, it is only necessary to determine whether the relation pattern in the question exists or not in the input text according to the question. We use the final hidden state H_i of [CLS] as the representation of X to predict the probability by a binary classifier. It can be calculated as

$$y_i = \sigma(\mathbf{w}_i^T \cdot \mathbf{h}_i + \mathbf{b}_i) \tag{1}$$

where $\sigma(\bullet)$ is a sigmoid function, $\mathbf{w}_i$ is a trainable parameter vector, $\mathbf{w}_i \in \mathbb{R}^{d_h \times 1}$, d_h is the dimension of the last hidden layer of BERT, and $\mathbf{b}_i$ is the bias. Given a confidence threshold $\alpha \in [0, 1][0, 1]$, $y_i > \alpha$ indicates that the text C contains pattern p_i, and transforms it to the downstream entity recognition module. A binary cross-entropy loss function is used to optimize the module.

3.4 Entity Recognition

According to the relation recognition module, there is pattern p contained in C. In this module, the query related with entity recognition is constructed guided by the pattern p. The entity recognition consists of two steps: first, using the query to extract the head entity; second, the other new query is generated according to the head entity and the relational pattern to extract the tail entity.

Head Entity Identification: Conventional approaches treat the information extraction task as a sequential labeling task, in which a label is assigned to each input by the BIO labeling rule or its variants. In this paper, two bi-classifiers are used for entity recognition, one is used to predict whether the token is the start index of an entity, the other is used to predict whether the token is the end index. The two bi-classifiers make it possible to output multiple start indexes and multiple end indexes for a given text C and question q. The entity recognition module matches the start index and end index by a matching strategy to generate the target entity.

Similar to the relation recognition module, the given new string $\{[CLS], q_1, q_2, ..., q_m, [SEP], c_1, c_2, ..., c_n\}$ is encoded using the BERT encoder to generate the encoding matrix $X \in \mathbb{R}^{n\times d}$, d is the vector dimension of the last layer of BERT.

Since entity recognition does not require to recognizing the query token, we drop the query representation generated from the BERT.

Given the output encoding matrix X, the model first predicts the probability that each token is the starting index.

$$p(start_i|c_i) = \sigma(\mathbf{w}_i^T \cdot \mathbf{h}_i + \mathbf{b}_i) \tag{2}$$

where $start_i$ is the starting token of the entity. $P_{start} \in \mathbb{R}^{n \times 2}$, given the query $p(start_i|c_i)$ is the probability distribution of each index being the start position of an entity.

The cross-entropy loss function is used to optimize the entity start position prediction module.

$$\ell(p(start_i|c_i), t) = -t \times \log(p(start_i|c_i)) - (1-t) \times \log(1 - p(start_i|c_i)) \tag{3}$$

where $t \in [0, 1]$ determines the token is the start index or not.

The end index prediction is same as the start index prediction, and slimily uses a binary classifier to predict the probability that each token is the end position of the entity. Unlike the start index prediction, the end index prediction yields a probability prediction matrix $P_{end} \in \mathbb{R}^{n \times 2}$.

Given text C, there will be multiple entities belonging to a certain type. This indicates that multiple start indexes and end indexes are generated in the entity prediction module. Due to the entity nested, the method of simply matching the start and end indexes by the proximity principle does not work. Therefore, we use argmax for each row of P_{start} and P_{end} respectively, and then uses the sigmoid function to make a determination.

Tail Entity Identification: After recognizing the head entity, it guides the construction of the tail entity recognition query and generates a new string $\{[CLS], q_1, q_2, ..., q_m, [SEP], c_1, c_2, ..., c_n\}$. The process is same to the head entity recognition except that the query guiding the tail entity recognition is different.

3.5 Triples Determination

The relation recognition and entity recognition step may generate incorrect candidate answers, and the purpose of the triples determination module is to eliminate the incorrect answers accumulated in the first two steps. Triples determination is similar to relation recognition, except that these two modules have different training data. We form a set of entities from all the correct entities, and after the de-duplication operation, we define one set as the head entity set H and the other as the tail entity set T. The Cartesian product between the head entity, the relation and the tail entity is calculated to generate the set of triples D. The set D^- is obtained by filtering out the triples in D that have the same head and tail entities. We denote the set of triples obtained from the relation identification and entity identification as F. We compare F with D^-, positive samples means the triples appear in both F and D^-, and negative samples means the triples appear only in F but not in D^-. We conduct a new query for f_i, where $f_i \in F$ and splice the new query with original text generating a new string $\{[CLS], q_1, q_2, ..., q_m, [SEP], c_1, c_2, ..., c_n\}$. Then perform the same operation as the relation recognition to determine whether the triple is correct or not by the binary classifier.

4 Experiment

4.1 Dataset

To validate the effectiveness of MRC-IE model, we conducted experiments on two widely used standard English datasets, ACE05 [28] and CoNLL03 [29], a Chinese dataset, OntoNotes4.0 [30], and a self-constructed dataset Recipe, respectively.

ACE05 is a multi-domain and multi-lingual dataset that defines seven entity types, including Person (Per), Organization (Org), Geographic Entity (Gpe), Location (LOC), Facility (Fac), Weapon (Wea), and Vehicle (Veh). For each pair of entities, there are six relation types, including Person-Society (Per-Soc), Agent-Art (Art), Geographic Entity-Association (Gpe-Aff), Physically (Phys), Whole-Part (Par-Who), and Employee-Organization (Emp-Org). These datasets were well labeled, and we divided the dataset into training data, validation data, and test data according to the data processing method of Lin et al. [31].

CoNLL03 is an English news dataset, defining four entity types, including Loc, Org, Peo, and Other. For each pair of entities there are five relation types, including Located-In, Work-For, OrgBased-In, Live-In, and Kill. In this study, the CoNLL03 dataset was divided into training, validation, and test data according to the data processing method of Miwa and Sasaki [32].

OntoNotes4.0 is a Chinese dataset from news texts, and there are 18 named entity types. In this paper, we only choose 4 common entity types among them, which are Loc, Org, Peo and Gpe.

In this paper, we use 3723 Chinese recipe documents as the original data and split these by sentence. This is a food safety-related complex entity-relation extraction dataset. It was annotated by two supervisors and 30 annotators. Each sentence is annotated back-to-back by two annotators, and in addition to the initial training of these annotators, a second annotation is performed by the supervisor when the annotated data are inconsistent during the annotation process.

Recipes were constructed for subsequent food sampling, which contains seven categories of entities and seven categories of relationships. The entity types include dish name (Dis), cuisine (Cui), ingredient (Ing), cooking method (Cme), nutrient (Nut), people (Peo) and efficacious (Eff). The relation types include belong to, ingredient is, cook by, contain, suitable for, efficacy is and overcome. Recipes dataset is divided into training, validation and testing sets according to the principle of no duplication between datasets. We list the question templates of the Recipes dataset in Table 1 and Table 2.

Detailed statistical information of the four datasets is shown in Table 3.

4.2 Baseline Model

In this paper, we selected some classical and SOTA models as baseline models and compared with the MRC-IE model on two language types, Chinese and English, respectively. The specific choices of baseline models are as follows:

For the English dataset, there are five baseline models in three categories:

Table 1. Templates of Recipes for relation recognition

Patter	Query	Answer
(Dis, Belong-To, Cui)	Does this context contains pattern: a dish belonged to one cuisine? 句子中是否存在三元组模式: 某菜名属于某菜系?	T/F
(Dis,Ingredient-Is,Ing)	Does this context contains pattern: a dish contains some ingredients? 句子中是否存在三元组模式: 某道菜包含某些原材料?	T/F
(Dis, Cook-By, Cme)	Does this context contains pattern: a dish cooked by a cooking method? 句子中是否存在三元组模式: 某道菜由某种工艺制作?	T/F
(Ing, Contain, Nut)	Does this context contains pattern: an ingredient contains some nutritionist? 句子中是否存在三元组模式: 某原材料包含某些营养物质?	T/F
(Ing,Suitable-For,Peo)	Does this context contains pattern: an ingredient suitable for some person? 句子中是否存在三元组模式: 某原材料适合某类人群?	T/F
(Ing, Efficacy-Is, Eff)	Does this context contains pattern: an ingredient has a certain efficacy? 句子中是否存在三元组模式: 某原材具有某种功效?	T/F
(Ing, Overcome, Ing)	Does this context contains pattern: two ingredients overcome with each other? 句子中是否存在三元组模式: 某两种原材相克?	T/F

Multi-turn QA model [26], which uses MRC as a framework to extract tail entities after extracting head entities; Patti model [33] is a pattern-first pipelined entity-relationship extraction model;

The tree-based LSTM model [34], ELMo labeling model [35] and BiLSTM-CNNs-CRF model [14] are the classical models for entity-relation extraction. Among them, the BiLSTM-CNNs-CRF model performs end-to-end entity-relation extraction based on adequate acquisition of word and lexical representations; the tree-based LSTM model fully acquires the dependencies between information; the ELMo tagging model jointly learns decaNLP (Natural Language Decathlon, a natural language decathlon) for all tasks efficiently improving model performance through multi-pointer decoders;

Table-filing model [32] converts the entity-relation extraction task into a table-filling task for joint modeling of entities and relations;

Table 2. Templates of Recipes for entity recognition

Patter	Query	Answer
(Dis, Belong-To, Cui)	Find the dish name which belongs to one cuisine 找到句子中属于某菜系的菜名。	e_i
	Find the cuisine that e_i belongs 找到句子中 e_i 所在的地方。	e_j
(Dis, Ingredient-Is, Ing)	Find the dish name which has some ingredients 找到句子中包含某原材料的菜名。	e_i
	Find the ingredient that e_i has 找到句子中 e_i 所包含的原材料。	e_j
(Dis, Cook-By, Cme)	Find the dish name which cooked by one cooking method 找到句子中使用某制作工艺的菜名。	e_i
	Find the cooking method that e_i cooked by 找到句子中 e_i 所用的烹饪工艺。	e_j
(Ing, Contain, Nut)	Find the ingredient which has some nutritionent 找到句子中包含营养物质的原材料。	e_i
	Find the nutritionent that e_i has 找到句子中 e_i 所包含的营养物质。	e_j
(Ing, Suitable-For, Peo)	Find the ingredient which suitable for some person 找到句子中适合某人群的原材料。	e_i
	Find the person that e_i suitable for 找到句子中 e_i 所适合的人群。	e_j
(Ing, Efficacy-Is, Eff)	Find the ingredient which has a certain efficacy 找到句子中包含某些功效的原材料。	e_i
	Find the efficacy that e_i has 找到句子中 e_i 所包含的功效。	e_j
(Eff, Overcome, Eff)	Find the efficacy which overcome with other efficacy 找到句子与某一原材料相克的原材料。	e_i
	Find the efficacy that has overcome relationship with e_i 找到句子与 e_i 相克的原材料。	e_j

PURE model [36] for the entity overlap problem based on a span representation using a pipelined approach for extraction;

For the Chinese dataset, we similarly set five baseline models in three categories:

BERT-MRC model [27] is also a MRC-based model and is the main comparative model in this paper.

Lattice-LSTM [37] encodes the input characters and all potential words matched with the lexicon. The tagging approach of the Bert-Tagger model [38] predicts the real values to solve the entity nesting problem. Glyce-BERT model [39] uses the glyph vectors of Chinese character representations to make full use of the glyph information.

Table 3. Statistical information of the datasets

Datasets	$\lvert\mathcal{E}\rvert$	$\lvert\mathfrak{R}\rvert$	Sentence		
			Train	Dev	Test
ACE05(English)	7	6	6,051	1,624	1,498
CoNLL03(English)	4	5	14,040	3,249	3,684
OntoNotes4(Chinese)	4		7,749	2,124	1,879
Recipes(Chinese)	7	7	8,732	2,318	2,334

SpIE model [24] is based on the span representation and makes full use of the category label information of entities for overlapping entity extraction and multi-relation classification.

4.3 Implementation Details

We adopt BERT-wwm-ext1 [40] as the core for our experiment. We employ the Adam Optimizer and a peak learning rate of 2e−5. The batch size is set as 32, dropout rate to 0.3. The processor was an Intel(R) Xeon(R) CPU E5–2603 v3 @ 1.60 GHz 1.60 GHz, experimented on Pytorch (python version 3.6, torch version 1.8.0) with GPU (model NIVIDA GeForce GTX 1060 6GB). For evaluation, we use precision (P), recall (R) and micro-F1 score (F1) as metrics in our experiments.

4.4 Result and Analysis

Table 4 shows the experimental results on the English dataset. For ACE05, compared with the SOTA model, the F1 values of the MRC-IE model on the two tasks are improved by 0.2% and 2.2%, respectively. For the CoNLL03 dataset, the F1 values of the MRC-IE model improved by 1.05% and 0.12%, respectively. Compared with the ordinary entities in the CoNLL03 dataset, the ACE05 dataset contains 22% of overlapping entities, and the greater performance improvement of the MRC-IE model on the ACE05 dataset illustrates that overlaying two binary classifiers on the MRC model to determine the starting and ending positions of entities separately can solve the recognition of nested entities. The improvement in relation recognition is higher than entity recognition on both English datasets because the MRC-IE model changes the traditional extraction order of the pipelined model in which entity recognition is performed first and then relationship recognition. The results show that relation-first approach can only alleviate but not eliminate the error propagation problem.

As shown in Table 5 the MRC-IE model achieves better or competitive results with the baseline model on the Chinese dataset. The Recipes dataset contains 29.45% of overlapping entities, and the experimental results similarly show that the MRC-IE model can effectively solve the overlapping entity problem. To verify the effectiveness of the relation-guided entity relationship extraction model for solving redundant entities, we selected four patterns for entity relationship extraction, including (Dis, Belong-To, Cui),

Table 4. Experiment results on English datasets

Datasets	Model	Entity recognition			Relation recognition		
		recall	precision	F1	recall	precision	F1
ACE05	Multi-turn QA	**84.7**	**84.9**	84.8	**64.8**	56.2	60.2
	Patti	70.25	72.81	71.51	61.35	**63.58**	62.45
	tree-based LSTM	82.9	83.9	83.4	57.2	54.0	55.6
	PURE	–	–	**89.7**	–	–	**69.0**
	MRC-IE	**89.6**	**90.2**	**89.9**	**70.3**	**72.1**	**71.2**
CoNLL03	Multi-turn QA	**89.0**	**86.6**	87.8	69.2	68.2	68.9
	Patti	77.26	80.15	78.67	70.92	**73.01**	**71.95**
	BiLSTM-CNNs-CRF	–	–	91.03	–	–	–
	ELMo	–	–	**92.22**	–	–	–
	Table-filing	81.2	80.2	80.7	**76.0**	50.9	61.0
	PURE	–	–	89.7	–	–	69.0
	MRC-IE	**91.6**	**93.1**	**92.34**	73.8	**72.3**	**73.0**

(Dis, Ingredient-Is, Ing), (Ing, Suitable-For, Peo), (Ing, Efficacy-Is, Eff). We treat (Dis, Cook-By, Cme), (Ing, Contain, Nut), and (Ing, Overcome, Ing) as noise. Considering the processed data as the Redundant-Recipes dataset. Compared with the Recipes dataset, there are 613, 156 and 203 redundant entities were added to the training set, validation set and test set, respectively. We conducted experiments on the Recipes dataset and the Redundant- Recipes dataset using the code provided in the original paper of the BERT-MRC model. The experimental results show that there is no significant difference on the Recipes and Redundant- Recipes datasets, but the F1 values of BERT-MRC decrease by 6.2% and 7.16%, respectively. This indicates that the MRC-IE model can effectively deal with the problem of entity redundancy, while BERT-MRC, which performs entity identification before relationship extraction, shows a degradation in model performance due to the influence of redundant entities.

The validity of the MRC-IE model has been demonstrated by comparing it with different models on different datasets. We further thoroughly analyze the effectiveness of the various components of MRC-IE model in three aspects: 1. Pre-trained language model vs. MRC model; 2. Entity first vs. relation first; and 3. The effect of different query styles.

There are two core aspects of the MRC-IE model. One is the use of a large-scale pre-trained language model and the other is encoding of prior knowledge in query by the MRC model. To verify the specific role of these two aspects in improving the performance of entity-relation extraction, we compare tree-based LSTM model with Multi-turn QA model. Tree-based LSTM model neither rely on large-scale pre-trained language models nor is a MRC-based model. Multi-turn QA rely on a large-scale pre-trained language models. Based on the use of pre-trained language models, the SpIE and MRC-IE

Table 5. Experiment results on Chinese datasets

Datasets	Model	Entity recognition			Relation recognition		
		recall	precision	F1	recall	precision	F1
OntoNotes4.0	BERT-MRC	**82.98**	81.25	**82.11**	–	–	–
	Lattice-LSTM	76.35	71.56	73.88	–	–	–
	Bert-Tagger	78.01	80.35	79.16	–	–	–
	Glyce-BERT	81.87	**81.40**	81.63	–	–	–
	MRC-IE	**83.1**	**82.63**	**82.86**	–	–	–
Recipes	SpIE	**97.28**	**96.52**	**96.90**	**96.72**	**97.21**	**96.96**
	BERT-MRC	96.23	95.84	96.03	94.57	95.32	94.94
	MRC-IE	96.78	**97.83**	**97.3**	**97.49**	**98.42**	**97.85**
Redundant-Recipes	BERT-MRC	87.63	92.15	89.83	81.26	86.47	87.78
	MRC-IE	**96.42**	**97.88**	**97.14**	**97.23**	**98.46**	**97.84**

models are using span-based representations and MRC architecture, respectively. As shown in Table 4 and 5, the MRC-based Multi-turn QA model significantly outperforms the tree-based LSTM model, and the MRC-IE model outperforms the SpIE for both entity recognition and relationship recognition tasks, which indicates that encoding prior knowledge plays an important role in entity relationship extraction. Compared with the MRC-IE model, the BERT-MRC model only does entity-relationship extraction and the MRC-IE outperforms the BERT-MRC model in entity recognition under the premise of extracting pre-defined relationships in text. The query used in the MRC-IE are generated from natural language templates, and can also be obtained by automatically generating pseudo-questions. We experimentally verify the effectiveness of these two question generation methods. The pseudo-questions of the triplet model are combinations of head entity types, relations, and tail entity types, e.g., name of dish, belong, cuisine. The entity pseudo-questions are generated based on different descriptions of entity types, e.g., for dish name the entity can be dish name and find dish name. The lack of auxiliary words containing entity structure information in pseudo-questions makes it difficult for the model to understand the pseudo-questions, while natural language can provide more semantic information makes the model recognition using natural language better.

5 Conclusion

Extracting entity and relations from recipe text is the basis of multi-tasking for the food safety domain. In this paper, we propose MRC-IE, a framework for entity-relation extraction based on machine reading comprehension, to solve entity nested and overlapping relation recognition. MRC-IE effectively transform unstructured recipe text to structured triadic information by generating question templates to achieve. We set up three types of baseline models MRC-based, traditional entity-relation extraction models and span-based models respectively and conduct extensive experiments on the English datasets ACE05 and CoNLL03, the Chinese dataset OntoNotes4.0 and the self-constructed dataset Recipes. Experimental results against the state-of-the-arts sufficiently demonstrate the effectiveness of MRC-IE model.

References

1. Zhao, X., Wu, W., Tao, C., et al.: Low-resource knowledge-grounded dialogue generation. In: International Conference on Learning Representations (2022)
2. Wang, P., Jiang, H., Xu, J., et al.: Knowledge graph construction and applications for web search and beyond. Data Intell. **1**(4), 333–349 (2019)
3. Deng, S., Zhang, N., Zhang, W., et al.: Knowledge-driven stock trend prediction and explanation via temporal convolutional network. In: Companion Proceedings of The 2019 World Wide Web Conference, pp. 678–685. ACM, San Francisco (2019)
4. Nadeau, D., Sekine, S.: A survey of named entity recognition and classification. Lingvisticae Investigationes **30**(1), 3–26 (2007)
5. Bach, N., Badaskar, S.: A review of relation extraction (2007)
6. Lin, Y., Shen, S., Liu, Z., et al.: Neural relation extraction with selective attention over instances. In: Proceedings of the 54th Annual Meeting of the Association for Computational Linguistics, vol. 1: Long Papers, pp. 2121–2133. Association for Computational Linguistics, Berlin (2016)
7. Nickel, M., Murphy, K., Tresp, V., et al.: A review of relational machine learning for knowledge graphs. Proc. IEEE **104**(1), 11–33 (2016)
8. Mikheev, A., Moens, M., Grover, C.: Named entity recognition without gazetteers. In: Ninth Conference of the European Chapter of the Association for Computational Linguistics, pp. 1–8. Association for Computational Linguistics, Bergen (1999)
9. Sekine, S., Nobata, C.: Definition, dictionaries and tagger for extended named entity hierarchy. In: Proceedings of the Fourth International Conference on Language Resources and Evaluation (LREC 2004). European Language Resources Association (ELRA), Lisbon (2004)
10. McNamee, P., Mayfield, J.: Entity extraction without language-specific resources. In: Proceedings of the 6th Conference on Natural Language Learning, vol. 20, pp. 1–4. Association for Computational Linguistics (2002)
11. Collins, M., Singer, Y.: Unsupervised models for named entity classification. In: 1999 Joint SIGDAT Conference on Empirical Methods in Natural Language Processing and Very Large Corpora (1999)
12. dos Santos, C., Guimarães, V.: Boosting named entity recognition with neural character embeddings. In: Proceedings of the Fifth Named Entity Workshop, pp. 25–33. Association for Computational Linguistics, Beijing (2015)
13. Collobert, R., Weston, J., Bottou, L., et al.: Natural language processing (almost) from scratch. J. Mach. Learn. Res. **12**, 2493–2537 (2011)
14. Ma, X., Hovy, E.: End-to-end sequence labeling via bi-directional LSTM-CNNs-CRF. In: Proceedings of the 54th Annual Meeting of the Association for Computational Linguistics, vol. 1: Long Papers, pp. 1064–1074. Association for Computational Linguistics, Berlin (2016)
15. Zheng, S., Wang, F., Bao, H., et al.: Joint extraction of entities and relations based on a novel tagging scheme. In: Proceedings of the 55th Annual Meeting of the Association for Computational Linguistics, vol. 1: Long Papers, pp. 1227–1236. Association for Computational Linguistics, Vancouver (2017)
16. Yang, Z.H., Cao, M., Luo, L.: Joint drug entity and relations extraction based on neural network. J. Comput. Res. Dev. **56**, 1432–1440 (2019)

17. Wei, Z., Su, J., Wang, Y., et al.: A novel cascade binary tagging framework for relational triple extraction. In: Proceedings of the 58th Annual Meeting of the Association for Computational Linguistics, pp. 1476–1488. Association for Computational Linguistics (2020)
18. Sui, D., Zeng, X., Chen, Y., et al.: Joint entity and relation extraction with set prediction networks. IEEE Trans. Neural Netw. Learn. Syst. 1–12 (2023)
19. Lee, K., He, L., Lewis, M., et al.: End-to-end neural coreference resolution. In: Proceedings of the 2017 Conference on Empirical Methods in Natural Language Processing, pp. 188–197. Association for Computational Linguistics, Copenhagen (2017)
20. Dixit, K., Al-Onaizan, Y.: Span-level model for relation extraction. In: Proceedings of the 57th Annual Meeting of the Association for Computational Linguistics, pp. 5308–5314. Association for Computational Linguistics, Florence (2019)
21. Luan, Y., Wadden, D., He, L., et al.: A general framework for information extraction using dynamic span graphs. In: Proceedings of the 2019 Conference of the North, pp. 3036–3046. Association for Computational Linguistics, Minneapolis (2019)
22. Wadden, D., Wennberg, U., Luan, Y., et al.: Entity, relation, and event extraction with contextualized span representations. In: Proceedings of the 2019 Conference on Empirical Methods in Natural Language Processing and the 9th International Joint Conference on Natural Language Processing (EMNLP-IJCNLP), pp. 5783–5788. Association for Computational Linguistics, Hong Kong (2019)
23. Eberts, M., Ulges, A.: Span-based joint entity and relation extraction with transformer pre-training. In: 24th European Conference on Artificial Intelligence - ECAI 2020, Santiago de Compostela, Spain, pp. 2006–2013 (2020)
24. Zhang, M., Ma, L., Ren, Y., et al.: Span-based model for overlapping entity recognition and multi-relations classification in the food domain. Math. Biosci. Eng. **19**(5), 5134–5152 (2022)
25. Levy, O., Seo, M., Choi, E., et al.: Zero-shot relation extraction via reading comprehension. In: Proceedings of the 21st Conference on Computational Natural Language Learning (CoNLL 2017), pp. 333–342. Association for Computational Linguistics, Vancouver (2017)
26. Li, X., Yin, F., Sun, Z., et al.: Entity-relation extraction as multi-turn question answering. In: Proceedings of the 57th Annual Meeting of the Association for Computational Linguistics, pp. 1340–1350. Association for Computational Linguistics, Florence (2019)
27. Li, X., Feng, J., Meng, Y., et al.: A unified MRC framework for named entity recognition. In: Proceedings of the 58th Annual Meeting of the Association for Computational Linguistics, pp. 5849–5859. Association for Computational Linguistics (2020)
28. Doddington, G., Mitchell, A., Przybocki, M., et al.: The automatic content extraction (ACE) program – tasks, data, and evaluation. In: Proceedings of the Fourth International Conference on Language Resources and Evaluation (LREC 2004). European Language Resources Association (ELRA), Lisbon (2004)
29. Tjong, K., Sang, E.F., De Meulder, F.: Introduction to the CoNLL-2003 shared task: language-independent named entity recognition. In: Proceedings of the Seventh Conference on Natural Language Learning at HLT-NAACL 2003, pp. 142–147 (2003)
30. Che, W., Wang, M., Manning, C.D., et al.: Named entity recognition with bilingual constraints. In: Proceedings of the 2013 Conference of the North American Chapter of the Association for Computational Linguistics: Human Language Technologies, pp. 52–62. Association for Computational Linguistics, Atlanta (2013)
31. Lin, H., Lu, Y., Han, X., et al.: Sequence-to-nuggets: nested entity mention detection via anchor-region networks. In: Proceedings of the 57th Annual Meeting of the Association for Computational Linguistics, pp. 5182–5192. Association for Computational Linguistics, Florence (2019)
32. Miwa, M., Sasaki, Y.: Modeling joint entity and relation extraction with table representation. In: Proceedings of the 2014 Conference on Empirical Methods in Natural Language

Processing (EMNLP), pp. 1858–1869. Association for Computational Linguistics, Doha (2014)
33. Chen, Z., Guo, C.: A pattern-first pipeline approach for entity and relation extraction. Neurocomputing **494**, 182–191 (2022)
34. Miwa, M., Bansal, M.: End-to-end relation extraction using LSTMs on sequences and tree structures. In: Proceedings of the 54th Annual Meeting of the Association for Computational Linguistics, vol. 1: Long Papers, pp. 1105–1116. Association for Computational Linguistics, Berlin (2016)
35. Peters, M.E., Neumann, M., Iyyer, M., et al.: Deep contextualized word representations. In: Proceedings of the 2018 Conference of the North American Chapter of the Association for Computational Linguistics: Human Language Technologies, vol. 1 (Long Papers), pp. 2227–2237. Association for Computational Linguistics, New Orleans (2018)
36. Zhong, Z., Chen, D.: A frustratingly easy approach for entity and relation extraction. In: Proceedings of the 2021 Conference of the North American Chapter of the Association for Computational Linguistics: Human Language Technologies, pp. 50–61. Association for Computational Linguistics (2021)
37. Zhang, Y., Yang, J.: Chinese NER using lattice LSTM. In: Proceedings of the 56th Annual Meeting of the Association for Computational Linguistics, vol. 1: Long Papers, pp. 1554–1564. Association for Computational Linguistics, Melbourne (2018)
38. Devlin, J., Chang, M.W., Lee, K., et al.: BERT: pre-training of deep bidirectional transformers for language understanding. In: Proceedings of the 2019 Conference of the North American Chapter of the Association for Computational Linguistics: Human Language Technologies, vol. 1 (Long and Short Papers), pp. 4171–4186. Association for Computational Linguistics, Minneapolis (2019)
39. Meng, Y., Wu, W., Wang, F., et al.: Glyce: glyph-vectors for Chinese character representations. arXiv (2020). https://arxiv.org/abs/1901.10125. 22 Mar 2023
40. Cui, Y., Che, W., Liu, T., et al.: Pre-training with whole word masking for Chinese BERT. IEEE/ACM Trans. Audio Speech Lang. Process. **29**, 3504–3514 (2021)

A Revamped Sparse Index Tracker Leveraging K–Sparsity and Reduced Portfolio Reshuffling

Yiu Yu Chan and Chi-Sing Leung(✉)

Department of Electrical Engineering, City University of Hong Kong, Kowloon, Hong Kong
yiuyuchan3-c@my.cityu.edu.hk, eeleungc@cityu.edu.hk

Abstract. In financial engineering, sparse index tracking (SIT) serves as a specialized and cost-effective passive strategy that seeks to replicate a financial index using a representative subset of its constituents. However, many existing SIT algorithms have two imperfections: (1) they do not allow investors to explicitly control the number of assets held in the portfolio, and (2) these algorithms often result in excess purchasing and selling activities during the rebalancing process. To address these deficiencies, this paper first proposes a practical constrained optimization problem. Afterwards, the paper develops the corresponding algorithm, termed the index tracker with four portfolio constraints via projected gradient descent (IT4-PGD). With IT4-PGD, investors can freely define the settings of a portfolio, including the number of holding assets, the maximum holding position, and the maximum turnover ratio of each constituent. Simulation results using real-world data demonstrate that IT4-PGD outperforms existing methods by its lower magnitude of daily tracking error (MDTE) and lower accumulative turnover ratio (ATR).

Keywords: Sparse index tracking (SIT) · projected gradient descent (PGD) · ℓ_0-norm constraint · turnover constraint · portfolio optimization

1 Introduction

Over the past decade, machine learning (ML) techniques have emerged as significant tools for tackling investment optimization challenges [1,2]. In the realm of passive portfolio management, index tracking (IT) has garnered attention [3,4]. An IT model aims to construct a portfolio vector, denoted as $\boldsymbol{x} = [x_1, \cdots, x_n]^\top$, based on the historical data of a financial index and its n-constituents. Here, x_i represents the percentage of investment allocated to the i-th constituent. The objective of an ideal IT model is that the returns of the portfolio align closely with the fluctuations of the index.

However, the full replication approach has some critical issues. First, management and transaction costs should be high when the portfolio comprises a large number of assets. Second, a fully replicating portfolio may incorporate many

B. Luo et al. (Eds.): ICONIP 2023, CCIS 1966, pp. 499–512, 2024.
https://doi.org/10.1007/978-981-99-8148-9_39

small-cap and illiquid stocks. Further, we assume that a small bucket of potential stocks leads the market index over a period of time. Accordingly, we consider applying a sparsity constraint to the IT formulation. Sparse index tracking (SIT) is regarded as a special form of the sparse regression problem. Additionally, a portfolio should adhere to several constraints. We formulate it as an ℓ_0-sparse optimization problem, with the squared ℓ_2-norm objective loss. This problem is subject to a capital budget constraint $\mathbf{1}^\top \boldsymbol{x} = 1$; a long-only constraint that mandates $x_i \geq 0$ for all constituents; and an ℓ_0-norm constraint that $\|\boldsymbol{x}\|_0 \leq K$ enforces a limited number of nonzero components.

Since the ℓ_0-norm problem is NP-hard, a variety of surrogate and ℓ_0-norm approximation functions have been studied. Among these, the least absolute shrinkage and selection operator (LASSO) [5] is commonly used but it has some limitations, such as ineffectiveness in the non-short-sale (i.e., $x_i \geq 0$), selection bias for large coefficient values, and sensitivity to multicollinearity [6]. The adaptive elastic net (Aenet) [7] has been developed. Aenet introduced the reweighted ℓ_1-norm and ℓ_2-norm regularizers together as a remedy to surmount the defects of LASSO.

Besides, several nonconvex approximations of the ℓ_0-norm have been proposed. For instance, a half thresholding operator [8] is designed to promote a sparser solution. Alternatively, a continuous and differentiable function [9] has been devised to replace the ℓ_0-norm, and the majorization-minimization (MM) method is used to tackle this nonconvex problem. In the aforementioned works, direct control over the number of holding constituents is not allowed. Also, no effective way is provided to guide the control of the sparseness of the portfolio and the effectiveness of regularization parameters.

A recent method, NNOMP-PGD [10], allows investors to define the upper bound of the sparsity level. However, NNOMP-PGD and existing works may reconstruct a portfolio with high turnover during rebalancing. Since they prioritize single-period optimization rather than maintaining a stable portfolio over multiple periods, their models may result in myopic strategies that necessitate frequent and substantial portfolio reshuffling. Note that "portfolio reshuffling" refers to the changes in the investment allocation within a portfolio, interpreted as sale and purchase activities.

This paper first introduces a revamped constrained portfolio optimization problem featuring the ℓ_0-norm and turnover constraints. Then, we propose a two-step iterative projection algorithm to solve this problem. The proposed algorithm is named the index tracker with four portfolio constraints via projected gradient descent (IT4-PGD). IT4-PGD incorporates gradient descent and nonconvex projection. Furthermore, the projection optimality and the convergence behavior of the objective function value are guaranteed. In experiments with real market data, our IT4-PGD exhibits superior performance compared to the existing methods, in terms of both the magnitude of daily tracking error (MDTE) and the accumulative turnover ratio (ATR).

The rest of this paper is organized as follows: Sect. 2 reviews the relevant literature and various SIT formulations. In Sect. 3, we derive the IT4-PGD algorithm and present a convergence analysis. Section 4 discusses numerical simulation results, and Sect. 5 concludes this paper.

2 Background

Consider a financial index composed of n constituents. Let $\boldsymbol{\mathcal{A}} \in \mathbb{R}^{T \times n}$ be the daily return matrix of the n constituents over the past T trading days, and let $\boldsymbol{\mathcal{Y}} \in \mathbb{R}^{T}$ be the daily returns of the index during these days. The SIT with explicit sparsity control is given by

$$\min_{\boldsymbol{x}} \frac{1}{T}\|\boldsymbol{\mathcal{A}}\boldsymbol{x} - \boldsymbol{\mathcal{Y}}\|_2^2 \text{ s.t. } \mathbf{1}^\top \boldsymbol{x} = 1,\ x_i \geq 0\ \forall i,\ \|\boldsymbol{x}\|_0 \leq K, \tag{1}$$

where $\boldsymbol{x} = [x_1, \cdots, x_n]^\top$ is the portfolio vector, x_i is the investment percentage in the i-th constituent, $\mathbf{1}$ is a constant vector of ones, and K is the desired number of constituents involved in the tracking portfolio. The constraint of $x_i \geq 0$ is known as the long-only constraint, which is implemented to mitigate the hidden risk of short-sale activities that may be infinitely high. Also, some financial management firms disallow the short-sale policies.

The LASSO penalty is adopted as the main alternative of the ℓ_0-norm. The LASSO-based index tracking (LASSO) [5] is formulated as

$$\min_{\boldsymbol{x}} \frac{1}{T}\|\boldsymbol{\mathcal{A}}\boldsymbol{x} - \boldsymbol{\mathcal{Y}}\|_2^2 + \lambda\|\boldsymbol{x}\|_1 \text{ s.t. } \mathbf{1}^\top \boldsymbol{x} = 1. \tag{2}$$

However, in LASSO, we cannot introduce a long-only constraint when the ℓ_1-norm is used. Since "$x_i \geq 0$" and "$\mathbf{1}^\top \boldsymbol{x} = 1$" imply $\|\boldsymbol{x}\|_1$ is equal to 1. Undoubtedly, LASSO lacks direct and explicit control over portfolio sparsity. We must tune λ to achieve the desired sparsity level.

To circumvent the downsides of LASSO, the SIT with the Aenet penalty has been proposed [7]. It formulated as

$$\min_{\boldsymbol{x}} \frac{1}{T}\|\boldsymbol{\mathcal{A}}\boldsymbol{x} - \boldsymbol{\mathcal{Y}}\|_2^2 + \lambda_1\|\boldsymbol{w}^\top \boldsymbol{x}\|_1 + \lambda_2\|\boldsymbol{x}\|_2^2 + \lambda_3\|\boldsymbol{x} - \overline{\boldsymbol{x}}\|_1 \text{ s.t. } \mathbf{1}^\top \boldsymbol{x} = 1, \boldsymbol{x} \geq 0 \tag{3}$$

where $\boldsymbol{w}$ and $\overline{\boldsymbol{x}}$ are the adaptive weight vector and the previous portfolio vector, respectively. In (3), while the term $\lambda_2\|\boldsymbol{x}\|_2^2$ improves the stability of the solution paths, $\lambda_1\|\boldsymbol{w}^\top \boldsymbol{x}\|_1$ addresses the inconsistency problem. Additionally, $\lambda_3\|\boldsymbol{x} - \overline{\boldsymbol{x}}\|_1$ is to promote low turnover during rebalancing. Obviously, managing Aenet involves a complex interplay of three regularization parameters, and, similar to LASSO, it does not provide a way to directly and explicitly control sparsity.

In [8], the $\ell_{1/2}$-regularization term is exploited to promote a sparser solution via a half thresholding operator. Also, a continuous differentiable function $\varphi_{p,\gamma}(x_i)$ [9] is designed where $\varphi_{p,\gamma}(x_i) = \log(1 + |x_i|/p)/\log(1 + \gamma/p)$ for $0 < p \ll 1$ and $\gamma > 0$.

Upon reviewing the existing works, we highlight some areas for enhancement, including (i) the tuning of sparse regularization parameters to achieve the target K-sparsity is a non-trivial task, and (ii) the lack of turnover control in portfolio reshuffling leads to a sharp and continuous increase in the accumulative turnover ratio, posing a critical management issue.

3 SIT Based on IT4-PGD Algorithm

In this section, we refine the sparse index tracking design and propose a two-step iterative algorithm to tackle the proposed optimization problem.

3.1 Logarithmic Return

In the major existing works, they use the simple return values to construct the daily return matrix $\mathcal{A}$ of the n constituents and the daily return vector $\mathcal{Y}$ of a financial index. In finance, analysts prefer to use "log return" rather than "simple return" because log return values have three main characteristics, including log-normality, approximation of raw-log equality, and time-additivity [3]. In particular, when considering accumulated returns over successive periods, the estimation errors of log returns are independent and additive. This means that the tracking error from a prior period would not carry over to influence the estimation of the present period. The log daily return matrix $\boldsymbol{A}$ and log daily return vector $\boldsymbol{y}$ are given by

$$[\boldsymbol{A}]_{ji} = \log(1 + [\mathcal{A}]_{ji}) \text{ and } [\boldsymbol{y}]_j = \log(1 + [\mathcal{Y}]_j). \tag{4}$$

It is worth noting that we convert back to the original return values when evaluating the performance.

3.2 The Proposed Portfolio Design

Section 2 highlights two potential directions for SIT improvements. To handle (i), we adopt the K-sparsity constraint to allow the portfolio more controllable. To address (ii), we introduce the turnover constraint to avoid the excessive activities of purchases and sales during rebalancing.

Let $\overline{\boldsymbol{x}}$ be the current portfolio vector before rebalancing. The turnover constraint aims to restrict the total reshuffling between the estimated portfolio $\boldsymbol{x}$ and $\overline{\boldsymbol{x}}$ by a controlled parameter ϵ (i.e., $\|\boldsymbol{x}-\overline{\boldsymbol{x}}\|_1 \leq \epsilon$). With the ℓ_1-norm turnover constraint, solving the optimization problem becomes complicated. Instead, we can apply the turnover constraint on each constituent (i.e., $|x_i - \overline{x}_i| \leq \epsilon \ \forall i$). In addition, we also set an upper bound h on the holding percentage of each constituent.

The proposed formulation is as follows:

$$\begin{aligned} &\min_{\boldsymbol{x}} \ f(\boldsymbol{x}) := \|\boldsymbol{A}\boldsymbol{x} - \boldsymbol{y}\|_2^2 \\ &\text{s.t.} \ \ \mathbf{1}^\top \boldsymbol{x} = 1, \ \|\boldsymbol{x}\|_0 \leq K, \ 0 \leq x_i \leq h, \text{ and } |x_i - \overline{x}_i| \leq \epsilon \ \ \forall i. \end{aligned} \tag{5}$$

Now, we use the penalty method, moving the sum-to-one constraint into the objective function. With the quadratic penalty term, we finalize the problem as

$$\begin{aligned} &\min_{\boldsymbol{x}} \ f(\boldsymbol{x}) := \|\boldsymbol{A}\boldsymbol{x} - \boldsymbol{y}\|_2^2 + \frac{\rho}{2}\left(\mathbf{1}^\top \boldsymbol{x} - 1\right)^2 \\ &\text{s.t.} \ \ \|\boldsymbol{x}\|_0 \leq K, \text{ and } \max\{0, \overline{x}_i - \epsilon\} \leq x_i \leq \min\{h, \overline{x}_i + \epsilon\} \ \ \forall i, \end{aligned} \tag{6}$$

where $\rho > 0$ controls the severity of the penalty for violating the constraint. For a sufficiently large ρ, the resultant $\boldsymbol{x}$ is close to the optimal $\boldsymbol{x}^*$ of the original problem. As suggested by [10,11], the recommended value of ρ is $\rho \geq 10$.

In (6), direct and explicit control of the sparsity K is allowed, while ϵ limits the turnover ratio for each constituent. We also set a maximum position h for each holding constituent. Typically, h is between 0.08 and 0.15, and for ϵ, the practical turnover ratio is set between 0.001 and 0.05.

To simplify the expression, $\max\{0, \overline{x}_i - \epsilon\} \leq x_i \leq \min\{h, \overline{x}_i + \epsilon\}\ \forall i$, we denote ζ_i's and υ_i's as the corresponding lower and upper bound values, respectively. After that, we rewrite (6) as

$$\min_{\boldsymbol{x}} f(\boldsymbol{x}) := \|\boldsymbol{A}\boldsymbol{x} - \boldsymbol{y}\|_2^2 + \frac{\rho}{2}\left(\mathbf{1}^\top \boldsymbol{x} - 1\right)^2 \text{ s.t. } \|\boldsymbol{x}\|_0 \leq K, \text{ and } \zeta_i \leq x_i \leq \upsilon_i\ \forall i. \quad (7)$$

3.3 Algorithm for IT4-PGD

In this subsection, we develop our projected gradient based algorithm for solving (7). Our two-step iterative scheme is divided into gradient descent and nonconvex projection. It entails the following two steps:

$$\boldsymbol{z}^{(t+1)} = \boldsymbol{x}^{(t)} - \eta\nabla f(\boldsymbol{x}^{(t)}), \text{ and } \boldsymbol{x}^{(t+1)} = \mathcal{P}_{\mathcal{C}}(\boldsymbol{z}^{(t+1)}), \quad (8)$$

where $\nabla f(\boldsymbol{x}^{(t)}) = 2\boldsymbol{A}^\top(\boldsymbol{A}\boldsymbol{x}^{(t)} - \boldsymbol{y}) + \rho(\mathbf{1}^\top \boldsymbol{x}^{(t)} - 1)\mathbf{1}$ is the gradient vector, $\eta > 0$ is the step size. For the choice of a learning rate η, $\eta \in (0, \alpha/L]$ and $0 < \alpha \leq 1$ are commonly used [12]. Constant L is the largest eigenvalue of $\nabla^2 f(\boldsymbol{x}^{(t)})$. It should be noted that $\nabla^2 f(\boldsymbol{x}^{(t)}) = 2\boldsymbol{A}^\top\boldsymbol{A} + \rho\mathbf{1}\mathbf{1}^\top$ does not depend on the iteration number t.

To improve legibility, we denote $\boldsymbol{x} = \boldsymbol{x}^{(t+1)}$ and $\boldsymbol{z} = \boldsymbol{z}^{(t+1)}$. In (8), the solution of $\mathcal{P}_{\mathcal{C}}(\boldsymbol{z})$ is obtained through the below optimization problem:

$$\min_{\boldsymbol{x}} \|\boldsymbol{x} - \boldsymbol{z}\|_2^2 \text{ s.t. } \|\boldsymbol{x}\|_0 \leq K, \text{ and } \zeta_i \leq x_i \leq \upsilon_i\ \forall i. \quad (9)$$

The details of $\mathcal{P}_{\mathcal{C}}(\boldsymbol{z})$ will be discussed in Sect. 3.4. Additionally, the convergence analysis of IT4-PGD will be presented in Sect. 3.5.

3.4 Algorithm for Nonconvex Projection $\mathcal{P}_{\mathcal{C}}(\boldsymbol{z})$

This subsection covers the operations for nonconvex projection $\mathcal{P}_{\mathcal{C}}(\boldsymbol{z})$.

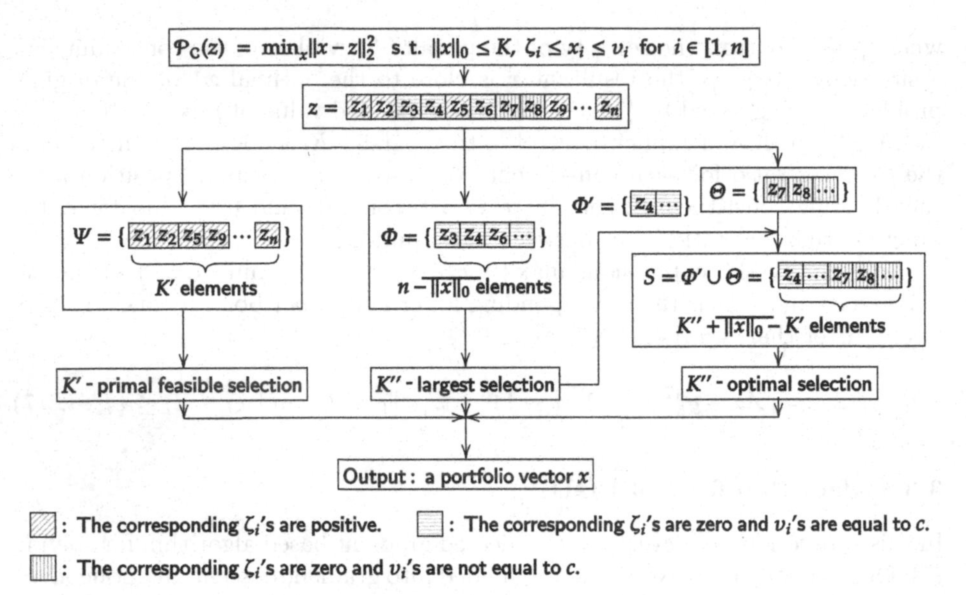

Fig. 1. The flowchart of the proposed projection $\mathcal{P}_C(z)$

Figure 1 provides an overview of the proposed projection. It consists of the following three parts: **(1) K'-primal feasible selection**, **(2) K''-largest selection**, and **(3) K''-optimal selection**.

Operation for (1) K'-primal feasible selection:
Let Ω be the index set for $\zeta_i > 0$, and Ψ be the set of elements: $\{z_i | i \in \Omega\}$. Since the constraint of $\zeta_i > 0$ depends on the nonzero elements of the previous portfolio $\overline{x}$ (i.e., $\overline{x}_i - \epsilon > 0$). Also, $\|\overline{x}\|_0 \le K$ holds, the number K' of ζ_i's with $\zeta_i > 0$ must be less than or equal to K (i.e., $K' \le K$). For each z_i in Ψ, its projected value of x_i is $x_i = \min\{\max\{\zeta_i, z_i\}, \upsilon_i\}$.

The rationale of selecting $\{z_i \in \Psi\}$ for the projection stated above is that the corresponding constraints, $x_i \ge \zeta_i > 0$ must be primally satisfied. Note that $x_i \ge \zeta_i \; \forall i \in \Omega$ is a necessary condition if $\zeta_i > 0$.

Remarks:
If $K' = K$, the projection is complete. For all z_i's not included in Ψ, their projected values of x_i's are zero. Otherwise, $K' \ll K$ due to the effect of a large ϵ, we denote $K'' = K - K'$ as the number of elements that will be projected in **(2)** and **(3)**.

Operation for (2) K''-largest selection:
Let Γ be the index set for $\upsilon_i = c$, and Φ be the set of elements: $\{z_i | i \in \Gamma\}$, where $c = \epsilon$ except $c = h$ when initializing the portfolio. Also, let n_Φ denote the number of elements in Φ, where $n_\Phi = n - \|\overline{x}\|_0$. For all z_i's in Φ, we consider the following projection:

$$\min_{x} \|x - z\|_2^2 \text{ s.t. } \|x\|_0 \le K'', \text{ and } 0 \le x_i \le c \text{ for } i \in \Gamma. \tag{10}$$

In Φ, we first sort the elements in descending order by value. After sorting, Φ is expressed as $\Phi = \{z_{\pi_1}, \cdots, z_{\pi_{n_\Phi}}\}$, where $\{\pi_1, \cdots, \pi_{n_\Phi}\}$ represents the sorted index list. Next, we define another element set Φ' containing the top K'' elements, where $\Phi' = \{z_{\pi_1}, \cdots, z_{\pi_{K''}}\}$. The projected values for all elements in Φ' are given by $x_i = \min\{\max\{0, z_i\}, c\}$ for each z_i in Φ'.

Remarks:
So far, elements that are not in either Ψ or Φ have not been considered yet. The projection will continue to **(3)**. Also, for all z_i's in Φ that are not included in Φ', their projected values of x_i's are zero. Note that since the size of Φ' is K'', the primary advantage of **(2)** is to reduce the problem size in **(3)** from $n - K'$ to $K'' + \|\overline{\boldsymbol{x}}\|_0 - K'$.

Operation for (3) K''-optimal selection:
Let Ξ be an element set, defined as $\Xi = \{z_1, \cdots, z_n\}$, and let Θ be another element set, $\Theta = \Xi \setminus (\Psi \cup \Phi)$, that is, Θ consists of elements that are in Ξ but not in either Ψ or Φ. The number of elements in Θ is $\|\overline{\boldsymbol{x}}\|_0 - K'$, and we denote it as n_Θ. Also, we represent Θ as $\Theta = \{z_{\tau_1}, \cdots, z_{\tau_{n_\Theta}}\}$, where $\{\tau_1, \cdots, \tau_{n_\Theta}\}$ indicates their corresponding positions in the original set.

To optimally select K'' elements of z_i's from a combination of Φ' and Θ, we define an element set $\mathcal{S} = \Phi' \cup \Theta$, and let $n_\mathcal{S}$ be the number of elements in $\mathcal{S}$. We then partition $\mathcal{S}$ into two non-overlapped sets: Λ and Λ^c, where Λ contains the K'' optimal elements of $\mathcal{S}$. Clearly, $\Lambda^c = \mathcal{S} \setminus \Lambda$.

Consider any switching operation between Λ and Λ^c. If $z_i \in \Lambda$ is swapped with any $z_j \in \Lambda^c$, the difference in squared ℓ_2-loss before and after the switch is given by

$$(x_i - z_i)^2 + z_j^2 - \left((x_j - z_j)^2 + z_i^2\right) = (x_i - z_i)^2 - z_i^2 - \left((x_j - z_j)^2 - z_j^2\right). \quad (11)$$

In (11), let δ_i denote $(x_i - z_i)^2 - z_i^2$ and δ_j denote $(x_j - z_j)^2 - z_j^2$. For $z_i \in \Lambda$ to be optimal, the loss difference for any switching cases must be non-increasing. That is, $\delta_i - \delta_j \leq 0 \Rightarrow \delta_i \leq \delta_j$. Then, we compute all δ's in $\mathcal{S}$ and sort them in ascending order. This leads to the inequality as follows,

$$\delta_{\psi_1} \leq \cdots \leq \delta_{\psi_{K''}} \leq \delta_{\psi_{K''+1}} \leq \cdots \leq \delta_{\psi_{n_\mathcal{S}}}. \quad (12)$$

where $\{\psi_1, \cdots, \psi_{n_\mathcal{S}}\}$ is the sorted index list.

As a result, we formulate $\Lambda = \{z_i | i \in [\psi_1, \psi_{K''}]\}$. For the elements in Λ, their projected values of x_i's are given by $x_i = \min\{\max\{0, z_i\}, v_i\}$. On the other hand, $\Lambda^c = \{z_i | i \in [\psi_{K''+1}, \psi_{n_\mathcal{S}}]\}$. For the elements in Λ^c, their projected values of x_i's are assigned to zero. Based on the above inequality (12), any switching elements between Λ and Λ^c would not further decrease the objective loss $\|\boldsymbol{x} - \boldsymbol{z}\|_2^2$. This is because there exists no counterexample that could show $\delta_i - \delta_j > 0$ when $z_i \in \Lambda$ and $z_j \in \Lambda^c$.

Remarks

- The formal proof of the optimality of $\mathcal{P}_\mathcal{C}(\boldsymbol{z})$ is lengthy. The main philosophy is to divide z_i's into two non-overlapping sets based on the projected values obtained from **(1)** to **(3)**. One set contains z_i's with the corresponding projected value x_i's that are greater than 0. The other set contains z_i's with the corresponding projected values x_i's that are equal to zero. Then, we must prove that there are no pairs of elements between the two sets that can be swapped in a way that reduces the objective loss value $\|\boldsymbol{x} - \boldsymbol{z}\|_2^2$.

- To showcase that our projection solver is efficient and fast, we note that from **(1)** to **(3)**, the average runtime measured on MATLAB with a 3.6 GHz Intel Core i7-12700K is $\lessapprox 4 \times 10^{-5}$ s for a problem with $n = 877$ and $K = 90$.

3.5 Convergence Analysis of IT4-PGD

The convergence behavior of our objective function is corroborated in Theorem 1.

Theorem 1. *In IT4-PGD, the objective function* $f(\boldsymbol{x}) = \|\boldsymbol{A}\boldsymbol{x} - \boldsymbol{y}\|_2^2 + \frac{\rho}{2}\left(\mathbf{1}^\top \boldsymbol{x} - 1\right)^2$ *satisfies the following properties:*

(i) $f(\boldsymbol{x})$ *is lower bounded, where* $f(\boldsymbol{x}) \geq 0$.
(ii) $\nabla f(\boldsymbol{x})$ *is L-Lipschitz, where L is the largest eigenvalue of* $\nabla^2 f(\boldsymbol{x}^{(t)})$.
(ii) IT4-PGD with step size $\eta \in (0, 1/L)$ *is a descent method such that the objective function value* $f(\boldsymbol{x}^{(t+1)}) \leq f(\boldsymbol{x}^{(t)})$. *The equality holds if and only if* $\boldsymbol{x}^{(t+1)} = \boldsymbol{x}^{(t)}$.

Therefore, the convergence of $\left\{f(\boldsymbol{x}^{(t)})\right\}_{t=1}^{\infty}$ *can be guaranteed with* $0 < \eta < 1/L$.

The main steps of proving Theorem 1 are as follows:

1. Proving that $f(\boldsymbol{x})$ is lower bounded, which is straightforward.
2. By $\nabla^2 f(\boldsymbol{x}^{(t)}) = 2\boldsymbol{A}^\top \boldsymbol{A} + \rho \mathbf{1}\mathbf{1}^\top$, we can show that $\nabla f(\boldsymbol{x})$ is Lipschitz continuous.
3. Utilizing the L-smoothness property, we can show the following inequality:

$$f(\boldsymbol{x}^{(t+1)}) \leq f(\boldsymbol{x}^{(t)}) + \frac{1}{2}\left(L - \frac{1}{\eta}\right)\|\boldsymbol{x}^{(t+1)} - \boldsymbol{x}^{(t)}\|_2^2. \tag{13}$$

4. We conclude that if $\eta \in (0, 1/L)$ and $\boldsymbol{x}^{(t+1)} \neq \boldsymbol{x}^{(t)}$, then $f(\boldsymbol{x}^{(t+1)}) < f(\boldsymbol{x}^{(t)})$. The objective function value $f(\boldsymbol{x}^{(t)})$ is monotonically decreasing with respect to t.

Note that all the subsequences of $\{\boldsymbol{x}^{(t)}\}_{t=1}^{\infty}$ are bounded and guaranteed to converge to the same limit because $\{f(\boldsymbol{x}^{(t)})\}_{t=1}^{\infty}$ is convergent and the descent property of $f(\boldsymbol{x}^{(t)})$ holds.

Figures 2(a)–(b) show the convergence behavior of the proposed IT4-PGD under different datasets. The details of them will be explained in the next section. Zooming into Fig. 2, we can observe that the objective function value converges within 250 iterations.

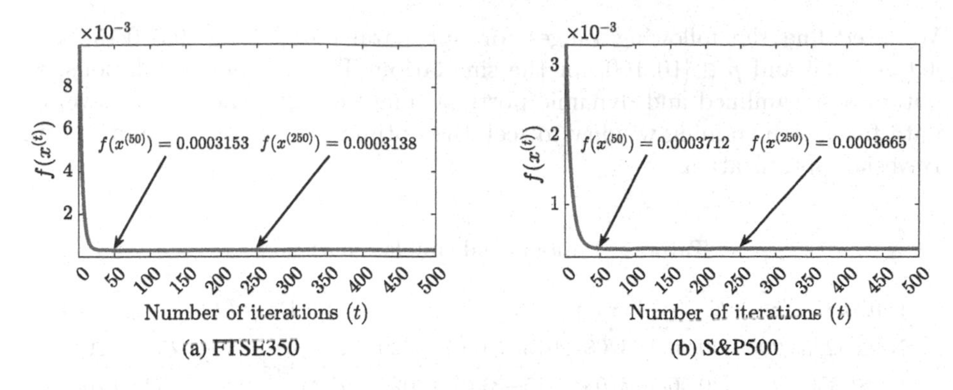

Fig. 2. Convergence behavior of our IT4-PGD ($K = 30$)

4 Numerical Experiments

4.1 Datasets and Settings

In time-series analysis, the rolling window is prevalently adopted for forecasting and backtesting purposes [10,13]. As depicted in Fig. 3, given the historical data, we first construct a portfolio from a training window T_{train} and secondly examine its performance with the associated window T_{test}. At the end of T_{test}, rebalancing occurs. The currently held portfolio is updated using the latest data from the subsequent training window.

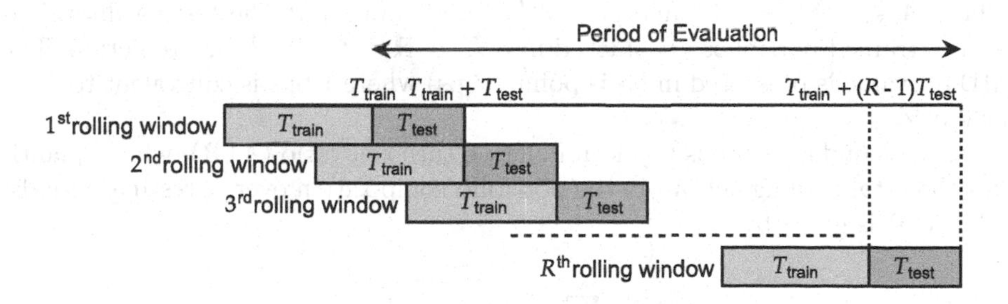

Fig. 3. Rolling window approach

Three leading financial market benchmarks are considered, including Nikkei225, FTSE350, and S&P500. In line with common practice, the stock tickers that are either suspended or newly enlisted during the simulation period are excluded [13,14]. Details of the settings are summarized in Table 1.

In the experiments, the algorithm stops when the criterion is satisfied, or the number of iterations reaches $t = 10^3$. The stopping criterion is given by

$$\frac{\|\boldsymbol{x}^{(t+1)} - \boldsymbol{x}^{(t)}\|_2^2}{\|\boldsymbol{x}^{(t+1)}\|_0} \leq 10^{-8}. \tag{14}$$

We predefine the following ranges for our parameters: $h \in [0.8, 0.15]$, $\epsilon \in [0.001, 0.05]$ and $\rho \in [10, 100]$ for the simulation. Through our simulations, we obtain a streamlined and dynamic portfolio management scheme. We leverage data from each training window to seek the optimal values for h, ϵ and ρ using Bayesian optimization.

Table 1. Datasets and simulation settings

Dataset	n	K	Period	Total Days (T)	T_{train}	T_{test}
Nikkei225	205	[30,90]	03/08/2015–31/07/2020	1212	252[a]	20[b]
FTSE350	232	[30,90]	03/08/2015–31/07/2020	1252	252	20
S&P500	437	[30,90]	03/08/2015–31/07/2020	1252	252	20

[a] 252 trading days per year
[b] Monthly rebalancing $\approx$ 20 days

4.2 Quantitative Measurements

Two metrics are adopted for evaluation. One is the magnitude of daily tracking error (MDTE) [3], which assesses the average performance of index replication. The MDTE is defined as

$$\text{MDTE} = \frac{1}{T - T_{\text{train}}} \|\text{diag}(\widetilde{\boldsymbol{A}}\widetilde{\boldsymbol{X}}) - \widetilde{\boldsymbol{y}}\|_2, \tag{15}$$

where $\widetilde{\boldsymbol{A}} \in \mathbb{R}^{(T-T_{\text{train}})\times n}$ and $\widetilde{\boldsymbol{y}} \in \mathbb{R}^{(T-T_{\text{train}})}$ contain all the testing data, and all the trained portfolios are stacked into $\widetilde{\boldsymbol{X}} \in \mathbb{R}^{n\times(T-T_{\text{train}})}$ for prediction. The MDTE value is presented in basis points (bps) where 1 bps is equivalent to 10^{-4} or 0.01%.

The second measure is the accumulative turnover ratio (ATR), which quantifies the total trading actions in the portfolio composition over R testing periods. The ATR is given by

$$\text{ATR} = \sum_{i=1}^{R} \|\boldsymbol{x}^{i\text{-th}} - \overline{\boldsymbol{x}}^{(i-1)\text{-th}}\|_1, \tag{16}$$

where R is the total number of rebalancing periods, $\boldsymbol{x}^{i\text{-th}}$ is the newly-constructed portfolio after rebalancing while $\overline{\boldsymbol{x}}^{(i-1)\text{-th}}$ is the currently held portfolio before rebalancing. When $i = 1$, we initialize the portfolio $\boldsymbol{x}^{i\text{-th}}$. Since $\overline{\boldsymbol{x}}^{(i-1)\text{-th}}$ does not exist, we assume it to be a zero vector. As the turnover ratio is positively proportional to transaction costs, a large ATR signifies that the total transaction fees are costly. A unit of ATR represents that the percentage change of investment on a portfolio is 100%.

4.3 Performance Comparison

This subsection compares our IT4-PGD with four existing algorithms. They are nonnegative orthogonal matching pursuit with projected gradient descent (NNOMP-PGD) [10], Lasso-based index tracking (LASSO) [5], specialized linear approximation for index tracking (SLAIT) [9], and hybrid half thresholding (HHT) [8]. Notably, only our IT4-PGD and NNOMP-PGD are capable of controlling sparsity explicitly. The SLAIT, LASSO, and HHT algorithms cannot directly control the sparsity level K. Hence, we adjust their respective regularization parameter values to ensure that the sparsity level meets the specified level. The simulation results are shown in Figs. 4, 5, 6 and 7.

Figures 4, 5 and 6 demonstrate that our IT4-PGD outperforms NNOMP-PGD, SLAIT, LASSO, and HHT in terms of MDTE and ATR.

Under the same sparsity, compared with other algorithms, our IT4-PGD obtains a significantly lower MDTE value. In Fig. 4, the MDTE of IT4-PGD is approximately 0.454 bps at $K = 30$. Among the comparison algorithms, NNOMP-PGD offers the best performance at $K = 30$ with an MDTE of 0.635 bps. Besides, S&P500 consists of around five hundred constituents. Remarkably, in Fig. 6, our IT4-PGD still obtains a lower MDTE value even when K is extremely small (i.e., $K \ll n$). For example, in Fig. 6, the MDTE value of our IT4-PGD is 0.498 bps at $K = 30$. Among the comparison algorithms, NNOMP-

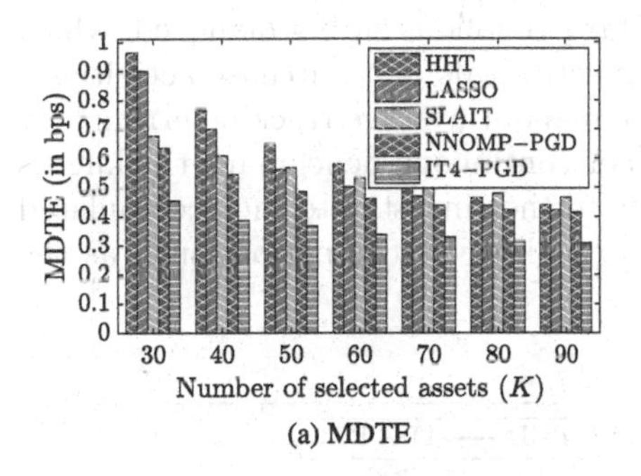

(a) MDTE

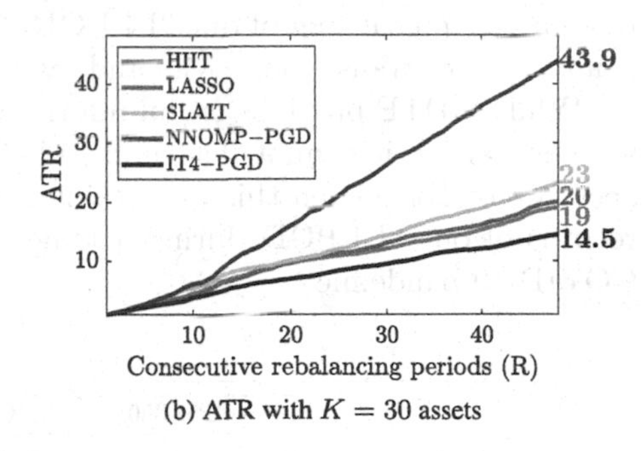

(b) ATR with $K = 30$ assets

Fig. 4. The performance of various algorithms on Nikkei225

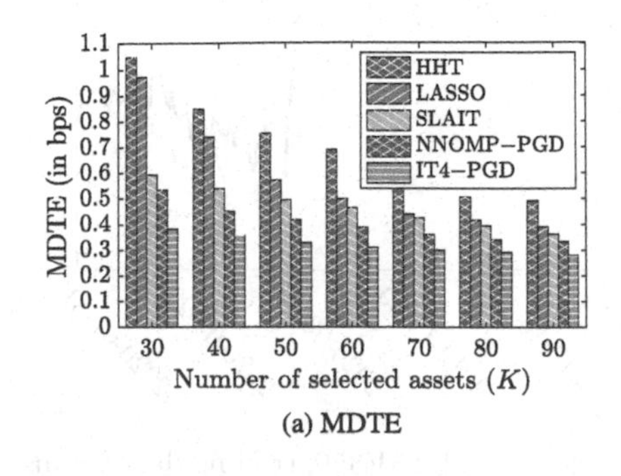

(a) MDTE

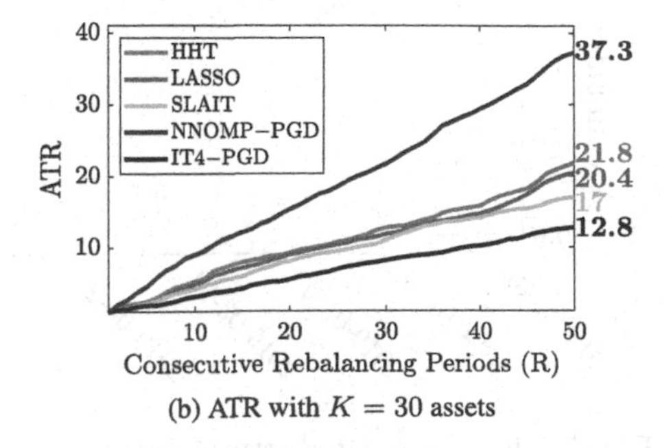

(b) ATR with $K = 30$ assets

Fig. 5. The performance of various algorithms on FTSE350

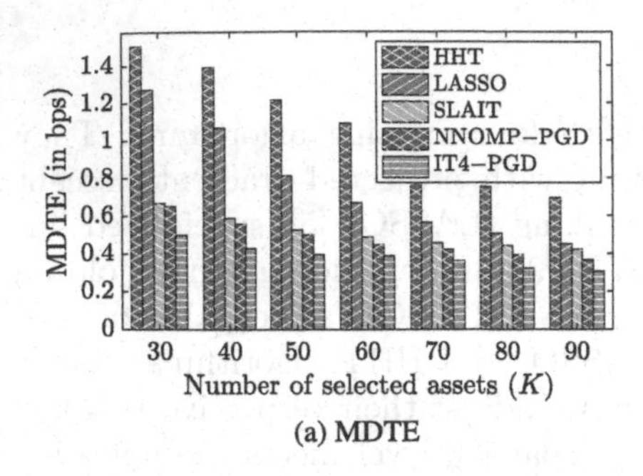

(a) MDTE

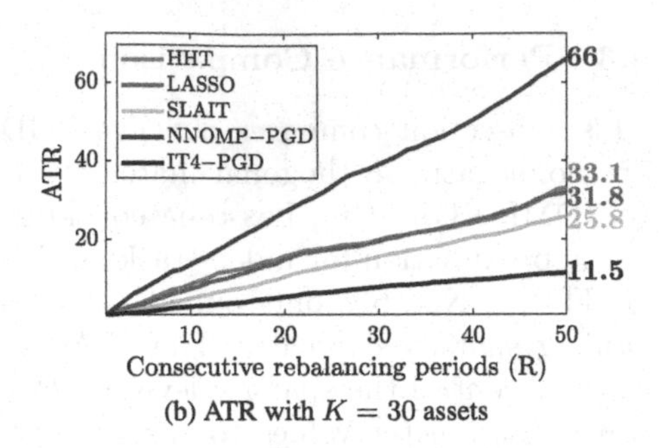

(b) ATR with $K = 30$ assets

Fig. 6. The performance of various algorithms on S&P500

PGD achieves an MDTE value of around 0.65 bps. It is worth pointing out that the MDTE values of our IT4-PGD with a sparser K are also better than those comparison algorithms with a larger K.

Simultaneously, our IT4-PGD also surpasses other methods in terms of ATR. In Figs. 4, 5 and 6, our IT4-PGD achieves lower ATR values than other methods on different datasets. In particular, the ATR of our IT4-PGD is 11.5 units in S&P500 (Fig. 6), whereas the next best-performing algorithm, SLAIT, registers an ATR value of 25.8 units. Therefore, the ATR of our IT4-PGD is noticeably lower than the ATR of SLAIT. This implies that lower turnover rates are guaranteed in the rebalancing of our IT4-PGD. Our portfolio is stably managed, which leads to lower transaction costs and avoids active sale and purchase activities.

While MDTE provides useful information about average tracking errors over a given period, it cannot determine whether continuous benchmark tracking is accurate or not. Given this uncertainty, we further investigate the accumulated returns of our IT4-PGD during a long and highly volatile period, such as the COVID-19 pandemic.

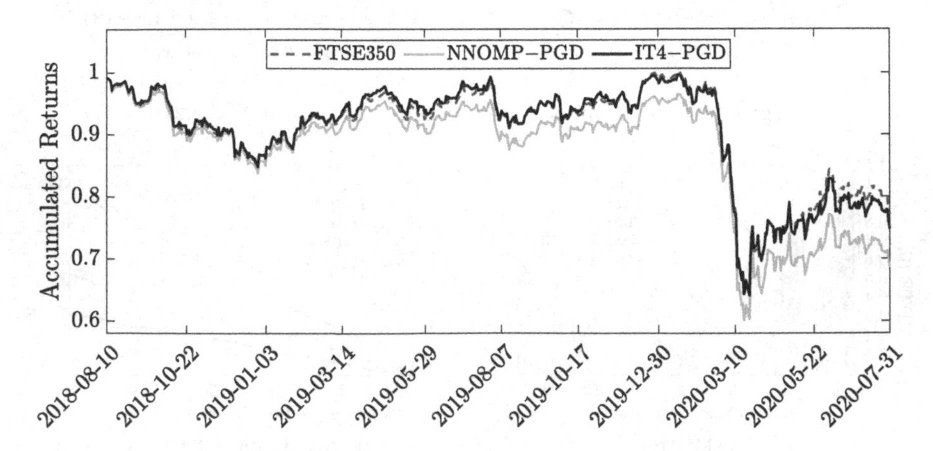

Fig. 7. The accumulated returns of tracking portfolios on FTSE350, trading days versus accumulated returns with $K = 30$ assets and rolling size = 500 days.

For clarity, we focus on our IT4-PGD and the second-best method (in terms of MDTE), NNOMP-PGD. Figure 7 reveals that our IT4-PGD with $K = 30$ could effectively track the trends of FTSE350 from August 2018 to July 2020. The accumulated returns of our IT4-PGD always stick close to the benchmark itself. In contrast, NNOMP-PGD starts diverging from January 2019. This shows that poor estimates during certain periods may lead to the propagation of accumulative errors. This behavior of NNOMP-PGD results in a significant deviation in accumulated returns relative to the benchmark target. In summary, the results of accumulated returns reaffirm the excellent performance of our IT4-PGD in long-term index tracking.

5 Conclusion

This paper derives a novel algorithm to deal with the SIT problem. The proposed IT4-PGD offers flexible portfolio management by providing direct and explicit control over the number of holding assets, the maximum holding position, and the maximum turnover ratio of each constituent. Moreover, the convergence of our IT4-PGD is guaranteed. Simulations on various real-world datasets are conducted. In our evaluation, we compared our IT4-PGD with existing SIT methods, and our IT4-PGD demonstrated its superior performance in terms of MDTE and ATR.

References

1. Perrin, S., Roncalli, T.: Machine learning optimization algorithms & portfolio allocation. In: Machine Learning for Asset Management: New Developments and Financial Applications, pp. 261–328 (2020)
2. Ma, Y., Han, R., Wang, W.: Portfolio optimization with return prediction using deep learning and machine learning. Expert Syst. Appl. **165**, 113973 (2021)
3. Benidis, K., Feng, Y., Palomar, D.P., et al.: Optimization methods for financial index tracking: From theory to practice. Found. Trends® Optim. **3**(3), 171–279 (2018)
4. Li, N., Niu, Y., Sun, J.: Robust sparse portfolios for index tracking based on m-estimation. In: Communications in Statistics-Simulation and Computation, pp. 1–13 (2022)
5. Sant'Anna, L.R., Caldeira, J.F., Filomena, T.P.: Lasso-based index tracking and statistical arbitrage long-short strategies. North Am. J. Econ. Finance **51**, 101055 (2020)
6. Kremer, P.J., Brzyski, D., Bogdan, M., Paterlini, S.: Sparse index clones via the sorted ℓ_1-norm. Quant. Finance **22**(2), 349–366 (2022)
7. Shu, L., Shi, F., Tian, G.: High-dimensional index tracking based on the adaptive elastic net. Quant. Finance **20**(9), 1513–1530 (2020)
8. Xu, F., Xu, Z., Xue, H.: Sparse index tracking based on $L_{1/2}$ model and algorithm. arXiv preprint arXiv:1506.05867 (2015)
9. Benidis, K., Feng, Y., Palomar, D.P.: Sparse portfolios for high-dimensional financial index tracking. IEEE Trans. Signal Process. **66**(1), 155–170 (2017)

10. Li, X.P., Shi, Z.L., Leung, C.S., So, H.C.: Sparse index tracking with K-sparsity or ε-deviation constraint via ℓ_0-norm minimization. IEEE Trans. Neural Netw. Learn. Syst. (2022)
11. Shi, Z.L., Li, X.P., Leung, C.S., So, H.C.: Cardinality constrained portfolio optimization via alternating direction method of multipliers. IEEE Trans. Neural Netw. Learn. Syst. (2022)
12. Parikh, N., Boyd, S., et al.: Proximal algorithms. Found. Trends® Optim. **1**(3), 127–239 (2014)
13. Leung, M.F., Wang, J.: Minimax and biobjective portfolio selection based on collaborative neurodynamic optimization. IEEE Trans. Neural Netw. Learn. Syst. **32**(7), 2825–2836 (2020)
14. Wu, L., Feng, Y., Palomar, D.P.: General sparse risk parity portfolio design via successive convex optimization. Signal Process. **170**, 107433 (2020)

Anomaly Detection of Fixed-Wing Unmanned Aerial Vehicle (UAV) Based on Cross-Feature-Attention LSTM Network

Li Xu, Yingduo Yang, Xiaoling Wen, Chunlong Fan, and Qiaoli Zhou(✉)

Shenyang Aerospace University, Shenyang, Liaoning, China
zhou_qiao_li@hotmail.com

Abstract. With the gradual penetration of unmanned aerial vehicle (UAV) technology and its related applications in people's lives, the safety of UAVs has become an important research focus. In this paper, we present an anomaly detection method based on cross-feature-attention LSTM neural networks. In an unsupervised setting, we use two types of networks to extract temporal and spatial features from flight data and predict future states to detect abnormal flight behavior. We conduct experiments on real flight data, the Air Lab Fault and Anomaly (ALFA) dataset, using multiple sets of different feature combinations. The results indicate that our method can maintain high performance across different feature combinations, achieving an average accuracy of 0.96 and a response time of 5.01 s.

Keywords: Autonomous aerial vehicles · Anomaly detection · Long short-term memory

1 Introduction

With the rapid development of unmanned aerial vehicle (UAV) technology, UAVs have gradually transitioned from military to civilian use. As of 2022, the global civiliDuean UAV market size has exceeded 30 billion, which is likely to exceed 80 billion by 2025. Therefore, UAV security has become an important research topic [1]. During UAV operations, safety accidents can occur due to improper operation, sensor malfunction, ground station failure, etc. [2]. Accordingly, UAV anomaly detection has become an integral approach for enhancing UAV reliability [3].

Unlike manned aircraft, UAVs are programmed to fly autonomously in many applications. During flight, UAVs send the data recorded by onboard sensors to the ground control station. This allows UAVs to collect a large amount of data of different types related to the flight for recording and analysis while executing the flight program. For the field of autonomous UAV anomaly detection, there are two main methods: model-based and data-driven [4].

Supported by Department of Education of Liaoning Province, LJKZ0209.

B. Luo et al. (Eds.): ICONIP 2023, CCIS 1966, pp. 513–527, 2024.
https://doi.org/10.1007/978-981-99-8148-9_40

Model-based methods are mainly based on known robot dynamic models for anomaly detection, exhibiting typically desirable response time and accuracy. Because in the case of a known model, the robot's behavior probability distribution can be easily obtained, and anomaly detection can be performed by comparing the probability distribution under the observed state and predicted probability distribution [5]. Chen et al. [6] modeled the full-scale 3D structure of a wing and used FLUENT and ANSYS software for finite element simulation analysis to determine the fault monitoring nodes. Melnyk et al. [7] represented multivariate time series using vector autoregressive exogenous models and constructed a distance matrix between objects based on the vector autoregressive exogenous model between objects, and finally performed anomaly detection based on object differences. Model-based methods are suitable for UAV systems with precise dynamic models, but these models are not universal, and modeling different types of robots is cumbersome. Moreover, signal noise also poses a challenge to the robustness of the model [8]. Therefore, model-based methods are primarily unable to perform anomaly detection tasks for complex structural UAVs with complex flight environments.

Data-driven methods mainly collect robot behavioral data and analyze and detect anomalies based on these data [9]. By analyzing the correlation between different types or periods of data collected from the robot, the operating status of the robot can be detected for the anomaly. Bronz M et al. [10] used UAV flight logs and SVM algorithm to classify UAV behavior during normal flight and fault stage. Das S et al. [11] proposed a multi-kernel anomaly detection algorithm that uses multiple kernels to combine multidimensional data source information and jointly detect anomalies in flight data using OCSVM algorithm. The advantage of Data-based lies in the lower computational complexity after the model training is completed, and it is more sensitive to irregularly distributed noisy data.

The anomaly detection task on the ALFA dataset can be divided into two categories: feature-based anomaly classification and prediction-based anomaly detection. The former mainly classifies features for each time step, while the latter detects anomalies for each flight data. Specifically, [12–14] use autoencoder to calculate the reconstruction error of feature information for classification tasks. Tao Yang et al. [15] convert feature information into images and train and classify them using CNN networks. These methods are all feature-based anomaly classification methods. In specific aircraft models or datasets, this method often achieves higher accuracy, but it usually requires a large number of features, which makes them less generalizable. In addition, these methods require high sampling frequency, which makes them unable to perform real-time anomaly detection tasks. [16,17] are prediction-based anomaly detection methods. Azarakhsh Keipour et al. estimate the model between related input-output signals online to perform anomaly detection. Jae-Hyeon Park et al. use stacked LSTM to predict future states and perform real-time anomaly detection based on the difference between predicted and real values. Prediction-based anomaly detection has a stronger generalization ability, requires fewer types of features, and is less computationally intensive.

In recent years, the attention mechanism has been widely used in deep learning [18]. It is similar to the human visual system and can focus on crucial information and ignore some redundant information to adjust the weights of neural networks [19]. In unmanned aerial vehicle flight, time-sequenced features have complex spatial-temporal relationships. To address this issue, we propose combining the Cross-Feature-Attention network with the LSTM network to model the spatial-temporal relationships among different UAV features and perform anomaly detection for different input types of features. Comparing our method with other prediction-based anomaly detection methods, our method achieves less response time and higher accuracy.

The organization of this article is as follows: Sect. 2 introduces the dataset and the proposed model. Section 3 describes the experimental settings. Section 4 describes the experimental results. Section 5 presents the conclusion based on the results.

2 Materials and Methods

2.1 Dataset

In this study, we used the publicly available AirLab Failure and Anomaly (ALFA) dataset produced by Keipour et al. [20]. This is a dataset of multiple fault types of fixed-wing unmanned aerial vehicles for anomaly detection and hazard detection research. The data collection was performed by a customized version of the Carbon Z T-28 model aircraft, which has a 2-meter wingspan, a front electric motor, and components such as ailerons, flaps, elevators, and rudders. The aircraft is equipped with a Holybro PX4 2.4.6 autopilot, a Pitot tube, a GPS module, and an Nvidia Jetson TX2 embedded computer.

The ALFA dataset contains a total of 47 processed flight data, including 10 normal flight scenarios, 23 engine failure scenarios, 4 rudder failure scenarios, 2 elevator failure scenarios, and 8 flap failure scenarios. In the fault flight scenarios, the UAV first flies normally, and then the firmware parameters are modified to simulate the fault state of the aircraft during flight. Therefore, in the 37 fault flight scenarios, there are a total of 3377 s of normal flight data and 777 s of fault flight data [20]. During the flight, the ROS Kinetic Kame in the onboard computer reads the flight status and state information from Pixhawk, and the data is recorded using the MAVROS package.

2.2 Data Processing

During the flight, different feature data are recorded at different sampling frequencies, because different sensors have different recording frequencies. The ALFA dataset records data from 20 topics and hundreds of feature types, with sampling frequencies ranging from 1 [Hz] to 50 [Hz]. As we aim to perform real-time anomaly detection with our model, we selected the feature with the lowest sampling rate of 4 [Hz] in our experiments, as the benchmark frequency without performing any data augmentation. To ensure consistency in data sampling, we used timestamps to organize the data, with each timestamp interval of 250 [ms]. Among features with different sampling frequencies, we selected the data closest to the timestamp for recording to maximize feature synchronization. Although the feature data recorded in the dataset have fixed sampling frequencies, the sampling frequencies of some features may change abruptly in certain cases. As shown in Fig. 1, the sampling rate of Feature 4 is low and unstable. Since the second data point is the closest point to both TS1 and TS2, it is recorded twice.

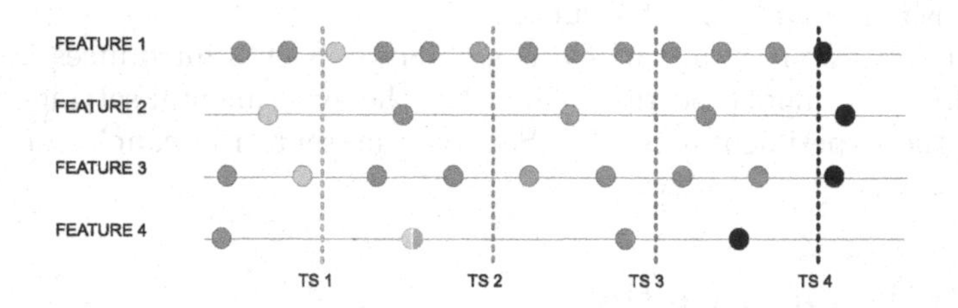

Fig. 1. Sample data according to timestamps.

To demonstrate the generalizability of the model, we selected data from three types of features: Position Status, including Position Status with multi-dimensional features, Command Status with multi-dimensional features, and System Status with one-dimensional features. Table 1 shows which topic these data came from and their initial sampling rate. In the following sections, we will explain how to use these data for training and testing.

2.3 Overview

Given the diversity of drones and their complex internal structures, it is difficult to use a single model or algorithm for anomaly detection tasks. Therefore, our method extracts spatial-temporal relationships from the flight feature data to model the flight process nonlinearly. In our method, we separate our method into three parts (Fig. 2).

Cross-Feature-Attention Network. It consists of an LSTM network and a multi-head attention mechanism [21]. Its main function is to blend different feature information and focus on different parts of the sequence to discover spatial information in the data.

LSTM Network. It consists of multiple LSTM layers and Dense layers [22]. Its main function is to control the flow of information in and out of the model using gated units, allowing for the transfer and processing of information across time steps, and focusing on the temporal information in the sequence.

Table 1. Selection and source of feature data

Category	Feature Name	Field Name	Average rate
Position Status	Local Position(x,y,z)	mavros_local_position_pose	4 Hz
	Local Velocity(x,y,z)	mavros_nav_info_velocity	20–25 Hz
	Local Altitude(quaternion)	mavros_nav_info	20–25 Hz
	Local Angular-Velocity(x,y,z)	mavros_local_position_velocity	4–5 Hz
	Local Linear-Acceleration(x,y,z)	mavros_imu_data_raw	10 Hz
Command Status	Command Velocity(x,y,z)	mavros_nav_info_velocity	20–25 Hz
	Command-Attitude(quaternion)	mavros_nav_info	20-25 Hz
	Command Linear-Acceleration(x,y,z)	mavros_setpoint_raw_target_global	20–50 Hz
System Status	Fluid Pressure	mavros_imu_atm_pressure	10 Hz
	Temperature	mavros_imu_temperature	10 Hz
	Altitude Error	mavros_nav_info_errors	20–25 Hz
	Airspeed Error	mavros_nav_info_errors	20–25 Hz
	Tracking Error (x)	mavros_nav_info_errors	20–25 Hz
	WP Distance	mavros_nav_info_errors	20–25 Hz

Detection Module. This part mainly compares the predicted state vector and the actual state vector by normalizing their values, quantifying their residual values, filtering the residual values, and comparing them to a threshold to determine if the flight status is anomalous [23].

2.4 Spatial-Temporal Features Processing Module

In typical robot state prediction problems, the robot's state prediction is often described as a state transition equation between the current state and the control signal as follows:

$$S_{t+1} = f(S_t, U_t) \tag{1}$$

where S represents the robot state, U represents the control signal, f is the state transition function, t represents the current time, and $t+1$ represents the next time step [24]. The motion and environmental interaction of robots can be viewed as a process of continuous state transitions, which can be described by dynamic equations. By using the state transition equation, a series of state predictions can be derived, which is useful for helping the robot make more accurate decisions.

However, in many cases, due to a series of problems such as sensor sensitivity, we may not be able to obtain effective control signals. Therefore, we need to use other types of features to predict the robot's state. In our method, we generalize the problem of state prediction to feature-based prediction as follows:

$$S_{t+1} = g(S_t, X_t) \tag{2}$$

where X represents a series of different features including control signals, and g represents the state transition function that is based on the current state and current feature. Similarly, this process is also continuous. Specifically, in our research, S represents the specific flight state of the drone, including position, linear velocity, angular velocity, attitude, etc. Furthermore, X represents the features other than the flight state, such as control signals, temperature, trajectory errors, and other features.

In our method, we use various types of features for non-linear feature fitting. Because in addition to control signals, other features can also be used to evaluate the robot's behavior and serve as inputs to the neural network to predict future states. To achieve this, the model must have a strong generalization ability for feature selection, which cannot be achieved only by LSTM network [25]. There are two reasons for this. First, as the data comes from different sensors, different features have errors, missing data, and redundancies, which cause insufficient data quality and result in unreasonable weight allocation during the training process. Second, for different feature selections, the same network parameters cannot be used. Otherwise, the network will have difficulty distinguishing the importance of different features due to the unreasonable distribution of parameters. To solve the various issues caused by different feature selections, we propose a Cross-feature-attention network, which consists of an LSTM layer, a multi-head attention layer, and a concatenate layer. Its function is to fuse and recombine different types of features, dynamically assign weights to the original features during the fusion process, and enable the network to fully extract the spatial relationships between features, thereby predicting the state of the drone for different types of input features (Fig. 3).

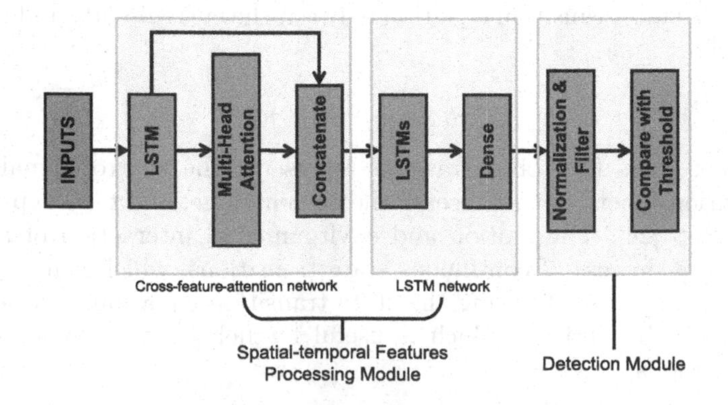

Fig. 2. Cross-feature-attention LSTM network architecture.

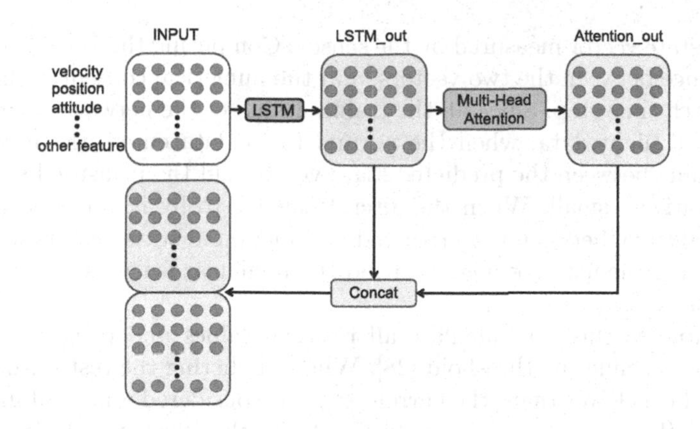

Fig. 3. Spatial data processing module.

Before entering the Cross-feature-attention network, the time series data are transformed into tensor format that can be received by the recurrent neural network, and each feature is normalized with the $Z - score$ method. As shown in the figure, we use LSTM-out as Q, K, and V respectively to perform linear transformations and input them into the scaled dot-product attention [21]. This step is performed 8 times. The specific formula for calculating the attention score is as follows:

$$Attention(Q, K, V) = \sum_{s=1}^{num=8} softmax(\frac{Q_s K_s^T}{\sqrt{d}})V_s \tag{3}$$

where s represents the order of the head, $\sqrt{d}$ is used to normalize the attention weights. The attention scores obtained here are used to calculate the weighted sum of the reference sequence to generate the output vector. This process allows the model to learn more spatially related information between different features in different subspaces. In essence, this process performs spatial feature extraction on each position in the input sequence, and the network automatically focuses on the important feature positions in the sequence [26].

We add a Concatenate layer in the network to concatenate the results of multihead-attention and the original LSTM output along the channel dimension, in order to obtain more comprehensive feature information. Although Cross-feature-attention can perform spatial feature extraction to a large extent, since these features are calculated based on previous input information, they may be disturbed by some noise or bias, resulting in some errors in the output feature vectors. In the subsequent network, we use a combination of LSTM network and linear layer to extract the temporal information of the features [27], and finally output the predicted state vector.

2.5 Detection Module

After the model completes the unsupervised learning iteration on the training set data, it predicts the state vector, which is then compared with the current measured state vector measured by the sensor. Computing the Euclidean norm of the difference between the two vectors, and the numerical content is the residual value at each time step. Since all the training data of the network is constructed from normal flight data, when the normal flight data enters the network, the residual value between the predicted state vector and the measured state vector will be relatively small. When the aircraft begins to fly in an abnormal state, the residual value between the predicted and measured state vectors will become larger because the network has not fitted the nonlinear features under abnormal conditions.

According to this, we calculate all residual values under the normal flight state and determine the threshold [28]. When predicting the test set data, if the residual value is lower than, the current state is considered a normal flight state. Conversely, if the residual value is higher than, the current state is considered an abnormal flight state.

Since we only use the uniform sampling rate according to the timestamp when using the data and do not perform denoising processing, and only simple normalization processing is done before the data enters the neural network training. This leads to our network model being unable to accurately capture the most real trend of data changes, which in turn leads to unstable residual values. If the direct residual values are used for anomaly detection, it will cause many false positives and reduce the model accuracy. Therefore, we need to smooth and denoise the residual values, remove the high-frequency oscillations caused by noise, and restore the overall trend and features of the residual values (Fig. 4).

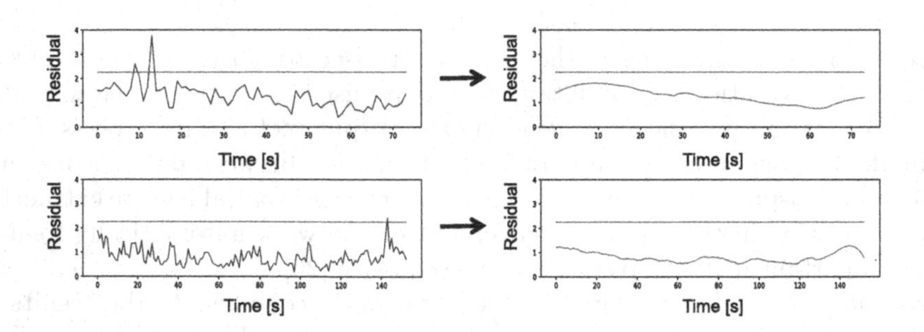

Fig. 4. The left side shows the residual values without filtering, while the right side shows the residual values after filtering. Before filtering, the residual values exceed the threshold.

To achieve this, we use the Savitzky-Golay filter [29], a filtering method based on local polynomial least squares fitting in the time domain. It can remove high-frequency noise while maintaining the shape and width of the signal. The basic algorithm is to select a filtering window size, and then use a polynomial of a certain order to fit the data within this window. The specific algorithm is as follows:

$$y = a_0 + a_1 x + a_2 x^2 + \cdots + a_{k-1} x^{k-1} \tag{4}$$

where x is the parameter to be fitted, y is the fitted data, and a is the parameter to be solved. Compared to low-pass filters, Savitzky-Golay filters can retain detailed information after smoothing, thereby more accurately capturing the features and trends of the signal and improving the response time of anomaly detection.

3 Experiment

In this section, we will explain how to use the method proposed in the previous section to detect anomalies on the ALFA dataset for fixed-wing drones. First, as in Table 2, we will divide the features introduced in Sect. 2 into three groups of feature combination.

Table 2. Feature combination and input/output dimensions

Combination	Category	Input Size	Output Size
feature combination 1	Position Status Command Status System Status	32	10
feature combination 2	Position Status System Status	26	10
feature combination 3	Position Status	22	10

To perform state prediction and performance evaluation, we divide the dataset into three parts:

Training Set: containing only normal flight data, used for network training.

Validation Set: containing only normal flight data, used for calculating threshold and other hyperparameters.

Test Set: containing normal and abnormal flight data, used for performance evaluation and calculation of response time.

Regardless of how the input data feature types change, they always contain state information(P, V, Θ, Ω, F), where P is position, V is velocity, Θ is attitude, Ω is angular velocity, and F is linear acceleration. In our input data, we used (V, Θ, Ω), and we applied $Z-score$ normalization in the data preprocessing stage:

Table 3. The structure and parameters of network

Layer(type)	Output Dim	Parameters	Return Sequences
Input layer	32/26/22	0	TRUE
LSTM	32	8320/7808/7808	TRUE
Multi-head Attention	32	8736	TRUE
Concatenate	64	0	TRUE
LSTM	64	33024	TRUE
LSTM	128	98816	TRUE
LSTM	64	49408	FALSE
Dense	32	2080/1888/1888	-
Dropout	32	0	-
Dense	32	1056	-
Dropout	32	0	-
Dense	10	330	-

$$z = (x - \mu)/\sigma \tag{5}$$

where x is the value of the data, μ is the mean of the data, and σ is the standard deviation of the data. In order to ensure that our anomaly detection method can be used as a real-time system, the mean and standard deviation used in $Z-score$ normalization for the training set, validation set, and test set are all derived from the training data. When position information P and linear acceleration information F are used as the predicted state during testing, there may be a large error leading to a deterioration of model performance, so they are not set as predicted outputs. In addition, all information about attitude is replaced by quaternions, which is beneficial for the stability of the model parameters.

Regarding data distribution, regardless of any feature combinations, we chose 23 flight data as the training set, 8 flight data as the validation set, and the remaining 15 flight data as the test set. The training set and validation set are composed of normal flight data and abnormal flight data of fault removal part, while the test set includes 3 rudder faults, 2 elevator faults, 2 aileron faults, and 8 engine faults.

During the experiment, we used three feature combinations to predict the state vector without modifying the network model parameters. Furthermore The return-sequence column indicates whether the output is a sequence. True represents that the layer outputs the complete sequence, while False represents that the layer outputs only the last value of the output sequence. We used the mean squared error loss function and the Adam optimizer. Table 3 contains a detailed record of the network structure and parameters.

4 Result

We used 15 fault flight scenarios as the test set to demonstrate the performance of our method. These scenarios are time series data of normal flights, followed by sudden control surface faults. We used the model trained in the third part to predict the state and calculate the residuals r, and used the SG filter to smooth it. We used the threshold T calculated from the validation set as the judgment criterion. If $r_{smooth} < T$, the UAV is in a normal state; if $r_{smooth} > T$, it is in an abnormal state. Once $r_{smooth} > T$ occurs, we consider that the whole flight scenario is an abnormal flight (Fig. 5 and 6).

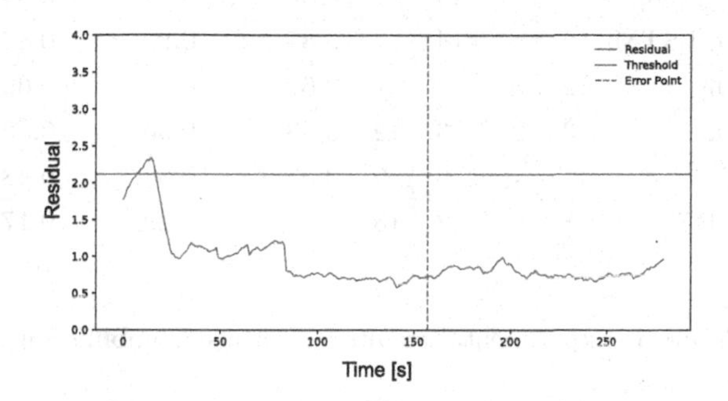

Fig. 5. False positive (FP) is considered when the fault is incorrectly diagnosed before the actual fault point, and false negative (FN) is considered when the fault is not detected after the fault point.

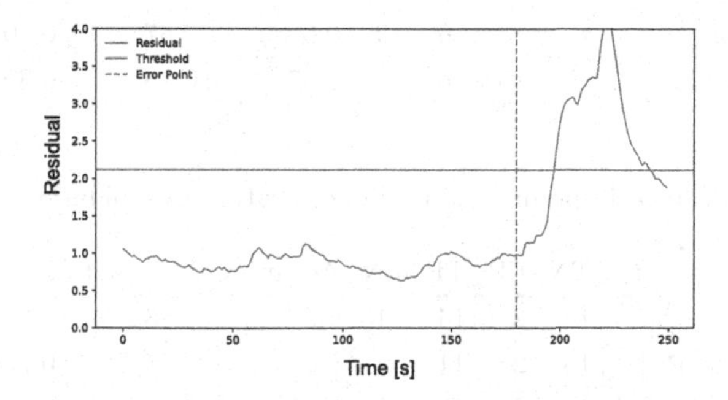

Fig. 6. True negative (TN) is considered when the fault is not detected before the actual fault point, and true positive (TP) is considered when the fault is successfully diagnosed after the fault point.

To compare the performance of different models, we compared our model with other prediction-based anomaly detection models. Using performance indicators that take into account the actual UAV flight in anomaly detection, including true positives, false positives, true negatives, and false negatives. Under faulty conditions, if the residual is always below the threshold, the number of true negatives is increased by 1; if the residual is greater than the threshold at least once, the number of false positives is increased by 1. We provide a more intuitive display of the comparison in the Table 4, 5 and 6.

Table 4. Experimental results of Feature Combination 1.

Method	TN	FP	TP	FN	Accuracy	Precision	F1-score	Recall
Our Method	15	0	14	1	0.97	1.00	0.97	0.93
Stacked LSTM [16]	12	3	13	2	0.83	0.81	0.84	0.87
Self-Attention-LSTM	12	3	14	1	0.87	0.82	0.87	0.93
x-velocity [30]	12	3	8	7	0.67	0.73	0.62	0.53
y-velocity [30]	13	2	3	12	0.53	0.60	0.30	0.20
z-velocity [30]	11	4	6	9	0.57	0.60	0.48	0.40
Linear mode [17]	8	7	2	13	0.33	0.22	0.17	0.13

Table 5. Experimental results of Feature Combination 2.

Method	TN	FP	TP	FN	Accuracy	Precision	F1-score	Recall
Our Method	15	0	12	3	0.9	1.00	0.89	0.8
Stacked LSTM [16]	12	3	13	2	0.83	0.81	0.84	0.87
Self-Attention-LSTM	13	2	11	4	0.8	0.85	0.79	0.73
x-velocity [30]	13	2	7	8	0.67	0.78	0.58	0.47
y-velocity [30]	13	2	5	10	0.60	0.71	0.45	0.33
z-velocity [30]	12	3	6	9	0.60	0.67	0.50	0.40
Linear model [17]	5	10	6	9	0.37	0.38	0.39	0.40

Table 6. Experimental results of Feature Combination 3.

Method	TN	FP	TP	FN	Accuracy	Precision	F1-score	Recall
Our Method	13	2	14	1	0.9	0.88	0.90	0.93
Stacked LSTM [16]	13	2	11	4	0.8	0.85	0.79	0.73
Self-Attention-LSTM	13	2	12	3	0.83	0.86	0.83	0.80
x-velocity [30]	10	5	7	8	0.57	0.58	0.52	0.47
y-velocity [30]	12	3	6	9	0.60	0.67	0.50	0.40
z-velocity [30]	13	2	5	10	0.60	0.71	0.45	0.33
Linear model [17]	6	9	7	8	0.43	0.44	0.45	0.47

From the experimental results, it is evident that our method has significant performance advantages. Besides comparing our method with stacked LSTM and self-attention LSTM, We also compared our method with Wang [30] which predicted a single variable. Since data error will lead to instability of output variables, our model outputs complete state vectors, which can reduce the impact of instability of single variable. Keipour [17]established a model for the relationship between relevant input and output and performed anomaly detection. Our model performs best in feature extraction and maintains good performance in different feature combinations.

The response time of our cross-feature-attention LSTM, stacked LSTM [16], Wang et al. [30], and Keipour et al. [17] are in Fig. 7. Response time refers to the time it takes for the UAV to detect a fault from the beginning of the fault, and this indicator can directly reflect the performance of different anomaly detection algorithms. We selected the two variables with the fastest response time for Wang et al.'s method, z-velocity, and pitch rate. We show the average response time and average response step size of different methods in a bar chart. The smaller these two values, the faster the anomaly detection speed. As can be seen in Fig. 7, our method has the shortest response time and can detect anomalies the earliest.

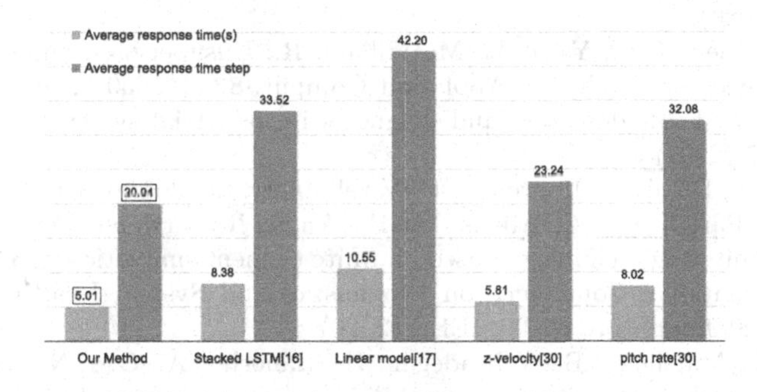

Fig. 7. The average response time and average response steps for different methods.

We also compared with the method proposed by Muhammad Waqas Ahmad [31], which uses an intelligent framework for UAV anomaly detection and has the earliest response time of 0.74 s among all existing methods. However, their method increases the sampling rate of the data to 60 Hz through interpolation, which means, on average, it takes 44.4 steps to detect anomalies, while our method only takes 20.04 steps to detect anomalies. Therefore, our method has the fastest response efficiency for the ALFA dataset.

5 Conclusions

In this paper, we proposed an anomaly detection method based on a cross-feature-attention LSTM neural network, a data-based unsupervised learning model. Without using the UAV kinematic model, we fused spatial-temporal features for different types of UAV feature combinations using unlabeled normal data, then predicted future state vectors, computed residuals, smoothed the residuals with filtering, and detected anomalies using thresholding. Our method was tested on the public dataset ALFA, demonstrating that our model has higher accuracy and faster response time. Future research directions include extending this method to anomaly detection tasks for other fixed-wing UAVs.

Acknowledgments. This research was funded by Department of Education of Liaoning Province, LJKZ0209.

References

1. Kopardekar, P., Rios, J., Prevot, T., Johnson, M., Jung, J., Robinson, J.E.: Unmanned aircraft system traffic management (UTM) concept of operations. In: AIAA Aviation and Aeronautics Forum (Aviation 2016), number ARC-E-DAA-TN32838 (2016)
2. González-Sieira, A., Cores, D., Mucientes, M., Bugarín, A.: Autonomous navigation for UAVs managing motion and sensing uncertainty. Robot. Auton. Syst. **126**, 103455 (2020)
3. Khan, S., Liew, C.F., Yairi, T., McWilliam, R.: Unsupervised anomaly detection in unmanned aerial vehicles. Appl. Soft Comput. **83**, 105650 (2019)
4. Ziegel, E.R.: Fault detection and diagnosis in industrial systems. Technometrics **44**(2), 197 (2002)
5. Caswell, G., Dodd, E.: Improving UAV reliability, no. 301, p. 7 (2014)
6. Chen, M., Pan, Z., Chi, C., Ma, J., Hu, F., Wu, J.: Research on UAV wing structure health monitoring technology based on finite element simulation analysis. In: 2020 11th International Conference on Prognostics and System Health Management (PHM-2020 Jinan), pp. 86–90. IEEE (2020)
7. Melnyk, I., Matthews, B., Valizadegan, H., Banerjee, A., Oza, N.: Vector autoregressive model-based anomaly detection in aviation systems. J. Aerosp. Inf. Syst. **13**(4), 161–173 (2016)
8. Albuquerque Filho, J.E., Brandão, L.C.P., Fernandes, B.J.T., Maciel, A.M.A.: A review of neural networks for anomaly detection. IEEE Access (2022)
9. Agrawal, S., Agrawal, J.: Survey on anomaly detection using data mining techniques. Procedia Comput. Sci. **60**, 708–713 (2015)
10. Bronz, M., Baskaya, E., Delahaye, D., Puechmore, S.: Real-time fault detection on small fixed-wing UAVs using machine learning. In: 2020 AIAA/IEEE 39th Digital Avionics Systems Conference (DASC), pp. 1–10. IEEE (2020)
11. Das, S., Matthews, B.L., Srivastava, A.N., Oza, N.C.: Multiple kernel learning for heterogeneous anomaly detection: algorithm and aviation safety case study. In: Proceedings of the 16th ACM SIGKDD International Conference on Knowledge Discovery and Data Mining, pp. 47–56 (2010)
12. Bell, V., Rengasamy, D., Rothwell, B., Figueredo, G.P.: Anomaly detection for unmanned aerial vehicle sensor data using a stacked recurrent autoencoder method with dynamic thresholding. arXiv preprint arXiv:2203.04734 (2022)
13. Park, K.H., Park, E., Kim, H.K.: Unsupervised fault detection on unmanned aerial vehicles: encoding and thresholding approach. Sensors **21**(6), 2208 (2021)

14. Tlili, F., Ayed, S., Chaari, L., Ouni, B.: Artificial intelligence based approach for fault and anomaly detection within UAVs. In: Barolli, L., Hussain, F., Enokido, T. (eds.) AINA 2022. LNNS, vol. 449, pp. 297–308. Springer, Cham (2022). https://doi.org/10.1007/978-3-030-99584-3_26
15. Yang, T., Chen, J., Deng, H., Yu, L.: UAV abnormal state detection model based on timestamp slice and multi-separable CNN. Electronics **12**(6), 1299 (2023)
16. Park, J.-H., Shanbhag, S., Chang, D.E.: Model-free unsupervised anomaly detection of a general robotic system using a stacked LSTM and its application to a fixed-wing unmanned aerial vehicle. In: 2022 IEEE/RSJ International Conference on Intelligent Robots and Systems (IROS), pp. 4287–4293. IEEE (2022)
17. Keipour, A., Mousaei, M., Scherer, S.: Automatic real-time anomaly detection for autonomous aerial vehicles. In: 2019 International Conference on Robotics and Automation (ICRA), pp. 5679–5685. IEEE (2019)
18. Chaudhari, S., Mithal, V., Polatkan, G., Ramanath, R.: An attentive survey of attention models. ACM Trans. Intell. Syst. Technol. (TIST) **12**(5), 1–32 (2021)
19. Enze, S., Cai, S., Xie, L., Li, H., Schultz, T.: Stanet: a spatiotemporal attention network for decoding auditory spatial attention from EEG. IEEE Trans. Biomed. Eng. **69**(7), 2233–2242 (2022)
20. Keipour, A., Mousaei, M., Scherer, S.: Alfa: a dataset for UAV fault and anomaly detection. Int. J. Robot. Res. **40**(2–3), 515–520 (2021)
21. Vaswani, A., et al.: Attention is all you need. In: Advances in Neural Information Processing Systems, vol. 30 (2017)
22. Yong, Yu., Si, X., Changhua, H., Zhang, J.: A review of recurrent neural networks: LSTM cells and network architectures. Neural Comput. **31**(7), 1235–1270 (2019)
23. Zhang, P., Ding, S.X.: A model-free approach to fault detection of continuous-time systems based on time domain data. Int. J. Autom. Comput. **4**, 189–194 (2007)
24. Usenko, V., Engel, J., Stückler, J., Cremers, D.: Direct visual-inertial odometry with stereo cameras. In: 2016 IEEE International Conference on Robotics and Automation (ICRA), pp. 1885–1892. IEEE (2016)
25. Sherstinsky, A.: Fundamentals of recurrent neural network (RNN) and long short-term memory (LSTM) network. Physica D **404**, 132306 (2020)
26. Li, B., et al.: Dense nested attention network for infrared small target detection. IEEE Trans. Image Process. **32**, 1745–1758 (2022)
27. En, F., Zhang, Y., Yang, F., Wang, S.: Temporal self-attention-based conv-LSTM network for multivariate time series prediction. Neurocomputing **501**, 162–173 (2022)
28. Bergmann, P., Batzner, K., Fauser, M., Sattlegger, D., Steger, C.: The MVTec anomaly detection dataset: a comprehensive real-world dataset for unsupervised anomaly detection. Int. J. Comput. Vision **129**(4), 1038–1059 (2021)
29. Press, W.H., Teukolsky, S.A.: Savitzky-Golay smoothing filters. Comput. Phys. **4**(6), 669–672 (1990)
30. Wang, B., Liu, D., Peng, Y., Peng, X.: Multivariate regression-based fault detection and recovery of UAV flight data. IEEE Trans. Instrum. Meas. **69**(6), 3527–3537 (2019)
31. Ahmad, M.W., Akram, M.U., Ahmad, R., Hameed, K., Hassan, A.: Intelligent framework for automated failure prediction, detection, and classification of mission critical autonomous flights. ISA Trans. **129**, 355–371 (2022)

Spatial and Frequency Domains Inconsistency Learning for Face Forgery Detection

Caili Gao, Peng Qiao(✉), Yong Dou, Qisheng Xu, Xifu Qian, and Wenyu Li

National University of Defense Technology, Changsha, China
{gaocl,pengqiao}@nudt.edu.cn

Abstract. With the rapid development of face forgery technology, it has attracted widespread attention. The current face forgery detection methods, whether based on Convolutional Neural Network (CNN) or Vision Transformer (ViT), are biased towards extracting local or global features respectively. These methods are relatively one-sided, and the extracted features are not robust and general enough. In this work, we exploit intra-frame inconsistency as well as inter-modal inconsistency between spatial and frequency domains to improve performance and generalization for face forgery detection. We efficiently extract intra-frame inconsistency by utilizing the capabilities of Swin Transformer. Its self-attention mechanism and attention mapping between patch embeddings naturally represent the inconsistency relations, allowing for simultaneous modeling of both local and global features, making it our ideal choice. Meanwhile, we also introduce frequency information to further improve detection performance, and design a Cross-Attention Feature Fusion (CAFF) module to exploit the inconsistency between spatial and frequency modalities to extract more general feature representations. Extensive experiments demonstrate the effectiveness of the proposed method.

Keywords: Face Forgery Detection · Inconsistency Learning · Swin Transformer · Feature Fusion

1 Introduction

The rapid advancement of face forgery technology makes fake faces more and more realistic in appearance and expression, and it is difficult to distinguish them from real faces. The abuse of this technology may lead attackers to easily forge fake videos of public figures and politicians, bringing serious trust crises to society and the country. Although some studies [1–4] have achieved good performance on known forgery techniques, existing detectors often show significant performance degradation when faced with new and unseen forgery methods. At the same time, new forgery techniques are emerging at an unprecedented pace, emphasizing the urgent need for a face forgery detection method with strong generalization capabilities.

B. Luo et al. (Eds.): ICONIP 2023, CCIS 1966, pp. 528–542, 2024.
https://doi.org/10.1007/978-981-99-8148-9_41

Forgery traces usually manifest as local texture, geometry, and illumination anomalies, which enable most convolutional neural network (CNN)-based methods [3,5–8] to achieve remarkable results. However, CNNs exhibit limited capabilities in modeling global context and long-range dependencies. To further improve the performance and generalization of forgery detection, some methods start to explore intra-frame inconsistency [9,10]. These approaches leverage the self-attention mechanism and attention maps between patch embeddings in Vision Transformers (ViTs) to capture consistent relationships and achieve satisfactory results. Nevertheless, ViTs tend to pay less attention to modeling local features, thereby limiting their ability to effectively capture subtle forgery traces.

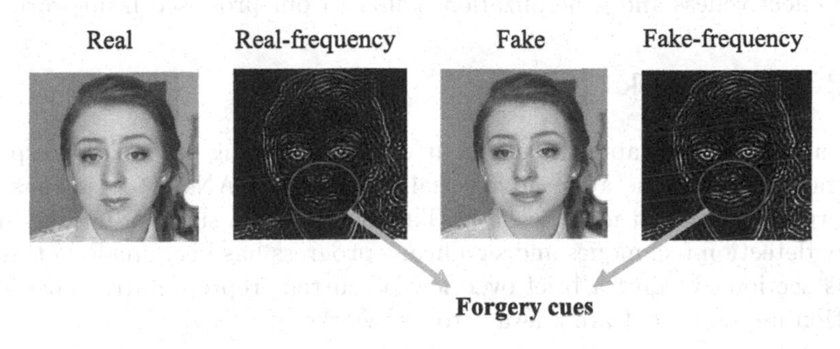

Fig. 1. The visualization results of a certain frequency band for real and fake faces, the frequency domain can be used as an important supplementary information.

Furthermore, subtle artifacts can induce anomalies in the frequency domain, especially in middle and high frequency bands, as illustrated in Fig. 1, many methods have extensively introduced the frequency domain as supplementary information [2,11–14]. However, these approaches often overlook the connections and inconsistencies between the spatial and frequency domains. For instance, the FDFL [15] method simply fuses the spatial stream and the frequency stream through a convolutional layer. This simple fusion method cannot make full use of the correlation between the spatial and frequency modalities, resulting in the extracted features not being general enough.

In this paper, we propose a Spatial and Frequency Domains Inconsistency Learning framework. Specifically, our framework focuses on two key types of inconsistencies: intra-frame inconsistency, and inter-modal inconsistency between the spatial and frequency domains. We leverage the Swin Transformer's capacity to efficiently extract intra-frame inconsistency. It's window-based self-attention mechanism and hierarchical structure enable the simultaneous modeling of both local and global features, making it an ideal choice for our purposes. Additionally, we use frequency as supplementary information and develop a Cross-Attention Feature Fusion (CAFF) module to discover inconsistencies between the two modalities. This module provides complementary information according to the inconsistency between the modalities, facilitating comprehensive interaction between the information from both spatial and frequency domains, thus enabling the extraction of more generalized features.

Our contributions can be summarized as follows:

- Our method can consider both local and global features, effectively extracting intra-frame inconsistency clues, and thereby enhancing the detection performance.
- We design a Cross-Attention Feature Fusion (CAFF) module to extract more general feature representations by exploiting the inconsistency between spatial and frequency modalities.
- We conduct extensive experiments on two benchmark datasets, demonstrating the effectiveness and generalization ability of our proposed framework.

2 Related Work

In recent years, the rapid development of deep learning technology, especially the emergence of generative adversarial networks (GANs) [16–19], has made face forgery easier and more realistic. This has led to a surge of interest in face forgery detection techniques and significant progress has been made in this field. In this section, we give a brief overview of current representative face forgery detection methods that are relevant to our work.

2.1 Face Forgery Detection

Spatial-Based. To improve the accuracy of fake face detection, various techniques have been proposed, with many of them focusing on spatial domain analysis. Early methods relied on identifying intuitive inconsistencies during the synthesis process to detect forgeries. For instance, Li et al. [20] observed that the blink rate of fake videos was lower than that of real individuals. Yang et al. [21] detect fake videos by checking for inconsistencies in 3D head poses. Similarly, Matern et al. [22] uses artificially designed visual features (eg. eyes, nose, teeth, etc.) to distinguish real and fake faces.

Recent methods exploit deep neural networks to capture high-level features from the spatial domain and achieve outstanding performance. Rossler et al. [5] achieved promising results on a variety of forgery algorithms using the Xception [23]. Face X-ray [6] mainly focuses on the fusion step present in most face forgeries, which remains effective even when dealing with forgeries produced using previously unseen face manipulation techniques. Both RFM [24] and SBIs [25] enable CNN classifiers to learn general and robust representations through data augmentation. Ni et al. [26] proposed a consistent representation learning framework based on data augmentation, explicitly constraining the consistency of different representations, thereby enabling traditional CNN models to achieve optimal performance. However, most of the current methods only use the spatial domain information and are not sensitive enough to the small forgery cues that are difficult to detect in the spatial domain. In our work, we combine the spatial and frequency domains to take full advantage of both.

Frequency-Based. In digital image processing, frequency domain analysis is widely used in tasks such as image classification, super-resolution, image enhancement, and denoising. In recent research, some methods utilize frequency domain information to capture forgery traces, they usually use Discrete Fourier Transform (DFT), Discrete Cosine Transform (DCT), or Wavelet Transform (WT) to convert the image to a frequency domain. F^3-Net [2] uses DCT to extract frequency domain information and analyze statistical features for face forgery detection. Two-branch [11] uses Laplacian of Gaussian for frequency enhancement to suppress image content in low-level feature maps. Liu et al. [13] observed that cumulative upsampling leads to significant changes in the frequency domain, especially the phase spectrum, and thus utilized the phase information to detect upsampling artifacts. While these methods extract more comprehensive features, there is still room for improvement in utilizing frequency information effectively. To learn more generalized features, we propose a cross-attention fusion mechanism that fully exploits the correlations between the spatial and frequency domains.

2.2 Vision Transformer

The introduction of the Transformer [27] model, widely used in the field of natural language processing, into computer vision, gave rise to Vision Transformer (ViT) [28]. Compared to traditional convolutional neural networks, ViT can capture more global and richer feature representations, while also exhibiting strong parallel computing capabilities and scalability. In the domain of face forgery detection, ViT has been successfully applied in several approaches. For example, Khan et al. [29] proposed a video transformer that leverages ViT to extract spatial features with temporal information, enabling effective forgery detection. Zheng et al. [30] designed a lightweight temporal transformer based on their full temporal convolutional network, which explores temporal coherence for general video manipulation detection. ICT [9] utilizes ViT to extract both internal and external identity features of faces, allowing for authenticity judgment based on identity consistency. UIA-ViT [10] designs a ViT-based unsupervised block consistency learning module that focuses on capturing intra-frame inconsistency without requiring pixel-level annotations. Recognizing some limitations of ViT in certain visual tasks, a multitude of variants quickly emerged [31–33]. Building on the work of Liu et al. [31], who demonstrated the Swin Transformer's excellent ability to capture both global and local information, we applied this variant to face forgery detection and achieved commendable performance.

3 Method

3.1 Overview

Current face image forgery techniques have made synthetic images indistinguishable from real photos, which also makes forgery detection extremely difficult.

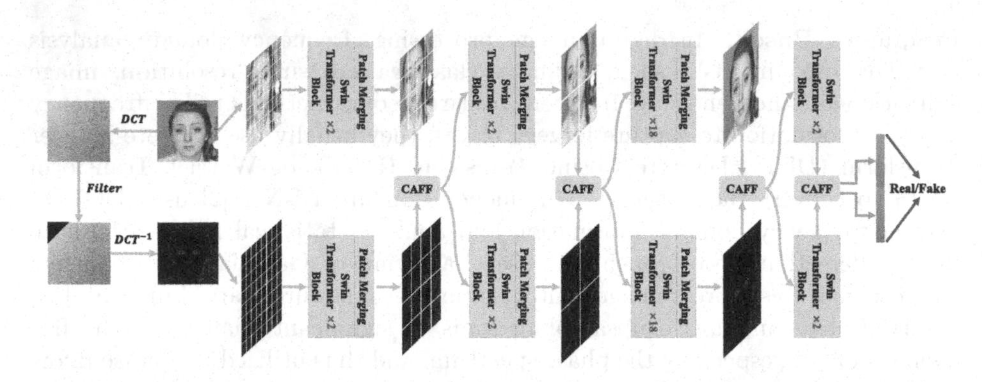

Fig. 2. The framework of the proposed Spatial and Frequency Domains Inconsistency Learning for Face Forgery Detection.

Although we can look for forgery traces in the spatial domain, some forgeries are not easy to detect in the spatial domain and may appear in the form of unusual frequency distributions. Therefore, we consider frequency domain information as a supplementary clue and approach forgery detection as a multimodal task, leveraging both spatial domain and frequency domain information as distinct modalities to uncover forgery clues.

To achieve this goal, as shown in Fig. 2, we adopt a two-stream architecture, use Swin Transformer to extract spatial domain and frequency domain features separately, and fuse the information of these two modalities through the cross-attention mechanism to identify intra-frame inconsistency and inter-modal inconsistency. This comprehensive approach enables us to more accurately detect traces of forgery. In the following sections, we will delve into the specific details of our approach.

3.2 Frequency Clue

Due to subtle artifacts that can cause anomalies in the frequency domain, we introduce a frequency information auxiliary network to capture complementary information from both the spatial and frequency domains. Leveraging the advantageous frequency distribution properties of the DCT, we utilize it to extract frequency cues in our approach.

Suppose the input RGB image is denoted as $x_1 \in \mathbb{R}^{H \times W \times 3}$, where H and W represent the height and width of the image, respectively. We first convert x_1 to a grayscale image:

$$x_1^g = G(x_1) \tag{1}$$

where $x_1^g \in \mathbb{R}^{H \times W \times 1}$ and G represents the grayscale image conversion. Next, we transform it from the spatial domain to the frequency domain using the DCT:

$$x_1^d = D(x_1^g) \tag{2}$$

where D represents the DCT operation. We then filter out low-frequency information to amplify tiny artifacts at high frequencies. This is achieved through a high-pass filtering operation:

$$x_1^f = F(x_1^d, \alpha) \tag{3}$$

where F represents the high-pass filtering function, and the parameter α is used to control the filtering of low-frequency components. Specifically, F sets the upper-left corner of x_1^d to 0 within a triangular region of length α.

Finally, we invert x_1^f back to the RGB color space by performing the inverse DCT transform D^{-1}, ensuring that the artifact locations in the spatial and frequency domains correspond. The entire process can be summarized as follows:

$$x_2 = D^{-1}(F(D(G(x_1)), \alpha)) \tag{4}$$

where $x_2 \in \mathbb{R}^{H \times W \times 3}$ represents the frequency domain image.

3.3 Intra-frame Inconsistency Feature Extraction

Intra-frame inconsistency has been shown to improve the performance and generalization of forgery detection [10], we leverage the window-based self-attention mechanism and hierarchical structure of the Swin Transformer [31] to capture inconsistencies across various spatial locations and layers. This is achieved through two main components: the Swin Transformer block and the patch merging layer.

The Swin Transformer block adopts a shifted window attention mechanism. By applying a local window on the feature map and using a shift operation to enable the window to cover the entire feature map, this mechanism can learn more representative intra-frame inconsistency. For the input feature F, the calculation formula of the Swin Transformer block is as follows:

$$\begin{aligned} q &= \mathrm{LN}(F)W_q \\ k &= \mathrm{LN}(F)W_k \\ v &= \mathrm{LN}(F)W_v \\ F' &= \mathrm{(S)W\text{-}MSA}(q, k, v) + F \\ F_o &= \mathrm{MLP}(\mathrm{LN}(F') + F') \end{aligned} \tag{5}$$

where LN represents the layer normalization, W_q, W_k, and W_v are the parameter matrices for linear projections, W-MSA and SW-MSA respectively denote window-based multi-head self-attention using regular and shifted window partitioning configurations, while MLP stands for multi-layer perceptron.

To generate a hierarchical representation, as the network deepens, the patch merging layer merges the extracted local features to obtain a more global feature representation. Specifically, the features of each group of 2×2 neighboring patches are concatenated to form a $4\times$ dimensional concatenated feature, and then a linear layer is applied to reduce the dimension to $2\times$, thus merging small-size patches into larger patches. Combined with the window-based self-attention mechanism, the network can capture both local details and global context simultaneously.

3.4 Cross-Attention Feature Fusion Module

For fake images, spatial domain information is useful for locating abnormal textures, while frequency information helps amplify tiny artifacts. Effectively fusing these two types of information can achieve a more comprehensive feature representation. Most of the current fusion methods use a simple structure at the end of the network [15], and we design multiple spatial and frequency domains Cross-Attention Feature Fusion (CAFF) modules in the backbone network to leverage the inter-modal relationship and capture inconsistencies, thereby enhancing the quality of extracted features. Experimental results have confirmed the effectiveness of this design.

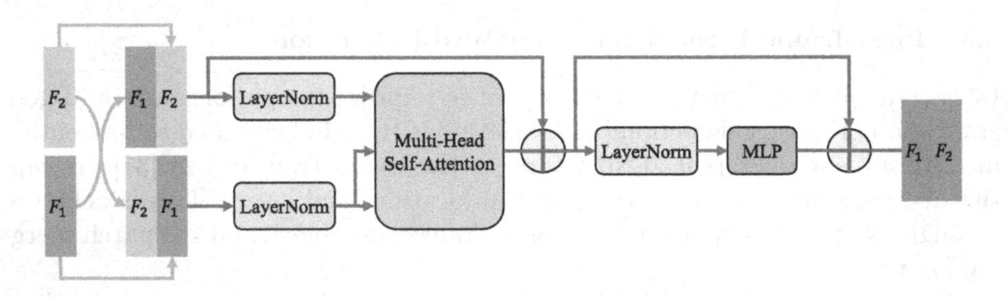

Fig. 3. Cross-Attention Feature Fusion Module.

Specifically, the structure of the CAFF is shown in Fig. 3. For the spatial domain features F_1 and frequency domain features F_2 extracted by the Swin Transformer block, where $F_1, F_2 \in \mathbb{R}^{H\times W\times D}$, with H, W, and D denoting the height, width, and feature dimension respectively, we employ cross-attention for feature fusion. Unlike self-attention, we make F_1 fuse F_2 information, linearly project F_1 to get q, and linearly project F_2 to get k and v. Conversely, the same is true for F_2 fused with F_1 information. To enable parallel computation of the cross-fusion process of F_1 and F_2, we concatenate F_1 and F_2. The specific calculation formula is as follows:

$$
\begin{aligned}
q &= \mathrm{LN}([F_1, F_2])W_q \\
k &= \mathrm{LN}([F_2, F_1])W_k \\
v &= \mathrm{LN}([F_2, F_1])W_v \\
[F_1, F_2]' &= \mathrm{MSA}(q, k, v) + [F_1, F_2] \\
[F_1, F_2]_o &= \mathrm{MLP}(\mathrm{LN}([F_1, F_2]') + [F_1, F_2]'
\end{aligned} \tag{6}
$$

where MSA represents the standard multi-head self-attention.

4 Experiments

4.1 Dataset

We evaluate our proposed method on two publicly available face forgery detection datasets, FaceForensics++ (FF++) [5], and Celeb-DF [34]. FF++ [5] is one of the most widely used datasets, containing 1000 real videos. And 4000 fake videos were synthesized using four forgery methods. In order to simulate the real situation, the original video was compressed with different compression rates: Raw, High Quality (HQ), and Low Quality (LQ), we use HQ for experiments. Celeb-DF [34] comprises 590 real videos of 59 celebrities collected from YouTube, along with 5639 high-quality fake videos synthesized using an enhanced Deepfake algorithm. Compared to previous datasets, Celeb-DF exhibits superior visual quality.

4.2 Experimental Setting

Data Pre-processing. We first extract image frames, then use RetinaFace [35] to detect faces and resize them to 256×256. For Celeb-DF, we sample 3 frames per second. For FF++, we take the first 270 frames for each training video, and the first 110 frames for each validation/testing video.

Implementation Details. We apply random and center cropping into training and testing, resizing images to 224×224, and additionally horizontally flip each image with 50% probability during training. We use the Swin Transformer-base architecture as the backbone network, where the patch size is 4×4 and the window size is 7×7. It contains a total of 4 stages, and each stage contains {2, 2, 18, 2} blocks, corresponding to the number of attention heads is {4,8,16,32}. The batch size is set to 16, each batch consists of 8 real and 8 fake face images, and the Adam optimizer with an initial learning rate of 3e-5 is used. For the filtering hyperparameter α, we conducted exploratory experiments and set it to 0.333.

Evaluation Metrics. We report the Area Under Curve (AUC) of ROC as the primary metric, and True Detect Rate (TDR) at False Detect Rate (FDR) of 0.01% (denoted as $\text{TDR}_{0.01\%}$) and 0.1% (denoted as $\text{TDR}_{0.1\%}$) as secondary metrics.

4.3 Comparison with Other Methods

Evaluation on FF++. We divide the existing methods into CNN, RNN, and Vision Transformer classes, as shown in Table 1. When comparing our proposed method with these existing methods, we can observe that:

1) Our proposed method outperforms most of the compared methods. Both our proposed method and the state-of-the-art method achieve an AUC value exceeding 99.50%. This result highlights the effectiveness of our proposed method in detecting face forgery.

Table 1. Quantitative results in terms of AUC(%) on the FaceForensics++ dataset. Our proposed approach achieves a similar performance to the SOTA method. We report the results from the original paper. **Bold** and underlined values correspond to the best and the second-best value, respectively.

Methods	Reference	Backbone	AUC
Face X-ray [6]	CVPR 2020	HRNet	87.40
Two-branch [11]	ECCV 2020	LSTM	88.70
F^3-Net [2]	ECCV 2020	Xception	98.10
PCL+I2G [36]	ICCV 2021	ResNet-34	99.11
Local-relation [14]	AAAI 2021	Xception	99.46
LTW [8]	AAAI 2021	EfficientNet-B0	99.17
DCL [37]	AAAI 2021	EfficientNet-B4	99.30
MADD [3]	CVPR 2021	EfficientNet-B4	99.29
FDFL [15]	CVPR 2021	Xception	99.30
CORE [26]	CVPR 2022	Xception	**99.66**
UIA-ViT [10]	ECCV 2022	Vision Transformer	99.33
Ours		Swin Transformer	99.51

Table 2. Performance (%) comparison among previous state-of-the-art methods on the Celeb-DF dataset.

Methods	Reference	AUC	$TDR_{0.1\%}$	$TDR_{0.01\%}$
Xception [23]	CVPR 2017	99.85	89.11	84.22
RE [38]	CVPR 2017	99.89	88.11	85.20
AE [39]	AAAI 2020	99.84	84.05	76.63
Patch [40]	ECCV 2020	99.96	91.83	86.16
RFM-Xception [24]	CVPR 2021	99.94	93.88	87.08
RFM-Patch [24]	CVPR 2021	**99.97**	93.44	89.58
CORE [26]	CVPR 2022	**99.97**	95.58	93.98
Ours		**99.97**	**99.34**	**95.67**

2) Compared to the two-stream approaches, Two-branch [11] and Local-relation [14], which consider both spatial and frequency domains simultaneously, our method achieves better performance. This demonstrates the effectiveness of our proposed CAFF.
3) Compared with the same Transformer-based method UIA-ViT [10], our method exceeds it by 0.18%. Instead of only using spatial domain information, our model integrates spatial domain and frequency domain information, so that it can more comprehensively capture traces of forgery.

Evaluation on Celeb-DF. As shown in Table 2, our method performs the best among the evaluated methods. Specifically, when compared to the previous state-of-the-art method CORE [26], our method achieves improvements of 3.76 $TDR_{0.1\%}$ and 1.69 $TDR_{0.01\%}$, proving that our method can achieve good performance on multiple datasets.

4.4 Cross-Dataset Evaluation

In this section, we evaluated the generalization ability of our proposed model and several state-of-the-art models, and the results of some methods were directly cited from [3].

Table 3. Cross-dataset evaluation in terms of AUC(%) on Celeb-DF by training on FaceForensics++. Our method achieves the best performance.

Methods	Train Set	DF	Celeb-DF
Two-stream [41]	FF++	70.10	53.80
Meso4 [42]	FF++	84.70	54.80
Mesolnception4 [42]	FF++	83.00	53.60
FWA [43]	FF++	80.10	56.90
Xception [23]	FF++	99.70	65.30
Multi-task [44]	FF++	76.30	54.30
Capsule [45]	FF++	96.60	57.50
DSP-FWA [43]	FF++	93.00	64.60
Two Branch [11]	FF++	93.18	73.41
F^3-Net [2]	FF++	98.10	65.17
EfficientNet-B4 [46]	FF++	99.70	64.29
MADD [3]	FF++	99.80	67.44
Finfer [1]	FF++	95.67	70.60
Ours	FF++	**99.93**	**74.18**
SOLA [4]	DF	-	76.02
Ours	DF	**99.95**	**81.55**

Referring to the settings of MADD [3], we used multiple forgery methods for training on FF++ and tested on Deepfakes class (In-dataset) and Celeb-DF (Cross-dataset). As shown in Table 3, our method achieves the best performance both in-dataset and cross-dataset, confirming the effectiveness and superior generalization of our approach.

In addition, we noticed that SOLA [4] only uses the Deepfakes class for training, and the cross-dataset performance of Celeb-DF is higher. This is because the Celeb-DF dataset is also synthesized using the Deepfake forgery method.

This setting is unfair. We also conducted corresponding experiments and found that our method greatly outperforms SOLA, which also combines spatial and frequency domains.

4.5 Ablation Study

Ablation Study on Low-Frequency Component Size α. We explored the setting of filtering out the low-frequency component size α on the Celeb-DF dataset, and the results are shown in Fig. 4. We found that the model with $\alpha = 0.333$ achieved the best performance.

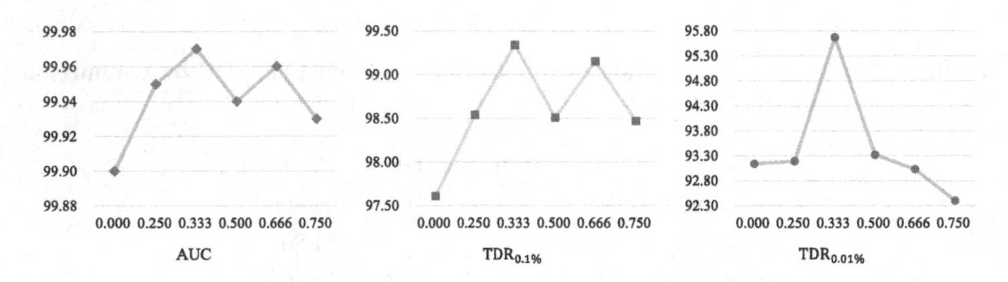

Fig. 4. Results (%) with different α on Celeb-DF.

Ablation Study on Fusion Methods. To demonstrate the effectiveness of Cross-Attention Feature Fusion, we compare it with the following approaches:

1) Spatial and Spatial (×2): One-stream structures in the spatial domain with single and double parameters, respectively.
2) FF (Feature Fusion): Feature fusion fuses the output features of each stream to classify real and fake. Each stream uses Swin Transformer to extract corresponding features, which are then concatenated together and fed into a fully connected layer to make the final decision.
3) SF (Score Fusion): Use the score fusion strategy to fuse the classification information of each stream. The score of each stream corresponds to the output of the last fully connected layer, which is essentially the result of passing the features extracted by the Swin Transformer through a fully connected layer. Let S_1 represent the score of the spatial stream and S_2 represent the score of the frequency stream. The final score value S of the model is defined as follows:

$$S = w_1 \cdot \text{Softmax}(S_1) + w_2 \cdot \text{Softmax}(S_2) \tag{7}$$

where w_1 and w_2 respectively represent the weight values assigned to the two streams, which are learnable variables.

Table 4. The ablation study results(%) on fusion methods.

Methods	AUC	$TDR_{0.1\%}$	$TDR_{0.01\%}$
Spatial	99.95	98.58	92.65
Spatial (×2)	99.94	99.03	91.80
Spatial+Frequency (FF)	99.96	99.32	94.04
Spatial+Frequency (SF)	99.85	95.12	87.51
Spatial+Frequency (CAFF)	**99.97**	**99.34**	**95.67**

The experimental results on the Celeb-DF dataset are shown in Table 4. Firstly, the one-stream structure in the spatial domain achieves satisfactory results, but the effect is slightly worse than that of the two-stream structure. We also observe that the two-stream structure is better than the one-stream double parameters structure, which shows that the performance improvement of the two-stream structure is not brought by the larger parameter scale.

Secondly, we noticed a degradation in performance when combining spatial and frequency domains through score fusion. This hints at the complexity of information fusion. Nonetheless, our proposed Cross-Attention Feature Fusion method shows the best performance.

5 Conclusion

In this paper, we propose a novel framework for face forgery detection, Spatial and Frequency Domains Inconsistency Learning, which effectively captures both intra-frame inconsistency and inter-modal inconsistency between the spatial and frequency domains. To capture intra-frame inconsistency, we employ the Swin Transformer model, while the CAFF module is utilized to capture inconsistency between spatial and frequency modalities. Through extensive experiments conducted on two publicly available datasets, our method exhibits exceptional performance and remarkable generalization capabilities.

References

1. Hu, J., Liao, X., Liang, J., Zhou, W., Qin, Z.: Finfer: frame inference-based deepfake detection for high-visual-quality videos. In: Proceedings of the AAAI Conference on Artificial Intelligence, vol. 36, pp. 951–959 (2022)
2. Qian, Y., Yin, G., Sheng, L., Chen, Z., Shao, J.: Thinking in frequency: face forgery detection by mining frequency-aware clues. In: Vedaldi, A., Bischof, H., Brox, T., Frahm, J.-M. (eds.) ECCV 2020. LNCS, vol. 12357, pp. 86–103. Springer, Cham (2020). https://doi.org/10.1007/978-3-030-58610-2_6
3. Zhao, H., Zhou, W., Chen, D., Wei, T., Zhang, W., Yu, N.: Multi-attentional deepfake detection. In: Proceedings of the IEEE/CVF Conference on Computer Vision and Pattern Recognition, pp. 2185–2194 (2021)

4. Fei, J., Dai, Y., Yu, P., Shen, T., Xia, Z., Weng, J.: Learning second order local anomaly for general face forgery detection. In: Proceedings of the IEEE/CVF Conference on Computer Vision and Pattern Recognition, pp. 20270–20280 (2022)
5. Rossler, A., Cozzolino, D., Verdoliva, L., Riess, C., Thies, J., Nießner, M.: Faceforensics++: learning to detect manipulated facial images. In: Proceedings of the IEEE/CVF International Conference on Computer Vision, pp. 1–11 (2019)
6. Li, L., et al.: Face X-ray for more general face forgery detection. In: Proceedings of the IEEE/CVF Conference on Computer Vision and Pattern Recognition, pp. 5001–5010 (2020)
7. Li, W., Qiao, P., Dou, Y.: Efficiently classify synthesized facial images generated by different synthesis methods. In: Twelfth International Conference on Graphics and Image Processing (ICGIP 2020), vol. 11720, pp. 250–256. SPIE (2021)
8. Sun, K., et al.: Domain general face forgery detection by learning to weight. In: Proceedings of the AAAI Conference on Artificial Intelligence, vol. 35, pp. 2638–2646 (2021)
9. Dong, X., et al.: Protecting celebrities from deepfake with identity consistency transformer. In: Proceedings of the IEEE/CVF Conference on Computer Vision and Pattern Recognition, pp. 9468–9478 (2022)
10. Zhuang, W., et al.: UIA-ViT: unsupervised inconsistency-aware method based on vision transformer for face forgery detection. In: Avidan, S., Brostow, G., Cissé, M., Farinella, G.M., Hassner, T. (eds.) ECCV 2022. LNCS, vol. 13665, pp. 391–407. Springer, Cham (2022). https://doi.org/10.1007/978-3-031-20065-6_23
11. Masi, I., Killekar, A., Mascarenhas, R.M., Gurudatt, S.P., AbdAlmageed, W.: Two-branch recurrent network for isolating deepfakes in videos. In: Vedaldi, A., Bischof, H., Brox, T., Frahm, J.-M. (eds.) ECCV 2020. LNCS, vol. 12352, pp. 667–684. Springer, Cham (2020). https://doi.org/10.1007/978-3-030-58571-6_39
12. Yu, M., Zhang, J., Li, S., Lei, J., Wang, F., Zhou, H.: Deep forgery discriminator via image degradation analysis. IET Image Proc. **15**(11), 2478–2493 (2021)
13. Liu, H., et al.: Spatial-phase shallow learning: rethinking face forgery detection in frequency domain. In: Proceedings of the IEEE/CVF Conference on Computer Vision and Pattern Recognition, pp. 772–781 (2021)
14. Chen, S., Yao, T., Chen, Y., Ding, S., Li, J., Ji, R.: Local relation learning for face forgery detection. In: Proceedings of the AAAI Conference on Artificial Intelligence, vol. 35, pp. 1081–1088 (2021)
15. Li, J., Xie, H., Li, J., Wang, Z., Zhang, Y.: Frequency-aware discriminative feature learning supervised by single-center loss for face forgery detection. In: Proceedings of the IEEE/CVF Conference on Computer Vision and Pattern Recognition, pp. 6458–6467 (2021)
16. Goodfellow, I., et al.: Generative adversarial networks. Commun. ACM **63**(11), 139–144 (2020)
17. Choi, Y., Choi, M., Kim, M., Ha, J.-W., Kim, S., Choo, J.: Stargan: unified generative adversarial networks for multi-domain image-to-image translation. In: Proceedings of the IEEE Conference on Computer Vision and Pattern Recognition, pp. 8789–8797 (2018)
18. Karras, T., Laine, S., Aila, T.: A style-based generator architecture for generative adversarial networks. In: Proceedings of the IEEE/CVF Conference on Computer Vision and Pattern Recognition, pp. 4401–4410 (2019)
19. Karras, T., Aila, T., Laine, S., Lehtinen, J.: Progressive growing of GANs for improved quality, stability, and variation. arXiv preprint arXiv:1710.10196 (2017)

20. Li, Y., Chang, M.-C., Lyu, S.: In ictu oculi: exposing AI created fake videos by detecting eye blinking. In: 2018 IEEE International Workshop on Information Forensics and Security (WIFS), pp. 1–7. IEEE (2018)
21. Yang, X., Li, Y., Lyu, S.: Exposing deep fakes using inconsistent head poses. In: ICASSP 2019-2019 IEEE International Conference on Acoustics, Speech and Signal Processing (ICASSP), pp. 8261–8265. IEEE (2019)
22. Matern, F., Riess, C., Stamminger, M.: Exploiting visual artifacts to expose deepfakes and face manipulations. In: 2019 IEEE Winter Applications of Computer Vision Workshops (WACVW), pp. 83–92. IEEE (2019)
23. Chollet, F.: Xception: deep learning with depthwise separable convolutions. In: Proceedings of the IEEE Conference on Computer Vision and Pattern Recognition, pp. 1251–1258 (2017)
24. Wang, C., Deng, W.: Representative forgery mining for fake face detection. In: Proceedings of the IEEE/CVF Conference on Computer Vision and Pattern Recognition, pp. 14923–14932 (2021)
25. Shiohara, K., Yamasaki, T.: Detecting deepfakes with self-blended images. In: Proceedings of the IEEE/CVF Conference on Computer Vision and Pattern Recognition, pp. 18720–18729 (2022)
26. Ni, Y., Meng, D., Yu, C., Quan, C., Ren, D., Zhao, Y.: Core: consistent representation learning for face forgery detection. In: Proceedings of the IEEE/CVF Conference on Computer Vision and Pattern Recognition, pp. 12–21 (2022)
27. Vaswani, A., et al.: Attention is all you need. In: Advances in Neural Information Processing Systems, vol. 30 (2017)
28. Dosovitskiy, A., et al.: An image is worth 16x16 words: transformers for image recognition at scale. arXiv preprint arXiv:2010.11929 (2020)
29. Khan, S.A., Dai, H.: Video transformer for deepfake detection with incremental learning. In: Proceedings of the 29th ACM International Conference on Multimedia, pp. 1821–1828 (2021)
30. Zheng, Y., Bao, J., Chen, D., Zeng, M., Wen, F.: Exploring temporal coherence for more general video face forgery detection. In: Proceedings of the IEEE/CVF International Conference on Computer Vision, pp. 15044–15054 (2021)
31. Liu, Z., et al.: Swin transformer: hierarchical vision transformer using shifted windows. In: Proceedings of the IEEE/CVF International Conference on Computer Vision, pp. 10012–10022 (2021)
32. Li, Y., et al.: Efficientformer: Vision transformers at mobilenet speed. Adv. Neural. Inf. Process. Syst. **35**, 12934–12949 (2022)
33. Chen, C.-F.R., Fan, Q., Panda, R.: Crossvit: cross-attention multi-scale vision transformer for image classification. In: Proceedings of the IEEE/CVF International Conference on Computer Vision, pp. 357–366 (2021)
34. Li, Y., Yang, X., Sun, P., Qi, H., Lyu, S.: Celeb-DF: a large-scale challenging dataset for deepfake forensics. In: Proceedings of the IEEE/CVF Conference on Computer Vision and Pattern Recognition, pp. 3207–3216 (2020)
35. Deng, J., Guo, J., Ververas, E., Kotsia, I., Zafeiriou, S.: Retinaface: single-shot multi-level face localisation in the wild. In: Proceedings of the IEEE/CVF Conference on Computer Vision and Pattern Recognition, pp. 5203–5212 (2020)
36. Zhao, T., Xu, X., Xu, M., Ding, H., Xiong, Y., Xia, W.: Learning self-consistency for deepfake detection. In: Proceedings of the IEEE/CVF International Conference on Computer Vision, pp. 15023–15033 (2021)
37. Sun, K., Yao, T., Chen, S., Ding, S., Li, J., Ji, R.: Dual contrastive learning for general face forgery detection. In: Proceedings of the AAAI Conference on Artificial Intelligence, vol. 36, pp. 2316–2324 (2022)

38. Zhong, Z., Zheng, L., Kang, G., Li, S., Yang, Y.: Random erasing data augmentation. In: Proceedings of the AAAI Conference on Artificial Intelligence, vol. 34, pp. 13001–13008 (2020)
39. Wei, Y., Feng, J., Liang, X., Cheng, M.M., Zhao, Y., Yan, S.: Object region mining with adversarial erasing: a simple classification to semantic segmentation approach. In: Proceedings of the IEEE Conference on Computer Vision and Pattern Recognition, pp. 1568–1576 (2017)
40. Chai, L., Bau, D., Lim, S.-N., Isola, P.: What makes fake images detectable? Understanding properties that generalize. In: Vedaldi, A., Bischof, H., Brox, T., Frahm, J.-M. (eds.) ECCV 2020. LNCS, vol. 12371, pp. 103–120. Springer, Cham (2020). https://doi.org/10.1007/978-3-030-58574-7_7
41. Zhou, P., Han, X., Morariu, V.I., Davis, L.S.: Two-stream neural networks for tampered face detection. In: 2017 IEEE Conference on Computer Vision and Pattern Recognition Workshops (CVPRW), pp. 1831–1839. IEEE (2017)
42. Afchar, D., Nozick, V., Yamagishi, J., Echizen, I.: Mesonet: a compact facial video forgery detection network. In: 2018 IEEE International Workshop on Information Forensics and Security (WIFS), pp. 1–7. IEEE (2018)
43. Li, Y., Lyu, S.: Exposing deepfake videos by detecting face warping artifacts. arXiv preprint arXiv:1811.00656 (2018)
44. Nguyen, H.H., Fang, F., Yamagishi, J., Echizen, I.: Multi-task learning for detecting and segmenting manipulated facial images and videos. In: 2019 IEEE 10th International Conference on Biometrics Theory, Applications and Systems (BTAS), pp. 1–8. IEEE (2019)
45. Nguyen, H.H., Yamagishi, J., Echizen, I.: Capsule-forensics: using capsule networks to detect forged images and videos. In: ICASSP 2019-2019 IEEE International Conference on Acoustics, Speech and Signal Processing (ICASSP), pp. 2307–2311. IEEE (2019)
46. Tan, M., Le, Q.: Efficientnet: rethinking model scaling for convolutional neural networks. In: International Conference on Machine Learning, pp. 6105–6114. PMLR (2019)

Enhancing Camera Position Estimation by Multi-view Pure Rotation Recognition and Automated Annotation Learning

Shuhao Jiang[1], Qi Cai[2(✉)], Yuanda Hu[3], and Xiuqin Zhong[1]

[1] School of Computer Science and Engineering, University of Electronic Science and Technology of China, Chengdu 611731, China
zhongxiuqin@uestc.edu.cn

[2] School of Electronic Information and Electrical Engineering, Shanghai Jiao Tong University, Shanghai 200240, China
qicaicn@gmail.com

[3] College of Design and Innovation, Tongji University, Shanghai 200092, China

Abstract. Pure rotational anomaly recognition is a critical problem in 3D visual computation, requiring precise recognition for reliable camera pose estimation and robust 3D reconstruction. Current techniques primarily focus on model selection, parallax angle, and intersection constraints within two-view geometric models when identifying pure rotational motion. This paper proposes a multi-view pure rotational detection method that draws upon two-view rotation-only recognition indicators to identify pure rotational views that cause pose estimation anomalies. An automatic data annotation and training strategy for rotation-only anomaly recognition in multi-view pose estimation data is also introduced. Our experiments demonstrate that our proposed model for rotation-only anomaly recognition achieves an accuracy of 91% on the test set and is highly effective in improving the precision, resilience, and performance of camera pose estimation, 3D reconstruction, object tracking, and other computer vision tasks. The effectiveness of our approach is validated through comparison with related approaches in simulated camera motion trajectory experiments and Virtual KITTI dataset.

Keywords: 3D Visual Computation · Position Estimation · Rotation-only Recognition · Video Understanding

1 Introduction

In 3D visual computation, cameras acquire a series of images or videos by traversing specific trajectories and capturing the surrounding environment. Visual geometry typically utilizes projection equations to define the relationship between the 3D scene and image observations. As a fundamental technology in Simultaneous Localization and Mapping (SLAM) and Structure from Motion (SfM), the Bundle Adjustment (BA) technique recovers camera poses and the surrounding scenery based on the obtained image observations and projection

B. Luo et al. (Eds.): ICONIP 2023, CCIS 1966, pp. 543–557, 2024.
https://doi.org/10.1007/978-981-99-8148-9_42

equations [2,9,16], which is a nonlinear optimization process that requires appropriate initial results for successful execution. Rotation-only motion is a persistent challenge in visual SLAM and SfM systems, where the camera only rotates without any translational motion. In fact, both the BA technique and mainstream camera position estimation methods (e.g., LUD [15] and LiGT [2] algorithm) may encounter anomalies when dealing with camera rotation-only movements [2]. When faced with rotation-only movements, the SLAM and SfM pipelines may generate low-quality initialization of pose and 3D scene for BA, which in turn leads to unreliable optimization outcomes. As a consequence, this results in a lack of robustness in camera position estimation and 3D scene reconstruction. To ensure stability, it is generally crucial to identify rotation-only movements, particularly those provoking estimation anomalies, before proceeding with camera pose estimation and BA. This paper primarily focuses on the following questions related to rotation-only movement issues in three-dimensional visual geometry computation:

- How can the two-view rotation-only recognition indicators be employed to recognize pure rotational views in multi-view estimation?
- Can pure rotational views and those causing position estimation anomalies be directly recognized from images before conducting 3D visual geometry computation?

Addressing above issues will help to improve the precision, resilience, and performance of camera pose estimation, 3D reconstruction, SLAM, SfM, object tracking, and other computer vision tasks.

Thus, the main contributions of this paper are two-fold:

1. Building upon the two-view rotation-only recognition indicators, a multi-view pure rotational indicator is devised to quantify the extent of rotation-only motion and is capable of identifying pure rotational views that cause pose estimation anomalies.
2. An automatic data annotation and training strategy for rotation-only anomaly recognition in multi-view pose estimation is designed. Through the trained neural network, images can be directly examined to recognize and eliminate pure rotational anomalies before visual geometry computation.

2 Related Work

Some previous work has demonstrated that handling rotation-only motion in 3D visual computation remains a challenging problem [1,2,4,17,21]. When depth-map information is readily established, Zhou et al. [22] combined a probabilistic depth map model based on Bayesian estimation with the main framework of LSD-SLAM, enabling it to track new pixels with low expectation values during rotation motion. Monocular SLAM can handle rotation-only motion if the camera observes image features with known 3D landmarks. Conversely, when landmark information is unavailable or real-time processing is necessary, identifying and managing rotation-only movements become crucial.

In cases where 3D landmark information was unknown, one approach identified rotation-only movements through geometric model selection. Civera et al. [4] utilized the interacting multiple model (IMM) approach to Bayesian model selection to identify the dominant motion model in the sequence, including pure rotation. The paper shows that the IMM approach can switch automatically between parameter sets in a dimensionless formulation of monocular SLAM. Gauglitz et al. [7,8] mentioned that SLAM systems did not support rotation-only camera motion and required that each feature be observed from two distinct camera locations. They tackled the recognition of rotation-only movements by calculating different models under various camera motion situations. Then, the geometric robust information criterion (GRIC) [20] was applied to choose the best representation of camera motion for the current camera motion and scene structure. When rotation-only movements were detected, the system switched to panorama tracking and map-making mode.

Alternatively, other approaches primarily employed pure rotational indicators to recognize such movements. In 2013, Pirchheim et al. [17] introduced a keyframe-based SLAM method for handling rotation-only motion. They suggested using the approximated parallax angle criterion to detect camera motion that allowed robust triangulation and 3D reconstruction. The system employed scale-independent thresholds for parallax and camera view angles to detect rotation-only motion. Building upon the theory of two view geometry, Herrera et al. [11] in 2014 utilized the characteristics of epipolar segments and proposed a real-time visual SLAM system that seamlessly transitions between pure rotation, essential matrix, and full 6DoF models by evaluating parallax. In 2019, Cai et al. [1] analyzed the geometric properties of rotation-only movements. They constructed the M_3 statistical indicator to identify two-view rotation-only movements, which is composed of all matched normalized image coordinates in images V_i and V_j, along with the rotation matrix $R_{i,j}$ between them. This indicator tends to be consistent with zero under rotation-only movements. Studies [1,21] demonstrate that utilizing the average m_3 statistic can identify whether there are rotation-only movements between two views. Additionally, [1] suggested an alternative pure rotational indicator for two-view scenarios by exploiting geometric properties, such as the observation matrix approaching zero, which is derived from intersection inequalities. Nevertheless, these indicators cannot determine which view may contribute to rotation-only movements in camera position estimation and 3D reconstruction from multiple views. Moreover, they cannot ascertain if any rotation-only movement demands attention in the current multi-view estimation, potentially resulting in challenges regarding robustness in multi-view pose recovery and 3D reconstruction.

3 Methodology

3.1 Pure Rotational Indicators in Multiple Views

Consider a set of 3D feature points $X = \{\boldsymbol{X}_1, \ldots, \boldsymbol{X}_m\}$ that are projected onto image feature observations within a multi-view set $V = \{V_1, \ldots, V_n\}$. Assume that camera is calibrated, rotation matrix and translation vector for each view V_i are represented by R_i, and $\boldsymbol{t}_i$, respectively. The Bundle Adjustment technology relies on the multi-view geometry representation, which utilizes projection equations as described by [9]. These projection equations of image observation set $\boldsymbol{x}$ can be expressed as follows:

$$\boldsymbol{x} = \{\boldsymbol{x}_{i,k} \mid \boldsymbol{x}_{i,k} \sim R_i(\boldsymbol{X}_k - \boldsymbol{t}_i), 1 \le i \le n, 1 \le k \le m\}, \tag{1}$$

where $\boldsymbol{x}_{i,k} = (x_{i,k}, y_{i,k}, 1)^T$, and $\sim$ represents equality in homogeneous coordinates.

In a two-view configuration denoted as (i, j) and established by V_i and V_j, the parallax angle associated with point $\boldsymbol{X}_k$ can be defined as $\alpha_{i,j,k} = \angle(\boldsymbol{t}_i\boldsymbol{X}_k\boldsymbol{t}_j)$. To simplify the description, assume that in the two-view (i, j), we can observe m' 3D feature points $\boldsymbol{X}' \in \boldsymbol{X}$, [17] used an approximate formula for the parallax angle, expressed as $\alpha_{i,j} = 2\arctan(d_{i,j}/(2f_{i,j}))$. Here, $f_{i,j}$ denotes the average depth of 3D points $\boldsymbol{X}'$ as seen in two views, and $d_{i,j}$ refers to the relative distance between the positions of $\boldsymbol{t}_i$ and $\boldsymbol{t}_j$. In the two-view (i, j), the M_3 pure rotational indicator provided by [1] is

$$\varphi_{i,j}^{M_3} = \frac{1}{m'} \sum_{k=1}^{m'} \frac{\left\| [\boldsymbol{x}_{j,k}]_\times R\boldsymbol{x}_{i,k} \right\|}{\|\boldsymbol{x}_{j,k}\| \, \|\boldsymbol{x}_{i,k}\|}. \tag{2}$$

We can use $\varphi_{i,j}^{M_3} < \delta_{M_3}$ to recognize the rotation-only relationship between two-view (i, j). Define the matrix $A = (\boldsymbol{x}_{i,1} \otimes \boldsymbol{x}_{j,1}, \ldots, \boldsymbol{x}_{i,m'} \otimes \boldsymbol{x}_{j,m'})^T$. Choose the singular vector $\boldsymbol{q}$ corresponding to the smallest singular value, and construct the matrix $Q_{3\times3}$ in column order from $\boldsymbol{q}$. Perform SVD decomposition on Q, we have $Q = U\Sigma V^T = (\boldsymbol{u}_1, \boldsymbol{u}_2, \boldsymbol{u}_3)\,\Sigma\,(\boldsymbol{v}_1, \boldsymbol{v}_2, \boldsymbol{v}_3)^T$. Then, we can define a vector

$$\boldsymbol{y} = \begin{pmatrix} \|\boldsymbol{x}_{j,1}\| \, \boldsymbol{x}_{i,1}^T & \|\boldsymbol{x}_{i,1}\| \, \boldsymbol{x}_{i,1}^T \\ \vdots & \vdots \\ \|\boldsymbol{x}_{j,m'}\| \, \boldsymbol{x}_{i,m'}^T & \|\boldsymbol{x}_{i,m'}\| \, \boldsymbol{x}_{i,m'}^T \end{pmatrix} \begin{pmatrix} -\det(UV)\boldsymbol{v}_3 \\ \boldsymbol{u}_3 \end{pmatrix}. \tag{3}$$

Then, the two-view PRI indicator provided in [1] is

$$\varphi_{i,j}^{PRI} = \frac{1}{m'} \sum_{k=1}^{m'} |y_i|, \tag{4}$$

where y_i is the i-th element in $\boldsymbol{y}$. If this indicator smaller than δ_{PRI}, it is considered to be pure rotational relationship for (i, j). The above-mentioned pure rotational indicators focus on relationships between two views, limiting their capacity to comprehend rotational motion in multi-view scenarios. Conventional methods only assess rotation-only between specific views, neglecting information from other views. As a result, these methods might produce an incomplete representation of the actual motion, which could potentially impact the accuracy of associated applications.

Algorithm 1. Calculate multiple-view pure rotational degree

Input: global rotation set $R = \{R_1, \ldots, R_n\}$, image observation set $\boldsymbol{x}$
Output: pure rotational degrees $\varphi = \{\varphi_1, \ldots, \varphi_n\}$
1: **for** $i = 1$ to n **do**
2: $V_{coviews} \leftarrow$ find all co-views of V_i from $\boldsymbol{x}$
3: $n_{coview} \leftarrow$ obtain the number of co-views
4: **for** $j = 1$ to $n_{coviews}$ **do**
5: $R_{i,j} \leftarrow$ calculate relative rotation from R
6: calculate and store two-view indicator $\varphi_{i,j}$ using eq.(2) or (4)
7: **end for**
8: $\varphi_i \leftarrow$ average of all $\varphi_{i,j}$ values for the current view V_i
9: **end for**

Algorithm 1 integrates information from multiple views, evaluating the pure rotational degree for each view. By combining image observations from various co-views, the algorithm achieves a more reliable estimation of rotational motion in the scene, improving representation and facilitating motion dynamics understanding. The algorithm iteratively processes each view, identifying all co-views sharing feature points with the current view. For every pair of co-views, it calculates relative rotation and evaluates the two-view indicator $\varphi_{i,j}$. Finally, it computes the average of all $\varphi_{i,j}$ values corresponding to the current view V_i, determining the degree of rotation-only for that view. Accounting for multiple views concurrently, Algorithm 1 enables a more accurate evaluation of pure rotational motion presence in each view. This approach identifies views with significant pure rotational motion while providing insights into the global structure and motion of the scene. Algorithm 1 can be regarded as a multiple-view pure rotational degree (PRD) function for a two-view pure rotational indicator, denoted by $prd()$. In the following experiments, $prd(M_3)$ and $prd(PRI)$ will be used to denote the respective statistical values derived from Algorithm 1. Furthermore, this algorithm can be effortlessly incorporated into existing computer vision workflows, thereby improving the robustness and performance of techniques such as SfM and Visual SLAM (Fig. 1).

3.2 Rotation-Only Anomaly Recognition Model

This section addresses the second question raised in the introduction: whether pure rotational views and those that cause position estimation anomalies can be

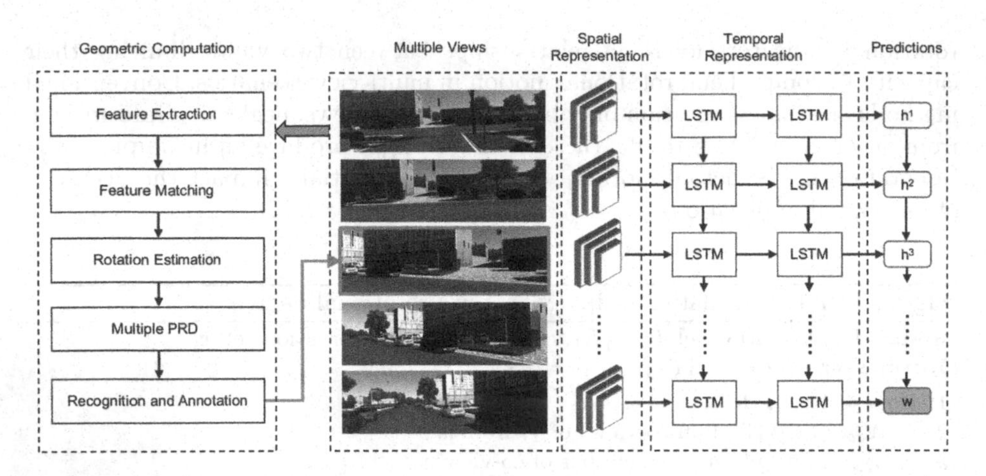

Fig. 1. Overview of automated annotation training scheme for rotation-only recognition.

directly recognized from images before conducting 3D visual geometry computations. Conventional approaches predominantly rely on pure rotation recognition indicators, which require the calculation of various visual geometric models or the determination of pure rotational indicators. Employing visual geometry methodologies, this section demonstrates the systematic annotation of an arbitrary open-source dataset for pure rotation recognition purposes. Subsequently, neural networks are utilized for training in order to ascertain the viability of automated annotation and the direct identification of pure rotation based solely on image data.

We extended prior research on video understanding [5,13,14,18,19] by implementing a baseline approach for identifying rotation-only anomaly in multi-view pose estimation. Specifically, we employed a network [5] combining CNNs and long short-term memory (LSTM) networks to extract the spatio-temporal features of the view sequences. Figure 6 provides an overview of the proposed architecture, which is designed to accurately identify and classify rotation-only anomalies within multi-view pose estimation data.

For a given i-th view within a multi-view set, we construct a sliding window of length w centered around V_i, denoted as

$$W_i = \{V_{i-[w/2]}, \dots, V_{i-1}, V_i, V_{i+1}, \dots, V_{i+[w/2]}\}. \tag{5}$$

Here, $[\cdot]$ represents the floor division, ensuring an integer value for the window size. By including the neighboring frames within the temporal context of V_i, this window enables a collaborative assessment to determine whether V_i represents a pure rotation view.

The sliding window is initially processed by a CNNs to extract spatial representations from each frame. The CNNs utilizes convolutional layers, pooling operations, and non-linear activations to identify visual patterns and features in the frames. These spatial representations are then fed into a two-layer LSTM, which captures the temporal dependencies within the sliding window and learns the sequential relationships between frames. The final hidden representation obtained from the LSTM, denoted as h^w, is further processed by passing it through a fully connected (FC) layer. The FC layer maps the learned temporal representation to a one-hot label, enabling classification of the input based on the assigned probability distribution across binary classes. The model is trained by binary cross entropy with logits loss (BCE with Logits Loss), as shown in Eq. 6.

$$Loss(y_i, \hat{y}_i) = -\frac{1}{N}\sum_{i=0}^{N} y_i \cdot \log(\hat{y}_i) + (1 - y_i) \cdot \log(1 - \hat{y}_i), \tag{6}$$

where N denotes the number of training samples.

This loss function has been shown to handle imbalanced datasets and missing labels effectively by adjusting class weights accordingly, making it a reliable choice for our tasks. During the optimization process, the model update its parameters based on the discrepancy between its predictions $\hat{y}_i$ and true labels y_i, thereby facilitating learning and continual enhancement of its accuracy over time. In conclusion, the systematic annotation training approach for rotation-only recognition can be comprehensively delineated through the ensuing steps:

Step 1: Perform feature extraction and matching for two-view estimation.
Step 2: Estimate relative and global rotations, taking into account image matching point pairs and global rotation matrix sets, and compute two-view pure rotational indicators.
Step 3: Employ Algorithm 1 to calculate the multi-view pure rotational degree, denoted as $prd(PRI)$.
Step 4: Given that $prd(PRI) > \delta_{prd}$, automatically annotate pure rotational views.
Step 5: Positive samples correspond to pure rotation views, while negative samples correspond to other views.
Step 6: Each view is processed by constructing a sliding window of length w centered around it, serving as the input for the model.

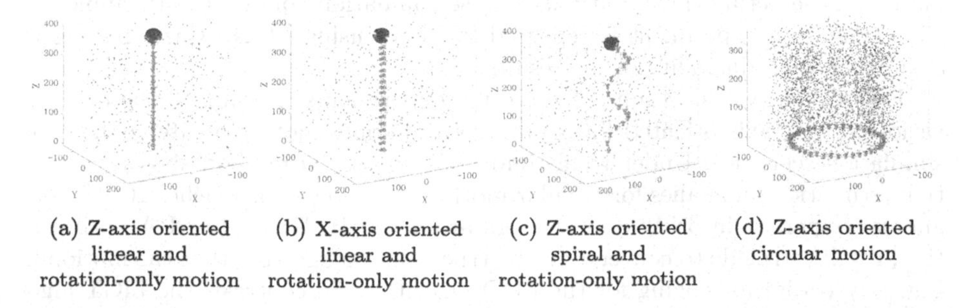

(a) Z-axis oriented linear and rotation-only motion (b) X-axis oriented linear and rotation-only motion (c) Z-axis oriented spiral and rotation-only motion (d) Z-axis oriented circular motion

Fig. 2. Comparison of various simulated motion types (Black points: 3D feature locations. Red boxes: regular camera motion positions. Blue boxes: camera rotation-only positions) (Color figure online)

4 Experiments

The experiments conducted in this paper have several primary objectives, including: (1) verifying that existing camera pose estimation methods display anomalies caused by specific pure rotational views, even under ideal zero-noise conditions; (2) utilizing simulation environment data to confirm the ability of the new method to accurately detect the presence of rotation-only anomaly views and identify the views responsible for these anomalies; and (3) validating the effectiveness of the proposed model for rotation-only anomaly recognition.

4.1 Experimental Comparison of Four Simulated Motion Scenarios

With the known global rotation set R obtained by [3], The simulation experiment utilizes the two camera position estimation methods employed in mainstream global SfM platforms: LUD [15] (Theia platform) and LiGT [2] (OpenMVG platform). As demonstrated in study [2], linear motion can easily lead to abnormal position estimation, with LUD exhibiting anomalies under co-linear motions. To enhance the challenge of detecting pose estimation recovery anomalies, the experiment combines linear motion with a substantial amount of pure rotational motion, thereby creating a relatively complex and anomaly-prone simulation environment. The effectiveness of the new method is validated through comparison with related approaches.

In particular, the simulated camera motion trajectory, as shown in Fig. 2, consists of 40 views, including 20 regular motion views and 20 pure rotational views. The pure rotational views have a fixed position corresponding to the camera position of the last normal view. As depicted in Fig. 2, the simulation experiment tests four different camera motions by adjusting the normal camera orientations and analyzing the impact of the rotation-only view set on multi-view pose estimation recovery. The initial number of 3D feature points in the experimental scene is 4,000, uniformly distributed in a cube around the center of the camera motion line. Imaging of feature points in the simulation environment includes constraints such as chirality, image plane dimensions, etc., to approximate real imaging scenarios. Since the primary goal of this experiment is to validate the presence of rotation-only anomalies and the effectiveness of the new method in detecting them, ideal zero-noise simulation conditions are employed. The simulation experiment is repeated 20 times using Monte Carlo testing to ensure the reproducibility of the results.

Figure 3 shows the results of the error experiments, leading to these observations: (1) X-axis oriented linear and rotation-only motion create challenging simulation scenarios. Under ideal conditions, the LUD method shows no position estimation anomalies for circular motion (position error reaches 10^{-13}), but anomalies persist in other cases (errors around 10^{0}). Even after BA optimization, results struggle to converge to the true value, suggesting that the motion in Fig. 2(a)–(c) is challenging for the LUD algorithm. In contrast, the LiGT algorithm displays significant anomalies only for X-axis oriented linear and rotation-only motion, with position error levels between 10^{-1} and 10^{0}. The experiment

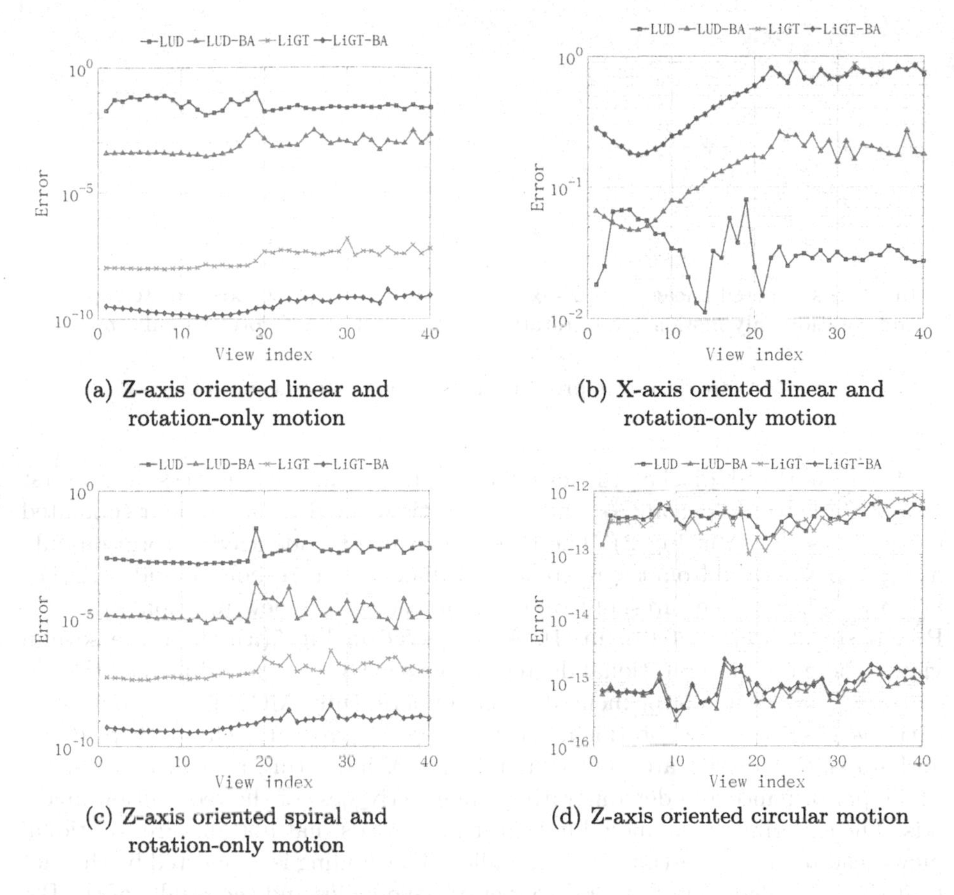

(a) Z-axis oriented linear and rotation-only motion

(b) X-axis oriented linear and rotation-only motion

(c) Z-axis oriented spiral and rotation-only motion

(d) Z-axis oriented circular motion

Fig. 3. Position error of various simulated motion types

confirms that X-axis oriented linear and rotation-only motion poses challenges for state-of-the-art algorithms; (2) The LiGT algorithm's robustness and accuracy allow it to maintain normal position estimation levels in Fig. 3(a), (c), and (d). Therefore, in subsequent rotation-only recognition experiments, comparing the LiGT algorithm results before and after rotation-only recognition is sufficient to verify the performance of the multi-view PRD indicator; (3) The LiGT algorithm can handle a certain degree of rotation-only motion.

We calculate the BA reprojection error in each Monte Carlo test. Figure 4 indicates that using BA reprojection error is incapable of identifying anomalies in camera position estimation. Specifically, taking the case of X-axis oriented linear and rotation-only motion as an example, Fig. 4 shows that BA technique can reduce the reprojection errors of LUD and LiGT algorithm results. However, after comparing the position errors in Fig. 3(a)–(c) with the results in Fig. 4 one by one, it can be observed that even after BA optimization, the position estimation remains abnormal (with error distribution around 10^{-1}). At this point, the BA reprojection errors are minimal, with optimized errors primarily around 10^{-10}.

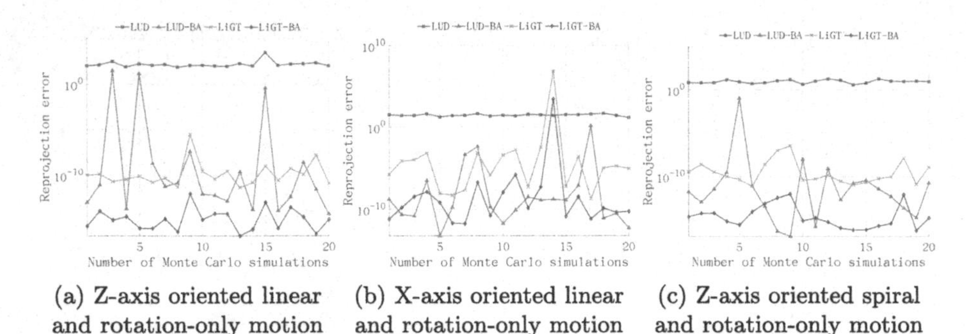

(a) Z-axis oriented linear and rotation-only motion (b) X-axis oriented linear and rotation-only motion (c) Z-axis oriented spiral and rotation-only motion

Fig. 4. Comparison of reprojection errors for different motion types

Figure 5(a)–(b) display the calculated pure rotational degrees in the 1st Monte Carlo test based on φ_{M_3} and φ_{PRI} statistical indicators for four simulated motions (as shown in Fig. 2). The threshold value for identifying rotation-only using M_3, is derived from the reference parallax angle threshold provided in [17], i.e., $\delta_{M_3} < 5$ deg. The threshold value for detecting two-view rotation-only with PRI is set at $\delta_{PRI} = 0.015$ in [1]. As observed in Fig. 5(a), there are several views where the pure rotational degree surpasses 0.8; thus, an additional threshold of $\delta_{prd} = 0.8$ was implemented in the experiments. After filtering out these extreme cases of pure rotational motion, Fig. 5(c) reveals that both $prd(M_3)$ and $prd(PRI)$ results are approximately 10^{-8}, indicating a return to normal LiGT performance and demonstrating the effectiveness of the recognition methods. The experiment conducted in this study reveals that not all pure rotational views lead to position estimation anomalies. This finding is supported by the fact that the LUD algorithm is unable to identify anomalies and the results of the BA reprojection error. The ability of the LiGT algorithm to handle a certain degree of pure rotational views is the theoretical explanation for this phenomenon.

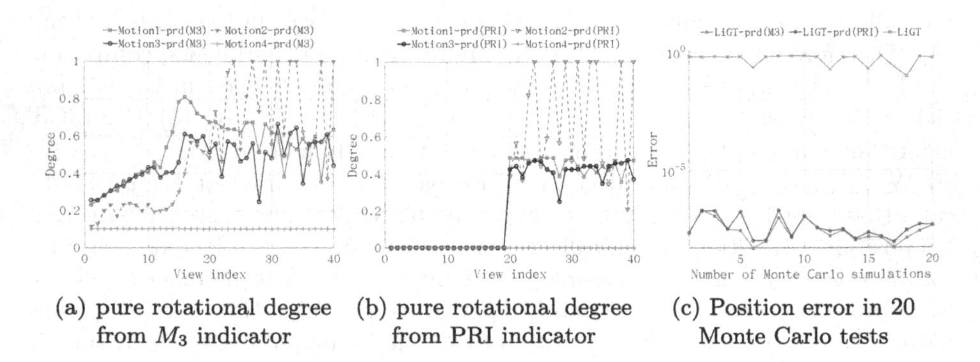

(a) pure rotational degree from M_3 indicator (b) pure rotational degree from PRI indicator (c) Position error in 20 Monte Carlo tests

Fig. 5. Performance of the rotation-only recognition

It is important to note that in the first simulated structure, the motion direction coincides with the baseline of relative camera displacement along the z-axis, causing the epipole to lie within the image plane and resulting in multiple views with low parallax angles. Consequently, $prd(M_3)$ does not effectively distinguish between views corresponding to pure rotational configurations. However, the results obtained using $prd(PRI)$ suggest that the PRI-based criterion can still accurately identify the set of pure rotational views $V_{rot} = \{V_{21}, \ldots, V_{40}\}$ specified in the simulation under such conditions.

Table 1. Comparison of position error before and after PRD detection.

Method	Noise in pixel				
	0	0.1	0.5	1	2
PRD-LiGT	3.87e−8	8.33e−5	7.24e−4	7.55e−4	3.06e−3
PRD-LiGT-BA	3.42e−9	1.14e−5	5.80e−5	1.59e−4	3.48e−4
LiGT	4.46e−1	6.44e−1	6.37e−1	6.56e−1	6.44e−1
LiGT-BA	6.44e−1	6.45e−1	6.38e−1	6.57e−1	6.43e−1

We present the position error before and after $prd(PRI)$ detection under varying levels of noise in pixel values, as illustrated in Table 1, ranging from 0 to 2 pixels. The results demonstrate the effectiveness of the PRD approach in identifying and correcting anomalies in camera position estimation. As seen from the table, the PRD-LiGT method significantly reduces the position errors compared to the original LiGT algorithm. Even under noiseless conditions, PRD-LiGT achieves a position error of 3.87×10^{-8}, which is orders of magnitude lower than the error observed for the LiGT algorithm (4.46×10^{-1}). Moreover, as the noise level increases, the PRD-LiGT method maintains its superior performance over the LiGT algorithm.

It is worth noting that the BA optimization applied to both methods further improves their accuracy. For instance, in the case of PRD-LiGT-BA, the position error decreases to 3.42×10^{-9} under zero-noise conditions, whereas the LiGT-BA method still exhibits a high position error of 6.44×10^{-1}. This observation suggests that the proposed PRD method not only effectively detects anomalies in camera position estimation but also benefits from the BA optimization.

Furthermore, the robustness of the PRD-LiGT method against increasing noise levels is evident. Although the position error increases as the noise level grows, it remains significantly lower than the errors observed for the LiGT and LiGT-BA methods across all noise levels. This indicates that the PRD approach can effectively handle noisy data while maintaining accurate camera position estimation.

4.2 Experimental Results in Virtual KITTI Dataset

We utilized the Virtual KITTI dataset [6] to evaluate the efficacy of our rotation-only anomaly recognition model. This dataset is a photo-realistic synthetic video dataset designed for a range of video understanding tasks and contains 50 high-resolution monocular videos generated from five virtual worlds under varying imaging and weather conditions. The videos are fully annotated and provide 3D camera pose information for each frame, making it an excellent resource for evaluating rotation-only anomaly recognition in multi-view pose estimation.

The geometric computation method outlined in Sect. 3.2 is utilized to automatically annotate pure rotational views. Positive samples are assigned to views exhibiting pure rotation, while negative samples are assigned to other views. Each view is processed by constructing a sliding window of length w centered around it, serving as the input for the model.

To enable comprehensive evaluation, we adopted random partitioning to segregate the dataset into training (80%), validation (10%), and testing (10%) sets. The ResNet50 [10] was selected as the backbone employed in our model. The training steps consisted of 10 epochs, utilizing the Adam optimization algorithm [12] with a learning rate of 0.001. A mini-batch size of 32 was employed throughout the training process. Through extensive training and evaluation, our experiments demonstrate that our proposed model for rotation-only anomaly recognition is highly effective. In particular, we achieved an accuracy of 91% on the test set, demonstrating the feasibility and potential of this approach in real-world applications.

The results of the qualitative visualization on the Virtual KITTI dataset are presented in Fig. 6. The correctly classified views are indicated by blue border images, while the misclassified views are indicated by red borders. These results demonstrate the effectiveness and robustness of the proposed model in detecting pure rotation views under different environmental and lighting conditions. However, it is worth noting that the model may face difficulty in capturing the subtle differences between adjacent views when the pure rotation is not significant. This limitation is illustrated in the 5-th row of Fig. 6, where the view is a pure rotation view, but the model fails to detect it as an anomaly. This observation suggests that the model is not perfect and may require further refinement to improve its accuracy in detecting pure rotation views.

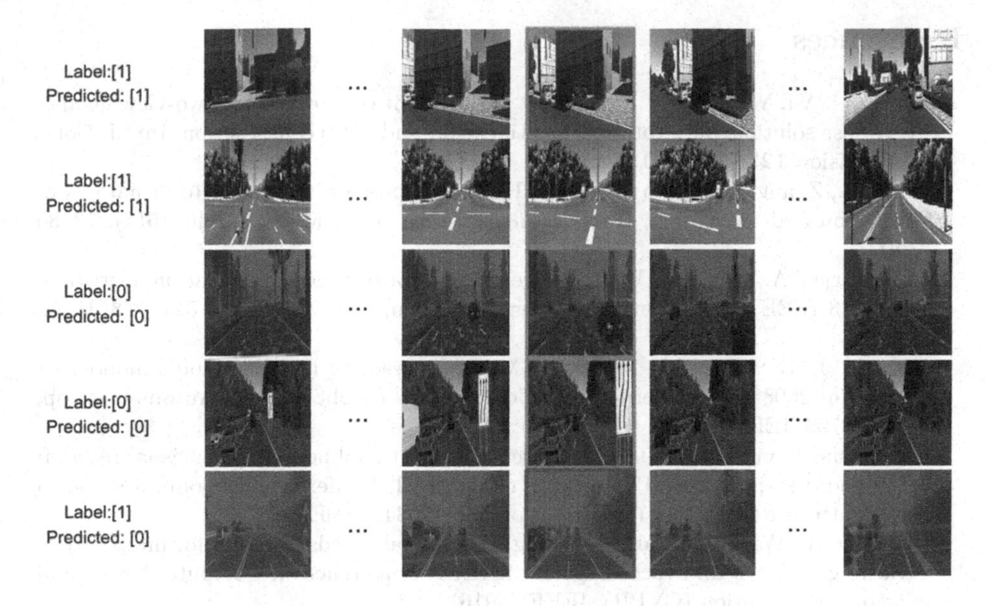

Fig. 6. Example visual results for the Virtual KITTI dataset.

5 Conclusion

In summary, the paper introduces a new method for detecting pure rotation anomalies in multi-view pose estimation data. The approach utilizes a multi-view pure rotational detection method inspired by related two-view approaches, along with an automated data annotation and training strategy using a CNNs combined with LSTM network for anomaly recognition. The experimental results demonstrate that the proposed model achieves an accuracy of 91% on the test set, indicating its high effectiveness. The systematic annotation training approach for pure rotation recognition involves feature extraction and matching for two-view estimation, estimating relative and global rotations, and computing two-view pure rotational indicators. The proposed approach has the potential to enhance the robustness and performance of techniques such as structure from motion and visual simultaneous localization and mapping by identifying pure rotation anomalies in multi-view pose estimation data. Future work could explore the application of our approach in real-world scenarios and investigate additional neural network models in the domain of video understanding for potential improvements in accuracy and efficiency.

Acknowledgements. This work was partially funded by the National Natural Science Foundation of China (61876034).

References

1. Cai, Q., Wu, Y., Zhang, L., Zhang, P.: Equivalent constraints for two-view geometry: pose solution/pure rotation identification and 3D reconstruction. Int. J. Comput. Vision **127**(2), 163–180 (2019)
2. Cai, Q., Zhang, L., Wu, Y., Yu, W., Hu, D.: A pose-only solution to visual reconstruction and navigation. IEEE Trans. Pattern Anal. Mach. Intell. **45**(1), 73–86 (2023)
3. Chatterjee, A., Govindu, V.M.: Efficient and robust large-scale rotation averaging. In: 2013 IEEE International Conference on Computer Vision, pp. 521–528. IEEE (2013)
4. Civera, J., Davison, A.J., Montiel, J.M.M.: Interacting multiple model monocular slam. In: 2008 IEEE International Conference on Robotics and Automation, pp. 3704–3709. IEEE (2008)
5. Donahue, J., et al.: Long-term recurrent convolutional networks for visual recognition and description. In: Proceedings of the IEEE Conference on Computer Vision and Pattern Recognition (CVPR), pp. 2625–2634. IEEE (2015)
6. Gaidon, A., Wang, Q., Cabon, Y., Vig, E.: Virtual worlds as proxy for multi-object tracking analysis. In: Proceedings of the IEEE Conference on Computer Vision and Pattern Recognition (CVPR). IEEE (2016)
7. Gauglitz, S., Sweeney, C., Ventura, J., Turk, M., Höllerer, T.: Live tracking and mapping from both general and rotation-only camera motion. In: 2012 IEEE International Symposium on Mixed and Augmented Reality (ISMAR), pp. 13–22. IEEE (2012)
8. Gauglitz, S., Sweeney, C., Ventura, J., Turk, M., Höllerer, T.: Model estimation and selection towards unconstrained real-time tracking and mapping. IEEE Trans. Visual Comput. Graphics **20**(6), 825–838 (2014)
9. Hartley, R., Zisserman, A.: Multiple View Geometry in Computer Vision. Cambridge University Press, Cambridge (2003)
10. He, K., Zhang, X., Ren, S., Sun, J.: Deep residual learning for image recognition. In: Proceedings of the IEEE Conference on Computer Vision and Pattern Recognition (CVPR), pp. 770–778. IEEE (2016)
11. Herrera, D.C., Kim, K., Kannala, J., Pulli, K., Heikkilä, J.: DT-slam: deferred triangulation for robust slam. In: 2014 2nd International Conference on 3D Vision, vol. 1, pp. 609–616. IEEE (2014)
12. Kingma, D.P., Ba, J.: Adam: a method for stochastic optimization (2017). https://arxiv.org/abs/1412.6980
13. Liu, C., Ma, X., He, X., Xu, T.: Hierarchical multimodal attention network based on semantically textual guidance for video captioning. In: Tanveer, M., Agarwal, S., Ozawa, S., Ekbal, A., Jatowt, A. (eds.) ICONIP 2022. LNCS, vol. 13625, pp. 158–169. Springer, Cham (2023). https://doi.org/10.1007/978-3-031-30111-7_14
14. Ma, C.Y., Chen, M.H., Kira, Z., AlRegib, G.: TS-LSTM and temporal-inception: exploiting spatiotemporal dynamics for activity recognition. Signal Process. Image Commun. **71**, 76–87 (2019)
15. Özyeşil, O., Singer, A.: Robust camera location estimation by convex programming. In: Proceedings of the IEEE Conference on Computer Vision and Pattern Recognition (CVPR). IEEE (2015)
16. Özyeşil, O., Voroninski, V., Basri, R., Singer, A.: A survey of structure from motion. Acta Numer. **26**, 305–364 (2017)

17. Pirchheim, C., Schmalstieg, D., Reitmayr, G.: Handling pure camera rotation in keyframe-based slam. In: 2013 IEEE International Symposium on Mixed and Augmented Reality (ISMAR), pp. 229–238. IEEE (2013)
18. Seo, Y., Defferrard, M., Vandergheynst, P., Bresson, X.: Structured sequence modeling with graph convolutional recurrent networks. In: Cheng, L., Leung, A.C.S., Ozawa, S. (eds.) ICONIP 2018. LNCS, vol. 11301, pp. 362–373. Springer, Cham (2018). https://doi.org/10.1007/978-3-030-04167-0_33
19. Simonyan, K., Zisserman, A.: Two-stream convolutional networks for action recognition in videos. In: Ghahramani, Z., Welling, M., Cortes, C., Lawrence, N., Weinberger, K. (eds.) Advances in Neural Information Processing Systems, vol. 27, pp. 68–576. Curran Associates, Inc. (2014)
20. Torr, P.H.: Bayesian model estimation and selection for epipolar geometry and generic manifold fitting. Int. J. Comput. Vision **50**(1), 35–61 (2002)
21. Zhao, J.: An efficient solution to non-minimal case essential matrix estimation. IEEE Trans. Pattern Anal. Mach. Intell. **44**(4), 1777–1792 (2022)
22. Zhou, Y., Yan, F., Zhou, Z.: Handling pure camera rotation in semi-dense monocular SLAM. Vis. Comput. **35**(1), 123–132 (2019)

Detecting Adversarial Examples via Classification Difference of a Robust Surrogate Model

Anjie Peng[1,2], Kang Deng[1], Hui Zeng[1](✉), Kaijun Wu[3], and Wenxin Yu[1]

[1] Southwest University of Science and Technology, Mianyang 621010, Sichuan, China
zengh5@mail2.sysu.edu.cn

[2] Engineering Research Center of Digital Forensics, Ministry of Education, Nanjing University of Information Science and Technology, Nanjing, China

[3] Science and Technology on Communication Security Laboratory, Chengdu 610041, Sichuan, China

Abstract. Existing supervised learning detectors may deteriorate their performance when detecting unseen adversarial examples (AEs), because they may be sensitive with training samples. We found that (1) the CNN classifier is modest robust against AEs generated from other CNNs, and (2) such adversarial robustness is rarely affected by unseen instances. So, we construct an attack-agnostic detector based on an adversarial robust surrogate CNN to detect unknown AEs. Specifically, for a protected CNN classifier, we design a surrogate CNN classifier and predict the image with different classification labels on them as an AE. In order to detect transferable AEs and maintain low false positive rate, the surrogate model is distilled from the protected model, aiming at enhancing the adversarial robustness (i.e., suppress the transferability of AE) and meanwhile mimicking the output of clean image. To defend the potential ensemble attack targeted at our detector, we propose a new adversarial training scheme to enhance the security of the proposed detector. Experimental results of generalization ability tests on Cifar-10 and ImageNet-20 show that our method can detect unseen AEs effectively and performs much better than the state-of-the-arts.

Keywords: CNN · adversarial example · detection · distillation · adversarial training

1 Introduction

Convolution neural networks (CNN) have been widely adopted as fundamental machine learning tools in artificial intelligence tasks, due to their superior performance in image classification [1], object detection [2], etc. However, adversarial example (AE) exploits the vulnerability of CNN and forces the CNN to give an error prediction label [4]. How to defend against AEs is of great interest, especially when deploying CNNs in safety-critical applications.

B. Luo et al. (Eds.): ICONIP 2023, CCIS 1966, pp. 558–570, 2024.
https://doi.org/10.1007/978-981-99-8148-9_43

Detecting AE in advance is an important defense to prevent the CNN from being fooled. Recent detection methods include input transformation induced inconsistency [5–8], nearest neighborhood clustering [9, 10], and feature-based supervised binary classifiers [11–17, 25]. Benefiting from the supervised training, feature-based methods perform well for detecting known AEs, but may not detect unseen AEs in the future applications. For example, the-state-of-art ESRM [16] trained from the AEs on BIM [33] failed to detect the AEs generated from a DDN attack [21].

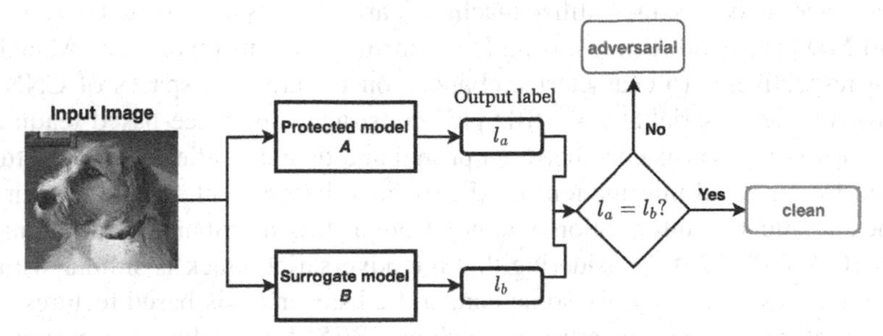

Fig. 1. The proposed detection workflow. For an input image, if the CNN model *A* and *B* output different classified labels, it is predicted as adversarial, otherwise clean.

In this paper, we try to propose a detector for detecting unseen AEs. Inspired from the detector TPD [18], we find that the current CNN is difficult to be fooled or be fooled with similar mistakes by weak transferable AEs generated from other CNNs. Such a property of CNN is called *adversarial robustness*. Furthermore, the adversarial robustness is rarely affected by the distribution of images, and thus is non-sensitive with training samples. Motivated by these observations, for a protected CNN classifier *A*, we design a surrogate CNN classifier *B* to detect AEs, as shown in Fig. 1. If an input image has different outputs on *A* and *B*, our method predicts it as adversarial otherwise clean.

However, our detector built from an initial surrogate model *B* cannot detect high transferable AEs, such as AEs with large-perturbation or transferability boosting [19, 24]. Besides, our initial detector also cannot resist the ensemble attack against both *A* and *B*. These problems are main drawbacks of the similar detectors TPD [18] and SID [13]. To address these issues, we design a combination of new distillation loss and adversarial training loss to train a robust and secure surrogate model *B*. The proposed detector equipped with the enhanced model *B* not only improves the detectability for high transferable AEs, but also maintains a low false positive rate. Experimental results on Cifar-10 and ImageNet-20 validate that our method can detect AEs from various adversarial attacks, including optimization-based attacks [21, 27, 29], gradient-based attacks [24, 33, 34], and their transferability-enhanced version DI [19], TI [23]. Our method performs much better than state-of-the-arts in the cross-attack tests and robustness tests.

2 Related Detection Methods

Many detection methods from different aspects have been proposed. Assuming that adversarial perturbations are easily disturbed by image transformations (e.g., de-nosing, filtering, quantization), input transformation-based methods, such as noise adding or reduction [6, 7], feature squeezing (FS) [5], first feed a questioned image and its transformed version into a protected CNN, then flag the image as adversarial if the CNN outputs different results. Feature-based bi-classifiers first extract effective feature pairs of clean images and AEs, then utilize machine learning tools to construct detectors. LID [11] and MD [12] build features using local intrinsic dimensionality and Mahalanobis distance respectively. To characterize changes on feature map spaces of CNN layers that caused by adversarial attacks, SID [13] constructs a distance-based feature based on the sensitivity inconsistency between primal and dual classifiers. Some features are exploited from related information of AE, such as the gradient stream used in PACA [14], the classification information obtained from a class-disentangled variational autoencoder (CD-VAE) [25]. Considering that the adversarial attack is similar with a sort of accidental steganography [4], some handcrafted steganalysis-based features are proposed to capture image space artifacts, such as ESRM [16], color co-occurrence [17]. From the point of embedding rate, the adversarial attack generally results in more statistical changes than the steganography does. So, steganalysis-based methods perform very well for large-perturbation attacks. However, they deteriorate for cross-databases, cross-attacks, and few-perturbation attacks, motivating us to develop a generalized detector against various adversarial attacks.

3 Methodology

Figure 1 depicts the workflow of the proposed method. To protect a fixed CNN classification model A, we select a surrogate model B and distill it from the model A to form the detector. For a test image, if the model A and B output different labels, it is detected as adversarial otherwise clean. The key issue of our method is how to distill the surrogate model B, making it output the same/different labels with A when detecting clean/adversarial images. In the following, we first analyze the adversarial robustness of an initial model B, then enhance its adversarial robustness via a distillation method.

3.1 Analyzing Adversarial Robustness of the Initial Surrogate Model

The adversarial robustness of CNN exists naturally. As pointed by PGD [30] and TPD [18], a weak adversarial example generated from the A model is difficult to fool the B model with different architecture or strong capacity. For the detection purpose, we relax the definition of traditional adversarial robustness. Under the proposed detection framework, *a surrogate model that outputs different labels with the protected model for AEs is viewed as adversarial robust.* This means that, an adversarial robust surrogate model can prevent an AE to fool it or to mislead it to output the same error label with the model A. For AEs that successfully fool the model A, we predict their labels by B, and then count the number of AEs with different predicted label on A to measure the adversarial robustness of *B*. This metric of adversarial robustness is equal with the true positive rate (TPR) of the proposed detector. The higher TPR means a stronger adversarial robust model.

Table 1. The metric of adversarial robustness for the initial model *B* (on rows) via TPR (%) of our detector. ResNet-50 is selected as the model *A*, and is attacked to generate adversarial examples. The classification accuracy of *B* on clean images are depicted in the bracket.

Initial surrogate model *B* (classification accuracy (%))	BIM 10-step, $\varepsilon = 1$	C&W 200-step, $\kappa = 0$	C&W 200-step, $\kappa = 10$
ResNet18(89.2)	94.4	97.1	96.2
EfficientNet-B0 (90.0)	**95.9**	**97.9**	**97.3**
EfficientNet-B5 (89.3)	95.5	97.9	96.9
RegNetY-800MF (88.0)	94.2	97.2	96.4
RegNetY-1.6GF (88.5)	95.3	97.7	96.7
MNASNet1–0 (89.4)	95.6	97.8	97.3
MNASNet1–3 (89.6)	95.7	97.9	97.1
MobileNetV3-small (85.7)	94.9	96.7	96.2
MobileNetV3-large (89.1)	95.7	97.9	96.7

We select ResNet-50 as the protected model *A* and evaluate the adversarial robustness of various surrogate models downloaded from PyTorch platform.To generate adversarial examples, BIM [33] and C&W [29] are employed to attack the model A on the proposed Imagenet-20. Please refer to Sect. 4.1 for details of the Imagenet-20. Please refer to Sect. 4.1 for details of the Imagenet-20. Only successfully attacked AEs are used for calculating TPR.As seen in Table 1, all TPRs are > 94%, indicating that the initial surrogate model is modest adversarial robust when facing with the adversarial attack. The results in Table 1 show that an accurate model *B* with different architecture of *A* tends to achieve high TPR. Because these models have high capacity and are difficult to be attacked by black-box AEs, as indicated by PGD [30]. For example, when taking EfficientNet-B0 as the surrogate model *B*, the detector achieves the best TPRs. These results motivate the criterion of selecting the surrogate model.

As shown in Table 1, increasing the confidence κ of C&W from 0 to 10 (i.e., increasing transferability of adversarial attack), the TPRs are decreasing. This indicates that the initial surrogate may be no longer robust for high transferable AEs. In the following, we will enhance adversarial robustness of the surrogate model to resist large-perturbation and transferability augmentation attacks.

3.2 Enhancing Adversarial Robustness of the Surrogate Model

In order to improve TPR for detecting high transferable AEs and meanwhile maintain a low false positive rate (FPR) for detecting clean images, the surrogate model B is designed to output different/same outputs with A for the AEs/clean images. To this end, a distillation scheme is employed to make B learn different/similar behaviors of A on AEs/clean images.

$$\min_{\theta} \ E_{(x,y)\sim D}[\alpha L_{\mathrm{KL}}(\theta) + (1-\alpha-\beta-\gamma)L_{\mathrm{CE}}(\theta) + \beta L_{\mathrm{MSE}}(\theta) + \gamma L_{\mathrm{CEAT}}(\theta)]$$

$$\text{where}\ L_{\mathrm{KL}}(\theta) = \mathrm{KL}(B_{\theta}^{t}(x), A^{t}(x)),\ L_{\mathrm{CE}}(\theta) = \mathrm{CE}(B_{\theta}(x), y),$$

$$L_{\mathrm{MSE}}(\theta) = \frac{1}{\mathrm{MSE}(B_{\theta}^{t}(x^{adv}), A^{t}(x^{adv}))},\ L_{\mathrm{CEAT}}(\theta) = \mathrm{CE}(B_{\theta}(x^{adv}), y). \quad (1)$$

We design a composite loss function in (1) and solve it to distill the surrogate model B. In (1), E(.) is an expectation function, x with label y is an input image under the distribution D, x^{adv} is an adversarial example generated from A, B_{θ}^{t} is a surrogate model with parameter θ, t is a temperature constant.

We now explain the role of each loss in (1). $L_{\mathrm{KL}}(\theta)$ is a commonly used KL divergence loss used for knowledge distillation, aiming at reducing FPR. During the distillation, the student B is trained from the teacher A to mimic outputs on clean images. Inspired by [28], the standard cross entropy loss $L_{\mathrm{CE}}(\theta)$ is added to improve natural accuracy of B on clean images for further reducing FPR. $L_{\mathrm{MSE}}(\theta)$ is the reciprocal of a MSE loss, which is used to enhance adversarial robustness of the surrogate model for improving TPR. Minimizing $L_{\mathrm{MSE}}(\theta)$ means maximizing the MSE loss between $B_{\theta}^{t}(x^{adv})$ and $A^{t}(x^{adv})$, which forces B to learn different outputs of A on AEs. When the model B and A are exposed to an adversary, an ensemble attack [22] may be utilized to fool A and B to output the same error labels for AEs. This attack will cause the proposed detector to fail to detect AEs. To resist the ensemble attack, the model B is adversarially trained by ensemble AEs generated from A and B. So, a cross entropy loss $L_{\mathrm{CEAT}}(\theta)$ of adversarial training is used to enhance the security of detector.

Algorithm 1: Distilling the surrogate model B
Inputs: a protected model A(teacher), initial model B with parameter θ (student), data set $\{x, y\}$, learning rate μ, the weight α, β, γ of loss in (1)
Outputs: the distilled model B

For Epoch=1, ..., N **do**
  For Batch =1, ..., M **do**
    For each x_i, generate an AE x_i' on A using k-step PGD;
    If γ=0 // *without ensemble attack* //
      Compute the gradients of loss in (1):
      $\nabla_\theta L(x,\theta) =$
      $\sum_i \nabla_\theta[\alpha L_{KL}(B_\theta, A, x_i) + (1\text{-}\alpha\text{-}\beta) L_{CE}(B_\theta, x_i) +$
      $\beta L_{MSE}(B_\theta, A, x_i')]$
      Update the model B by: $\theta \leftarrow \theta - \mu \nabla_\theta L(x,\theta)$;
    ELSE // *With ensemble attack against A and B* //
      For each x_i, generate the AE x_i'' using k-step PGD ensemble attack as [22] on B and A, and compute the gradients of loss in (1):
      $\nabla_\theta L(x,\theta) =$
      $\sum_i \nabla_\theta[\alpha L_{KL}(B_\theta, A, x_i) + (1\text{-}\alpha\text{-}\beta\text{-}\gamma) L_{CE}(B_\theta, x_i) +$
      $\beta L_{MSE}(B_\theta, A, x_i') + \gamma L_{CEAT}(B_\theta, A, x_i'')]$
      Update the model B by: $\theta \leftarrow \theta - \mu \nabla_\theta L(x,\theta)$;
    End If
  End For
End For

We allocate a weight parameter within [0, 1] to each loss function. Users can set weights according to their application requirements. In experiments, we empirically set $\alpha = 0.1$, $\beta = 0.6$, $\gamma = 0/0.2$ under the case that the attacker cannot/can access to the surrogate model. The temperature constant t is set to be 1 in $L_{KL}(\theta)$ and $L_{MSE}(\theta)$, due to our detection results depended on hard labels.

The distillation of the surrogate model is summarized in Algorithm 1. We find that $L_{MSE}(\theta)$ is too small, so the loss $L_{MSE}(\theta)$ will be multiplied with the size of a mini-batch in distillation. We utilize the 10-steps PGD [30] with $\varepsilon = 8$ to generate training AEs for distillation. That's because PGD runs fast and has strong transferability. We expect that the distilled surrogate model is robust against other adversarial attack to improve the generalization power of our detector.

Figure 2 visually demonstrates that the proposed distillation enhances the adversarial robustness of surrogate model from the view of decision boundary similarity [20, 26]. Dist $(A, B, x) = |d(B, x)\text{-}d(A, x)|$ is a decision boundary similarity metric between B and A for the image x, which gives a perturbation distance that transfers x^{adv} (adversarial example of x) from A to attack B. As shown in Fig. 2, d (A, x) is a metric of the decision boundary of A for x along the positive x-axis. The larger Dist (A, B, x), the more difficult to transfer x^{adv} from A to fool B. As seen, the proposed distillation enlarges the purple decision spaces of the "sea lion", and thus has a larger Dist (A, B, x) than the non-distilled

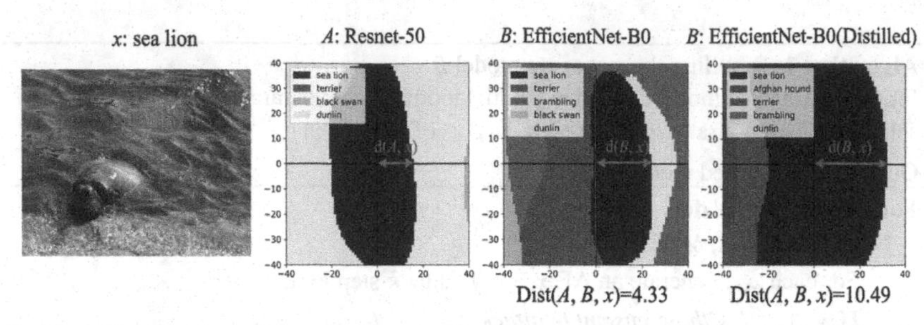

Fig. 2. Visualization the enhanced adversarial robustness of surrogate model via the decision boundary similarity as [20, 26] for an example "sea lion". As seen, the distilled model B enlarges the decision boundary similarity Dist (A, B, x) from 4.33 to 10.49, which enlarges the attack distance for the adversarial example x^{adv} from A to fool B, and thus enhancing the adversarial robustness of distilled B.

case. This means x^{adv} generated from A are more difficult to deceive B. We also inspect Dist (A, B, x) for all images on ImageNet-20. Similarly, after distillation on B, the average Dist (A, B, x) calculated on ImageNet-20 increases from 5.93 to 14.27. These results empirically verify that the proposed distillation enhances adversarial robustness of the surrogate model.

3.3 The Proposed Detector

After distilling the surrogate model B with Algorithm 1, the proposed detector is completed. Please refer to Fig. 1 to see the workflow of our detector. For a fixed protected classification model, we suggest that the model with different architecture and high classification capacity is selected as the initial surrogate model. Because such surrogate model naturally owns strong adversarial robustness, and also can be easily enhanced in the distillation phase. Once obtaining the distilled model B, our adversarial robustness-based detector no longer needs to train. This is the main reason that our detector can identify unseen adversarial examples. Furthermore, our detector needs not cost time to extract feature nor set a threshold. This is our advantage compared with the popular threshold-based detectors [5–7, 13, 18].

4 Experimental Results

4.1 Setup

Cifar-10 and ImageNet-20, a subset of the validation dataset of ImageNet (ILSVRC2012) are selected as database. To reduce the computation resources for distillation, as Liang *et al.* [6] done, we choose 20 categories[1] having $20 \times 900 = 18000$ images of ILSVRC 2012 to construct the proposed ImageNet-20. All images in ImageNet-20 will be resized to $224 \times 224 \times 3$ before further processing.

[1] 20 categories: Tench, harvestman, dunlin, brambling, black grouse, sea lion, water ouzel, lorikeet, Afghan, bullfrog, black swan, anole, wolfhound, flatworm, pit bull terrier, alligator, fiddler crab, Sealyham, night snake, flamingo.

The light-weight MobileNetV2 [31] on Cifar-10 and ResNet-50 [1] on ImageNet-20 are selected as attacked models to generate adversarial examples. Only clean images with correct classification label are used to generate AEs. Various attacks are considered, including gradient-based attacks: 10-step BIM [33], 10-step NI-FGSM (NI) [24], Auto attack (AA) [34] and optimization-based attacks: L_2 200-step C&W [29], EAD [27], DDN [21]. The few-perturbation and large-perturbation versions of each attack are also considered. Other attack parameters are set with default values.

For the proposed detector, ResNet-50 and EfficientNet-B0 [32] are selected as the protected model for MobileNetV2 [31] and ResNet-50, respectively. For distillation, the Adam optimizer is used with a mini-batch of $M = 128$ shuffled images. The learning rate μ is 0.001 and dropped by a factor of 0.1 at 80 and 120 epochs, with a budget of $N = 200$ epochs. The protected model ResNet-50 is distilled on the training set of Cifar-10 (50000 images), which costs about 5 h. EfficientNet-B0 is distilled using $20 \times 800 = 16000$ images on ImageNet-20, which also costs about 5 h.

The state-of-the-arts: LID [11], MD [12], SRM [16], TPD [18], CDVAE built on MD [25], are used for comparisons. Except our method and TPD, the rest are supervised-learning detectors which trained from the same training set. For the clean training set, 10000 images from the training set of Cifar-10 and $20 \times 800 = 16000$ images from ImageNet-20 are randomly selected respectively. For the adversarial training set, we uniformly select 10-step BIM using $\varepsilon = 8$ to generate AEs from the clean training set. Other training parameters are set with default values for all supervised-learning detectors. All supervised-learning detectors trained on clean images and BIM adversarial examples will be tested for detecting unseen AEs generated by unknown attacks or the BIM attack with different attack strength ε.

All methods are evaluated on the test set of Cifar-10 with 10000 images and the remaining $20 \times 100 = 2000$ images on ImageNet-20. Only the AEs which successfully fool the CNN model are considered for evaluation. Because of equal number of clean images and AEs for testing, we use $Acc = (\text{TPR} + 1 - \text{FPR})/2$ for evaluation. We focus on the tests for unknown attacks including cross-attacks and attacks with unknown image operations. These test results are informative when deploying the AE detector in practical applications.

4.2 Evaluation Under the Case of Un-aware Detections

In this sub-section, we suppose an adversary cannot access to the detector and will not attack the detector to escape from the detection. Under this case, we set $\gamma = 0$ in (1) for our method.

Test on Cross-Attack. In this test, the detector has no prior knowledges of testing attacks. The results in Table 2 and Table 3 show that our method achieves stable *Acc*s of 92.2%–95.2% on Cifar-10 and 94.4%–97.3% ImageNet-20 respectively. In all tests, the FPR of our method is less than 4%. These results indicate that our detector is probably generable when facing with unknown attacks. The supervised detectors SRM, MD and LID performs well for known BIM ($\varepsilon = 8$) used for generating training AEs and some homogeneous attacks, such as NI ($\varepsilon = 8$), AA ($\varepsilon = 8$). In contrast, the supervised detectors deteriorate for detecting unseen heterogeneous attacks. For example, the *Acc*s of

Table 2. The *Acc* (%) of different detectors (on column) on **Cifar-10** for cross-attack test. The tested adversarial examples are generated from MobileNetV2 using attacks with different attack strength shown in the bracket. The "*" denotes the known attack used for generating training samples for the supervised detector. Notice that our detector and TPD needs not training.

Attack method (attack strength)	SRM	MD	LID	CDVAE	TPD	Ours
BIM($\varepsilon = 1$)	53.9	55.6	54.7	75	92.1	**93.6**
BIM($\varepsilon = 4$)	87.2	81.7	85.3	74.9	85.5	**94.3**
BIM($\varepsilon = 8$)*	**97.9**	92.2	94.8	75.3	74.7	93.9
NI($\varepsilon = 1$)	54.4	56.9	54.9	74.9	92.0	**93.8**
NI ($\varepsilon = 4$)	97.1	85.9	89.1	75.3	79.1	**94.0**
NI ($\varepsilon = 8$)	**98.6**	95.0	96.9	74.9	61.7	92.2
AA($\varepsilon = 1$)	63.2	61.7	61.4	74.9	93.1	**94.7**
AA($\varepsilon = 4$)	**98.2**	95.8	98.4	75.3	80.8	94.9
AA($\varepsilon = 8$)	**98.6**	99.4	99.9	74.9	62.2	94.8
C&W($\kappa = 0$)	52.6	52.1	67.4	74.9	94.9	**95.1**
C&W($\kappa = 10$)	60.8	58.6	65.3	74.9	93.8	**95.2**
DDN($steps = 100$)	53.6	52.7	67.8	75.0	**94.6**	94.5
DDN($steps = 200$)	53.2	52.3	67.6	74.9	**92.9**	92.8
EAD($\kappa = 0$)	53.2	51.7	67.4	74.9	94.7	**95.1**
EAD($\kappa = 10$)	61.8	59.6	65.4	74.9	91.2	**95.0**
Average *Acc*	72.3	70.1	75.8	75.0	85.6	**94.3**

SRM for detecting C&W, DDN and EAD are $<62\%$, due to very low TPR. For weak transferable AEs from gradient-based attacks ($\varepsilon = 1$) and optimization-based attacks, our method obtains similar performance with TPD, a simple version of our method. However, for high transferable AEs (such as large-perturbation attacks with $\varepsilon = 8$), our method performs much better than TPD, attributing to that the proposed surrogate model B suppress the transferability of AEs.

Test of Robustness. Motivated by digital image forensics, we test the robustness of detector for detecting adversarial examples with some unknown extra-operations. The popular Gaussian noise (zero mean, standard variance $\sigma = 5$) is selected as a pre-processing operation. After adding the noise, the adversarial example is generated on the noisy image. The composite transferability augmented operation DI-TI (DT) as done in [19] is used to test whether our method can detect high transferable AEs. The Gaussian noise cannot enhance the transferability of AE, and thus rarely affects *Acc* of our method as shown in Table 4. Since our method decreases the transferability, DT slightly reduces the *Acc* of our method up to 4%. DT disturbs the co-occurrence relationships among neighbor pixels, so it makes SRM almost invalid (TPR $< 5\%$) for detecting AEs.

Table 3. The *Acc* (%) of different detectors (on column) on **ImageNet-20**. The tested adversarial examples are generated from ResNet-50 using attacks with different attack strength. The "*" denotes the known attack used for generating training samples for the supervised detector. Notice that our detector and TPD needs not training.

Attack method (attack strength)	SRM	MD	LID	CDVAE	TPD	Ours
BIM($\varepsilon = 1$)	50.1	66.8	75.4	71.1	94.4	**96.6**
BIM($\varepsilon = 4$)	71.3	90.5	94.3	71.5	86.1	**97.2**
BIM($\varepsilon = 8$)*	**98.9**	97.8	97.9	71.9	74.6	97.1
NI($\varepsilon = 1$)	50.1	65.7	67.9	71.6	92.7	**94.4**
NI ($\varepsilon = 4$)	52.1	82.5	87.9	71.2	92.2	**97.2**
NI ($\varepsilon = 8$)	**97.9**	92.9	95.2	70.6	82.5	97.1
AA($\varepsilon = 1$)	50.4	73.6	81.6	71.2	95.3	**96.9**
AA($\varepsilon = 4$)	98.3	98.4	98.3	77.3	86.3	**97.1**
AA($\varepsilon = 8$)	98.7	**99.9**	99.6	72.6	70.9	97.1
C&W($\kappa = 0$)	50.2	62.2	65.0	70.5	97.0	**97.3**
C&W($\kappa = 10$)	50.1	67.3	80.3	70.7	96.2	**97.3**
DDN(*steps* = 100)	50.1	62.5	64.5	70.4	95.4	**95.5**
DDN(*steps* = 200)	50.0	62.4	64.4	70.4	95.1	**95.3**
EAD($\kappa = 0$)	50.0	62.2	65.8	70.6	97.0	**97.1**
EAD($\kappa = 10$)	50.2	66.2	80.9	70.5	94.4	**97.2**
Average *Acc*	64.6	76.7	81.3	71.5	90.0	**96.7**

4.3 Discussion of the Security

In this sub-section, we discuss the security of our method when the protected model *A* and the surrogate model *B* are both exposed to the attacker. One possible adaptive attack is an ensemble attacking on *A* and *B* [22] to force them outputting the same error labels. To defend against such attack, we enable the loss $L_{\text{CEAT}}(\theta)$ in (1) with setting $\gamma = 0.2$ and adversarially train (AT) [3] the model *B* using AEs generated from the ensemble 10-step PGD attack ($\varepsilon = 8$) on *A* and *B*. The results in Table 5 show that the security concern loss $L_{\text{CEAT}}(\theta)$ helps to improve TPR from 0.3%, 2.1% to 77.0%, 81.6% for the attack BIM and NI with $\varepsilon = 8$. For the large-perturbation ensemble attack (such as BIM with $\varepsilon = 8$), the performance of our method still needs to be improved, which is our future work.

Table 4. The *Acc* (%) of robustness test against the pre-processed gaussian noise and transferability augmentation DT. We use the same detectors in Table 3 for testing.

	ImageNet-20, attacked model: ResNet-50					
	SRM	MD	LID	CDVAE	TPD	Ours
DT-BIM($\varepsilon = 1$)	50.0	63.6	57.2	71.6	92.3	**94.5**
DT-BIM($\varepsilon = 4$)	50.5	79.6	75.1	71.5	85.4	**96.3**
DT-BIM($\varepsilon = 8$)	55.6	90.7	90.3	72.0	70.1	**95.4**
DT-NI($\varepsilon = 1$)	49.9	62.7	60.8	71.6	87.3	**90.3**
DT-NI($\varepsilon = 4$)	50.3	73.1	83.3	71.6	90.3	**96.1**
DT-NI($\varepsilon = 8$)	51.2	83.1	91.5	71.3	79.9	**94.8**
Noise + C&W($\kappa = 0$)	84.0	59.8	64.8	68.6	**97.1**	96.9
Noise + DDN($step = 100$)	83.8	59.7	65.4	68.7	95.8	**97.1**
Noise + EAD($\kappa = 0$)	81.5	60.1	64.2	68.7	**97.1**	95.8
Average *Acc*	61.9	70.3	72.5	70.6	88.4	**95.2**

Table 5. The TPR (%) of our method on ImageNet-20 for detecting AEs of different ensemble attacks (column) on ResNet-50(A) and EfficientNet-B0 (B).

Our detector	BIM $\varepsilon = 1, 4, 8$	NI $\varepsilon = 1, 4, 8$	C&W $\kappa = 10$
Train without AT loss ($\gamma = 0$)	54.2/3.5/0.3	77.3/12.5/2.1	58.8
Trian with AT loss ($\gamma = 0.2$)	93.8/87.5/77.0	94.9/89.4/81.6	94.6

5 Conclusion

In order to detect adversarial examples that fool a protected CNN classifier, we design a surrogate model and proposed a detector based on the classification difference between them. A new distillation and adversarial training are designed to enhance the adversarial robustness of the surrogate model, which enables us to develop an effective and secure detection-based defense against unseen adversarial attacks. This is our main contribution. Experimental results show that our method achieves at least 6.7% higher of average detection accuracy in the generalization ability test than state-of-the-arts.

References

1. He, K., Zhang, X., Ren, S., Sun, J.: Deep residual learning for image recognition. In: Proceedings of the IEEE Conference on Computer Vision and Pattern Recognition, pp. 770–778 (2016)
2. Redmon, J., Farhadi, A.: Yolov3: an incremental improvement. arXiv preprint arXiv:1804.02767 (2018)

3. Jia, X., Zhang, Y., Wu, B., Ma, K., Wang, J., Cao, X.: LAS-AT: adversarial training with learnable attack strategy. In: Proceedings of the IEEE/CVF Conference on Computer Vision and Pattern Recognition, pp. 13398–13408 (2022)
4. Goodfellow, I.J., Shlens, J., Szegedy, C.: Explaining and harnessing adversarial examples. arXiv preprint arXiv:1412.6572 (2014)
5. Xu, W., Evans, D., Qi, Y.: Feature squeezing: detecting adversarial examples in deep neural networks. arXiv preprint arXiv:1704.01155 (2017)
6. Liang, B., Li, H., Su, M., Li, X., Shi, W., Wang, X.: Detecting adversarial image examples in deep neural networks with adaptive noise reduction. IEEE Trans. Depend. Secure Comput. **18**(1), 72–85 (2018)
7. Hu, S., Yu, T., Guo, C., Chao, W.L., Weinberger, K.Q.: A new defense against adversarial images: Turning a weakness into a strength. Adv. Neural Inf. Process. Syst. **32** (2019)
8. Roth, K., Kilcher, Y., Hofmann, T.: The odds are odd: a statistical test for detecting adversarial examples. In: International Conference on Machine Learning, pp. 5498–5507. PMLR (2019)
9. Feinman, R., Curtin, R.R., Shintre, S., Gardner, A.B. Detecting adversarial samples from artifacts. arXiv preprint arXiv:1703.00410 (2017)
10. Papernot, N., McDaniel, P.: Deep k-nearest neighbors: towards confident, interpretable, and robust deep learning. arXiv preprint arXiv:1803.04765 (2018)
11. Ma, X., et al.: Characterizing adversarial subspaces using local intrinsic dimensionality. arXiv preprint arXiv:1801.02613 (2018)
12. Lee, K., Lee, K., Lee, H., Shin, J.: A simple unified framework for detecting out-of-distribution samples and adversarial attacks. Adv. Neural Inf. Process. Syst. **31** (2018)
13. Tian, J., Zhou, J., Li, Y., Duan, J.: Detecting adversarial examples from sensitivity inconsistency of spatial-transform domain. In: Proceedings of the AAAI Conference on Artificial Intelligence, vol. 35, no. 11, pp. 9877–9885 (2021)
14. Chen, K., et al.: Adversarial examples detection beyond image space. In: ICASSP 2021–2021 IEEE International Conference on Acoustics, Speech, and Signal Processing (ICASSP), pp. 3850–3854 (2021)
15. Fan, W., Sun, G., Su, Y., Liu, Z., Lu, X.: Integration of statistical detector and Gaussian noise injection detector for adversarial example detection in deep neural networks. Multimedia Tools Appl. **78**, 20409–20429 (2019)
16. Liu, J., et al.: Detection based defense against adversarial examples from the steganalysis point of view. In: Proceedings of the IEEE/CVF Conference on Computer Vision and Pattern Recognition, pp. 4825–4834 (2019)
17. Peng, A., Deng, K., Zhang, J., Luo, S., Zeng, H., Yu, W.: Gradient-based adversarial image forensics. In: Neural Information Processing: 27th International Conference, ICONIP 2020, Bangkok, Thailand, 23–27 November 2020, Proceedings, Part II, vol. 27, pp. 417–428 (2020)
18. Guo, F., et al.: Detecting adversarial examples via prediction difference for deep neural networks. Inf. Sci. **501**, 182–192 (2019)
19. Xie, C., et al.: Improving transferability of adversarial examples with input diversity. In: Proceedings of the IEEE/CVF Conference on Computer Vision and Pattern Recognition, pp. 2730–2739 (2019)
20. Waseda, F., Nishikawa, S., Le, T.N., Nguyen, H.H., Echizen, I.: Closer look at the transferability of adversarial examples: how they fool different models differently. In: Proceedings of the IEEE/CVF Winter Conference on Applications of Computer Vision, pp. 1360–1368 (2023)
21. Rony, J., Hafemann, L.G., Oliveira, L.S., Ayed, I.B., Sabourin, R., Granger, E.: Decoupling direction and norm for efficient gradient-based l2 adversarial attacks and defenses. In: Proceedings of the IEEE/CVF Conference on Computer Vision and Pattern Recognition, pp. 4322–4330 (2019)

22. Dong, Y., et al.: Boosting adversarial attacks with momentum. In: Proceedings of the IEEE Conference on Computer Vision and Pattern Recognition, pp. 9185–9193 (2018)
23. Dong, Y., Pang, T., Su, H., Zhu, J. Evading defenses to transferable adversarial examples by translation-invariant attacks. In: Proceedings of the IEEE/CVF Conference on Computer Vision and Pattern Recognition, pp. 4312–4321 (2019)
24. Lin, J., Song, C., He, K., Wang, L., Hopcroft, J.E.: Nesterov accelerated gradient and scale invariance for adversarial attacks. arXiv preprint arXiv:1908.06281 (2019)
25. Yang, K., Zhou, T., Zhang, Y., Tian, X., Tao, D.: Class-disentanglement and applications in adversarial detection and defense. Adv. Neural Inf. Process. Syst. **34**, 16051–16063 (2021)
26. Tramèr, F., Papernot, N., Goodfellow, I., Boneh, D., McDaniel, P.: The space of transferable adversarial examples. arXiv preprint arXiv:1704.03453 (2017)
27. Chen, P.Y., Sharma, Y., Zhang, H., Yi, J., Hsieh, C.J.: EAD: elastic-net attacks to deep neural networks via adversarial examples. In: Proceedings of the AAAI Conference on Artificial Intelligence (2018)
28. Hinton, G., Vinyals, O., Dean, J.: Distilling the knowledge in a neural network. arXiv preprint arXiv:1503.02531 (2015)
29. Carlini, N., Wagner, D.: Towards evaluating the robustness of neural networks. In: 2017 IEEE Symposium on Security and Privacy, pp. 39–57 (2017)
30. Madry, A., Makelov, A., Schmidt, L., Tsipras, D., Vladu, A.: Towards deep learning models resistant to adversarial attacks. arXiv preprint arXiv:1706.06083 (2017)
31. Sandler, M., Howard, A., Zhu, M., Zhmoginov, A., Chen, L.C.: Mobilenetv2: inverted residuals and linear bottlenecks. In: Proceedings of the IEEE Conference on Computer Vision and Pattern Recognition, pp. 4510–4520 (2018)
32. Tan, M., Le, Q.: Efficient-net: rethinking model scaling for convolutional neural networks. In: International Conference on Machine Learning, pp. 6105–6114 (2019)
33. Kurakin, A., Goodfellow, I.J., Bengio, S.: Adversarial examples in the physical world. In: Artificial Intelligence Safety and Security, pp. 99–112 (2018)
34. Croce, F., Hein, M.: Reliable evaluation of adversarial robustness with an ensemble of diverse parameter-free attacks. In: International Conference on Machine Learning, pp. 2206–2216. PMLR (2020)

Minimizing Distortion in Steganography via Adaptive Language Model Tuning

Cheng Chen[1(✉)], Jinshuai Yang[2], Yue Gao[2], Huili Wang[1], and Yongfeng Huang[2(✉)]

[1] Institute for Network Sciences and Cyberspace, Tsinghua University, Beijing, China
chenchen21@mails.tsinghua.edu.cn

[2] Department of Electronic Engineering, Tsinghua University, Beijing, China
yfhuang@mail.tsinghua.edu.cn

Abstract. Linguistic steganography, a technique that hides secret information within normal text, possesses tremendous potential in various applications such as protecting user privacy. However, previous research in linguistic steganography has primarily focused on adjusting the probability distribution of steganographic text (stegotext) to minimize the difference with text generated by language models, thereby achieving indistinguishability between the two. Nonetheless, the significant gap between real text and generated text has often been overlooked. To address this issue, this paper proposes an innovative method: using an adaptive model tuning strategy, the generated stegotext becomes statistically closer to real text. We leverage a well-trained classifier in conjunction with a fundamental generative language model to produce stegotext that aligns closely with the distribution of real text. Consequently, we gain better control over the distortion between the stegotext and real text, while effectively embedding secret information. Compared to traditional methods, our approach reduces Kullback-Leibler divergence and steganography detection rates, demonstrating its enhanced effectiveness.

Keywords: Natural language processing · Linguistic steganography · Generative steganography

1 Introduction

In an era of ever-evolving information technology and network communications, the importance of information security has become widely recognized. For individuals, companies, and government agencies, it is crucial to protect information from unauthorized access, use, or disclosure. Against this backdrop, steganography has emerged as an important tool for secure communication. Steganography is a technique that hides secret information within innocent natural carriers such as images [1,2,8,10,41], audio [19,32,39], video [17,21], text [7,12,22,25,27], etc., to avoid being intercepted. Unlike encryption techniques that merely alter the readability of information through cryptographic algorithms, steganography

B. Luo et al. (Eds.): ICONIP 2023, CCIS 1966, pp. 571–584, 2024.
https://doi.org/10.1007/978-981-99-8148-9_44

hides information within everyday, seemingly harmless data carriers, making the existence of the information itself unknown to others, giving steganography advantages under certain circumstances.

Steganography can be classified into different categories depending on the type of carrier used during the steganographic process. However, steganography using images, audio, video and so on as carriers often faces distortions such as compression, rotation, and cropping during transmission, which can damage the secret information hidden within. In contrast, linguistic steganography can resist common distortions due to the highly encoded nature of text [9], and its ubiquity and universality make its application scenarios broader. Meanwhile, with the development of deep learning and NLP technologies, we are now able to generate more natural and fluent text, thus improving the stealthiness and embedding capacity of steganography.

The generative linguistic steganography is an important branch of linguistic steganography, and its core idea is to directly transform secret information into seemingly harmless stegotext without the need for any covertext [20,28,30,37]. The advantage of this approach is that by leveraging advanced deep learning and natural language processing (NLP) techniques, such as Variational Autoencoders (VAE) [11,35] and transformer [29], fluent and natural text can be generated, allowing for the selection of appropriate symbols to encode secret information at each time step, greatly increasing the embedding capacity. However, generative text steganography also faces some challenges.

The main challenge is how to minimize the distortion of the generated text while ensuring the integrity and security of the secret information. On the one hand, steganographyers need to ensure that the secret information is not damaged or lost during the encoding, transmission, and decoding processes. On the other hand, steganographyers need the generated stegotext to be as close as possible to real natural text in terms of semantics, syntax, and style, to avoid arousing suspicion. In this context, minimizing the distortion between the stegotext and the real text becomes a key issue.

Early methods such as Huffman coding [34] and arithmetic coding [42] were purely source coding algorithms, and although they hid secret information to some extent, they did not control the distortion of the stegotext. Subsequent methods such as Adaptive Dynamic Grouping (ADG) [40] and the Iterative Minimum Entropy Coupling approach (iMEC) [3] began to address the distortion between the generated stegotext and the generated text. However, not only did they overlook the distortion between the real text and the stegotext, but existing steganography encoding algorithms also struggle to effectively handle this issue. Therefore, controlling the distortion between the real text and the stegotext while ensuring high embedding efficiency is a problem that needs to be solved in the field of generative linguistic steganography.

However, under the existing generative linguistic steganography framework, the distribution of real text is usually learned by the language model (LM), and all the above encoding methods simply optimize the difference between the statistical distribution of the stegotext and the generated text produced by the

LM, without considering the difference between the generated text and the real text. In fact, Transformer LMs always systematically underestimate the probability of sequences drawn from the target language and tend to overestimate the probability of rare perturbed sequences [14]. Therefore, we attempt to adjust the probability distribution of the existing LM to be closer to the distribution of real text. By doing so, we aim to enhance the security of steganography, making the embedded content harder to detect.

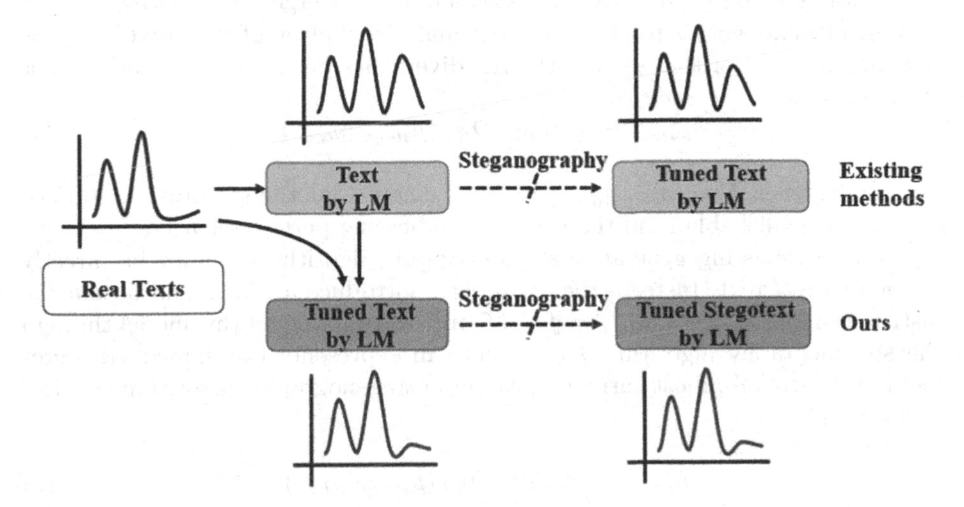

Fig. 1. The gray box above in the figure is the process of traditional generation-based linguisitc steganography, and the orange part below is the process of our proposed method. Compared with the traditional method, our method makes the probability distribution of stegotext closer to that of real text. (Color figure online)

In this paper, we proposes an innovative method aimed at minimizing the distortion of the generated text while ensuring the integrity and security of the secret information, as illustrated in Fig. 1. We delve into the probability distribution of the LM and attempt to adjust its distribution so that the generated text is statistically closer to the real text. To this end, we combine a trained classifier with the basic generative LM to generate stegotext closer to the real text distribution. Through this method, we can better control the distortion between the generated stegotext and the real text, while efficiently embedding secret information, achieving a more reliable steganography method.

2 Proposed Framework

2.1 Problem Definition and Optimization Goal

In this research, we focus on the problem of steganography in text generated by LMs. Linguistic steganography is an information hiding technique with the main goal of embedding secret information into seemingly ordinary text to evade unwanted attention or detection. The key challenge of this technique lies in the necessity of making the generated stegotext statistically indistinguishable from real text, ensuring the security of the information.

To quantify this goal, we first define the probability distribution of real text p_{true}, the probability distribution of text generated by the LM p_{LM}, and the probability distribution of stegotext p_{stego} generated by applying the embedding function f_{emb} to the LM $\mathcal{L}$. We measure the difference between the probability distributions p and q using Kullback-Leibler (KL) divergence $D_{KL}(p||q)$.

Our ultimate goal is to obtain the optimal distribution of stegotext p^*_{stego} by steganography algorithm so that the KL divergence between p_{stego} and p_{true} is minimized, i.e.,

$$p^*_{stego} = \underset{p_{stego}}{\arg\min}\ D_{KL}(p_{stego}||p_{true}). \tag{1}$$

Ideally, when $D_{KL}(p_{stego}||p_{true}) = 0$, it means that the stegotext is statistically indistinguishable from the real text, indicating perfect security.

However, existing generative steganographic algorithms cannot be directly applied to real text. Instead, they indirectly introduce a LM $\mathcal{L}$ and obtain the distribution of stegotext $p_{stego} = f_{emb}(\mathcal{L}, \mathbf{m})$ by sampling on this model through the steganography algorithm f_{emb}, where $\mathbf{m}$ represents the embedded secret message. Therefore, most current research on steganographic algorithms focuses on making

$$p^*_{stego} = \underset{p_{stego}}{\arg\min}\ D_{KL}(p_{stego}||p_{LM}), \tag{2}$$

which aims to find an ideal steganography algorithm f^*_{emb} such that

$$f^*_{emb} = \underset{f_{enb}}{\arg\min}\ D_{KL}(f_{emb}(\mathcal{L}, \mathbf{m})||p_{LM}). \tag{3}$$

It is important to note that theoretically, if a model can perfectly model the distribution of real text, then this approach is optimal as it ensures that the distribution of the generated text does not change too much after steganography. Yet, existing LMs cannot achieve this level of perfection, which serves as our first reason for shifting focus to the difference between real text and text generated by LMs.

Furthermore, while the optimization goal of steganography algorithms usually focuses on the difference between generated text and stegotext, in practice, steganalysis often detects the gap between real text and stegotext [16,23,24,31,33,36,38,43]. This detection method proves advantageous as it can be applied across different LMs and parameters and pays more attention to the differences between characteristics of real text and stegotext. This constitutes our second reason for this focus shift.

For these two reasons, we focus more on optimizing the KL divergence between the distribution of text generated by the LM p_{LM} and the distribution of real text p_{true}. That is,

$$p^*_{LM} = \underset{p_{LM}}{\arg\min}\ D_{KL}(p_{LM}||p_{true}). \tag{4}$$

In summary, our optimization goal is to find a modification method for the LM that minimizes the $D_{KL}(p_{LM}||p_{true})$. This minimization, in turn, ensures that $D_{KL}(p_{stego}||p_{true})$ is sufficiently small.

Intuitively, if the KL divergence between the generated text and stegotext is constrained, that is, $D_{KL}(p_{LM}||p_{stego}) < \delta$, then reducing $D_{KL}(p_{true}||p_{LM})$ will help to reduce $D_{KL}(p_{true}||p_{stego})$.

Although KL divergence does not satisfy the triangle inequality, we can still prove that when the distribution difference between stegotext and generated text is small enough, that is, the noise ϵ introduced by f_{emb} is small enough, considering the Taylor expansion of the KL divergence around the generated text probability distribution p_{LM}, we have

$$\begin{aligned} D_{KL}(p_{true}||p_{stego}) &= D_{KL}(p_{true}||p_{LM} + \epsilon) \\ &\approx D_{KL}(p_{true}||p_{LM}) \\ &+ \epsilon \cdot \nabla D_{KL}(p_{true}||p_{LM}) + O(\epsilon^2). \end{aligned} \tag{5}$$

This implies that if the noise ϵ is small enough, $D_{KL}(p_{true}||p_{stego})$ is primarily determined by $D_{KL}(p_{true}||p_{LM})$. Therefore, reducing $D_{KL}(p_{true}||p_{LM})$ can effectively reduce $D_{KL}(p_{true}||p_{stego})$.

In fact, many existing steganographic algorithms, such as ADG and iMEC, claim to be provably secure, ensuring that $D_{KL}(p_{stego}||p_{LM})$ is small enough, i.e., the noise introduced during the steganographic process is small enough.

In this context, we emphasize that our optimization objective has shifted to minimizing $D_{KL}(p_{LM}||p_{true})$, rather than minimizing $D_{KL}(p_{stego}||p_{LM})$. The reason behind this shift is that while existing steganography algorithms can effectively ensure that $D_{KL}(p_{stego}||p_{LM})$ is small enough, this does not directly imply that $D_{KL}(p_{stego}||p_{true})$ is also sufficiently small. As per our previous discussion, to achieve optimal steganographic security, we need to ensure a minimal difference between the stegotext and the real text. Therefore, optimizing $D_{KL}(p_{true}||p_{LM})$ better serves this purpose. Under the assumptions made earlier, we have theoretically proven the effectiveness of this approach.

2.2 Adaptive Language Model Tuning

To address the previously mentioned challenges, we reassess the problem from a Bayesian perspective, modeling and analyzing it using the Bayesian formula $p(\boldsymbol{x}|a) \propto p(a|\boldsymbol{x})p(\boldsymbol{x})$. To effectively calculate $p(a|x)$, we adopt the idea of Plug-and-Play Language Model (PPLM) [4] to design a backpropagation method based on classifiers. The fundamental objective of PPLM is to design the probabilistic model $p(x|a)$, such that Natural Language Generation (NLG) can generate text $\boldsymbol{x}$ that matches a given attribute a.

Our strategy uniquely adjusts the history matrix of a Transformer-based LM, pushing the distribution of the text produced closer to the actual text's distribution. We follow a few key steps to accomplish this:

First, we employ a pre-trained Transformer-based LM. Next, we introduce an update vector ΔH_t to modify the history matrix H_t, yielding a new logic vector $\tilde{o}_{t+1}$:

$$\tilde{o}_{t+1}, H_{t+1} = \text{LM}(w_t, H_t + \Delta H_t). \tag{6}$$

Lastly, we utilize a classifier $p(a|\boldsymbol{x})$, and update ΔH_t via gradient descent, driving the text generated closer to the actual text:

$$\Delta H_t \leftarrow \Delta H_t + \alpha \frac{\nabla_{\Delta H_t} \log p(a|H_t + \Delta H_t)}{||\nabla_{\Delta H_t} \log p(a|H_t + \Delta H_t)||^{\gamma}}, \tag{7}$$

where α is the step size, and γ is the scalar coefficient of the regularization term.

However, our approach goes beyond using PPLM to generate text with specific attributes. We found that the Bayesian formula can also control the probabilistic distribution of the LM's output. In this process, attribute 'a' acts as an adjustable factor that influences the conditional probability model $p(x|a)$, thereby making the distribution of the text generated more similar to that of the real text.

Therefore, our aim is to use this method to draw the distribution of the text generated closer to the real text distribution, thus achieving our optimization goal of minimizing $D_{KL}(p_{LM}||p_{true})$. This requires consideration of how to guide the generated text to be closer to the real text in terms of probability distribution when training the classifier.

2.3 Embeding and Extraction

Algorithm 1. Embedding Algorithm

1: **Input:** Language Model $\mathcal{L}$, Classifier $\mathcal{C}$, Message to Embed m, Prompt *prompt*
2: **Output:** Stegotext S
3: $S \leftarrow prompt$
4: **while** not the end of m **do**
5: $o_{t+1}, H_{t+1} \leftarrow \mathcal{L}(S, H_t)$
6: // Next, adjust H_t so that the distribution corresponding to o_{t+1} is closer to the real distribution
7: $\Delta H_t \leftarrow 0$
8: **for** $iteration \leftarrow 0, iteration <$ max_iterations **do**
9: // Update the history matrix's update value using the classifier $\mathcal{C}$
10: $Loss \leftarrow \mathcal{C}(o_{t+1})$
11: $\Delta H_t \leftarrow$ update_matrix($Loss$)
12: $o_{t+1}, H_{t+1} \leftarrow \mathcal{L}(S, H_t + \Delta H_t)$
13: **end for**
14: $P_{t+1} =$ generate_adjusted_distribution(o_{t+1})
15: $w_{t+1} \leftarrow f_{\text{emb}}(P_{t+1}, m)$ // Embed secret information using an embedding algorithm
16: $S \leftarrow S||w_{t+1}$
17: **end while**

In our approach, we focus on adjusting the LM to bring the distribution of its generated text closer to that of the real text. This section succinctly delineates the embedding and extraction process for better comprehension.

In the embedding process, our method is somewhat similar to traditional steganography methods [34], which also encode information through probabilities. However, the characteristic of our method lies in that we use the probabilities generated by the adjusted LM for embedding. Specifically, we first use a well-trained classifier to guide the LM to make its generated text distribution closer to the real text in terms of probability distribution. Therefore, in the embedding process, the generated text not only encodes the information we need to hide, but also has a probability distribution closer to the real text. A concise description of this algorithmic procedure can be found in Algorithm 1.

In the extraction process, we first obtain the text generated in the embedding process. Since we share the same LM, the generated probability distribution is the same under the same context. This allows us to get the corresponding secret information based on the currently selected candidate word and its corresponding probability encoding algorithm, thereby extracting the hidden content. A brief overview of the methodology is presented in Algorithm 2.

Algorithm 2. Extraction Algorithm

Require: Language Model $\mathcal{L}$, Classifier $\mathcal{C}$, Stegotext S, Prompt *prompt*
Ensure: Message to Embed m
1: $T \leftarrow prompt$
2: **while** not the end of m **do**
3: $\quad o_{t+1}, H_{t+1} \leftarrow \mathcal{L}(T, H_t)$
4: // Next, adjust H_t so that the distribution corresponding to o_{t+1} is closer to the real distribution
5: $\quad \Delta H_t \leftarrow 0$
6: $\quad$ **for** iteration = 0, iteration ¡ max_iterations **do**
7: $\quad\quad$ Loss $\leftarrow \mathcal{C}(o_{t+1})$
8: $\quad\quad \Delta H_t \leftarrow$ update_matrix(Loss)
9: $\quad\quad o_{t+1}, H_{t+1} \leftarrow \mathcal{L}(T, H_t + \Delta H_t)$
10: $\quad$ **end for**
11: $\quad P_{t+1} \leftarrow$ generate_adjusted_distribution(o_{t+1})
12: $\quad m \leftarrow m || f_{\text{ext}}(P_{t+1}, w_{t+1})$ // Extract secret information using an extraction algorithm
13: $\quad T \leftarrow T || w_{t+1}$
14: **end while**

In conclusion, our method adjusts the probability distribution of the LM, making the generated text closer to the real text in terms of probability distribution, while also ensuring the compatibility of our method with all previous generative text steganography methods.

3 Experiment and Analysis

3.1 Experiment Settings

In this research, our goal is to adjust the trained LMs to bring them closer to real texts and thereby enhance the security of text steganography. We will evaluate the steganographic performance from multiple perspectives, such as KL divergence, anti-steganalysis ability, sentence quality, and embedding rate. All experiments were conducted on a machine equipped with a GeForce GTX 1080 GPU. The experimental process can be roughly divided into three steps: training LMs, training discriminators and adjusting LMs, and performing steganography and evaluating the results. Here are more detailed experimental settings:

Training of Language Models. We chose three commonly used natural text corpora, Tweet [5], Movie [18], and News[1], as our datasets. Using these datasets, we trained three types of LMs: GPT-2 [26], BART [15], and Long Short-Term Memory networks (LSTM) [6], resulting in a total of 9 LMs. In the training process of GPT-2 and BART models, the learning rate was set to 0.0001, the warm-up ratio was set to 0.06, the weight decay was set to 0.01, the training rounds were 10, each batch contained 16 sentences, and the length of the sentences was 100. The LSTM model was set to 2 layers, with a dimension of 512, and trained using the AdamW optimizer. The learning rate was set to 0.0001, each batch contained 32 sentences, and the length of the sentences was 100.

Training of Discriminator and Adjustment of Language Models. We used the trained LMs to generate a certain number (2,000) of sentences, with sentence lengths between 5 and 200. Then, these generated texts and natural texts from the datasets were mixed for training the discriminator. During the training process of the discriminator, the learning rate was set to 0.0001, the training rounds were 10, and the batch size was set to 16. We used the trained discriminator to adjust the original LMs. The number of repetitions m for the backward gradient was set to 3, the KL divergence constraint λ_{KL} between the adjusted LM and the original model was set to 0.1, the step size was 0.2, and the intensity γ_{gm} was 0.3.

Steganography and Results Evaluation. Finally, we used the original LMs and adjusted LMs for steganography, with the steganography algorithm being the provably secure steganography algorithm, i.e., the ADG algorithm [40]. We will generate 1,000 sentences for subsequent testing.

3.2 Metrics

In this section, we introduce four key metrics for evaluating the performance of steganographic algorithms. These metrics comprehensively assess the performance of steganographic algorithms from different perspectives.

[1] https://www.kaggle.com/datasets/snapcrack/all-the-news.

Effective Embedding Rate (EER). The Embedding Rate (ER) denotes the average quantity of concealed information carried by each word, quantified in bits per word (bpw). In this study, we employ the Effective Embedding Rate [40] to evaluate the text's security in a more comprehensive manner. The Effective Embedding Rate is calculated using the formula:

$$\mathrm{EER} = 2 \times (1 - \mathrm{Acc}) \times \mathrm{ER}. \tag{8}$$

Here, Acc stands for detection accuracy. This metric reflects that if a stegotext has a probability of being detected, the average quantity of concealed information actually transmitted should correspondingly be adjusted. In an extreme scenario where stegotext is detected with 100

KL Divergence (KLD). The KL Divergence between stegotext and actual text directly characterizes the security of a steganographic algorithm. Generally speaking, the smaller the statistical distribution difference between stegotext and actual text, the more difficult it is to distinguish between them. We usually use third-party tools [13] to calculate the fixed-length density vectors $\boldsymbol{v_x}$ and $\boldsymbol{v_y}$ for steganographic and actual text, and then assume that these two vectors follow an isotropic Gaussian distribution, therefore, the corresponding KLD can be calculated by

$$\begin{aligned} &D_{KL}(p(\boldsymbol{v_x})||p(\boldsymbol{v_y})) \\ &\approx \sum \left(\log \frac{\boldsymbol{\sigma_x}}{\boldsymbol{\sigma_y}} + \frac{\boldsymbol{\sigma_x}^2 + (\boldsymbol{\mu_x} - \boldsymbol{\mu_y})^2}{2\boldsymbol{\sigma_y}^2} - \frac{1}{2} \right), \end{aligned} \tag{9}$$

where μ and σ are the mean and standard deviation of $\boldsymbol{v_x}$ and $\boldsymbol{v_y}$.

In theory, the smaller the KLD, the more difficult it is to distinguish statistical differences between steganographic and actual text, making it harder for steganalysis algorithms to detect whether a text contains hidden information.

Perplexity (PPL). Perplexity mainly reflects the quality of the generated sentences. The lower the Perplexity, the better the quality of the sentences generated, and the more they conform to human language habits.

Detection Accuracy (Acc). Steganalysis is a detection technique used to discover possible hidden information in a text, allowing for the practical testing of the security of steganographic algorithms. In this paper, we choose TS-CSW [36], a fast, lightweight, and highly accurate detection algorithm, as our detection method. Obviously, the lower the detection accuracy, the more difficult it is for the steganalysis algorithm to detect the stegotext in practice, and the better the security performance of the steganographic algorithm.

Through these metrics, we can comprehensively evaluate and compare the performance and security of various steganographic algorithms from different aspects. Please note that these metrics are not absolute but relative; that is, under the same test conditions and environments, these metrics can be used to compare the strengths and weaknesses of various algorithms.

3.3 Evaluation Results and Analysis

Table 1. Performance Comparison of ADG Steganography with Adaptive Fine-Tuning

Method		PPL↓		EER↑		KLD↓		Acc↓	
Dataset	Model	Baseline	Our	Baseline	Our	Baseline	Our	Baseline	Our
movie	GPT-2	51.236	**50.796**	3.386	**3.578**	0.668	**0.658**	0.516	**0.514**
	BART	64.142	**59.818**	3.365	**3.415**	0.678	**0.519**	**0.500**	0.522
	LSTM	**294.596**	297.448	5.228	**5.658**	7.670	**7.334**	0.540	**0.502**
news	GPT-2	56.594	**54.928**	3.385	**3.393**	0.407	**0.364**	0.520	**0.504**
	BART	88.751	**84.272**	3.682	**3.764**	0.286	**0.247**	0.518	**0.501**
	LSTM	**465.598**	469.489	**6.192**	5.982	1.580	**1.514**	0.508	**0.507**
tweet	GPT-2	**63.693**	64.115	2.740	**3.138**	**2.496**	2.864	0.581	**0.521**
	BART	73.163	**71.081**	2.612	**2.992**	3.195	**3.068**	0.591	**0.530**
	LSTM	**284.746**	301.687	4.804	**5.152**	0.918	**0.891**	0.544	**0.517**

Our evaluation results are summarized in Table 1, which shows a performance comparison of the ADG steganography method with and without our adaptive fine-tuning strategy on various datasets and models.

Firstly, it can be observed that our method enhances the EER in most cases, which signifies an improvement in the capacity of information hiding. Secondly, our method generally results in a decrease in the KLD, indicating that the stegotext generated by our method is closer to real text. This validates the effectiveness of our approach.

In terms of Perplexity, we observe some interesting patterns. The perplexity of LSTM slightly increases when applying our method, whereas it decreases for GPT-2 and BART. This might be due to the fact that our adaptive fine-tuning strategy, which utilizes PPLM, is inherently better suited to models based on the Transformer architecture, such as GPT-2 and BART. Transformer models are more capable of capturing long-distance dependencies in the text, which might be particularly beneficial for steganography. On the other hand, LSTM, with a different architecture, might not benefit as much from the PPLM-based fine-tuning, leading to a slight increase in perplexity.

From the perspective of steganalysis detection accuracy (Acc), our method shows a promising trend in decreasing the detection rate, thus enhancing the security of the steganography.

Lastly, it's worth noting that the results vary across different datasets, indicating that the nature of the text might have an influence on the performance of the steganography. For instance, our method does not improve KLD as significantly on the tweet dataset as it does on the movie and news datasets. This suggests that the language style of tweets, which tends to be more casual and abbreviated, might pose unique challenges to text steganography. This is a potential area for further investigation.

In summary, our approach has demonstrated promising results in improving the performance of steganographic algorithm, particularly on Transformer-based models. However, as with any method, there are opportunities for further optimizations and adjustments, especially when dealing with different model architectures and text styles.

4 Conclusion

This research paper primarily discusses the issue of steganography in text generated by LMs. The key challenge lies in making the generated stegotext statistically indistinguishable from real text, ensuring the security of the information. The paper proposes an innovative method aimed at minimizing the distortion of generated text while ensuring the integrity and security of the hidden information. By delving into the probability distribution of LMs and attempting to adjust this distribution, the generated text becomes statistically closer to real text. To achieve this, we combined a well-trained classifier with a basic generative LM to produce stegotext that closely aligns with the distribution of real text. This method provides better control over the distortion between the generated stegotext and real text and effectively embeds secret information, thus offering a more reliable steganography approach. Experimental results show that our method generally improves the EER in most cases, indicating enhanced information hiding capability. Furthermore, Experimental results show that our method typically results in a reduction of KLD, and a decrease in detection accuracy, proving that the stegotext generated with our method is closer to real text, thereby enhancing the security of the stegotext.

Acknowledgements. We would like to express our deepest gratitude to Professor Yongfeng Huang for his guidance, patience, and support throughout the course of this research. Additionally, we are grateful to our families and friends for their encouragement and support. Lastly, the authors thank anonymous reviewers for their insightful suggestions.

References

1. Cao, Y., Zhou, Z., Sun, X., Gao, C.: Coverless information hiding based on the molecular structure images of material. Comput. Mater. Continua **54**(2), 197–207 (2018)
2. Cao, Y., Zhou, Z., Wu, Q.M.J., Yuan, C., Sun, X.: Coverless information hiding based on the generation of anime characters. EURASIP J. Image Video Process. **2020**(1), 1–15 (2020). https://doi.org/10.1186/s13640-020-00524-4
3. Christian, Sokota, S., J, Foerster, J., Strohmeier, M.: Perfectly secure steganography using minimum entropy coupling. arXiv pre-print server (2023). https://doi.org/Nonearxiv:2210.14889, https://arxiv.org/abs/2210.14889

4. Dathathri, S., et al.: Plug and play language models: a simple approach to controlled text generation. arXiv preprint arXiv:1912.02164 (2019)
5. Go, A., Bhayani, R., Huang, L.: Twitter sentiment classification using distant supervision. CS224N Project Report, Stanford **1**(12), 2009 (2009)
6. Hochreiter, S., Schmidhuber, J.: Long short-term memory. Neural Comput. **9**(8), 1735–1780 (1997)
7. Huang, D., Yan, H.: Interword distance changes represented by sine waves for watermarking text images. IEEE Trans. Circ. Syst. Video Technol. **11**(12), 1237–1245 (2001)
8. Hussain, M., Wahab, A.W.A., Idris, Y.I.B., Ho, A.T., Jung, K.H.: Image steganography in spatial domain: a survey. Signal Process. Image Commun. **65**, 46–66 (2018)
9. Inui, K., Jiang, J., Ng, V., Wan, X.: Proceedings of the 2019 conference on empirical methods in natural language processing and the 9th international joint conference on natural language processing (EMNLP-IJCNLP). In: Proceedings of the 2019 Conference on Empirical Methods in Natural Language Processing and the 9th International Joint Conference on Natural Language Processing (EMNLP-IJCNLP) (2019)
10. Kaur, S., Singh, S., Kaur, M., Lee, H.N.: A systematic review of computational image steganography approaches. Arch. Comput. Meth. Eng. **29**, 4775–4797 (2022)
11. Kingma, D.P., Welling, M.: Auto-encoding variational bayes. arXiv preprint arXiv:1312.6114 (2013)
12. Krishnan, R.B., Thandra, P.K., Baba, M.S.: An overview of text steganography. In: 2017 Fourth International Conference on Signal Processing, Communication and Networking (ICSCN), pp. 1–6. IEEE (2017)
13. Le, Q., Mikolov, T.: Distributed representations of sentences and documents. In: International Conference on Machine Learning, pp. 1188–1196. PMLR (2014)
14. LeBrun, B., Sordoni, A., O'Donnell, T.J.: Evaluating distributional distortion in neural language modeling. arXiv preprint arXiv:2203.12788 (2022)
15. Lewis, M., et al.: BART: denoising sequence-to-sequence pre-training for natural language generation, translation, and comprehension. arXiv preprint arXiv:1910.13461 (2019)
16. Li, S., Wang, J., Liu, P.: Detection of generative linguistic steganography based on explicit and latent text word relation mining using deep learning. IEEE Trans. Dependable Secure Comput. **20**, 1476–1487 (2022)
17. Liu, Y., Liu, S., Wang, Y., Zhao, H., Liu, S.: video steganography: a review. Neurocomputing **335**, 238–250 (2019)
18. Maas, A., Daly, R.E., Pham, P.T., Huang, D., Ng, A.Y., Potts, C.: Learning word vectors for sentiment analysis. In: Proceedings of the 49th Annual Meeting of the Association for Computational Linguistics: Human Language Technologies, pp. 142–150 (2011)
19. Mishra, S., Yadav, V.K., Trivedi, M.C., Shrimali, T.: Audio steganography techniques: a survey. In: Bhatia, S.K., Mishra, K.K., Tiwari, S., Singh, V.K. (eds.) Advances in Computer and Computational Sciences. AISC, vol. 554, pp. 581–589. Springer, Singapore (2018). https://doi.org/10.1007/978-981-10-3773-3_56
20. Moraldo, H.H.: An approach for text steganography based on Markov chains. arXiv preprint arXiv:1409.0915 (2014)
21. Mstafa, R.J., Younis, Y.M., Hussein, H.I., Atto, M.: A new video steganography scheme based on Shi-Tomasi corner detector. IEEE Access **8**, 161825–161837 (2020)

22. Murphy, B., Vogel, C.: The syntax of concealment: reliable methods for plain text information hiding. In: Security, Steganography, and Watermarking of Multimedia Contents IX, vol. 6505, pp. 351–362. SPIE (2007)
23. Niu, Y., Wen, J., Zhong, P., Xue, Y.: A hybrid R-BILSTM-C neural network based text steganalysis. IEEE Sig. Process. Lett. **26**(12), 1907–1911 (2019)
24. Peng, W., Zhang, J., Xue, Y., Yang, Z.: Real-time text steganalysis based on multi-stage transfer learning. IEEE Signal Process. Lett. **28**, 1510–1514 (2021)
25. Por, L.Y., Ang, T., Delina, B.: WhiteSteg: a new scheme in information hiding using text steganography. WSEAS Trans. Comput. **7**(6), 735–745 (2008)
26. Radford, A., Wu, J., Child, R., Luan, D., Amodei, D., Sutskever, I.: Language models are unsupervised multitask learners. OpenAI Blog **1**(8), 9 (2019)
27. Satir, E., Isik, H.: A Huffman compression based text steganography method. Multimedia Tools Appl. **70**(3), 2085–2110 (2014)
28. Shniperov, A.N., Nikitina, K.: A text steganography method based on Markov chains. Autom. Control. Comput. Sci. **50**(8), 802–808 (2016)
29. Vaswani, A., et al.: Attention is all you need. In: Advances in Neural Information Processing Systems, pp. 5998–6008 (2017)
30. Wolf, T., et al.: Transformers: state-of-the-art natural language processing. In: Proceedings of the 2020 Conference on Empirical Methods in Natural Language Processing: System Demonstrations, pp. 38–45 (2020)
31. Wu, H., Yi, B., Ding, F., Feng, G., Zhang, X.: Linguistic steganalysis with graph neural networks. IEEE Signal Process. Lett. **28**, 558–562 (2021)
32. Wu, J., Chen, B., Luo, W., Fang, Y.: Audio steganography based on iterative adversarial attacks against convolutional neural networks. IEEE Trans. Inf. Forensics Secur. **15**, 2282–2294 (2020)
33. Xiang, L., Liu, Y., You, H., Ou, C.: Aggregating local and global text features for linguistic steganalysis. IEEE Signal Process. Lett. **29**, 1502–1506 (2022)
34. Yang, Z.L., Guo, X.Q., Chen, Z.M., Huang, Y.F., Zhang, Y.J.: RNN-Stega: linguistic steganography based on recurrent neural networks. IEEE Trans. Inf. Forensics Secur. **14**(5), 1280–1295 (2018)
35. Yang, Z.L., Zhang, S.Y., Hu, Y.T., Hu, Z.W., Huang, Y.F.: VAE-Stega: linguistic steganography based on variational auto-encoder. IEEE Trans. Inf. Forensics Secur. **16**, 880–895 (2020)
36. Yang, Z., Huang, Y., Zhang, Y.J.: TS-CSW: text steganalysis and hidden capacity estimation based on convolutional sliding windows. Multimedia Tools Appl. **79**(25), 18293–18316 (2020)
37. Yang, Z., Jin, S., Huang, Y., Zhang, Y., Li, H.: Automatically generate steganographic text based on Markov model and Huffman coding. arXiv preprint arXiv:1811.04720 (2018)
38. Yang, Z., Wang, K., Li, J., Huang, Y., Zhang, Y.J.: TS-RNN: text steganalysis based on recurrent neural networks. IEEE Signal Process. Lett. **26**(12), 1743–1747 (2019)
39. Yi, X., Yang, K., Zhao, X., Wang, Y., Yu, H.: AHCM: adaptive Huffman code mapping for audio steganography based on psychoacoustic model. IEEE Trans. Inf. Forensics Secur. **14**(8), 2217–2231 (2019)
40. Zhang, S., Yang, Z., Yang, J., Huang, Y.: Provably secure generative linguistic steganography. arXiv preprint arXiv:2106.02011 (2021)
41. Zhou, Z., Cao, Y., Wang, M., Fan, E., Wu, Q.J.: Faster-RCNN based robust coverless information hiding system in cloud environment. IEEE Access **7**, 179891–179897 (2019)

42. Ziegler, Z., Deng, Y., Rush, A.: Neural linguistic steganography. In: Proceedings of the 2019 Conference on Empirical Methods in Natural Language Processing and the 9th International Joint Conference on Natural Language Processing (EMNLP-IJCNLP), pp. 1210–1215. Association for Computational Linguistics, Hong Kong, China (2019). https://doi.org/10.18653/v1/D19-1115. https://aclanthology.org/D19-1115
43. Zou, J., Yang, Z., Zhang, S., Rehman, S., Huang, Y.: High-performance linguistic steganalysis, capacity estimation and steganographic positioning. In: Zhao, X., Shi, Y.-Q., Piva, A., Kim, H.J. (eds.) IWDW 2020. LNCS, vol. 12617, pp. 80–93. Springer, Cham (2021). https://doi.org/10.1007/978-3-030-69449-4_7

Efficient Chinese Relation Extraction with Multi-entity Dependency Tree Pruning and Path-Fusion

Weichuan Xing, Weike You(✉), Linna Zhou, and Zhongliang Yang

School of Cyberspace Security, Beijing University of Posts and Telecommunications, Beijing 100083, China
{xing_weichuan,ywk,zhoulinna,yangzl}@bupt.edu.cn

Abstract. Relation Extraction (RE) is a crucial task in natural language processing that aims to predict the relationship between two given entities. In recent years, a large majority of approaches utilized syntactic information, particularly dependency trees, to enhance relation extraction by providing superior semantic guidance. Compared with other fields, Chinese texts are more semantic complex, and contain multiple pairs of entities. However, many studies only focus on removing extraneous information from the dependency tree that pertains to a single entity pair. We hypothesis that preserving the semantic and structural interaction between multiple entity pairs in the tree is more conducive to the identification of the current entity pair relationship. Therefore, we propose a new pruning strategy called Multi-entity dependency Tree Pruning and path-Fusion (MTPF), which preserves the ancestor nodes of each entity pair to their lowest common ancestor, as well as the shortest path from that node to each entity. Then we introduce A-GCN as the encoder for the syntax tree obtained above, and the idea of multi-classification sequence as the decoder. Experimental results on two Chinese benchmark datasets, the financial dataset constructed by ourselves and DUIE1.0, demonstrate the effectiveness of our pruning strategy for CRE, where our approach outperforms strong dependency-tree baselines and achieve state-of-the-art results on both datasets.

Keywords: Chinese RE · Multi-entity pairs · Dependency Tree pruning Strategy · Attentive graph convolutional network

1 Introduction

Relation Extraction, as an important subtask in information extraction, aims to detect relations among entities in the text, and can also be regarded as a relation classification problem. It plays an important role in various downstream

This research was supported by the National Nature Science Foundation of China 62172053, National Key R&D Program of China 2021YFC3340700, 2022YFC3303300, 2021YFC3340600, 2022YFC3300800 and the Fundamental Research Funds for the Central Universities 2023RC30.

B. Luo et al. (Eds.): ICONIP 2023, CCIS 1966, pp. 585–598, 2024.
https://doi.org/10.1007/978-981-99-8148-9_45

tasks of natural language processing (NLP), such as question answering [1], summarization [2] and text mining [3]. Relational facts are usually represented as triples containing two entities (subject and object) and the relationship. Chinese relation extraction (CRE) refers to the process of identifying the relationship between entities in a given Chinese text.

Most current relation extraction (RE) models employ deep neural networks such as LSTM and CNN [4,5] to extract and analyze the semantic features of text sequences, or pre-trained language models like BERT [6] and XLNet [7] learn richer semantic knowledge, which greatly enhances the performance of RE. Although sequence-based models are prevalent, dependency tree-based models have shown to be effective for extracting non-local syntactic relationships that are challenging to obtain solely from the sentence surface [8]. However, research has indicated that feeding the entire syntax tree as input can introduce redundant noise to the encoder, leading to a significant decrease in the F1 value of the extracted results [9]. Therefore, it is critical to prune unnecessary noise from the syntactic tree. Some papers propose to integrate traditional fixed pruning strategies, such as Shortest Dependency Path (SDP) [10] and the Lowest Common Ancestor (LCA) [8] between one pair of entities, before entering the encoder. However, these rule-based pruning strategies can exclude crucial information. To address this issue, most existing approaches use complicated network architectures or heavy feature engineering (Roth and Lapata et al., 2016 [11]; Marcheggiani and Titov et al., 2017 [12]; Zhang, 2018 [9]; Guo et al.,2019 [13]; Yuanhe Tian et al., 2020 [14]; Bowen Yu et al., 2020 [15]) to soft-prune the dependency tree.

Compared with other fields, Chinese sentences are often semantically complex and can contain multiple relation triples within a single sentence [16]. To extract entity relations, pruned syntactic tree features and strong neural network models from the general domain are commonly utilized [17], but these methods do not fully consider the influence of other entity pairs on the current entity relation extraction.

Based on the above analysis, we argue that the main limitation is that current dependency tree models only pay attention to the single pair entity pruning strategy, and this limits the information interaction effect of other entities in the tree on the current entity pair if there are multiple entities in one tree. Hence, we assert that there is a pressing need to investigate the impact of other entities on the identification of the relationship between the current entity pair.

In this paper, we introduce a novel pruning strategy called MTPF for dependency syntactic trees that contain multiple entity pairs. This strategy enables us to remove extraneous information while preserving the structural and semantic interaction between the entity pairs. We then introduce an efficient dependency-driven graph convolutional network to encode and soft-prune the resulting dependency tree, followed by a multi-class sequence labeling method for decoding the tree. We evaluate our model on the widely used Chinese dataset, DUIE1.0, as well as on a financial Chinese dataset we created ourselves. Our experimental results demonstrate the effectiveness of our new pruning strategy in retaining the structural and semantic interaction information between multiple entity pairs. Notably, our model achieves state-of-the-art performance on both Chinese datasets.

To recap, our main contributions are:

- We firstly propose a pruning and path-fusion strategy which can preserve the structural and semantic interaction path of multiple entities.
- Our model introduces dependency-driven Attentive GCN to encode and soft-prune the syntactic tree, and adapts the idea of multi-classification sequence labeling to identify the relationship of entity pairs in a sentence simultaneously, which effectively improves the accuracy of each pair of entity relationship recognition.
- We have conducted a lot of experiments and analyses to prove the validity of our pruning strategies. Experiment results show that our framework consistently achieves superior performance over previous dependency-tree models and the other relation models.

2 Related Work

2.1 Pruning Strategy

One common approach to utilizing dependency information in relation extraction is to reduce the parse tree to the shortest dependency path between the entities, as proposed by Xu et al. [10]. Another popular approach is to perform bottom-up or top-down computation along the parse tree or the subtree below the lowest common ancestor of the entities, as suggested by Miwa [8].

However, current methods for extracting entity pairs often overlook the importance of information interaction between multiple entities and the entity of interest. If we simply transfer existing strategies to extract multiple entity pairs, we encounter certain issues. 1) When two entity pairs are situated at the beginning and end of a text, there may be no path intersection between them. If we were to apply the existing strategies like SDP and calculate the shortest path for each pair individually, the impact of the pruning strategy would be almost negligible. For example, in the first syntax tree shown in Fig. 1, the entity pair of "To the Happiest People - Gigi Leung" at the beginning of the sentence cannot interact with the entity pair of "The Great Entertainer - Gigi Leung" at the end of the sentence, so it is difficult to judge the third relationship between the song and the album; 2) Imply taking the lowest common ancestor (LCA) path between two entities captures all the subtree information, but it also introduces a lot of irrelevant noise. In the second syntax tree shown in Fig. 1, the rest of the irrelevant noise contained under the minimum common ancestor, such as dates like "2004年12月15日", will also be included. As shown in the Sect. 3.1 below, our approach, on the other hand, leverages the information interaction paths between multiple entity pairs in the syntactic dependency tree and offers a more effective solution compared to existing strategies.

2.2 RE Dependency-Based Models

RE is conventionally performed as a typical classification task. Therefore, in addition to leveraging advanced text encoders (e.g., Transformer [18], BERT [6]) to capture contextual information, the dependency information, as one kind of knowledge, of the running text has been widely used as an effective resource to improve RE. Zhang et al. [9] adopted GCN to model the dependencies and proposed a trade-off pruning strategy in between Xu et al. [10] and Miwa and Bansal [8]. To further emphasize the target-relevant information, Bowen Yu [15] proposes a novel dynamic Pruned Graph Convolutional Network (DP-GCN) architecture, which learns pruned dependency trees through reflection. Besides, there are other graph-based models for RE that utilize layers of multi-head attentions [13], dynamic pruning [19], and additional attention layers [20] to encode dependency trees. And there is another track to leverage it by pre-training dependency-based word embeddings through an auxiliary module to learn the dependency information by treating the dependencies as additional input during pre-training [14].

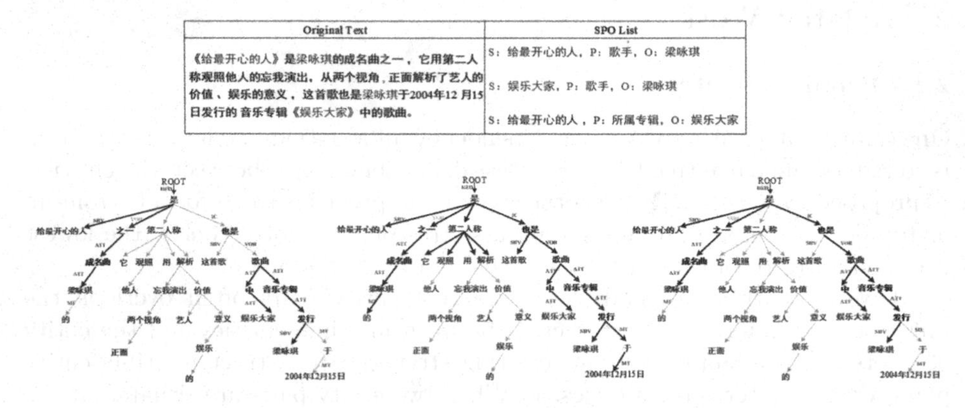

Original Text	SPO List
《给最开心的人》是梁咏琪的成名曲之一，它用第二人称观照他人的忘我演出，从两个视角，正面解析了艺人的价值、娱乐的意义，这首歌也是梁咏琪于2004年12 月15日发行的 音乐专辑《娱乐大家》中的歌曲。	S：给最开心的人，P：歌手，O：梁咏琪 S：娱乐大家，P：歌手，O：梁咏琪 S：给最开心的人，P：所属专辑，O：娱乐大家

Fig. 1. An example modified from the DUIE Dataset. The three syntactic trees are the results of pruning multi-entity trees by using traditional SDP strategy, LCA strategy and MTPF strategy respectively. The original tree is parsed by the DDParser tool and contains 2 pairs of entities—"给最开心的人-梁咏琪", "娱乐大家-梁咏琪". Our pruning strategy MTPF (as described in Sect. 3.1) maximizes the interaction between the two pairs of entities, and the path highlighted in black is the path that is ultimately preserved after pruning.

3 Methods

In this section, we first detail our pruning strategy for Chinese multi-entity pairs in Sect. 3.1 and then introduce how to use the graph convolutional neural network integrated with dependency coding to drive further soft pruning of the dependency syntactic tree obtained above, and finally decode it using the multi-classification sequence annotation method in Sect. 3.2.

3.1 Pruning Strategy

Dependency trees provide valuable structural knowledge for relation extraction. Previous research has used the shortest path SDP and the least common ancestor LCA to prune and delete unnecessary information. However, these methods only consider the syntactic tree where a pair of entities is located, ignoring the influence of other entities on the semantic and structural characteristics of the current entity pair when there are multiple pairs of entities in a sentence. We propose a pruning strategy for multi-pair entity syntactic trees and integrate paths based on traditional strategies to enhance the interaction between entity pairs.

To preserve the path where multiple entity pairs are located on a dependency tree, we propose the following pruning strategy, which we name MTPF. Through observation of the syntax tree, we find that the lowest common root of the current entity pair extracted is often the trigger word for the relationship between the entities, and the words on the path connecting the two entities and the root provide key information. By combining the advantages of SDP and LCA, we optimize the integration of the root nodes and critical paths of multiple entities, leading to the following pruning Algorithm 1:

Algorithm 1. Multi-entity Dependency Tree Pruning and Path-Fusion.

Input:
Dependency syntax tree containing multiple entity pairs;
Output:
Pruned syntactic tree;
1: Identify the common ancestor of each pair of entities, denoted as T_x (where $x =$ 1,2,...,n);
2: Determine the minimum common ancestor, denoted as L^*, by finding the common ancestor that is furthest from the root of the tree;
3: Keep the shortest path from L^* to each child node of the entities;
4: **return** Pruned syntactic tree;

The experiments have shown that the shortest path in Step 3 must include the root node, which is the key root node that needs to be preserved. To illustrate, let's consider the syntax tree parsed by the tool as shown by the third syntactic tree in Fig. 1. Firstly, we need to locate the lowest common ancestor of "给最开心的人，梁咏琪" and "娱乐大家，梁咏琪" which are "是" and "歌曲" respectively. Secondly, we need to find the smallest common ancestor, L*, of "是" and "歌曲", which is "是". Finally, we preserve the shortest path from "是" to all known entities.

3.2 The Whole Architecture of Our Model

With the pruning strategy for the multi-entity pair tree described above, we can obtain the syntax tree for multi-entity pair information interaction. To encode the tree and perform soft pruning, we employ a dependency-driven graph convolutional neural network that captures deep features within the tree. Then, our decoder uses the idea of multi-classification sequence predict the maximum possible category of relationship between a given entity token and other tokens. The overall architecture of our model is depicted in Fig. 2.

Encoding. The input of the training model based on words is seen as $\mathbf{s} = \{\mathbf{w}_1, \mathbf{w}_2, ..., \mathbf{w}_n\}$, where $\mathbf{w}_i$ is the i-th word in a sentence. Due to features of Long-Short Term Memory (LSTM) [21], BiLSTM is applied to $\mathbf{w}_i$ which is a combination of forward LSTM layers and backward LSTM layers, and has been proven to be effective in capturing semantic information between each word in a sentence. As shown in Eq.(1), the hidden vector representations of input sentence are denoted as $\mathbf{h}^{(0)} = \{\mathbf{h}_1^{(0)}, \mathbf{h}_2^{(0)}, ..., \mathbf{h}_3^{(0)}\}$.

$$\mathbf{h}_i^{(0)} = \overrightarrow{\mathbf{LSTM}}\left(\mathbf{x}_i, \overrightarrow{\mathbf{h}}_{i-1}\right) \oplus \overleftarrow{\mathbf{LSTM}}\left(\mathbf{x}_i, \overleftarrow{\mathbf{h}}_{i+1}\right) \tag{1}$$

Generally, Graph convolutional networks (GCN) is a widely used architecture to encode the information in a graph. Given a graph with n nodes, we can represent the graph structure as an n * n adjacency matrix, $A_{ij} = 1$ and $A_{ji} = 1$ if there is an edge from node i to node j, otherwise $A_{ij} = 0$ and $A_{ji} = 0$. However, it is worth noted that in standard GCN, each edge of the path in the pruned syntactic tree is treated equally. In addition, dependency types associated with the dependency connections are often omitted, in fact they can provide highly useful information. Therefore, we use A-GCN to distinguish the weights of different connections using the attention mechanism, dependency coding drives dynamic soft pruning of the previously obtained syntactic trees. We firstly construct the dependency type matrix. For example, d_{ij} = ATT indicates that the oriented dependency between words i and j is Deterministic relation (ATT), and then it is calculated as the corresponding Embedding matrix. Then the weight A_{ij} in formula is reexpressed as:

$$P_{i,j}^{(l)} = \frac{A_{i,j} \cdot \exp\left(\mathbf{t}_i^{(l)} \cdot \mathbf{t}_j^{(l)}\right)}{\sum_{j=1}^{n} A_{i,j} \cdot \exp\left(\mathbf{t}_i^{(l)} \cdot \mathbf{t}_j^{(l)}\right)} \tag{2}$$

where $\mathbf{t}_i^{(l)}$ and $\mathbf{t}_j^{(l)}$ respectively are the representation after the introduction of dependency syntax coding vector:

$$\mathbf{s}_i^{(l)} = \mathbf{h}_i^{(l-1)} \oplus \mathbf{e}_{i,j}^t \tag{3}$$

$$\mathbf{s}_j^{(l)} = \mathbf{h}_j^{(l-1)} \oplus \mathbf{e}_{i,j}^t \tag{4}$$

And $\mathbf{e}_{i,j}^t \in E$, the AGCN convolution operation at each layer is expressed as:

$$h_i^{(l)} = \sigma\left(\sum_{j=1}^{n} A_{ij} W^{(l)} h_j^{(l-1)} + b^{(l)}\right) \tag{5}$$

The sentence we expect to obtain after iteration of L layers GCN is expressed as:

$$\mathbf{h}_i^{(l)} = \sigma\left(\sum_{j=1}^{n} P_{i,j}^{(l)} \left(\mathbf{W}^{(l)} h_j^{(l-1)} + \mathbf{b}^{(l)}\right)\right) \tag{6}$$

Compared with standard GCN, A-GCN uses numerical weighting (i.e. $\mathbf{p}_{i,j} \in [0,1]$) rather than a binary choice for A_{ij} , to distinguish the importance of different connections so as to leverage them accordingly.

Decoding. In the last decoder, in order to output multiple pairs of entity relations in a sentence at the same time, we convert it into a multi-classification sequence annotation problem, that is, the last word of the head entity is annotated as the position of the tail entity and the corresponding relation category. The input is the output of AGCN coding network. Finally, we use a multiple sigmoid model to classify any kind of relationship between each token. The probability of token w_i choses token w_j as the tail of relation label r_k is:

$$P(w_j, r_k | w_i) = \text{sigmoid}\left(s(\hat{\text{h}}_i, \hat{h}_j, r_k)\right) \tag{7}$$

Finally, we minimize the objective function, which is the cross-entropy loss function:

$$L_{\text{rel}} = -\sum_{i=1}^{l}\sum_{j=1}^{m} \log P\left(y_{i,j}, r_{i,j} \mid w_i\right) \tag{8}$$

where $\mathbf{e}_j \in W$ and $\mathbf{r}_i \in R$ are the correct tail entities and relations, and m is the number of tail entities of w_i.

4 Experiments

4.1 Datasets

To test the performance of the proposed method, we perform experiments on two Chinese datasets. The first dataset is a financial dataset obtained by crawling the financial company announcement text from the website[1], consisting of 1.7w articles and 11 types of relationships, such as subsidiaries, employment, equity relationship transfer, pledge and so on. The second dataset is the DUIE1.0 dataset [16], which is a large schema-based Chinese relation extraction dataset with over 430,000 triples, 210,000 Chinese sentences, and 50 predefined relation types. The data in the DuIE1.0 dataset is obtained from various sources such as Baidu Baike, Baidu Post Bar, and Baidu Information Flow Text. The distribution of data categories is relatively uniform in the DuIE1.0 dataset.

[1] https://www.10jqka.com.cn/.

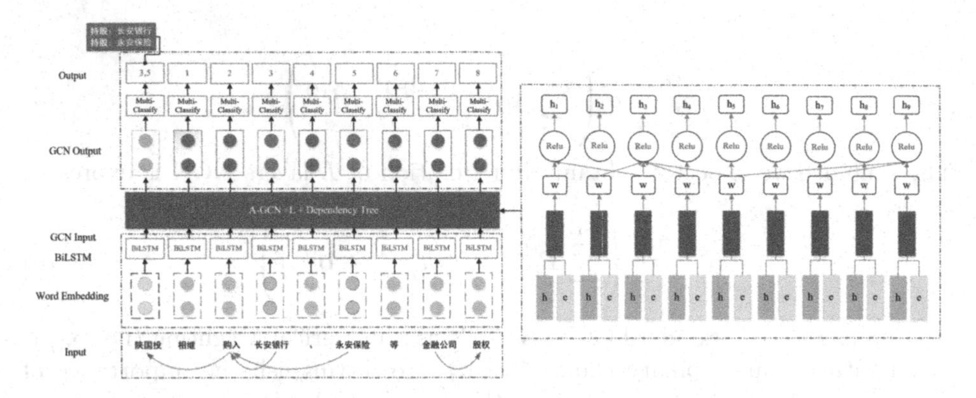

Fig. 2. The architecture of our model. The left side shows the overall architecture, and the syntactic tree where the multi-entity pairs are located obtained after 3.1 Pruning strategy is encoded by dependency-driven graph convolution. While on the right side, we only show the detailed A-GCN computation.

4.2 Dependency Graph Construction

Dependency syntax trees are a crucial feature of dependency extraction. To construct the dependency trees for multiple entity pairs and create the adjacency matrix needed for the graph convolutional neural network, we utilize the DDParser dependency tree construction tool, which was developed and published by Baidu. DDParser is a deep learning-based dependency parsing tool that uses PaddlePaddle[2], a deep learning platform developed by Baidu, and large-scale annotated data. The training data covers various Chinese scenarios such as news and forums, making it suitable for parsing complex sentences in large-scale Chinese domains. Additionally, this tool also allows the inclusion of longer proper nouns that frequently appear in Chinese text through the use of custom dictionaries.

4.3 Experimental Setup

In our main experiments, we compared our model with the previous relation extraction models, especially some different dependency-based models. For example, 1) Yuhao Zhang [9] proposed a path center pruning strategy and utilized GCN for feature extraction. 2) Zhijiang Guo [13] developed AGGCN, which uses a"soft pruning" strategy to transform the original dependency tree into a fully connected edge-weighted graph. 3) Yuanhe Tian [14] introduced AGCN, which dynamically learns different dependency connections to intelligently trim dependency syntax trees. Since the above models only see one pair of entities when encoded by LSTM or BERT, our tree can see multiple pairs, we wanted to make a fairer comparison by allowing the convolutional network encoder to

[2] https://github.com/PaddlePaddle/Research.

observe all entity pairs equally. To accomplish this, we followed the method outlined in Chen Danqi's paper [22] and independently input multiple entity pairs into the network by inserting entity marks for both subject and object.

4.4 Implementation Details

Effective text representation is crucial for achieving outstanding performance in many NLP tasks. To improve the quality of our experiments, we leverage pre-trained language models, specifically BERT, as the base encoder for syntax induction with dependency masking. Our BERT model consists of 12 layers and has a dimensionality of 768. Additionally, the batch size is set to 128, BSGD is used for optimization with a learning rate of 0.001, momentum 0.5, weight decay rate of 0.0001.

To evaluate the performance of our model, we will use the standard evaluation scheme and employ micro-F1 as our evaluation metric. Additionally, we will evaluate the boundaries of tail entities accurately, and predict the relationship type of two spans correctly, which we refer to as boundaries evaluation. We will consider the predicted relationship as correct if it matches the ground truth relationship.

5 Results and Analysis

5.1 Overall Results

Table 1 presents the results of our approach on the test sets of the Financial Data set and DUIE1.0. We compared our approach with dependency-tree baselines described in 4.3. Furthermore, we also use different encoders trained on one, two, and three layers of AGCN (e.g., "2-AGCN" denotes our approach with 2 layers of AGCN), with or without BERT. There are several following observations. Firstly, our pruning strategy effectively preserves the information interaction between the multi-entity pairs, and thus performs better than the other relational extraction studies (Zeng, 2014 [23]; M.Miwa, 2016 [8]; Wang, 2016 [24]; Yu, 2020 [19]; Wang and Lu, 2020 [25]), especially those using dependency trees as the main feature input (Zhang, 2018 [9]; Guo, 2019 [13]; Tian, 2021 [14]). It is shown that our model is 2.1% higher than the best dependency syntax baseline model (AGCN). This further demonstrates the effectiveness of our strategy MTPF in leveraging the information interaction between multi-entity pairs in the syntax tree. Additionally, our approach can input multiple pairs of entities in a sentence at once and identify the relationship between them simultaneously. Compared to previous research that could only input a single pair of entities at a time, our model runs more efficiently. Secondly, among the encoding models in different layers of AGCN, the ones with 3 layers of AGCN had the best performance 95.04% F1 on the financial data set and 77.33% on DUIE and could capture the most structural and semantic information from the tree. Thirdly, our approach

Table 1. Main results on Chinese Financial Dataset and DUIE1.0.

Models	Financial Dataset			DuIE1.0		
	Pre	Rec	F1	Pre	Rec	F1
Zeng, 2014(CNN)	87.93	82.39	85.06	-	-	-
M. Miwa, 2016(LCA+Tree-LSTM)	-	-	-	76.23	66.24	70.89
Wang, 2016(Multi-Level Attention CNN)	86.85	67.25	75.80	-	-	-
Zhang, 2018(k-LCA+GCN)	89.56	90.06	89.81	86.85	67.25	75.80
Guo, 2019(AGGCN)	90.31	91.59	90.95	83.77	70.29	76.44
Yu, 2020(Rethinking)	91.24	89.15	90.18	-	-	-
Wang and Lu, 2020(Joint Model)	92.76	90.35	91.54	-	-	-
Tian, 2021(A-GCN(L+G))	93.02	92.88	92.94	80.56	71.96	76.02
MTPF+3-AGCN	**94.83**	**93.96**	**94.39**	**76.31**	**76.98**	**76.64**
BERT-base+MTPF+1-AGCN	94.02	92.67	93.33	75.87	76.86	76.36
BERT-base+MTPF+2-AGCN	94.67	93.02	93.83	76.23	77.60	76.91
BERT-base+MTPF+3-AGCN	**95.32**	**94.76**	**95.04**	**76.74**	**77.92**	**77.33**

performed 0.65% better than the normal model when BERT was used as the initial encoder. This confirms that the pre-trained language model is excellent in understanding the complex long sentences in Chinese.

Overall, the results indicate that MTPF is effective in capturing the interactions between multi-entity pairs and can outperform the baseline dependency tree models. The three-layer AGCN model provides the best performance and can capture the most information from the syntax tree.

5.2 The Effect of Pruned Strategy

To further investigate the impact of our designed multi-entity dependency tree pruning strategy MTPF on CRE tasks, we conducted experiments using dependency trees constructed with various pruning strategies, including No pruning, traditional SDP, and PDT (optimized LCA), with all using AGCN as the graph encoder.

Figure 3 presents the experimental results of these traditional pruning strategies and our proposed novel multi-entity pair co-pruning strategy on GCN networks. We find that removing the pruning further hurts the result by another 4.77 F1. This is because the entire tree as input introduces too much unnecessary noise in the lengthy Chinese text, such as the title content of a company's protocol. Furthermore, When we use SDP and PDT strategies, F1 values drops by 1.85% and 1.64%, respectively. This confirms that using the SDP strategy is challenging to capture the interaction information between entity pairs that are far apart. Besides, pruning with the LCA strategy results in preserving path nearly covers the entire tree.

As demonstrated, our multi-entity pair pruning strategy refines the advantages of traditional strategies by preserving the information on the critical path and the subtree root that may serve as the relationship trigger word, providing more semantic and structural information interaction on other entity pairs for the RE of the current entity pairs. This results in 95.04% F1 on the Chinese financial data set.

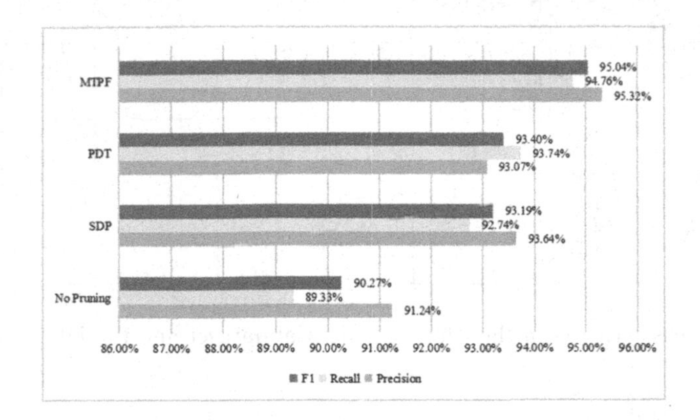

Fig. 3. Experimental results of different pruning strategies.

5.3 Case Study

Further, we compare the prediction results of MTPF and other dependency-base models or strategies (K.Xu, 2015 [10]; Zhang, 2018 [9]; Tian, 2021 [14]) in Fig. 4.

For the first example shown in Fig. 4, only our method correctly predicts the relationship between "中国平安" and "中国中药控股有限公司" is "持股", and all other methods predict the relationship as "子公司". We believe the phenomenon can be attributed that our pruning strategy takes into account information from another entity pair triples "国药集团-子公司-中国中药控股有限公司", but other models are misled by semantic information such as "旗下" appearing nearby. In the second example, because the traditional strategy cannot capture the interaction path between the first and last two entity pairs over long distances, besides, the neural network does not fully understand the "这首歌" pronoun and therefore does not correctly judge the relationship between the song "给最开心的人" and the album "娱乐大家". Our MTPF fuses the syntactic tree paths of these two entity pairs, and uses AGCN to further soft pruning, finally successfully identifying the correct relationship.

Original Text	Entity pairs	Ground Truth Relation	Prediction Results			
			K.Xu (SDP)	Zhang (LCA +GCN)	Tian (AGCN)	Ours
2018年3月，中国平安入股国药集团旗下中国中药控股有限公司，成为中药控股第二大股东，从此开启了与国药集团全面战略合作的进程。	<中国平安，中国中药控股有限公司>	持股	子公司	子公司	子公司	**持股**
	<国药集团，中国中药控股有限公司>	子公司	子公司	子公司	子公司	**子公司**
《给最开心的人》是梁咏琪的成名曲之一，它用第二人称观照他人的忘我演出，从两个视角，正面解析了艺人的价值，娱乐的意义，这首歌也是梁咏琪于2004年12月15日发行的音乐专辑《娱乐大家》中的歌曲。	<给最开心的人，娱乐大家>	所属专辑	-	-	-	**所属专辑**

Fig. 4. Two case studies on the Chinese Financial dataset and the DUIE respectively.

6 Conclusions

Relation classification is a basic block of natural language processing tasks. While most recent work leverage dependency grammar information of sentences, the effect of other entities interact with the information of current entities in the dependency tree is not fully explored. In this paper, we propose a new pruning strategy MTPF that preserves the structural and semantic interaction information between the entity pairs. By preserving the ancestor nodes of each entity to their lowest common ancestor, as well as the shortest path from that node to each entity, it enhances the information exchange between other entities and the current entity, and improves the efficiency of identifying multiple entity pairs relationships at once. Then, we encode the syntactic tree using a dependency -driven convolutional neural network, and decode it using the idea of multi-class sequence labeling. The experimental results showed a improvement over the baseline model on both Chinese datasets.

References

1. Xu, K., Reddy, S., Feng, Y., Huang, S., Zhao, D.: Question answering on freebase via relation extraction and textual evidence, pp. 2326–2336 (2016)
2. Lu, W., Cardie, C.: Focused meeting summarization via unsupervised relation extraction. In: Association for Computational Linguistics, pp. 304–313 (2016)
3. Distiawan, B., Weikum, G., Qi, J., Zhang, R.: Neural relation extraction for knowledge base enrichment. In: Proceedings of the 57th Annual Meeting of the Association for Computational Linguistics, pp. 229–240 (2019)
4. Wang, L., Cao, Z., De Melo, G., Liu, Z.: Relation classification via multi-level attention CNNs, pp. 1298–1307 (2016)

5. Zeng, D., Liu, K., Lai, S., Zhou, G., Zhao, J.: Relation classification via convolutional deep neural network. In: International Conference on Computational Linguistics, pp. 2335–2344 (2014)
6. Devlin, J., Chang, M. W., Lee, K., Toutanova, K.: BERT: pre-training of deep bidirectional transformers for language understanding, pp. 4171–4186 (2018)
7. Yang, Z., Dai, Z., Yang, Y., Carbonell, J., Salakhutdinov, R.R., Le, Q.V.: XLNet: generalized autoregressive pretraining for language understanding, pp. 5753–5763 (2019)
8. Miwa, M., Bansal, M.: End-to-end relation extraction using LSTMs on sequences and tree structures, pp. 1105–1116 (2016)
9. Zhang, Y., Qi, P., Manning, C.D.: Graph convolution over pruned dependency trees improves relation extraction, pp. 2205–2215 (2018)
10. Xu, K., Feng, Y., Huang, S., Zhao, D.: Semantic relation classification via convolutional neural networks with simple negative sampling. Comput. Sci. **71**(7), 941–9 (2015)
11. Roth, M., Lapata, M.: Neural semantic role labeling with dependency path embeddings, pp. 1192–1202 (2016)
12. Marcheggiani, D., Titov, I.: Encoding sentences with graph convolutional networks for semantic role labeling, pp. 1506–1515 (2017)
13. Guo, Z., Zhang, Y., Lu, W.: Attention guided graph convolutional networks for relation extraction, pp. 241–251 (2019)
14. Tian, Y., Chen, G., Song, Y., Wan, X.: Dependency-driven relation extraction with attentive graph convolutional networks. In: Proceedings of the 59th Annual Meeting of the Association for Computational Linguistics and the 11th International Joint Conference on Natural Language Processing (Volume 1: Long Papers), pp. 4458–4471 (2021)
15. Yu, B., Mengge, X., Zhang, Z., Liu, T., Yubin, W., Wang, B.: Learning to prune dependency trees with rethinking for neural relation extraction. In Proceedings of the 28th International Conference on Computational Linguistics, pp. 3842–3852. Barcelona, Spain (Online) (2020). International Committee on Computational Linguistics
16. Shi, Y., Li. S., He, W.: DuIE: a large-scale Chinese dataset for information extraction, pp. 791–800 (2019)
17. Li, Z., Ding, N., Liu, Z., Zheng, H., Shen, Y.: Chinese relation extraction with multi-grained information and external linguistic knowledge. In: Proceedings of the 57th Annual Meeting of the Association for Computational Linguistics, pp. 4377–4386 (2019)
18. Parmar, N., et al.: Attention is all you need, pp. 5998–6008 (2017)
19. Zhang, Z., Liu, T., Yubin, W., Yu, B., Xue, M., Wang, B.: Learning to prune dependency trees with rethinking for neural relation extraction. In: Proceedings of the 28th International Conference on Computational Linguistics, pp. 3842–3852 (2020)
20. Mandya, A., Bollegala, D., Coenen, F.: Graph convolution over multiple dependency sub-graphs for relation extraction. In: Proceedings of the 28th International Conference on Computational Linguistics, pp. 6424–6435 (2020)
21. Peng, Z., Wei, S., Tian, J., Qi, Z., Bo, X.: Attention-based bidirectional long short-term memory networks for relation classification. In: Proceedings of the 54th Annual Meeting of the Association for Computational Linguistics (Volume 2: Short Papers), pp. 207–212 (2016)
22. Chen, D., Zhong, Z.: A frustratingly easy approach for joint entity and relation extraction, pp. 50–61 (2021)

23. Zeng, D., Liu, K., Lai, S., Zhou, G., Zhao, J.: Relation classification via convolutional deep neural network. In: International Conference on Computational Linguistics, pp. 2335–2344 (2014)
24. Wang, L., Cao, Z., De Melo, G., Liu, Z.: Relation classification via multi-level attention CNNs, pp. 1298–1307 (2016)
25. Wang, J., Lu, W.: Two are better than one: joint entity and relation extraction with table-sequence encoders, pp. 1706–1721 (2020)

A Lightweight Text Classification Model Based on Label Embedding Attentive Mechanism

Fan Li, Guo Chen(✉), Jia Wen Yi, and Gan Luo

School of Automation, Central South University, Changsha 410000, Hunan, China
224612134@csu.edu.cn, guochen@ieee.org

Abstract. This paper presents a lightweight model based on the self-attention mechanism for text classification tasks. In our model, we incorporate auxiliary information of the label through the label embedding method, enabling the model to capture the contextual language variations of the same word. Furthermore, we address the issue of misclassification of similar texts by introducing the contrastive loss function, in conjunction with the traditional cross-entropy loss function. Experimental evaluations are conducted on multiple datasets, comparing our model against others with similar parameter scales, thus demonstrating the effectiveness of the proposed approach.

Keywords: Text Classification · Label Embedding · Metric Learning

1 Introduction

Text classification tasks are considered classical tasks in the field of natural language processing and have found extensive applications across various domains, including sentiment classification [9], news classification [21], and public opinion analysis [20]. Enhancing classification accuracy has emerged as a prominent research focus in recent years.

In recent years, the application of attention mechanisms in the field of text classification has gained significant popularity. The attention mechanism assigns diverse attention scores to individual words in the text vector, enabling the model to prioritize more pertinent feature words and enhance its effectiveness. Several studies have integrated the attention mechanism to enhance the feature perception capabilities of the original models, such as LSTMAtt and ATAE-LSTM [17,19], resulting in more flexible feature learning. However, despite the incorporation of the attention mechanism, there are still inherent limitations. Specifically, it fails to account for the fact that the same word should carry different weights across different classification tasks.

Furthermore, when dealing with a large number of classification categories, it is common to encounter semantic similarities among certain categories. For instance, within the THUCNews dataset, samples belonging to the finance and stocks categories often exhibit high textual similarity. In such cases, the attention

B. Luo et al. (Eds.): ICONIP 2023, CCIS 1966, pp. 599–609, 2024.
https://doi.org/10.1007/978-981-99-8148-9_46

mechanism alone is inadequate and unable to effectively handle this issue, leading to potential misclassification by the classifier.

In order to tackle the aforementioned challenges, this study presents LabelML, a lightweight text classification model. In the main structure of the model, the label embedding technique is introduced to assign different attention weights to the same word in different contexts, capturing the importance of each word in various classification tasks. Additionally, to tackle the problem of misclassification of similar texts, the model incorporates a contrastive loss function during the training process. The objective of this loss function is to minimize the distance between text vectors within the same category in the vector space, while maximizing the distance between text vectors from different categories. This approach enhances the discriminative ability of the vectors and improves the overall classification performance of the model.

To assess the performance of the proposed model across various text classification datasets, a comprehensive set of experiments was conducted. The experimental findings substantiate the efficacy of the proposed model in achieving desirable outcomes in text classification tasks.

2 Related Work

2.1 Text Classification

Traditional text classification algorithms based on machine learning often use one-hot encoding, bag-of-words model, TF-IDF, and similar methods to construct vector representations of texts. These representations are then used with Text classification methods based on machine learning techniques to classify the obtained text feature vectors.However, these methods require substantial time and effort to set up the features. On the other hand, deep learning-based text classification algorithms typically do not require manual feature engineering. Instead, they utilize word embedding models to obtain semantically enriched text vector representations. Models such as [2–5, 15] are considered classical deep learning approaches for text classification.

2.2 Label Embedding

The utilization of category label information was initially applied in image classification tasks. [1] improved the classification performance by encoding labels as vectors and incorporating them as additional information during model training. In the field of Natural Language Processing (NLP), [16,18]combined label embedding with attention mechanisms to learn attention weights related to labels during the model training process. These weights were then applied to weight the text representations, resulting in more precise text representations. Furthermore, [12] and others embedded labels and word vectors in the same space, ensuring the flexibility of adaptive convolution while fully leveraging the auxiliary role of label information. [7]jointly encoded text and labels in their mutually participating representations, allowing the model to perceive the interconnected parts of both.

2.3 Metric Learning

Metric learning provides an effective solution for handling the issue of a large number of sample categories in domains such as retrieval and face recognition. Its purpose is to learn a metric function that enlarges the distance between dissimilar samples and reduces the distance between similar samples. Due to its advantage of enhancing feature representation capability, This approach has found extensive applications in the domain of text classification as well [11,13,22]. Contrastive loss and triplet loss [8] are two common methods in metric learning. Contrastive loss distinguishes samples from different categories by setting a threshold for their distance. The similarity among samples belonging to the same class should be higher than a predefined threshold, while the dissimilarity between samples from different classes should be greater. On the other hand, the triplet loss function selects an anchor sample randomly from the training dataset and then constructs triplets by selecting another sample from the same class as the anchor and a sample from a different class. Its objective is to train samples with smaller differences and theoretically reduces the constraints on the embedding space.

3 Model

3.1 Model Structure

The proposed LabelML model in this paper consists of three modules: the label embedding-based weight matrix layer, the feature extraction layer, and the classification layer. The network architecture diagram is shown in Fig. 1. Firstly, the input is the text vector $X = \{x_1, x_2 \cdots x_n\}$, and it is passed through an embedding layer to obtain the original word vectors $E = \{e_1, e_2, \cdots e_n\}$. Then, it undergoes an Bi-LSTM layer to obtain the feature vector $H = \{h_1, h_2, \cdots h_n\}$ that encapsulates the latent semantics of the text. After the fusion of this feature vector with the weight matrix, it is passed through a self-attention module and then normalized. Next, it undergoes regularization and scaling to obtain a vector to be classified. Finally, it is input to a linear layer to obtain the final classification result.

Defining $M = \{m_1, m_2, \ldots, m_n\}$ as the weight matrix corresponding to the text under each category label, where m_i represents the attention score of x_i under different label semantics. The calculation formula for the weight matrix based on label embedding is as follows:

$$Y_1 = HM^T \tag{1}$$

Then, we apply a maximum pooling operation to obtain the highest weight value for each word across all labels, representing the weight in the current text semantics:

$$Y_2 = \max\_\text{pooling}\,(Y_1) \tag{2}$$

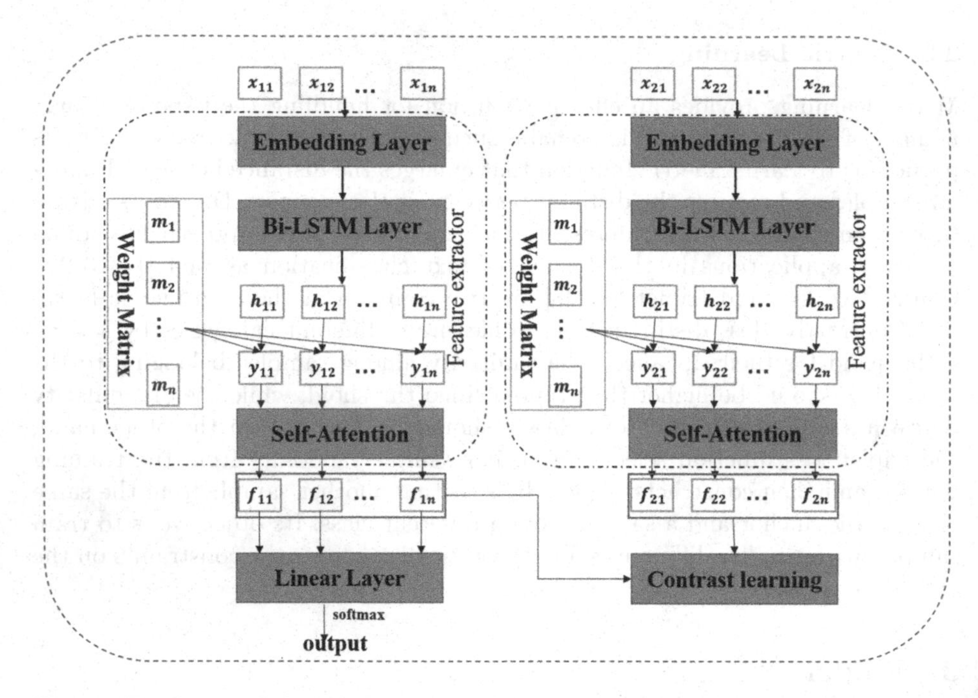

Fig. 1. The design of the LabelML model. The LabelML model consists of three main components: feature extraction, weight matrix, and metric learning. The LSTM Layer employs Bi-LSTM for feature extraction from the Embedding Vector.

Finally, we apply L2 regularization to the obtained vector and use softmax normalization to ensure that the sum of attention weights for all n words equals 1, resulting in the text semantic attention weight value:

$$Y = \text{softmax}\left(L_2\left(Y_2\right)\right) \tag{3}$$

Next, combine it with the original context vector, and to address the issue of a significant decrease in data magnitude caused by vector multiplication,we introduce a correction factor:

$$f = \alpha \cdot H \cdot Y \tag{4}$$

The correction factor is usually set to 100. To the best of our knowledge, previous works [6,11,13,22] seldom multiplied by this correction factor, while multiplying it greatly improves the model's performance. In the experimental section, we demonstrate this point.

Subsequently, the vector f undergoes a self-attention mechanism to fuse the features. It is then subjected to L2 regularization, followed by amplification, before entering a linear classification layer to obtain the final prediction results.

3.2 Metric Learning

To address the problem of misclassifying similar texts from different categories, this study introduces a metric learning auxiliary task to enhance the discriminative capacity of the text features acquired by the classifier, thereby improving the classification performance of the model. To achieve this while ensuring the effectiveness of the auxiliary task and maximizing efficiency, a strategy similar to that described in reference [14] is employed. The specific workflow is as follows:

(1) Randomly divide the samples into batches of size m, where m is the same as the batch size used during model training. Then, Obtain the text feature vector f by randomly selecting a data batch as input for the model.
(2) Randomly select two text vectors for each class from all text feature vectors f with different class labels to form a new set $S = \{s_1, s_2, \cdots, s_{2n}\}$, satisfying $i \neq j$ for each pair in S. Combine them to create a triplet (s_i, s_j, c), where c is the label. If the original labels corresponding to s_i and s_j are the same, assign 1; otherwise, assign 0. Then, obtain a new set V, where v_i represents the triplet (s_i, s_j, c).
(3) Calculate the Euclidean distance d between the s_i and s_j vectors for each $v_i = (s_i, s_j, c)$ in the set V. Use the following loss function to calculate the contrastive loss, where margin is a threshold hyperparameter:

$$Loss_2 = \frac{1}{n} \sum_{k=1}^{n} \left(cd^2 + (1 - c) \max (0, margin - d)^2 \right) \tag{5}$$

(4) Finally, we adopt a similar approach to [10] to construct the final training loss:

$$L = (1 - \lambda) \cdot Loss_1 + \lambda \cdot Loss_2 \tag{6}$$

The $Loss_1$ here refers to the cross-entropy loss function. By randomly selecting only one vector per class, it greatly reduces the computational complexity while ensuring the effectiveness of the model. During the experimental process, we observed that setting $\lambda = 10^{-3}$ to a specific value yielded good results. Therefore, we used this parameter in all our experiments.

4 Experiments

4.1 Experimental Setup

To evaluate the performance of the proposed LabelML model on various text classification tasks, we conducted experiments on five datasets and compared them with five different baseline models.

Firstly, we employed bert-base tokenizer provided by the transformer community to perform tokenization. We set the appropriate length for generating vectors based on the text length of each dataset. For texts that were shorter than this length, we padded them with <pad> tokens,Table 1 provides a detailed description of the datasets used in this study.

Table 1. Detailed Information of Experimental Datasets

Dataset	Train set	Val set	Test set	Input length	Category_num
THUNews	180000	10000	10000	32	10
Toutiaonews	183684	43291	76534	128	15
R8	4937	547	2189	128	8
TREC	4952	500	500	512	6
CR	2997	349	425	128	2

Experimental Datasets

THUNews Dataset: This dataset is derived from the historical data of the Sina News RSS feed channel between 2005 and 2011, filtered and selected by the Natural Language Processing Laboratory at Tsinghua University. For this study, we extracted 200,000 news headlines with text lengths ranging from 20 to 30 to test the model's performance on short texts. There are a total of 10 categories, with 20,000 samples in each category.

Toutiao Dataset: This dataset is collected from the client-side of the Today's Headlines application. It includes 15 categories such as current affairs, culture, entertainment, and sports. There are 382,688 data samples in total, with approximately 25,500 samples per category.

R8 Dataset: This dataset focuses on document topic classification and consists of 8 categories. It contains a total of 7,973 data samples.

TREC Dataset: This dataset is designed for question classification and consists of 6 categories: description, entity, person, location, numeric, and abbreviation. It includes 5,952 data samples in total.

CR Dataset: The CR dataset is a user comment dataset that includes two sentiment categories: positive and negative. It is a binary classification problem with a total of 3,771 data samples.

Introduction of Baseline Models

In this paper, we selected five deep learning models with the same parameter scale as our proposed model as baseline models:

TextCNN: It utilizes convolutional layers to extract N-gram features from the text. In our study, we used 256 convolutional kernels with sizes of 2, 3, and 4.

DPCNN: It is the first widely effective deep convolutional network for text classification. By deepening the network, it can capture long-distance textual dependencies without significantly increasing computational cost. We used 250 convolutional kernels in our study.

RNN-Att: It first employs Bi-LSTM to obtain feature vectors of the input text and then captures key information using an attention module.

Transformer: It utilizes an encoder module to encode the input text and obtain feature vectors. Then, a linear layer is applied to obtain the final model output.

TextRCNN: It first learns contextual information through Bi-LSTM, then concatenates the forward and backward hidden layer states with the word embedding layer, and finally applies max-pooling to extract key information.

To reduce the impact of the embedding layer, we did not use pre-trained embeddings for training but instead used randomly initialized weights. In all our experiments, we used a batch size of 128. During the entire training phase, we utilized the AdamW optimizer. The dimension of the Bi-LSTM is set to 150, and the dropout rate is 0.5. Due to variations in dataset sizes, we selected appropriate learning rates for different datasets.

The models' performance in the experiments was evaluated using accuracy, which measures the proportion of correctly predicted samples out of the total number of samples.

4.2 Experimental Result

The classification prediction results of different text classification models on each dataset are shown in Table 2. It can be observed that the proposed model in this study performs remarkably well. Among the selected baseline models, Transformer does not exhibit outstanding results on all datasets, likely due to its utilization of only an encoder and a simple linear classifier. The TextCNN model shows relatively good performance, benefiting from its N-Gram-like processing mechanism, which enables it to perform well on datasets with varying text lengths.

Table 2. Experimental results of different models on all datasets.

Model	THUNews	Toutiaonews	R8	TREC	CR
TextCNN	89.47%	84.09%	96.07%	97.20%	80.24%
DPCNN	89.64%	82.15%	95.31%	75.53%	79.06%
RNN-Att	90.15%	83.31%	95.02%	97.20%	79.76%
Transformer	87.38%	78.36%	95.84%	96.80%	75.53%
TextRCNN	89.81%	83.14%	95.66%	97.60%	77.88%
LabelML	**90.23%**	**84.32%**	**96.57%**	**98.60%**	**81.65%**

Compared to the baseline models, our model utilizes weighted embedding layers and employs metric learning for training, resulting in text features with higher discriminability. This is evident in the significant advantage demonstrated by LabelML, particularly on datasets with a large number of categories.

We also conducted ablation experiments to demonstrate the contributions of the Metric Learning and the Label Embedding module in enhancing classification accuracy. As shown in Table 3, the experimental results indicate that when our model does not incorporate Label Embedding technology and the auxiliary task of metric learning, it exhibits significant performance degradation compared to

LabelML. Furthermore, when only Label Embedding technology is employed, the performance drop is at most 2.49% compared to LabelML, further illustrating the beneficial effect of metric learning in text classification. These experimental results align with our expectations.

Table 3. Experimental results of ablation study. - means "without" here.ml: Metric Learning, le:Label Embedding.

Model	THUNews	Toutiaonews	R8	TREC	CR
LabelML	**90.23%**	**84.32%**	**96.57%**	**98.60%**	**81.65%**
-ml	89.78%	82.15%	96.48%	96.60%	79.06%
-le	89.60%	82.17%	96.25%	96.60%	79.17%

The aforementioned experimental results provide evidence that our proposed method significantly enhances text classification by integrating attention weights and employing metric learning. Furthermore, it is evident that our method exhibits favorable performance even without the incorporation of metric learning and the enhancement of attention weights.

Impact of Adjustment Factors. In Eq. 4, we assign weights to the vectors and multiply them by a scaling factor to compensate for the scaling error introduced by the multiplication of weights. Taking the R8 dataset as an example, while keeping the other model parameters constant, we vary the size of the adaptive factor. Table 4 presents the performance of LabelML with different adjustment factors. It is evident that the scaling factor has a significant impact on the model's performance. A value of 100 yields better results, while selecting a larger value can even lead to a decrease in performance.

Table 4. Classification performance of LabelML with different adjustment factors.

Alpha value	R8 result
$\alpha = 1$	95.80%
$\alpha = 10$	95.93%
$\alpha = 100$	**96.57%**
$\alpha = 1000$	95.70%

Visualization Analysis. To provide a more intuitive illustration of the effectiveness of LabelML in text classification tasks and to demonstrate the enhancement of model discrimination among different classes through Metric learning, this study takes the r8 dataset as an example. Firstly, the test dataset is transformed into corresponding text feature vectors using LabelML. Then, the feature

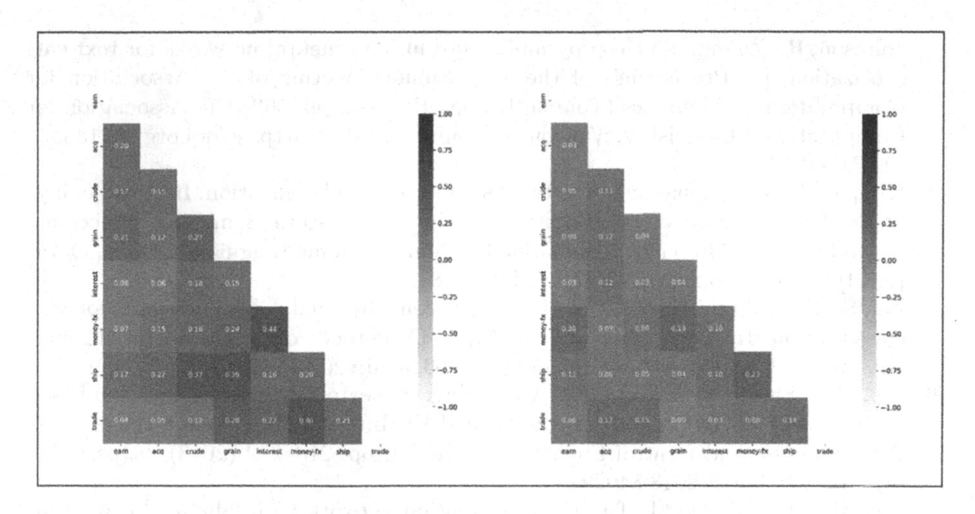

Fig. 2. Comparison of average vector cosine similarity heatmaps obtained by LabelML on the R8 dataset. The left figure represents the results without using metric learning, while the right figure represents the results with metric learning.

vectors of the same class are averaged to obtain the average vector for each class. Subsequently, The cosine similarity is computed between these average vectors, and the results are visualized through a confusion matrix, as depicted in Fig. 2. It can be observed that the addition of metric learning further reduces the feature vectors' dissimilarity between different classes.

5 Conclusion

In this study, we have developed a lightweight text classification model based on label embedding. It effectively utilizes label information to assign varying weights to the same words based on their semantic meanings. Additionally, during the training process, we have incorporated a metric learning auxiliary task to encourage the model to extract feature vectors with larger distances between different classes and smaller distances within the same class. Experimental results demonstrate the superiority of the model proposed in this study compared to other models with similar parameter scales. The results analysis further confirms the effectiveness of our model. In future research, we will focus on exploring methods to improve the performance of lightweight models.

References

1. Akata, Z., Perronnin, F., Harchaoui, Z., Schmid, C.: Label-embedding for image classification. IEEE Trans. Pattern Anal. Mach. Intell. **38**(7), 1425–1438 (2016). https://doi.org/10.1109/TPAMI.2015.2487986
2. Bai, S., Kolter, J.Z., Koltun, V.: An empirical evaluation of generic convolutional and recurrent networks for sequence modeling (2018)

3. Johnson, R., Zhang, T.: Deep pyramid convolutional neural networks for text categorization. In: Proceedings of the 55th Annual Meeting of the Association for Computational Linguistics (Volume 1: Long Papers), pp. 562–570. Association for Computational Linguistics, Vancouver, Canada (2017). https://doi.org/10.18653/v1/P17-1052
4. Kim, Y.: Convolutional neural networks for sentence classification. In: Proceedings of the 2014 Conference on Empirical Methods in Natural Language Processing (EMNLP), pp. 1746–1751. Association for Computational Linguistics, Doha, Qatar (2014). https://doi.org/10.3115/v1/D14-1181
5. Lai, S., Xu, L., Liu, K., Zhao, J.: Recurrent convolutional neural networks for text classification. In: Proceedings of the AAAI Conference on Artificial Intelligence, vol. 29. no. 1 (2015). https://doi.org/10.1609/aaai.v29i1.9513
6. Liu, C.Z., Sheng, Y.x., Wei, Z.Q., Yang, Y.Q.: Research of text classification based on improved TF-IDF algorithm. In: 2018 IEEE International Conference of Intelligent Robotic and Control Engineering (IRCE), pp. 218–222 (2018). https://doi.org/10.1109/IRCE.2018.8492945
7. Liu, M., Liu, L., Cao, J., Du, Q.: Co-attention network with label embedding for text classification. Neurocomputing **471**, 61–69 (2022). https://doi.org/10.1016/j.neucom.2021.10.099
8. Musgrave, K., Belongie, S., Lim, S.-N.: A metric learning reality check. In: Vedaldi, A., Bischof, H., Brox, T., Frahm, J.-M. (eds.) ECCV 2020. LNCS, vol. 12370, pp. 681–699. Springer, Cham (2020). https://doi.org/10.1007/978-3-030-58595-2_41
9. Peng, S., et al.: A survey on deep learning for textual emotion analysis in social networks. Digital Commun. Netw. **8**(5), 745–762 (2022). https://doi.org/10.1016/j.dcan.2021.10.003
10. Schick, T., Schütze, H.: Exploiting cloze-questions for few-shot text classification and natural language inference. In: Proceedings of the 16th Conference of the European Chapter of the Association for Computational Linguistics: Main Volume, pp. 255–269. Association for Computational Linguistics, Online (2021). https://doi.org/10.18653/v1/2021.eacl-main.20. https://aclanthology.org/2021.eacl-main.20
11. Song, P., Geng, C., Li, Z.: Research on text classification based on convolutional neural network. In: 2019 International Conference on Computer Network, Electronic and Automation (ICCNEA), pp. 229–232. IEEE, Xi'an, China (2019). https://doi.org/10.1109/ICCNEA.2019.00052
12. Tan, C., Ren, Y., Wang, C.: An adaptive convolution with label embedding for text classification. Appl. Intell. **53**(1), 804–812 (2023). https://doi.org/10.1007/s10489-021-02702-x
13. Tian, H., Wu, L.: Microblog emotional analysis based on TF-IWF weighted word2vec model. In: 2018 IEEE 9th International Conference on Software Engineering and Service Science (ICSESS), pp. 893–896. IEEE, Beijing, China (2018). https://doi.org/10.1109/ICSESS.2018.8663837
14. Tu Zhenchao, M.J.: Item categorization algorithm based on improved text representation. Data Anal. Knowl. Discovery **6**, 34–43 (2022). https://doi.org/10.11925/infotech.2096-3467.2021.0958
15. Vaswani, A., et al.: Attention is all you need (2017)
16. Wang, G., et al.: Joint embedding of words and labels for text classification (2018)
17. Wang, Y., Huang, M., Zhu, X., Zhao, L.: Attention-based LSTM for aspect-level sentiment classification. In: Proceedings of the 2016 Conference on Empirical Methods in Natural Language Processing, pp. 606–615. Association for Computational Linguistics, Austin, Texas (2016). https://doi.org/10.18653/v1/D16-1058

18. Xu Yuemei, Fan Zuwei, C.H.: A multi-task text classification model based on label embedding of attention mechanism. Data Anal. Knowl. Discovery, **6**(2/3) pp. 105–116 (2022). https://doi.org/10.11925/infotech.2096-3467.2021.0912
19. XUAN, W.: Logistics service quality sentiment analysis with deeper attention lstm model with aspect embedding. Tehnicki vjesnik - Technical Gazette **30**(2), 634–641 (2023). https://doi.org/10.17559/TV-20221018031450
20. Yan, C., Liu, J., Liu, W., Liu, X.: Research on public opinion sentiment classification based on attention parallel dual-channel deep learning hybrid model. Eng. Appl. Artif. Intell. **116**, 105448 (2022). https://doi.org/10.1016/j.engappai.2022.105448
21. Yao, L., Mao, C., Luo, Y.: Graph convolutional networks for text classification. CoRR abs/1809.05679 (2018). http://arxiv.org/abs/1809.05679
22. Zhao, W., Zhu, L., Wang, M., Zhang, X., Zhang, J.: WTL-CNN: a news text classification method of convolutional neural network based on weighted word embedding. Connect. Sci. **34**(1), 2291–2312 (2022). https://doi.org/10.1080/09540091.2022.2117274

[illegible] and Zhang, H.: [illegible] a joint model based on label [illegible] (2) pp. 105-[illegible] (2022). [illegible] 10.1007/s[illegible]-[illegible]-2021-0[illegible]
[illegible] and [illegible]: [illegible] (2020) [illegible]
[illegible] (2022). [illegible] 10.1109/[illegible]
[illegible] Classification [illegible] (2023). [illegible]
[illegible] Zhang, [illegible] CNN [illegible] word [illegible] (2021). [illegible] 10.1016/j.[illegible]

Author Index

B. Luo et al. (Eds.): ICONIP 2023, CCIS 1966, pp. 611–613, 2024.
https://doi.org/10.1007/978-981-99-8148-9

Printed in the United States
by Baker & Taylor Publisher Services